Cancer Stem Cells

Cancer Stem Cells

Editor
Vinagolu K. Rajasekhar

WILEY Blackwell

Published by John Wiley & Sons, Inc., Hoboken, New Jersey

Published simultaneously in Canada

Library of Congress Cataloging-in-Publication Data:

Cancer stem cells (2014)
Cancer stem cells / editor, Vinagolu K. Rajasekhar.
  1 online resource.
  Includes index.
  Description based on print version record and CIP data provided by publisher; resource not viewed.
  ISBN 978-1-118-35617-3 (ePub) – ISBN 978-1-118-35618-0 (Adobe PDF) – ISBN 978-1-118-35616-6 (cloth)
I. Rajasekhar, Vinagolu K., editor of compilation. II. Title.
  [DNLM: 1. Neoplastic Stem Cells. QZ 202]
  RC269.7
  616.99′402774–dc23

2013043558

Cover images: Top row: Alternate bright field and fluorescent (green or red protein) images of primary sphere-forming cancer stem cells that are isolated from human patient derived xenografts (PDX) of prostate tumor specimens (V.K. Rajasekhar and John H. Healey, Unpublished data). Bottom row: Immunohistochemical (bright field) and immunofluorescent (green) detection of nuclear localization of NF-κB in the sphere-forming prostate cancer stem cells isolated from the human PDXs and nuclei counterstained by DAPI (blue) (Rajasekhar et al., Nat Commun., 2011 January 18; 2: 162). Background: iStock 26050470, © luismmolina.

Cover designer: Nicole Teut
Printed and bound in Malaysia by Vivar Printing Sdn Bhd

1 2014

Dedication

I sincerely dedicate this book to:

- All Physicians, Basic Research Scientists, Physician Scientists, Other Employees at All the Levels and Patients at Memorial Sloan-Kettering Cancer Center, New York.
- My Family: My parents, Late Mr. Vinagolu Krishnamachari and Mrs. V.K. Suseela (on her 75th Birthday), my wife, Mrs. Birgit Luise Baur, and my children Julia Ruby Vinagolu-Baur and Jessica Pallavi Vinagolu-Baur, my parents-in-law, Mr. Hugo Baur (on his 90th Birthday) and Mrs. Erika Baur, my late sister and late brother V.K. Bhanu and V.K. Suresh respectively, and other brothers, cousins (especially A.R. Thulasi Krishna and. Kumbakonam Prasaad), uncles (particularly Hari Gopal, Raghunath, A. Venkatachari, Mahi, K. Jagadeshan, and K. Chandra), aunts (particularly Jaya Aunty), nephews, nieces, and last but not least my childhood friends (R. Dharmarao, A. Ramesh Reddy, E. Lokeshwar Reddy, S. Gopi, and S. Venkateswara Prasad).
- And, many others not limited to Drs. Irv Weissman, Nahum Sonenberg, Bjørn Nicolaissen, Alan Trounson, John Dick, Max Wicha, Andreas Trumpp, David Lyden, Shahin Rafii, Laurie Glimcher, Marc Tessier-Lavigne, Elaine Fuchs, David Allison, Nancy Colburn, Knud Nierhaus, Aboulghassem Shahdadfar, Jesintha Navaratnam, Sally Temple, Craig Jordan, Inder Verma, Leonard Zon, Fred Gage, Shinya Yamanaka, Alex Meissner, Kevin Eggan, Konrad Hochedinger, Chad Cowan, Karla Kim, Douglas Melton, David Scadden, Bert Vogelstein, Tyler Jacks, Ronald DePinho, Stephen Elledge, Martin Pera, Oven Witte, Micheal Shen, George Daley, Stuart Orkin, Rudolf Jaenisch, Kornelia Polyak, Eric Holland, Harold Varmus, Robert Wittes, Stephen Nimer, Eric Lander, Craig Venter, Michael Karin, Sanjay Tyagi, Maitradas Panicker, Fay Betsou, Pasquale De Blasio, Timothy Osborne, the late William Gerald, and also to plant biology colleagues not limited to Professors Sudhir Sopory, Hans Mohr, Krishna Tewari, Michael Mulligan, Donald Fosket, Winslow Briggs, Wilbur Campbell, Brent Nielsen, Ralf Oelmuller, L. Vijayamohan Rao, M.K. Reddy, as well as The Samuel Roberts Noble Foundation, Ardmore, OK and the late: V.S. Ramadas, Sipra-Guha Mukherjee, and Christopher Lamb.

Contents

About the Editor

Dr. Vinagolu K. Rajasekhar, MSc, MPhil, PhD, is a Senior Research Scientist at Memorial Sloan-Kettering Cancer Center, New York. Dr. Rajasekhar and his research team have purified human prostate CSCs, discovered new biomarkers, and revealed a clinically relevant signaling pathway distinct from that found in bulk tumor analysis (www.Genomeweb.com). This study has also opened up a field of live biobanking of patient CSC-derived xenograft (PDX^{CSC}) tumor models amenable for individualized therapeutic testing. His publications accomplished an impressive number of citations, previews in lead journals, and exceptional post-publication reviews by Faculty of 1,000.

Dr. Rajasekhar is the senior editor of Regulatory Networks in Stem Cells, one of the top 25% most downloaded ebooks in the relevant Springer eBook Collection in 2012. His graduate work at the Jawaharlal Nehru University, New Delhi, fetched him the internationally competitive research fellowship from the Alexander von Humboldt Foundation, Germany. He has researched and/or also taught courses in other leading institutions including the University of Freiburg, Germany, Texas Tech Univeristy, Lubbock, TX, Michigan Tech University, Houghton, MI, University of California, Irvine, CA, Humboldt University, Berlin, Germany, The Samuel Roberts Noble Foundation, Ardmore, OK, MD Anderson Cancer Center, Houston, TX, and the University of Medicine and Dentistry of New Jersey, NJ. He has been a reviewer for many lead scientific journals and an invited speaker and chair/discussion leader in many national and international conferences related to Biobanking. Recently, Dr. Rajasekhar has been nominated to ISBER Biospecimen Working Group.

Dr. Rajasekhar is currently integrating the live biobanking approach to human tumor specimens from the generous pre-consented patients with the recently emerging mouse hospital concept for mechanistic studies. In principle, these approaches are aimed at combating cancers at their roots and looking forward to share the reagents worldwide for harmonization of research materials towards facilitating patient-specific translational research and also enhancing clinical cancer outcomes with fidelity.

Contributors

Hinrich Abken
Center for Molecular Medicine Cologne
University of Cologne
Department I Internal Medicine
University Hospital Cologne
Cologne, Germany

Sheela A. Abraham
Paul O'Gorman Leukaemia Research Centre
Institute of Cancer Sciences
College of Medical, Veterinary and Life Sciences
University of Glasgow
Scotland, UK
and
Stem Cell and Leukaemia Proteomics Laboratory
School of Cancer and Enabling Sciences
Manchester Academic Health Science Centre
University of Manchester
Manchester, UK

Aamir Ahmad
Department of Pathology
Barbara Ann Karmanos Cancer Institute
Wayne State University School of Medicine
Hudson Webber Cancer Research Center
Detroit, MI, USA

Elianna M. Amin
Molecular Pharmacology and Chemistry Program
and Center for Cell Engineering
Memorial Sloan-Kettering Cancer Center
New York, NY, USA

Mark A. Badeaux
Department of Molecular Carcinogenesis
University of Texas M.D Anderson Cancer Center
Science Park
Smithville, TX, USA

Inmaculada Bañón-Rodríguez
Centro de Biología Molecular Severo-Ochoa
Consejo Superior de Investigaciones Científicas (CSIC)
Madrid, Spain

Bin Bao
Department of Pathology
Barbara Ann Karmanos Cancer Institute
Wayne State University School of Medicine
Hudson Webber Cancer Research Center
Detroit, MI, USA

Jeremy Bastid
Orega Biotech
L'espace Européen
Ecully, France

M. Baumann
OncoRay
National Center for Radiation Research in Oncology
Faculty of Medicine Carl Gustav Carus
Dresden, Germany

Ilenia Bernascone
Centro de Biología Molecular Severo-Ochoa
Consejo Superior de Investigaciones Científicas (CSIC)
Madrid, Spain

Ravi Bhatia
Division of Hematopoietic Stem Cell
and Leukemia Research
City of Hope National Medical Center
Duarte, CA, USA

Scott V. Bratman
Department of Radiation Oncology
Stanford University
Stanford, CA, USA

Jacqueline Bromberg
Department of Medicine
Memorial Sloan-Kettering Cancer Center
New York, NY, USA

Monika Brüggemann
2nd Medical Department
University Hospital Schleswig-Holstein
Campus Kiel
Kiel, Germany

Nadia P. Castro
Tumor Growth Factor Section
Laboratory of Cancer Prevention
Frederick National Laboratory for
Cancer Research
Frederick, MD, USA

Derryn Xin Hui Chan
Institute of Medical Biology
Agency for Science, Technology
and Research (A*STAR)
Singapore

Ling Chen
Department of Chemistry and
Biotechnology
Graduate School of Natural Science
and Technology
Okayama University
Okayama, Japan
and
Department of Pathology
Tianjin Central Hospital of Gynecology
Obstetrics
Tianjin, People's Republic of China

Stephen S. Chung
Leukemia Service
Department of Medicine
Human Oncology and Pathogenesis Program
Memorial Sloan-Kettering Cancer Center
New York, NY, USA

Anne T. Collins
YCR Cancer Research Unit
Departnent of Biology
University of York
Heslington, York, UK

Bikul Das
Medicine/Oncology Stanford Institute for
Stem Cell Biology and Regenerative Medicine
Stanford University
Stanford, CA, USA
and
Forsyth Institute
Harvard School of Dental Medicine
Cambridge, MA, USA

Piero Dalerba
Instructor of Medicine and Siebel
Scholar
Stanford Institute for Stem Cell Biology
and Regenerative
Medicine
Stanford University
Lokey Stem Cell Research Building
Stanford, CA, USA

Filemon S. Dela Cruz
Columbia University Medical Center
Department of Pediatrics
Division of Pediatric Oncology
New York, NY, USA

Maximilian Diehn
Department of Radiation Oncology
Institute for Stem Cell Biology and
Regenerative Medicine and Cancer
Institute
Stanford University
Stanford, CA, USA

A. Dubrovska
OncoRay
National Center for Radiation Research
in Oncology
Faculty of Medicine Carl Gustav Carus
Dresden, Germany

Nicole Facompre
Tumor Microenvironment and Metastasis
Program
Melanoma Research Center
The Wistar Institute
Philadelphia, PA, USA

Li Fu
Department of Breast Cancer Pathology and Research Laboratory
Center Hospital of Tianjin Medical University
Tianjin, People's Republic of China

Idoia García-Ramírez
Experimental Therapeutics and Translational Oncology Program
Instituto de Biología Molecular y Celular del Cáncer
CSIC/University of Salamanca and Institute of Biomedical Research of Salamanca (IBSAL)
Salamanca, Spain

Susana García-Silva
Molecular Pathology Programme
Stem Cells & Cancer Group
Spanish National Cancer Research Center
Madrid, Spain

Shanchun Guo
Microbiology, Biochemistry and Immunology
Morehouse School of Medicine
Atlanta, GA, USA

Pamela Jo Harris
Cancer Therapy Evaluation Program
Division of Cancer Treatment and Diagnosis
National Cancer Institute
National Institute of Health
Rockville, MD, USA

Christopher Heeschen
Molecular Pathology Programme
Stem Cells & Cancer Group
Spanish National Cancer Research Center
Madrid, Spain

Lisa Hopcroft
Paul O'Gorman Leukaemia Research Centre
Institute of Cancer Sciences
College of Medical, Veterinary and Life Sciences
University of Glasgow
Scotland, UK

Meenhard Herlyn
Tumor Microenvironment and Metastasis Program
Melanoma Research Center
The Wistar Institute
Philadelphia, PA, USA

Tessa L. Holyoake
Paul O'Gorman Leukaemia Research Centre
Institute of Cancer Sciences
College of Medical, Veterinary and Life Sciences
University of Glasgow
Scotland, UK

Joerg Huelsken
ISREC (Swiss Institute for Experimental Cancer Research)
and National Center of Competence in Research (NCCR)
Molecular Oncology
Ecole Polytechnique Fédérale de Lausanne (EPFL)
Lausanne, Switzerland

J.M. Iglesias
Regulation of Cell Growth Laboratory
Fundacion Inbiomed
San Sebastian, Spain

S. Percy Ivy
Cancer Therapy Evaluation Program
Division of Cancer Treatment and Diagnosis
National Cancer Institute
National Institute of Health
Rockville, MD, USA

Hideaki Karasawa
Tumor Growth Factor Section
Laboratory of Cancer Prevention
Frederick National Laboratory for Cancer Research
Frederick, MD, USA

Tomonari Kasai
Department of Chemistry and Biotechnology
Graduate School of Natural Science and Technology
Okayama University
Okayama, Japan

Michael G. Kharas
Molecular Pharmacology and Chemistry Program
and Center for Cell Engineering
Memorial Sloan-Kettering Cancer Center
New York, NY, USA

Yutaka Kondo
Division of Epigenomics
Aichi Cancer Center Research Institute
Chikusa-ku, Nagoya, Japan

Dejuan Kong
Department of Pathology
Barbara Ann Karmanos Cancer Institute
Wayne State University School of Medicine
Hudson Webber Cancer Research Center
Detroit, MI, USA

Steffen Koschmieder
Department of Medicine Oncology, Hematology, and Stem Cell Transplantation
University Medical Center of Aachen
Aachen, Germany

Takayuki Kudoh
Department of Chemistry and Biotechnology
Graduate School of Natural Science and Technology
Okayama University
Okayama, Japan

I. Kurth
OncoRay
National Center for Radiation Research in Oncology
Faculty of Medicine Carl Gustav Carus
Dresden, Germany

Caterina A.M. La Porta
Laboratory of Molecular Oncology
Department of Bioscience
University of Milan
Milan, Italy

Li Li
Laboratory of Translational Cancer Research
Ochsner Clinic Foundation
New Orleans, LA, USA

Yiwei Li
Department of Pathology
Barbara Ann Karmanos Cancer Institute
Wayne State University School of Medicine
Hudson Webber Cancer Research Center
Detroit, MI, USA

James W. Lillard
Microbiology, Biochemistry and Immunology
Morehouse School of Medicine
Atlanta, GA, USA

Keigo Machida
Department of Molecular Microbiology and Immunology
University of Southern California School of Medicine
Los Angeles, CA, USA
and
Southern California Research Center for ALPD and Cirrhosis
Los Angeles, CA, USA

Jessica Maddox
Department of Pathology
University of Utah School of Medicine
Salt Lake City, UT, USA

Norman J. Maitland
YCR Cancer Research Unit
Department of Biology
University of York
Heslington, York, UK

Jennifer Makalowski
Center for Molecular Medicine Cologne
University of Cologne
Department I Internal Medicine
University Hospital Cologne
Cologne, Germany

Aneet Mann
Faculty of Health Sciences
McMaster University Stem Cell and Cancer Research Institute
McMaster University
Hamilton, ON, Canada

Branavan Manoranjan
Faculty of Health Sciences
McMaster University Stem Cell and
Cancer Research Institute
McMaster University
Hamilton, ON, Canada

David A. Margolin
Department of Colon and Rectal Surgery
Ochsner Clinic Foundation
New Orleans, LA, USA
and
Department of Surgery
Ochsner Clinical School
The University of Queensland School
of Medicine
Queensland, Australia

A.G. Martin
Regulation of Cell Growth Laboratory
Fundacion Inbiomed
San Sebastian, Spain

Fernando Martín-Belmonte
Centro de Biología Molecular
Severo-Ochoa
Consejo Superior de Investigaciones
Científicas (CSIC)
Madrid, Spain

Alberto Martín-Lorenzo
Experimental Therapeutics and
Translational Oncology Program
Instituto de Biología Molecular y Celular
del Cáncer
CSIC/ University of Salamanca and
Institute of Biomedical Research of
Salamanca (IBSAL)
Salamanca, Spain

Igor Matushansky
Columbia University Medical Center
Department of Medicine
Division of Medical Oncology
New York, NY, USA

Robyn McClelland
Regenerative Medicine Program
Department of Biochemistry and Medical
Genetics
University of Manitoba
Winnipeg, Manitoba, Canada

Sean P. McDermott
Department of Internal Medicine
Division of Hematology and Oncology
University of Michigan
Comprehensive Cancer Center
Ann Arbor, MI, USA

Nicole McFarlane
Faculty of Health Sciences
McMaster University Stem Cell and
Cancer Research Institute
McMaster University
Hamilton, ON, Canada

Giovanna Merchand-Reyes
Unit for Research and Development in
Bioprocesses (UDIBI)
National School of Biological Sciences
National Polytechnic Institute
Mexico

Akifumi Mizutani
Department of Chemistry and
Biotechnology
Graduate School of Natural Science
and Technology
Okayama University
Okayama, Japan

Hiroshi Murakami
Department of Chemistry and
Biotechnology
Graduate School of Natural Science
and Technology
Okayama University
Okayama, Japan

Abhilasha Nair
Division of Cancer Treatment
and Diagnosis
National Cancer Institute
National Institute of Health
Rockville, MD, USA

Srikanth Nama
Institute of Medical Biology
Agency for Science, Technology
and Research (A*STAR)
Singapore

Randy van Ommeren
Faculty of Health Sciences
McMaster University Stem Cell and Cancer Research Institute
McMaster University
Hamilton, ON, Canada

Christopher Y. Park
Human Oncology and Pathogenesis Program
Departments of Pathology and Clinical Laboratory Medicine
Memorial Sloan-Kettering Cancer Center
New York, NY, USA

Lenin Pavón
Department of Psychoimmunology
National Institute of Psychiatry "Ramón de la Fuente"
Mexico

C. Peitzsch
OncoRay
National Center for Radiation Research in Oncology
Faculty of Medicine Carl Gustav Carus
Dresden, Germany

Rosana Pelayo
Oncology Research Unit
Mexican Institute for Social Security
Mexico City, Mexico

Richard G. Pestell
Department of Cancer Biology
Kimmel Cancer Center
Thomas Jefferson University
Philadelphia, PA, USA

Thorsten Raff
2nd Medical Department
University Hospital Schleswig-Holstein Campus Kiel
Kiel, Germany

Vinagolu K. Rajasekhar
Department of Medicine
Memorial Sloan-Kettering Cancer Center
New York, NY, USA

Pradeep S. Rajendran
David Geffen School of Medicine
University of California—Los Angeles
Los Angeles, CA, USA

Maria Cristina Rangel
Tumor Growth Factor Section
Laboratory of Cancer Prevention
Frederick National Laboratory for Cancer Research
Frederick, MD, USA

Lucia Ruiz-Roca
Experimental Therapeutics and Translational Oncology Program
Instituto de Biología Molecular y Celular del Cáncer
CSIC/University of Salamanca
Institute of Biomedical Research of Salamanca (IBSAL)
Salamanca, Spain

David S. Salomon
Tumor Growth Factor Section
Laboratory of Cancer Prevention
Frederick National Laboratory for Cancer Research
Frederick, MD, USA

Prabha Sampath
Institute of Medical Biology
Agency for Science, Technology and Research (A*STAR)
Singapore
and
Department of Biochemistry
Yong Loo Lin School of Medicine
National University of Singapore
Singapore

Isidro Sanchez-Garcia
Experimental Therapeutics and Translational Oncology Program
Instituto de Biología Molecular y Celular del Cáncer
CSIC/University of Salamanca
Institute of Biomedical Research of Salamanca (IBSAL)
Salamanca, Spain

Pasquale Sansone
Department of Medicine
Memorial Sloan-Kettering Cancer Center
New York, NY, USA

Albert Santamaria i Martínez
ISREC (Swiss Institute for Experimental Cancer Research)
and National Center of Competence in Research (NCCR)
Molecular Oncology
Ecole Polytechnique Fédérale de Lausanne (EPFL)
Lausanne, Switzerland

Fazlul H. Sarkar
Department of Pathology and Oncology
Barbara Ann Karmanos Cancer Institute
Wayne State University School of Medicine
Hudson Webber Cancer Research Center
Detroit, MI, USA

Hideyuki Saya
Division of Gene Regulation
Institute for Advanced Medical Research
School of Medicine, Keio University
Tokyo, Japan

Masaharu Seno
Department of Chemistry and Biotechnology
Graduate School of Natural Science and Technology
Okayama University
Okayama, Japan

Keshav K. Singh
Departments of Genetics, Pathology, Environmental Health
Center for Free Radical Biology
Center for Aging
and
University of Alabama at Birmingham Comprehensive Cancer Center
University of Alabama at Birmingham
Birmingham, AL, USA

Sheila Singh
Faculty of Health Sciences
McMaster University Stem Cell and Cancer Research Institute
McMaster University
Hamilton, ON, Canada

Rajasekharan Somasundaram
Tumor Microenvironment and Metastasis Program
Melanoma Research Center
The Wistar Institute
Philadelphia, PA, USA

Gopinath Sundaram
Institute of Medical Biology
Agency for Science, Technology and Research (A*STAR)
Singapore

Naoko Takebe
Cancer Therapy Evaluation Program
Division of Cancer Treatment and Diagnosis
National Cancer Institute
National Institute of Health
Rockville, MD, USA

Dean G. Tang
Department of Molecular Carcinogenesis
University of Texas MD Anderson Cancer Center
Science Park
Smithville, TX, USA

Dean Tantin
Department of Pathology
University of Utah School of Medicine
Salt Lake City, UT, USA

Marco Velasco-Velázquez
Department of Pharmacology
School of Medicine
National Autonomous University of Mexico
Mexico City, Mexico

L. Vellon
Cell Reprogramming Unit
Fundacion Inbiomed
San Sebastian, Spain

Chitra Venugopal
Faculty of Health Sciences
McMaster University Stem Cell and Cancer Research Institute
McMaster University
Hamilton, ON, Canada

Parvez Vora
Faculty of Health Sciences
McMaster University Stem Cell and Cancer Research Institute
McMaster University
Hamilton, ON, Canada

Tamra Werbowetski-Ogilvie
Regenerative Medicine Program
Department of Biochemistry and Medical Genetics
University of Manitoba
Winnipeg, Manitoba, Canada

Anthony D. Whetton
Stem Cell and Leukaemia Proteomics Laboratory
School of Cancer and Enabling Sciences
Manchester Academic Health Science Centre
University of Manchester
Manchester, UK

Max S. Wicha
Department of Internal Medicine
Division of Hematology and Oncology
University of Michigan
Comprehensive Cancer Center
Ann Arbor, MI, USA

Lily Yang
Department of Surgery
Emory University School of Medicine
Atlanta, GA, USA

Stefano Zapperi
CNR-IENI
Milano, Italy
and
ISI Foundation
Torino, Italy

Dov Zipori
Department of Molecular Cell Biology
Weizmann Institute of Science
Rehovot, Israel

Margot Zöller
Department of Tumor Cell Biology
University Hospital of Surgery
Heidelberg, Germany

Foreword

During the last decade, the conceptual themes of stem cell biology have been re-applied, with a new vigor, to the field of oncology. The idea that, similar to normal tissues, tumors can be viewed as "complex societies," where different cell types are generated as the result of multi-lineage differentiation processes, and organize themselves in hierarchical structures, has now entered the realm of solid tumor biology, and altered the way we think of cancer as a disease. Most importantly, the possibility that tumor tissues, similar to normal ones, might be sustained in their long-term growth by a subset of cancer cells endowed with stem cell properties (i.e., a mutated "cancer stem cell" population capable of both aberrant self-renewal as well as differentiation) has important implications for the future development of targeted therapies. In this beautiful book, Dr. Vinagolu K. Rajasekhar (Memorial Sloan Kettering Cancer Center, New York) thoughtfully weaved together the perspectives and contributions from several of the leading scientists in the field. This book is both an elegant review and a practical guide to the exciting, and still largely uncharted, world of "cancer stem cells," I praise the editor and the authors for this wonderful endeavor, rich of provocative ideas and challenging concepts, not only for a better understanding of basic cancer biology, but also for the future development of new, more effective, anti-tumor treatments.

– Michael F. Clarke, MD,
Stanford University, Stanford, CA, USA

"The cancer stem cell (CSC) concept posits that not all cells in tumors are equal, but that dedicated cells fuel tumor growth. A major attraction of the CSC concept rests in the explanations it provides for several poorly understood clinical phenomena. The CSCs are built to last a lifetime, to be resilient to electromagnetic and chemical insults, to be able to slumber for prolonged periods of time, and to colonize other parts of the body. Thus, the CSC hypothesis explains why a cancer patient should never be considered cured, even when the initial response to radiation or chemotherapy is encouragingly robust. The concept guides the development of more effective treatments, targeting the 'beating heart' of the tumor: the CSC. This authoritative book, written by a range of world-leading cancer researchers, provides a comprehensive overview of the cancer stem cell, its micro-environment, and how these insights will lead to novel clinical strategies."

– Hans Clevers, MD, PhD,
Hubrecht Institute, Utrecht,
The Netherlands

"The nature and clinical relevance of cancer stem cells are timely topics covered with an appropriately broad and insightful brush in this comprehensive book devoted entirely to this subject. Chapters include emerging provocative evidence that a cancer stem cell, although still necessarily defined operationally, actually refers to a molecular state that may be unstable or altered reversibly. In this respect, the cancer stem cell field has entered a new era of complexity building on discoveries of concurrent intrinsic and extrinsic regulators of the stem cell state in normal tissues. Nevertheless, in spite of this evolution, many investigations in specific types of malignancies have proven useful and more are expected. For those wanting to stay abreast of the field from a basic as well as a clinical perspective, this book will be a welcome read and resource."

– Connie J. Eaves, PhD, FRSC,
Terry Fox Laboratory,
Vancouver, Canada

"Cancer stem cells have moved onto center stage for those who are interested in the behavior of solid tumors. In the context of carcinomas, these cells hold the prospect of explaining many aspects of the malignant behavior of high-grade tumor cells, including their metastatic dissemination and their responsiveness to a variety of therapies. Those who are interested in developing novel therapeutic strategies for treating solid tumors can no longer afford to ignore these important subpopulations of cancer cells, which increasingly appear to be critical determinants of the success or failure of existing treatments. This volume reports on many aspects of these cells in a variety of human tumors, justifying the notion that CSCs are likely to be important players in virtually all types of human tumors."

– Robert A. Weinberg, PhD,
Whitehead Institute, Massachusetts
Institute of Technology,
Cambridge, MA, USA

Preface

How Could Individual Cancer Treatments Be Improved Going Forward?

This provocative and challenging question provides an ideal framework for introducing this book, "Cancer Stem Cells". It broadly echoes the questions posed during many of my encounters with researchers, clinicians, and patients at Memorial Sloan-Kettering Cancer Center (MSKCC) in New York, as well as members of the general public not limited to the co-pedestrians in the streets and avenues of the Upper East Side of Manhattan to and fro MSKCC. After my first identification of a plant gene, which was previously unknown in any database and whose human homolog was subsequently characterized as the cancer related Jab 1, I first initiated studies to complement this plant gene in human system at the MD Anderson Cancer Center, Houston, Texas. Deeply touched by the patients pouring in from all over the globe and to help their strong will to survive against cancers, I began to shift my research focus from biotic and abiotic stress signaling in plants originally aimed at increasing global plant productivity towards targetable signaling in human cancers for benefitting the patients' life quality worldwide. Inspired by the tremendous progress in patient focused research performed for over 100 years, and also a new and challenging opportunity to be a part of it, I started at the MSKCC as a beginner to cancer biology and clinical translational research.

My alternative perspective steered me away from conventional cancer research studies towards the poorly understood origins of cancer and the undiscovered layers of molecular control in oncogenesis. This book will introduce cancer stem cells (CSCs) to scientists unfamiliar with this area of cancer research and to clinicians interested in developing careers as physician-scientists. To provide this audience with an appropriate context for the discussions in this book, I will present my own casual appraisal of CSCs from the conclusions of prior literature, which is too extensive to list here with my sincere apologies, that laid the foundation for our current work on this topic. The goal is to synthesize a coherent set of queries so that inquisitive readers will be provoked to go online, investigate further, conceive more research themes, and even amend my thoughts here and those of the authors of the book chapters. Often, after analyzing the complexity of the questions raised by my own translationally oriented basic research and contemplating the general topic of CSCs, I find myself returning to the same fundamental issue: to improve cancer treatments we must unearth the roots of cancer and understand the soil nourishing them as much as we prune and fight against the more easily accessible, wild bushy branches of the disease. These roots include the ability to initiate, sustain and metastasize tumor growth. These are the properties of CSCs that drive tumor initiation and possibly thrive minimal residual disease in cancers, which are the focus of the chapters to follow.

Why have many human cancers remained largely incurable?

I will begin by discussing the above recurring theme: While we continue to make progress on extending the list of curable cancers, many cancer types are still associated with extremely high mortality rates and short survival. Over the last century, we have significantly advanced our understanding of cancer, beginning with the earliest microscopic observations of transformed cancer cells and progressing to current in vitro techniques of functional interrogation and manipulation of established cancer cell lines. Through these efforts, we have learned a great deal about the cellular properties of cancer, knowledge which has undoubtedly influenced developments in clinical treatment. Extensive progress has also been made in the modeling of human cancers in animals, as in the many genetically engineered mouse (GEM) models that provide the tumors for testing drug toxicity and treatment strategies. However, studies in these animal models can only inform us to a certain degree, and the substantial gap that separates these model tumors from those of cancer patients means many of the treatment strategies fail to make the jump to real-world efficacy.

Undoubtedly, there has been significant progress in improving our understanding of oncogenesis. But until recently, most researchers in the field generally interpreted their data within a broader paradigm in which any elevated or inhibited signaling pathway intermediates were correlated to a presumed linear functional representation of the relevant genes in the bulk tumors. Oncogenes (tumor-causing) and tumor suppressor (tumor-inhibiting) genes that are represented by mutations etc., in otherwise normal developmental genes have been extensively pursued as the true targets of cancer treatment, employing several related GEM models irrespective of the fact that those models are unlikely to reflect the clinical heterogeneity of actual patient or the behavior and interaction of these tumor cells within the body. The overwhelming confidence in this paradigm continues, even though we are now realizing, based on the ongoing human cancer genome sequencing initiatives, that mutated oncogenes may be present in the mature cells of a healthy person's body, in which malignant disease does not develop for exceptionally long times, or perhaps ever.

I will not even dare to delve extensively into the subject of chromothripsis, a chromosome catastrophe characterized by several gene copy numbers and cataclysmic genome disruptions, even within a single chromosome, that occurs within at least 2–3% of cancer genomes and 25% of all bone cancers, or any of the other unknown mechanisms that are challenging the thus far believed conventional model of sequential accumulation of mutations in the biogenesis of cancers. Moreover, such new observations would raise even more questions if the often overlooked layers of regulatory control and additional feedback mechanisms may impinge into other unexplored signaling cascades within cancer cells. Some of such over looked layers of control could include: (i) distinct signals in CSCs versus bulk tumor cells, (ii) an aberrant control in the early steps of polyribosome recruitment of oncogenic transcripts, (iii) altered metabolism in the cytoplasm of the CSCs involving the intracellular organelles like mitochondria, (iv) epigenetic modifiers in the functional genomic loci, etc. Nevertheless, most studies still continue to concentrate largely on tumor shrinkage rather than the biology of the disease and the impact of therapeutics on the functional cells that actually initiate tumors. It is important to note that, clinically, there is now an increasing understanding that decreasing the tumor burden is not necessarily a functional criterion for cancer cure; because of this, we must relentlessly identify and track the mechanisms of metastasis and treat the true tumor-initiating CSCs.

From Simple Concepts to Complex Signaling Networks

Unfortunately, concomitant with the persistently high proportion of incurable cancers, the list of new oncogenes and tumor suppressors continues to grow with the increasing amounts of data provided through the efforts of the Cancer Genome Atlas (TCGA) and the International Cancer Genome Consortium (ICGC). These data suggest a complex web in which the majority of human genes and known signaling pathways are related in one way or another to cancer development in humans. Due to the dynamic nature of intra-/inter-tumor heterogeneity in human cancers, it is very important to note that the tumors utilized to generate cancer genome sequencing data or the global gene expression data can never fully represent the biologically, functionally, genetically, epigenetically and metabolically harmonized tumors in a living patient.

Therefore, the important and perhaps most relevant driver genes and their functional transcripts in the rare populations of CSCs could become nearly undetectable because of the "noise" of other gene transcripts that reflect the heterogeneity of bulk tumor tissues. Also, caution must be exercised in interpreting any single miRNA as the regulator of an entire oncogenesis program using bulk tumor tissues, especially when a small subset of distinct and detached miRNA networks are known to function in cell type–specific ways in cancers as against a large and organized tree-like network of up to hundreds of miRNAs in regulation of healthy cells. For example, the CSCs typically represent about 0.1% of the bulk tumor cell population, although that proportion may be increased under certain circumstances (e.g., hypoxia). Unfortunately, the presence of a mere 1,000 CSCs among a million bulk tumor cells, when subjected to genome sequencing, would yield data that are normally excluded as part of the accepted error rate of the instrument measurements. Thus, to pinpoint the driver gene mutations responsible for the origin of human tumors, we may perhaps need to step back and carefully look into the above lacuna in the present general race to sequence as many cancer genomes as possible. We first need cooperation and generous support from federal and state agencies, private companies, or philanthropic/nonprofit foundations for accomplishing the difficult and expensive endeavor of identification, prospective purification, functional validation, and modeling of the minor populations of CSCs from freshly resected patient tumor specimens. Then, the tumor-initiating CSC population may be subjected to comparative genome, transcriptome, proteome, and metabolome analysis. This approach is expected to identify the true driver mutations that could be functionally traced to unveil real therapeutic targets, which would nudge us much closer to durable cancer treatments.

The issues discussed above are a sample of the many reasons why the precise treatment and cure of human cancers is still not on the horizon. Another intriguing puzzle is determining why the same set of activated oncogenic growth factor receptors and their downstream signaling components play roles in other diseases, such as degenerative and metabolic diseases. Moreover, given that more than 300 different signaling proteins and about 900 signaling relationships representing feedback/feedforward cross-talks potentially operate in cancers, how can we design drugs that target these cancers without affecting the normal cells in the body and thereby causing treatment-associated morbidities? If we have not adequately resolved these bulk tumor related issues, can we effectively define personalized cancer therapies?

Thus, many targeted therapies for cancers such as melanomas and medulloblastomas have provided only transient beneficial effects and were followed by aggressive, uncontrollable relapses that resulted in faster-than-anticipated mortality. At least for now, we may have to re-evaluate the study of how a single activated growth factor receptor may influence different downstream signaling cascades in different cell types, or even within one focus of a multifocal tumor. For example, primary tumors of the prostate, kidney, etc., are known to present as multifocal cancers, even within the same organ. For example, each of the foci may be characterized by a distinct genetic defect, and yet various cell types found in the same focus might respond differently to the same gene mutation depending on their cellular context within each focus.

Similarly, breast cancers also present multiple tumor subtypes (e.g., estrogen receptor positive to negative) within the same breast, albeit to different extents. Another important example is that in case of colon cancers, the normal epithelium, with an enriched signaling pathway, proceeds through early, intermediate and late adenoma stages before culminating into a colorectal carcinoma, switching through the activation of different signaling pathways mediated by epidermal growth factor receptor, transforming growth factor receptor β, a loss of p53 function, and many more. As a result, when a patient presents to a clinician with polyps representing these varied developmental stages, it is unclear which signaling pathways should be targeted for treatment, especially since we know that there are numerous feedback controls among these signaling pathways. Thus, these intellectually challenging and not easily ignored facts, which are associated with real-world clinical situations, ultimately take us back to the important question of how to identify the cell of origin that drives tumor initiation.

From Tumor Heterogeneity to Its Cell of Origin

How many and how much of each of the above mechanisms are influenced through dynamic epigenetic changes in the affected gene loci and are relevant to the overall manifestation of the disease phenotype? Which of these are causes or consequences of changes in the tumor microenvironment of any given single tumor or single focus within a multifocal tumor? How can we attempt to solve these critical and complex puzzles in patients with the present-day basic research information? And how will personalized targeted therapies resolve these issues unless they also address tumor heterogeneity, which dynamically manifests itself three-dimensionally in a tumor, whether from center to periphery, from one side to the other, or from top to bottom? Molecular heterogeneity must also be addressed, not only among different tumors of the same type (inter-tumor heterogeneity), but also among different regions within the same tumor (intra-tumor heterogeneity). Increasingly, there is recognition that only small populations of undifferentiated cells in bulk tumor tissue have the capacity to recreate the full tumor heterogeneity of the original tumor upon transplantation into appropriate animal models. Thus, examining the data derived from heterogeneous bulk tumor tissues is much like digging through a gold mine and expecting to find lumps of pure gold ready to be picked up. Just as raw materials from the mine must be processed to extract the refined gold, we need to purify tumor initiating CSCs from heterogeneous bulk tumor tissues.

The CSCs are largely undifferentiated cells with stem-like cell characteristics, namely self-renewal and multipotent differentiation, of adult stem cells in the same body organs (Figure 1, A–D). Adult stem cells are activated only when required, as with skin stem cell activation after skin damage to initiate the necessary tissue repair processes; upon completion, the skin stem cells return to quiescence. A widely held contention is that, analogous to adult stem cells of any given organ, the CSCs also maintain their number with self-renewal through asymmetrical cell divisions and increase their numbers as needed by proliferation through symmetrical cell divisions, thereby maintaining the stem cell number homeostasis (Figure 1A: Normal tissue). But unlike adult stem cells, the CSCs appear to exhibit a prolonged proliferation phase, undergoing more symmetrical cell divisions resulting in increased stem cell numbers and their aberrant differentiation (Figure 1A: Tumor tissue). Although it remains to be proven, the theory is plausible if, once a tumor reaches to a definite size, perhaps due to the levels of tumor-secreting cytokines and growth factors, a feedback process inhibiting the proliferating CSCs sets in; this would be consistent with observations that the treatments successful in inhibiting tumor

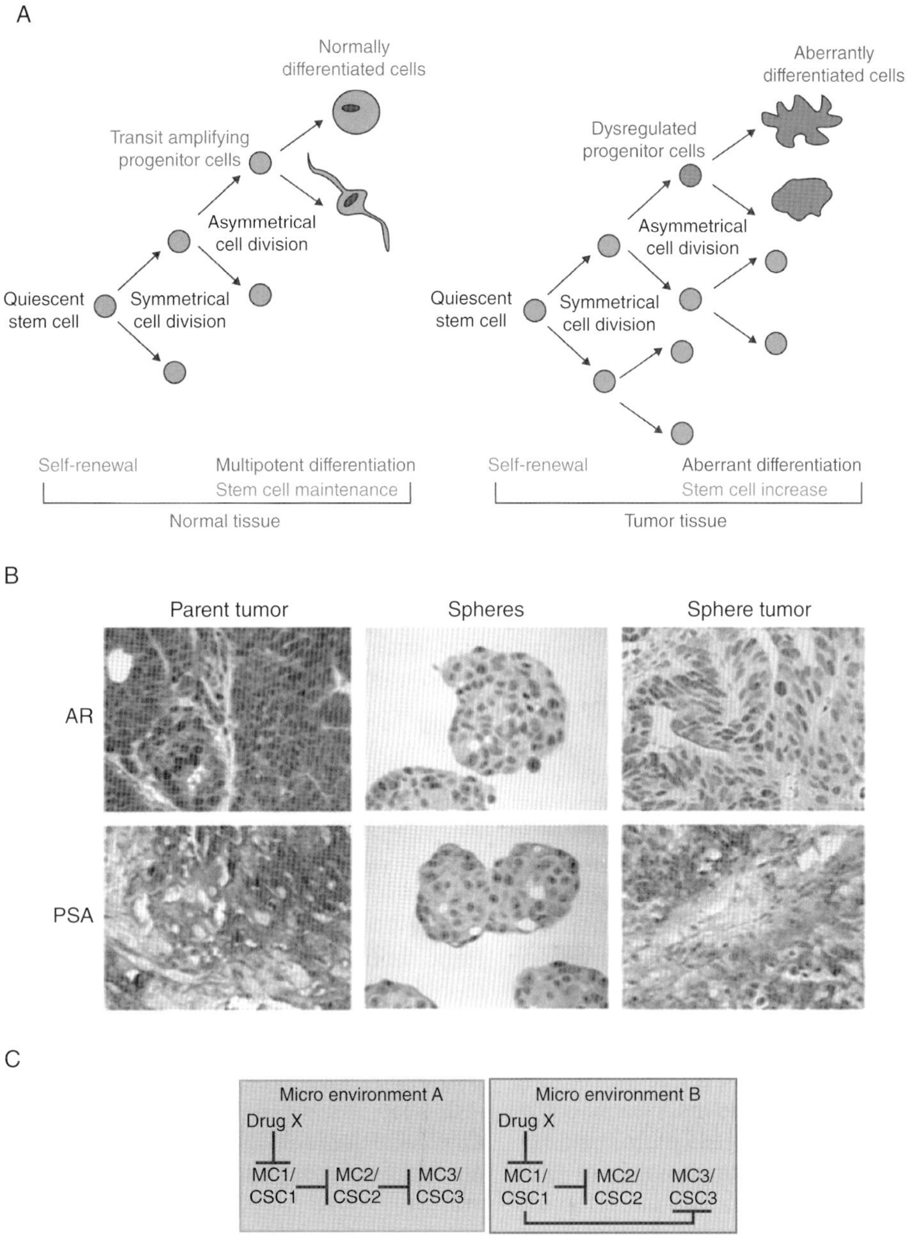

Figure 1. Cancer Stem Cells. **A**. Schematic representations of adult stem cell number maintenance and their differentiation in the normal tissues versus cancer stem cell number increase and their aberrant differentiation in the tumor tissues. **B**. Comparative immunohistochemistry of androgen receptor (AR) and prostate specific antigen (PSA) in the human primary prostate tumor (parent tumor), the tumor-derived primary spheres (spheres), and the sphere-derived tumor (sphere tumor) respectively [see http://www.nature.com/ncomms/journal/v2/n1/full/ncomms1159.html]. **C**. Hypothetical representation of hierarchical arrangement of multiple mutant clones (MC1/2/3) or progenies of different CSC pools (CSC1/2/3). For color detail, please see color plate section.

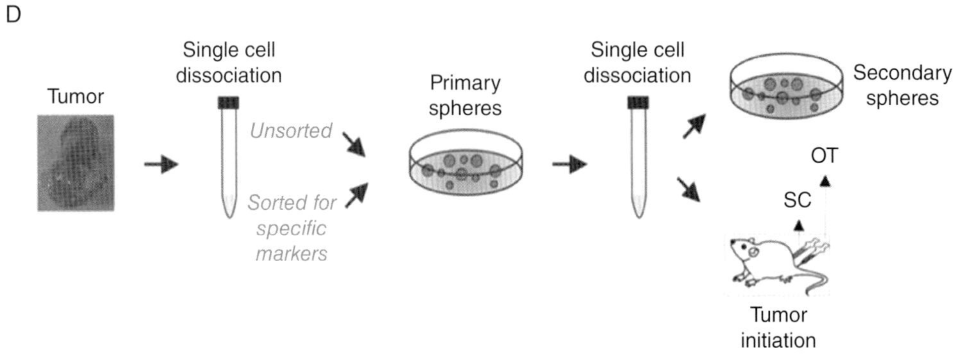

Figure 1. (*Continued*). **D**. Schematic representation of a rapid and novel mini-screen strategy for identifying the conventional therapy resistant CSCs and screening small molecule inhibitors in bulk tumor tissues [see http://www.genomeweb.com/proteomics/sloankettering-team-ids-non-psa-producing-cells-potentially-linked-prostate-can]. For color detail, please see color plate section.

burden often also increase the numbers of CSCs in many cancers.

Many other questions that are not yet fully answered have also arisen from the above experimental data: (1) What is the trigger for the origin of CSCs? (2)What is the cell of origin for CSCs. Is the CSC a mutated adult stem cell, progenitor cell, stem/progenitor cell with differentiation arrest, dedifferentiated adult cell, or a combination of these? (3) How do CSCs originate in the body by genetic or epigenetic mechanisms, or both? (4) How are CSCs persistently maintained as minor pools of cells, even in well differentiated tumors? (5) Are there any biological feedback controls that maintain a homeostasis among different pools of CSCs in a given bulk tumor? While it is known that genotoxic and chemotherapeutic stresses contribute to the serial development of hypoxia, aneuploidy, and cancer like genomic alterations in single-cell eukaryotes, it remains an open question whether similar cascades trigger the normal stem cells, progenitors, and the differentiated cell progeny to become tumor-initiating CSCs.

It is well known that untreated infection or long-term ulcerated wounds resulting from viral and bacterial infections can cause human cancers such as cervical and colon cancers, respectively. Infection, inflammation and the consequent reprogramming of target cells to become CSCs is one of the most plausible hypotheses, but has never been directly tested by researchers. It is also known that the terminally differentiated host cells can be reprogrammed into stem-like progenitor cells by bacterial pathogens. Because of this an altered paracrine signaling may result in the microenvironment at the infection site and this unexpected stress may destabilize the cellular homeostasis in that area resulting in possible cell-specific additional mutations and thereby the aneuploidy in adult stem cells and other competent cells in that area as they undergo reprogramming and dedifferentiation. Thus, depending on the differentiation state or cell type, and the competence of the cell, it is possible to have more than one subtype of CSCs. There is also evidence in the literature for the existence of more than one subtype of CSCs in some cancers such as brain tumors.

Importantly, it is known from animal modeling studies that when the same oncogenic mutation occurs in somatic cells as well as in stem cells, the tumor-like growths may initially result from both the cell types if the cellular proliferation was the only consequence of that mutation. But, tumor-like growth if originated from somatic cells may not be sustained for longer periods or the tumors may even regress spontaneously, as happens indeed in some cancer patients (sometimes viewed as 'cancer miracles'). At the same time, the tumor growths from mutated CSCs which could result in an incurable disease as evidenced with some engineered colon cancer animal model systems by other groups. These observations reiterate the importance of the tumor microenvironment and perhaps the niche specific factors in tumor-initiation and maintenance.

Typically, once the CSCs are prospectively purified by exploiting known cell-surface markers, they are first verified for their characteristic self-renewal ability by examining in vitro sphere formation, which is a surrogate assay for their in vivo self-renewal capacity (see cover page upper strip, which shows primary spheres expressing green or red fluorescent proteins). Next, these CSCs are subjected to a gold standard experimental validation of their characteristic multipotency via transplantation of small numbers of these purified cells into animal models by exploiting limiting cell dilution approach. Most animal models of human cancer are considered to demonstrate treatment benefit if the overall tumor size decreases following a drug treatment or by target gene knockout/transcript knockdown of the predicted oncogene(s). But, the same treatments rarely cure the cancers in human patients with the same oncogene activation, raising the obvious question: WHY? Most of the above approaches are also undermined if the in vivo effects of the drug were indeed a consequence of a specific effect on target cells or a combined effect on the mobilized host cells in the animal model.

Identification of CSCs is not a simple endeavor, not only because they represent a minor population in bulk tumor tissues, but also due to dynamic changes in their markers expression profiles during tumor development. Thus, handling cell type specificity within a dynamic context of cells isolated at given stage of patient tumor development becomes the most important challenge for the basic researchers and the physician scientists. Interestingly, minor pools of undifferentiated CSCs are always maintained by still-unknown mechanisms even after the apparently fullest differentiation of CSCs into the original parent-like aberrant tumor tissues during sequential derivation and xeno-transplantation of serially derived CSCs over many passages of sphere-to-tumor -to-sphere cycles in animal models. This data suggests that the CSCs are indeed tumor driving cells with tumor-initiation and maintenance functions. Thus, the CSC-sensitive tumor targeting may be an effective strategy in conferring the greatest therapeutic benefit to patients, especially when we can delineate clinically relevant novel signaling cascades that are specifically prominent in the CSCs.

Invoking PDX^{CSC} Cancer Models

Moreover, compared to cancer cell lines or bulk tumor cells that are normally transplanted in multiples of millions for inducing tumors in immunocompromised animals, CSCs are required only on the order of hundreds, or sometimes less. This confirms that bulk tumors contain only a minor population of CSCs and, therefore, that bulk tumor analysis may largely identify tumor-associated or differentiated cell targets rather than targets specific to the tumor-driving function of CSCs. Because the primary tumor tissue volumes obtained from surgical resections are highly heterogeneous, with tissue composition that is variable depending on the tumor stage, size, necrosis, etc., the ability to isolate sufficient numbers of CSCs may also be inconsistent and challenging. Although it is by no means an alternative to conventional banking of flash frozen tissues used to identify the physiological status of disease at the time of biopsy or surgery, the live biobanking approach offers a renewable resource of functional tumor tissue for real-time basic and preclinical studies. The well-characterized patient-derived xenograft (PDX) cancer model systems may therefore be far superior to any other contextual models and form a harmonized, reproducible, and renewable platform for addressing original patient tumor heterogeneity. However, it must be kept in mind that most human tumor xenografts, including the PDX tumors, consist of about 40–50% mobilized host animal cells. Because of differences in human versus mouse cytokinins as well as their signaling cross talks, it will be important to investigate the human CSC-derived PDX (PDX^{CSC}) cancer models. These models will be useful in developing individually tailored patient CSC-specific therapeutics for successfully treating patient specific cancers. Importantly, it is necessary to frequently validate PDX^{CSC} cancer models with comparisons to the parental PDX cancer models by testing for the expression of known functional genes, and by short tandem repeat analysis for confirming the maintenance of gross genomic integrity over many passages during the renewal of the tumor tissues. Nevertheless, the invoking PDX^{CSC} cancer models offer an excellent opportunity to obtain sufficient numbers of functional CSCs in a faithfully renewable fashion for biochemical and molecular analysis and for the discovery of novel and useful biomarkers.

For example, by exploiting an established subcutaneous xenograft model of primary human prostate carcinoma [PDX prostate cancer model], it was found that all the differentiated cells in the primary human prostate tumor xenografts have expressed androgen receptor (AR) and its downstream target, prostate specific antigen (PSA), which have been the most widely employed clinical markers for prostate cancer (Figure 1B). However, the sphere-forming prostate CSCs (spheres) isolated from the patient tumor or the primary PDX (patient tumor) tumor were largely undifferentiated, expressing low levels or none of these markers. Upon orthotopic transplantation of these CSCs into mouse prostate, the resulting $PDX^{\text{prostate CSC}}$ cancer model represents a well-differentiated parent-like tumor tissue re-expressing the markers AR and PSA, as can be seen in the sphere-derived tumor (sphere tumor). Through the limiting cell dilution experiments, the prospectively purified tumor-initiating CSCs were specifically found to be enriched with clinically relevant and also self-perpetuating NF-kB – IL-6 signaling. Please refer to the cover page, lower strip for an example of immunohistochemical staining of nuclear NF-kB shown in bright field followed by green immunofluorescence in the blue background of nuclear staining with DAPI. A subset of human prostate cancer patients that had undergone the radical prostatectomy and thus had negative margins for expression of AR and PSA, succumbed to death more quickly than others. Interestingly, the above negative margins of this moribund subset of patients have been found to contain an increased levels and activation of NF-kB.

Thus, we propose that the enriched NF-kB signaling is evolved as a possible anti-apoptotic mechanism in the CSCs, conferring resistance to antiandrogen or AR antagonist treatments. Increased NF-kB signaling is being found in the CSCs isolated from a variety of other organ tumors, and more recently, an involvement of activated NF-kB has also been suggested in the expression of characteristic markers of CSCs such as Sox9. Our and others' hypothesis that the NF-kB signaling, which occurs at heightened levels in CSCs and which plays an important role in tumor initiation, could persist in CSCs and regenerate tumors over many generations, may be verified through serial transplantation assays by employing different PDX^{CSC} cancer models.

With respect to preclinical and translational studies, as outlined in Figure 2, it will also be very informative to generate the PDX and PDX^{CSC} cancer models in a longitudinal fashion, beginning at a patient's primary diagnosis, through therapy resistance, metastasis, and all the way until a patient's death, followed by warm autopsy. Such strategies will significantly augment our understanding of the evolution of cancer development in the clinical setting of therapy resistance. In collaboration with and under the leadership of Dr. John Healey, Chief of the Orthopaedic Surgical Service at MSKCC, as well as with many intellectual consultations with Dr. Irving Weissman of Stanford University, we have been developing live biobanking of rare bone primary tumors and all other metastatic tumor specimens, as the majority of

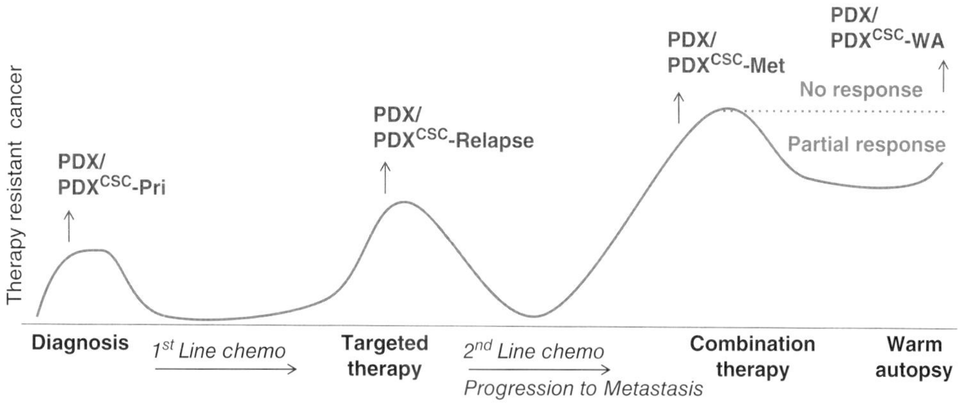

Figure 2. A schematic for strategic development of live biobanking of patient-specific and disease-stage specific tissue, including therapy-resistant PDX and PDX^{CSC} cancer models, starting from patient diagnosis and proceeding through warm autopsy. For color detail, please see color plate section.

mortality associated with the most common cancers result from tumor metastasis to distant organs. Out of an expected 700,000 new invasive cancers per year, about 300,000 cases were associated with bone cancers, which are expected to double by the year 2020 in the United States. Moreover, 7–15% of the total metastatic bone cancers were derived from unknown and uncharacterized rare primary cancers. Normally, the metastasized tumor cells and CSCs will be in limited numbers during their initial dormancy period, which could last for decades in a given patient before emerging as a detectable metastatic tumor. Thus, the patient-specific development of PDX and PDX^{CSC} cancer models is expected to facilitate the discovery of novel CSC markers that could aid in the early clinical detection of cancer metastasis in patients who have presented with primary tumors.

The overall tumor growth in the animal models incorporates a cooperative paracrine signaling between the host animal cells and the transplanted guest human tumor cells, including the CSCs. It is also unknown if the crosstalk between cells of mouse stroma and the human tumor cells in the PDX cancer models are comparable to that seen in real patient tumors with only human stroma in reference to the patient-specific CSCs. Moreover, since murine and human cytokines are not the same, any observed results related to signaling in the animal models must be carefully verified with the tumor-derived cells from an actual patient as soon as they are available. Thus, it may be important to employ animal models that are humanized to express a relevant set of human cytokines which are known to participate in the dysregulated proliferation and aberrant differentiation of CSCs. Nevertheless, the live biobanking of tumor tissues through the proposed PDX models unveils a paradigm shift in the development of patient-specific individual cancer treatment strategies, as well as in the field of next-generation tissue biobanking, which is Live Tumor Tissue Banking. Recently emerging next-generation targeted genome editing approaches involving Talens, CRISPR/Cas, etc., may also be extended to the above-mentioned PDX^{CSC} cancer models for validating the targets and perhaps for correcting the cancers, at least in the in vivo model systems. But the challenge still remains as to how to target the CSCs specifically in solid tumors, which brings us back to the important question of what are the CSC-specific cell surface markers that may be exploited in this important goal.

Finally, the live biobanking approach described above for generating viable tumor tissues also has the potential for a broader applicability well beyond the realm of its suitability to cancers, and it may eventually open up "cure models" of diseases in general.

Therapy Resistance—How to Understand the Problem?

Currently, the major challenge in treating cancers hinges upon whether the minimal residual disease (residual cancer) and the development of therapy resistant disease were the consequence of context-dependent emergence of therapy resistant clones by acquiring fresh mutations or were due to therapy-dependent activation of definite subsets of cancer stem cell pools that were otherwise dormant and now resulting in generation of parent-like differentiated heterogeneity. Thus, it is expected from the current treatment strategies that we could encounter a sort of heterogeneity even within an observed therapy resistance (resistance heterogeneity). Targeted therapies developed to date against activation of a given signaling pathway intermediate (eg., mutant clone 1; MC1) are of only transient benefit to patients, because they invariably develop resistance followed by an emergence of new mutant clones (mutant clones: MC2 and MC3). It is unknown how exactly such therapy-resistant clones emerge, bypassing the same pathway intermediates during tumorigenesis. Most importantly, it is unknown why the new clones (MC2 and MC3) are prevented from emerging before targeted therapy against MC1. It also remains unknown whether the new MCs emerge as a result of newly acquired mutations or if they proliferate afresh from the pre-existing mutant clone. In the former scenario, any other non-tumor stromal cells could also acquire such new mutations, whereas in the latter scenario, the newly emerged clones MC2 and/or MC3, etc., may have been inhibited by the functional MC1 as modeled in Figure 1C.

Thus, it will be interesting to discover if one mutant clone affects the proliferation of other mutant clones and thereby impact overall tumor development in a dynamic fashion. This is likely to contribute to a possible altered tumor heterogeneity and thereby a sort of resistance heterogeneity. For example, if the MC1 is exerting a hierarchical control on the expression of MC2 and then MC3, targeting the MC1 will only increase the levels of MC2, but not MC3 as depicted in the microenvironment A of Figure 1C. On the other hand, if the MC1 is exerting a parallel control on the expression of MC2 and MC3, targeting MC1 will increase both the levels of MC2 and MC3, as depicted in the microenvironment B of Figure 1C. Alternatively, it also may be possible that newly emerged targeted therapy-resistant clones (MC2 and MC3) are derivatives of different subsets of CSC pools (CSC2 and CSC3) that were otherwise dormant until the CSC1 progeny were functionally inactivated or eliminated during primary targeted therapy. Because the targeted therapy-resistant clones eventually re-create the tumor heterogeneity comparable to the original parent tumor, which is also the characteristic of stem cell multipotency, the therapy resistance may be of the stem cell origin. Thus, it becomes important to delineate the hierarchy of CSC subsets and their combined complex interactions associated with the functional development of tumor within the context of its microenvironment (Figure 1C; Compare Microenvironment A versus Microenvironment B). One can execute more intelligent experiments if one subset indeed exerts control over the other subsets of CSCs.

Finally, it will also be important to understand the effects of targeting individual mutant clones on the overall metabolism of the CSCs in their respective microenvironments. For example, it is unknown if targeting CSCs in their original tumor microenvironment will require different ammunition from what would be required after targeting the predominant MC1-3/CSC1-3 in the bulk tumors. The reason I highlight this point is to emphasize that we should not exclude any potential direct cross talks between CSCs and their differentiated progeny, irrespective of the tumor stroma.

Thus, many targeted therapies developed to date, based solely on bulk tumor analysis, may have to be re-examined if we hope to develop durable therapies against the cells of cancer origin in the actual clinical settings of the patients.

I have also developed a simple small-scale protocol to screen for small molecule inhibitors against CSCs. The assay uses the stem cell self-renewing characteristics such as sphere formation *in vitro* to identify and functionally test the self-renewing properties of single cells harvested from bulk tumor samples (Figure 1D). This assay can be used to screen various candidate molecule inhibitors in search of compounds that specifically disrupt the sphere-forming capacity. We have previously verified this principle in human CSCs isolated from primary spheres to secondary sphere formation assays and/or to direct functional tumor-initiating transplantation studies. Thus, the lack of sphere-formation and/or tumor-initiation in xenografts will be the preliminary test of efficacy for any compounds that specifically target CSCs. Moreover, the drugs that affect CSCs can be distinguished from those that act on bulk tumor without interfering with tumor relapse. The targets of such CSC specific drugs would likely have long-term clinical benefit in patients, and we have demonstrated their effect directly on tumor initiation in the human patient prostate CSC-derived xenograft ($PDX^{prostateCSC}$) cancer model system (http://www.nature.com/ncomms/journal/v2/n1/full/ncomms1159.html), along with our ongoing studies in human patient breast CSC-derived xenograft ($PDX^{breatstCSC}$) cancer model system.

Nevertheless, it may not be purely hypothetical to consider that comparable to bacterial quorum sensing, whereby the bacteria use concentration of signaling molecules released in their environment to sense the number of bacteria and function, the unaffected CSCs or MCs may orchestrate their function depending on their altered microenvironment in the tumors undergoing targeted therapies (Figure 1C). Importantly, clinical data exist suggesting that, in addition to the mutational status of a tumor cell, the age of the patient also plays a role in manifesting the cancer development. Thus, identifying the role for age-dependent changes in the tumor microenvironment and its consequences on the function of CSCs will be other important avenues to pursue further investigations for developing successful and durable therapies to cancer patients.

Integrating CSCs with the Mouse Hospital Concept and the Prospects

The above critical appraisal is aimed at targeting CSCs in vivo, which relies on precise identification of unique cell surface markers associated with the CSCs (http://www.genomeweb.com/proteomics/sloan-kettering-team-ids-non-psa-producing-cells-potentially-linked-prostate-can); this, in turn, facilitates their prospective purification, which eventually paves the way for discovery of novel biomarkers and/or also functional therapeutics. Furthermore, once the unique cell surface markers in the functional CSCs are identified, the patient's own T cells can also be reprogrammed to target the CSCs. This recently developed approach may be another worthy avenue to expedite, especially in the context of the dynamic nature of cell surface markers expression in CSCs. Thus, debulking of the bulk tumors (largely by surgery) followed by the therapeutic targeting of CSCs by one or more methods, like those suggested above, would be a productive approach to the prospective and successful treatment of cancer patients. Single cell genome sequencing, and analyses of the epigenome, kinome, proteome, and metabolome in CSCs and in PDX^{CSC} cancer models are expected to provide a hitherto of unrecognized landscape of tumor initiation mechanisms and lead to treatment strategies that can successfully and predictably eliminate patient tumors.

All the above discussion calls for the functionally focused specialization in the existing bulk tumor based-mouse hospital concept, which may be referred to as 'PDX^{CSC} specialized -mouse hospital'. Here, each type of PDX^{CSC} cancer models could be a representative of the given type of cancer patient and the overall infrastructure being live tumor tissue resource of human cancers and their CSCs for translational studies. In the bulk tumor tissue based mouse hospital approach, the same structure and procedures as in phase 1 and phase 2 human clinical trials are performed simultaneously in mouse models and humans. As the clinical trials in mice are expected to move faster, integrating such data as a predictive of human response needs many other careful considerations. For example, a group of independently developed PDX and PDX^{CSC} models of a given cancer type may apparently reflect the comparable inter-tumor heterogeneity as expected among the real life patients of given cancer types; but, the mouse tissue heterogeneity is also added up into the same tumor models. Thus, the signaling interplays between the mouse and the human cells, their influence on the overall PDX development, and their therapy response may have to be carefully considered before exploiting the mouse clinic conclusions for clinical trials in human patients. Otherwise, we may continue to cure cancers in mice without similar impact in human patients especially when we are convinced of patient specific individualized therapies for cancers that did not have contribution from animal signaling cross talks. A simple take-home lesson would be to employ functionally validated CSCs to generate harmonized PDX^{CSC} cancer models, engineered to express relevant human cytokines and make them available to all researchers globally as a common platform for verifying the consensus and for making new discoveries. Any preclinical treatment data derived from these humanized mouse models of PDX^{CSC} may guide expectations for treatment outcomes in patients and, thus, the models serve as extremely valuable tools for achieving successful clinical outcome. We have just began state of the art preliminary studies standardizing the PDX and PDX^{CSC} cancer models for non-invasively determining the growth, permeability, and metabolic status of the tumors growing in orthotopic sites and their metastatic spreads as means to gauge therapy responses. The therapy response characteristics of the PDX^{CSC} models of a particular cancer type to a particular patient CSC-specific therapy protocol in the mouse hospital would enhance the chances for developing precise, successful, and timely treatment options to the respective patients in the companion human clinics. Thus, the above proposed studies and ideas on the structurally integrated and yet parallel settings of human patient clinic and PDX^{CSC}-mouse hospital could bring our clinical decisions much closer to accomplishing a cure for all cancer patients.

Classification of the Chapters

The following chapters have been written by an excellent group of researchers and scholars who discussed many of the queries and topics mentioned above. I hope that many readers will be also inspired by the provocative hypotheses and novel ideas in the following chapters. For the convenience of readers, I have classified the 35 chapters in this edition into following 6 sections:

1. Essentials of Cancer Stem Cells and Conceptual Modeling (Chapters 1–10),
2. Stem Cells in Liquid Tumors (Chapters 11 and 12),
3. Stem Cells in Solid Tumors(Chapters 13–18),
4. Cancer Stem Cells in Tumor Metastasis Perspective(Chapters 19–22),
5. Novel and Potential Targets in Cancer Stem Cells(23–31), and
6. Clinical Relevance of Cancer Stem Cells in Patients (Chapters 32–35).

Thank you for your interest in the book and your appraisals are welcome.

Vinagolu K. Rajasekhar
Email: Dr.VKRajasekhar@gmail.com or Vinagolr@mskcc.org

Acknowledgments

Due to space constraints, I deeply apologize to all of the authors of tens of thousands of excellent cancer stem cell papers for my deliberately being unable to cite their works in the preface and the following chapters. I sincerely appreciate Mathew Jones, Yildirim Dogan, Rosalind Simmons, and Sarah Murphy for their efficient scientific text editing and Laura Daly and Julia Vinagolu-Baur for the help with artwork.

I am indeed indebted to:

Drs. John Healey and his staff; all members of the laboratories of Lorenz Studer, Eric Holland, Mark Ptashne, Howard Scher, Jackie Bromberg, John Petrini, Paul Marks, Andrew Koff, Gary Schwartz, Kitai Kim, Hans Guido Wendel, Daniel Heller, Vivaine Tabar, Alan Hall, Niel Rosen, Kenneth Marians, Gabriella Chiosis, Malcolm Moore, Robert Benezra, and Jose Baselga for generous gift of reagents; David Scheinberg, Victor Reuter, Raju Chaganti, Christopher Park, Prasad Adusumilli, Sarat Chandarlapaty, Johanna Joyce, David Solit, Omar Abdel Waheb, Philip Paty, Bhvanesh Singh, Pattrick Bolland, Ingo Millinghoff, Brett Carver, Jaspreet Sandhu, Nick Socci, Timothy Chan, Michael Berger, William Tap, Peter Besmer, David Klimstra, Larry Norton, Monica Morrow, Steve Larson, Lisa De Angelis, John Petrini, Robert Benezra, Jatin Shah, Harry Herr, Bernard Bochner, Jim Faggin, Steven Leach, David Abramson, Richard O'Reillly, Roger Wilson, Bayard Clarkson, Jerard Hurwitz, Scott Lowe, Scott Armstrong, Eric Cottington, Peter Scardino, George Bosl, Charles Sawyers, Joan Massague, Jose Baselga, Craig Thompson, and also many other tri-institute faculties and leaders for scientific advice/motivation;

Drs. Yildirim Dogan, Santosh Narayan, Sandhya Dharmarao Pasarakonda, Dharmarao Thapi, Pasquale Sansone, Nagavarakishore Pillarsetty, Daniel Thorek, Maria Skamagki,Diane Tabarini, Rebecca Sherwood, John Maciejowski, Veronica Rodriguez Bravo, Sonja Kriks, Tamara Major, Zehra Dincer, Kosuke Funato, Bikul Das, Annamalai Selvakumar, Armugam Jayakumar, Brahmam Reddy, Devarajan Eswaran, Swarnali Acharyya, Aarti Santanam, Lindy Barrett, Svetlana Pavlovic, HuiYong Zhao, Anuradha Gowtham, Lakshmi Ratnakumar Bugga, Surya Pardha Namuduri, Gayatri Rao, Sangadala Sreedhar, Lakshmi Narasaiah Gavini, Liat Iftachy, David Thaler, Martin Begemann, Zaki Qureshi, Julie Cerrato, Tahira Khan, Sarah Kwak, Kumaran Manoharan, Rama Shankar Varma, Kotha Subbaramaiah, A. Premkumar, Velvizhi Gokuladoss, Soroush Tahmasebi, Nora Ghandour, Anne-Marie Tasse, and many colleagues for help with many helpful discussions;

Nancy Middleton, Jesse Galle, Kerith Luchins, Diana Champhers, Ana Rojas, Donna Gibson, Sarah Jewell, Linda Wilkins, Diane Mejia, Desiree Ehleiter, Lynne Rudwick, Daniel Spagna, Shivangi Shah, Mesruh Turkekul, Ning Fan, Yvette Chin, Elgilda Musi, Sadia Rahman, Christie Park, Tarsha Barton, Diane Domingo, Gisela Venta-Perez, Meghan Choy, Frank Fusaro, Lang Ngo, Becky Liu, and all members of all the core facilities of small animal imaging, antitumor, molecular cytology, molecular cytogenetics, genomics, flow cytometry, animal facility, the IACUC, PDX task force committee, and all the non-scientific staff at the center for their generous help and care in many ways.

Section I

Essentials of Cancer Stem Cells and Conceptual Modeling

Chapter 1

Theoretical and Experimental Foundations of the "Cancer Stem Cell" Model

Pradeep S. Rajendran[1] and Piero Dalerba[2]
[1] *David Geffen School of Medicine, University of California—Los Angeles, Los Angeles, CA, USA*
[2] *Stanford Institute for Stem Cell Biology and Regenerative Medicine, Stanford University, Stanford, CA, USA*

Definition of Stem Cells

Many human tissues are in a state of constant flux, in which large numbers of mature differentiated cells are continuously eliminated and replaced. Despite the short-lived nature of many of their cells, tissues maintain their size and structure over time through a tightly regulated process of proliferation and differentiation. In each tissue, this process is orchestrated by a remarkable population of long-lived cells, called "stem cells" (Weissman, 2000).

Historically, the word "stem cell" originated to identify an ancestor/progenitor cell that stands at the "stem" of a genealogic tree, used to depict either evolutionary (i.e., phylogenetic) or developmental (i.e., ontogenetic) processes (Ramalho-Santos and Willenbring, 2007). Its meaning remains essentially intact today, where it is mainly used in developmental biology, to identify a cell capable to generate and sustain over time a specific set of diversified cell populations whose aggregate interaction leads to the formation of either an entire living organism (e.g., embryonic stem cells with pluripotent capacity) or a specific subset of its organs and tissues (e.g., adult stem cells with oligopotent or multipotent capacity) (Weissman, 2000).

Traditionally, stem cells are defined by three main properties:

1. Differentiation: the ability to give rise to a heterogeneous population of daughter cells, which then progressively diversify and specialize according to a hierarchical process, thereby replenishing tissues of fully mature, differentiated elements.
2. Self-renewal: the ability to form identical copies of themselves, other stem cells that retain intact potential for long-term proliferation, expansion, and differentiation, thereby maintaining a constant stem cell pool.
3. Homeostatic control: the ability to balance self-renewal and differentiation, and modulate the frequency of these two developmental outcomes based on environmental needs, and within the limits of a specific set of genetic constraints.

The "Cancer Stem Cell" (CSC) Model

Similar to normal tissues, tumor tissues are composed of a multiplicity of cell types, including various subpopulations of cancer cells. Traditionally, the cellular heterogeneity

Cancer Stem Cells, First Edition. Edited by Vinagolu K. Rajasekhar.

observed within an individual tumor is explained using a random or "stochastic" model. In this model, phenotypic differences observed between different subsets of cancer cells are explained as the result of the progressive and divergent accumulation of independent somatic mutations, which in turn lead to the formation of distinct genetic subclones. An alternative explanation is provided by the "cancer stem cell" (CSC) model. This model envisions tumors as "pathological organs," sustained in their aberrant growth by a mutated population of stem cells, in which normal homeostatic controls on tissue expansion have been lost. According to this model, phenotypic differences among cancer cells can also originate as the result of epigenetic diversity, secondary to multi-lineage differentiation processes akin to those sustained by normal stem cells (Dalerba et al., 2007a; Reya et al., 2001). The two models are not mutually exclusive, as they point to two independent sources of variability, both of which can act simultaneously and contribute to the observed diversity (Gerlinger et al., 2012; Lagasse, 2008).

Theoretical Considerations in Support of the CSC Model

The CSC model is supported by several considerations of theoretical nature. The concept that tumors can be viewed as "caricatures" of normal tissues, aberrant versions of normal developmental processes, is very old, and dates back to the beginnings of anatomic pathology (von Hansemann, 1897). A rapid look at a tumor's histology on a simple tissue-section usually reveals a complex three-dimensional structure, whose architecture and cellular composition usually mirror, though often aberrantly, those of the normal tissue from which the tumor originates. For instance, in colorectal carcinomas, cancer cells frequently organize themselves in crypt-like gland structures, and contain cellular subpopulations, such as mucus-producing goblet cells, that are reminiscent of normal colon tissues (Figure 1.1). In their daily practice, pathologists routinely rely on tissue-specific morphological features, such as tissue-architecture, cell composition, and expression of tissue-specific differentiation markers, to guide their diagnostic approach to tumor lesions, and for instance, to identify the anatomical origin of a metastatic tumor when the primary site is unknown (Pavlidis and Pentheroudakis, 2012). This approach presupposes that neoplastic tissues, even distant metastases, do usually retain, at least in part, the developmental patterns and differentiation programs of their tissue of origin. The degree by which a tumor retains the capacity to undergo differentiation, a feature usually referred to as "grading," is used as an important prognostic variable in many types of human cancer (Carriaga and Henson, 1995).

From a more mechanistic perspective, ample experimental evidence indicates that, in order to undergo malignant transformation, cells must progressively acquire a set of multiple and independent genetic mutations, frequently over a period of many years (Fearon and Vogelstein, 1990). Given their self-renewal capacity and long life span, stem cells are ideal targets for malignant transformation, as they have the opportunity to accumulate several mutations over long periods of time. On the contrary, their differentiated progeny is frequently short-lived, and thus less likely to survive long enough to complete the transformation process. Finally, one of the critical steps in tumorigenic transformation is the acquisition of the capacity to escape the limits on proliferation imposed by cellular senescence, a property often referred to as "immortality" (Hanahan and Weinberg, 2011). Given their capacity to self-renew, stem cells are naturally endowed with the potential for long-term growth and proliferation. Therefore, as compared to other cell types, stem cells appear better poised to undergo malignant transformation, as they do not need to artificially acquire immortal traits de novo. Taken together, these considerations suggest that, at least in the early stages of neoplastic transformation, a first set of oncogenic mutations might accumulate in the stem cell compartment of normal tissues, and that this population of mutated stem

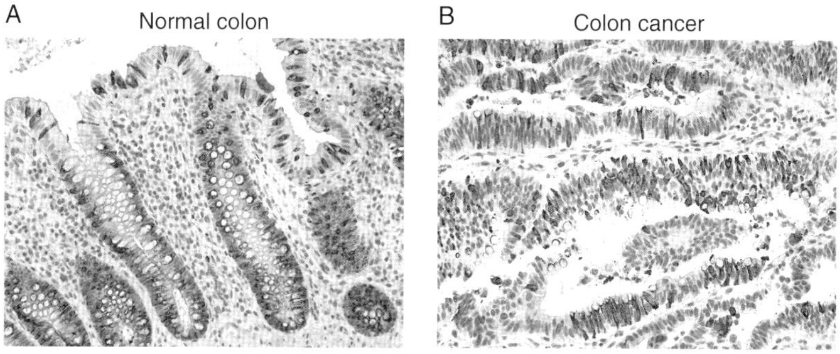

Figure 1.1. Tumors act as "*developmental caricatures*" of normal tissues. **A**. Normal colon epithelium analyzed by immunohistochemistry for expression of the MUC2 protein, a marker of mucus-producing, differentiated goblets cells (brown staining). **B**. Malignant colon cancer tissue analyzed by immunohistochemistry for expression of the MUC2 protein (brown staining). When compared to normal colon epithelia, human colon cancer tissues often reveal close similarities in terms of three-dimensional architecture (e.g., formation of crypt-like gland structures) and cell composition (e.g., presence of mucus-producing goblet-like cells). For color detail, please see color plate section.

cells might sustain the growth of a pathological tissue, which frequently retains the capacity for multi-lineage differentiation. This scenario, however, does not exclude that, over time, the progressive accumulation of additional mutations might ultimately disable the molecular constraints that prevent self-renewal outside of the stem cell compartment, and eventually "unleash" a fully malignant phenotype also in the more differentiated progeny of the neoplastic clone (Weissman, 2005).

Development of the CSC Model in Human Tumors

The experimental foundations of the CSC model, and especially of its more modern formulations, have been laid by studies that investigated the relationship between hematopoietic stem cells (HSC) and human myeloid types of leukemia. In the late 1970s, Philip Fialkow and colleagues (1977) initiated a set of seminal investigations on the monoclonal nature of hematological malignancies, using glucose-6-phosphate dehydrogenase (G6PD) as a marker. The gene encoding for G6PD is located on the X-chromosome. Secondary to random X-inactivation processes in female patients, one of the two X-chromosomes becomes inactivated in somatic cells. As a result, female patients who are heterozygous at the G6PD locus contain two different populations of somatic cells, each expressing either one allele or the other, and each found approximately at 50% frequency. The authors investigated the relative frequency of the two G6PD alleles in heterozygous female patients affected by different forms of hematological malignancies, including acute myeloid leukemia (AML), chronic myeloid leukemia (CML), essential thrombocythemia and polycythemia vera (Adamson et al., 1976; Fialkow et al., 1981a; Fialkow et al., 1977; Fialkow et al., 1981b). In all of these cases, while cells from healthy skin tissues were found to express either one or the other of the two G6PD alleles, and at similar frequency (50:50), circulating tumor cells were found to homogeneously express always the same allele, strongly pointing to a monoclonal origin of the disease. Most remarkably, the authors also showed that, in all of these disorders, the monoclonal expansions encompassed multiple independent lineages of mature circulating blood cells (e.g., granulocytes, erythrocytes, platelets) thus suggesting that the disease might have originated from a transformed hematopoietic stem cell (HSC) that preserved the capacity for in vivo multilineage differentiation.

In the 1980s, Irving Weissman and colleagues led a systematic effort to identify surface markers that could allow for the live isolation, and thus the prospective functional validation, of HSCs. Through a momentous sequence of studies, they progressively dissected the cellular hierarchy and the lineage relationships that underpin normal hematopoietic development in mammals (Weissman and Shizuru, 2008). Building on this experimental template, in the 1990s, John Dick and colleagues sought to understand whether human blood malignancies preserved not only the cell composition, but also the functional hierarchy observed in normal hematopoiesis (Bonnet and Dick, 1997; Lapidot et al., 1994). Working on human acute myeloid leukemia (AML), they used flow cytometry to sort different phenotypic subsets of the leukemic clone from individual human patients, and compared their functional properties side-by-side, in a prospective way, by transplantation into NOD/SCID immunodeficient mice (Bonnet and Dick, 1997; Lapidot et al., 1994). Remarkably, they observed that not all leukemic cells were equally tumorigenic (i.e., capable to initiate tumor growth upon transplantation), and that tumorigenic cells appeared enriched in the $CD34^{+}/CD38^{neg}$ subset of the neoplastic clone, whose phenotype corresponds to that of the immature stem/progenitor compartments of the normal human hematopoietic lineage. Importantly, leukemic expansions that originated in mice from $CD34^{+}/CD38^{neg}$ cells contained multiple phenotypic subsets and reconstituted the full phenotypic diversity that was characteristic of the original AML in corresponding patients. These studies provided the first evidence for the existence, within human AML, of a hierarchical organization reminiscent of a stem cell system.

In the last decade, the CSC model has been progressively applied to the study of many other tumor types, especially solid epithelial carcinomas. Progress has been slow, due to limited knowledge on the developmental hierarchy of many normal human tissues, and the frequent lack of known cell-surface markers for differential isolation of cellular subsets. In 2003, Michael Clarke and colleagues developed a flow cytometry method for the differential purification of distinct phenotypic subsets within human breast cancer epithelia and showed that $CD44^{+}/CD24^{neg/low}$ cells were substantially enriched for tumorigenic capacity in NOD/SCID mice (Al-Hajj et al., 2003). Importantly, tumors originated from $CD44^{+}/CD24^{neg/low}$ cells contained mixed populations of cancer cells, fully re-creating the phenotypic heterogeneity of the parent tumors. This study provided the first experimental evidence of a functional hierarchy reminiscent of a stem cell system in a human solid tumor (Clarke, 2005). Similar findings have since been obtained in many other human solid tumors, including glioblastoma multiforme (Galli et al., 2004; Singh et al., 2004), colon cancer (Chu et al., 2009; Dalerba et al., 2007b; O'Brien et al., 2007; Ricci-Vitiani et al., 2007), head and neck cancer (Prince et al., 2007), pancreatic cancer (Li et al., 2007), bladder cancer (Chan et al., 2009), prostate cancer (Collins et al., 2005; Patrawala et al., 2006; Rajasekhar et al., 2011), and cholangiocarcinoma (Wang et al., 2011).

Taken together, these studies contributed to define a common methodological framework for the experimental validation of the CSC model. As a general rule, a tumor is usually described as following a CSC model based on the fulfillment of three criteria:

1. Within tumor tissues, only a subset of cancer cells is endowed with tumorigenic capacity when transplanted into immunodeficient mice. These cells are characterized by a distinctive repertoire of surface markers and can be reproducibly purified from other cells that do not have this capacity.
2. Tumors grown from tumorigenic cells contain mixed populations of both tumorigenic and non-tumorigenic cancer cells, thus re-creating the phenotypic heterogeneity of the parent tumor.

3. Tumors grown from tumorigenic cells can be serially passaged multiple times.

According to this definition, the phenotypic subsets of cancer cells endowed with tumorigenic capacity are defined as "cancer stem cells" (CSC), as they display two of the hallmark properties of stem cells: self-renewal (i.e., the capacity to form other CSC with intact long-term proliferation and expansion potential), and differentiation (i.e., the capacity to form a variety of other cell types, thus re-creating the cellular diversity of the parent tissue) (Dalerba et al., 2007a).

Origin, Identity, and Evolution of CSCs During Disease Progression

It is important to note that, in its current use, the term "cancer stem cell" (CSC) entails a purely operational definition (i.e., it is based not on the display of a descriptive trait but on the experimental fulfillment of a specific set of functional properties). As a result, the term CSC indicates a tumorigenic cell endowed with self-renewal and differentiation capacity, but the term does not necessarily indicate a cell whose molecular or phenotypic identity can always be matched with that of a normal stem cell. From a theoretical perspective, stem cells represent ideal targets for malignant transformation, because they are naturally endowed with self-renewal properties, and because their long life span allows for the progressive accumulation of multiple rounds of oncogenic mutations. However, it is also possible that, over time, as more and more mutations accumulate in the neoplastic clone, some of the differentiated elements of the mutated stem cell's progeny might aberrantly acquire the capacity of self-renew (Weissman, 2005). As a result, during disease progression, the neoplastic clone could acquire additional CSC populations, some of whom are characterized by a more differentiated phenotype, and sometimes by higher proliferation rates.

Once again, these concepts are best illustrated by investigations initially performed in hematological malignancies. In human CML, for instance, in the initial "chronic phase" of the disease, the malignant clone appears sustained by a mutated population of HSCs, which usually maintain the capacity for multilineage differentiation (Takahashi et al., 1998). However, as CML progresses toward its terminal "blast crisis," the leukemic clone acquires a second population with CSC properties, whose surface marker phenotype mirrors that of a more mature granulocyte-macrophage progenitor cell (GMP) (Jamieson et al., 2004). Conceptually similar findings hold true also for AML, where cells with CSC properties can be found in populations whose surface marker phenotype is not always strictly characteristic of hematopoietic stem-cells (HSC) but can encompass also multipotent progenitors (MPP) (Blair et al., 1997; Eppert et al., 2011; Miyamoto et al., 2000), and sometimes more differentiated oligo-lineage precursors, such as GMPs (Goardon et al., 2011).

Taken together, these studies suggest that the hierarchical organization of tumor tissues is likely to "evolve" during the natural history of each disease, and that the phenotypic and molecular identity of its CSC populations might change over time and across different patients (Jan et al., 2012). The molecular basis of these events is the subject of intense investigations. Experiments in mice have shown that aberrant acquisition of self-renewal capacity can result from the inactivation of classic tumor-suppressor genes, as observed in the case of mice bearing simultaneous inactivation of the Tp53, $p16^{INK4a}$ and $p19^{ARF}$ genes, a genetic lesion able to confer aberrant self-renewal capacity to normal blood MPPs, which are usually devoid of such property (Akala et al., 2008). In human CML, aberrant acquisition of self-renewal capacity by neoplastic GMP-like cells during progression to "blast crisis" is associated to abnormal mis-splicing of the GSK3β gene (Abrahamsson et al., 2009).

Validation of the CSC Model in Animal Tumors

Due to obvious ethical reasons, transplantation experiments using human cancer cells can be performed only in xenogeneic hosts (e.g., in immunodeficient animals). This experimental limitation raises the question of whether the lack of tumorigenicity observed for specific subsets of human cancer cells (i.e., their inability to form tumors when transplanted in immunodeficient mice) is caused by an incompatibility with the mouse tissue microenvironment rather than an intrinsic deficiency in long-term proliferation capacity (Quintana et al., 2008). Examples of cross-species barriers that could limit the in vivo engraftment of cancer cells in these experimental settings include the rejection of transplanted cells by residual elements of the host's innate immune system (e.g., macrophages, NK cells), or the inability of cancer cells to cross-talk with stromal populations of the host (e.g., fibroblasts, endothelial cells) due to a reduced capacity of specific growth factors to activate their receptor counterparts across different species (Quintana et al., 2008). To circumvent these problems, several authors set out to investigate whether the CSC model can be applied also to the study of animal tumors, where transplantation experiments can be performed in syngeneic systems (i.e., between genetically identical individuals), thus by-passing all cross-species barriers to engraftment. Studies on mouse epithelial tumors, such as the MMTV-Wnt1 transgenic mouse model of breast cancer and the DMBA/TPA-induced mouse model of skin squamous cell carcinoma, have shown that several forms of mouse cancer contain multiple phenotypic subsets of cancer cells, which differ substantially in their tumorigenic capacity (Cho et al., 2008; Malanchi et al., 2008). Importantly, in these animal models, the tumorigenic populations fulfilled the definition for CSC, as they were able to sustain the formation of tumor tissues that recapitulated the phenotypic diversity and hierarchical organization of the parent tissues, and included the presence of non-tumorigenic cancer cells. Similar to what observed in many human tumors, CSC from mouse tumors are frequently defined by a surface marker phenotype characteristic of more immature stem/progenitor cell compartments of the corresponding normal tissues (i.e., $CD24^{+}/Thy1^{+}$ in the mouse breast epithelium, $CD34^{+}/KRT10^{neg}$ in the mouse skin epidermis).

Experimental Proof of Multi-lineage Differentiation in Solid Tumors

In hematological malignancies, such as CML, leukemic populations can encompass multiple lineages of differentiated circulating blood elements (e.g., granulocytes, monocytes, B-cells), likely originated from multi-lineage differentiation of a mutated hematopoietic stem/progenitor cell. In solid tumors, however, the possibility that the phenotypic diversity observed among cancer cells could originate as the result of a multi-lineage differentiation process, similar to that normally observed in corresponding healthy tissues, remained historically controversial (Shackleton et al., 2009). Experimental efforts to address this question have traditionally been focused on human colon cancer, as malignant colon epithelia are frequently composed of multiple cell populations that mirror those of the normal tissue (Figure 1.1). Several pioneering studies provided evidence in support of multi-lineage differentiation (Kirkland, 1988; Odoux et al., 2008; Vermeulen et al., 2008), but could not take advantage of a comprehensive molecular characterization of the various malignant populations. Recently, analysis by single-cell PCR of the cell composition of human colon cancer tissues revealed that, indeed, malignant colon epithelia contain a diversity of multiple cell subsets, whose transcriptional identity closely mirrors that of the various lineages and differentiation stages of the normal colon (e.g., goblet cells, enterocytes, $Lgr5^{+}$ columnar basal cells) (Dalerba et al., 2011). Formal demonstration of multi-lineage differentiation, however, requires a prospective experiment, in which the fate of a single (n = 1) cell is followed over time, together with careful molecular tracing of the monoclonal nature of its progeny (Figure 1.2). This experiment was performed by injection a single colon CSC ($EpCAM^{high}/CD44^{+}$), purified by flow cytometry from a xenograft line, and injected into a $NOD/SCID/IL2R\gamma^{-/-}$ mouse after infection with a lentivirus encoding for the enhanced green fluorescent protein (EGFP) (Dalerba et al., 2011). The resulting tumor's monoclonality was confirmed by the presence of a unique lentivirus DNA integration site in cancer cells, all of which expressed EGFP. Importantly, the monoclonal tumor generated in this experiment re-created the full phenotypic heterogeneity of the malignant parent tissue, in terms of both tissue histology and repertoire of cell populations. Most strikingly, a single-cell PCR gene-expression analysis of the monoclonal tumor demonstrated its heterogeneous lineage composition, again closely mirroring the cellular diversity of normal colon tissues.

Although transplantation experiments provided formal proof of the multi-lineage differentiation capacity of CSCs in human colon cancer, the artificial nature of transplantation procedures left open the question of whether this actually occurs in primary tumors, in the natural environment where cancer cells are normally found. To better clarify this point, a recent set of studies addressed the question of multi-lineage differentiation in mouse tumors, taking advantage of transgenic mouse strains genetically engineered to allow for in vivo lineage-tracing experiments. A classic example of such strains is represented by mice engineered to carry simultaneously: a) a copy of the gene encoding for the chimeric Cre/ER recombinase, placed under the transcriptional control of a lineage-specific promoter, such as the KRT14 promoter or the Lgr5 promoter, and b) a gene encoding for a fluorescent reporter protein (e.g., the yellow fluorescence protein or YFP), placed under the transcriptional control of a constitutive promoter (e.g., the Rosa26 promoter), but rendered inactive by the introduction of a poli-adenylation cassette flanked by LoxP sites between the promoter and the gene. In these mice, a pulsed injection of tamoxifen is able to cause transient activation of the Cre/ER recombinase but only in the specific subset of cells where the lineage-specific promoter is active (i.e., $KRT14^{+}$ or $Lgr5^{+}$ cells). Transient activation of the Cre/ER recombinase leads to excision of the poly-adenylation cassette and irreversible activation of the fluorescent reporter gene, resulting in the permanent in vivo labeling of individual $KRT14^{+}$ or $Lgr5^{+}$ cells and, most importantly, of all of their subsequent in vivo progeny. Experiments performed using this system were performed both in the DMBA/TPA-induced mouse model of skin papilloma, using a mouse strain engineered with a KRT14-Cre/ER construct active in the basal layer of the skin's epidermis (Driessens et al., 2012), and in the $APC^{fl/fl}$ mouse model of intestinal adenoma, using a strain engineered with the Lgr5-Cre/ER construct active in a subset of progenitor cells at the bottom of colonic crypts (Schepers et al., 2012). These studies allowed genetic labeling and prospective tracing of individual tumor cells in vivo, in a context of unperturbed tumor growth. In both models, this strategy revealed that a minority subset of neoplastic cells had both the capacity to persist long-term in tumor tissues (i.e., to self-renew) and to generate a diverse cellular progeny that encompassed all other mature cell types normally found in the original tissue (i.e., to undergo multi-lineage differentiation).

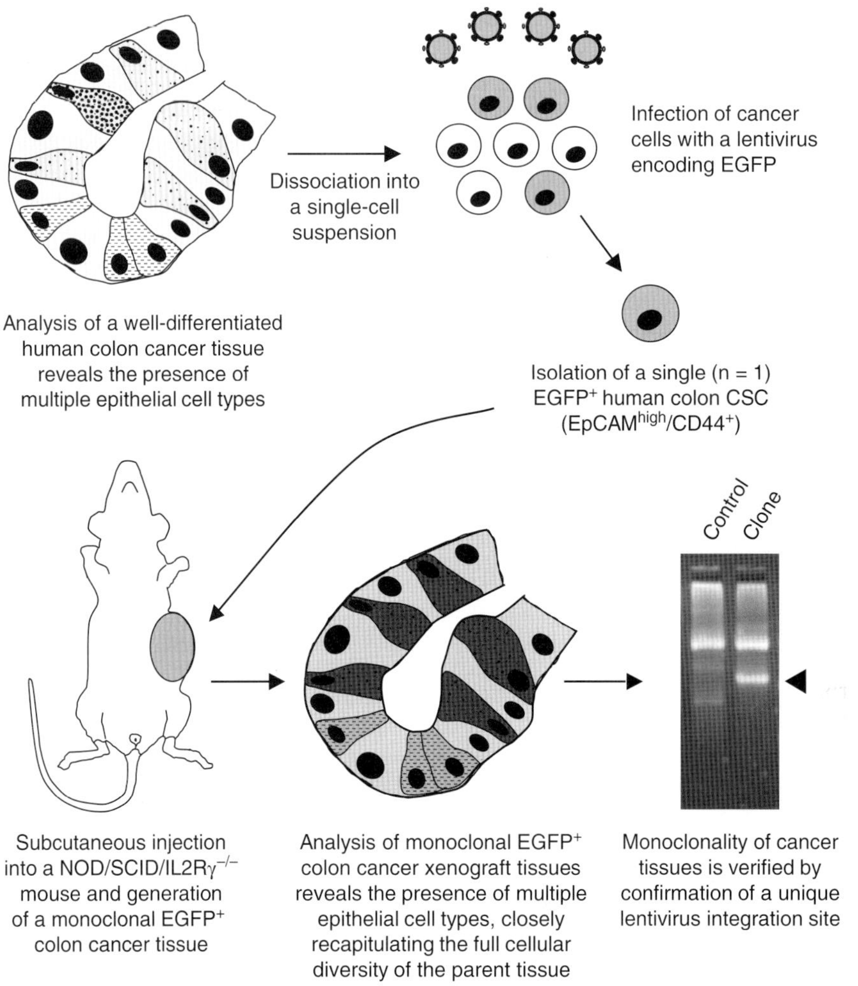

Figure 1.2. Prospective evaluation of multi-lineage differentiation capacity using transplantation experiments. Tumor tissues are frequently composed of a heterogeneous population of cancer cells, containing multiple cellular subtypes. According to the CSC model, the cellular diversity of tumor tissues can originate as the result of multi-lineage differentiation processes, akin to those sustained by normal stem cells. Formal demonstration of multi-lineage differentiation requires a prospective experiment, in which the fate of a single ($n = 1$) cancer cell is followed over time, together with careful molecular tracing of the monoclonal nature of its progeny. A possible way to perform this experiment is to inject into an immunodeficient animal host a single ($n = 1$) CSC purified from a tumor tissue known to contain a heterogeneous population of cancer cells (e.g., a single $EpCAM^{high}/CD44^{+}$ human cancer cell from a well-differentiated human colon cancer xenograft line) after infection with a lentivirus encoding for the enhanced green fluorescence protein (EGFP). The monoclonal nature of resulting tumors can be demonstrated by showing the presence of a unique lentivirus DNA integration site in the genome of cancer cells, all of which must also express EGFP (*green cells*). The observation that monoclonal tumor tissues generated by transplantation of a single cancer cell are able to recapitulate the phenotypic diversity and lineage composition of parent tumors (*various shades of green cells*) provides formal evidence of the multi-lineage differentiation capacity of the transplanted "*cancer stem cell.*" For color detail, please see color plate section.

Clinical Implications of the CSC Model

Beyond its intellectual appeal for the theoretical modeling of cancer biology, the CSC theory has profound clinical implications in medical oncology, especially for the discovery of novel prognostic biomarkers and for the design and validation of novel anti-tumor treatments. For instance, the CSC model suggests that variations in the cellular composition of tumor tissues, including both variations in the relative content of distinct subsets of mature cell types (e.g., produced by perturbations in multi-lineage differentiation processes), and variations in the frequency and phenotype of CSC populations (e.g., produced by aberrant acquisition of self-renewal capacity by mature cell types), might have substantial impact in disease biology. Furthermore, these variations might underpin changes in growth rate and differential sensitivity to individual drugs.

Given the ominous capacity of CSCs to self-renew and form tumors in mice, the CSC model suggests that tumors characterized by a higher content of CSCs could be characterized by a more aggressive natural history, and worse clinical outcomes. This concept was explored in a study that investigated whether the gene-expression profile of purified human breast CSCs ($CD44^{+}/CD24^{neg/low}$) could be used to generate a gene signature for prognostic applications (Liu et al., 2007). Indeed, the study showed that breast cancer patients whose whole tumor gene-expression profile displayed higher similarity to the CSC-derived signature were characterized by higher incidence of tumor recurrence and lower overall survival rates. Very similar results have recently been obtained also in the case of AML (Eppert et al., 2011; Gentles et al., 2010) and colon cancer (Merlos-Suarez et al., 2011). Microarray-derived gene-expression signatures are often difficult to implement for routine clinical use, but they can be frequently reduced to more simplified marker sets, focusing on a few robust genes (Gentles and Alizadeh, 2011). In the case of CSC signatures, a detailed understanding of multi-lineage differentiation processes can lead to the identification of reference markers for individual cell types, which in turn can be exploited as a measure of tissue cell composition. For example, a careful dissection of the differential gene-expression profile of the various epithelial cell lineages of the colon epithelium led to the development a two-gene classifier system (i.e., KRT20 versus CA1, SLC26A3, MS4A12, or CD177) that allowed the stratification of large cohorts of human colorectal carcinomas into discrete biological subgroups, characterized by very different clinical outcomes (Dalerba et al., 2011). Once again, tumors characterized by a more immature, undifferentiated phenotype (e.g., $KRT20^{neg}/CA1^{neg}$) were associated with worse survival outcomes. Remarkably, the prognostic power of this two-gene classifier system appeared superior to that of pathological grade, which is currently one of the few parameters used in the design of treatment algorithms for colon cancer. A similar strategy was successfully applied also to bladder cancer, using a limited set of gene-expression markers known to be differentially expressed during the various differentiation stages of the normal bladder epithelium (e.g., KRT14, KRT5, KRT20) (Volkmer et al., 2012).

Another important implication of the CSC model is related to the design and evaluation of anti-tumor treatments. According to the CSC model, eradication of tumor tissues is strictly dependent upon elimination of their CSC component, as CSCs are endowed with self-renewal capacity and hold the potential to sustain tumor recurrence (Figure 1.3). Historically, many conventional cancer treatments, such as many forms of chemotherapy and radiation, have been selected based on their capacity to induce tumor response (i.e., to rapidly reduce tumor size). As a result, these treatments are usually capable of killing a majority of cancer cells, and frequently target rapidly proliferating cells, but are not specifically designed to target CSC populations. Importantly, several studies suggest that CSCs might be preferentially

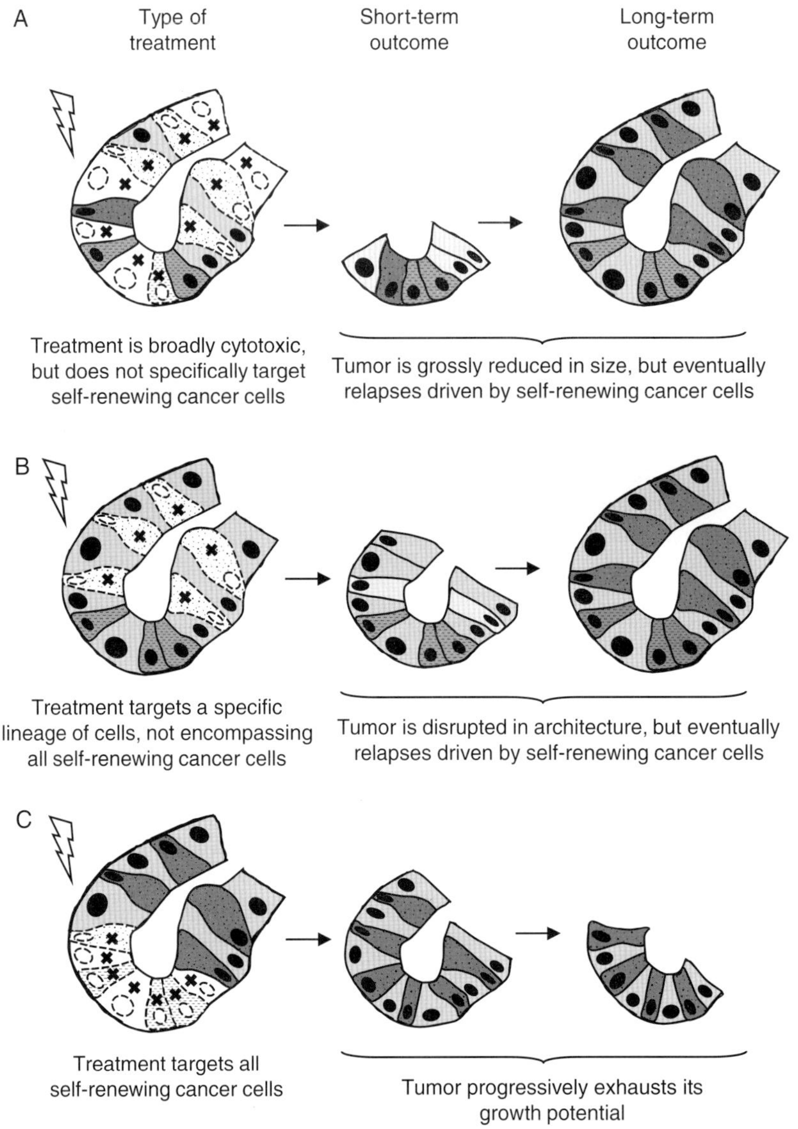

Figure 1.3. Implications of the CSC model for the design and evaluation of anti-tumor treatments. The heterogeneous cell composition of cancer tissues implies the possibility of a differential sensitivity of different cell types to various treatments. **A**. Anti-tumor treatments endowed with broad cytotoxic activity (e.g., treatments targeting rapidly proliferating cancer cells) might be able to hit multiple cell types (*yellow cells with red crosses*) and induce substantial reductions in tumor size. However, if cancer cells endowed with self-renewal capacity (*dark grey cells*) are spared, they hold the potential to cause tumor relapse. **B**. Anti-tumor treatments targeted to a specific lineage of cancer cells (*black cells*) might induce tissue disruption and temporary reduction of tumor volume. However, if the targeted lineage does not encompass all cells with self-renewal capacity, tumor tissues are likely to relapse, sustained by self-renewing cancer cells. At relapse, cells of the targeted lineage might be re-created as a result of multi-lineage differentiation processes. **C**. Anti-tumor treatments able to target and eliminate all cancer cells with self-renewal capacity, although not necessarily designed to induce rapid reductions in tumor volume, could cause long-term exhaustion of tissue growth, and hold the potential for permanent tumor eradication. For color detail, please see color plate section.

resistant to several standard anti-tumor treatments, including both chemotherapy (Dylla et al., 2008; Guzman et al., 2001) and radiotherapy (Bao et al., 2006; Diehn et al., 2009), and that they might be responsible for fueling tumor recurrence after initial response (Chen et al., 2012). Based on these findings, a new wave of studies is currently being initiated, aimed at a detailed molecular characterization of CSCs across the full spectrum of human tumors. Hopefully, a deeper understanding of the biochemical pathways that define CSC properties will identify novel therapeutic targets, and pave the way to new, more effective therapies.

Future Perspectives

During the last 10 years, the CSC model progressively established itself as a powerful research tool in cancer biology, endowed with substantial heuristic value. It provided us with a better understanding of intra-tumor heterogeneity and of the hierarchical organization that exists within cancer tissues. According to this model, the cellular heterogeneity observed within tumors originates, at least in part, as the result of multi-lineage differentiation, an epigenetic diversification process characteristic of normal stem cell systems. Perturbations in multi-lineage differentiation can translate into variations in the cellular composition of malignant tissues, and likely underpin important differences in tumor behavior across patients. As a general rule, for instance, tumors charaterized by a gene-expression profile characteristic of more immature, undifferentiated cells display a more aggressive clinical course and associate with worse survival. Identification of genes differentially expressed across cell populations holds the potential for the discovery of both novel prognostic biomarkers and highly specific cellular targets for drug development. Ultimately, a deep molecular understanding of the similarities and differences between CSCs and normal stem cells will allow the development of a new generation of therapies, able to target cancer tissues specifically, while leaving normal tissues unharmed.

Acknowledgments

The authors have been supported by grants and scholarships from the UCLA-Caltech MSTP (NIH T32GM008042), the Thomas and Stacey Siebel Foundation, and the California Institute for Regenerative Medicine (CIRM).

References

Abrahamsson, A.E., Geron, I., Gotlib, J., Dao, K.H., Barroga, C.F., Newton, I.G., Giles, F.J., Durocher, J., Creusot, R.S., Karimi, M., et al. (2009). Glycogen synthase kinase 3beta missplicing contributes to leukemia stem cell generation. Proc Natl Acad Sci USA 106, 3925–3929.

Adamson, J.W., Fialkow, P.J., Murphy, S., Prchal, J.F., and Steinmann, L. (1976). Polycythemia vera: stem-cell and probable clonal origin of the disease. N Engl J Med 295, 913–916.

Akala, O.O., Park, I.K., Qian, D., Pihalja, M., Becker, M.W., and Clarke, M.F. (2008). Long-term haematopoietic reconstitution by Trp53−/− p16Ink4a−/−p19Arf−/− multipotent progenitors. Nature 453, 228–232.

Al-Hajj, M., Wicha, M.S., Benito-Hernandez, A., Morrison, S.J., and Clarke, M.F. (2003). Prospective identification of tumorigenic breast cancer cells. Proc Natl Acad Sci USA 100, 3983–3988.

Bao, S., Wu, Q., McLendon, R.E., Hao, Y., Shi, Q., Hjelmeland, A.B., Dewhirst, M.W., Bigner, D.D., and Rich, J.N. (2006). Glioma stem cells promote radioresistance by preferential activation of the DNA damage response. Nature 444, 756–760.

Blair, A., Hogge, D.E., Ailles, L.E., Lansdorp, P.M., and Sutherland, H.J. (1997). Lack of expression of Thy-1 (CD90) on acute myeloid leukemia cells with long-term proliferative ability in vitro and in vivo. Blood 89, 3104–3112.

Bonnet, D., and Dick, J.E. (1997). Human acute myeloid leukemia is organized as a hierarchy that originates from a primitive hematopoietic cell. Nat Med 3, 730–737.

Carriaga, M.T., and Henson, D.E. (1995). The histologic grading of cancer. Cancer 75, 406–421.

Chan, K.S., Espinosa, I., Chao, M., Wong, D., Ailles, L., Diehn, M., Gill, H., Presti, J., Jr., Chang, H.Y., van de Rijn, M., et al. (2009). Identification, molecular characterization, clinical prognosis, and therapeutic targeting of human bladder tumor-initiating cells. Proc Natl Acad Sci USA 106, 14016–14021.

Chen, J., Li, Y., Yu, T.S., McKay, R.M., Burns, D.K., Kernie, S.G., and Parada, L.F. (2012). A restricted cell population propagates glioblastoma growth after chemotherapy. Nature 488, 522–526.

Cho, R.W., Wang, X., Diehn, M., Shedden, K., Chen, G.Y., Sherlock, G., Gurney, A., Lewicki, J., and Clarke, M.F. (2008). Isolation and molecular characterization of cancer stem cells in MMTV-Wnt-1 murine breast tumors. Stem Cells 26, 364–371.

Chu, P., Clanton, D.J., Snipas, T.S., Lee, J., Mitchell, E., Nguyen, M.L., Hare, E., and Peach, R.J. (2009). Characterization of a subpopulation of colon cancer cells with stem cell-like properties. Int J Cancer 124, 1312–1321.

Clarke, M.F. (2005). A self-renewal assay for cancer stem cells. Cancer Chemother Pharmacol 56 Suppl 1, 64–68.

Collins, A.T., Berry, P.A., Hyde, C., Stower, M.J., and Maitland, N.J. (2005). Prospective identification of tumorigenic prostate cancer stem cells. Cancer Res 65, 10946–10951.

Dalerba, P., Cho, R.W., and Clarke, M.F. (2007a). Cancer stem cells: models and concepts. Annu Rev Med 58, 267–284.

Dalerba, P., Dylla, S.J., Park, I.K., Liu, R., Wang, X., Cho, R.W., Hoey, T., Gurney, A., Huang, E.H., Simeone, D.M., et al. (2007b). Phenotypic characterization of human colorectal cancer stem cells. Proc Natl Acad Sci USA 104, 10158–10163.

Dalerba, P., Kalisky, T., Sahoo, D., Rajendran, P.S., Rothenberg, M.E., Leyrat, A.A., Sim, S., Okamoto, J., Johnston, D.M., Qian, D., et al. (2011). Single-cell dissection of transcriptional heterogeneity in human colon tumors. Nat Biotechnol 29, 1120–1127.

Diehn, M., Cho, R.W., Lobo, N.A., Kalisky, T., Dorie, M.J., Kulp, A.N., Qian, D., Lam, J.S., Ailles, L.E., Wong, M., et al. (2009). Association of reactive oxygen species levels and radioresistance in cancer stem cells. Nature 458, 780–783.

Driessens, G., Beck, B., Caauwe, A., Simons, B.D., and Blanpain, C. (2012). Defining the mode of tumour growth by clonal analysis. Nature 488, 527–530.

Dylla, S.J., Beviglia, L., Park, I.K., Chartier, C., Raval, J., Ngan, L., Pickell, K., Aguilar, J., Lazetic, S., Smith-Berdan, S., et al. (2008). Colorectal cancer stem cells are enriched in xenogeneic tumors following chemotherapy. PloS One 3, e2428.

Eppert, K., Takenaka, K., Lechman, E.R., Waldron, L., Nilsson, B., van Galen, P., Metzeler, K.H., Poeppl, A., Ling, V., Beyene, J., et al. (2011). Stem cell gene expression programs influence clinical outcome in human leukemia. Nat Med 17, 1086–1093.

Fearon, E.R., and Vogelstein, B. (1990). A genetic model for colorectal tumorigenesis. Cell 61, 759–767.

Fialkow, P.J., Faguet, G.B., Jacobson, R.J., Vaidya, K., and Murphy, S. (1981a). Evidence that essential thrombocythemia is a clonal disorder with origin in a multipotent stem cell. Blood 58, 916–919.

Fialkow, P.J., Singer, J.W., Adamson, J.W., Vaidya, K., Dow, L.W., Ochs, J., and Moohr, J.W. (1981b). Acute nonlymphocytic leukemia: heterogeneity of stem cell origin. Blood 57, 1068–1073.

Fialkow, P.J., Jacobson, R.J., and Papayannopoulou, T. (1977). Chronic myelocytic leukemia: clonal origin in a stem cell common to the granulocyte, erythrocyte, platelet and monocyte/macrophage. Am J Med 63, 125–130.

Galli, R., Binda, E., Orfanelli, U., Cipelletti, B., Gritti, A., De Vitis, S., Fiocco, R., Foroni, C., Dimeco, F., and Vescovi, A. (2004). Isolation and characterization of tumorigenic, stem-like neural precursors from human glioblastoma. Cancer Res 64, 7011–7021.

Gentles, A.J., and Alizadeh, A.A. (2011). A few good genes: simple, biologically motivated signatures for cancer prognosis. Cell Cycle 10, 3615–3616.

Gentles, A.J., Plevritis, S.K., Majeti, R., and Alizadeh, A.A. (2010). Association of a leukemic stem cell gene expression signature with clinical outcomes in acute myeloid leukemia. JAMA 304, 2706–2715.

Gerlinger, M., Rowan, A.J., Horswell, S., Larkin, J., Endesfelder, D., Gronroos, E., Martinez, P., Matthews, N., Stewart, A., Tarpey, P., et al. (2012). Intratumor heterogeneity and branched evolution revealed by multiregion sequencing. N Engl J Med 366, 883–892.

Goardon, N., Marchi, E., Atzberger, A., Quek, L., Schuh, A., Soneji, S., Woll, P., Mead, A., Alford, K.A., Rout, R., et al. (2011). Coexistence of

LMPP-like and GMP-like leukemia stem cells in acute myeloid leukemia. Cancer Cell 19, 138–152.

Guzman, M.L., Neering, S.J., Upchurch, D., Grimes, B., Howard, D.S., Rizzieri, D.A., Luger, S.M., and Jordan, C.T. (2001). Nuclear factor-κB is constitutively activated in primitive human acute myelogenous leukemia cells. Blood 98, 2301–2307.

Hanahan, D., and Weinberg, R.A. (2011). Hallmarks of cancer: the next generation. Cell 144, 646–674.

Jamieson, C.H., Ailles, L.E., Dylla, S.J., Muijtjens, M., Jones, C., Zehnder, J.L., Gotlib, J., Li, K., Manz, M.G., Keating, A., et al. (2004). Granulocyte-macrophage progenitors as candidate leukemic stem cells in blast-crisis CML. N Engl J Med 351, 657–667.

Jan, M., Snyder, T.M., Corces-Zimmerman, M.R., Vyas, P., Weissman, I.L., Quake, S.R., and Majeti, R. (2012). Clonal evolution of preleukemic hematopoietic stem cells precedes human acute myeloid leukemia. Sci Transl Med 4, 149ra118.

Kirkland, S.C. (1988). Clonal origin of columnar, mucous, and endocrine cell lineages in human colorectal epithelium. Cancer 61, 1359–1363.

Lagasse, E. (2008). Cancer stem cells with genetic instability: the best vehicle with the best engine for cancer. Gene Ther 15, 136–142.

Lapidot, T., Sirard, C., Vormoor, J., Murdoch, B., Hoang, T., Caceres-Cortes, J., Minden, M., Paterson, B., Caligiuri, M.A., and Dick, J.E. (1994). A cell initiating human acute myeloid leukaemia after transplantation into SCID mice. Nature 367, 645–648.

Li, C., Heidt, D.G., Dalerba, P., Burant, C.F., Zhang, L., Adsay, V., Wicha, M., Clarke, M.F., and Simeone, D.M. (2007). Identification of pancreatic cancer stem cells. Cancer Res 67, 1030–1037.

Liu, R., Wang, X., Chen, G.Y., Dalerba, P., Gurney, A., Hoey, T., Sherlock, G., Lewicki, J., Shedden, K., and Clarke, M.F. (2007). The prognostic role of a gene signature from tumorigenic breast-cancer cells. N Engl J Med 356, 217–226.

Malanchi, I., Peinado, H., Kassen, D., Hussenet, T., Metzger, D., Chambon, P., Huber, M., Hohl, D., Cano, A., Birchmeier, W., et al. (2008). Cutaneous cancer stem cell maintenance is dependent on beta-catenin signalling. Nature 452, 650–653.

Merlos-Suarez, A., Barriga, F.M., Jung, P., Iglesias, M., Cespedes, M.V., Rossell, D., Sevillano, M., Hernando-Momblona, X., da Silva-Diz, V., Munoz, P., et al. (2011). The intestinal stem cell signature identifies colorectal cancer stem cells and predicts disease relapse. Cell Stem Cell 8, 511–524.

Miyamoto, T., Weissman, I.L., and Akashi, K. (2000). AML1/ETO-expressing nonleukemic stem cells in acute myelogenous leukemia with 8;21 chromosomal translocation. Proc Natl Acad Sci USA 97, 7521–7526.

O'Brien, C.A., Pollett, A., Gallinger, S., and Dick, J.E. (2007). A human colon cancer cell capable of initiating tumour growth in immunodeficient mice. Nature 445, 106–110.

Odoux, C., Fohrer, H., Hoppo, T., Guzik, L., Stolz, D.B., Lewis, D.W., Gollin, S.M., Gamblin, T.C., Geller, D.A., and Lagasse, E. (2008). A stochastic model for cancer stem cell origin in metastatic colon cancer. Cancer Res 68, 6932–6941.

Patrawala, L., Calhoun, T., Schneider-Broussard, R., Li, H., Bhatia, B., Tang, S., Reilly, J.G., Chandra, D., Zhou, J., Claypool, K., et al. (2006). Highly purified CD44+ prostate cancer cells from xenograft human tumors are enriched in tumorigenic and metastatic progenitor cells. Oncogene 25, 1696–1708.

Pavlidis, N., and Pentheroudakis, G. (2012). Cancer of unknown primary site. Lancet 379, 1428–1435.

Prince, M.E., Sivanandan, R., Kaczorowski, A., Wolf, G.T., Kaplan, M.J., Dalerba, P., Weissman, I.L., Clarke, M.F., and Ailles, L.E. (2007). Identification of a subpopulation of cells with cancer stem cell properties in head and neck squamous cell carcinoma. Proc Natl Acad Sci USA 104, 973–978.

Quintana, E., Shackleton, M., Sabel, M.S., Fullen, D.R., Johnson, T.M., and Morrison, S.J. (2008). Efficient tumour formation by single human melanoma cells. Nature 456, 593–598.

Rajasekhar, V.K., Studer, L., Gerald, W., Socci, N.D., and Scher, H.I. (2011). Tumour-initiating stem-like cells in human prostate cancer exhibit increased NF-κB signalling. Nat Commun 2, 162.

Ramalho-Santos, M., and Willenbring, H. (2007). On the origin of the term "stem cell." Cell Stem Cell 1, 35–38.

Reya, T., Morrison, S.J., Clarke, M.F., and Weissman, I.L. (2001). Stem cells, cancer, and cancer stem cells. Nature 414, 105–111.

Ricci-Vitiani, L., Lombardi, D.G., Pilozzi, E., Biffoni, M., Todaro, M., Peschle, C., and De Maria, R.

(2007). Identification and expansion of human colon-cancer-initiating cells. Nature 445, 111–115.

Schepers, A.G., Snippert, H.J., Stange, D.E., van den Born, M., van Es, J.H., van de Wetering, M., and Clevers, H. (2012). Lineage tracing reveals Lgr5+ stem cell activity in mouse intestinal adenomas. Science 337, 730–735.

Shackleton, M., Quintana, E., Fearon, E.R., and Morrison, S.J. (2009). Heterogeneity in cancer: cancer stem cells versus clonal evolution. Cell 138, 822–829.

Singh, S.K., Hawkins, C., Clarke, I.D., Squire, J.A., Bayani, J., Hide, T., Henkelman, R.M., Cusimano, M.D., and Dirks, P.B. (2004). Identification of human brain tumour initiating cells. Nature 432, 396–401.

Takahashi, N., Miura, I., Saitoh, K., and Miura, A.B. (1998). Lineage involvement of stem cells bearing the Philadelphia chromosome in chronic myeloid leukemia in the chronic phase as shown by a combination of fluorescence-activated cell sorting and fluorescence in situ hybridization. Blood 92, 4758–4763.

Vermeulen, L., Todaro, M., de Sousa Mello, F., Sprick, M.R., Kemper, K., Perez Alea, M., Richel, D.J., Stassi, G., and Medema, J.P. (2008). Single-cell cloning of colon cancer stem cells reveals a multi-lineage differentiation capacity. Proc Natl Acad Sci USA 105, 13427–13432.

Volkmer, J.P., Sahoo, D., Chin, R.K., Ho, P.L., Tang, C., Kurtova, A.V., Willingham, S.B., Pazhanisamy, S.K., Contreras-Trujillo, H., Storm, T.A., et al. (2012). Three differentiation states risk-stratify bladder cancer into distinct subtypes. Proc Natl Acad Sci USA 109, 2078–2083.

von Hansemann, D. (1897). Die mikroskopische Diagnose der bösartigen Geschwülste. [The microscopic diagnosis of malignant tumors.] (Berlin, Verlag von August Hirschwald).

Wang, M., Xiao, J., Shen, M., Yahong, Y., Tian, R., Zhu, F., Jiang, J., Du, Z., Hu, J., Liu, W., et al. (2011). Isolation and characterization of tumorigenic extrahepatic cholangiocarcinoma cells with stem cell-like properties. Int J Cancer 128, 72–81.

Weissman, I. (2005). Stem cell research: paths to cancer therapies and regenerative medicine. JAMA 294, 1359–1366.

Weissman, I.L. (2000). Stem cells: units of development, units of regeneration, and units in evolution. Cell 100, 157–168.

Weissman, I.L., and Shizuru, J.A. (2008). The origins of the identification and isolation of hematopoietic stem cells, and their capability to induce donor-specific transplantation tolerance and treat autoimmune diseases. Blood 112, 3543–3553.

Chapter 2

The Hallmarks of Prostate Cancer Stem Cells

Norman J. Maitland and Anne T. Collins
YCR Cancer Research Unit, Department of Biology, University of York, Heslington, York, UK

Introduction

The concept of cancer stem cells (CSC) is now 15 years old in the modern era, which probably began with cell isolation studies in acute myeloid leukemia (Bonnet and Dick, 1997). Historically the existence of a unique, tumor-initiating cell with a distinctive phenotype, relative to the bulk of a tumor, can be traced back to the very earliest times of cancer research, where a 'root' for cancers was surmised rather than proven, based on the embryonic rest hypothesis (Conheim, 1875). During the 1960s and 1970s, as cell culture expertise expanded, clonogenicity assays identified individual populations of cells with both higher clonogenicity and cancer-causing ability in animal models (Wodinsky et al., 1968; Hamburger and Salmon, 1977). A fuller account of the history of the cancer stem cell hypothesis has been covered elsewhere (Maitland and Collins, 2010; Sell, 2010). A lack of purity in the original cell preparations undermined the positive testing of the hypothesis, but now, despite a number of more recent controversies (Quintana et al., 2008), it is clear that within the inherent tumor heterogeneity there is a ready source of cells with the canonical properties of a cancer stem cell (Baker, 2012). The identification, purification, and characterization of these cells are necessary prerequisites for developing CSC-based therapies. The number of papers in this area has also increased, providing a great deal of data but also a great deal of collateral 'noise' in terms of our understanding.

A Systematic Approach to Etiology: Lessons from Koch's Postulates

The current situation is perhaps similar to studies in the last century, linking 'agents' to infectious disease. It was only when the great German microbiologist, Robert Koch, produced a series of postulates in the nineteenth century that the study of such infectious agents was systematized. Initially Koch proposed four postulates (Koch, 1884) shown in Table 2.1, which were modified 100 years later to include data from the nucleic acids age (Table 2.2A) (Fredericks and Relman, 1996) and further modified to discuss the molecular basis of disease, as our understanding and results of clinical trials became more systematic (Table 2.2B).

Cancer Stem Cells, First Edition. Edited by Vinagolu K. Rajasekhar.

Table 2.1. Koch's Postulates

	Postulate
1	Can the organism be isolated from every case of the disease?
2	Can the organism be propagated in pure culture in vitro?
3	Can the disease be reproduced by injecting the pure organism into a suitable recipient?
4	Can the organism be re-isolated from the secondary disease?

The postulates (published in 1884) formed a logical framework for the association of specific organisms with common diseases, such as anthrax and smallpox, and were devised at a time when the concept of microorganisms and disease was still in its infancy.

Table 2.2. Updated Koch's Postulates

2.2A. Reformulated for the Nucleic Acid Era

	Postulate
1	A nucleic acid sequence belonging to a putative pathogen should be present in most cases of an infectious disease. Viral nucleic acids should be found preferentially in those organs or gross anatomic sites known to be diseased, and not in those organs that lack pathology.
2	Fewer, or no, copies of viral nucleic acid sequences should occur in hosts or tissues without disease.
3	With resolution of disease, the copy number of viral nucleic acid sequences should decrease or become undetectable. With clinical relapse, the opposite should occur.
4	When sequence detection predates disease, or sequence copy number correlates with severity of disease or pathology, the sequence-disease association is more likely to be a causal relationship.
5	The nature of the virus inferred from the available sequence should be consistent with the known biological characteristics of that group of organisms.
6	Tissue-sequence correlates should be sought at the cellular level; efforts should be made to demonstrate specific in situ hybridization of viral sequence to areas of tissue pathology and to visible viral particles or to areas where the viruses are presumed to be located.
7	These sequence-based forms of evidence for viral causation should be reproducible.

Note: This (longer) version of the original postulates was published in 1996 by Fredericks and Relman, focusing on virally induced disease. It lacks the specific questioning of the original postulates.

2.2B. Reformulated for the Molecular Basis of Disease

	Postulate
1	Which gene (or gene product) is responsible for virulence?
2	Is the gene present in strains of virus that cause the disease?
3	Is this gene not present in avirulent strains?
4	Does disruption of the gene reduce virulence?
5	Does introduction of cloned genes into an avirulent strain confer virulence?
6	Is this gene expressed in vivo in diseased patient samples?
7	Does a specific immune response to this gene protect against the disease?

This version restores the questioning as a basis for enquiry, while still retaining the infectious etiology as its core, and introducing the importance of prevention/treatment to augment the original postulates.

Table 2.3. Koch's Postulates Modified for the Viral Etiology of Human Cancer

	Postulate	Experimental Approaches
1	Do patients with the tumor have evidence of viral infection?	Sero-epidemiology
2	Are viral genes present (and active) in the tumor?	Viral nucleic acid and antigen detection
3	Can the tumor be linked to the presence of an active viral gene product?	Mutagenesis Tumor antigen/ oncogene
4	Does prevention of the infection stop the tumor?	Vaccination in experimental animals and ultimately man

Note: This modified group, originally developed in the 1970s as the concept of viral oncogenes emerged, provided a logical framework for the study viral etiology of human cancer. Only recently have vaccination studies (postulate 4) begun; concept to fact has therefore taken more than 30 years.

The same was true for studies of the infectious etiology of cancer in the 1970s and 1980s, where Koch's postulates were again invoked (Table 2.3), for example, in studies of the link between human papillomaviruses and cervical cancer (zur Hausen, 2002; Moore and Chang, 2010).

The modified postulates, and in particular those for viral etiology of cancer (Table 2.3), have not as yet, been completely fulfilled. We are in the strong position of having vaccines against a number of these infectious agents—and here the key postulate is number 4—which states that medical intervention or prevention of the infection should result in a drop off in tumor incidence coincident with that of the infection. For this we shall have to wait some 10 to 20 years beyond the first vaccinations, which really only became widespread within the last 5 to 6 years (Markman, 2012). However, an unwavering adherence to Koch's postulates has also been criticized (Hanson, 1988), particularly for multifactorial disease and unculturable pathogens, but the application of such tests still has many advantages in the complex study of disease etiology.

Could the Study of Cancer Stem Cells Benefit from Koch's Framework for Investigation?

Is it possible to systematize the study of cancer stem cells in the same way, to remove the potential noise from the system, and to focus attention on the essential features for study? A series of such postulates is listed in Table 2.4. Some of these have already been fulfilled, but the final postulate is pivotal. If we can identify CSCs and isolate them, not only from cell lines but also from primary human tumors, then what is the outcome for a tumor from which CSCs have been eliminated? The first of these modified postulates, for instance,"does a patient's tumor contain cells with CSC properties?" is rather more complex than it seems. There is a considerable body of literature on stem-like cells from established cancer-derived cell lines. The various phenotypes used to identify and isolate cells from cell lines and xenografts with some or all of the hallmarks of a CSC in prostate are shown in Table 2.4. However, the microenvironment in which these cell lines are cultured will quite clearly change the cells' phenotypes. Therefore, the phenotype of cells from cell lines, advanced xenograft models, or indeed mouse models may not be the same as that found within the complex microenvironment in a primary or indeed metastatic human tumor.

The second postulate deals with the 'isolation and propagation of the cells' (similar to the isolation and propagation of a microbiological agent in disease). There are a number of methods described for this, but they are only rarely applied to primary prostate cancers (Collins et al., 2005; Shepherd et al., 2008). Cultural expansion has proven to be more difficult, unlike embryonal carcinoma cells, which can be induced to proliferate in high serum and specialized media. The same conditions may promote the growth but induce differentiation of adult SC from primary normal and malignant sources. Equally, long-term culture of tumor cells relative to the normal equivalents in the inevitable mixture one

Table 2.4. Proposed Koch's Postulates Linking Cancer Stem Cells to a Key Tumor Initiating Role in Human Cancer

	Postulate	Experimental Approaches
1	Does a patient's tumor contain cells with CSC properties?	Fractionation of human prostate cancer tissues (cf cell lines and established xenografts)
2	Can we isolate and propagate these cells?	In vitro culture and xenograft in immunocompromised mice
3	If we reintroduce these cells into a recipient does it cause an identical disease?	Serial transplantability of tumor-initiating fraction
4	Do these cells contain and express specific gene products that give them the properties of CSCs?	Phenotypic analysis of CSCs relative to normal tissue SC (and differentiated cell products within the normal tissue and cancers)
5	If we disrupt these genes, do the cells lose the properties of CSCs?	Specific inhibition of core properties, i.e., tumor induction by blocking the actions of specific genes and pathways with small molecules, siRNA, antibodies
6	If we eliminate the CSCs, do we eliminate the cancer?	Xenotransplantation experiments to measure tumor induction after ex vivo treatment, or the response and cell content of established tumours after CSC elimination.

finds within a clinical biopsy, remains a frequent technical problem.

The third postulate, that is, reintroduction of cells into a recipient, is also technically challenging for prostate cancers. The ideal recipient is the original human patient, but ethical restrictions prevent these experiments. Hence, there is a reliance on mouse models for many lineage studies (Wang et al., 2009; Choi et al., 2012). The act of cell purification clearly breaks many of the microenvironmental controls, which establish epithelial phenotypes within a primary tissue. A xenograft of these cells into immunocompromised mice almost inevitably results in a tumor (van Weerden et al., 2008; Maitland et al., 2011), which is usually of a less differentiated type than the original prostate cancers. Since such grafts are often carried out in the presence of embryonal stroma to promote/preserve the SC content, this is perhaps not surprising. Co-culture of prostatic fibroblasts with epithelial cells induces a more complete form of luminal differentiation (Lang et al., 2006), but the ability to reconstruct such tissues in a mouse, in the absence of substantial invasion by murine stroma is challenging, particularly for serial xenografts. However, the tumors obtained retain the ability to undergo luminal differentiation and express many of the features of a prostate cancer, including expression of the androgen receptor (AR) from an initial input, which is AR− (i.e., a primitive basal type cell) (Maitland et al., 2011). Xenografts most closely model the induction of metastasis, which principally occurs with tumors of Gleason patterns 4 and 5, paralleling the clinical situation (Chan et al., 2000).

This result links to the fourth of the modified Koch's postulates, which concerns the expression of specific gene products. A number of laboratories have phenotyped both cell cultures and xenografts from prostate cancer, seeking a fingerprint of the gene expression profile. (See Table 2.5.)

Several common features have emerged from this, but comparisons of this less differentiated population with the bulk of a prostate cancer, which has a largely luminal, androgen receptor and prostate-specific antigen [PSA] expressing phenotype, are likely to be 'contaminated' with signatures for cellular differentiation (Birnie et al., 2008; Oldridge et al., 2011). Sequence Tag experiments from the 1990s (Adams et al., 1995) confirm that

Table 2.5. Phenotypes of Stem-Like Cells from Human Prostate Cancer (Cell Lines and Established Xenografts)

Source of Prostate Cells	CSC Selection Phenotype	Reference
Cell Lines		
Multiple	$CD44^+$	Patrawala et al., 2006
DU145	$CD133^+$/a2b1 integrin$^+$/ $CD44^+$	Chen et al., 2012c
LNCaP	$CD44^+/CD24^-$	Hurt et al., 2008
PC3-MM2	$CD133^+$/ $CD44^+$/ $CD166^+$	Rowehl et al., 2008
PC3	$Fam65B^{High}$/ $Mfl2^{low}/LEF1^{low}$	Zhang and Waxman, 2010
PC3	$CD133^+$/ $CD44^+$	Fan et al., 2012
RC165N/ RC-92a	$CD133^+$/ CXCR4	Miki et al., 2008
Xenografts		
CWR22	TRA-1-60$^+$/ CD151/CD166	Rajasekhar et al., 2011
LAPC-9	Hoechst 33342 effluxing side population	Patrawala et al., 2005
DU145, LAPC4, LAPC9	Integrin a2b1High/ $CD44^+$	Patrawala et al., 2007
22RV1	$CD117^+$/ $ABCG2^+$	Liu et al., 2010

luminal-specified genes are overexpressed and are expressed at an order of magnitude higher compared to basal genes. Many such whole tumor signatures and indeed whole genome sequencing studies, for example those by True (2006) and Wu and others (2012) have now been performed, almost inevitably reporting the potent upregulation of luminal markers seen in prostate cancer.

What markers are linked directly to prostate tumors? One common emerging feature appears to be the upregulation of genes normally involved in inflammatory processes. This is also a feature of the bulk of prostate cancers (De Marzo et al., 2007) but is particularly striking within the CSC population. For example, overexpression of activated NF-kB is frequently observed, as is activated IL-6 signaling (Birnie et al., 2008; Rajasekhar et al., 2011). It is unlikely that CSC gene expression can be used as a reliable marker for disease, given the relative rarity of the cancer stem cell population within the tumor mass (around 0.1%). The functional significance of these individual genes and the signaling pathways to which they contribute are of more importance. One strategy would be to knock down specific gene expression or inhibit protein function, and to monitor the effects on the CSC population (Figure 2.1). If such knockdowns/ inhibitors result in a loss of tumor initiation, self renewal, or tumor spread then it is clear the 'marker' has more than just a phenotypic significance for the CSC population (see below).

The fifth of the new postulates considers the effects of disrupting the disease-inducing genes (postulate 4) and how this can be related to the loss of properties of CSCs. The last (sixth) postulate is equivalent to the vaccination approach for pathogens, which may indeed also be applicable for the destruction of cancer stem cells (i.e., that elimination of the CSCs results in elimination of the tumor). In a clinical specialty where oncology 'leaps forward' are measured in terms of extra months of survival (Scher et al., 2012), it is clear that agents targeted against the large mass of tumor cells are unlikely to provide lasting treatment in a tumor where the CSCs have a variety of resistance mechanisms in-built as part of the SC phenotype, such as rapid drug efflux and radiation resistance. This is not surprising since normal SC are required to remain unaltered phenotypically throughout the life span of the organism. Genomic changes and mutations may well be advantageous to escape from, for example, the immune system. The consideration of Koch's postulates perhaps leads us inevitably toward defining hallmarks of CSCs, which describe in molecular detail the essential features not only in terms of biological activity but also for therapeutic targeting.

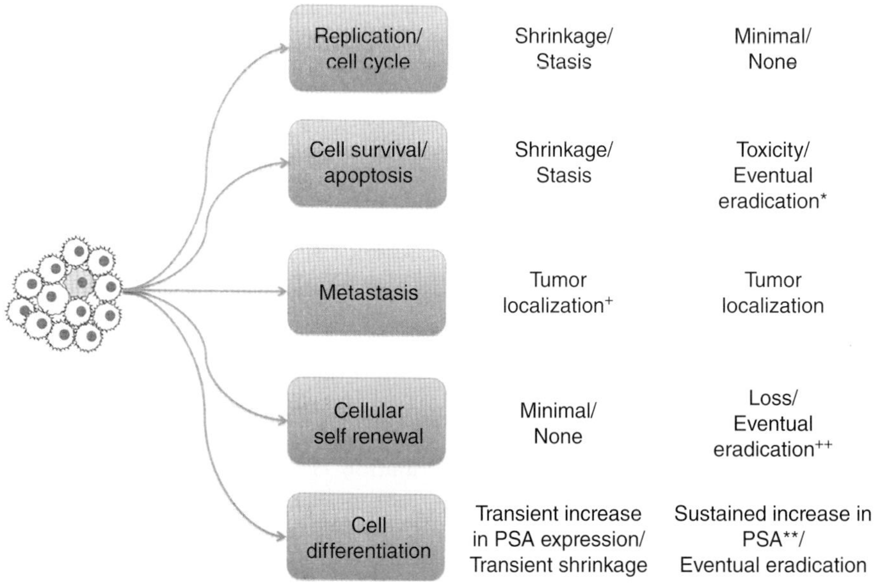

Figure 2.1. Outcomes of treatment strategies using therapies targeting bulk tumor cells and stem cells in prostate cancer. Five treatment strategies and their outcomes for a heterogeneous prostate cancer are illustrated.

* For apoptotic agents, targeted only to CSCs, toxicity in the bulk tumor would NOT be observed.

** For agents that induce differentiation, it is likely that there would be a transient increase in common markers such as PSA, when targeting the whole tumor. This increase is likely to be more sustained after treatment with CSC-differentiating agents, as the SC population shifts toward a more luminal type.

+Inhibitors of metastatic spread in the bulk tumor may transiently inhibit spread but will not offer permanent treatment unless they affect the CSC fraction.

++According to the CSC hypothesis, it is only by elimination of the CSC fraction that tumor extinction can be achieved. The timing to achieve this (i.e., how long a tumor is self sustaining in the absence of a SC/TA fraction) is currently unknown.

A Systematic Approach to Prostate Cancer Stem Cell Biology: Learning from the 'Hallmarks of Cancer'

This means of systematizing the study of cancer was first described by Hanahan and Weinberg (2000). In 2011 the hallmarks were updated (Hanahan and Weinberg, 2011) to now include 10 key features of cancer, derived mainly from studies of cell lines in vitro and of our increasingly knowledge of tumor tissue masses in the era of total genome studies. These hallmarks also provided a framework for the design and study of new cancer treatments, most of which are designed to reduce the replicating cells that form the bulk tumor, as a result of de-regulated cell cycle and rapid growth. Given the distinctive phenotype of CSCs, can these same hallmarks be applied in a systematic study of a cell with such different biological characteristics?

As shown in Table 2.6A, the various hallmarks of cancer are indeed rather limited in either their application to or their relevance in CSCs. For example, sustained proliferative signaling is at best unlikely in the CSC fractions in studies of several different tumor types (including prostate) has indicated that the stem cells are largely quiescent. As such they provide a 'reserve cell' capable of re-populating

Table 2.6. Hallmarks of Cancer (after Hanahan and Weinberg, 2011)

2.6A. Hallmarks of Total Tumor Cells and Tissues

	Biological Function – Hallmark	Inhibition Strategy	Relevance to CSCs*
1	Sustained proliferative signalling	Growth factor inhibitors	No/Limited
2	Evasion of growth suppressors	Cyclin-dependent kinase inhibitors	No/Limited
3	Evasion of immune destruction	Blockade of immune response, e.g., by anti-CTLA4	Important
4	Induction of replicative immortality	Telomerase inhibitors	Depends on tumor stage and grade
5	Induction of tumor-promoting inflammation	Anti-inflammatories (e.g., aspirin)	Important
6	Ability to invade/metastasise	Inhibitors of EMT or HGF/c-met	Important
7	Ability to induce angiogenesis	Inhibitors of VEGF signalling	Unknown
8	Presence of genome instability and mutations	Inhibitors of defective DNA repair, e.g., PARP inhibitors	Unknown
9	Resistance to cell death mechanisms	Promotion of apoptosis or inhibition of anti-apoptotic genes such as Bcl2	Probably important, although effects on normal SC unknown
10	Deregulation of cellular energetics controls	Inhibitors of aerobic glycolysis	Unknown

*These hallmarks (Hanahan and Weinberg, 2011) are designed to apply to a growing tumor, and are therefore less likely to universally apply to a reserve/quiescent cell population as suggested by current experimental evidence for CSCs.

2.6B. Hallmarks of Cancer Stem Cells

	Biological Function – Hallmark
1	Isolated CSCs can be used to initiate new tumor growth
2	Isolated CSCs are responsible for tumor invasion and spread
3	Isolated CSCs are present in minimal residual disease after conventional treatments
4	Residual CSCs are responsible for tumor recurrence after therapy
5	CSCs can differentiate from a more primitive phenotype to produce the recognized differentiated phenotype of bulk tumor cells
6	CSCs can differentiate from a primitive phenotype to produce the recognized phenotype of bulk tumor cells
7	CSCs have a distinctive phenotype compared to the bulk of the tumor
	Other properties commonly investigated
1	CSCs express sets of 'stem-like' genes
2	CSCs activate 'stem cell signaling'
3	CSCs have a stem cell-like epigenetic imprint
4	CSCs divide asymmetrically, self-renew
5	CSCs grow in semi-solid support media

after insult or damage to the tumor mass (Li and Bhatia, 2011; Chen et al., 2012b). Essentially the same can be said to be true of growth suppressors, where many of the cyclin-dependent kinase inhibitors that have been used in cancer treatment have almost no effect on an SC population (unpublished data). In fact, had treatment with these sophisticated and targeted drugs been effective against CSCs, they would probably be capable of eliminating tumors, if recurrence is indeed a matter of CSC resistance.

Evasion of immune destruction is a more interesting property, which may well be the Achilles heel of the stem cells. Normal stem cells in a tissue situation should have an inherent resistance to destruction by autoimmunity, and the antigens that are required by a SC to survive outside tissues could prove to be an exploitable weakness, if the CSCs does not possess immune suppressive properties, for which at the moment there is no evidence.

Telomerase inhibitors, such as those used for replicative immortality, have also been claimed to be effective against prostate CSCs (Marian and Shay, 2009; Xu et al., 2011). This is not universally the case however, as in a quiescent cell, activation of telomerase is not a sine qua non for survival and the extended life span attributed to cancers (Oldridge et al., 2011). It is however an important part of the life cycle of rapidly dividing cells, where telomere shorting at each apoptotic cycle could be a limit on the overall survival of the cells. However, the evidence in prostate cancer is that the initial events have probably occurred in the absence of telomerase activation, since the tumors are known to exhibit considerably shorter (but stable) telomeres, compared to the surrounding normal tissue (Meeker et al., 2002). This was an intriguing result, which was also mirrored in pre-malignancies (PIN) in prostate cancer, suggesting there should be a telomerase-negative stage in tumor development, which we would propose occurs at stem cell level (Rane et al., manuscript in preparation).

Excessive production of proteins associated with inflammation (Birnie et al., 2008; Rajasekhar et al., 2011) is another important feature of prostate CSCs.

Excessive production of the IL6 cytokine and downstream signaling through STAT proteins and more importantly NF-kappa B (Birnie et al. 2008; Rajasekhar et al. 2011), indicate that prostate cancers may well arise from a stem cell, in which inflammation addiction has been promoted through the multiple cycles of inflammatory stimuli present during prostatitis (Maitland and Collins, 2008; Sfanos and De Marzo, 2012). Patients with frequent prostatitis also have higher levels of prostate cancer (Roberts et al., 2004; Sfanos and De Marzo, 2012), while anti-inflammatories such as aspirin are known (with long-term use) to suppress the more malignant types of prostate cancer (Bosetti et al., 2012).

The ability to induce angiogenesis is a common feature of many tumors, and is associated with the combating of metabolic stress and hypoxia experienced as a tumor expands. Since many CSCs are known to exist in a hypoxic environment (Ma et al., 2011) then angiogenic stimuli may not be an important concept in the hallmarks of CSCs. However, angiogenic factors have many alternative roles in cell survival.

The frequency of genome instability (i.e., mutations in SCs) is something that at the moment is relatively unknown. Whole-genome, deep-sequencing studies indicate considerable heterogeneity in a tumor mass with regard to mutation (Shah et al., 2012), and considerable attempts have been made to divide these mutations into those that have a phenotypic significance (the founder mutations) from those that are merely passengers and part of the noise of a bulk tumor cell population, which is growing sufficiently quickly to accumulate random mutations. Founder mutations, and particularly those that allow an SC to escape from niche controls are more likely to be found in the CSC population. We would propose that one such founder mutation is the TMPRSS2-ERG translocation in prostate cancer, which we have detected in stem cells from prostate cancer (Birnie et al., 2008; Polson et al., 2013) Activation of ERG and other members of the ETS families are key regulators of SC fate in hematopoietic systems (Taoudi et al., 2011). The existence of genome instability (i.e., a propensity to develop new mutations and a tolerance for these mutations) remains controversial (Bodmer and Loeb, 2008; Loeb et al., 2008; Welch et al., 2012). Such instability has been observed in many tumor types, but its relevance to prostate cancer development remains in question. The evidence for enhanced DNA repair defects in prostate cancer is still controversial (Hällström and Laiho, 2008), and in our most recent data, the CSCs, when studied on an individual basis, have shown, if anything, an increased resistance to mutagenic damage, and an ability to repair that is also reflected in their normal SC equivalents (Frame et al., 2013). After low dose nonlethal irradiation, the stem cell component of a tumor is remarkably resistant, as also seen in normal tissue SC, thus promoting the survival of normal tissues through a lifetime of low-level radiation exposure. The key feature of studies in this area is whether the resistance is the result of an active mechanism (i.e., production of blockers of apoptosis such as BCL2), or whether it is a passive process involving the quiescent nature of the SC. If normal SCs are indeed the source of the cancer, it is not unreasonable to suggest that these properties would be retained within the CSC phenotype, offering considerable resistance to treatment by conventional chemotherapeutics.

Lastly, and a relatively new addition to the hallmarks of prostate cancer, is deregulation of cellular energetic controls. In many ways this again is related to the increased proliferation rates of the bulk of the tumor. Energetic controls are an important part of maintaining a homeostatic relationship, which is lost within tumor mass. However, in a cancer where the SC is indeed quiescent, the situation may be reversed. The shutdown of many energetic controls mirroring those that are required to maintain a cycling cell (since it is not a completely energy-free process) may well turn out to be unique within the CSC fraction and are again worthy of future study.

Hallmarks of Prostate Cancer Stem Cells

Perhaps a set of CSC hallmarks may help us to understand the processes, which are essential to identify SCs, rather than those that may be desirable, or that may perhaps be more established in the thinking of molecular cell biologists. Despite the current limitations in our knowledge of the nature of CSCs, Table 2.6B lists some alternative hallmarks, which describe the behavior of CSCs in a way and that can be used to study their functions in more detail. Like the original hallmarks proposed in 2000 (Hanahan and Weinberg, 2000), these hallmarks are likely to change and be enhanced as we learn more of the properties of the tumor-initiating cell fractions. They are based around two distinct but overlapping levels of classification.

The Relative Importance of CSC Phenotype and Function

The first classification (Table 2.6B, upper panel) is mainly related to function whereas the second (lower panel) considers the individual phenotype of the CSCs. Many published studies have focused in the opposite direction (i.e., phenotype then function), frequently employing known phenotypes to identify particular cell populations, which are subsequently tested for function. For the most part, these initial studies were carried out on cell lines rather than primary tumor tissues, and it is clear from the original hallmarks of cancer, that the situation within a tissue with respect to many of the properties in the 10 more recently delineated by Hanahan and Weinberg (2011) are quite different. To take one example: angiogenesis is inconsequential in cell culture but absolutely vital within a growing tissue. Equally, the ability to evade immune destruction is clearly present in tissues, but is not required in many animal and xenograft models of cancer. Essentially the hallmarks are a reductionist mechanism, which will allow us to study a biological system in more detail. If our modeling is accurate, it can then be applied in the in vivo situation. No model can completely encapsulate human disease, and a great danger in experimental studies is to overinterpret inadequate models.

In Vivo Modeling: A Necessary Series of Compromises

With respect to properties that can define the CSC, for many years the gold standard has been the ability of isolated CSCs to initiate new tumor growth in xenograft experiments with high efficiency relative to the bulk cells within the tumor, although there have been a number of contradictory reports of the efficacy of the xenograft model depending on the degree of immune-insufficiency present in the experimental animal (Quintana et al., 2008; Ishizawa et al., 2010). More modern double knockout models that lack not only T and B cells but also natural killer cells have a higher take rate for a variety of different tumors, compared to the more established nude mice, which contain measurable natural killer activity. Equally, should the murine hosts be injected with highly purified epithelial cells dissociated from their normal micro-environment, or should it be a co-inoculum with stromal cells (ideally from prostate but also perhaps from embryonal sources) to promote the formation of a new SC niche in an alien environment such as a subcutaneous or sub-renal site? The stromal requirement is perhaps not so essential if one limits the interpretation of the model to the ability of a CSC to establish extra-prostatic growth, for example in bone where the influences of the bone stroma are rather different from those within the prostate (van der Pluijm, 2011). Also at issue is the number and purity of the CSC inoculum. Do CSCs migrate and initiate new tumor growth as single cells, or do they migrate as a transformed epithelial-mesenchymal transition (EMT)-transformed cell population, which then re-establishes a normal epithelial phenotype at the metastatic site? Should our level of precision be 1 cell, 10 cells, 100 cells, or 1,000 cells instead of the 10^6 cells normally used in tumor induction experiments? The phenotype of many CSCs strongly suggests that they are responsible for tumor invasion and spread. Elements of EMT, high levels of matrix modification, and altered adhesion molecules are all part of this particular phenotype (Birnie et al., 2008; Rajasekhar et al., 2011), but rather it is the biological behavior, the ability to invade in vitro assays, and to metastasize in animals from an initial inoculation site, which are more important here as a hallmark.

CSC as Residual Disease After Treatment

Another feature of the CSC hypothesis is the contention that these cells represent minimal residual disease after successful treatment and shrinkage of a bulk tumor. This is a difficult hallmark to establish because such cells are likely to be relatively rare and may even be contaminated with other cells, which are more resistant to the treatment. For example, in castration experiments, fully competent luminal cells susceptible to the loss of androgenic stimuli are often still present for up to 12 months after the initial treatment. Are these the CSC? A re-application at this stage of the initial hallmark would undoubtedly decide this one way or the other. Since CSCs are probably residual after therapy, the fourth hallmark (i.e., the ability of the SC to repopulate and regenerate the tumor after therapy) may be viewed as a trivial extension of the previous hallmark. However, its significance is that the CSC has the ability to emerge after therapy, and to survive in the post-therapy environment: (i) post castration in prostate cancer, (ii) post radiotherapy in many other tumors, and (iii) post chemotherapy in multiple tumor types where cytotoxic drugs are targeting fast-replicating cells.

Here, the CSC would exist as the reservoir of the inactivating mutations (e.g., in detoxifying enzymes and the gene amplifications such as those seen after the castration therapy in prostate cancer) (Bubendorf et al., 1999; Li et al., 2002; Chen et al., 2003; Taplin et al., 2003). The hallmark predicts that markers of therapy resistance should be present within the CSC population. However, this is not an absolute requirement, as the CSCs can still provide a reservoir for an expanding population, which may well carry the mutations and be maintained with the essential tumor DNA mutations to repopulate after further rounds of therapy. Commitment to a particular mutagenic pattern should perhaps be seen as both a limited and un-stem like response to treatment based on natural selection/survival. Rather, a potent epigenetic silencing (Feinberg et al., 2006) of susceptibility genes (while accompanied by activation of enhancer genes) is probably more likely and has been seen in some experimental models of prostate cancer with cell lines (Pellacani et al., 2011; Yan et al., 2011).

Differentiation Capacity of CSCs

A key feature of the CSC hypothesis is that the CSC can regenerate the more differentiated cell types seen in the bulk of a tumor, for example in prostate cancer a more androgen-receptor-expressing luminal phenotype. It is clear from phenotyping carried out more than 10 years ago that the initial responses to castration of the LAPC9 xenograft is indeed an enrichment of a more primitive basal type (Craft et al., 1999), but the tumors that emerge after castration therapies are more frequently a replicating luminal, androgen receptor positive, PSA-secreting tumor (hence the successful use of PSA for monitoring), although a small minority of cases seem to lack this property. What is more intriguing in terms of differentiation is the response to castration/chemotherapy in prostate cancer, where the resistant lesion is very frequently not luminal but has a much more neuroendocrine phenotype (Abrahamsson, 1998). Is this the product of a trans-differentiation of a luminal cell to a more neuroendocrine cell or is it a re-awakening of the neuroendocrine pathway in a more basal undifferentiated cell type, allowing this to repopulate the tumor? Lastly and by implication from the previous section, the CSCs should have a more primitive and also a distinctive phenotype than the rest of the tumor. This is important because it provides the basis for therapy resistance based on the particular differentiated cell products commonly selected as therapeutic targets, such as prostate-specific antigen, prostatic acid phosphatase, prostate-specific membrane antigen, and prostate stem cell antigen. A primitive phenotype also provides the possibility of identifying new CSC targets for therapy, but will impose new requirements for the output of CSC-specific treatments, as the industry standard assays for cell proliferation and tumor growth are unlikely to be applicable (as discussed earlier).

Hallmarks of the CSC Phenotype

In the lower panel of Table 2.6B, a number of properties are also included that appear in many papers on prostate CSCs. These properties are desirable and are commonly investigated but do not address the functional significance of the cancer and therefore can be misleading, particularly in a tumor context where phenotype can be radically altered without affecting the behavior of the cell. The hypothesis of a pseudo-stem cell, which has activated a few genes, or another cell type, which one or two markers of EMT have been activated, is a very common output from this research. Hallmarks should therefore focus on function as allied to phenotype, rather than a phenotype which can be fitted into function. These subsidiary hallmarks include expression of stem-like genes and activation of stem cell signaling. Interestingly, CSCs appear to have a very particular epigenetic imprint, in a very similar way to embryonic and tissue stem cells. Of all cells, a SC must retain a degree of flexibility to allow it to be maintained through the various selective pressures imposed upon the whole organism through its lifetime. A rigidly programmed (or mutated) stem cell is unlikely to survive and therefore would lead to the extinction of the animal. In an essential organ for reproduction such as the prostate, such redundancy and survival characteristics are even more likely to have been preserved.

An understudied property in both SC and CSCs is the ability of the stem cell to divide asymmetrically to provide another daughter SC (self-renewal) and a transit-amplifying (TA) cell capable of helping to repopulate the tissue. Such asymmetric division is vital for tissue maintenance. A SC must retain the ability to divide symmetrically, in order to expand the pool in response to damage, but also to increase the amplifying population by providing two daughter cells again in terms of rapid regeneration of the tissue. Elements of asymmetric division have already been studied in prostate cell lines by quantifying the production of differentiated progeny (Qin et al., 2012) and also in our own experiments on prostate tissues by relating the presence of the TMPRSS 2-ERG to its expression within prostate cancers and normal tissues (Polson et al., 2013).

Several investigators have used prostaspheres (Duhagon et al., 2010), although our own data (Lang et al., 2010) and a more recent report suggests that this is not feasible from primary tissues in prostate (Chen et al., 2012a). It was for this reason that we grow our prostate CSCs in a two-dimensional feeder system, on a collagen matrix similar to that used for embryonic stem cells (Collins et al., 2005), which maintains the SCs or at least allows a modest (arithmetic) expansion, compared to the more rapid geometric expansion of the cycling transit amplifying and committed basal cells, which depletes the SC pool. Induction of differentiation to luminal cells also results in a depletion of the SC pool in vitro.

Drug Development for CSC Therapeutics: Good Drug-Bad Test or Vice Versa?

Assuming that we are now able to identify CSCs in prostate cancers, and know something of their properties and phenotype, how can we exploit this knowledge to develop treatments? Based on the biological properties, it is clear that traditional drug discovery methods will not yield significant data.

What biological systems are available to test the CSC hypothesis and perhaps sustain a therapeutic development program? These are listed in Table 2.7, emphasizing both strengths and weaknesses. In the time of Koch (postulate 2), the ability to culture and propagate an organism was very limited, and even now some of the world's most dangerous pathogens resist laboratory propagation. To what extent can we actually propagate the true CSC, and if we do so, have we irrevocably altered its phenotype, given the quiescent nature of CSCs in vivo, as discussed in the previous section? The high throughput strategies of the pharmaceutical industry require large cell numbers, but apart from some of the established cell lines, these are unlikely to be available for CSCs. It is perhaps better to screen the stem cells in their normal environment, as part of a colony/prostasphere or indeed a primary tumor, and to subsequently screen for SC content before and after treatments. It is equally important to measure the content of other cell types since the CSCs could differentiate, as well as die, as a result of the treatment. To overcome this restriction, we have employed 'near patient' xenografts from high Gleason grade and CRPC patients as outlined in Figure 2.2.

Although xenografted tumors lack the prostatic stromal environment, they can be viewed as a test for metastatic disease. As illustrated, this system permits the analysis of the following: (i) tumor induction after an ex vivo treatment of intact or disaggregated tumor, (ii) sustained inhibition of tumor

Table 2.7. Model Systems for CSC Therapeutic Development

Model System	Advantages	Disadvantages
Human tumors	Closest to reality	Difficult to manipulate (patient consent) Heterogeneous material
Primary Xenografts of human tumors	Best phenocopy of primary disease Provides increased amounts of CSCs for study Can be manipulated	Difficult to establish Heterogeneous Role of mouse tissues No immune response
Established xenografts of human tumors	Mimic original tumors No limitations in material for study Can be manipulated ex vivo	Heterogeneous Contamination (retroviruses) No immune system Role of mouse tissues Genomic instability
Primary cell cultures of human tumors	Mimic original tumors Ease of manipulation Fewer limitations on material	Heterogeneous (cancer origins?) Culture induced changes Growth medium (microenvironment) induced changes
Human cell lines	No limitations on study material Ease of manipulation 'Industry standard'	Heterogeneous (clonal variations in culture) Growth condition-dependence of phenotype Genetic instability Relevance to primary human cancers
Autochthonous mouse models	In vivo tumor initiation Mimic human disease in premalignant stages Intact immune responses	Mice do not spontaneously develop prostate cancer Mouse prostate differs both anatomically and in cell content from human gland Gene knockout versus haploinsufficiency
Transgenic oncogene-driven mouse models	In vivo tumor initiation Mimic human disease in premalignant stages Intact immune responses	Mice do not spontaneously develop prostate cancer Mouse prostate differs both anatomically and in cell content from human gland Multiple effects of viral oncogenes

induction by conducting a secondary graft, (iii) treatment of established tumors, which is unlikely to have a gross anatomical effect, followed by disaggregation and counting cell proportions, and finally (iv) whether CSC treatment of an intact tumor, in contrast to (i), can prevent tumor relapse.

The other major concern in these experiments is the effect of treatments on normal SCs, both in the prostate and in other tissues. Many of the commonly targeted developmental signaling pathways (e.g., notch, hedgehog, and wnt) are essential for tissue homeostasis elsewhere. These effects are unlikely to be seen in short-term assays either in vivo or in vitro. Undesirable side effects such as the induction of hemangiomas after long-term inhibition of the DLL4 component of notch signaling (Yan et al., 2010), may not represent a major risk in elderly male prostate cancer patients, and could be acceptable in older men with castration-resistant prostate cancer with a life expectancy of around 2 years. One argument against the application of CSC therapy in CRPC patients is the likelihood that there are multiple CSCs, all clonally derived, but with different phenotypes in response to first line treatments. Currently, we simply do not know enough to predict this. It would therefore be more practical and scientifically logical to treat earlier stage disease (with a predictably more

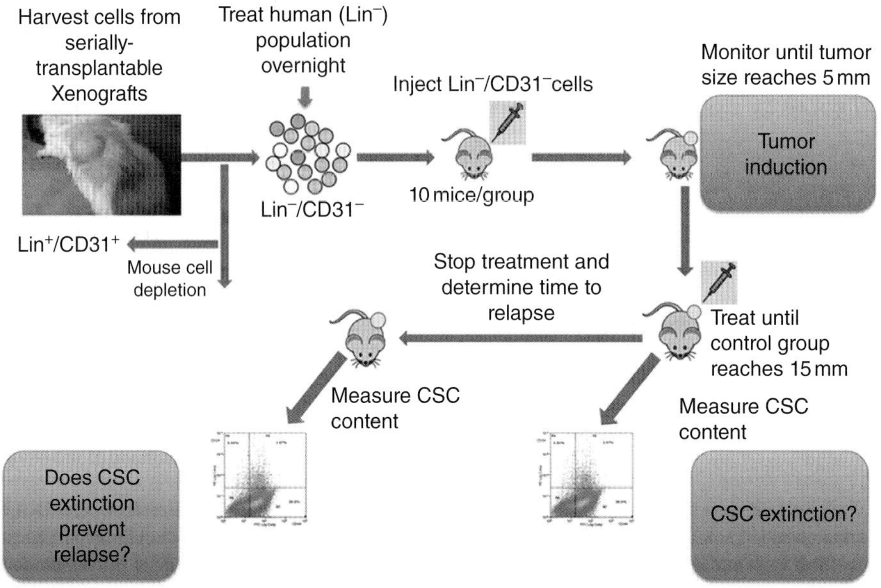

Figure 2.2. Xenograft-based strategies to test the cancer stem cell hypothesis. Four different levels of analysis are illustrated:

1. By ex vivo treatment of the constituent cells from a pre-formed primary xenograft with an agent that targets CSCs, regrafting of the CSC-depleted population of *live* bulk tumor cells should be unable to initiate new tumor growth. If the bulk tumor cells can regenerate CSC-like cells, this assay will fail.
2. If a tumor initiates in these experiments, there is the possibility that it is derived from a stem-like transit-amplifying cell, which should be unable to sustain secondary grafting.
3. Direct treatment intratumorally (or systemically) with an anti-CSC agent is unlikely to have any short-term effect of tumor bulk or growth rate. However, retrieval of the treated tumor and cell phenotype analysis (by FACS, for example) would measure efficacy, and regrafting should indicate whether an effect upon tumor initiation has been achieved.
4. The same effects are seen after termination of anti-CSC treatment. If (3.) indicates elimination of the CSC phenotype in the tumor, what is the long-term outcome of a secondary (conventional treatment)? Frequent relapse would negate the hypothesis.

Table 2.8. Experimental Approaches to Cancer Stem Cell Therapies

Treatment	Primary Assay	Comments
Direct killing of CSCs	Induction of apoptosis, necrosis autophagy Detected by FACS or colony forming/tumor induction assays	Preventative, long term Stem cell resistance to small molecules No/little effect on tumor size Access for agents in vivo
Prevention of CSC-mediated tumor spread	Boyden chamber Xenograft	Not permanent, SC can develop variants over time No effect on tumor size
Induction of CSC differentiation	FACS analysis of cell populations (and proportions) in vitro and in vivo Tumor induction/colony formation/ sphere formation	Depletes SC pool More differentiated progeny can be killed by other means No/little/negative primary effect on tumor size
Combination therapies	Tumor induction and recurrence	Life extension and tumor shrinkage

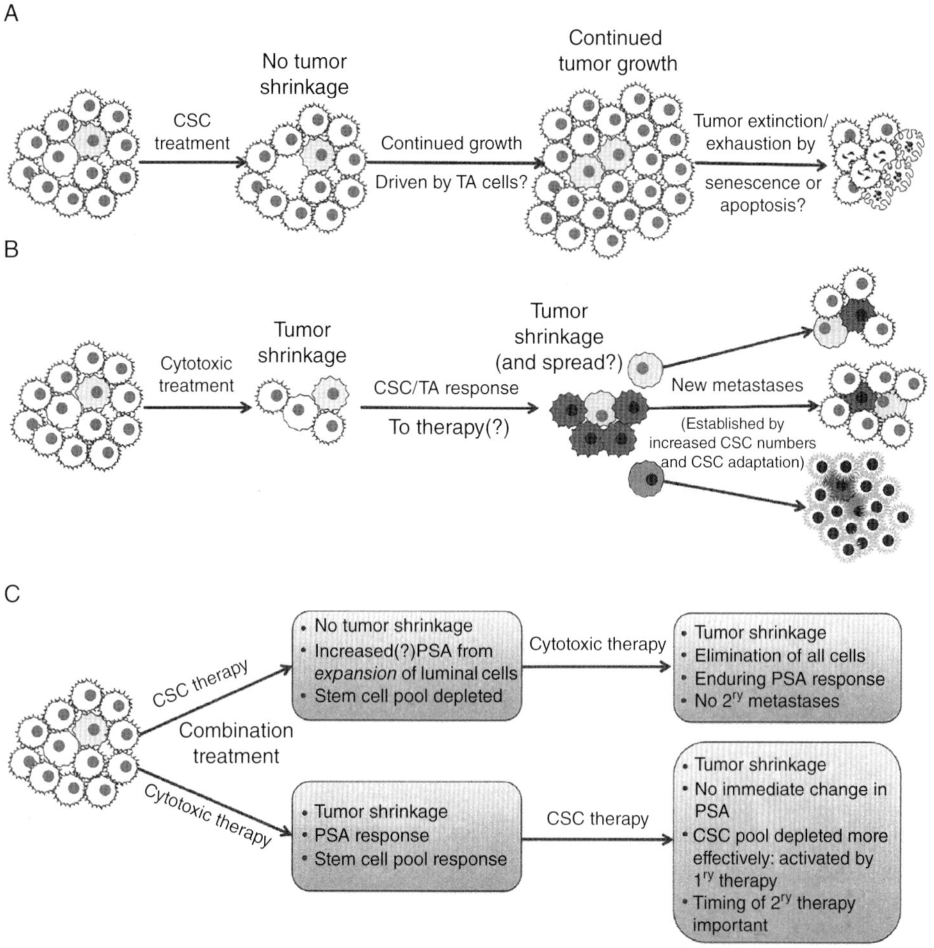

Figure 2.3. Predicted outcomes of CSC and conventional treatments on heterogeneous tumors, either as single agents or in combination. **A**. Effects of a single CSC treatment leads to unsustainable tumor growth and eventual tumor extinction. The time to tumor extinction is unknown and could vary between tumors even from the same tissue of origin. **B**. Conventional chemotherapy, which kills the replicating tumor bulk tissues, results either in enrichment of the CSC/TA fraction, or indeed to a 'wound healing' amplification of CSCs, potentially increased metastatic spread. Successful new tumors have probably undergone a new mutation/selection resulting in the emergence of new clonal variants. We do not know whether this occurs in the SC or in a TA fraction. **C**. The order and timing of a combination therapy to eliminate both the CSC and bulk (replicating tumor cells) is likely to be crucial to achieve the desired effects. The potential outcomes are illustrated. For color detail, please see color plate section.

homogeneous CSC population), but this is unlikely to be the patient population of choice for early stage clinical trials.

The readouts of the efficacy of anti-CSC agents will therefore differ from those normally used by the pharmaceutical industry. As shown in Table 2.8, direct killing, prevention of spread, and perhaps induction of differentiation from the resistant stem cell phenotype are all realistic possibilities, assuming that inherent SC resistance can be overcome (Frame and Maitland, 2013). How then, apart from cell fractionation and quantification, can the effects of a drug on CSCs be monitored? A return to the hallmarks might be advantageous here. If the CSC hypothesis is correct, then the primary function of a CSC is to induce the growth of a new tumor, either in the original organ, or with higher-grade cancers, at metastatic sites. Therefore, the readout of such treatment assays should not rely on proliferation rates (which would actually measure the progeny of the CSC: TA and other cells) but rather the ability to found new growth in vitro in the form of primary (and secondary) colony formation, in either 2D or 3D. Ultimately, the assay of choice for preclinical studies must be in a relevant in vivo model. It is now evident that simplistic xenografts of 30-year-old cell lines are not adequate. Various arguments concerning the applicability of xenograft versus autochthonous mouse models are also summarized in Table 2.7. Perhaps we do need to consider whether the very different anatomical structure of the rodent prostate is indeed relevant to human disease (Pienta et al., 2008; Shen and Abate-Shen, 2010), or take into account the lack of a viable immune system in immunocompromised mouse xenografts, which might overestimate or underestimate agent effectiveness. However, in both cases, perhaps the greatest consideration is the grade of prostate cancer under study. As discussed above, it is unlikely that early stage, low Gleason grade prostate tumors would be used in the testing and targeting of new CSC therapeutics. Therefore, an immunecompromised mouse carrying castration-resistant prostate cancer (CRPC) tumors, independent of the graft site, remains a viable model of late stage disease. As illustrated in Figure 2.3, CSC treatments are unlikely to shrink such cancers when used in isolation, and the bulk of the tumor should be able to retain a proliferative capacity in the course of such experiments (see Hallmarks of Cancer, 2000 and 2012). In fact, it may be that the serum PSA levels of a prostate cancer patient on a CSC trial would rise, rather than fall initially. However, if the tumor-inducing capacity has been eliminated, then the secondary grafting capacity (in both autochthonous and xenograft systems) would be compromised. There is also the prospect (Figure 2.3A) that if the CSCs provide a renewing core for the cancer that the tumor will become exhausted with time (which could however be many years later) and with resulting morbidity. At its most simplistic level, if CSCs are indeed responsible for initiation of new metastatic growth, then their elimination could restrict the development of the fatal multiple metastases.

Prospects for Therapeutic Applications: Combinations Are Better

The lack of an immediate response and tumor shrinkage in the presence of an active CSC agent (Figure 2.3C) is not a desirable output, either from a late-stage preclinical test, and more importantly in human clinical trials. In the past, such slow or negative responses would terminate a drug development program. The answer to CSC therapeutics may therefore lie in the reject bin from tests of combinatorial compound libraries, already carried out during the boom in the chemical biology.

It would seem more sensible therefore to combine the CSC effect with another agent capable of tumor bulk shrinkage. This was achieved by combining irinotecan with an anti-DLL4 antibody by Hoey and others (2009). This promising study could be applied in other situations such as prostate tumors; however, the timing of the dual treatments is still far from optimized (Figure 2.3). For example, should the SC treatment precede the debulking, or should the overt cytotoxicity be used to 'expose' the SC compartment of the tumors, perhaps enhancing the CSC treatment? Adjuvant hormone therapy to debulk tumors before radiotherapy has been remarkably successful in this regard (Payne and Mason, 2011) and may inadvertently have achieved this aim. Since prostate CSCs are less radiosensitive compared to the tumor mass (Frame et al., 2013), perhaps the response to adjuvant hormone therapy is achieved by increasing the CSC killing efficiency by radiation.

Equally, what would the effects of such treatments be on the tumor microenvironment? If CSCs constitute a reservoir for relapse and alternative differentiation, would dual treatment simply eliminate the original phenotype and permit a new more aggressive therapy resistant variant to emerge? There is perhaps evidence of a negative CSC effect (as outlined in Figure 2.3B) in the data from the Reduction by Dutasteride of Prostate Cancer Events (REDUCE) trial and Prostate Cancer Prevention Trial (PCPT) (Andriole et al., 2010), where the expected result (i.e., a 25% reduction in the incidence of low grade [Gleason <6] prostate cancer rate) was seen. However, of the cancers detected, many more aggressive, high-grade (Gleason 8–10) tumors were seen compared to placebo treatment. While androgen blockade would be expected to change the histo-pathological phenotype of the cancers to that observed in the PCPT trial, the significantly increased frequency of aggressive, ultimately fatal disease could clearly have arisen as a result of CSC enrichment. Thus, treatment with an inhibitor of androgen response had changed the natural history of the disease.

Ultimately, the presence of CSCs in prostate cancer is teaching us that the cells in a cancer are not homogeneous in their responses to therapy, and just because the majority of the population is susceptible to a particular treatment, this does not imply that the effects of this treatment on a minority, tumor-initiating population are zero, and may even be detrimental to the survival of the patient. We are already aware that prostate cancer treatment changes the natural history of the disease; CRPC overall has a more neuroendocrine phenotype (Abrahamsson, 1998; Matei et al., 2012), but is this reflected in the stem cell population of CRPC as well?

Conclusion: Proceed With Caution

There is still much to understand before the first CSC efficacy trial in man, and premature application could be faced with the same disastrous results produced in the early days of gene therapy (Buckley, 2002). To quote Sir Peter Medawar, the pioneering British immunologist, "I cannot give any scientist of any age better advice than this: The intensity of a conviction that a hypothesis is true has no bearing over whether it is true or not." A systematic approach to hypothesis testing, as advocated by Koch, but without some of the rigidity that limited pathogenicity studies, rather than a headlong rush to the clinic, should begin to answer the outstanding issues regarding cancer stem cells as a therapeutic target.

Acknowledgments

The research underpinning this review is supported by Yorkshire Cancer Research. We thank Michelle Scaife for assistance in producing the manuscript.

References

Abrahamsson, P.A. (1998). Neuroendocrine differentiation and hormone-refractory prostate cancer. Prostate. 8, 37–42.

Adams, M.D., Kerlavage, A.R., Fleischmann, R.D., Fuldner, R.A., Bult, C.J., Lee, N.H., Kirkness, E.F., Weinstock, K.G., Gocayne, J.D., and White, O. (1995). Initial assessment of human gene diversity and expression patterns based upon 83 million nucleotides of cDNA sequence. Nature 377, 3–174.

Andriole, G.L., Bostwick, D.G., Brawley, O.W., Gomella, L.G., Marberger, M., Montorsi, F., Pettaway, C.A., Tammela, T.L., Teloken, C., Tindall, D.J., et al. (2010). Effect of dutasteride on the risk of prostate cancer. N Engl J Med 362, 1192–1202.

Baker, M. (2012). Cancer stem cells tracked. Nature 488, 13–14.

Birnie, R., Bryce, S.D., Roome, C., Dussupt, V., Droop, A., Lang, S.H., Berry, P.A., Hyde, C.F., Lewis, J.L., Stower, M.J., et al. (2008). Gene expression profiling of human prostate cancer stem cells reveals a pro-inflammatory phenotype and the importance of extracellular matrix interactions. Genome Biol. 9, R83.

Bodmer, W., and Loeb, L.A. (2008). Genetic Instability Is Not a Requirement for Tumor Development. Cancer Research 68, 3558–3561.

Bonnet, D., and Dick, J.E. (1997). Human acute myeloid leukemia is organized as a hierarchy that originates from a primitive hematopoietic cell. Nat Med 3, 730–737.

Bosetti, C., Rosato, V., Gallus, S., and La Vecchia, C. (2012). Aspirin and urologic cancer risk: an update. Nat Rev Urol 9, 102–110.

Bubendorf, L., Kononen, J., Koivisto, P., Schraml, P., Moch, H., Gasser, T.C., Willi, N., Mihatsch, M.J., Sauter, G., and Kallioniemi, O.P. (1999). Survey of gene amplifications during prostate cancer progression by high-throughout fluorescence in situ hybridization on tissue microarrays. Cancer Research 59, 803–806.

Buckley, R.H. (2002). Gene therapy for SCID—a complication after remarkable progress. Lancet 360, 1185–1186.

Chan, T.Y., Partin, A.W., Walsh, P.C., and Epstein, J.I. (2000). Prognostic significance of Gleason score 3 + 4 versus Gleason score 4 + 3 tumor at radical prostatectomy. Urology 56, 823–827.

Chen, C.D., Welsbie, D.S., Tran, C., Baek, S.H., Chen, R., Vessella, R., Rosenfeld, M.G., and Sawyers, C.L. (2003). Molecular determinants of resistance to antiandrogen therapy. Nat Med 10, 33–39.

Chen, S., Principessa, L., and Isaacs, J.T. (2012a). Human prostate cancer initiating cells isolated directly from localized cancer do not form prostaspheres in primary culture. Prostate 72, 1478–1489.

Chen, Y., Li, D., Wang, D., Liu, X., Yin, N., Song, Y., Lu, S.H., Ju, Z., and Zhan, Q. (2012b). Quiescence and attenuated DNA damage response promote survival of esophageal cancer stem cells. J. Cell. Biochem.

Chen, W., and Wang, G.-M. (2012c). Gene expression profiling of cancer stem cells in the Du145 prostate cancer cell line. Oncol Lett. 3(4): 791–796.

Choi, N., Zhang, B., Zhang, L., Ittmann, M., and Xin, L. (2012). Adult murine prostate basal and luminal cells are self-sustained lineages that can both serve as targets for prostate cancer initiation. Cancer Cell 21, 253–265.

Collins, A.T., Berry, P.A., Hyde, C., Stower, M.J., and Maitland, N.J. (2005). Prospective identification of tumorigenic prostate cancer stem cells. Cancer Research 65, 10946–10951.

Conheim J. Congenitales, quergestreiftes muskelsarkon der nireren. Virchows Arch. 1875; 65:64.

Craft, N., Chhor, C., Tran, C., Belldegrun, A., DeKernion, J., Witte, O.N., Said, J., Reiter, R.E., and Sawyers, C.L. (1999). Evidence for clonal outgrowth of androgen-independent prostate cancer cells from androgen dependent tumors through a two-step process. Cancer Research 59, 5030–5036.

De Marzo, A.M., Platz, E.A., Sutcliffe, S., Xu, J., Grönberg, H., Drake, C.G., Nakai, Y., Isaacs, W.B., and Nelson, W.G. (2007). Inflammation in prostate carcinogenesis. Nat Rev Cancer 7, 256–269.

Duhagon, M.A., Hurt, E.M., Sotelo-Silveira, J.R., Zhang, X., and Farrar, W.L. (2010). Genomic profiling of tumor initiating prostatospheres. BMC Genomics 11, 324.

Fan, X., Liu, S., Su, F., Pan, Q., and Lin, T. (2010). Effective enrichment of prostate cancer stem cells from spheres in a suspension culture system. Urol Oncol. 30(3):314–8.

Feinberg, A.P., Ohlsson, R., and Henikoff, S. (2006). The epigenetic progenitor origin of human cancer. Nature Reviews Genetics. 7, 21–33.

Frame, F. M., and Maitland, N.J. (2013) Cancer stem cells provide new insights into the therapeutic responses of human prostate cancer. S Cramer (Editor), Stem Cells and Prostate Cancer, DOI 10.1007/978-1-4614-6498-3_4 Springer Science + Business Media.

Frame, F.M., Pellacani, D., Collins, A.T., Simms, M.S., Mann, V.M., Jones, G., Meuth, M., Bristow, R.G., and Maitland, N.J. (2013). HDAC inhibitor confers radiosensitivity to prostate stem-like cells. Br J Cancer Nov 12. doi: 10.1038/bjc.2013.691. [Epub ahead of print]

Fredericks, D.N., and Relman, D.A. (1996). Sequence-based identification of microbial pathogens: a reconsideration of Koch's postulates. Clin. Microbiol. Rev. 9, 18–33.

Hällström, T.M.K.-A., and Laiho, M. (2008). Genetic changes and DNA damage responses in the prostate. Prostate 68, 902–918.

Hamburger, A.W., and Salmon, S.E. (1977). Primary Bioassay of Human Tumor Stem Cells. Science 197, 461–463.

Hanahan, D., and Weinberg, R.A. (2000). The hallmarks of cancer. Cell.100, 57–70.
Hanahan, D., and Weinberg, R.A. (2011). Hallmarks of cancer: the next generation. Cell.144, 646–74.
Hanson, R.P. (1988). Koch is dead. J.Wildl. Dis. 24, 193–200.
Hoey, T., Yen, W.-C., Axelrod, F., Basi, J., Donigian, L., Dylla, S., Fitch-Bruhns, M., Lazetic, S., Park, I.-K., Sato, A., et al. (2009). DLL4 Blockade Inhibits Tumor Growth and Reduces Tumor-Initiating Cell Frequency. Stem Cell 5, 168–177.
Hurt, E.M., Kawasaki, B.T., Klarmann, G.J., Thomas, S.B., Farrar, W.L. (2008). CD44+ CD24(−) prostate cells are early cancer progenitor/stem cells that provide a model for patients with poor prognosis. British Journal of Cancer 98(4), 756–765.
Ishizawa, K., Rasheed, Z.A., Karisch, R., Wang, Q., Kowalski, J., Susky, E., Pereira, K., Karamboulas, C., Moghal, N., Rajeshkumar, N.V., et al. (2010). Tumor-Initiating Cells Are Rare in Many Human Tumors. Stem Cell 7, 279–282.
Koch, R. (1884). Die Aetiologie der Tuberkulose. Mitt Kaiser Gesundh 2, 1–88.
Lang, S.H., Smith, J., Hyde, C., Macintosh, C., Stower, M., and Maitland, N.J. (2006). Differentiation of prostate epithelial cell cultures by matrigel/stromal cell glandular reconstruction. In Vitro Cell Dev Biol Anim 42, 273–280.
Lang, S.H., Anderson, E., Fordham, R., and Collins, A.T. (2010). Modeling the Prostate Stem Cell Niche: An Evaluation of Stem Cell Survival and Expansion In Vitro. Stem Cells and Development 19, 537–546.
Li, L., and Bhatia, R. (2011). Stem cell quiescence. Clinical Cancer Research 17, 4936–4941.
Li, P., Yu, X., Ge, K., Melamed, J., Roeder, R.G., and Wang, Z. (2002). Heterogeneous expression and functions of androgen receptor co-factors in primary prostate cancer. Am J Pathol 161, 1467–1474.
Liu, T., Xu, F., Du, X., Lai, D., Liu, T., Zhao, Y., Huang, Q. (2010). Establishment and characterization of multi-drug resistant, prostate carcinoma-initiating stem-like cells from human prostate cancer cell lines 22RV1. Mol Cell Biochem. 2010 Jul;340(1–2):265–73.
Loeb, L.A., Bielas, J.H., Beckman, R.A., and Bodmer, I.W. (2008). Cancers Exhibit a Mutator Phenotype: Clinical Implications. Cancer Research 68, 3551–3557.
Ma, Y., Liang, D., Liu, J., Axcrona, K., Kvalheim, G., Stokke, T., Nesland, J.M., and Suo, Z. (2011). Prostate cancer cell lines under hypoxia exhibit greater stem-like properties. PLoS ONE 6, e29170.
Maitland, N.J., and Collins, A.T. (2008). Inflammation as the primary aetiological agent of human prostate cancer: a stem cell connection? J. Cell. Biochem. 105, 931–939.
Maitland, N.J., and Collins, A.T. (2010). Cancer stem cells: A therapeutic target? Curr. Opin. Mol. Ther. 12, 662–673.
Maitland, N.J., Frame, F.M., Polson, E.S., Lewis, J.L., and Collins, A.T. (2011). Prostate cancer stem cells: do they have a basal or luminal phenotype? Horm Cancer 2, 47–61.
Marian, C.O., and Shay, J.W. (2009). Prostate tumor-initiating cells: A new target for telomerase inhibition therapy? BBA—Molecular Basis of Disease 1792, 289–296.
Markman, M. (2012). Risk of Cervical Cancer after HPV Vaccination. Curr Pharm Des. (EPUB AHEAD OF PRINT)
Matei, D.-V., Renne, G., Pimentel, M., Sandri, M.T., Zorzino, L., Botteri, E., De Cicco, C., Musi, G., Brescia, A., Mazzoleni, F., et al. (2012). Neuroendocrine differentiation in castration-resistant prostate cancer: a systematic diagnostic attempt. Clin Genitourin Cancer 10, 164–173.
Meeker, A.K., Gage, W.R., Hicks, J.L., Simon, I., Coffman, J.R., Platz, E.A., March, G.E., and De Marzo, A.M. (2002). Telomere Length Assessment in Human Archival Tissues Combined Telomere Fluorescence in Situ Hybridization and Immunostaining. Am J Pathol 160, 1259–1268.
Miki, J., Furusato, B., Li, H., Gu, Y., Takahashi, H., Egawa, S., Sesterhenn, I.A., McLeod, D.G., Srivastava, S., Rhim, J.S. (2007). Identification of putative stem cell markers, CD133 and CXCR4, in hTERT-immortalized primary nonmalignant and malignant tumor-derived human prostate epithelial cell lines and in prostate cancer specimens. Cancer Res. 2007 67(7):3153–61.
Moore, P.S., and Chang, Y. (2010). Why do viruses cause cancer? Highlights of the first century of human tumour virology. 1–12. Nat. Rev. Cancer 10, 878–89.
Oldridge, E.E., Pellacani, D., Collins, A.T., and Maitland, N.J. (2012). Prostate cancer stem cells: Are they androgen-responsive? Mol. Cell. Endocrinol. 360, 14–24.
Patrawala, L., Calhoun, T., Schneider-Broussard, R. (2005). Side population is enriched in

tumorigenic, stem-like cancer cells, whereas ABCG2+ and ABCG2− cancer cells are similarly tumorigenic. Cancer Research. Jul 15; 65(14):6207–19.

Patrawala, L., Calhoun, T., Schneider-Broussard, R., Li, H., Bhatia, B., Tang, S., Reilly, J.G., et al. (2006). Highly purified CD44+ prostate cancer cells from xenograft human tumors are enriched in tumorigenic and metastatic progenitor cells. Oncogene, 25(12), 1696–1708.

Patrawala, L., Calhoun-Davis, T., Schneider-Broussard, R., Tang, D.G. (2007). Hierarchical organization of prostate cancer cells in xenograft tumors: the CD44+alpha2beta1+ cell population is enriched in tumor-initiating cells. Cancer Research, 67(14), 6796–6805.

Payne, H., and Mason, M. (2011). Androgen deprivation therapy as adjuvant/neoadjuvant to radiotherapy for high-risk localised and locally advanced prostate cancer: recent developments. Br J Cancer 105, 1628–1634.

Pellacani, D., Packer, R.J., Frame, F.M., Oldridge, E.E., Berry, P.A., Labarthe, M.-C., Stower, M.J., Simms, M.S., Collins, A.T., and Maitland, N.J. (2011). Regulation of the stem cell marker CD133 is independent of promoter hypermethylation in human epithelial differentiation and cancer. Mol. Cancer 10, 94.

Pienta, K.J., Abate-Shen, C., Agus, D.B., Attar, R.M., Chung, L.W.K., Greenberg, N.M., Hahn, W.C., Isaacs, J.T., Navone, N.M., Peehl, D.M., et al. (2008). The current state of preclinical prostate cancer animal models. Prostate 68, 629–639.

Polson, E.S., Lewis J.L., Celik H., Mann V.M., Stower M.J., Simms M.S., Rodrigues G., Collins A.T., and Maitland N.J. (2013). Monoallelic expression of TMPRSS2/ERG in prostate cancer stem cells. Nature Communications, 4:1623. doi: 10.1038/ncomms2627.

Qin, J., Liu, X., Laffin, B., Chen, X., Choy, G., and Jeter, C.R. (2012). The PSA-/lo Prostate Cancer Cell Population Harbors Self-Renewing Long-Term Tumor-Propagating Cells that Resist Castration. Cell Stem Cell. 4, 556–569.

Quintana, E., Shackleton, M., Sabel, M., Fullen, D., Johnson, T., and Morrison, S. (2008). Efficient tumor formation by single human melanoma cells. Nature 456, 593–598.

Rajasekhar, V.K., Studer, L., Gerald, W., Socci, N.D., and Scher, H.I. (2011). Tumour-initiating stem-like cells in human prostate cancer exhibit increased NF-κB signalling. Nature Communications 2, 162.

Roberts, R.O., Bergstralh, E.J., Bass, S.E., Lieber, M.M., and Jacobsen, S.J. (2004). Prostatitis as a Risk Factor for Prostate Cancer. Epidemiology 15, 93–99.

Rowehl, R.A., Crawford, H., Dufour, A., Ju, J., Botchkina, G.I. (2008). Genomic analysis of prostate cancer stem cells isolated from a highly metastatic cell line. Cancer genomics & proteomics, 5(6), 301–310.

Scher, H.I., Fizazi, K., Saad, F., and Taplin, M.E. (2012). Increased survival with enzalutamide in prostate cancer after chemotherapy. N.E. J Med. 367, 1187–97.

Sell, S. (2010). On the stem cell origin of cancer. Am J Pathol.176, 2584–94.

Sfanos, K.S., and De Marzo, A.M. (2012). Prostate cancer and inflammation: the evidence. Histopathology 60, 199–215.

Shah, S.P., Roth, A., Goya, R., Oloumi, A., Ha, G., Zhao, Y., Turashvili, G., Ding, J., Tse, K., Haffari, G., et al. (2012). The clonal and mutational evolution spectrum of primary triple-negative breast cancers. Nature 486, 395–399.

Shen, M.M., and Abate-Shen, C. (2010). Molecular genetics of prostate cancer: new prospects for old challenges. Genes & Development 24, 1967–2000.

Shepherd, C.J., Rizzo, S., Ledaki, I., Davies, M., Brewer, D., Attard, G., de Bono, J., and Hudson, D.L. (2008). Expression profiling of CD133+ and CD133- epithelial cells from human prostate. Prostate 68, 1007–1024.

Taoudi, S., Bee, T., Hilton, A., Knezevic, K., Scott, J., Willson, T.A., Collin, C., Thomas, T., Voss, A.K., Kile, B.T., et al. (2011). ERG dependence distinguishes developmental control of hematopoietic stem cell maintenance from hematopoietic specification. Genes & Development 25, 251–262.

Taplin, M.-E., Rajeshkumar, B., Halabi, S., Werner, C.P., Woda, B.A., Picus, J., Stadler, W., Hayes, D.F., Kantoff, P.W., Vogelzang, N.J., et al. (2003). Androgen receptor mutations in androgen-independent prostate cancer: Cancer and Leukemia Group B Study 9663. J. Clin. Oncol. 21, 2673–2678.

True, L., Coleman, I., Hawley, S., Huang, C.Y., Gifford, D., Coleman, R., Beer, T.M., Gelmann, E., Datta, M., Mostaghel, E., Knudsen, B., Lange, P., Vessella, R., Lin, D., Hood, L., Nelson, P.S. (2006). A molecular correlate to the Gleason grading system for prostate adenocarcinoma. Proceedings of the National Academy of Sciences 103, 10991–10996.

van der Pluijm, G. (2011). Epithelial plasticity, cancer stem cells and bone metastasis formation. Bone 48, 37–43.

van Weerden, W., Bangma, C., and de Wit, R. (2008). Human xenograft models as useful tools to assess the potential of novel therapeutics in prostate cancer. Br J Cancer 100, 13–18.

Wang, X.I., Julio, M.K.-D., and Economides, K.D., Walker, D., Yu, H., Halili, M.V., Hu, Y.P., Price, S.M., Abate-Shen, C., Shen, M.M. (2009). A luminal epithelial stem cell that is a cell of origin for prostate cancer. Nature. 461, 495–500.

Welch, J.S., Ley, T.J., Link, D.C., Miller, C.A., Larson, D.E., Koboldt, D.C., Wartman, L.D., Lamprecht, T.L., Liu, F., Xia, J., et al. (2012). The origin and evolution of mutations in acute myeloid leukemia. Cell 150, 264–278.

Wodinsky, I., Swiniarski, J., and Kensler, C.J. (1968). Spleen colony studies of leukemia L1210. 3. Differential sensitivities of normal hematopoietic and resistant L1210 colony-forming cells to 6-mercaptopurine (NSC-755). Cancer Chemotherapy Reports Part 1 52, 251–255.

Wu, C., Wyatt, A.W., Lapuk, A.V., McPherson, A., McConeghy, B.J., Bell, R.H., Anderson, S., Haegert, A., Brahmbhatt, S., Shukin, R., et al. (2012). Integrated genome and transcriptome sequencing identifies a novel form of hybrid and aggressive prostate cancer. J. Pathol. 227, 53–61.

Xu, T., He, K., Wang, L., and Goldkorn, A. (2011). Prostate tumor cells with cancer progenitor properties have high telomerase activity and are rapidly killed by telomerase interference. Prostate 71, 1390–1400.

Yan, H., Chen, X., Zhang, Q., Qin, J., Li, H., Liu, C., Calhoun-Davis, T., Coletta, L.D., Klostergaard, J., Fokt, I., et al. (2011). Drug-tolerant cancer cells show reduced tumor-initiating capacity: depletion of CD44 cells and evidence for epigenetic mechanisms. PLoS ONE 6, e24397.

Yan, M., Callahan, C.A., Beyer, J.C., Allamneni, K.P., Zhang, G., Ridgway, J.B., Niessen, K., and Plowman, G.D. (2010). Chronic DLL4 blockade induces vascular neoplasms. Nature 463, E6–7.

Zhang, K., and Waxman, D.J. (2010). PC3 prostate tumor-initiating cells with molecular profile FAM65Bhigh/MFI2low/LEF1low increase tumor angiogenesis. Molecular cancer, 9, 319.

zur Hausen, H. (2002). Papillomaviruses and cancer: from basic studies to clinical application. Nat Rev Cancer. 2, 342–50.

Chapter 3
Self-Renewal, Induced Proliferation, and Autonomous Cell Growth Represent Distinct Modes of Cell Multiplication: Relevance to the Cancer Stem Cell Theory

Dov Zipori
Department of Molecular Cell Biology, Weizmann Institute of Science, Rehovot, Israel

Introduction

Hematopoietic stem cells (HSC) have been denoted as such because they give rise to the variety of blood cell lineages and to the multitudes of cell types within these lineages. In addition to this remarkable capacity, HSCs are supposed to self-renew (i.e., produce upon their division cells identical to the mother cell). This renewal property has traditionally been regarded as a bona fide stem cell trait. Self-renewal divisions may be symmetrical, leading to the formation of two identical daughter stem cells. An alternative is an asymmetrical division, wherein one daughter remains a stem cell while the other is a progenitor committed to differentiation. It has recently been suggested that the hemotopoietic system harbors yet another stem cell type: the leukemia stem cell, or as generally referred to, the cancer stem cell (CSC). The original reasons for denoting this latter cell type as a stem cell were that it harbors cell surface markers similar to those of HSCs. More importantly, upon transplantation into appropriate hosts, this cell re-creates a tumor, leukemia in this case, that resembles the disease from which the so-called leukemia stem cells were derived. I shall therefore refer to these cells herein as leukemia-initiating cells (LIC). It is noteworthy that leukemia, like other tumors, does not differentiate effectively, which is one of the major characteristics of tumorous tissues in general. Indeed, many tumors present an undifferentiated or non-functional phenotype. Therefore, a major difference between HSCs and the cells within tumors that initiate tumor formation in transplanted mice, the LICs, is the poor, or absent, differentiation capacity of the latter. This issue, that casts doubt on the cancer stem cell theory, has been reviewed in great detail along side other inconsistencies in the cancer stem cell theory (Zipori, 2009). LICs are often characterized using cell surface markers common with HSCs. One example is the $cKit^+$ $lineage^-$ chronic myelogenous leukemia (CML) LIC (Naka et al., 2010). However, the use of cell surface markers as a means of identifying TICs has been challenged (Clement et al., 2010). As an example, CD133 has been suggested to be a marker of neuronal stem cells and

Cancer Stem Cells, First Edition. Edited by Vinagolu K. Rajasekhar.

neuronal CSC, but CD133 negative cells are also tumor-initiating cells (TIC) (Lottaz et al., 2010). An additional trait of the so-called CSC is its rarity. Yet, this apparent property seems to result from the use of models that do not correctly predict tumor cell incidences (Bonnefoix and Callanan, 2010), as well as to animal models that restrict the growth of transplanted human cells. Similarly, low incidences of LICs are obtained when inappropriate culture conditions, which reduce the number of clonogenic cell proliferation in culture, are being employed. Indeed, improved animal models show that primary human prostatic tumors have a TIC incidence of 25% and probably higher (Quintana et al., 2008). In the colorectal tumor cell line, HCT116, almost all cells seem to be TICs (Yeung et al., 2010). The inconsistencies in the CSC theory drove many to use the term "stem cell-like" (Gao et al., 2010; Charles et al., 2010), "stem and progenitor" (Jeselsohn et al., 2010) or "stem/progenitor" (Fomchenko and Holland, 2005) with reference to the cellular entity otherwise called CSC. Notably, stem cells and progenitors are distinct cellular entities and differ markedly in their properties. Whereas stem cells rarely divide within their niches and are strictly niche dependent, progenitors are transiently expanding populations that proliferate extensively and are often niche independent (reviewed by Zipori, 2009). Progenitors readily grow in vitro, whereas HSCs are devoid of this capacity and die in vitro unless supported by stromal cells. It has been shown in the past and also more recently, that committed progenitors, rather then HSCs are the origin of LICs (Guibal et al., 2009; Bereshchenko et al., 2009; Bruns et al., 2009). Thus, several major aspects of the CSC theory remain doubtful.

Here, I focus on one major trait attributed to HSCs (i.e., their self-renewal capacity within their bone marrow niche). The question addressed is whether similarities exist between the mode by which HSCs and LICs reproduce themselves. Furthermore, a comparison is made between the mechanisms that regulate HSC self-renewal, within the bone marrow microenvironment, and the mode by which LICs are regulated. It is shown below that there are little or no similarities between these mechanisms. Therefore, the correlation drawn between HSCs and LICs is invalid. This together with other considerations, that I summarized in other publications (Zipori, 2009; Zipori, 2010) cast doubt on the cancer stem cell theory.

Do HSCs Self-Renew or Are They Just Maintained Throughout Life Span?

Upon transplantation of bone marrow cells into irradiated hosts, in either unpurified or otherwise as a fraction highly enriched with long-term repopulating (LTR)-HSCs, reconstitution of the recipients occurs. Enumeration of HSCs, indicates that this population is maintained, and increases in size over the life span of the transplanted animals. Such experiments clearly showed that new HSCs are generated from some precursors and were often interpreted as demonstrating the self-renewal capacity of HSCs. Nevertheless, this is not necessarily the case. Ample evidence exists in different progenitor cell systems, indicating that cells at later stages of differentiation, may "turn back" to gain stemness. The most obvious of all is the example of gonadal stem cells in *Drosophila* (Brawley and Matunis, 2004). Yet similar findings were also reported in mammalians in which spermatogonia, that descended the differentiation ladder, returned to an actual stem cell phenotype (Nakagawa et al., 2007). There is no unequivocal proof for HSC self-renewal resulting from symmetrical or asymmetrical divisions at this point. It is equally possible that generation of new HSCs occurs from the dedifferentiation of later stages in the maturation cascade. Questions regarding the self-regenerating capabilities of HSCs have been raised in the past (Chertkov, 1984) as well as in more recent (Metcalf, 2007) publications. There is a clear need to provide unequivocal proof

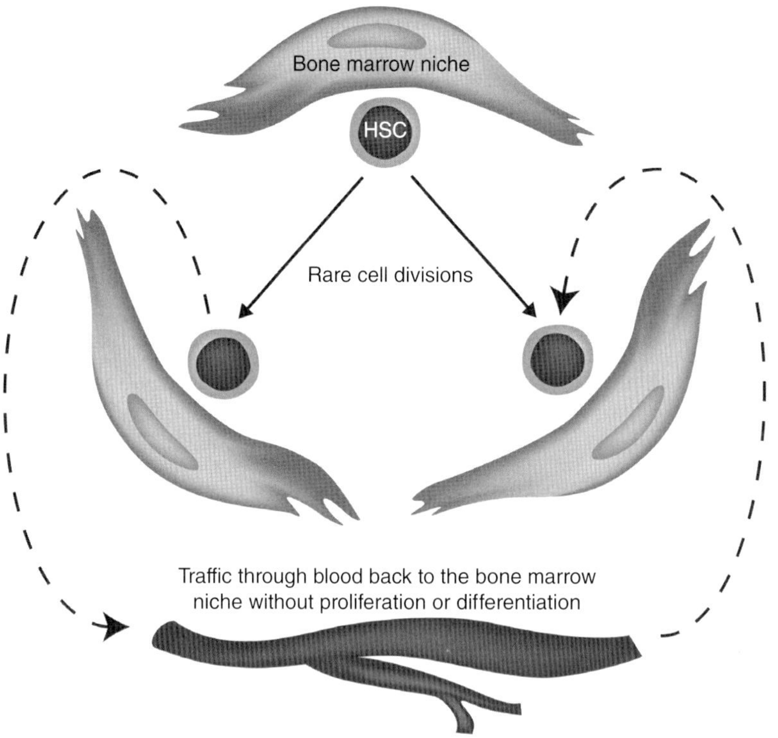

Figure 3.1. The HSC niche. HSCs depend on their bone marrow niche for survival and for the maintenance of their undifferentiated phenotype. These cells may divide rarely (*solid arrows*) to give rise to cells identical to them. HSCs may return to their niche, undifferentiated even following traffic through the blood (*dashed arrows*).

for the capacity of HSCs to maintain their numbers through their own divisions.

Even if HSCs do self-renew by division, how often does this occur? To put it differently, how robust is the self-renewal capacity of HSCs? Do these cells proliferate extensively? One estimate indicates that HSCs divide only five times throughout the mouse's life span (Wilson et al., 2008). Another study showed that the traffic of HSC, from the bone marrow into the blood and back into bone marrow niches, does not include cell division (Bhattacharya, et al., 2009) (Figure 3.1). In fact, a calculation of the number of HSCs present in the organism and their involvement in the differentiation processes shows that there are sufficient HSCs to cover the entire duration of a mammalian life span, without an urgent need for production of new stem cells (reviewed in Zipori, 2009). Thus, the fact that it has not been possible, to date, to cause HSC "expansion", even in the most sophisticated culture conditions, may simply mean that HSCs do not have this capacity at all. In fact, in vivo passage of HSCs by serial transplantation causes an abrupt decline in their numbers and subsequent exhaustion within two to three passages. The determining role of the bone marrow stem cell niche is also evident in aging. When gradual decline in stem cell functions is eventually observed during aging, it is related to damage caused to the stem cell niche by systemic signals (Mayack et al., 2010) rather than by stem cell autonomous processes.

It is therefore obvious from the information above that HSCs are mostly quiescent cells. Comparative analysis of HSC gene signatures, at an uninduced versus induced state, have been used to identify molecules relevant to the maintenance of the quiescent state of these cells. A list of molecules emerged, but it is still too early to conclude which of the molecules are the most relevant to this process (Forsberg et al., 2010).

LIC Proliferate Extensively

Before turning to the discussion on the properties of LICs, it is important to make the following distinction among several biological processes that include cell division. Self-renewal/maintenance of HSCs is enforced on the cells by their environment (i.e., the stem cell niche). This process is often confused in the literature with the general term "proliferation" being used. In the latter case, two options exist that do not relate to self-renewal: one is the division of transiently proliferating progenitors, triggered by cytokines, that ends in terminal differentiation with no further division. The second option is long-term proliferation, due to defects in the cellular division machinery, and diminished (or absent) responsiveness to external cues. These three forms of cell division (i.e., self-renewal, transient progenitor proliferation versus unlimited extensive proliferation) represent very different and distinct biological phenomena.

In sharp contrast to HSCs, LICs proliferate extensively, this being their most characteristic in vitro and in vivo property (Figure 3.2). A commonly used assay for TICs is colony

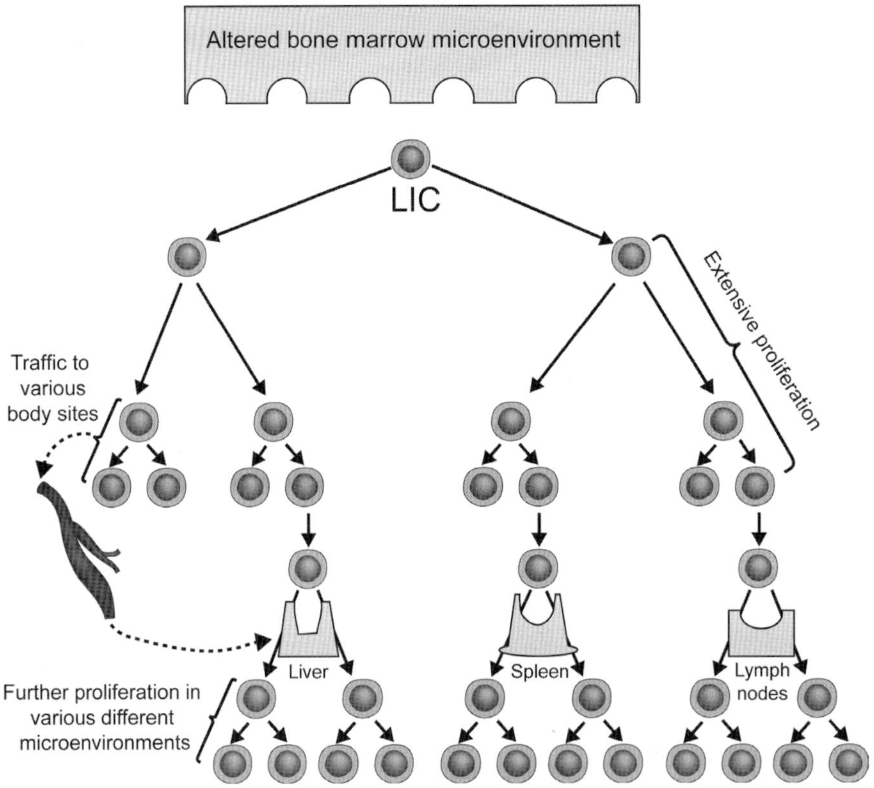

Figure 3.2. The microenvironment of leukemia cells. LICs proliferate in an altered microenvironment, different in nature from the normal one. These cells are endowed with a cell autonomous capacity of extensive proliferation, which is almost limitless. The traffic of LICs to distant organs leads to the formation of new sites of LIC proliferation.

formation in culture. Upon seeding in semisolid medium, a fraction of the tumor cells form large colonies, while others either form smaller size clusters (Yeung et al., 2010) or do not proliferate at all. The seeding efficiency of TICs varies between one type of tumor to another. In vitro colony formation, by normal bone marrow cells, is due to the proliferation of transiently expanding progenitor populations that cease proliferating after a period of a few days. They proliferate while gradually differentiating into mature progeny devoid of proliferation capacity. The end result is the formation of colonies that are reclonable only to a limited extent, if at all. In contrast, colonies produced by LICs are relatively easy to reclone, since they are a consequence of proliferation of a cell that does not differentiate well and therefore maintains its almost unlimited proliferation capacity.

Although colony formation in culture is often used as an assay system that measures the incidence of TICs, the more rigorous method is the analysis of the capacity of cells to form tumors in vivo. It is possible to inject LICs into susceptible animals and upon eruption of leukemia in the recipients, extract cells and inject them into a secondary recipient. This procedure can be repeated numerous times by passaging of cells to tertiary, quaternary recipients, etc. Clearly, HSCs cannot be passaged in such a manner. Upon two to three passages of HSCs, a decline process occurs and no rescue of the transplanted recipients happens (Eckner et al., 1982). In contrast, infection with a leukemia virus allows for many more passages. This demonstrates the fundamental difference between HSCs and LICs. The latter are endowed with a cell autonomous capacity to proliferate. In contrast, HSCs have a highly restricted capacity to maintain themselves and are certainly devoid of extensive proliferation potential. Thus, LICs grow in number as the leukemia spreads upon transplantation, while under similar conditions HSCs disappear.

MicroRNA-29a is normally expressed in HSCs and is downregulated in committed myeloid progenitors. The constitutive expression of this molecule in HSCs resulted in unlimited proliferation of the committed progenitors (Han et al., 2010). This again demonstrates the fundamental difference between HSC self-renewal, which is an inherently slow process, as compared to unrestricted growth of progenitors, which acquire a capacity to express a gene that they normally downregulate. Designating extensive proliferation, driven by genetic defects, as "self-renewal," is misleading.

The Nature of the HSC Niche

As detailed above, it is not entirely clear whether HSCs self-renew or are otherwise produced from cells at later stages of differentiation. If, for the moment we put aside, the issue of self-renewal and adhere to the question of maintenance, then the discussion becomes simpler. It is clear that HSCs are effectively maintained in the bone marrow niche throughout the life span of healthy individuals. This occurs due to the activity of the HSC niche, within the bone marrow microenvironment. The importance of this niche had been realized decades ago. However, the mode by which this specific microenvironment operates both at the cellular and at the molecular levels, is only partially understood. Nevertheless, much information has been gathered since the 1980s. First, it is clear that the cells, which are not descendents of HSCs, constitute a major component of the HSC niche. Collectively, these cells are called stroma. The latter is however an extremely heterogeneous population as cells that have been ascribed stromal functions are fibroblastoid cells, osteoblasts, adventitial reticular (ARC) cells, and pericytes, to mention just a few. Therefore, it is not possible, with any degree of confidence, to ascertain that a particular cell combination is the minimal structure of the HSC niche. A method of specific cell ablation is needed to determine unequivocally the cellular structure of the HSC niche. In this context, it is important to mention that mesenchymal stromal cells (MSC) are thought to contribute to the formation of the HSC niche since they support in vitro survival and growth of HSCs in a long-term manner.

In addition, MSCs have been shown to be identical to cells such as ARCs and pericytes (reviewed by Cai et al., 2009). These cells have been demonstrated to participate in the formation of the subendothelial niche of HSCs. Attempts to increase/reduce the incidence of some of these cells, such as osteoblasts, resulted in fluctuations in incidences of HSCs (Calvi et al., 2003). Although these experiments clearly implicate osteoblasts in contributing to the niche, in no way do they prove an absolute requirement of osteoblasts to HSC self-renewal. The mechanism by which the HSC niche operates, involves the need for close proximity between the HSC and the stromal cell (Zipori, 1981), as well as the activity of differentiation restraining factors that keep the stem cell undifferentiated (Zipori and Sasson, 1980 and reviewed by Zipori, 1990). A long list of molecules that the stromal elements produce and secrete have been suggested to contribute to niche formation (reviewed by Lane, 2009). These include mainly molecules that keep HSCs quiescent.

None of the cells or molecules that form the HSC niche has conclusively been shown to contribute to LIC survival and growth. One important distinction that should be emphasized here is between the terms "niche" and "microenvironment." The latter is a general term that indicates the closeness of any cell to its immediate neighbors. Thus, the microenvironment of differentiating erythroid progenitors is a central macrophage, and that of a differentiating thymocyte may be a thymic epithelial cell, widely represented within the thymic stroma. On the other hand, when the term "stem cell niche" is used, it denotes a specific, very rare, and highly specialized site within the bone marrow microenvironment, in which HSCs are maintained undifferentiated. The rarity of such niches matches the similarly low incidence of LTR-HSCs.

The Nature of the LIC Microenvironment

The question now raised is whether LIC are located in specific sites: Are these sites separate from the sites of localization of cells within tumors that do not perform as LIC? If LICs reside in specific sites, do these sites resemble the HSC niche? First and foremost, LTR-HSC, have been demonstrated found particularly in the bone marrow. Therefore, the stem cell niche in adult mouse is mainly in the bone marrow, and in the adult human, it is exclusively in the bone marrow. On the other hand, LICs can be found in diverse body sites including most tissues and organs. Therefore, the likelihood that the niche for LIC resembles that of HSCs is low. In fact it appears that the tumor cells which spread within the bone marrow microenvironment destroy normal HSC niches.

Leukemia cells spread outside of the bone marrow, and these cells are therefore not restricted to HSC niches in any case (Figure 3.2). The "latent niche" idea (McGovern et al., 2009) was suggested as an explanation for the spread of tumors and for metastasis formation. It is proposed that latent niches are formed prior to tumor cell seeding. Nevertheless, even if this is the case, such LIC niches would be formed in nonhematopoietic tissues. Therefore signals emanating from the hematopoiteic cells themselves, as well as from other components of the bone marrow niche, such as osteogenic and osteoclastogenic cells would be absent. (Kollet et al., 2006). Furthermore, if the latent niches are identical to HSC niches, these metastatic sites should have become a reservoir of HSCs. No evidence to this effect is available.

Ample evidence points to the fact that the tumor microenvironment is basically different from the normal one, making it unlikely that the LIC microenvironment resembles that of the HSC. A study of MSCs from infants with acute B lymphocytic leukemia showed that these cells express the MLL-AF4 fusion gene. Clearly, these MSCs that constitute part of the bone marrow microenvironment differ from their normal counterparts (Menendez et al., 2009). Similarly, mutations in MSCs have been suggested to predict development of malignancy (Houghton et al., 2010). Osteoblast dysfunction due to lack of Dicer 1 resulted in a microenvironmental defect that eventually led to myelodysplasia and subsequent acute myelogenous leukemia (AML) (Raaijmakers et al., 2010). Such findings clearly show that the emergence of AML requires a new microenvironment, different from the one supporting normal hematopoiesis. Similarly, intrinsic pathways that are essential for the growth of AML LICs, such as the Wnt/β-catenin pathway, are inessential for HSCs (Wang et al., 2010). In fact, many therapeutic approaches are based on the marked difference between normal progenitor cells and TICs in their intracellular signaling cascades (Jin et al., 2010; Hjelmeland et al., 2010). Clearly, divergent microenvironments would evoke correspondingly different intracellular responses in target cells. The dependence of the rare HSCs on their bone marrow niche is total. Conversely, it is acknowledged by supporters of the CSC concept that CSCs may not always be rare and may not relay on niche interactions (Mikkola et al., 2010). In addition to the differences in niche dependence, HSCs also seem to respond differently from LICs to cytokine challenges (Saito et al., 2010).

It is not surprising, and is actually expected, that malignant cells would share the need for particular cellular interaction with their normal counterparts, for example in terms of cell adhesion molecules. However, tumorigenesis and the survival of TICs may depend on modified/mutated cues rather than on the normal ones, meaning that the tumor microenvironment is inherently different from that of normal tissues.

Summary

LICs are essentially autonomously proliferating cells. Their growth potential is dictated cell autonomously, by damaged receptors and signaling cascades and/or unregulated cell cycle machineries. The aberrant malignant growth should not be confused with the process of stem cell maintenance. The latter process is enforced on stem cells by their immediate microenvironment (i.e., the stem cell niche). Stem cells in general and HSCs in particular do not have an autonomous in-built machinery that drives their proliferation. Rather, they are forced to maintain their undifferentiated phenotype by the surrounding cells that produce differentiation-restraining signals. This major difference between HSCs and LICs puts these two cell types within separate categories having little in common. Designating LICs as "stem cells" blurs the meaning of stemness and does not contribute in any why to a better understanding of the nature of malignancies, or otherwise, to the cure of cancer.

A growing body of evidence points to the possibility that differentiation processes are not unidirectional. The stem state theory proposes that differentiating cells may be capable of turning back and acquiring stemness (Zipori, 2004; Lander, 2009). A stem state can be achieved artificially by overexpression of transcription factors that lead to the formation of induced pluripotent stem cells (iPSC) (Okita et al., 2007). Cellular plasticity may also be partially maintained in cancer. Indeed, leukemia cells can be turned into iPSCs (Carette et al., 2010). By the same token, TICs may emerge from other cells in the tumor that do not exhibit tumor-forming ability at the time point of their examination by transplantation into susceptible recipients (Gupta et al., 2009). One proposition is that the presence of stem cells is characteristic of only one type of tumor, and others are clonal expansions (Shackleton et al., 2009). The above discussion focuses on the types of leukemia that do seem to possess a hierarchical organization. Nevertheless, this structure of leukemia types does not conform to the stem cell concept. It is undeniable that leukemia harbors a subpopulation of cells endowed with a greater capacity to form tumors in transplanted animals. However, these cells differ fundamentally from stem cells. Therefore, all types of leukemia are stem cell deficient.

References

Bereshchenko, O., Mancini, E., Moore, S., et al. (2009). Hematopoietic stem cell expansion precedes the generation of committed myeloid leukemia-initiating cells in C/EBPalpha mutant AML. Cancer Cell. 16:390–400.

Bhattacharya, D., Czechowicz, A., Ooi, A.G., et al. (2009). Niche recycling through division-independent egress of hematopoietic stem cells. J Exp Med. 206:2837–2850.

Bonnefoix, T., Callanan, M. (2010). Challenging the adult T-cell leukemia/lymphoma (ATL) stem cell concept based on limiting dilution transplantation assay. Blood. 115:2117–2118.

Brawley, C., Matunis, E. (2004). Regeneration of male germline stem cells by spermatogonial dedifferentiation in vivo. Science. 304:1331–1334.

Bruns, I., Czibere, A., Fischer, J.C., et al. (2009). The hematopoietic stem cell in chronic phase CML is characterized by a transcriptional profile resembling normal myeloid progenitor cells and reflecting loss of quiescence. Leukemia. 23:892–899.

Cai, X., Lin, Y., Friedrich, C.C., et al. (2009). Bone marrow derived pluripotent cells are pericytes which contribute to vascularization. Stem Cell Rev. 5:437–445.

Calvi, L.M., Adams, G.B., Weibrecht, K.W., et al. (2003). Osteoblastic cells regulate the haematopoietic stem cell niche. Nature. 425:841–846.

Carette, J.E., Pruszak, J., Varadarajan, M., et al. (2010). Generation of iPSCs from cultured human malignant cells. Blood.

Charles, N., Ozawa, T., Squatrito, M., et al. (2010). Perivascular nitric oxide activates notch signaling and promotes stem-like character in PDGF-induced glioma cells. Cell Stem Cell. 6:141–152.

Chertkov, IL. (1984). Does the hematopoietic stem cell exist? Mol Biol (Mosk). 18:565–573.

Clement, V., Marino, D., Cudalbu, C., et al. (2010). Marker-independent identification of glioma-initiating cells. Nat Methods. 7: 224–228.

Eckner, R.J., Hettrick, K.L., Greenberger, J.S., et al. (1982). Extended self-renewal capacity of pluripotent hematopoietic stem cells: association with persistent Friend spleen focus-forming virus. Cell. 31:731–738.

Fomchenko, E.I., Holland, E.C. (2005). Stem cells and brain cancer. Exp Cell Res. 306: 323–329.

Forsberg, E.C., Passegue, E., Prohaska, S.S., et al. (2010). Molecular signatures of quiescent, mobilized and leukemia-initiating hematopoietic stem cells. PLoS One. 5:e8785.

Gao, M.Q., Choi, Y.P., Kang, S., et al. (2010). CD24(+) cells from hierarchically organized ovarian cancer are enriched in cancer stem cells. Oncogene.

Guibal, F.C., Alberich-Jorda, M., Hirai, H., et al. (2009). Identification of a myeloid committed progenitor as the cancer-initiating cell in acute promyelocytic leukemia. Blood. 114:5415–5425.

Gupta, P.B., Chaffer, C.L., Weinberg, R.A. (2009). Cancer stem cells: mirage or reality? Nat Med. 15:1010–1012.

Han, Y.C., Park, C.Y., Bhagat, G., et al. (2010). microRNA-29a induces aberrant self-renewal capacity in hematopoietic progenitors, biased myeloid development, and acute myeloid leukemia. J Exp Med. 207:475–489.

Hjelmeland, A.B., Wu, Q., Wickman, S., et al. (2010). Targeting A20 decreases glioma stem cell survival and tumor growth. PLoS Biol. 8:e1000319.

Houghton, J.M., Li, H., Fan, X., et al. (2010). Mutations in bone marrow derived stromal stem cells unmask latent malignancy. Stem Cells Dev.

Jeselsohn, R., Brown, N.E., Arendt, L., et al. (2010). Cyclin D1 kinase activity is required for the self-renewal of mammary stem and progenitor cells that are targets of MMTV-ErbB2 tumorigenesis. Cancer Cell. 17:65–76.

Jin, Y., Lu, Z., Ding, K., et al. (2010). Antineoplastic mechanisms of niclosamide in acute myelogenous leukemia stem cells: inactivation of the NF-kappaB pathway and generation of reactive oxygen species. Cancer Res. 70:2516–2527.

Kollet, O., Dar, A., Shivtiel, S., et al. (2006). Osteoclasts degrade endosteal components and promote mobilization of hematopoietic progenitor cells. Nat Med. 12:657–664.

Lander, A.D. (2009). The 'stem cell' concept: is it holding us back? J Biol. 2009;8:70.

Lane, S.W., Scadden, D.T., Gilliland, D.G. (2009). The leukemic stem cell niche: current concepts and therapeutic opportunities. Blood. 114:1150–1157.

Lottaz, C., Beier, D., Meyer, K., et al. (2010). Transcriptional profiles of CD133+ and CD133− glioblastoma-derived cancer stem cell lines suggest different cells of origin. Cancer Res. 70:2030–2040.

Mayack, S.R., Shadrach, J.L., Kim, F.S., et al. (2010). Systemic signals regulate ageing and rejuvenation of blood stem cell niches. Nature. 463:495–500.

McGovern, M., Voutev, R., Maciejowski, J., et al. (2009). A "latent niche" mechanism for tumor initiation. Proc Natl Acad Sci USA. 106: 11617–11622.

Menendez, P., Catalina, P., Rodriguez, R., et al. (2009). Bone marrow mesenchymal stem cells from infants with MLL-AF4+ acute leukemia harbor and express the MLL-AF4 fusion gene. J Exp Med. 206:3131–3141.

Metcalf, D. (2007). Concise review: hematopoietic stem cells and tissue stem cells: current concepts and unanswered questions. Stem Cells. 25:2390–2395.

Mikkola, H.K., Radu, C.G., Witte, O.N. (2010). Targeting leukemia stem cells. Nat Biotechnol. 28:237–238.

Naka, K., Hoshii, T., Muraguchi, T., et al. (2010). TGF-beta-FOXO signalling maintains leukaemia-initiating cells in chronic myeloid leukaemia. Nature. 463:676–680.

Nakagawa, T., Nabeshima, Y., Yoshida, S. (2007). Functional identification of the actual and potential stem cell compartments in mouse spermatogenesis. Dev Cell. 12:195–206.

Okita, K., Ichisaka, T., Yamanaka, S. (2007). Generation of germline-competent induced pluripotent stem cells. Nature. 448:313–317.

Raaijmakers, M.H., Mukherjee, S., Guo, S., et al. (2010). Bone progenitor dysfunction induces myelodysplasia and secondary leukaemia. Nature.

Saito, Y., Uchida, N., Tanaka, S., et al. (2010). Induction of cell cycle entry eliminates human leukemia stem cells in a mouse model of AML. Nat Biotechnol. 28:275–280.

Shackleton, M., Quintana, E., Fearon, E.R., et al. (2009). Heterogeneity in cancer: cancer stem cells versus clonal evolution. Cell. 138:822–829.

Quintana, E., Shackleton, M., Sabel, M.S., et al. (2008). Efficient tumour formation by single human melanoma cells. Nature. 456:593–598.

Wang, Y., Krivtsov, A.V., Sinha, A.U., et al. (2010). The Wnt/beta-catenin pathway is required for the development of leukemia stem cells in AML. Science. 327:1650–1653.

Wilson, A., Laurenti, E., Oser, G., et al. (2008). Hematopoietic stem cells reversibly switch from dormancy to self-renewal during homeostasis and repair. Cell. 135:1118–1129.

Yeung, T.M., Gandhi, S.C., Wilding, J.L., et al. (2010). Cancer stem cells from colorectal cancer-derived cell lines. Proc Natl Acad Sci USA. 107:3722–3727.

Zipori, D. (2010). The stem cell niche, the stem state and multiple myeloma. Cancer Microenvironment.

Zipori, D. (2009). Biology of stem cells and the molecular basis of the stem state.

Zipori, D. (2004). The nature of stem cells: state rather than entity. Nat Rev Genet. 5:873–878.

Zipori, D. (1990). Regulation of hemopoiesis by cytokines that restrict options for growth and differentiation. Cancer Cells. 2:205–211.

Zipori, D. (1981). Cell interactions in the bone marrow microenvironment: role of endogenous colony-stimulating activity. J Supramol Struct Cell Biochem. 17:347–357.

Zipori, D., Sasson, T. (1980). Adherent cells from mouse bone marrow inhibit the formation of colony stimulating factor (CSF) induced myeloid colonies. Exp Hematol. 8:816–817.

Chapter 4

Human Embryonic Stem Cells and Cancer: Modeling Disease in a Dish

Tamra Werbowetski-Ogilvie and Robyn McClelland
Regenerative Medicine Program, Department of Biochemistry and Medical Genetics, University of Manitoba, Winnipeg, Manitoba, Canada

Abbreviations

ATM	Ataxia telangiectasia mutated
CSC	cancer stem cell
hESCs	human embryonic stem cells
iPSC	induced pluripotent stem cells
MB	medulloblastoma
NP	neural precursor
PTEN	phosphatase and tensin homologue
TPC	tumor propagating cell
Trans-hESC	transformed human embryonic stem cell

Introduction

Human embryonic stem cells (hESC) are derived from the inner cell mass of the pre-implantation embryo (Thomson et al., 1998) and are defined by two properties: (1) unlimited self-renewal capacity or the ability to indefinitely maintain an undifferentiated state and (2) pluripotent potential or the ability to generate cell types from all three germ layers (ectoderm, mesoderm, and endoderm) that comprise the body. First isolated and cultured in 1998 (Thomson et al., 1998), hESCs can be manipulated in vitro through variations in culture conditions to generate differentiated cell types. This includes everything from dopaminergic neurons (Kriks et al., 2011; Perrier et al., 2004; Roy et al., 2006), to cells of the hematopoietic system (Chadwick et al., 2003; Wang et al., 2004) as well as pancreatic beta cells (Nostro and Keller, 2012) and more recently intestinal cells (Wang and Sander, 2012) (Figure 4.1) Thus, hESCs hold great promise for regenerative medicine and cell replacement therapy.

In vivo, the pluripotent potential of hESCs, is defined by their ability to give rise to a teratoma, a benign tumor composed of

Cancer Stem Cells, First Edition. Edited by Vinagolu K. Rajasekhar.

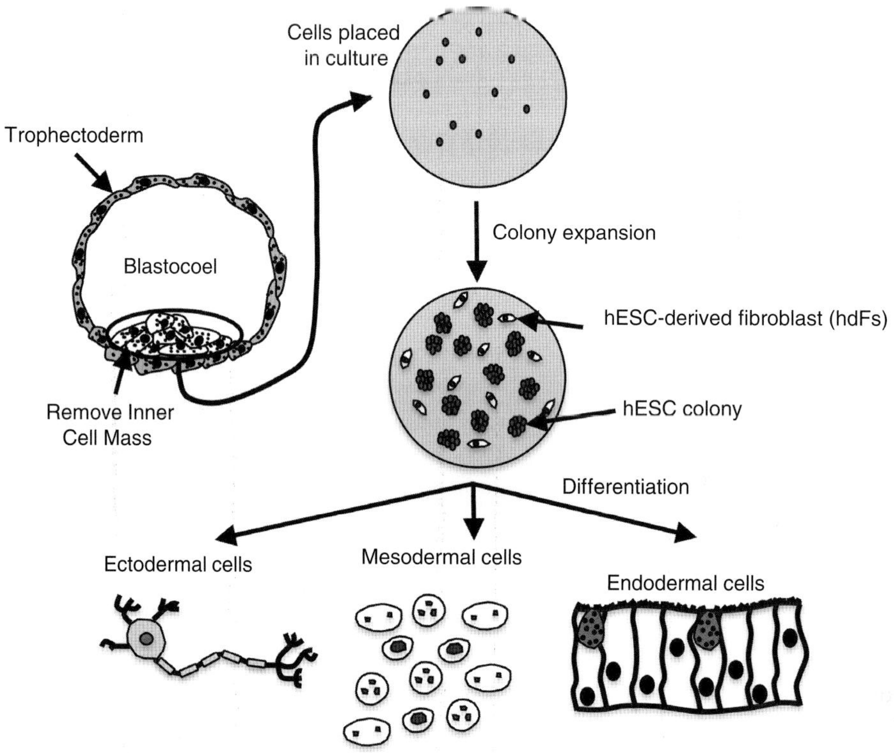

Figure 4.1. Generation and culture of hESC lines. hESCs are isolated from the inner cell mass of the pre-implantation blastocyst. These cells are then typically placed onto a feeder layer such as mouse embryonic fibroblasts or an extracellular matrix like Matrigel (depicted here) for support and expansion. In feeder-free conditions, hESCs expand as colonies of pluripotent cells. These cells spontaneously generate a more differentiated layer of fibroblast-like cells (hdFs) that serve as a supportive niche and provide the necessary factors to keep the hESCs in a self-renewing state. Following expansion, hESCs can be differentiated toward cell types associated with all three germ layers: ectoderm, mesoderm, and endoderm. Defined cell culture conditions and growth factors have enabled researchers to generate a multitude of cell types including differentiated neurons, hematopoietic cells, and even intestinal cells.

tissues derived from all three germ layers, when injected into immunodeficient mice (Thomson et al., 1998). Although this stringent test for pluripotency has become the standard in hESC research laboratories around the world, the tumorigenic capacity of hESCs still hinders their clinical utility. In fact, hESCs share cellular and molecular features with cancer cells including a rapid proliferation rate, high telomerase activity, elevated expression of oncogenes, and genomic instability (Ben-David and Benvenisty, 2011). Several studies have reported karyotypic changes in hESCs and long-term adaptation in culture (Baker et al., 2007). These studies identified major chromosomal abnormalities or genetic hotspots in hESCs; however, not all hESCs that exhibit transformed features display an abnormal karyotype by standard cytogenic techniques (Werbowetski-Ogilvie et al., 2009). Notably, phenotypic changes that accompany hESC transformation include an elevated capacity for self-renewal, a lowered dependence on exogenous growth factors, and aberrant differentiation both in vitro and in vivo compared with "normal" hESCs (Werbowetski-Ogilvie et al., 2009). Interestingly, these cellular features mirror the properties that accompany cellular transformation, neoplastic progression and the acquisition of cancer stem cell (CSC) properties. hESCs that harbor karyotypic abnormalities and exhibit neoplastic features cannot be utilized in the clinic. The self-renewal capacity and pluripotent potential thus serve as both tremendous assets and potent drawbacks for the use of hESCs in regenerative medicine. Although solutions include complete differentiation and removal of residual pluripotent stem cells in culture, the risks associated with hESC tumorigenicity still remain.

The discovery of induced pluripotent stem cells (iPSC) in 2006 revolutionized pluripotent stem cell research (Park et al., 2008; Takahashi et al., 2007; Takahashi and Yamanaka, 2006; Yu et al., 2007). The demonstration that somatic cells could be "reprogrammed" to the undifferentiated or embryonic state underscored the plasticity of differentiated cell types and offered alternatives to hESCs that had been plagued by ethical considerations surrounding the use of pre-implantation embryos. Like hESCs, they too have been shown to exhibit tumorigenic features (Ben-David and Benvenisty, 2011). Despite the obvious risks associated with both types of human pluripotent stem cells, the idea of "modeling disease in a dish" has emerged as an attractive and exciting alternative use for both iPSCs and hESCs. A large number of studies have been published demonstrating the generation of disease-specific iPSCs for investigating a variety of human disorders (Zhu et al., 2011). However, this chapter specifically focuses on hESCs and their more recent utilization as models for early stages of tumorigenesis as well as recent applications to CSC biology.

The vast cellular and molecular differences between mouse and human have led to the increasing demand for more biologically relevant human disease models. The discovery of mouse embryonic stem cells (Evans and Kaufman, 1981; Martin, 1981) revolutionized the field of mouse genetics and greatly enhanced our understanding of both normal and aberrant development through the generation of transgenic and knockout models. However, for some genetic disorders and cancers, mouse models are insufficient in recapitulating the human counterpart or fail to define the early events that underlie the disease. The manipulation of hESCs therefore serves as an alternative system for understanding the early developmental events that lead to disease or tumorigenic progression.

hESCs as Early Developmental Models of Human Cancers

Molecular control of self-renewal is well established in embryonic stem cells (ESC) and is governed by a core set of transcription factors that maintain the undifferentiated state (Boyer et al., 2006). These factors, which include Oct4 and Nanog as well as other ESC-associated genes, have also recently been associated with highly aggressive adult tumors (Ben-Porath et al., 2008; Somervaille et al., 2009; Wong et al., 2008). In particular, targets of Oct4, Sox2, Nanog, and c-myc are frequently overexpressed in poorly differentiated tumors such as high-grade ER-negative breast cancer, bladder carcinoma, and glioblastoma (Ben-Porath et al., 2008). In fact, core transcription factors, such as Nanog, have also been linked to Hedgehog signaling specifically in malignant brain tumors (Po et al., 2010; Zbinden et al., 2010). Conversely, recent studies have demonstrated that differentiated derivatives from both hESCs and human iPSCs exhibit more transcriptional overlap with cancer cells then corresponding normal primary cells (Ghosh et al., 2011).

Additional work has capitalized on the relationship between pluripotency and malignant cancers by knocking out common tumor suppressor genes in hESCs. For example, studies have shown that knockout of p53, the most commonly mutated tumor suppressor in human cancers, in hESCs leads to genetic instability (Song et al., 2010). These cells represent a valuable resource for the study of human cancers that are not recapitulated by p53 −/− mouse models (Donehower et al., 1992; Jacks et al., 1994; Song et al., 2010). In particular, p53−/− mice typically develop sarcomas and lymphomas (Donehower et al., 1992; Jacks et al., 1994); whereas, human patients suffer from mostly carcinomas. p53 has also been shown to play an important role in the G2/M checkpoint of human cells; however, it is not required at the G2/M checkpoint in mouse cells (Sancar et al., 2004; Song et al., 2007).

In addition to p53, Song and others (2010) also knocked out the Ataxia telangiectasia mutated (ATM) gene in hESCs. ATM is a protein kinase mutated in Ataxia telangiectasia syndrome characterized by cerebellar degeneration, radiation sensitivity, immunodeficiency, and an increased risk for cancer (Shiloh, 1995). While ATM −/− hESCs maintain a normal karyotype, they do exhibit defects in DNA repair and cell cycle checkpoints (Song et al., 2010). Along with the p53 −/− hESCs, these cells represent a valuable resource for the study of human tumorigenesis.

In addition to p53 and ATM, studies have also examined the role of another tumor suppressor gene, phosphatase and tensin homologue (PTEN), in hESCs (Alva et al. 2011). Knockdown of PTEN in hESCs increased tumorigenic properties including enhanced self-renewal, survival, and proliferation relative to control cells (Alva et al., 2011). This was coupled with continued expression of the pluripotent transcription factors Oct4 and Nanog following differentiation and an enhancement of neuroectoderm differentiation in teratoma assays (Alva et al., 2011). Similar to the work conducted in hESCs, studies have also shown knockdown of PTEN in murine embryonic stem cells leads to an accumulation of a small group of highly aggressive teratoma-initiating embryonal carcinoma cells (Lindgren et al., 2011).

hESC Markers and CSC Populations

The cancer stem cell hypothesis suggests that a rare subpopulation of cells exhibit stem cell properties such as self-renewal capacity and multilineage differentiation (Rosen and Jordan, 2009; Visvader, 2011). These cancer stem cells (CSC) or tumor-propagating cells (TPC) would ultimately drive tumorigenesis and must therefore be specifically targeted to abrogate malignant progression. Although this work originated in leukemia (Bonnet and Dick, 1997), a large number of studies have since been published on TPCs in solid tumors (Rosen and Jordan, 2009; Visvader, 2011).

In addition to studying the expression patterns of ES cell-associated factors in malignant cancers, more recent studies have moved into evaluating a role for pluripotent factors specifically in TPC subpopulations from multiple tumor subtypes. For example, the pluripotent transcription factor Nanog is required for maintaining self-renewal of both mouse and human ES cells (Chambers et al., 2003; Pan and Thomson, 2007). Recently, studies have shown that Nanog1, together with high Wnt/β-catenin signaling functionally define the TPC population in colorectal cancers (Ibrahim et al., 2012).

Oct4 is essential for the maintenance of stem cell properties in glioma TPC populations through regulation of Sox2 expression (Ikushima et al., 2011). Sox2 is regulated by a self-regulatory loop in fetal neural progenitors demonstrating that different mechanisms underlie stem cell properties in glioma TPCs relative to normal counterparts. Oct4 overexpression in melanoma cells is also accompanied by the acquisition of cancer stem cell phenotypes such as enhanced capacity for self-renewal and multipotent differentiation (Kumar et al., 2012). Similarly, Oct4 overexpression in primary human mammary epithelial breast cells led to the isolation of colonies with unlimited self-renewal capacity and tumor-initiation and metastasis in nude mice (Beltran et al., 2011). Combined ectopic expression of Oct4 and Nanog in lung adenocarcinoma cells increased expression of the putative CSC marker CD133, promoted epithelial to mesenchymal transition (EMT) and enhanced tumor-initiating capacity in vivo (Chiou et al., 2010).

The abovementioned studies demonstrate a role for pluripotent transcription factors such as Oct4 and Nanog in tumor-initiating cell subpopulations, however these factors cannot be used to selectively isolate TPCs from bulk culture, as this requires utilization of cell surface markers. Taking the relationship between CSCs and ESCs one step further, Rajasekhar and others (2011) have recently used the cell surface pluripotent marker tumor rejection antigen Tra-1-60 in conjunction with other markers to selectively identify and isolate a small subpopulation of stem-like human prostate cancer cells (Rajasekhar et al., 2011). These cells displayed enhanced sphere-forming ability in vitro and tumor-initiation in vivo thus underscoring the value in utilizing hESC cell surface markers as novel tools for the isolation of putative TPC populations. This will become increasingly important as we move away from using markers with minimal specificity such as CD133 towards identification and utilization of markers that better represent the more primitive embryonic-like CSC state.

hESCs and Pediatric Cancers

The most frequently diagnosed forms of childhood cancer are leukemia, brain tumors, and lymphomas. Researchers have made substantial progress on the molecular mechanisms that govern these highly malignant tumors. However, because these cancers typically manifest during infancy or early childhood, it seems that the genetic events required for tumorigenesis take place prenatally. Although murine models exist for many pediatric cancers, they do not always faithfully recapitulate the human disease counterpart. In addition, many researchers utilize fully transformed, highly malignant patient samples for analysis. Although primary tissue typically provides the most biologically relevant data, the earliest events leading to tumor initiation and progression cannot be examined with these samples. Thus, hESCs have recently become an alternative and complementary system for the modeling of the early aspects of neoplastic progression and the acquisition of CSC phenotypes. Gain and loss of function studies with a particular gene of interest followed by lineage specification allows for comparative analysis of abnormal versus normal hESC derivatives for further study. Examples of hESC manipulation to study the effects of specific genes on the early developmental stages of pediatric tumorigenesis are outlined below.

hESCs and Infant Pro-B Acute Lymphoblastic Leukemia (ALL)

Infant pro-B acute lymphoblastic leukemia (ALL) is a malignant cancer of the lymphoid progenitor cells (Pui, 2000). The presence of the MLL-AF4 fusion gene is a hallmark of high-risk, poor prognosis in ALL infant patients (Pui, 2000). High concordance rates of this leukemia in monozygotic twin studies strongly suggests that the early events required for aberrant hematopoietic differentiation and subsequent tumorigenesis take place in utero (Greaves et al., 2003). Unfortunately, murine models of ALL do not fully recapitulate human disease in terms of both latency and phenotype (Chen et al., 2006; Krivtsov et al., 2008; Metzler et al., 2006). To address this issue, Bueno and others (2012) have recently utilized hESCs and their hematopoietic derivatives to delineate a role for MLL-AF4 during embryonic hematopoiesis. Interestingly, overexpression of MLL-AF4 resulted in fate decision changes during the stages of hematopoiesis (Bueno et al., 2012). MLL-AF4 enhanced hemogenic precursor specification from hESCs; however, further commitment revealed that cells were primed to differentiate toward endothelial maturation at the expense of hematopoiesis (Bueno et al., 2012). Moreover, overexpression of MLL-Af4 did not induce tumorigenic properties in vitro or in vivo demonstrating that this particular "hit" is insufficient for transformation toward neoplastic features (Bueno et al., 2012). This novel model system has therefore provided tremendous insight into how overexpression of MLL-AF4 may lead to impaired embryonic blood formation. In addition, this work demonstrates that additional, cooperating genetic events are necessary for tumorigenesis or perhaps another cell type, either more primitive or differentiated, along this lineage must be targeted for oncogenic transformation.

hESCs and Medulloblastoma

Over the past decade, accumulating evidence has suggested a transformed neural stem cell origin for malignant brain tumors (Hadjipanayis and Van Meir, 2009; Singh et al., 2004), as it has been proposed that a subset of cancer cells, termed brain tumor stem (or propagating) cells are responsible for brain tumor initiation and maintenance (Singh et al., 2004). However, these features may be acquired during the transformation process, and may not actually indicate a "cell of origin" (Visvader, 2011). Indeed, recent work has shown that differentiated neural progenitors are the "cells of origin" even though neural stem cells sustain the initial oncogenic hits (Liu et al., 2011; Persson et al., 2010; Read et al., 2009; Schuller et al., 2008; Yang et al., 2008). These studies emphasize the importance of tracing the entire tumorigenic process, however they focus mainly on cells derived from transgenic and knockout mouse models, and the direct relevance to human brain tumors is unclear.

Medulloblastoma (MB) is the most common malignant primary pediatric brain tumor (Louis et al., 2007). Following surgery and treatment with chemotherapy or radiation, MB tumors can recur as a consequence of massive tumor cell infiltration into the brain and frequent metastasis through the cerebral spinal fluid (CSF) (Mehta et al., 2011; Wu et al.). Despite numerous advances in understanding the molecular basis for the malignant state and a 5-year survival rate of 60 to 70% (Louis et al., 2007), children with MB typically suffer from permanent cognitive and physical dysfunction attributed to the impact of harsh toxic therapies on their developing nervous system (Louis et al., 2007).

Most current MB mouse models recapitulate the Shh-pathway variant (Hatten and Roussel, 2011). Even though they have provided tremendous insight into molecular mechanisms underlying this subtype, these murine MB models typically require a loss of p53 to generate a tumor, and p53 mutations are infrequent in human MB (Thompson et al., 2006). Elucidating the oncogenic drivers associated with early-stage disease will be critical to specifically target and eradicate human brain tumors before they reach malignancy. These early oncogenic events are difficult to assess using late-stage tumors from primary patient samples or derived human cell lines that have acquired numerous karyotypic abnormalities or extensive cellular heterogeneity, and they often fail to recapitulate the invasive phenotype in xenograft models (Visvader, 2011). More relevant human models are needed to identify the molecular drivers of MB.

Major research efforts are focused on generating cancer-specific iPSCs from patient samples. However, "reprogramming" somatic cancer cells is more difficult amidst a background of major karyotypic abnormalities (Sun and Liu, 2011). To date, iPSCs have not been generated from malignant brain tumors. Similar to the ALL studies described, hESCs serve as a potential alternative model to study the oncogenic drivers of pediatric brain tumorigenesis.

hESCs have recently been identified that, in contrast to normal hESCs, have spontaneously "transformed" and acquired neoplastic or CSC features during culture (Werbowetski-Ogilvie et al., 2009). These properties include enhanced self-renewal, proliferation, and tumor-initiating capacity, as well as aberrant neural lineage specification (Werbowetski-Ogilvie et al., 2009). Neural precursors (NP) derived from neoplastic or transformed hESCs (N-trans-hESCs) also exhibit features of human brain tumors both in vitro and in vivo (Figure 4.2A) (Werbowetski-Ogilvie et al., 2012). NPs derived from trans-hESCs have enhanced cell proliferation and an inability to mature toward the astrocytic lineage, compared with progeny derived from normal embryonic stem cells (N-hESC). Intracranial transplantation of N-trans-hESCs and N-hESCs into nonobese diabetic severe combined immunodeficient (NOD SCID) mice revealed neuroectodermal tumors only in mice injected with the N-trans-hESCs

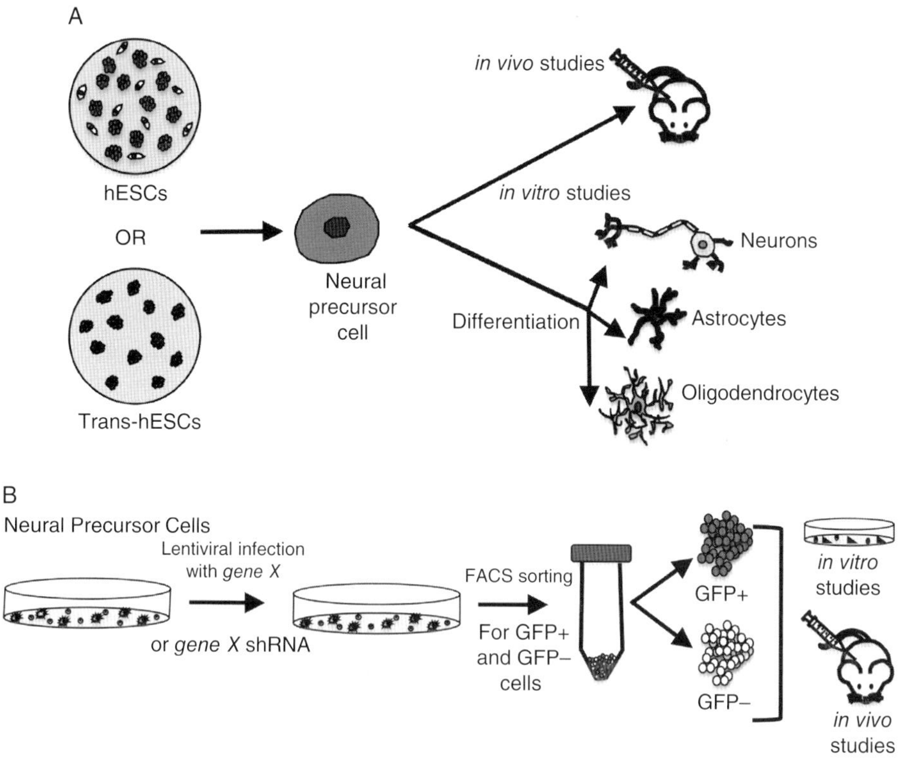

Figure 4.2. Using neural derivatives from hESCs to investigate the role of specific genes in the early stages of pediatric brain tumorigenesis. **A**. Two-stage differentiation procedure employed to generate neural derivatives from hESCs or trans-hESCs for use in both in vitro and in vivo studies. Pluripotent hESCs or trans-hESCs are placed into neural precursor conditions to generate a heterogeneous cell population consisting of neural stem cells and more differentiated progenitors. Culture conditions vary but typically include basic fibroblast growth factor (bFGF) and/or epidermal growth factor (EGF), known to be essential for neurosphere growth and maintenance. Following neural precursor differentiation, cells can be further differentiated toward the three cell types of the central nervous system: neurons, oligodendrocytes, and astrocytes. In turn, these cells as well as their more primitive derivatives can be used for further characterization in vitro as well as in vivo using mouse xenograft models. **B**. Summary of lentiviral infection and sorting procedures currently utilized for examining the role of specific genes during normal and aberrant neural development from hESCs. Genes known to play an oncogenic role are overexpressed in neural precursor cells from normal hESCs (N-hESCs) using lentiviral infection. Alternatively, these same genes can be knocked down in N-trans-hESCs using lentiviral infection with short-hairpin RNA (shRNA). Coupled with a GFP fluorescent tag as an indicator of transduction efficiency, cultures are then sterile sorted to physically separate infected (GFP+) and non-infected (GFP−) cells. Sorted populations are then used for self-renewal and invasion assays both in vitro and in vivo. Collectively, these studies enable researchers to determine whether a single gene is necessary and sufficient to induce tumor initiation in vivo or neoplastic properties in vitro such as enhanced self-renewal, invasion, and proliferation. Tumor development in mouse xenograft models can be further characterized for properties associated with known pediatric brain tumors such as medulloblastoma.

NPs (Werbowetski-Ogilvie et al., 2012). These tumors infiltrated the cerebellum and the ventricles of recipient mice, and displayed morphological, phenotypic, and molecular features associated with classic MB, a medulloblastoma variant. Collectively, these data provide the necessary foundation to test the hypothesis that comparative studies of NPs from normal and neoplastic human pluripotent stem cells can be used to identify early oncogenic events leading to human MB formation or to validate known MB regulators.

Current studies are combining existing technologies in hESC differentiation and brain tumor biology to address a research question that cannot be evaluated using patient samples and cell lines that acquire numerous karyotypic abnormalities and phenotypic heterogeneity: What are the oncogenic drivers of early human MB development? This involves both gain and loss of function studies with either genes already known to be involved with MB progression or genes with currently unknown roles in brain tumorigenesis. Gain of function studies with normal hESCs or their neural derivatives would enable one to determine whether dysregulation of a single gene could induce tumorigenic properties both in vitro and in vivo (Figure 4.2B). Additionally, loss of function studies in either undifferentiated trans-hESCs or their neural precursors derivatives could determine whether a single gene is necessary and sufficient to maintain tumorigenic properties (Figure 4.2B). In this way, one can test whether a gene is important for brain tumor initiation and/or progression amidst a clean genetic background. This innovative integrated approach underscores a previously unappreciated role for hESCs as early developmental models of human malignant brain tumors.

Elucidating the oncogenic drivers of MB will be critical for the design of new strategies to detect and target premalignant cells. This will benefit not only patients with sporadic tumors but also those suffering from familial cancer syndromes such as Turcot or Gorlin syndrome (Louis et al., 2007), who are predisposed to MB.

Although initial work has focused on classic MB, neural derivatives of hESCs can also be manipulated to examine early events in a variety of other pediatric neural cancers including gliomas and the most common extracranial tumor neuroblastoma that is derived from neural crest cells of the sympathetic nervous system (Maris, 2010). The pluripotent potential and general utility of hESCs and their transformed derivatives makes these cells a highly advantageous and alternative model system for the study of human malignant neural tumor development and progression.

hESCs and Drug Discovery: Taking It to the Next Level

In addition to the utility of hESCs and their neoplastic derivatives in gain and loss of function studies, recent investigations have also used these cells in high throughput screens to identify chemical compounds that selectively target "cancer stem cell" populations while leaving their hypothetical normal equivalent intact. For example, Sachlos and others (2012) have demonstrated that by conducting parallel chemical screens with transformed and normal hESCs, they could identify compounds that selectively differentiate the neoplastic hESC population while having no effect on normal hESCs. Interestingly, thioridizine, an antipsychotic drug, selectively targeted the transformed hESCs, and this was validated using leukemic models of CSCs both in vitro and in vivo. These studies underscore the general utility of neoplastic hESCs versus normal hESCs in anticancer screens, as they provide a reliable and scalable in vitro surrogate for human somatic cell CSC populations. Furthermore, they enable researchers to conduct large-scale experiments with more readily available, large quantities of cells, prior to moving into more specific drug-testing studies with precious patient samples.

Conclusions

The idea of using hESCs as a model system to study human cancer initiation and progression is a relatively new concept. Although originally intended for regenerative medicine purposes, these cells have been the center of much controversy since their initial isolation in 1998. This is mainly attributed to the ethical considerations regarding the use of pre-implantation embryos for derivation, and of course, the general properties of these cells that make them highly tumorigenic. Even though the discovery and generation of iPSCs from somatic sources has alleviated some ethical concerns, nevertheless, the tumor-initiating capacity of both types of human pluripotent cells remains an issue. Therefore, alternative uses for human pluripotent stem cells have gained momentum. With recent published studies demonstrating the utilization of hESCs as models for both pediatric blood and solid tumors, it will be interesting to see whether other malignant cancers can be modeled using this unique system and whether the use of pluripotent cell surface markers to isolate putative CSC subpopulations will become much more common. These advances with hESCs will not replace existing animal models or studies with primary patient samples; however, they will definitely provide a complementary system, as we move forward in trying to delineate the early molecular mechanisms responsible for all human malignancies and those that sustain tumor-propagating or cancer stem cell subpopulations.

References

Alva, J.A., Lee, G.E., Escobar, E.E., and Pyle, A.D. (2011). Phosphatase and tensin homolog regulates the pluripotent state and lineage fate choice in human embryonic stem cells. Stem Cells 29, 1952–1962.

Baker, D.E., Harrison, N.J., Maltby, E., Smith, K., Moore, H.D., Shaw, P.J., Heath, P.R., Holden, H., and Andrews, P.W. (2007). Adaptation to culture of human embryonic stem cells and oncogenesis in vivo. Nat Biotechnol 25, 207–215.

Beltran, A.S., Rivenbark, A.G., Richardson, B.T., Yuan, X., Quian, H., Hunt, J.P., Zimmerman, E., Graves, L.M., and Blancafort, P. (2011). Generation of tumor-initiating cells by exogenous delivery of OCT4 transcription factor. Breast Cancer Res 13, R94.

Ben-David, U., and Benvenisty, N. (2011). The tumorigenicity of human embryonic and induced pluripotent stem cells. Nat Rev Cancer 11, 268–277.

Ben-Porath, I., Thomson, M.W., Carey, V.J., Ge, R., Bell, G.W., Regev, A., and Weinberg, R.A. (2008). An embryonic stem cell-like gene expression signature in poorly differentiated aggressive human tumors. Nat Genet 40, 499–507.

Bonnet, D., and Dick, J.E. (1997). Human acute myeloid leukemia is organized as a hierarchy that originates from a primitive hematopoietic cell. Nature Medicine 3, 730–737.

Boyer, L.A., Mathur, D., and Jaenisch, R. (2006). Molecular control of pluripotency. Current Opinion in Genetics & Development 16, 455–462.

Bueno, C., Montes, R., Melen, G.J., Ramos-Mejia, V., Real, P.J., Ayllon, V., Sanchez, L., Ligero, G., Gutierrez-Aranda, I., Fernandez, A.F., *et al.* (2012). A human ESC model for MLL-AF4 leukemic fusion gene reveals an impaired early hematopoietic-endothelial specification. Cell Res 22, 986–1002.

Chadwick, K., Wang, L., Li, L., Menendez, P., Murdoch, B., Rouleau, A., and Bhatia, M. (2003). Cytokines and BMP-4 promote hematopoietic differentiation of human embryonic stem cells. Blood 102, 906–915.

Chambers, I., Colby, D., Robertson, M., Nichols, J., Lee, S., Tweedie, S., and Smith, A. (2003). Functional expression cloning of Nanog, a pluripotency sustaining factor in embryonic stem cells. Cell 113, 643–655.

Chen, W., Li, Q., Hudson, W.A., Kumar, A., Kirchhof, N., and Kersey, J.H. (2006). A murine Mll-AF4 knock-in model results in lymphoid and myeloid deregulation and hematologic malignancy. Blood 108, 669–677.

Chiou, S.H., Wang, M.L., Chou, Y.T., Chen, C.J., Hong, C.F., Hsieh, W.J., Chang, H.T., Chen, Y.S., Lin, T.W., Hsu, H.S., *et al.* (2010). Coexpression of Oct4 and Nanog enhances malignancy in lung adenocarcinoma by inducing cancer stem cell-like properties and epithelial-mesenchymal transdifferentiation. Cancer Res 70, 10433–10444.

Donehower, L.A., Harvey, M., Slagle, B.L., McArthur, M.J., Montgomery, C.A., Jr., Butel, J.S., and Bradley, A. (1992). Mice deficient for p53 are developmentally normal but susceptible to spontaneous tumors. Nature 356, 215–221.

Evans, M.J., and Kaufman, M.H. (1981). Establishment in culture of pluripotential cells from mouse embryos. Nature 292, 154–156.

Ghosh, Z., Huang, M., Hu, S., Wilson, K.D., Dey, D., and Wu, J.C. (2011). Dissecting the oncogenic and tumorigenic potential of differentiated human induced pluripotent stem cells and human embryonic stem cells. Cancer Research 71, 5030–5039.

Greaves, M.F., Maia, A.T., Wiemels, J.L., and Ford, A.M. (2003). Leukemia in twins: lessons in natural history. Blood 102, 2321–2333.

Hadjipanayis, C.G., and Van Meir, E.G. (2009). Brain cancer propagating cells: biology, genetics and targeted therapies. Trends Mol Med 15, 519–530.

Hatten, M.E., and Roussel, M.F. (2011). Development and cancer of the cerebellum. Trends Neurosci 34, 134–142.

Ibrahim, E.E., Babaei-Jadidi, R., Saadeddin, A., Spencer-Dene, B., Hossaini, S., Abuzinadah, M., Li, N., Fadhil, W., Ilyas, M., Bonnet, D., *et al.* (2012). Embryonic NANOG Activity Defines Colorectal Cancer Stem Cells and Modulated through AP1- and TCF-dependent Mechanisms. Stem Cells.

Ikushima, H., Todo, T., Ino, Y., Takahashi, M., Saito, N., Miyazawa, K., and Miyazono, K. (2011). Glioma-initiating cells retain their tumorigenicity through integration of the Sox axis and Oct4 protein. J Biol Chem 286, 41434–41441.

Jacks, T., Remington, L., Williams, B.O., Schmitt, E.M., Halachmi, S., Bronson, R.T., and Weinberg, R.A. (1994). Tumor spectrum analysis in p53-mutant mice. Curr Biol 4, 1–7.

Kriks, S., Shim, J.W., Piao, J., Ganat, Y.M., Wakeman, D.R., Xie, Z., Carrillo-Reid, L., Auyeung, G., Antonacci, C., Buch, A., *et al.* (2011). Dopamine neurons derived from human ES cells efficiently engraft in animal models of Parkinson's disease. Nature 480, 547–551.

Krivtsov, A.V., Feng, Z., Lemieux, M.E., Faber, J., Vempati, S., Sinha, A.U., Xia, X., Jesneck, J., Bracken, A.P., Silverman, L.B., *et al.* (2008). H3K79 methylation profiles define murine and human MLL-AF4 leukemias. Cancer Cell 14, 355–368.

Kumar, S.M., Liu, S., Lu, H., Zhang, H., Zhang, P.J., Gimotty, P.A., Guerra, M., Guo, W., and Xu, X. (2012). Acquired cancer stem cell phenotypes through Oct4-mediated dedifferentiation. Oncogene.

Liu, C., Sage, J.C., Miller, M.R., Verhaak, R.G., Hippenmeyer, S., Vogel, H., Foreman, O., Bronson, R.T., Nishiyama, A., Luo, L., *et al.* (2011). Mosaic analysis with double markers reveals tumor cell of origin in glioma. Cell 146, 209–221.

Louis, D., Ohgaki, H., Wiestler, O.D., and Cavenee, W.K. (2007). WHO Classification of Tumors of the Central Nervous System, Fourth Edition.

Maris, J.M. (2010). Recent advances in neuroblastoma. N Engl J Med 362, 2202–2211.

Martin, G.R. (1981). Isolation of a pluripotent cell line from early mouse embryos cultured in medium conditioned by teratocarcinoma stem cells. Proc Natl Acad Sci USA 78, 7634–7638.

Mehta, M., Chang S., Newton H., Guha A., and M., V. (2011). Principles and Practice of Neuro-oncology: A multidisciplinary approach (Demos Medical Publishing; 1 edition, Oct 21 2010).

Metzler, M., Forster, A., Pannell, R., Arends, M.J., Daser, A., Lobato, M.N., and Rabbitts, T.H. (2006). A conditional model of MLL-AF4 B-cell tumourigenesis using invertor technology. Oncogene 25, 3093–3103.

Nostro, M.C., and Keller, G. (2012). Generation of beta cells from human pluripotent stem cells: Potential for regenerative medicine. Semin Cell Dev Biol.

Pan, G., and Thomson, J.A. (2007). Nanog and transcriptional networks in embryonic stem cell pluripotency. Cell Res 17, 42–49.

Park, I.H., Zhao, R., West, J.A., Yabuuchi, A., Huo, H., Ince, T.A., Lerou, P.H., Lensch, M.W., and Daley, G.Q. (2008). Reprogramming of human somatic cells to pluripotency with defined factors. Nature 451, 141–146.

Perrier, A.L., Tabar, V., Barberi, T., Rubio, M.E., Bruses, J., Topf, N., Harrison, N.L., and Studer, L. (2004). Derivation of midbrain dopamine neurons from human embryonic stem cells. Proc Natl Acad Sci USA 101, 12543–12548.

Persson, A.I., Petritsch, C., Swartling, F.J., Itsara, M., Sim, F.J., Auvergne, R., Goldenberg, D.D., Vandenberg, S.R., Nguyen, K.N., Yakovenko, S., *et al.* (2010). Non-stem cell origin for oligodendroglioma. Cancer Cell 18, 669–682.

Po, A., Ferretti, E., Miele, E., De Smaele, E., Paganelli, A., Canettieri, G., Coni, S., Di Marcotullio, L., Biffoni, M., Massimi, L., *et al.* (2010). Hedgehog controls neural stem cells through p53-independent regulation of Nanog. EMBO J 29, 2646–2658.

Pui, C.H. (2000). Acute lymphoblastic leukemia in children. Curr Opin Oncol 12, 3–12.

Rajasekhar, V.K., Studer, L., Gerald, W., Socci, N.D., and Scher, H.I. (2011). Tumour-initiating stem-like cells in human prostate cancer exhibit increased NF-kappaB signalling. Nat Commun 2, 162.

Read, T.A., Fogarty, M.P., Markant, S.L., McLendon, R.E., Wei, Z., Ellison, D.W., Febbo, P.G., and Wechsler-Reya, R.J. (2009). Identification of CD15 as a marker for tumor-propagating cells in a mouse model of medulloblastoma. Cancer Cell 15, 135–147.

Rosen, J.M., and Jordan, C.T. (2009). The increasing complexity of the cancer stem cell paradigm. Science 324, 1670–1673.

Roy, N., Cleren, C., Singh, S., Yang, L., Beal, M., and Goldman, S. (2006). Functional engraftment of human ES cell-derived dopaminergic neurons enriched by coculture with telomerase-immortalized midbrain astrocytes. Nature Medicine 12, 1259–1268.

Sachlos, E., Risueño, R.M., Laronde, S., Shapovalova, Z., Lee, J.H., Russell, J., Malig, M., McNicol, J.D., Fiebig-Comyn, A., Graham, M., Levadoux-Martin, M., Lee, J.B., Giacomelli, A.O., Hassell, J.A., Fischer-Russell, D., Trus, M.R., Foley, R., Leber, B., Xenocostas, A., Brown, E.D., Collins, T.J., Bhatia, M. (2012). Identification of drugs including a dopamine receptor antagonist that selectively target cancer stem cells. Cell. 149(6):1284–97.

Sancar, A., Lindsey-Boltz, L.A., Unsal-Kacmaz, K., and Linn, S. (2004). Molecular mechanisms of mammalian DNA repair and the DNA damage checkpoints. Annu Rev Biochem 73, 39–85.

Schuller, U., Heine, V.M., Mao, J., Kho, A.T., Dillon, A.K., Han, Y.G., Huillard, E., Sun, T., Ligon, A.H., Qian, Y., *et al.* (2008). Acquisition of granule neuron precursor identity is a critical determinant of progenitor cell competence to form Shh-induced medulloblastoma. Cancer Cell 14, 123–134.

Shiloh, Y. (1995). Ataxia-telangiectasia: closer to unraveling the mystery. Eur J Hum Genet 3, 116–138.

Sikkema, A.H., den Dunnen, W.F., Hulleman, E., van Vuurden, D.G., Garcia-Manero, G., Yang, H., Scherpen, F.J., Kampen, K.R., Hoving, E.W., Kamps, W.A., *et al.* (2012). EphB2 activity plays a pivotal role in pediatric medulloblastoma cell adhesion and invasion. Neuro Oncol.

Singh, S.K., Hawkins, C., Clarke, I.D., Squire, J.A., Bayani, J., Hide, T., Henkelman, R.M., Cusimano, M.D., and Dirks, P.B. (2004). Identification of human brain tumour initiating cells. Nature 432, 396–401.

Somervaille, T.C., Matheny, C.J., Spencer, G.J., Iwasaki, M., Rinn, J.L., Witten, D.M., Chang, H.Y., Shurtleff, S.A., Downing, J.R., and Cleary, M.L. (2009). Hierarchical maintenance of MLL myeloid leukemia stem cells employs a transcriptional program shared with embryonic rather than adult stem cells. Cell Stem Cell 4, 129–140.

Song, H., Chung, S.K., and Xu, Y. (2010). Modeling disease in human ESCs using an efficient BAC-based homologous recombination system. Cell Stem Cell 6, 80–89.

Song, H., Hollstein, M., and Xu, Y. (2007). p53 gain-of-function cancer mutants induce genetic instability by inactivating ATM. Nat Cell Biol 9, 573–580.

Sun, C., and Liu, Y.K. (2011). Induced pluripotent cancer cells: progress and application. J Cancer Res Clin Oncol 137, 1–8.

Takahashi, K., Tanabe, K., Ohnuki, M., Narita, M., Ichisaka, T., Tomoda, K., and Yamanaka, S. (2007). Induction of pluripotent stem cells from adult human fibroblasts by defined factors. Cell 131, 861–872.

Takahashi, K., and Yamanaka, S. (2006). Induction of pluripotent stem cells from mouse embryonic and adult fibroblast cultures by defined factors. Cell 126, 663–676.

Thompson, M.C., Fuller, C., Hogg, T.L., Dalton, J., Finkelstein, D., Lau, C.C., Chintagumpala, M., Adesina, A., Ashley, D.M., Kellie, S.J., *et al.* (2006). Genomics identifies medulloblastoma subgroups that are enriched for specific genetic alterations. J Clin Oncol 24, 1924–1931.

Thomson, J.A., Itskovitz-Eldor, J., Shapiro, S.S., Waknitz, M.A., Swiergiel, J.J., Marshall, V.S., and Jones, J.M. (1998). Embryonic stem cell lines derived from human blastocysts. Science 282, 1145–1147.

Visvader, J.E. (2011). Cells of origin in cancer. Nature 469, 314–322.

Wang, A., and Sander, M. (2012). Generating cells of the gastrointestinal system: current approaches and applications for the differentiation of human pluripotent stem cells. J Mol Med (Berl) 90, 763–771.

Wang, L., Li, L., Shojaei, F., Levac, K., Cerdan, C., Menendez, P., Martin, T., Rouleau, A., and Bhatia, M. (2004). Endothelial and hematopoietic cell fate of human embryonic stem cells originates from primitive endothelium with hemangioblastic properties. Immunity 21, 31–41.

Werbowetski-Ogilvie, T.E., Morrison, L.C., Fiebig-Comyn, A., and Bhatia, M. (2012). In vivo generation of neural tumors from neoplastic pluripotent stem cells models early human pediatric brain tumor formation. Stem Cells 30, 392–404.

Werbowetski-Ogilvie, T.E., Bosse, M., Stewart, M., Schnerch, A., Ramos-Mejia, V., Rouleau, A., Wynder, T., Smith, M.J., Dingwall, S., Carter, T., *et al.* (2009). Characterization of human embryonic stem cells with features of neoplastic progression. Nat Biotechnol 27, 91–97.

Wong, D.J., Liu, H., Ridky, T.W., Cassarino, D., Segal, E., and Chang, H.Y. (2008). Module map of stem cell genes guides creation of epithelial cancer stem cells. Cell Stem Cell 2, 333–344.

Wu, X., Northcott, P.A., Dubuc, A., Dupuy, A.J., Shih, D.J., Witt, H., Croul, S., Bouffet, E., Fults, D.W., Eberhart, C.G., *et al.* (2012). Clonal selection drives genetic divergence of metastatic medulloblastoma. Nature 482, 529–533.

Yang, Z.J., Ellis, T., Markant, S.L., Read, T.A., Kessler, J.D., Bourboulas, M., Schuller, U., Machold, R., Fishell, G., Rowitch, D.H., *et al.* (2008). Medulloblastoma can be initiated by deletion of Patched in lineage-restricted progenitors or stem cells. Cancer Cell 14, 135–145.

Yu, J., Vodyanik, M., Smuga-Otto, K., Antosiewicz-Bourget, J., Frane, J., Tian, S., Nie, J., Jonsdottir, G., Ruotti, V., Stewart, R., *et al.* (2007). Induced pluripotent stem cell lines derived from human somatic cells Science.

Zbinden, M., Duquet, A., Lorente-Trigos, A., Ngwabyt, S.N., Borges, I., and Ruiz i Altaba, A. (2010). NANOG regulates glioma stem cells and is essential in vivo acting in a cross-functional network with GLI1 and p53. EMBO J 29, 2659–2674.

Zhu, H., Lensch, M.W., Cahan, P., and Daley, G.Q. (2011). Investigating monogenic and complex diseases with pluripotent stem cells. Nat Rev Genet 12, 266–275.

Chapter 5

Cancer Stem Cell as a Result of a Reprogramming-Like Mechanism: Implications in Tumor Development and Treatment

J.M. Iglesias[1], Idoia García-Ramírez[2], Alberto Martín-Lorenzo[2], L. Vellon[3], Lucia Ruiz-Roca[2], A.G. Martin[4], and Isidro Sanchez-Garcia[2]

[1] *Regulation of Cell Growth Laboratory, Fundacion Inbiomed, San Sebastian, Spain*
[2] *Experimental Therapeutics and Translational Oncology Program, Instituto de Biología Molecular y Celular del Cáncer, CSIC/University of Salamanca and Institute of Biomedical Research of Salamanca (IBSAL), Salamanca, Spain*
[3] *Cell Reprogramming Unit, Fundacion Inbiomed, San Sebastian, Spain*
[4] *Regulation of Cell Growth Laboratory, Fundacion Inbiomed, San Sebastian, Spain*

The Cancer Stem Cell Theory

The cancer stem cell (CSC) theory hypothesizes that a cancer maintains a hierarchical organization similar to a normal tissue. Thus, the tumor mass is the result of differentiated progeny of rarer CSCs with self-renewal capacity (Sanchez-Garcia et al., 2007). Chronic myeloid leukemia (CML) is universally regarded as providing the strongest evidence in support of the CSC concept. Fialkow and his colleagues first suggested that CML arose from rare transformed hematopoietic stem cells (HSC) nearly 40 years ago, when they showed that both granulocytes and red blood cells from CML patients were derived from a common cell (Fialkow et al., 1977). However, the term tumor/cancer stem cell was first coined nearly 40 years ago to highlight the observation that only a minority of multiple myeloma cells were capable of clonogenic growth (Hamburger and Salmon, 1977). The last decade has witnessed an increasing re-appreciation of the role of these heterogenous cellular cues in cancer development and therapy. This re-evaluation represents a rather crucial detour from the widely held view that the neoplastic phenotype resulted from uncontrolled proliferation of tumor cells. The CSC concept would explain not only the low clonogenic capacity of most malignancies but also why complete treatment responses translate into cures in only a minority of cancer patients. Initial responses in cancer represent therapeutic effectiveness against the bulk cancer cells, while rarer resistant CSCs could be responsible for relapse. Accordingly, improving the results of cancer therapy would require identification and better understanding of the biology of CSC (Perez-Caro et al., 2009; Saito et al., 2010) (Figure 5.1).

Within this framework, fundamental determinants of neoplastic disease are to be found within the CSC, thus the role of CSC regarding cancer biology, management, and therapy needs to be evaluated (Sanchez-Garcia, 2009). It should be noted that partial tumor responses to therapy mean little if CSCs are the major cells determining outcome (Sanchez-Garcia, 2009).

Cancer Stem Cells, First Edition. Edited by Vinagolu K. Rajasekhar.

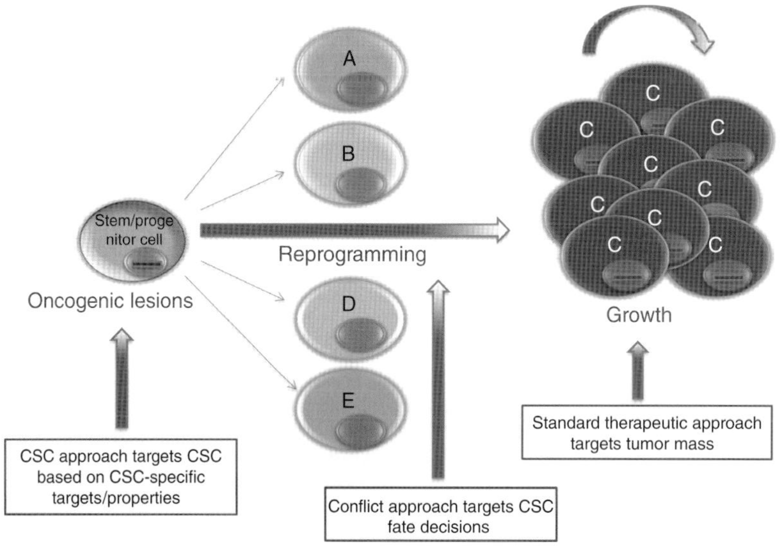

Figure 5.1. Potential approaches to target cancer stem cells (CSC).

Because of the difficulty of assessing the effects of therapies on the rare CSCs responsible for cancer maintenance and relapse, the development of new clinical approaches will require new clinical paradigms and methodologies that should rely heavily on preclinical modeling, using novel preclinical assays to evaluate the fate of CSCs (Sanchez-Garcia et al., 2007). Preclinical studies should assess the effects of therapies on CSC and differentiated cancer cell populations. This could allow us to take directly to the patient a fully functional new approach (Figure 5.1).

A related concept is that the exact definition of "stemness" is elusive, and stemness may be more of a continuum or a property that may be regained in cancer, which would suggest that neither the hierarchical nor the stochastic model are exclusively right.

Furthermore, we must call attention to the fact that the fundamental concept essential to the CSC hypothesis does not have anything to do with the absolute frequency of these cells within the tumor; indeed, what the model states is that there is a functional heterogeneity within the tumor cellular components, and that there is only a defined population of cells that can initiate/maintain malignant growth in vivo while the remaining cells cannot. Thus, the therapeutic implications of the CSC concept are equally important whatever their frequency is within each tumor type: they are the cells that must be effectively targeted to achieve a definitive cure on the long run (Perez-Caro et al., 2009; Saito et al., 2010) (Figure 5.1).

Stem Cells and Cancer Initiation

The nature of the cell in which the initiating mutation occurred in human cancer has received little attention during the last decades. Since the process of carcinogenesis needs to accumulate a number of oncogenic events during long periods of time, only cells with self-renewal capacity would be in the tissue enough time to accumulate the oncogenic alterations necessary for the complete cell transformation. This fact seems to be particularly evident in tumors originated in tissues with high cellular turnover, as the skin, the intestine or the breast, where normal stem cells should be the target for the oncogenic initiation event (Al-Hajj et al., 2003; Jacques et al., 2010; Singh et al., 2004; Wang et al., 2009). For more differentiated cells to originate epithelial cancer, it would be necessary that the first oncogenic event to induce a fully tumor phenotype, or at least be able to trigger a partial stem cell-like program that permit the differentiated progenitor to acquire surviving and self-renewal capabilities. In recent years, there is growing evidence that stem cells are the cells of origin for several types of cancer (Sanchez-Garcia et al., 2007; Vicente-Duenas et al., 2009). An example is provided by the chronic myelogenous leukemia (CML), a granulocytic disease (Melo and Barnes, 2007). However, the *BCR-ABL* translocation does not arise in a granulocyte, but rather in a cell at the beginning of the hematopoietic differentiation tree (Jamieson et al., 2004).

Expression of Pluripotency-Related Genes in Cancer

A growing body of evidence has been reported supporting the notion that tumors are organized in a hierarchy of heterogeneous cell populations where the ability to initiate and sustain growth is restricted to a small population of tumor cells, named cancer stem cells (CSC). Evidence that tumors arise from stem/progenitor cells has been obtained from leukemia (Bonnet and Dick, 1997), breast (Al-Hajj et al., 2003), brain (Singh et al., 2003), colon (Ricci-Vitiani et al., 2007), and most other tumors (pancreas, melanoma, glioblastoma, ovary, liver, prostate). However, the target cell for transformation that originates CSCs remains an unknown issue. This could be a tissue stem cell, a progenitor cell, or a terminally differentiated cell that acquires, through mutations and epigenetic changes, the stem cell self-renewing property. Nowadays, the direct reprogramming of somatic cells into a pluripotent state is a common methodology used to generate the termed induced pluripotent stem cells (iPSC). This discovery offers a unique experimental tool to address key questions regarding cell fate determination and epigenetic regulation and has attracted a great deal of attention. This process requires the overexpression of just four transcription factors, Oct4, Sox2, Klf4, and c-Myc (Takahashi et al., 2007), although Klf4 and c-Myc can be replaced by Lin28 and Nanog, and may even be dispensable (Kim et al., 2009; Meissner et al., 2007; Nakagawa et al., 2008; Okita et al., 2008). The efficiency of this reprogramming process is extremely low and remains so far as an in vitro phenomenon, since there is no evidence that it can naturally occur in vivo. Cell reprogramming is an inefficient and stochastic process, in which the cells must overcome many roadblocks to reach pluripotency. However, much of the nature of these barriers remains unknown.

Although the mechanisms underlying the reprogramming process are not well understood yet, the three main transcription factors—Oct4, Sox2, and Nanog, called master regulators of pluripotency (Jaenisch and Young, 2008)—have proved responsible for maintaining the undifferentiated state (Yamanaka, 2008). Both c-Myc and Klf4 are well-known oncogenes, c-Myc is an oncogenic transcription factor frequently dysregulated in human cancer cells by amplification, mutation, overexpression, or protein stabilization (Soucek and Evan, 2010); KLF4 is a transcription factor that shows context-dependent activity as a tumor suppressor or an oncogene (Rowland et al., 2005). On the other hand, Oct4, Sox2, Nanog, or Lin28 are expressed in several forms of cancer, linking acquisition of self-renewal traits with cancer. For example, Sox2 and Oct4 expression was found to have independent prognostic value in gastric cancer (Matsuoka et al., 2012). Oct4 overexpression leads to the generation of dysplastic lesions in epithelial tissues due to aberrant expansion of early progenitor cells in mice (Nakagawa et al., 2008). Besides, transduction of Oct4 into primary human mammary epithelial breast cell preparations from reduction mammoplasties was sufficient to generate cell lines possessing tumor-initiating and colonization capacities (Kim and Nam, 2011). Lin28, which promotes malignancy by inhibiting the family of tumor-suppressor miRNAs *let-7* (often repressed in human cancers), is a gatekeeper of the pluripotent state (Viswanathan et al., 2009). An interesting auto-regulatory loop involving NF-kB, Lin28, and IL6 links inflammation to transformation and cancer (Iliopoulos et al., 2009). Both Oct4 and lin28 are co-expressed in human ovarian cancer samples and cell lines, identifying a subpopulation of cells with enhanced cell growth and survival (Peng et al., 2010).

Breast cancer patients with strong expression of Nanog were shown to present significantly lower disease-free survival (DFS) and overall survival rates than those with weak expression of Nanog, suggesting a possible use of Nanog as a prognostic marker. In contrast, patients with strong expression of Klf4 had better DFS (Nagata et al., 2012). Jeter and others (2011) reported that Nanog overexpression in a prostate cancer cell line promoted both drug resistance and castration resistance.

Several forms of lung cancer have shown expression of Oct4, Nanog, and Sox2 (Karoubi et al., 2010; Nirasawa et al., 2009; Yuan et al., 2010). In fact, Sox2 was recently shown to control the oncogenic transcriptional network in lung cancer (Chen et al., 2012). As the neural system is the only tissue that expresses Sox2 in the adult, Sox2 expression has been associated with cancer stem cell features in brain tumors, such as glioblastoma (Alonso et al., 2011). Sox2 has been identified as an oncogene in breast cancer, through upregulation of CKND1 (Chen et al., 2008). Others as well as our team have shown that Sox2 is expressed in the initial phases of breast tumorigenesis and then is lost as the tumor develops (Leis et al., 2012). This behavior resembles the expression of those genes in pluripotent stem cells that are turned off as the cells commit to specific lineages.

Recently, the processes of reprogramming and tumorigenesis have been linked (Kawamura et al., 2009; Li et al., 2009; Marion et al., 2009; Utikal et al., 2009), as the p53 tumor suppressor, one of the main gatekeepers of oncogenic transformation, which prevents the initiation of tumor formation by inducing cell cycle arrest, senescence, DNA repair and apoptosis, controls the induction of pluripotency. Loss of p53 confers stem-like properties in committed progenitors in the hematopoietic system (Akala et al., 2008) and loss of p53 increases by tenfold the size of the mammary repopulating compartment in mice, suggesting that p53 may act as a regulator of stem cell self-renewal in vivo (Cicalese et al., 2009). It is then possible that the acquisition of the self-renewal ability of cancer stem cells arises from a reprogramming-like mechanism in cells that harbor permissive mutations, such as inactivation of the p53 pathway. This hypothesis implies that CSCs could be any cell harboring the right genetic changes, which undergoes an altered reprogramming-like process. Interestingly, the process of iPSC generation is highly inefficient, and most cells become trapped into partially reprogrammed states characterized by immortality but not full pluripotency (Mikkelsen et al., 2008).

The ability of non-stem cells to convert to cancer stem cells implies intrinsic plasticity at the level of the cell of origin of the tumors. In this regard, transformation of an immortalized breast epithelial cell line (MCF10A) by the oncogene Src-generated cells with features of CSCs only when IL6 was added to the medium, suggests that CSCs can be originated from non-CSC transformed cells (Iliopoulos et al., 2011). Similar results were obtained attending to in vitro culture of floating human mammary epithelial cells where cells with stem-like features arise from FACS-sorted differentiated epithelial cells (Chaffer et al., 2011). We propose that spontaneous de-differentiation in vivo may involve reactivation of one or more pluripotency associated factors. What the mechanisms for re-activation are, how are they different from a full pluripotency program, and how contextual signals from the microenvironment (such as induction of EMT [Mani et al., 2008] participate in this process), will shed light into how tumors are initiated and how to control this process.

EMT, MET, and Cellular Reprogramming

The vast majority of spontaneous human tumors arise from epithelial-type cells. In normal homeostasis conditions, epithelial cells possess apical-basolateral polarity, are attached to the basement membrane, form regular layers, and neighbouring elements are adjoined by means of specialized junctions. In contrast, mesenchymal cells lack apical-basal polarity, do not form regular layers of cells, and can form adhesions focally with neighboring cells (Cannito et al., 2010).

During development, two critical processes have been shown to mediate the interconversion between epithelial and mesenchymal cells, namely EMT for epithelial to mesenchymal transition and MET for mesenchymal to epithelial transition. EMT defines a process paradigmatic of the concept of cell plasticity that was first described in embryonic development, where cell migration and tissue remodeling have a primary role in regulating morphogenesis in multicellular organisms. The initial step of EMT involves the orchestrated disruption of specialized junctional structures of epithelial cells following exposure to biological signals (growth factors, cytokines, changes in ECM, etc.) following a time-dependent "kinetics" of changes that usually requires the dissociation of tight junctions and then the dissociation of adherens junctions and desmosomes (Cannito et al., 2010).

During the reprogramming of fibroblasts toward induced-pluripotency, they are transformed into tight clusters of rounded cells through what appears to be an MET, the opposite to EMT. Indeed, mouse embryonic fibroblasts (MEF) undergoing reprogramming displayed well-defined epithelial-like intercellular junctions as well as increased expression of epithelial markers such as cytokeratins, E-Cadherin, and Epithelial Cell Adhesion Molecule (Ep-CAM). Concomitantly, the mesenchymal markers fibronectin (Fn) and N-Cadherin are reduced. Accordingly, reprogramming of MEFs also decreased the expression of a panel of classic EMT mediators including Snail and Slug (Li et al., 2010). Interestingly, TGF-β, one of the mayor players in EMT, appears to be deeply involved in the reprogramming process, since the addition of TGB-β1 to MEFs undergoing reprogramming inhibited the upregulation of epithelial markers and the downregulation of Snail by day 6 post-infection. Conversely, overexpression of Snail also prevented the formation of intercellular junctions and inhibited the appearance of alkaline phosphatase (AP) positive colonies (a typical marker of pluripotency). These results suggest that TGF-β1 and Snail abort reprogramming at an early stage (Li et al., 2010). Indeed, pro-EMT cytokines contained in serum (TGF-β1 and 2) or secreted by the fibroblasts (TGF-β1 and low TGF-β2) may impair reprogramming by repressing MET. This would be counteracted by the reprogramming factors, since the OSKM reprogramming factors progressively decreased TGF-β1 in the conditioned medium of transduced MEFs.

The two genes commonly changed in all forms of EMT are E-cadherin and Snail. By expressing each of the reprogramming factors individually, it was observed that Sox2 and Oct4 downregulate Snail whereas Klf4 drastically upregulated E-cadherin, thus providing an additional manner to shut down EMT during reprogramming (Li et al.,

2010). In accordance with this notion, gene expression studies in secondary iPSCs generated from chimeric mice revealed three phases during the direct reprogramming of MEFs referred as initiation, maturation, and stabilization (Samavarchi-Tehrani et al., 2010). No pluripotency-associated genes are expressed during the initiation phase, but there is a large induction of the epithelial associated genes Cdh1, Cldns −3, −4, −7, and −11, Occludin, and Ep-CAM. The beginning of the maturation phase is marked by the expression of Nanog, Sall4, Rex1, Cripto and Nodal at 8 days post-infection, and the stabilization phase starts around day 21 and was marked by the induction of Dnmt3l, Lin28, Utf1, Pecam, Stella, and Dppa4 (Samavarchi-Tehrani et al., 2010). During the initiation phase MET takes place, as shown by the gain of epithelial markers and the downregulation of mesenchymal regulators such as Snail, Slug, Zeb1, and Zeb2 (Samavarchi-Tehrani et al., 2010). This same study identified the bone morphogenic proteins (BMP), a family of TGF-β related factors, as major players during cell reprogramming. The miR-200 and miR-205 family of miRNAs also play a critical role in the induction of MET by downregulation of Zeb1 and Zeb2, since constitutive expression of the miR-200 family members miR-141, miR-200, and miR-429 promote MET and reprogramming even in the absence of BMP. In summary, the early phase of the reprogramming of MEFs involves the reversal of the EMT program. The pluripotency factors take over later, during the maturation phase (Houghton, 2005).

In apparent contrast with these findings, EMT has been shown to induce stem cell properties in human and murine mammary epithelial cells (Mani et al., 2008). Indeed, upon the induction of EMT immortalized human mammary epithelial cells acquire a $CD44^{high}/CD24^{low}$ phenotype commonly associated with breast cancer stem cells (CSC) and normal mammary epithelial stem cells. Moreover, these cells were able to form mammospheres in vitro in response to TGF-β1 or forced expression of Snail or Twist. These mammospheres contained mammosphere-forming cells that could be passaged serially, thus suggesting that these mammosphere-forming cells were capable of self-renewal. Moreover, these $CD44^{high}/CD24^{low}$ cells behaved like bipotential precursor cells, since they were able to generate heterogeneous populations of cells displaying either luminal or basal/myo-epithelial markers and a small proportion expressing both (Mani et al., 2008). More recent evidence supports these results since two transcription factors, Slug and Sox9, act cooperatively to induce and maintain the mammary stem cell (MaSCs) state (Guo et al., 2012). Indeed, co-expression of Slug and Sox9 resulted in the conversion of murine differentiated luminal cells of the mammary gland into MaScs. These MaScs acquired functional properties of stem cells as shown by the mammary reconstitution assay. Moreover, Slug and Sox9 were able to induce macro-metastasis formation capability in nonmetastatic human breast cancer models. This would suggest that normal tissue-specific stem cells and cancer stem cells share some key transcriptional regulators and that cells are maintained in their differentiated or stem cell states in a metastable fashion (Guo et al., 2012). The demonstration that EMT confers stemness properties to terminally differentiated normal mammary epithelial or breast cancer cells does not challenge the role of MET in reprogramming into pluripotency, since EMT produces tissue-specific stem cells. To challenge this notion, a recent report shows that induction of a mesenchymal program in prostate and bladder cancer models impairs major properties associated to cancer stem cells, such as anchorage-independent cell growth and metastatic potential (Celia-Terrassa et al., 2012). Conversely, knock-down of E-cadherin and pluripotency-associated genes in CSCs-enriched epithelial subpopulations resulted in loss of its epithelial properties, inhibition of anchorage-independent cell growth, enhanced invasiveness and inhibition of its capacity to colonize distant organs, indicating that CSCs features would be linked to an epithelial phenotype (Celia-Terrassa et al., 2012). Thus, it appears that the mesenchymal program would be regulating the invasive capacity, whereas cell growth at distant sites would be allowed by the reversion to an epithelial state and activation of self-renewal genes.

From the body of evidence presented above, it is tempting to propose once again that either in normal or tumor cells, the stem-cell state is a metastable condition depending on the cell genetic and epigenetic conditions and is "switched on and off" by cues from the microenvironment.

Organ Development and Cancer

For a long time, cancer and developmental scientists have realized that signaling pathways and genes relevant to cancer overlap with organ development, as reflected in reactivation of embryonic genes during tumor progression. At the same time that scientists realized this correlation, the question arose whether tumors could originate from transformation of tissue stem cells or "retro-differentiation" (now named reprogramming) of more differentiated cells (Uriel, 1976). Nearly 40 years later these ideas and questions are still hot spots in cancer research (Hu and Shivdasani, 2005; Li et al., 2012; Visvader, 2011).

To illustrate the links between organ development and cancer, we will focus on the mammary gland, for convenience, although our conclusions may be widely applied to tumor formation. The mammary gland is a complex secretory organ whose main function is the production of milk to feed the pups in mammals. The mammary gland epithelium form a tree-like structure embedded into the fat pad that consists of two main cell types: luminal and myoepithelial cells. The luminal cells line the ducts and secretory alveoli and produce milk during lactation, whereas the myoepithelial cells surround the luminal cells, are in close contact with the basal membrane and contract in response to oxytocin to secrete the milk through the nipple. The fat pad is made of adipocytes but also contains vascular endothelial cells and stromal cells, including fibroblast and immune cells. The interactions among all of these cell types, coordinated through the action of systemic hormones, and the local paracrine action of different growth factors during development, shape the organ.

The development of the mammary gland (for a review see (Howard and Gusterson, 2000; Howlin et al., 2006; Richert et al., 2000; Watson and Khaled, 2008) starts in the utero, but most of it takes place after birth and it is linked to sexual development and reproduction. At puberty, a burst of proliferation takes place and the rudimentary epithelial tree expands and gets ready for the next stages of development that take place during pregnancy, lactation and involution. Pregnancy is characterized by extensive proliferation and the functional differentiation of the milk-producing alveolar luminal cells. During lactation the mammary gland reaches its full maturation, and after weaning the mammary gland undergoes involution, a complex remodeling process that involves extensive apoptosis of the epithelial structures, resulting in a mammary gland that resembles its pre-pregnant state. This very dynamic developmental process, with rapid expansions of the epithelial populations and involution, is supported by the hierarchical organization of the tissue, with the mammary stem cells on the apex and differentiated myoepithelial, ductal, and alveolar luminal cells at the bottom. The different levels in this cellular hierarchy are expanded or reduced depending on the time point of development, and this fine tuning is controlled by the action of systemic hormones and the paracrine action of local factors secreted by the stroma and neighboring epithelial cells.

Changes in the size of the stem cell compartment, or other progenitor populations, may have a huge impact on tumor development, generating putative targets for cell transformation. For example, maximum blood levels of progesterone during the luteal dioestrus phase are associated with the expansion of the stem cell compartment (Joshi et al., 2010), and it has been shown that women on hormonal replacement therapy treated with estrogen and progesterone have a higher incidence of breast cancer than women treated just with estrogen (Narod, 2011). Another example is expansion of the luminal restricted progenitor population within the mammary gland epithelium in healthy BRCA1 germ line mutation carriers (Lim et al., 2009). Besides, using a mouse model of BRCA1 associated breast cancer, a subpopulation of luminal progenitor cells was identified as the likely origin of these tumors (Molyneux et al., 2010).

After puberty, the concerted action of ovarian and pituitary hormones drive ductal elongation, recurrent menstrual cycles promote branching throughout adulthood, and alveogenesis completes during pregnancy (Brisken and O'Malley, 2010). The powerful effect of these systemic hormones is mediated by the action of multiple secondary paracrine effectors. Studies with knock out mice have shown that the mammary epithelial cells are the primary target cells of estradiol action during puberty, with estrogen receptor alpha −ERα− expressing cells acting on their neighboring cells through the release of paracrine factors like AREG (Ciarloni et al., 2007; Mallepell et al., 2006). In the adult mammary gland epithelium, the ERα positive cells are distributed along ducts and control the proliferation of the neighboring epithelial cells, where most of the proliferating cells are ER negative. A similar mechanism of action had been proposed to explain the progesterone receptor −PR− induced proliferation (Beleut et al., 2010; Brisken et al., 1998). This sequential hormonal mechanism of action ensures the amplification of the systemic stimulus on the target organ, and must be disrupted in tumors where ERα positive and PR positive cells are actively proliferating (Clarke et al., 1997). The expression of the nuclear receptors for these ovarian hormones is deregulated in breast cancer, with ERα expressed in more than 60% of the tumors, and PR is expressed in about 40% of the tumors. Currently, the endocrine therapy targeting ERα signaling with tamoxifen or aromatase inhibitors is the most effective way of treating these hormone-dependent tumors (Obiorah and Jordan, 2011).

AREG is a key mediator of ERα function in the mammary gland driving the massive epithelial cell proliferation characteristic of the pubertal ductal elongation (Ciarloni et al., 2007). AREG is a member of the EGF family of growth factors, and it has been found that other members of this family and its receptors are involved in mammary gland development and cancer (Eccles, 2011). The family member attracting more attention is the orphan ErbB2/HER2/Neu receptor, because it is amplified in approximately 20% of human breast tumors (Eccles, 2011). This receptor is expressed in stromal and mammary epithelial cells during puberty and in adult virgin mice, as well as in the alveolar epithelium from pregnant and lactating mice. Studies using transgenic animals overexpressing a dominant negative form of the receptor in the mammary gland show defects in differentiation of the alveolar structures at the end of pregnancy, and these mice show impaired lactation (Jones and Stern, 1999). On the other hand, knock out and transplantation assays show that ErbB2 also plays a role in ductal elongation during puberty (Jackson-Fisher et al., 2004). This relationship shows the effect of altering organ development and tumor formation.

Microenvironment interactions play a major role in organ development. Tissue recombination and mammary reconstitution experiments in mice have shown that mammary gland microenvironment is able to reprogram spermatogenic and neural stem cells to contribute to a functional mammary gland epithelium (Booth et al., 2008; Boulanger et al., 2007). NSCs alone, without cooperation from co-injected mammary epithelial cells, fail to reconstitute the mammary gland epithelium, therefore signals from bona fide mammary epithelial cells are necessary to produce the chimeric epithelial tree. NSC progeny produce myoepithelial and fully functional luminal cells capable of milk production. A nice example showing how important these interactions are is the essential role that macrophages play during development of the mammary epithelial tree at puberty. In γ-irradiated mice mammary gland development is impaired, though it can be restored through bone marrow transplantation providing the macrophage population around the ducts. Furthermore, in the CSF-1 knock out mice, the macrophage population is severely diminished in the mammary gland and show restricted outgrowth and branching of the ductal tree at puberty. It is not yet completely clear how macrophages mediate these effects. On one hand, macrophages promote collagen fibrillogenesis around terminal end buds helping ductal elongation (Ingman et al., 2006), but on the other hand, it is possible that macrophages play a trophic role for epithelial cells secreting growth factors and cytokines to facilitate ductal invasion (Ingman et al., 2006). During tumor progression, cancer cells could take advantage of these very same properties. In fact, in breast tumors, overexpression of CSF-1 correlates with high grade, poor prognosis, and high leukocytic infiltration (Scholl et al., 1994).

It is interesting to note that the mammary gland microenvironment is able to reprogram cancer cells to mammary epithelial cell fates. When cell lines derived from mouse MMTV-*neu* tumors are mixed with normal mammary epithelial cells and injected in mouse empty fat pads, they are able to contribute to the development of normal mammary gland outgrowths (Booth et al., 2011). The progeny of this tumor cell lines produce both luminal and myoepithelial cells, and no tumor formation is observed even after 18 months post-surgery in either first- or second-generation outgrowths. So, some MMTV-*neu* initiated tumor cells are able to interact with normal mammary epithelial cells and contribute to normal mammary gland development and regeneration, evidencing the plasticity necessary to accommodate both organ development and tumor formation.

Caveats for Identification of CSC in Human Cancer

In human cancer the definition of the identity of CSCs comes from experiments of serial transplantation of flow cytometry-sorted cell populations into immunocompromised mice. The CSC-containing population should recapitulate the cellular heterogeneity present in the primary human cancer and must have the capacity for self-renewal on serial passaging (Cobaleda and Sanchez-Garcia, 2009). However, there are many technical issues concerning the isolation and determination of CSC capabilities from human cancer samples, ranging from the methods of selection of the cells themselves to the choice of the recipient animals where the cells can reveal their potential and to the injection site within the recipient (Cobaleda and Sanchez-Garcia, 2009). To avoid these caveats, an alternative way to study the CSC population is to use mice as a system model.

Identification of CSC in Mouse Models of Human Cancer

Much of our current conceptualization of how tumorigenesis occurs in humans is strongly influenced by mouse models of cancer development (Perez-Losada et al., 2002; Perez-Mancera et al., 2005a; Perez-Mancera et al., 2005b; Sanchez-Martin et al., 2002). But studies in mice in which the oncogenic alteration(s) is not directed to the specific cells of origin, as it normally occurs in most current mouse models, should be interpreted cautiously (Vicente-Duenas et al., 2010).

The genetic alterations found in human cancer seem to occur during specific periods of time and are restricted to a few specific cells. In several cases, like in the case of CML, the cancer cell-of-origin is a stem/progenitor cell, and this explains the stem properties that allow the CSCs to maintain the tumor mass. However, there are also many cancers where most probably the cancer cell-of-origin would be a more differentiated cell (Cobaleda et al., 2007). In these cases, the combination of the reprogramming capabilities of the oncogenic alteration and the intrinsic plasticity of the target cell (i.e., its susceptibility to the reprogramming) determine the final outcome of a CSC. Since not all of the cells present the same susceptibility to reprogramming, and not all the oncogenes possess the same reprogramming capacities (i.e., the ability to confer stem cell features to the target cell), the targeting of the oncogenic alteration to the wrong cellular compartment is a likely cause of failure in the generation of accurate mouse models of human cancer. Considering these facts, three independent groups have already shown that the genotype-phenotype correlations found in human cancer can be established in mice by specific targeting of stem cells (Barker et al., 2009; Perez-Caro et al., 2009; Zhu et al., 2009).

Cancer as a Reprogramming-Like Disease: Therapeutic Implications

In a normal stem cell-driven tissue, genetic programming of stem cells is all that is required to (re)constitute all differentiated cells forming the tissue, and the genetic information responsible for the stem cell programming is not expressed within the differentiated cells that form the tissue. As we have mentioned before, in the last years, much evidence has been accumulated indicating that cancers are also hierarchically organized tissues that can be created and maintained like a normal stem-cell-based tissue (Etzioni et al., 2003; Jemal et al., 2009; Sanchez-Garcia et al., 2007). The most challenging arena in which to prove this concept is in those tumors whose main cellular components are terminally differentiated cells. A clear example of this kind of tumor is the chronic phase of CML. To elucidate if CML is a stem cell-driven tissue, we developed mice limiting *BCR-ABL* expression to the Sca1+ cells (Sca1-BCRABL mice) (Sanchez-Garcia, 2009). Thus, our Sca1-BCRABL is a very suitable in vivo model to study the consequences of ectopic expression of *BCR-ABL* targeted to stem cells. However, in human CML and in most animal models of cancer, the oncogenic alteration(s) is (are) present in all the cellular types that compose the tumoral tissue, from the cancer cell-of-origin to the terminal differentiated granulocytes. In our stem cell-driven Sca1-BCRABL model, the expression of the oncogene is restricted to the stem/progenitor compartment but is nevertheless capable of generating a full-blown CML with all its differentiated cellular components. Of course, the demonstration that CML development can be established in mice by limiting oncogene expression to Sca1$^+$ cells implies that abolishing oncogene function does not interfere with the formation of differentiated tumor cells, and suggests that the oncogene imposes a gene regulatory state in stem cells that somehow persists during hematopoiesis and that imposes a tumor phenotype reflective of the usual CML, an observation that seems to apply to other cancer-initiating gene defects (Sanchez-Garcia, 2009). Therefore, we hypothesize that the oncogene mediates tumorigenesis through epigenetic/genetic modification of target genes that remain in this modified state in the mature tumor even in the absence of *BCR-ABL* in agreement with a reprogramming role for *BCR-ABL* in regulating CML formation. Supporting these observations, it has been recently shown that only stem cells, but not astrocytes, gave rise to brain tumors, independent of their location. This suggests a cell-autonomous mechanism that enables stem cells to generate brain tumors, underlining an important role of stem cells and the relevance of initial genetic mutations in the pathogenesis and phenotype of brain tumors.

Recent breakthroughs have shown that reprogramming of differentiated cells can be achieved by the transient expression of a limited number of transcription factors that can "reset" the epigenetic status of the cells and allow them to adopt a new plethora of possible fates. Several of these reprogramming factors were previously known for their oncogenic activity, already connecting the role of oncogenes with tumoral cell fate reprogramming. Furthermore, it has recently been shown that the elimination of p53, whose function is to prevent the survival and expansion of cells with genetic damage, greatly enhances the reprogramming efficiency in the generation of induced pluripotent cells (iPSC) (Castellanos et al., 2010). These p53-null reprogrammed cells carry, however, several types of mutations (Castellanos et al., 2010). These results confirm the fact that the absence of the tumor suppressor does not have an instructive role in tumorigenesis, but just a permissive one, so p53 would prevent cells with damage from being successfully terminally reprogrammed. This indicates that the driving force of the reprogramming process are the reprogramming factors themselves, and that just the necessity of maintaining genetic integrity prevents the reprogrammed cells with any kind of damage to progress along the newly programmed pathway. As a logical consequence, it has recently been proposed that cancer stem cells might arise through a reprogramming-like mechanism and that, if this is the case, perhaps the oncogenes that initiate tumor formation might be dispensable for tumor progression (Castellanos et al., 2010). In addition, it has also been shown in the haematopoietic and nervous systems that the susceptibility of cells to reprogramming is inversely proportional to their degree of differentiation, and that hematopoietic stem cells (HSC) are 300 times more prone to be reprogrammed than B or T cells (Castellanos et al., 2010). Our results show that this stem cell reprogramming is indeed possible in the case of some oncogenes (Perez-Caro et al., 2009; Vicente-Duenas et al., 2012a; Vicente-Duenas et al., 2012b). But perhaps the most crucial question is whether these hands-off regulation mechanisms can be found in other cancer types, especially tumors of epithelial origin, which represent the bulk of human cancers. Importantly, a small subset of *Sca1-BCR-ABL* mice develops additional solid tumors. Considering that Sca1 has been identified as an almost universal stem cell marker in many different tissues, these data would suggest that the view of cancer as a reprogramming-like disease is not specific to only hematopietic tissues, but rather represents a broader mechanism for deregulation of stem cell differentiation, providing a paradigm that can be applied to solid-organ cancers and, together with all the above discussed findings, provide enough experimental evidence to support the view of cancer as a reprogramming-like disease (Castellanos et al., 2010).

This model of cancer (Figure 5.1) is very informative with respect to the fact that the oncogenic mutations can have different roles in CSC versus differentiated cancer cells, and explains why targeted therapies like imatinib can eliminate the latter without affecting the former. However, we should be cautious in interpreting the data as a mimicking of human disease because mouse cells are more prone to transformation than human cells and thus one mutation can lead to full blown cancer in the mouse transgenic model but not in humans. Furthermore, the regulation of certain genes/pathways might differ between mouse and human.

Much evidence now suggests that human cancer could be considered as a reprogramming-like disease. If the potential growth of cancer depends on CSCs and on oncogenes that can function in a hands-off manner, it would be important to know how to eradicate these cells and/or inactivate the reprogramming mechanism (Castellanos et al., 2010) (Figure 5.1).

Association of Embryonic Surface Markers and Cancer Stem Cells

Because the CSC hypothesis could imply that cancer results from the activation of dormant embryonal-rest cells (Cohnheim, 1867; Virchow, 1855), we examined in CSCs the expression of embryonic surface markers that have been previously identified in undifferentiated mouse embryonic stem cells, and we could show that CSCs in our mouse models present embryonic figures (Perez-Caro et al., 2009; Vicente-Duenas et al., 2009; Vicente-Duenas et al., 2012b). Similarly, an association of embryonic surface markers and CSCs has been identified in human cancers (Ben-Porath et al., 2008; Rajasekhar et al., 2011; Schoenhals et al., 2009). Although overall these observations identify potential attractive targets for selective CSC removal (Figure 5.1), they may also just reflect the function of the oncogene at the stem cell level. One possible hypothesis would be that the oncogene activity might be promoting the aging of the targeted stem cells. We examined the expression of genes associated to physiological HSC aging in CSCs from our mouse models and showed that they shared several features with aged stem cells. These data support a model in which stem cells may be affected in similar ways by aging and by oncogenes. Because young stem cells are more resistant to chemotherapy than aged stem cells, these findings suggest a therapeutic window of opportunity to destroy the CSCs while keeping intact the normal stem cell compartment.

Conclusions

Much evidence now suggests that human cancer could be considered as a reprogramming-like disease (Castellanos et al., 2010). The major questions that arise in light of these findings are how oncogene reprogramming impacts the target cell, and what are the qualitative and/or quantitative figures that make stem/progenitor target cells more vulnerable to malignancy. However, the technical approach used so far to demonstrate that the stem cell compartment drives the pathogenesis of some type of cancers is not sufficiently stringent to conclusively rule out that expression from the transgene is limited to stem cells, because it would be difficult to definitely exclude that low levels of expression at certain transitional stages in the mature compartments were at the root of the malignancy. Mouse models of cancer where precise control of the timing of the oncogene exposure is possible will be instrumental to address these and other questions to understand the complexity of stem-associated cancers.

If the potential growth of cancer depends on CSCs and on oncogenes that can function in a hands-off manner, it would be important to know how to eradicate these cells and/or inactivate the reprogramming mechanism (Figure 5.1). The coming years will show whether this optimism is well founded, or whether the immense complexity of this disease will continue to confound our best endeavors to tackle cancer.

Acknowledgments

Research in ISG group was partially supported by FEDER and by MICINN (SAF2009-08803 and SAF2012-32810 to ISG), by MEC OncoBIO Consolider-Ingenio 2010 (Ref. CSD2007-0017), by NIH grant (R01 CA109335-04A1), by Sandra Ibarra Foundation, by Group of Excellence Grant (GR15) from Junta de Castilla y Leon, and the ARIMMORA project (FP7-ENV-2011, European Union Seventh Framework Program), and by proyecto en red de investigación en celulas madre tumorales en cancer de mama supported by Obra Social Kutxa y Conserjería de Sanidad de la Junta de Castilla y Leon. All Spanish funding is co-sponsored by the European Union FEDER program. Research in AGM group is supported by grants from Obra Social Kutxa, Gobierno Vasco (Saiotek program, Departamento de Educación PI2010-25 and Departamento de Sanidad 2011111141), Diputación Foral de Gipuzkoa OF 53/2011 and Instituto de Salud Carlos III Acción Estratégica en Salud PI2010-01035.

References

Akala, O.O., Park, I.K., Qian, D., Pihalja, M., Becker, M.W., and Clarke, M.F. (2008). Long-term haematopoietic reconstitution by Trp53−/−p16Ink4a−/−p19Arf−/− multipotent progenitors. Nature *453*, 228–232.

Al-Hajj, M., Wicha, M.S., Benito-Hernandez, A., Morrison, S.J., and Clarke, M.F. (2003). Prospective identification of tumorigenic breast cancer cells. Proc Natl Acad Sci USA *100*, 3983–3988.

Alonso, M.M., Diez-Valle, R., Manterola, L., Rubio, A., Liu, D., Cortes-Santiago, N., Urquiza, L., Jauregi, P., Lopez de Munain, A., Sampron, N., *et al.* (2011). Genetic and epigenetic modifications of Sox2 contribute to the invasive phenotype of malignant gliomas. PLoS One *6*, e26740.

Barker, N., Ridgway, R.A., van Es, J.H., van de Wetering, M., Begthel, H., van den Born, M., Danenberg, E., Clarke, A.R., Sansom, O.J., and Clevers, H. (2009). Crypt stem cells as the cells-of-origin of intestinal cancer. Nature *457*, 608–611.

Beleut, M., Rajaram, R.D., Caikovski, M., Ayyanan, A., Germano, D., Choi, Y., Schneider, P., and Brisken, C. (2010). Two distinct mechanisms underlie progesterone-induced proliferation in the mammary gland. Proc Natl Acad Sci USA *107*, 2989–2994.

Ben-Porath, I., Thomson, M.W., Carey, V.J., Ge, R., Bell, G.W., Regev, A., and Weinberg, R.A. (2008). An embryonic stem cell-like gene expression signature in poorly differentiated aggressive human tumors. Nat Genet *40*, 499–507.

Bonnet, D., and Dick, J.E. (1997). Human acute myeloid leukemia is organized as a hierarchy that originates from a primitive hematopoietic cell. Nat Med *3*, 730–737.

Booth, B.W., Boulanger, C.A., Anderson, L.H., and Smith, G.H. (2011). The normal mammary microenvironment suppresses the tumorigenic phenotype of mouse mammary tumor virus-neu-transformed mammary tumor cells. Oncogene *30*, 679–689.

Booth, B.W., Mack, D.L., Androutsellis-Theotokis, A., McKay, R.D., Boulanger, C.A., and Smith, G.H. (2008). The mammary microenvironment alters the differentiation repertoire of neural stem cells. Proc Natl Acad Sci USA *105*, 14891–14896.

Boulanger, C.A., Mack, D.L., Booth, B.W., and Smith, G.H. (2007). Interaction with the mammary microenvironment redirects spermatogenic cell fate in vivo. Proc Natl Acad Sci USA *104*, 3871–3876.

Brisken, C., and O'Malley, B. (2010). Hormone action in the mammary gland. Cold Spring Harb Perspect Biol *2*, a003178.

Brisken, C., Park, S., Vass, T., Lydon, J.P., O'Malley, B.W., and Weinberg, R.A. (1998). A paracrine role for the epithelial progesterone receptor in mammary gland development. Proc Natl Acad Sci USA *95*, 5076–5081.

Cannito, S., Novo, E., di Bonzo, L.V., Busletta, C., Colombatto, S., and Parola, M. (2010). Epithelial-mesenchymal transition: from molecular mechanisms, redox regulation to implications in human health and disease. Antioxid Redox Signal *12*, 1383–1430.

Castellanos, A., Vicente-Duenas, C., Campos-Sanchez, E., Cruz, J.J., Garcia-Criado, F.J., Garcia-Cenador, M.B., Lazo, P.A., Perez-Losada, J., and Sanchez-Garcia, I. (2010). Cancer as a reprogramming-like disease: implications in tumor development and treatment. Semin Cancer Biol *20*, 93–97.

Celia-Terrassa, T., Meca-Cortes, O., Mateo, F., de Paz, A.M., Rubio, N., Arnal-Estape, A., Ell, B.J., Bermudo, R., Diaz, A., Guerra-Rebollo, M., *et al.* (2012). Epithelial-mesenchymal transition can suppress major attributes of human epithelial tumor-initiating cells. The Journal of clinical investigation *122*, 1849–1868.

Chaffer, C.L., Brueckmann, I., Scheel, C., Kaestli, A.J., Wiggins, P.A., Rodrigues, L.O., Brooks, M., Reinhardt, F., Su, Y., Polyak, K., *et al.* (2011). Normal and neoplastic nonstem cells can spontaneously convert to a stem-like state. Proc Natl Acad Sci USA *108*, 7950–7955.

Chen, S., Xu, Y., Chen, Y., Li, X., Mou, W., Wang, L., Liu, Y., Reisfeld, R.A., Xiang, R., Lv, D., *et al.* (2012). SOX2 gene regulates the transcriptional network of oncogenes and affects tumorigenesis of human lung cancer cells. PLoS One *7*, e36326.

Chen, Y., Shi, L., Zhang, L., Li, R., Liang, J., Yu, W., Sun, L., Yang, X., Wang, Y., Zhang, Y., *et al.* (2008). The molecular mechanism governing the oncogenic potential of SOX2 in breast cancer. J Biol Chem *283*, 17969–17978.

Ciarloni, L., Mallepell, S., and Brisken, C. (2007). Amphiregulin is an essential mediator of estrogen receptor alpha function in mammary gland development. Proc Natl Acad Sci USA *104*, 5455–5460.

Cicalese, A., Bonizzi, G., Pasi, C.E., Faretta, M., Ronzoni, S., Giulini, B., Brisken, C., Minucci, S., Di Fiore, P.P., and Pelicci, P.G. (2009). The tumor suppressor p53 regulates polarity of self-renewing divisions in mammary stem cells. Cell *138*, 1083–1095.

Clarke, R.B., Howell, A., Potten, C.S., and Anderson, E. (1997). Dissociation between steroid receptor expression and cell proliferation in the human breast. Cancer research *57*, 4987–4991.

Cobaleda, C., Jochum, W., and Busslinger, M. (2007). Conversion of mature B cells into T cells by dedifferentiation to uncommitted progenitors. Nature *449*, 473–477.

Cobaleda, C., and Sanchez-Garcia, I. (2009). B-cell acute lymphoblastic leukaemia: towards understanding its cellular origin. Bioessays *31*, 600–609.

Cohnheim, J. (1867). Ueber entzundung und eiterung. In Path Anat Physiol Klin Med, pp. 1–79.

Eccles, S.A. (2011). The epidermal growth factor receptor/Erb-B/HER family in normal and malignant breast biology. Int J Dev Biol *55*, 685–696.

Etzioni, R., Urban, N., Ramsey, S., McIntosh, M., Schwartz, S., Reid, B., Radich, J., Anderson, G., and Hartwell, L. (2003). The case for early detection. Nat Rev Cancer *3*, 243–252.

Fialkow, P.J., Jacobson, R.J., and Papayannopoulou, T. (1977). Chronic myelocytic leukemia: clonal origin in a stem cell common to the granulocyte, erythrocyte, platelet and monocyte/macrophage. Am J Med *63*, 125–130.

Guo, W., Keckesova, Z., Donaher, J.L., Shibue, T., Tischler, V., Reinhardt, F., Itzkovitz, S., Noske, A., Zurrer-Hardi, U., Bell, G., *et al.* (2012). Slug and Sox9 cooperatively determine the mammary stem cell state. Cell *148*, 1015–1028.

Hamburger, A., and Salmon, S.E. (1977). Primary bioassay of human myeloma stem cells. J Clin Invest *60*, 846–854.

Houghton, F.D. (2005). Role of gap junctions during early embryo development. Reproduction *129*, 129–135.

Howard, B.A., and Gusterson, B.A. (2000). Human breast development. J Mammary Gland Biol Neoplasia *5*, 119–137.

Howlin, J., McBryan, J., and Martin, F. (2006). Pubertal mammary gland development: insights from mouse models. J Mammary Gland Biol Neoplasia *11*, 283–297.

Hu, M., and Shivdasani, R.A. (2005). Overlapping gene expression in fetal mouse intestine development and human colorectal cancer. Cancer research *65*, 8715–8722.

Iliopoulos, D., Hirsch, H.A., and Struhl, K. (2009). An epigenetic switch involving NF-kappaB, Lin28, Let-7 MicroRNA, and IL6 links inflammation to cell transformation. Cell *139*, 693–706.

Iliopoulos, D., Hirsch, H.A., Wang, G., and Struhl, K. (2011). Inducible formation of breast cancer stem cells and their dynamic equilibrium with non-stem cancer cells via IL6 secretion. Proc Natl Acad Sci USA *108*, 1397–1402.

Ingman, W.V., Wyckoff, J., Gouon-Evans, V., Condeelis, J., and Pollard, J.W. (2006). Macrophages promote collagen fibrillogenesis around terminal end buds of the developing mammary gland. Dev Dyn *235*, 3222–3229.

Jackson-Fisher, A.J., Bellinger, G., Ramabhadran, R., Morris, J.K., Lee, K.F., and Stern, D.F. (2004). ErbB2 is required for ductal morphogenesis of the mammary gland. Proc Natl Acad Sci USA *101*, 17138–17143.

Jacques, T.S., Swales, A., Brzozowski, M.J., Henriquez, N.V., Linehan, J.M., Mirzadeh, Z., C, O.M., Naumann, H., Alvarez-Buylla, A., and Brandner, S. (2010). Combinations of genetic mutations in the adult neural stem cell compartment determine brain tumour phenotypes. Embo J *29*, 222–235.

Jaenisch, R., and Young, R. (2008). Stem cells, the molecular circuitry of pluripotency and nuclear reprogramming. Cell *132*, 567–582.

Jamieson, C.H., Ailles, L.E., Dylla, S.J., Muijtjens, M., Jones, C., Zehnder, J.L., Gotlib, J., Li, K., Manz, M.G., Keating, A., *et al.* (2004). Granulocyte-macrophage progenitors as candidate leukemic stem cells in blast-crisis CML. N Engl J Med *351*, 657–667.

Jemal, A., Siegel, R., Ward, E., Hao, Y., Xu, J., and Thun, M.J. (2009). Cancer statistics, 2009. CA Cancer J Clin *59*, 225–249.

Jeter, C.R., Liu, B., Liu, X., Chen, X., Liu, C., Calhoun-Davis, T., Repass, J., Zaehres, H., Shen, J.J., and Tang, D.G. (2011). NANOG promotes cancer stem cell characteristics and prostate cancer resistance to androgen deprivation. Oncogene *30*, 3833–3845.

Jones, F.E., and Stern, D.F. (1999). Expression of dominant-negative ErbB2 in the mammary gland of transgenic mice reveals a role in lobuloalveolar development and lactation. Oncogene *18*, 3481–3490.

Joshi, P.A., Jackson, H.W., Beristain, A.G., Di Grappa, M.A., Mote, P.A., Clarke, C.L., Stingl, J., Waterhouse, P.D., and Khokha, R. (2010). Progesterone induces adult mammary stem cell expansion. Nature *465*, 803–807.

Karoubi, G., Cortes-Dericks, L., Gugger, M., Galetta, D., Spaggiari, L., and Schmid, R.A. (2010). Atypical expression and distribution of embryonic stem cell marker, OCT4, in human lung adenocarcinoma. J Surg Oncol *102*, 689–698.

Kawamura, T., Suzuki, J., Wang, Y.V., Menendez, S., Morera, L.B., Raya, A., Wahl, G.M., and Izpisua Belmonte, J.C. (2009). Linking the p53 tumour suppressor pathway to somatic cell reprogramming. Nature *460*, 1140–1144.

Kim, J.B., Greber, B., Arauzo-Bravo, M.J., Meyer, J., Park, K.I., Zaehres, H., and Scholer, H.R. (2009). Direct reprogramming of human neural stem cells by OCT4. Nature *461*, 649–643.

Kim, R.J., and Nam, J.S. (2011). OCT4 Expression Enhances Features of Cancer Stem Cells in a Mouse Model of Breast Cancer. Lab Anim Res *27*, 147–152.

Leis, O., Eguiara, A., Lopez-Arribillaga, E., Alberdi, M.J., Hernandez-Garcia, S., Elorriaga, K., Pandiella, A., Rezola, R., and Martin, A.G. (2012). Sox2 expression in breast tumours and activation in breast cancer stem cells. Oncogene *31*, 1354–1365.

Li, D., Duell, E.J., Yu, K., Risch, H.A., Olson, S.H., Kooperberg, C., Wolpin, B.M., Jiao, L., Dong, X., Wheeler, B., *et al.* (2012). Pathway analysis of genome-wide association study data highlights pancreatic development genes as susceptibility factors for pancreatic cancer. Carcinogenesis *33*, 1384–1390.

Li, H., Collado, M., Villasante, A., Strati, K., Ortega, S., Canamero, M., Blasco, M.A., and Serrano, M. (2009). The Ink4/Arf locus is a barrier for iPS cell reprogramming. Nature *460*, 1136–1139.

Li, R., Liang, J., Ni, S., Zhou, T., Qing, X., Li, H., He, W., Chen, J., Li, F., Zhuang, Q., *et al.* (2010). A mesenchymal-to-epithelial transition initiates and is required for the nuclear reprogramming of mouse fibroblasts. Cell Stem Cell *7*, 51–63.

Lim, E., Vaillant, F., Wu, D., Forrest, N.C., Pal, B., Hart, A.H., Asselin-Labat, M.L., Gyorki, D.E., Ward, T., Partanen, A., *et al.* (2009). Aberrant luminal progenitors as the candidate target population for basal tumor development in BRCA1 mutation carriers. Nature medicine *15*, 907–913.

Mallepell, S., Krust, A., Chambon, P., and Brisken, C. (2006). Paracrine signaling through the epithelial estrogen receptor alpha is required for proliferation and morphogenesis in the mammary gland. Proc Natl Acad Sci USA *103*, 2196–2201.

Mani, S.A., Guo, W., Liao, M.J., Eaton, E.N., Ayyanan, A., Zhou, A.Y., Brooks, M., Reinhard, F., Zhang, C.C., Shipitsin, M., *et al.* (2008). The epithelial-mesenchymal transition generates cells with properties of stem cells. Cell *133*, 704–715.

Marion, R.M., Strati, K., Li, H., Murga, M., Blanco, R., Ortega, S., Fernandez-Capetillo, O., Serrano, M., and Blasco, M.A. (2009). A p53-mediated DNA damage response limits reprogramming to ensure iPS cell genomic integrity. Nature *460*, 1149–1153.

Matsuoka, J., Yashiro, M., Sakurai, K., Kubo, N., Tanaka, H., Muguruma, K., Sawada, T., Ohira, M., and Hirakawa, K. (2012). Role of the stemness factors sox2, oct3/4, and nanog in gastric carcinoma. J Surg Res *174*, 130–135.

Meissner, A., Wernig, M., and Jaenisch, R. (2007). Direct reprogramming of genetically unmodified fibroblasts into pluripotent stem cells. Nat Biotechnol *25*, 1177–1181.

Melo, J.V., and Barnes, D.J. (2007). Chronic myeloid leukaemia as a model of disease evolution in human cancer. Nature reviews Cancer *7*, 441–453.

Mikkelsen, T.S., Hanna, J., Zhang, X., Ku, M., Wernig, M., Schorderet, P., Bernstein, B.E., Jaenisch, R., Lander, E.S., and Meissner, A. (2008). Dissecting direct reprogramming through integrative genomic analysis. Nature *454*, 49–55.

Molyneux, G., Geyer, F.C., Magnay, F.A., McCarthy, A., Kendrick, H., Natrajan, R., Mackay, A., Grigoriadis, A., Tutt, A., Ashworth, A., *et al.* (2010). BRCA1 basal-like breast cancers originate from luminal epithelial progenitors and not from basal stem cells. Cell Stem Cell *7*, 403–417.

Nagata, T., Shimada, Y., Sekine, S., Hori, R., Matsui, K., Okumura, T., Sawada, S., Fukuoka, J., and Tsukada, K. (2012). Prognostic significance of NANOG and KLF4 for breast cancer. Breast Cancer.

Nakagawa, M., Koyanagi, M., Tanabe, K., Takahashi, K., Ichisaka, T., Aoi, T., Okita, K., Mochiduki, Y., Takizawa, N., and Yamanaka, S.

(2008). Generation of induced pluripotent stem cells without Myc from mouse and human fibroblasts. Nat Biotechnol *26*, 101–106.

Narod, S.A. (2011). Hormone replacement therapy and the risk of breast cancer. Nat Rev Clin Oncol *8*, 669–676.

Nirasawa, S., Kobayashi, D., Tsuji, N., Kuribayashi, K., and Watanabe, N. (2009). Diagnostic relevance of overexpressed Nanog gene in early lung cancers. Oncol Rep *22*, 587–591.

Obiorah, I., and Jordan, V.C. (2011). Progress in endocrine approaches to the treatment and prevention of breast cancer. Maturitas *70*, 315–321.

Okita, K., Nakagawa, M., Hyenjong, H., Ichisaka, T., and Yamanaka, S. (2008). Generation of mouse induced pluripotent stem cells without viral vectors. Science *322*, 949–953.

Peng, S., Maihle, N.J., and Huang, Y. (2010). Pluripotency factors Lin28 and Oct4 identify a sub-population of stem cell-like cells in ovarian cancer. Oncogene *29*, 2153–2159.

Perez-Caro, M., Cobaleda, C., Gonzalez-Herrero, I., Vicente-Duenas, C., Bermejo-Rodriguez, C., Sanchez-Beato, M., Orfao, A., Pintado, B., Flores, T., Sanchez-Martin, M., *et al.* (2009). Cancer induction by restriction of oncogene expression to the stem cell compartment. Embo J *28*, 8–20.

Perez-Losada, J., Sanchez-Martin, M., Rodriguez-Garcia, A., Sanchez, M.L., Orfao, A., Flores, T., and Sanchez-Garcia, I. (2002). Zinc-finger transcription factor Slug contributes to the function of the stem cell factor c-kit signaling pathway. Blood *100*, 1274–1286.

Perez-Mancera, P.A., Gonzalez-Herrero, I., Perez-Caro, M., Gutierrez-Cianca, N., Flores, T., Gutierrez-Adan, A., Pintado, B., Sanchez-Martin, M., and Sanchez-Garcia, I. (2005a). SLUG in cancer development. Oncogene *24*, 3073–3082.

Perez-Mancera, P.A., Perez-Caro, M., Gonzalez-Herrero, I., Flores, T., Orfao, A., de Herreros, A.G., Gutierrez-Adan, A., Pintado, B., Sagrera, A., Sanchez-Martin, M., *et al.* (2005b). Cancer development induced by graded expression of Snail in mice. Hum Mol Genet *14*, 3449–3461.

Rajasekhar, V.K., Studer, L., Gerald, W., Socci, N.D., and Scher, H.I. (2011). Tumour-initiating stem-like cells in human prostate cancer exhibit increased NF-kappaB signalling. Nat Commun *2*, 162.

Ricci-Vitiani, L., Lombardi, D.G., Pilozzi, E., Biffoni, M., Todaro, M., Peschle, C., and De Maria, R. (2007). Identification and expansion of human colon-cancer-initiating cells. Nature *445*, 111–115.

Richert, M.M., Schwertfeger, K.L., Ryder, J.W., and Anderson, S.M. (2000). An atlas of mouse mammary gland development. J Mammary Gland Biol Neoplasia *5*, 227–241.

Rowland, B.D., Bernards, R., and Peeper, D.S. (2005). The KLF4 tumour suppressor is a transcriptional repressor of p53 that acts as a context-dependent oncogene. Nat Cell Biol *7*, 1074–1082.

Saito, Y., Uchida, N., Tanaka, S., Suzuki, N., Tomizawa-Murasawa, M., Sone, A., Najima, Y., Takagi, S., Aoki, Y., Wake, A., *et al.* (2010). Induction of cell cycle entry eliminates human leukemia stem cells in a mouse model of AML. Nat Biotechnol *28*, 275–280.

Samavarchi-Tehrani, P., Golipour, A., David, L., Sung, H.K., Beyer, T.A., Datti, A., Woltjen, K., Nagy, A., and Wrana, J.L. (2010). Functional genomics reveals a BMP-driven mesenchymal-to-epithelial transition in the initiation of somatic cell reprogramming. Cell Stem Cell *7*, 64–77.

Sanchez-Garcia, I. (2009). The crossroads of oncogenesis and metastasis. N Engl J Med *360*, 297–299.

Sanchez-Garcia, I., Vicente-Duenas, C., and Cobaleda, C. (2007). The theoretical basis of cancer-stem-cell-based therapeutics of cancer: can it be put into practice? Bioessays *29*, 1269–1280.

Sanchez-Martin, M., Rodriguez-Garcia, A., Perez-Losada, J., Sagrera, A., Read, A.P., and Sanchez-Garcia, I. (2002). SLUG (SNAI2) deletions in patients with Waardenburg disease. Human molecular genetics *11*, 3231–3236.

Schoenhals, M., Kassambara, A., De Vos, J., Hose, D., Moreaux, J., and Klein, B. (2009). Embryonic stem cell markers expression in cancers. Biochem Biophys Res Commun *383*, 157–162.

Scholl, S.M., Pallud, C., Beuvon, F., Hacene, K., Stanley, E.R., Rohrschneider, L., Tang, R., Pouillart, P., and Lidereau, R. (1994). Anti-colony-stimulating factor-1 antibody staining in primary breast adenocarcinomas correlates with marked inflammatory cell infiltrates and prognosis. J Natl Cancer Inst *86*, 120–126.

Singh, S.K., Clarke, I.D., Terasaki, M., Bonn, V.E., Hawkins, C., Squire, J., and Dirks, P.B. (2003). Identification of a cancer stem cell in human brain tumors. Cancer Res *63*, 5821–5828.

Singh, S.K., Hawkins, C., Clarke, I.D., Squire, J.A., Bayani, J., Hide, T., Henkelman, R.M., Cusimano, M.D., and Dirks, P.B. (2004). Identification of human brain tumour initiating cells. Nature *432*, 396–401.

Soucek, L., and Evan, G.I. (2010). The ups and downs of Myc biology. Curr Opin Genet Dev *20*, 91–95.

Takahashi, K., Tanabe, K., Ohnuki, M., Narita, M., Ichisaka, T., Tomoda, K., and Yamanaka, S. (2007). Induction of pluripotent stem cells from adult human fibroblasts by defined factors. Cell *131*, 861–872.

Uriel, J. (1976). Cancer, retrodifferentiation, and the myth of Faust. Cancer research *36*, 4269–4275.

Utikal, J., Polo, J.M., Stadtfeld, M., Maherali, N., Kulalert, W., Walsh, R.M., Khalil, A., Rheinwald, J.G., and Hochedlinger, K. (2009). Immortalization eliminates a roadblock during cellular reprogramming into iPS cells. Nature *460*, 1145–1148.

Vicente-Duenas, C., Cobaleda, C., Perez-Losada, J., and Sanchez-Garcia, I. (2010). The evolution of cancer modeling: the shadow of stem cells. Dis Model Mech *3*, 149–155.

Vicente-Duenas, C., Fontan, L., Gonzalez-Herrero, I., Romero-Camarero, I., Segura, V., Aznar, M.A., Alonso-Escudero, E., Campos-Sanchez, E., Ruiz-Roca, L., Barajas-Diego, M., *et al.* (2012a). Expression of MALT1 oncogene in hematopoietic stem/progenitor cells recapitulates the pathogenesis of human lymphoma in mice. Proc Natl Acad Sci USA *109*, 10534–10539.

Vicente-Duenas, C., Perez-Caro, M., Abollo-Jimenez, F., Cobaleda, C., and Sanchez-Garcia, I. (2009). Stem-cell driven cancer: "hands-off" regulation of cancer development. Cell Cycle *8*, 1314–1318.

Vicente-Duenas, C., Romero-Camarero, I., Gonzalez-Herrero, I., Alonso-Escudero, E., Abollo-Jimenez, F., Jiang, X., Gutierrez, N.C., Orfao, A., Marin, N., Villar, L.M., *et al.* (2012b). A novel molecular mechanism involved in multiple myeloma development revealed by targeting MafB to haematopoietic progenitors. Embo J.

Virchow, R. (1855). Editorial. In Virchows Arch Pathol Anat Physiol Med, p. 23.

Visvader, J.E. (2011). Cells of origin in cancer. Nature *469*, 314–322.

Viswanathan, S.R., Powers, J.T., Einhorn, W., Hoshida, Y., Ng, T.L., Toffanin, S., O'Sullivan, M., Lu, J., Phillips, L.A., Lockhart, V.L., *et al.* (2009). Lin28 promotes transformation and is associated with advanced human malignancies. Nat Genet *41*, 843–848.

Wang, X., Kruithof-de Julio, M., Economides, K.D., Walker, D., Yu, H., Halili, M.V., Hu, Y.P., Price, S.M., Abate-Shen, C., and Shen, M.M. (2009). A luminal epithelial stem cell that is a cell of origin for prostate cancer. Nature *461*, 495–500.

Watson, C.J., and Khaled, W.T. (2008). Mammary development in the embryo and adult: a journey of morphogenesis and commitment. Development *135*, 995–1003.

Yamanaka, S. (2008). Induction of pluripotent stem cells from mouse fibroblasts by four transcription factors. Cell Prolif *41 Suppl* 1, 51–56.

Yuan, P., Kadara, H., Behrens, C., Tang, X., Woods, D., Solis, L.M., Huang, J., Spinola, M., Dong, W., Yin, G., *et al.* (2010). Sex determining region Y-Box 2 (SOX2) is a potential cell-lineage gene highly expressed in the pathogenesis of squamous cell carcinomas of the lung. PLoS One *5*, e9112.

Zhu, L., Gibson, P., Currle, D.S., Tong, Y., Richardson, R.J., Bayazitov, I.T., Poppleton, H., Zakharenko, S., Ellison, D.W., and Gilbertson, R.J. (2009). Prominin 1 marks intestinal stem cells that are susceptible to neoplastic transformation. Nature *457*, 603–607.

Chapter 6

A Cancer Stem Cell Model: An Insight into the Conversion of Induced Pluripotent Stem Cells to Cancer Stem-Like Cells

Akifumi Mizutani[1], Ling Chen[1,2], Tomonari Kasai[1], Takayuki Kudoh[1], Hiroshi Murakami[1], Li Fu[3], and Masaharu Seno[1]

[1] *Department of Chemistry and Biotechnology, Graduate School of Natural Science and Technology, Okayama University, Okayama, Japan*

[2] *Department of Pathology, Tianjin Central Hospital of Gynecology Obstetrics, Tianjin, People's Republic of China*

[3] *Department of Breast Cancer Pathology and Research Laboratory, Center Hospital of Tianjin Medical University, Tianjin, People's Republic of China*

Introduction

The early concept of the cancer stem cell proposed by Pierce and Speers (1988) follows: a concept of neoplasms, based upon developmental and oncological principles, states that carcinomas are caricatures of tissue renewal, in that they are composed of a mixture of malignant stem cells, which have a marked capacity for proliferation and a limited capacity for differentiation under normal homeostatic conditions, and of the differentiated, possibly benign, progeny of these malignant cells. Nowadays, a concept of cancer stem cells/tumor-initiating cells is widely recognized because of accumulation of experimental evidence. Since various independent experimental systems are being carried out by a number of researchers, we need to define the property of cancer stem cells. A cancer stem cell is defined as a cell within a tumor that possesses the capacity to self-renew and to cause the heterogeneous lineages of cancer cells that comprise the tumor. Cancer stem cells can thus only be defined experimentally by their ability to recapitulate the generation of a continuously growing tumor (Clarke et al., 2006). The AACR Cancer Stem Cells Workshop reached the consensus in 2006.

Generally, cancer stem cells characterized by the cell surface markers, such as CD44

Cancer Stem Cells, First Edition. Edited by Vinagolu K. Rajasekhar.

and CD133 (O'Brien et al. 2010), are isolated from clinical samples. However, the biological significance of these cell surface markers still remains unclear, especially in the aspect of cancer stem cell transformation and/or maintenance. Thus, further analyses are necessary to understand the nature of cancer stem cells and to propose the new therapeutic approach to target cancer stem cells. However, rarity of cancer stem cells in tumors and difficulty of isolation as homogenous population of cancer stem cells from clinical samples make characterization and analysis of these cells limited.

Appropriate model cell lines recapitulating cancer stem cell properties would accelerate not only investigation of cancer stem cells but also development of new clinical cancer therapy by establishing a screening system for anti-cancer stem cell agents. In this chapter, we introduce our recent works and others to generate cells with cancer stem cell properties in vitro. We also will discuss the concept of cancer stem cells with results obtained from originally established cancer stem-like cells. Finally, we will describe future applications of the model to basic and clinical studies.

Generation of Cancer Stem-Like Cells In Vitro

Cancerous Niche: Microenvironment to Develop Malignancy

In the field of regeneration therapy, the pluripotent stem cells such as embryonic stem cells (ESC) and induced pluripotent stem cells (iPSC) are promising sources of differentiated cells for transplantation. When exposed to the appropriate environment, the stem cells should be directed to the progenitor cells such as hematopoietic cells, neural stem cells, and so on, which are destined to differentiate into each mature cell such as macrophage, monocyte, neural cell, cardiac cell, and pancreatic β-cell. The behavior of stem cells is tightly regulated by the signals from surrounding microenvironment, so called 'niche,' which supports the self-renewal of stem cells and controls the stem cell number. Also, a niche regulates the differentiation, and in turn, maintains tissue homeostasis (Moore and Lemischka, 2006). Thus, the cell fate is determined by the events and factors present in the range of a niche.

Taking the pluripotency of stem cells into consideration, it should be hypothesized that malignant neoplasm is one of the phenotypes differentiated from stem cells. In this context, a cancer stem cell could be described as a progenitor cell that is destined to differentiate into a cancer cell. This might be called 'canceration' rather than tumor initiation. And the niche that directs stem cells into cancer stem cells might be called a 'cancerous niche' (Figure 6.1). But a cancerous niche could hardly be defined because normal tissue or the body should have a normal niche. The question is how and where we can find the cancerous niche. We designed experiments to convert pluripotent stem cells into the cells, which have the characteristic properties of cancer stem cells, using established cancer cell lines.

A Model of Cancer Stem Cells Derived From Mouse iPSCs

We have recently reported that mouse iPSCs (miPSC) could acquire characters of cancer stem cells when miPSCs were cultured in the presence of conditioned medium prepared from various cancer cell lines (Figure 6.2) (Chen et al., 2012). The established cancer stem-like cells derived from miPSCs (miPS-CSCs) formed spheroids when they were cultured in non-adherent condition, suggesting their possession of self-renewal capacity. The genes associated with stem cell properties and an undifferentiated state such as *Nanog*, *Rex1*, *Eras*, *Esg1*, and *Cripto1* were kept expressed in miPS-CSCs. Most importantly, miPS-CSCs exhibited high tumorigenecity with rapid growth in Balb/c nude mice in contrast to parental miPSCs that provided only teratomas. Among established miPS-CSCs, miPS-LLCcm cells, which were derived from miPSCs cultured in the media containing conditioned medium

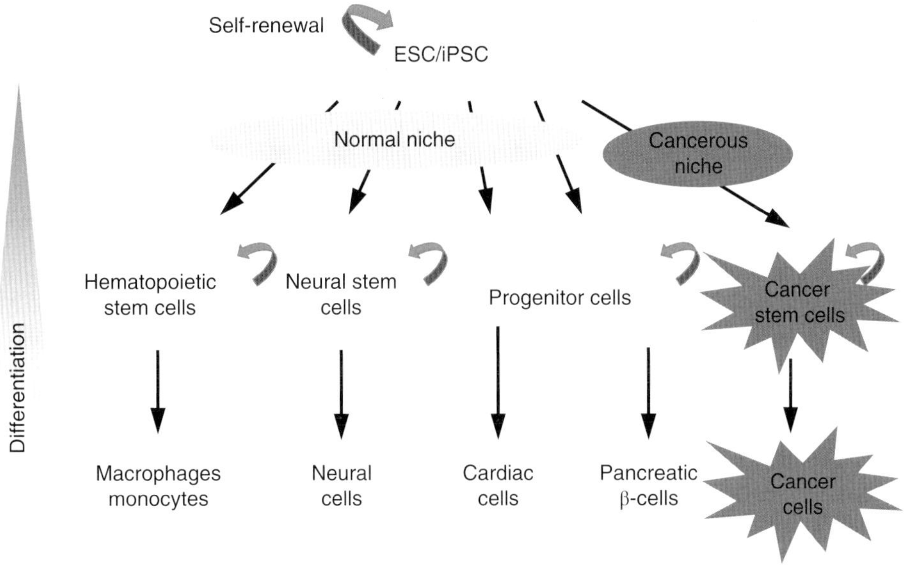

Figure 6.1. Concept of 'cancerous niche' to develop malignancy. In the appropriate niche, the stem cells should differentiate into progenitor cells and further into each mature cell. When a malignant neoplasm is considered to be a phenotype differentiated from stem cell, a cancer stem cell could be a progenitor cell giving rise to cancer cells. The niche contributing conversion of stem cells into cancer stem cells is hypothesized to be a 'cancerous niche' in this review. For color detail, please see color plate section.

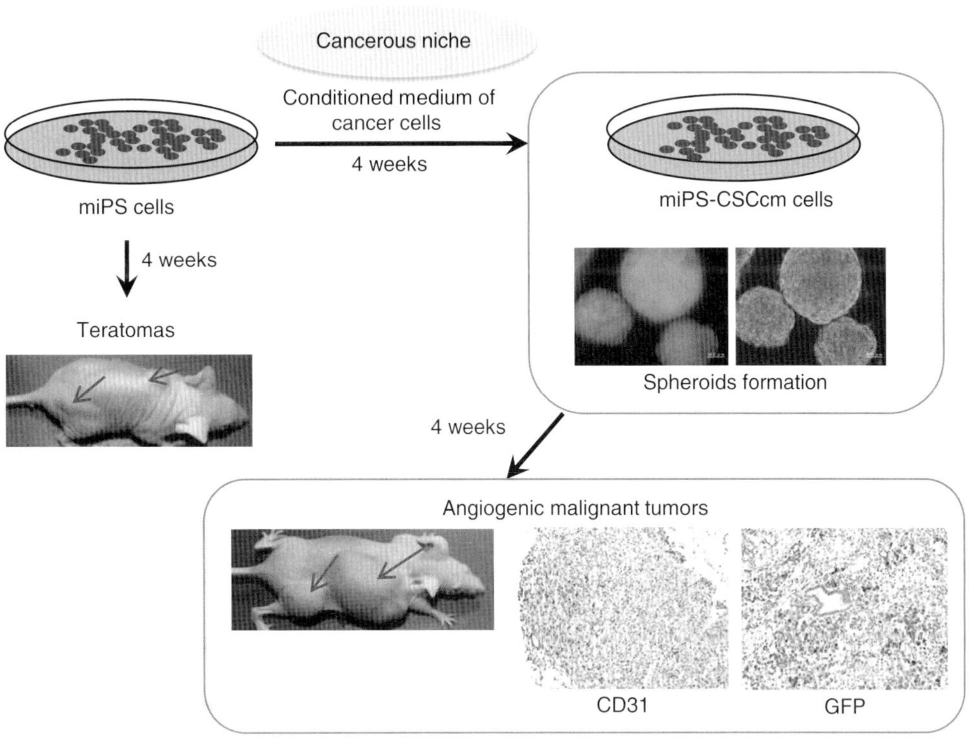

Figure 6.2. Generation of cancer stem-like cells from mouse iPS cells. miPS cells were cultured in the media containing the conditioned medium of various mouse cancer cell lines regarded as a cancerous niche. Resultant cells showed cancer stem cell properties such as sphere formation and rapid malignant tumorigenicity with extensive angiogenesis in vivo. For color detail, please see color plate section.

of Lewis lung carcinoma (LLC), typically showed highly angiogenic malignant phenotype after transplantation into nude mice. The procedure of establishment and assignment of this cell to a cancer stem cell is briefly described below.

Based on our hypothesis that the 'cancerous niche' could generate cancer stem cells by transforming or differentiating normal stem cells (Figure 6.1), we cultured miPSCs with conditioned medium of LLC cells in the absence of feeder cells and LIF. After 4 weeks of culture, survived cells were expanded in the normal medium without feeder cells and LIF. Interestingly, in this condition, approximately 30 to 50% of cells retained GFP expression, which was under the control of *Nanog* promoter (Okita et al. 2007), indicating those cells were undifferentiated. This implies that the mechanism(s) for maintenance of undifferentiated state of miPS-LLCcm cells should not depend on exogenous LIF or feeder cells. To evaluate the self-renewal capacity of the cells, we examined the growth of the cells in suspension culture. The formation of GFP-positive spheroids was observed. When the spheroids were dissociated, individual cells formed new spheroids during serial passage in the suspension culture.

miPS-LLCcm cells in either adherent or suspension culture formed adenocarcinomas in nude mice exhibiting cells with high nuclear to cytoplasmic ratio, nuclear pleomorphism, aberrantly high mitotic rates, and multiple pathological mitotic figures. Furthermore, CD31 positive staining in the tumor showed multiple vascular vessels, indicative of angiogenesis. Totally, histology of the tumor formed by miPS-LLCcm cells revealed malignant phenotype. It is noteworthy that 30 to 50% of the cells were GFP-positive in the tumors derived from miPS-LLCcm cells.

The characters of self-renewal capacity and tumorigenicity observed in miPS-LLCcm cells are consistent with the definition of cancer stem cells (Clarke et al., 2006). Thus, we are proposing miPS-LLCcm cell as a model of cancer stem cell. Furthermore, the section from miPS-LLCcm-derived tumor showed that more than half of the mesenchymal cells in the stroma of tumor turned out to be GFP negative. Also the gland like tissue was extensively stained with anti-CK antibodies. From these observations, we concluded that miPS-LLCcm cells have the potential of differentiation, showing the heterogeneous lineage in the tumor. As for the differentiation potential of miPS-LLCcm cells, we will discuss again later together with previously reported glioblastoma cancer stem cells differentiation (Ricci-Vitiani et al., 2010; Wang et al., 2010; Soda et al., 2011).

Cancer Stem-Like Cells in Other Literatures

Several cell lines that exhibit cancer stem-like features have been reported. Scaffidi and Misteli (2011) established a human cell line in vitro with cancer stem cell properties. In contrast to the niche-induced conversion demonstrated in our miPS-CSCs, they transformed human primary skin fibroblasts by retroviral introduction of the genes for telomerase, oncogenic H-RasV12, simian virus 40 large T, and small T antigens. After transfection, they isolated the stage-specific embryonic antigen (SSEA-1) positive cells as in vitro-generated cancer stem cells. These cells were assessed by the abilities of multipotency, self-renewal, and tumor initiation. SSEA-1 is known as a marker for human ESCs at the early differentiation stage, as well known as a cancer stem cell marker in brain tumors (Son et al., 2009).

To establish the safety standard in regenerative therapies, Werbowetski-Ogilvie and others (2009) have isolated and characterized variants of hESCs. During long-term cultivation, hESCs acquired genetic and/or epigenetic abnormalities, which lead the cells to cancerous transformation. The variants of hESCs showed some features of neoplastic progression such as high proliferation, growth factor independence, increased frequency of tumorigenicity, and aberrant lineage specification. However, the authors concluded that these variants were not malignant because there was no obvious metastasis in vivo. The enhanced expression of stem cell markers was observed in the variants of hESCs.

Forced expression of Yamanaka factors or microRNA to reprogram differentiated cell was applied in a series of generation of cancer-specific iPSCs, so called induced pluripotent cancer (iPC) cells (Miyoshi et al., 2010; Carette et al., 2010; Lin et al., 2008; and reviewed by Sun et al., 2011). Besides the similarity between iPC cells and iPSCs/ESCs in the points of pluripotency/multipotency and teratoma formation, iPC cells possess some oncogenic memories or surface markers as their parental cancer cells do. Although the iPC cell can form teratomas in immunodeficient mice, due to their oncogenic memories or reactivation of oncogenes, some iPC cells acquired malignancy after long-term cultivation in vitro (Nagai et al., 2010). Thus, iPC cells appear to be cancer stem-like cells, in mimicry of cancer stem cell's capacity of differentiation into cancer cells.

The cancer stem-like cells established in vitro including our miPS-CSCs were derived from stem cells or reprogrammed cells. Interestingly, these cancer stem-like cells described above were not required to be introduced with the genes for cell surface markers that are commonly used to characterize and isolate cancer stem cells. In contrast, Su and others (2011) introduced the CD44 gene into the $CD44^-$ subpopulation isolated from several $CD44^+$-colon cancer cell lines. The resultant cells showed cancer stem cell phenotype. CD44 is well known as a marker for several cancer stem cells, including breast, prostate, pancreatic, and colorectal cancers (reviewed by O'Brien et al., 2010). Although the cell surface markers are widely used for isolation of cancer stem cells, the functional role of the markers is still unclear. Especially, the relationship between their expression and transformation/maintenance of cancer stem cell is under investigation. It might be noteworthy that CD44 was translocated into nuclei functioning as a transcription factor, and led reprogrammed cancer cells into cancer stem cells in the case of colon cancer cell lines. CD44 was demonstrated to bind the promoter region of *twist1*, one of the markers for epithelial-mesenchymal transition, which is associated with the acquisition of the cancer stem cell properties (Mani et al., 2008).

Future Investigations and Applications of Cancer Stem-Like Cells

Differentiation Potential and Tumor Angiogenesis

The capacity of cancer stem cell differentiation should be considered with tumor angiogenesis. One of the definitive characters of cancer stem cell is the multipotency to create heterogeneous lineages in tumor. From 2010 to early 2011, three groups reported the origin of vessels in tumor (Ricci-Vitiani et al., 2010; Wang et al., 2010; Soda et al., 2011). These reports clarified that, in glioblastoma, a subpopulation of endothelial cells carried the same somatic mutation as found in tumor cells, indicating these endothelial cells were raised from the neoplastic origin. A series of analyses indicate that glioblastoma stem-like cells can differentiate into vascular endothelial cells in tumors. Because of obvious angiogenesis in the tumor derived from our miPS-LLCcm cells, it would be critical to investigate about tumor angiogenesis using the cancer stem cell model. Our recent analysis shows miPS-LLCcm cells could differentiate into endothelial cells forming tubular structure in vitro (Matsuda et al., 2013). The results from our study with miPS-CSCs will shed light on the molecular mechanisms of cancer stem cell differentiation and tumor angiogenesis. In addition, physiological significance of the differentiation potency and self-renewal capacity of cancer stem cells could be clarified through the analysis because endothelial cells are considered to create a stem cell niche promoting self-renewal of cancer stem cells (Krishnamurthy et al., 2010; Zhu et al., 2011).

Screening Platform for Anti-Cancer Stem Cell Agents

It is widely known that cancer stem cells show resistance toward the conventional chemo- and radiation-therapy. The resistance of cancer stem cells is considered to be one of the reasons for recurrence in patients after clinical treatment. The mechanisms involved in this resistance include expression of ABC drug pumps, expression of anti-apoptotic proteins, resistance to DNA damage, and so on (reviewed by Zhou et al., 2009). In vitro-generated cancer stem-like cells could be useful as tools to ask why and how they acquired those resistances. They also could be useful for screening new drugs that would eliminate cancer stem cells by restricting their survival and/or by promoting their differentiation to be more sensitive to traditional drugs, since this kind of screening should require a large number of cancer stem cells.

Cancer Stem Cell Vaccination Platform

Despite an attractive theory, cancer vaccination in clinical trials has not been satisfactory. A reason of failure might be a presence of cancer stem cells. Very recently, it has been shown that cancer vaccination induced the expression of *Nanog* in the tumor cells, and raised immune-resistant and stem-like cells (Noh et al., 2012). Although further investigation is required to elicit the molecular mechanisms for selection of *Nanog* expressing cells, this report implies the contribution of cancer stem cells to immuno-tolerance/-escape of tumor. In the mean time, Ning and others (2012) showed that enriched cancer stem cells were immunogenically more effective than the unselected tumor cells in inducing protective antitumor immunity. Their results proposed the novel type of cancer immunotherapy against cancer stem cells. Enough amounts of cancer stem cells will be required again to generate vaccine in good quality and quantity. Thus, cancer stem-like cells generated in vitro would have a great advantage as a good source of antigen.

Closing Remarks

Several studies have recently demonstrated that the ESC niche could have significant influence on the phenotype of aggressive cancer cells (Tzukerman et al., 2006; Postvit et al., 2008; Costa et al., 2009). These results indicate that the malignant phenotype of cancer cells could be suppressed in embryonic niche, accompanied by alternative expression of miRNAs and by change in epigenesis such as DNA methylation. The cancer microenvironment is supposed to play important roles in the initiation, progression, and metastasis of cancer (Hu and Polyak, 2008; Laconi, 2007). It has been reported that tumor cells can inhibit p53 induction, one of the most famous tumor suppressor, in adjacent fibroblasts. This suppression should depend on the factor secreted from tumor cells (Bar et al., 2009), which raises the possibility that cancer cells might convert adjacent stem cells by secreting factors.

Our miPS-CSCs were obtained from the culture in the conditioned medium of cancer-derived cells but were hardly obtained in co-culture with cancer cells (Chen et al., 2012). Collectively, there appears bidirectional communications between cancer cells and ESCs (Figure 6.3). Both communications should be regulating each other, so that the loss of either communications should make unbalanced regulations that would result in converting stem cells into cancer stem cells, and vice versa.

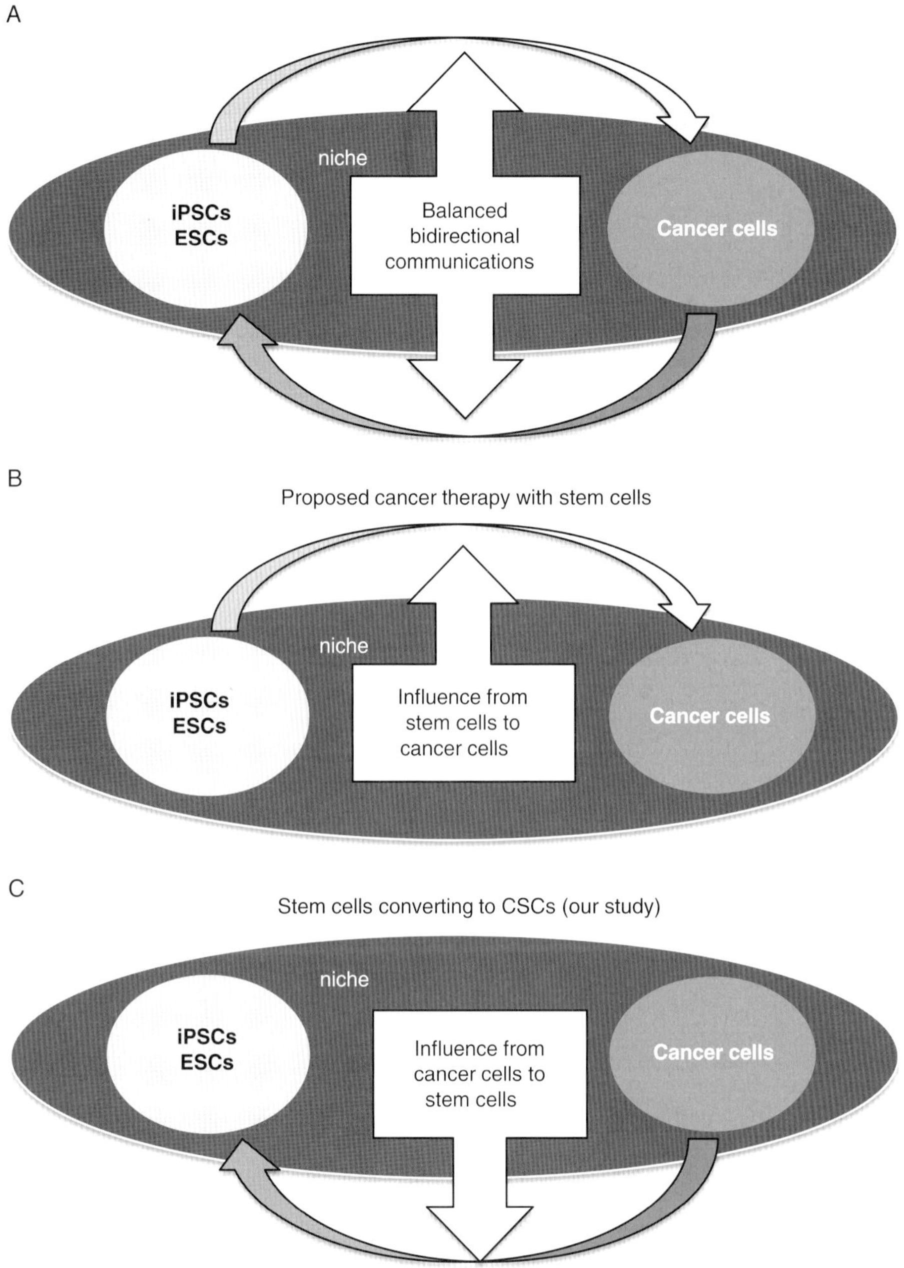

Figure 6.3. Proposed bidirectional communications between stem cells and cancer cells. **A**. Bidirectional communications between stem cells and cancer cells should be present to regulate each other in balance. **B**. Excessive unidirectional communication from stem cells to cancer cells suppresses the growth of cancer cells. **C**. Excessive unidirectional communication from cancer cells to stem cells converts the stem cells into cancer stem cells.

References

Bar, J., R. Feniger-Barish, N. Lukashchuk, H. Shaham, N. Moskovits, N. Goldfinger, D. Simansky, M. Perlman, M. Papa, A. Yosepovich, G. Rechavi, V. Rotter, and M. Oren. 2009. Cancer cells suppress p53 in adjacent fibroblasts. Oncogene. 28:933–936.

Carette, J.E., J. Pruszak, M. Varadarajan, V.A. Blomen, S. Gokhale, F.D. Camargo, M. Wernig, R. Jaenisch, and T.R. Brummelkamp. 2010. Generation of iPSCs from cultured human malignant cells. Blood. 115:4039–4042.

Chen, L., T. Kasai, Y. Li, Y. Sugii, G. Jin, M. Okada, A. Vaidyanath, A. Mizutani, A. Satoh, T. Kudoh, M.C. Hendrix, D.S. Salomon, L. Fu, and M. Seno. 2012. A model of cancer stem cells derived from mouse induced pluripotent stem cells. PLoS one. 7:e33544.

Clarke, M.F., J.E. Dick, P.B. Dirks, C.J. Eaves, C.H.M. Jamieson, D.L. Jones, J. Visvader, I.L. Weissman, and G.M. Wahl. 2006. Cancer stem cells-perspectives on current status and future directions: AACR workshop on cancer stem cells. Cancer Res. 66:9339–9344.

Costa, F.F., E.A. Seftor, J.M. Bischof, D.A. Kirschmann, L. Strizzi, K. Arndt, M.F. Bonaldo, M.B. Soares, and M.J. Hendrix. 2009. Epigenetically reprogramming metastatic tumor cells with an embryonic microenvironment. Epigenomics. 1:387–398.

Hu, M., and K. Polyak. 2008. Microenvironmental regulation of cancer development. Curr. Opin. Genet. Dev. 18:27–34.

Krishnamurthy, S., Z. Dong, D. Vodopyanov, A. Imai, J.I. Helman, M.E. Prince, M.S. Wicha, and J.E. Nor. 2010. Endothelial cell-initiated signaling promotes the survival and self-renewal of cancer stem cells. Cancer Res. 70:9969–9978.

Laconi, E. 2007. The evolving concept of tumor microenvionments. BioEssays. 29:738–744.

Lin, S.L., D.C. Chang, S. Chang-Lin, C.H. Lin, D.T. Wu, D.T. Chen, and S.Y. Ying. 2008. Mir-302 reprograms human skin cancer cells into a pluripotent ES-cell-like state. RNA. 14:2115–2124.

Mani, S., W. Guo, M.J. Liao, E.N. Eaton, A. Ayyanan, A.Y. Zhou, M. Brooks, F. Reinhard, C.C. Zhang, M. Shipitsin, L.L. Campbell, K. Polyak, C. Brisken, J. Yang, and R.A. Weinberg. 2008. The epithelial-mesenchymal transition generates cells with properties of stem cells. Cell. 133:704–715.

Matsuda, S., T. Yan, A. Mizutani, T. Sota, Y. Hiramoto, M. Prieto-Vila, L. Chen, A. Satoh, T. Kudoh, T. Kasai, H. Murakami, L. Fu, D.S. Salomon, and M. Seno. 2013. Cancer stem cells maintain a hierarchy of differentiation by creating their niche. Int. J. Cancer. doi:10.1002/ijc.28648 [Epub ahead of print].

Miyoshi, N., H. Ishii, K. Nagai, H. Hoshino, K. Mimori, F. Tanaka, H. Nagano, M. Sekimoto, Y. Doki, and M. Mori. 2010. Defined factors induce reprogramming of gastrointestinal cancer cells. Proc. Natl. Acad. Sci. USA 107:40–45.

Moore, K.A., and I.R. Lemischka, 2006. Stem cells and their niches. Science. 311:1880–1885.

Nagai, K., H. Ishii, N. Miyoshi, H. Hoshino, T. Saito, T. Sato, Y. Tomimaru, S. Kobayashi, H. Nagano, M. Sekimoto, Y. Doki, and M. Mori. 2010. Long-term culture following ES-like gene-induced reprogramming elicits an aggressive phenotype in mutated cholangiocellular carcinoma cells. Biochem. Biophys. Res. Commun. 395:258–263.

Ning, N., Q. Pan, F. Zheng, S. Teitz-Tennenbaum, M. Egenti, J. Yet, M. Li, C. Ginestier, M.S. Wicha, J.S. Moyer, M.E.P. Prince, Y. Xu, X.-L. Zhang, S. Huang, A.E. Chang, and Q. Li. 2012. Cancer stem cell vaccination confers significant antitumor immunity. Cancer Res. 72:1853–1864.

Noh, K.H., Y.-H. Lee, J.-H. Jeon, T.H. Kang, C.-P. Mao, T.-C. Wu, and T.W. Kim. 2012. Cancer vaccination derives Nanog-dependent evolution of tumor cells toward an immune-resistant and stem-like phenotype. Cancer Res. 72: 1717–1727.

O'Brien, C.A., A. Kreso, and C.H.M. Jamieson. 2010. Cancer stem cells and self-renewal. Clin. Cancer Res. 16:3113–3120.

Okita, K., T. Ichikawa, S. Yamanaka. 2007. Generation of germline-conpetent induced pluripotent stem cells. Nature. 448:313–317.

Pierce, G.B., and W.C. Speers. 1988. Tumors as caricatures of the process of tissue renewal: Prospects for therapy by directing differentiation. Cancer Res. 48:1996–2004.

Postovit, L.-M., N.V. Margaryan, E.A. Seftor, D.A. Kirschmann, A. Lipavsky, W.W. Wheaton, D.E. Abbott, R.E. Seftor, and M.J. Hendrix. 2008. Human embryonic stem cell microenvironment suppresses the tumorigenic phenotype of aggressive cancer cells. Proc. Natl. Acad. Sci. USA. 105:4329–4334.

Ricci-Vitiani, L., R. Pallini, M. Biffoni, M. Todaro, G. Invernici, T. Cenci, G. Maira, E.A. Parati, G. Stassi, L.M. Larocca, and R. De Maria. 2010. Tumour vascularization via endothelial differentiation of glioblastoma stem-like cells. Nature. 468:824–828.

Scaffidi, P., and T. Misteli. 2011. *In vivo* generation of human cells with cancer stem cell properties. Nat. Cell Biol. 13:1051–1061.

Soda, Y., T. Marumoto, D. Friedmann-Morvinski, M. Soda, F. Liu, H. Michiue, S. Pastorino, M. Yang, R.M. Hoffman, S. Kesari, and I.M. Verma. 2011. Transdifferentiation of glioblastoma cells into vascular endothelial cells. Proc. Natl. Acad. Sci. USA. 108: 4274–4280.

Son, M.J., K. Woolard, D-H. Nam, J. Lee, and H.A. Fine. 2009. SSEA-1 is an enrichment marker for tumor-initiating cells in human glioblastoma. Cell Stem Cell. 4:440–452.

Su, Y-J., H-M. Lai, Y-W. Chan, G-Y. Chen, and J-L. Lee. 2011. Direct reprogramming of stem cell properties in colon cancer cells by CD44. EMBO J. 30:3186–3199.

Sun, C., and Y.K. Liu. 2011. Induced pluripotent cancer cells: progress and application. J. Cancer Res. Clin. Oncol. 137:1–8.

Tzukerman, M., T. Rosenberg, I. Reiter, S. Ben-Eliezer, G. Denkberg, R. Coleman, Y. Reiter, and K. Skorecki. 2006. The influence of a human embryonic stem cell-derived microenvironment on targeting of human solid tumor xenografts. Cancer Res. 66:3792–3801.

Wang, R., K. Chadalavada, J. Wilshire, U. Kowalik, K.E. Hovinga, A. Geber, B. Fligelman, M. Leversha, C. Brennan, and V. Tabar. 2010. Glioblastoma stem-like cells give rise to tumour endothelium. Nature. 468:829–833.

Werbowetski-Ogilvie, T.E., M. Bossé, M. Stewart, A. Schnerch, V. Ramos-Mejia, A. Rouleau, T. Wynder, M. J. Smith, S. Dingwall, T. Carter, C. Williams, C. Harris, J. Dolling, C. Wynder, D. Boreham, M. Bhatia. 2009. Characterization of human embryonic stem cells with features of neoplastic progression. Nat Biotechnol. 27:91–97.

Zhou, B.-B.S., H. Zhang, M. Damelin, K.G. Geles, J.C. Grindley, and P.B. Dirks. 2009. Tumour-initiating cells: challenges and opportunities for anticancer drug discovery. Nat. Rev. Drug Discov. 8:806–823.

Zhu, T.S., M.A. Costello, C.E. Talsma, C.G. Flack, J.G. Crowley, L.L. Hamm, X. He, S.L. Hervey-Jumper, J.A. Heth, K.M. Muraszko, F. DiMeco, A.L. Vescovi, and X. Fan. 2011. Endothelial cells create a stem cell niche in glioblastoma by providing NOTCH ligands that nurture self-renewal of cancer stem cell-like cells. 71:6061–6072.

Chapter 7

Altruistic Stem Cells and Cancer Stem Cells

Bikul Das

Medicine/Oncology, Stanford Institute for Stem Cell Biology and Regenerative Medicine, Stanford University, Stanford, CA, USA and Forsyth Institute, Harvard School of Dental Medicine, Cambridge, MA, USA

Introduction

The initial period of stem cell biology research field evolved through our clinical experiences of managing disease processes including cancer. In the middle of the twentieth century, clinicians treating cancer had noted self-renewal and embryonal features (anaplasia) of aggressive tumors, particularly pediatric tumors including germ cell tumors, blastomas, and leukemic blast cells. The isolation of embryonic stem cells, and their in vitro propagation was achieved through clinico-pathological understanding of germ cell tumors including teratocarcinoma (Andrews, 2002). Now, in this early part of the twenty-first century, the rapidly evolving field of stem cell biology is providing new promises to understand the mechanism of cancer. Particularly, we may be able to address a critical question: Why do stem cell and cancer cell share certain phenotypic traits including self-renewal, hierarchical organization, and heterogeneity? In fact, during my early years of clinical practice in India, I was greatly disturbed by a similar question: Why should a cancer cell, toward the end of its evolution to become an aggressive, metastatic cancer cell exhibit features of embryonic stem cells? After all, embryonic stem cells create life, whereas aggressive cancer cells kill life. One of the potential answers to this intriguing question is that cancer hijacks the stem cell's survival/self-renewal mechanism to evolve into the aggressive, metastatic state. What then is the stem cell's survival/self-renewal mechanism that is relevant for cancer evolution? I suspect that stem cells evolved distinct mechanisms to maintain stemness in the microenvironment of hypoxia/oxidative stress and inflammation, which is hijacked by cancer cells to organized in a hierarchical manner having a few cancer cells, i.e. cancer stem cells maintaining the ability to self-renew. To investigate this possibility, I have utilized the hESC self-renewal during hypoxia-oxidative stress, with a particular focus on the p53 and its role in hESC self-renewal during stress. Colleagues and I recently reported that p53 is suppressed in a small population of heterogeneous hESC during hypoxia/oxidative stress (Das et al., 2012). This low p53 hESC population is transient, acquires certain cancer stem cell properties including self-sufficiency, and secrete growth factors/antioxidants that

Cancer Stem Cells, First Edition. Edited by Vinagolu K. Rajasekhar.

protect the rest of the hESCs from differentiation/apoptosis. I suspect this cytoprotection to be an altruistic mechanism, reminiscent of a bacterial altruistic defense mechanism. Although hESCs do not represent adult stem cells, the p53 system being highly active in stem cells, it could be presumed that a similar p53-related altruistic mechanism might also be involved in the self-renewal of adult stem cells during hypoxia/oxidative stress. The p53 protein is a common target of mutation/inactivation in almost all cancer. Hence, one of the potential answers to the question of why stem cells and cancer cells share similar phenotypic characters is that cancer evolution may involve the p53-related altruistic mechanism.

Hence, in this review, I will discuss a hypothesis that the molecular mechanisms might involve a unique reprogramming mechanism that switch the stem cell (stemness) to a low p53 state, the "enhanced stemness" reprogramming (Figure 7.1) (Das et al., 2009; Das et al., 2012). Note that stemness is defined here as a genetic network architecture (Furusawa et al., 2012) that corresponds to stem cell phenotype (the undifferentiated phenotypic state of stem cell). I will then discuss that the enhanced stemness reprogramming mechanism may correspond to the putative altruistic defense mechanism similar to microbial altruistic defense.

The p53 Activity During Stem Cell Self-Renewal in Hypoxia/Oxidative Stress Microenvironment

One of the ways to mimic the in vivo microenvironmental stress in the inflammatory tissues is to employ an in vitro assay of hypoxia (<0.1% O2) followed by reoxygenation. Hypoxia prevalent in the area of injury and inflammation is in dynamic states of flux with both chronic and intermittent components (Durand, 2001). The latter state being an in vivo process of hypoxia-reoxygenation may generate reactive oxygen species (ROS). It is possible that this microenvironment of intermittent hypoxia could also exist in the adult stem cell niche during inflammatory process, and might play an important role in the expansion of adult stem cells in their niche, as a response to inflammation. Although direct evidence of this speculation is yet to be reported, Suda and others (2005) showed that hematopoietic stem cells (HSC) in the hypoxic niche migrate to the angiogenic niche during stress-injured mobilization.

The hypoxia/re-oxygenation microenvironment could generate ROS in the stem cell population. The ROS thus generated could activate p53 system leading to apoptosis/differentiation of the stem cells. Hence, it

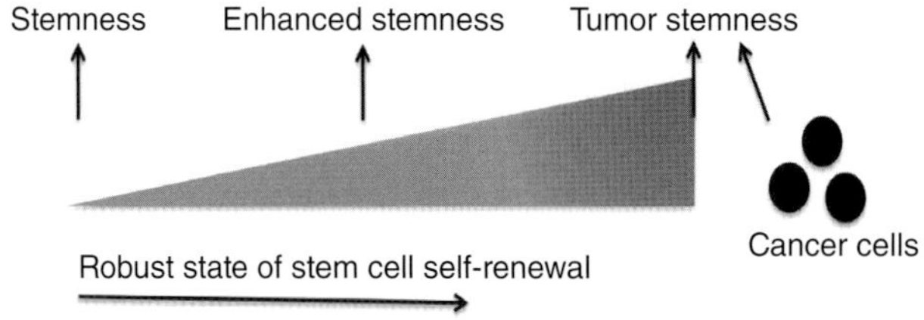

Figure 7.1. A schematic diagram of stem cell reprogramming to robust state of stemness under microenvironmental stress. The three states of stemness represent three robust states in the context of nongenetic, microenvironmental pressure. The stemness state is the resting, homeostatic state of quiescent stem cells in their niche. The enhanced stemness state is the robust state of stemness in the microenvironment of extreme oxidative stress described in Figure 7.2. The tumor stemness state is the malignant state, a robust state that tolerates genomic instability. The robustness is defined as the state of a biological system that continues to function despite genetic or nongenetic perturbations (Andreas, 2007, page 1). For color detail, please see color plate section.

could be speculated that stem cells evolved some kind of mechanism to protect itself from ROS-induced p53 activation. Is it possible that a few stem cells among the heterogeneous population of stem cells would suppress p53 activity, and then secrete growth factors/antioxidants to reduce ROS-induced stress in the rest of the stem cell population?

This possibility was investigated by using an in vitro acute hypoxia-reoxygenation model that I characterized during my PhD research (Das, 2007; Das et al., 2005a). In this model, hESCs were exposed to intense hypoxia (0.1% oxygen) for 24 hours and then re-exposed to normoxia for the next 24 hours (Figure 7.2). Using a human embryonic stem cell (hESC) model system, colleagues and I recently reported the reprogramming of a few hESCs (the SSEA3+/ABCG2+ cell fraction) to a low p53 state (Das et al., 2012). The SSEA3+/ABCG2+ cells acquired very high level of HIF-2alpha and low level of p53, as well as low ROS level. The p53/MDM2 system is a genetic circuit where p53 activates MDM2, which in turn degrades p53; the activity of this circuit is characterized by oscillation of p53 and MDM2 protein levels (Figure 7.3) (Geva-Zatorsky et al., 2006). A close observation of the SSEA3+/ABCG2+ cell p53/MDM2 oscillation pattern (Figure 7.3) indicated that the oscillation was arrested in a **reversible,** low p53 state for a 2-week period. Taken together, when exposed to the microenvironment of extreme hypoxia/oxidative stress, a rare fraction of hESCs, the SSEA3+/ABCG2+ cells exhibited an altered state of p53/MDM2 oscillation, and thus resisted the ROS-induced apoptosis/differentiation (Das et al., 2012).

Considering that the SSEA3+/ABCG2+ cells exhibited high HIF-2alpha and low p53 activity, the question arises whether they underwent phenotypic transformation. This question was investigated by performing several assays. Results described in Figure 7.4 showed that the post-hypoxia SSEA3+/ABCG2+ hESCs fraction exhibited more than fourfold increase of Nanog activity that was sustained for a 2-week period without any growth factor supplementation, thus indicating self-sufficiency (Das et al., 2012). When the SSEA3+/ABCG2+ cells were injected into immunocompromised mice to perform the serial transplantation assay, this hESC cell fraction exhibited more than a fifteen-fold increase in tumorigenicity compared to ES cells under normoxia. These results indicated that the post-hypoxia SSEA3+/ABCG2 hESCs exhibited phenotypic reprogramming to a highly self-sufficient state. This state, the "enhanced stemness state" (Figure 7.1), thus supports the hypothesis that the hESC's molecular mechanisms of survival during hypoxia/oxidative stress might involve a unique reprogramming mechanism that switch the stem cell (stemness) to a low p53 state, the "enhanced stemness" state.

Stem Cell Altruism: The Defense of Clonal Heterogeneity During Hypoxia/Oxidative Stress

Stem cells are a heterogeneous population. For example, the Sca-1 population in mouse bone marrow (BM) contain both Sca-1 + high and Sca-1 + low population. The flow cytometry sorted Sca-1 + high population, when allowed to expand, the heterogeneity of Sca-1 expression return to the parental state. Similar when the high Nanog fraction of hESCs is allowed to expand, the distribution of Nanog-expressing cells return to the parental state of Nanog expression (Furusawa et al., 2012). This clonal heterogeneity could be considered as the clonal identity or part of stemness of a particular stem cell type. In the experiment described in Figure 7.2, the SSEA3+/ABCG2− cells, when grown in the hESC media returned to the parental distribution of Nanog (Das et al., 2012). In contrast, the SSEA3+/ABCG2+ cells underwent apoptosis/differentiation. Interestingly, the SSEA3+/ABCG2+ cells exhibited altruistic behavior that contributed to the SSEA3+/ABCG2− self-renewal and maintenance of clonal heterogeneity. This aspect of the experiment in Figure 7.2 is described below, and the idea of stem cell altruism is introduced.

The SSEA3+/ABCG2+ cells acquired very high levels of HIF-2alpha and low levels of p53, as well as low ROS levels. These cells also secreted very high levels of glutathione antioxidant. In contrast, the SSEA3+/ABCG2− cells exhibited high p53 activity but low ROS, and a very high level of GSH. This active p53 state and a low ROS state in the SSEA3+/ABCG2− cells indicated that this cell type was probably protected by the high level of GSH secreted by the SSEA3+/ABCG2+ cells. This interpretation is further supported by the findings that when the SSEA3+/ABCG2− cells were flow sorted and then cultured in fresh media, the cells exhibited high ROS and underwent apoptosis. The result indicated that SSEA3+/ABCG2− cells required the cytoprotective activity of the SSEA3+/ABCG2+ cells to survive and maintain stemness (self-renewal and undifferentiated state). Indeed, the addition of conditioned media obtained from the SSEA3+/ABCG2+ cells helped to restore the survival and stemness of the flow cytometry sorted SSEA3+/ABCG2− cells. Thus, the SSEA3+/ABCG2+ cells provided cytoprotective mechanism to protect the SSEA3+/ABCG2− cells during oxidative stress.

When both the SSEA3+/ABCG2+ and the SSEA3+/ABCG2− cells were allowed to expand, most of the SSEA3+/ABCG22+ cells underwent apoptosis and/or differentiation after a 2-week period. Only a few cells survived and remained ES cells at the end of 2 weeks of culture. **These results support the notion of an altruistic mechanism, where the SSEA3+/ABCG2+ cells mostly underwent apoptosis and/or differentiation (loss of fitness) in order to protect the SSEA3+/ABCG2− cells that survived (gain of fitness).** This SSEA3+/ABCG2− cell fraction, when expanded, exhibited the parental heterogeneity of the hESCs, thus gain of fitness. Taken together, in an in vitro assay described in Figure 7.2, a preliminary finding on potential existence of altruistic stem cell phenotype has been observed.

I would like to propose that the altruistic mechanism exhibited by the SSEA3+/ABCG2+ cells represent a defense mechanism of clonal heterogeneity, because the mechanism contributed to maintaining the clonal heterogeneity of the hESCs in the microenvironment of hypoxia/oxidative stress. This putative altruistic defense mechanism in hESCs would be henceforth known as the altruistic stem cell defense (ASD) system.

What could be the underlying molecular mechanism that permits the SSEA3+/ABCG2+ cells to lose fitness, or in other words, to exhibit altruistic behavior? The underlying mechanism might be the reversibility of the p53/MDM2 oscillation system. The experimental analysis of the SSEA3+/ABCG2+ cell p53/MDM2 oscillation pattern (Figure 7.3) indicated that the oscillation was arrested in a **reversible,** low p53 state for a 2-week period. The differentiation and apoptosis of the SSEA3+/ABCG2+ fraction was preceded by the return of the p53/MDM2 oscillation to its basal state (Das et al., 2012). Thus, it appears that the fitness of the SSEA3+/ABCG2+ cells was reduced by the return of the p53/MDM2 oscillation to the basal state. Does the reversible arrest of the p53/MDM2 oscillation indicate a molecular mechanism of altruism in stem cells? Indeed, there are a few aspects of this oscillation mechanism that could make it an ideal system to exhibit altruism. First, the oscillation would allow temporary, that is, reversible suppression of p53 for a prolonged period of time. Second, the temporary suppression of p53 would allow the ES cells to reprogram to a higher state of stemness, including high secretion of antioxidant glutathione.

However, the problem arises of how to explain the mechanism of a reversible alteration of the p53/MDM2 oscillation system. It could be explained as the stochastic process of a biological feed back loop, known as the Hopf-bifurcation (Turcotte et al., 2008). Simply put, a genetic-circuit characterized by variable high amplitude occurring at a fixed interval is called Hopf-bifurcation (Rinzel et al., 1998). The p53/MDM2 system is a genetic circuit where p53 activates MDM2, which in turn degrades p53; the activity of this circuit is characterized by oscillation of p53 and MDM2 protein levels

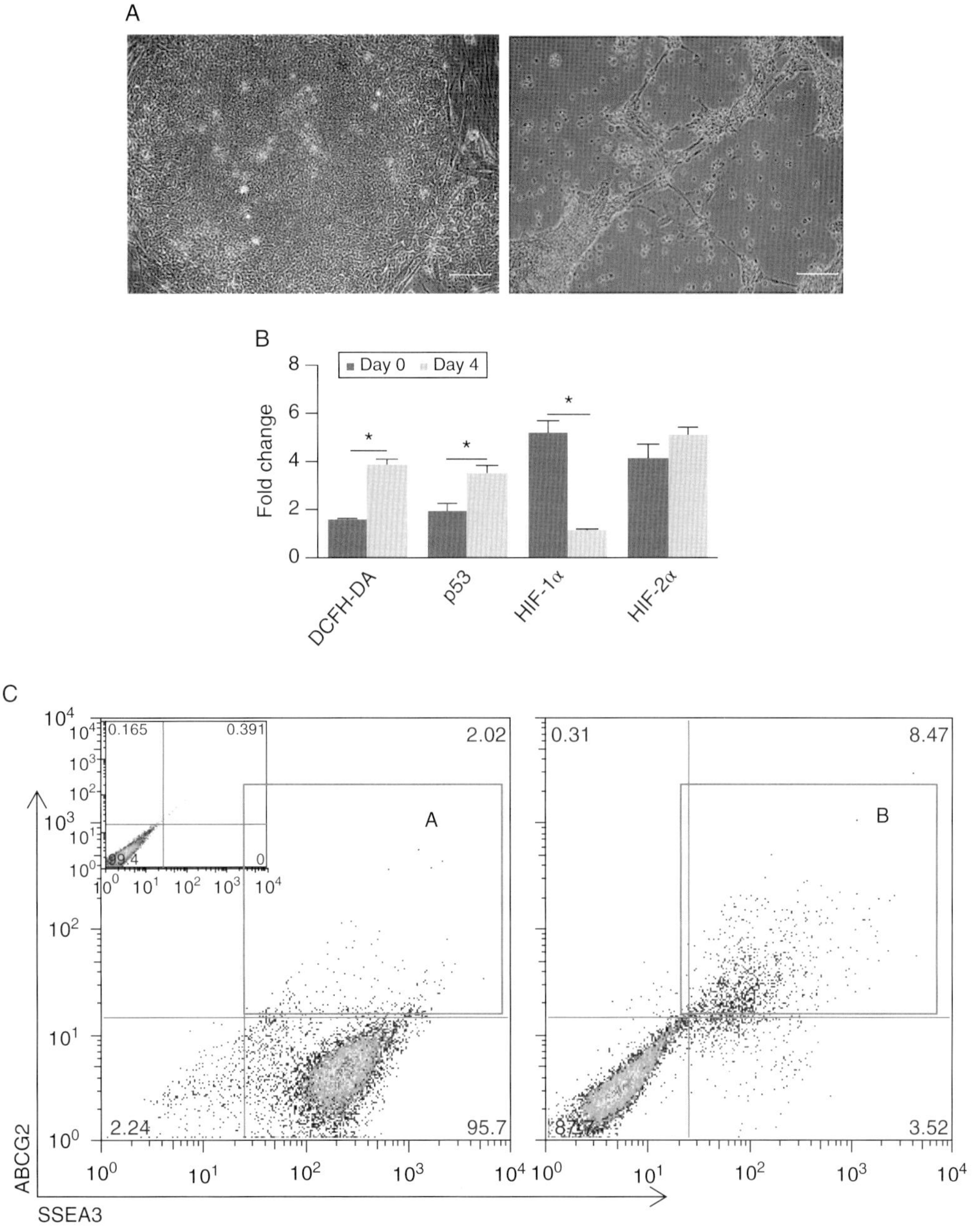

Figure 7.2. In vitro assay to study human ES cell altruism. (From Das B et al. Stem Cells, 2012). **A**. Morphological evidence of BGO1 hESC differentiation and death (image to the right) following exposure to extreme hypoxia followed by 4 days of reoxygenation. The hypoxia and re-oxygenation treated cells showed elongated neurite processes including fibroblastic-like features, as well as evidence of cell fragmentation. **B**. The fold change in ROS (DCFH-DA), p53, and HIF-2α in the BGO1 cells shown in **A**, following day 4 post-hypoxia/reoxygenation. **C**. Flow cytometry showing expansion (A versus B compartments) of the SSEA3+/ABCG2+ fraction on the day 4 post-hypoxia/reoxygenation. Insert shows the profile of isotype control. For color detail, please see color plate section.

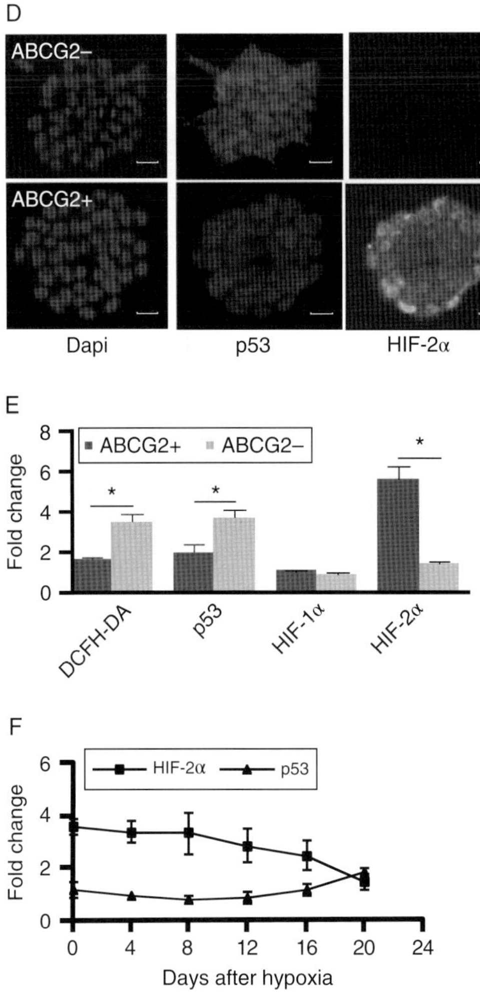

Figure 7.2. **D**. Immunofluorescence labeling of SSEA3+/ABCG2- and SSEA3+/ABCG2+ fractions on day 4 post-hypoxia/reoxygenation showing a high HIF-2α and low p53 signal in the SSEA3+/ABCG2+ fraction (Dapi, nuclear stain). **E**. The corresponding fold-change in DCFH-DA, p53, and HIFs in the SSEA3+/ABCG2+ versus SSEA3+/ABCG2− fractions on the day 4 post-hypoxia/reoxygenation. **F**. Protein levels of HIF-2α and p53 in the SSEA3+/ABCG2+ fraction were measured by ELISA. Note high HIF-2α versus low p53 until day 20. $*P<0.05$; $n = 4$. Scale bar (a) 40 μm and (d) 10 μm. For color detail, please see color plate section.

(Figure 7.3) (Geva-Zatorsky et al., 2006). Hence, it is possible that the p53/MDM2 genetic circuit could exhibit Hopf-bifurcation leading to the enhanced stemness reprogramming and so the appearance of the altruistic stem cell phenotype.

The Hopf-bifurcation would also allow the return of the p53/MDM2 oscillation to its basal state of active p53. The active p53 state could then eliminate the enhanced stemness reprogrammed cells exhibiting genomic damage. In this manner, the altruistic stem cells could be eliminated after a certain period of their cytoprotective function. Thus, the p53/MDM2 oscillation could provide a molecular mechanism for stem cells to exhibit altruistic phenotype to protect the stemness and heterogeneity of the ES cell population,

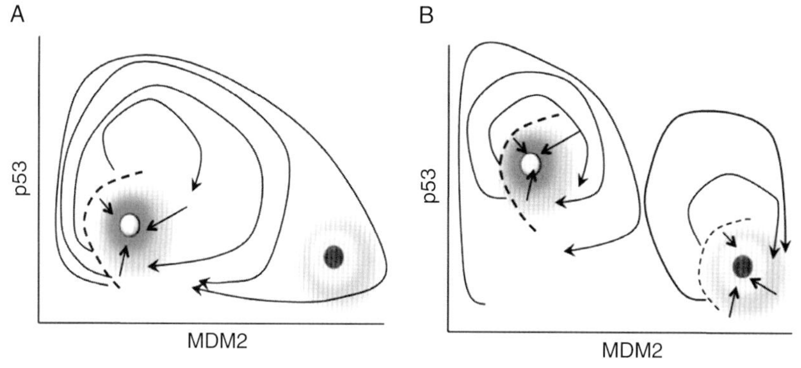

Figure 7.3. **A.** Model of p53/MDM2 oscillation during basal state. Minor fluctuation (*red zone*) away from the stable state (*yellow dot*) is immediately damped out (*short arrows*). Larger fluctuation beyond the threshold (*dashed line*) may generate unstable state of high MDM2 and low p53 (*blue dot*), which eventually bounces back to the stable state. **B.** Stabilization (Turcotte et al., 2008). Increase noise in the circuit might stochastically stabilize the unstable state. Figure **A** top panel was partially adopted from Tyson (1998). For color detail, please see color plate section.

without compromising the genetic integrity of its population.

The larger question is whether a similar mechanism may also exist in adult stem cells. This question is now under intense investigation. What is known so far is that the hypoxia/oxidative stress microenvironment that was used to perform the experiment described in Figure 7.2 could be highly relevant for adult stem cell population. Adult stem cells are heterogeneous and reside in their niches in a quiescent state (Cottler-Fox et al., 2003; Fuchs et al., 2004; Li and Xie, 2005). During tissue injury/inflammation, stem cell population undergoes symmetrical and asymmetrical self-renewal, a process commonly known as mobilization. The mobilization allows stem cells to rapidly increase their population, and participate in the repair/regeneration process. However, it is not yet clear about how the stem cell population maintains their heterogeneity during inflammation and oxidative stress, when ROS (reactive oxygen species) generated in such a microenvironment could activate the p53 system, leading to differentiation/death of stem cells (Prockop, 2009; Ito et al., 2004). Thus, it is possible that adult stem cells evolved sophisticated mechanisms to avoid p53-induced death/differentiation to serve their regenerative task as well as maintain the population heterogeneity in the oxidative stress microenvironment. The altruistic defense system of clonal heterogeneity, including the p53/MDM2 oscillation, could be such as a potential sophisticated defense mechanism that evolved in adult stem cells to maintain their population pool in the niches. In summary, although the data so far are obtained from ES cells, which does not represent adult stem cell population, it could be speculated that if adult stem cells employ the "enhanced stemness" reprogramming, then this altruism associated with the reprogramming mechanism could be considered to be a defense strategy common to stem cells cytoprotection at the site of injury.

Therefore, I suspect that the altruistic stem cells might play an important physiological role to protect adult stem cells in their niche, when exposed to hostile microenvironments. If so, it is expected that the altruistic HSCs will play an important cytoprotective role during the chemotherapy-induced mobilization. This possibility could be studied using various available animal models of HSC mobilization following chemotherapy, and therefore, are now under active investigation. Along a complementary line, the wound-healing process might also exhibit the ASD system. These possibilities are now being tested using both the in vitro and in vivo experiments.

From the Idea of Stemness to Altruistic Stem Cell Defense (ASD) System

I deem it important to briefly review and define the idea of the degree of stemness state, before continuing to discuss the enhanced stemness (Figure 7.1) and ASD system. The idea of stemness initially evolved as shared genetic, epigenetic, and molecular mechanisms among various types of stem cells to maintain self-renewal and undifferentiated state (Ramalho-Santos et al., 2002). However, such a shared genetic/epigenetic program had not yet been described (Cai et al., 2004; Mikkers and Frisen, 2005). Taking a different perspective on stemness, during my PhD work, I proposed stemness as a spectrum (Figure 7.1) of phenotypic states of stem cell's autocrine regulation of self-renewal and undifferentiated state (Das, 2007; Das et al., 2009). In this view, stem cells having a higher degree of autocrine regulation of self-renewal and undifferentiated state would be self-sufficient, and therefore, they would be considered to have a high degree of stemness state. Thus, highly self-sufficient stem cells would be able to maintain their de-differentiated state without help from paracrine signaling. On the opposite end of the spectrum is the lower degree of stemness state. Stem cells having a lower degree of stemness state would thus require the paracrine growth signaling of niche cells or paracrine growth factors to maintain their self-renewal and undifferentiated state. This perspective on stemness incorporates the variation in stemness, and so, allows us to consider stemness as a spectrum of stem cell states in a Waddington's epigenetic landscape metaphor (Das et al., 2009). In this spectrum, the stemness state of quiescent stem cells in their niche is considered to have a low degree of stemness (small hill in the epigenetic landscape), whereas the cancer stem cells having the ability to maintain a stemness state, without support of niche cells or paracrine growth factors, is considered to have the highest degree of stemness, the tumor stemness state (steep hill in the epigenetic landscape). Therefore, the tumor stemness state could also be considered as a state of **extreme self-sufficiency**. This idea is supported by findings that highly teratocarcinomagenic hESCs exhibit self-sufficiency (i.e., these hESCs could be cultured in vitro similar to embryonal carcinoma [EC] cells without the addition of growth factors and or feeder cell layer) (Andrews, 2002). Thus, highly self-sufficient hESCs could maintain an undifferentiated state even without addition of FGF-2 or ROCK-inhibitor, the latter being found to increase the survival of hESCs in the in vitro culture. Two intracellular proteins involved in resisting apoptosis, Bcl-2 and FGF-2, are associated with hESC self-sufficiency. Additionally, the core-transcription factors Oct-4 and Nanog are highly expressed in the self-sufficient hESCs. Thus, the EC like hESCs represents the tumor stemness state in the spectrum of stemness.

In this idea of stemness spectrum (Figure 7.1), the reprogramming phenomenon in ES cell self-renewal, where the HIF-2α transcription factor transiently reprograms ES cells to a robust state of stemness (the enhanced stemness state) (Figure 7.1) having low p53 activity could be the mechanism of how the stemness state switches from enhanced stemness to tumor stemness state.

I have already discussed that the enhanced stemness reprogramming mechanism describes the altruistic stem cell behavior. Hence, we could consider that the enhanced stemness reprogramming as a component of the stem cell altruistic defense system. In this defense mechanism, a few stem cells reprogram to an "enhanced stemness" phenotype, and exhibit an altruistic behavior to protect the rest of the stem cell population from stress-induced apoptosis/differentiation by secreting antioxidants and growth factors.

One of the characteristics of heterogeneity of stem cell population is the simultaneous existence of both quiescent and self-renewing population. The mechanism of how stem cells maintain such quiescent and self-renewing phenotype within the same population is not yet clear. It is considered that the heterogeneous state is dynamic and stochastic (i.e., at a given time, some cells will be at quiescent state, and then, the same cell population will exhibit self-renewal at a later time). It is tempting that such a heterogeneous strategy of population distribution would favor stem cells to exhibit ASD mechanism when exposed to microenvironmental stress. In Figure 7.2, the SSEA3+/ABCG2+ cells, which were rare and therefore probably quiescent, underwent enhanced stemness reprogramming to protect the self-renewing SSEA3+/ABCG2- cells. Based on this very preliminary data, it could be suggested that the quiescent cell fraction might be the candidate cells that undergo transient reprogramming to protect the rest of the population. Thus, maintaining a quiescent population could be part of the ASD mechanism. If so, we do have an explanation on the heterogeneity of stem cells. This heterogeneity might have evolved as a part of the evolution of the ASD mechanism.

Altruism is considered as an important evolutionary defense mechanism of microbial existence in hostile microenvironments (Youk and Qudenaarden, 2010). Therefore, I suspect that the ASD is probably a part of a primitive defense system that evolved in early unicellular Eukaroytes to protect their own heterogeneous population during microenvironmental stress. While writing a book on the cellular anti-oxidant defense system (Das, 2000), I proposed the potential existence of a primitive ancestral cytoprotective cell defense system. I suggested that a primitive unicellular system that potentially included the ancestral stem cell population during the evolution of unicellular to multicellular organism might have developed robust cytoprotective mechanism to protect their population heterogeneity from the prevalent microenvironment of low oxygen and high radiation of the Precambrian era. My observation that enhanced stemness reprogramming involves altruistic ES cell phenotypes provides preliminary support to my earlier speculation of the existence of a primitive cytoprotective defense system. (Das, 2000).

ASD System and Tumor Stemness Reprogramming

The ASD system described above could be used as a novel framework to address the intriguing question that I had decades ago: Why should a cancer cell, toward the end of its evolution to become an aggressive, metastatic cancer cell, exhibit features of stem cells? After all, stem cells support life, whereas aggressive cancer cells kill life. One of the potential answers to this intriguing question is that cancer hijacks stem cell's ASD mechanism to defend the clonal heterogeneity of a given cancer cell population. Let us first consider if the ASD system could itself be the source of the malignant transformation of normal stem cell to cancer stem cell. This consideration is schematically depicted in Figure 7.1, where the enhanced stemness state could further reprogram into a higher degree of stemness state, the tumor stemness state. The potential mechanism of tumor stemness reprogramming could be described in ES cells as follows.

In the pluripotency state, Nanog is maintained within a threshold level (Pan and Thomson, 2007) probably, because the high Nanog expression state might lead to reprogramming of the ES cell state to an EC cell-like state (Draper et al., 2004; Silva and Smith, 2008; Singh et al., 2007). This potential reprogramming of ES to an EC-like state could be experimented in the following ways: (1) by examining the Nanog expression in the SSEA3+/ABCG2+ cells, (2) by performing an in vitro self-sufficiency assay (e.g., testing the ability of ES cells to maintain stemness without the addition of growth factors in the media) (Silva and Smith, 2008), and (3) by performing in vivo serial transplantation assay of teratoma generation. In the study described in

Figure 7.2 and Figure 7.4, the above-mentioned experiments were attempted. Results showed that the post-hypoxia SSEA3+/ABCG2+ hESCs fraction exhibited more than a fourfold increase of Nanog activity and the ability to sustain stemness in media without any growth factor supplementation, thus indicating self-sufficiency (Das et al. 2012). When the SSEA3+/ABCG2+ cells were injected into immunocompromised mice to perform the serial transplantation assay, this hESC cell fraction exhibited more than a fifteen-fold increase in tumorigenicity compared to ES cells under normoxia. These results indicated that the post-hypoxia SSEA3+/ABCG2 hESCs exhibited self-sufficiency, one of the features of highly anaplastic cancer cells. The self-sufficiency could be considered as the phenotypic feature of tumor stemness reprogramming.

Considering that suppressing of p53 is one of the hallmarks of malignant cells, the potential trade-off of p53 suppression in the enhanced stemness and tumor stemness phenotype could potentially facilitate malignant transformation. Interestingly, the tumor stemness state of SSEA3+/ABCG2+ cells was transient (Figure 7.4), because the post-hypoxia SSEA3+/ABCG2+ xenograft-derived cells did not form secondary teratomas, and therefore, did not exhibit the full-blown EC-like malignant phenotype of teratocarcinoma. Thus, in Figure 7.1, although the enhanced stemness state could reprogram into the tumor stemness state as a part of the ASD system, this itself may not be sufficient for irreversible malignant transformation of stem cells to cancer cell state. The transient nature of the tumor stemness state probably functions as a possible safety mechanism that is built into ES cells to mitigate against malignant transformation. How is this safety mechanism then overridden leading to malignant transformation? What could be the potential mechanism of the transformation of tumor stemness from a transient to an irreversible permanent state?

I speculate that the tumor stemness state could achieve an irreversible state due to the stochastic stabilization of p53/MDM2 oscillation in a state of low p53 and high MDM2 state. One of the potential mechanisms of the p53/MDM2 genetic-circuit stabilization with an abnormal state of low p53 is a stochastic stabilization process of the Hopf-bifurcation (Figure 7.3). This possibility could be tested by repeating the experiment described in the Figure 7.2, and using a large number of ES cells (>50 million cells) and exposing them to serial hypoxia/oxidative stress. The reprogrammed cells could then be studied for the evidence of the stabilization of the p53/MDM2 genetic-circuit. I am currently working on this project, and hopefully will be able to find evidence for such a mechanism of abnormal stabilization of the p53/MDM2 oscillation that could underlie the initiating event of malignant transformation. The experimental results will indicate if the tumor stemness reprogramming mechanism, a part of the ASD system, could be the initial event of malignant transformation.

How do we incorporate the oncogenes in this process of tumor stemness reprogramming mediated malignant transformation? Two most commonly studied oncogenes, MYC and Ras, are also known to be involved in the wound-healing process. For example, recent reports suggest the role of the Ras oncogene in the repair process of wound re-epithelization (Tcharntke et al., 2005) and gastric ulcer (Pai et al., 1999). It is likely that during the wound-healing process, oncogenes are involved in the self-renewal of stem cells and therefore, could be involved in the tumor stemness reprogramming.

We do not have yet any direct evidence that chronic inflammation including hypoxia/oxidative stress microenvironment could transform stem cells to cancer stem cells. However, there are a few hints in the literature that provide support. For example, to explain the malignant transformation of tissue-specific stem cells (TSSC), Bissel and Radisky (2001) proposed a tissue injury model, where TSSC, after being damaged either by mutation or another mechanism, modify the microenvironment, which in turn further damages the transformed cells, leading to an escalation in

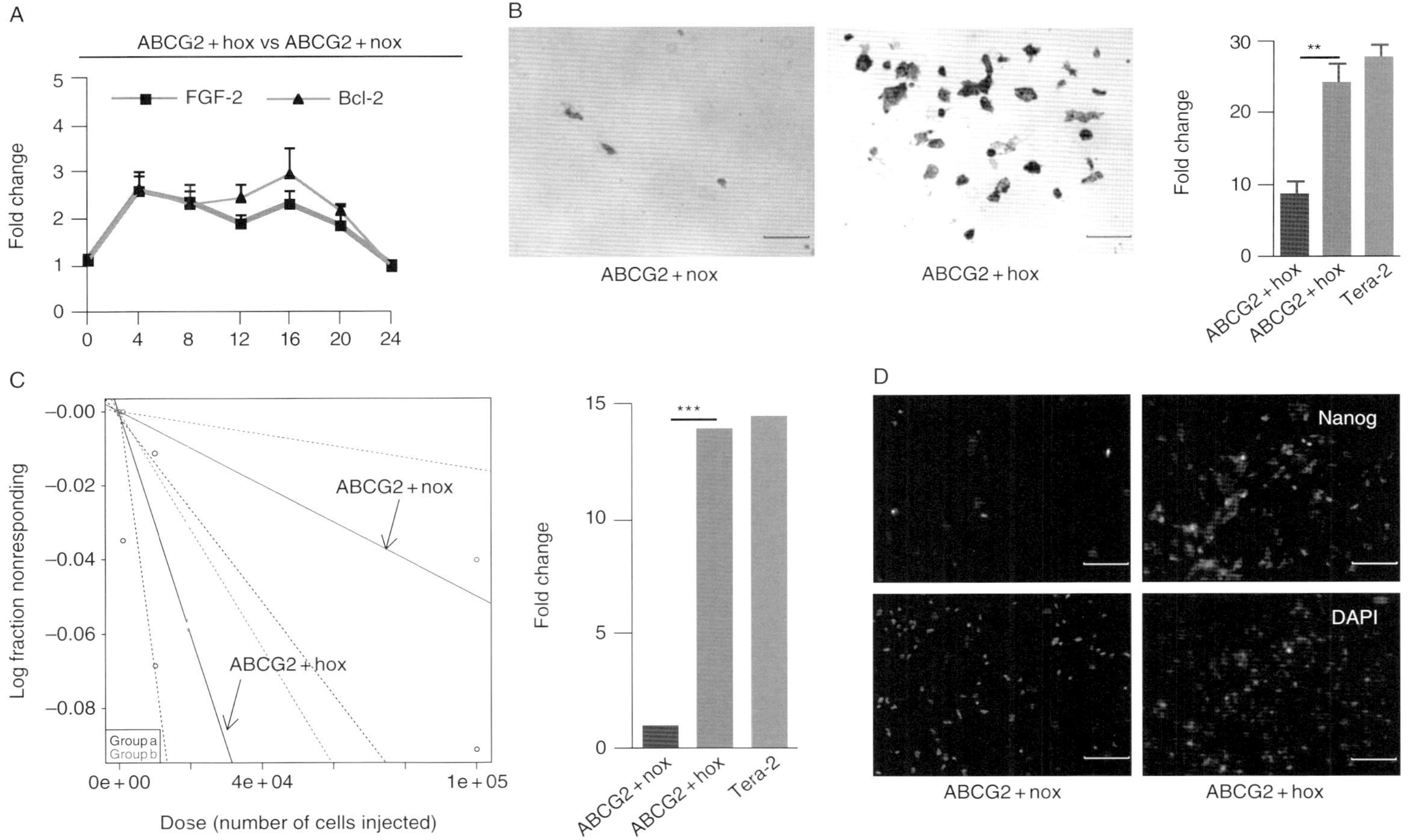

Figure 7.4. ABCG2 + hox fraction exhibits a phenotype of self-sufficiency (From Das et al., 2012). **A**. The fold change in FGF-2 and Bcl-2 in ABCG2 + hox cells relative to ABCG2 + nox cells. **B**. The single cell clonogenic assay showing the alkaline phosphatase staining of abundant colony formation in the ABCG2 + hox cell fraction. The histogram shows the significant fold increase in colonies derived from the ABCG2 + hox fraction. Tera-2 derived colonies were used as a control. **C**. The in vivo limiting dilution assay showing higher teratomagenic formation efficiency of ABCG2 + hox versus ABCG2 + nox. Tera-2 teratoma formation was used as a control. **D**. Immunofluorescence staining on a frozen section of ABCG2+ cells containing matrigel plug showing expression of Nanog positive cells. The quantification of the Nanog expressing cells is given in the Supplementary Figure 6 (Das et al., 2012). $*p<0.05$; $**p<0.001$; $***p<0.00001$, $n=4$. Scale bar **B** and **D** 40 μm. For color detail, please see color plate section.

pathological behavior and eventual tumor formation. This pathological behavior is likely to be the "repair/regeneration" response of the damaged TSSC to its surrounding "injured" environment. This repair/regeneration response could involve the ASD system, which then could provide a rare chance of malignant transformation of a few stem cells through the stochastic stabilization of the p53/MDM2 oscillation system.

Revisiting the Cancer Stem Cell Hypothesis

The cancer cells, whether it is the result of malignant transformation of stem cells or progenitor cells, exhibit mainly two phases of growth. The first stage is the dormant state, and the second state is the growth/metastasis (Nowell, 1986). For clinical importance, a third stage, the stage of relapse after cytotoxic therapy, could also be considered (Kim and Tannock, 2005).

For approximately the last about 3 decades, a serious attempt has been made to understand tumor growth and relapse as a function of cancer stem cell (CSC) quiescence and self-renewal (Trott, 1994; Mackillop et al., 1983; Reya et al., 2001; Pardal et al., 2003; Das et al., 2005b). The CSC hypothesis indicates the presence of a quiescence population of tumor-initiating cells in the heterogeneous population of tumor cells (Bonnet and Dick, 1997; Reya et al., 2001; Wicha et al., 2006; Dick, 2003; Al-Hajj et al., 2003; Kim et al., 2005). The quiescent state of CSC could be considered as the hallmark of CSC hypothesis, because it is the quiescent state that allows these cells to escape from the cytotoxic therapy-mediated elimination (Rich, 2007; Behbod and Rosen, 2005). It is thought that the quiescent CSC re-establishes the heterogeneous tumor following cytotoxic therapy (Pardal et al., 2003).

It is to be noted here that various studies on CSCs questioned the quiescence state of CSCs (Hill, 2006). For example, Singh and others (2004) found approximately 20% CD133+ CSCs in primary brain tumors. The large number of CSCs could reflect the heterogeneity of CSC population, where a small fraction constitutes quiescent, whereas the rest of the population are self-renewing CSCs. Our studies in a neuroblastoma SK-N-BE (2) SP cells indicated that SP cells expressing Oct-4 SPm (hox) cells (Das et al., 2008), when expanded, returned to the parental distribution of Oct-4, where only a small fraction of SP cells (0.1% of the SP cell population) maintained Oct-4 expression. Interestingly, the Oct-4 expressing SP cells were quiescent. Thus, CSCs could maintain both quiescent and self-renewing phenotypes. It is not yet clear how heterogeneous CSCs could maintain both quiescent and self-renewing CSCs.

I would like to suggest that CSCs incorporate an ASD-like defense system to maintain both quiescent and self-renewal cells. Considering that tumor microenvironment is often hypoxic/oxidative stress, it is indeed possible that CSCs hijack the ASD system of stem cells to defend clonal heterogeneity (i.e., maintaining both quiescent and self-renewing phenotypes).

Are there any experimental evidences that cancer cells could exhibit ASD-like features? My previous work on tumor side-population cells suggests that the answer is yes. Earlier, a colleague and I reported that when exposed to a hypoxia/oxidative stress microenvironment, the tumor side population (SP) (Hirschmann-Jax et al., 2004; Patrawala et al., 2005) cells underwent reprogramming to acquire ES cell-like gene expressions including high expression of Oct-4 and Nanog (Das et al., 2008). The post-hypoxia SP cells also exhibited significant tumorigenic activities compared to SP cells not exposed to hypoxia. We concluded that SP cells, during the expansion process, underwent a reprogramming to acquire a higher degree of tumor stemness (undifferentiated state and ability to initiate tumors in serial transplantation assay). In another model of stress, the chemotherapeutic-induced stress tumor SP cells exhibited tumor stemness phenotype including the expression of Oct-4, Nanog, and Bmi-1 are increases in the tumorigenic and metastatic activities. Briefly, in an osteosarcoma cell line HOS, which is non-tumorigenic, 3days of exposure to cisplatin expanded a quiescent but highly tumorigenic SP fraction (Tsuchida and Das et al., 2008).

Now, from an ASD point of view, it is possible to re-interpret the SP cell study data and the tumor stemness switch idea in a new way. The tumor stemness switch could be an ASD system of tumor to maintain its heterogeneous population during microenvironmental stress. A tumor may also be considered an injurious tissue, where hypoxic/necrotic zones may act as the area of injury (Cook et al., 2004). So, similar to the hESCs-derived SSEA3+/ABCG2+ cells undergoing enhanced stemness reprogramming during hypoxia/oxidative stress microenvironment (Figure 7.2), an CSC fraction within the heterogeneous CSC population might also undergo a reversible reprogramming process.

Indeed, the work on SP cells indicated that the post-hypoxia SP cells exhibited an ASD-like mechanism. Only a few SP cells (the ABCG2+/CXCR + cell fraction) among the heterogeneous SP cell population exhibited the tumor stemness phenotype (i.e., higher invasive, metastatic, tumorigenic, and ES cell gene expression phenotype). The phenotype was reversible. The phenotype also exhibited high secretion of angiogenic growth factors including VEGF that was probably instrumental in protecting the rest of the CSC population from the hypoxia/oxidative and or chemotherapy-induced stress. The ABCG2+/CXCR + fraction of SP cells again returned back to their quiescent state and resided in hypoxic niche (Das et al., 2008). Similar hypoxia-induced reversible phenotypic switch of cancer cells to immature cancer cell phenotype was also reported (Jogi et al., 2002).

The reversible nature of tumor stemness switch (i.e., the return of CSCs from their tumor stemness state, the state of active self-renewal or expansion) to quiescent state (Das et al. 2008) could explain why the heterogeneous CSC population remains both quiescent and actively self-renewing CSCs. Although further work is required to confirm if the findings described in the SP cell study (Das et al., 2008; Tsuchida and Das et al., 2008) may support the existence of the ASD-like system in cancer cells, these preliminary findings are encouraging, and they provide an experimental model system to investigate the potential role of ASD in CSC-mediated regeneration of tumors during microenvironmental stress including cytotoxic therapy.

Thus, the ASD mechanism may provide a novel way to understand the CSC hypothesis, including a possible explanation of how quiescent CSCs could be maintained in the heterogeneous population of cancer cells. The potential hijacking of the ASD system by cancer cells to adapt to microenvironmental stress may also provide an answer to the intriguing question of why cancer cells, toward the end of their evolution to become an aggressive, metastatic cancer cell exhibit features of stem cells. The cancer cell population probably hijacks stem cell's ASD mechanism to defend the clonal heterogeneity during microenvironmental stress.

It is proposed that CSC hypothesis of the hierarchical organization of cancer cell population is not compatible with the Darwinian natural selection process. Instead, a model of natural selection of competitive cells could explain why not all cancer cells are tumorigenic (Maenhaut et al. 2010). In this regard, the ASD mechanism supports the CSC hypothesis, because a hierarchical stemness pathway that underlies the ASD mechanism could be selected as the dominant mechanism of competitive and tumor-propagating cells of a given tumor population. The hierarchical stemness pathway could be a form of robust, bow-tie type architecture (Zhao et al., 2006), where the core would be formed by stemness genetic network. If so, the ASD system in a tumor would be highly dependent on a few core genes that maintain the bow-tie architecture of tumor stemness; inhibition of these candidate genes would lead to collapse of such bow-tie type tumor stemness architecture.

The ASD mechanism could also explain why metastasis is a reversible process. The CSCs undergoing tumor stemness switch will return back to its ground state of quiescence when the metastatic niche is established. In this manner, a reversible process of metastasis will be adequate enough for tumor cells to spread in the host body to establish seed population to ensure survival from an external threat. This is a strategy similar to what the microbial pathogens employ through their own ASD mechanism (Youk and Qudenaarden, 2010).

In a similar manner, understanding the ASD mechanism could also help to gain further insight about the regeneration/repopulation of a tumor following cytotoxic therapy, which is the main reason for tumor relapse (Fowler, 1991; Kim and Tannock, 2005). For example, cancer cells might hijack the ASD mechanism to maintain a quiescent fraction of cells, the CSCs among the heterogeneous population of cancer cells. During cytotoxic therapy, CSCs could undergo tumor stemness switch to re-establish the heterogeneous population of a given tumor.

Taken together, a new approach of an ASD-based model of tumor initiation, progression and relapse might provide us with some valuable insight about why it is so difficult to treat cancer, a vexing question in a globalized world, where cancer now stand as one of the serious global health issues.

ASD System: Future Direction

To follow Dobzhansky's warning, "Nothing in biology makes sense except in the light of evolution," it is crucial to develop a conceptual evolutionary framework to generate testable hypotheses about the idea of ASD described in the text. Earlier I made a crude attempt in this direction by proposing the idea that oxidative stress served as an evolutionary pressure in ancestral stem cells for the development of an antioxidant defense system. I proposed that molecular and metabolic architectures evolved to maintain a state of disequilibrium to maintain a high antioxidant state for cytoprotection in primitive ancestral stem cells (Figure 7.5). In this manner, the antioxidant defense system evolved to maintain the state of disequilibrium of a high antioxidant state, an important cytoprotective mechanism of survival. I further proposed that the state of disequilibrium is fluctuating, and the underlying architecture of the state could be a self-organization system (Das, 2000). Considering that an ASD system is primarily an antioxidant defense system, it is possible that the stemness genetic and metabolic networks that probably regulate ASD, initially evolved to maintain the state of antioxidant disequilibrium state. Hence, the evolution of stemness could be intricately linked to the evolution of antioxidant defense system in the early phase of life's origin in earth. A dynamical-system approach to study the robustness aspect of stemness might provide a conceptual framework of the robustness architecture of ASD. Because Darwinian selection might contribute to the evolution of robustness of a biological system (Andrreas, 2007), the study of the robustness aspect of stemness could generate some testable hypothesis to test the idea of ASD described in the text. Such a robustness of stemness study approach might also provide insight about tumor stemness and its fragility. Targeting the tumor stemness fragility would greatly contribute to the future discovery of novel anti-cancer strategies.

Conclusions

The ASD system incorporates the mechanism of stem cell and CSC self-renewal during microenvironmental stress. At the core of the defense system is a reprogramming mechanism of stem cells to higher stemness states of self-sufficiency, the enhanced stemness, and tumor stemness reprogramming. The former exhibit altruistic behavior, whereas the latter exhibit the ability to form a tumor by managing niche interaction. The tumor stemness state could be the ground state of malignant transformation of normal stem cell to cancer stem cell. The ASD system might be hijacked by cancer cells to grow, spread, and relapse following cytotoxic treatment.

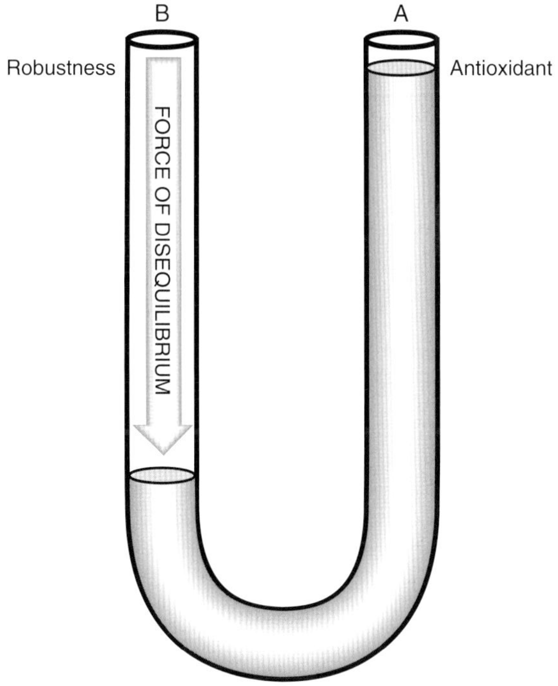

Figure 7.5. The biological state of disequilibrium: the model of antioxidant defense system architecture. The molecular and metabolic architectures of ancestral stem cells evolved to maintain a state of disequilibrium to maintain a high antioxidant state for cytoprotection in the pre-Cambrian earth, when the environment was mostly oxidant in nature. (Adapted from Science Behind Squalene, Das 2000, page 140.)

Acknowledgments

I thank Dr. Herman Yeger, Hospital for Sick Children, Toronto; Dr. Dean W. Felsher, Stanford University Medical School; and Dr. Peter W. Andrew, University of Sheffield, UK for helpful discussion. This work is supported by grants from the Grand Challenge Exploration Grant, Bill & Melinda Gates Foundation, Canadian Cancer Society, Laurel Foundation, USA, and KaviKrishna Foundation, Sualkuchi, Assam, India.

References

Al-Hajj M, Wicha MS, Benito-Hernandez A, Morrison SJ, Clarke MF. (2003). Prospective identification of tumorigenic breast cancer cells. Proc Natl Acad Sci USA. 100(7):3983–8.

Andreas W. (2007). Robustness and evolvability in living systems. Princeton University Press, NJ, page 277.

Andrews PW. (2002). From teratocarcinomas to embryonic stem cells. Philos Trans R Soc Lond B Biol Sci 357:405–417.

Bonnet D, Dick JE. (1997). Human acute myeloid leukemia is organized as a hierarchy that originates from a primitive hematopoietic cell. Nat Med 3(7):730–7.

Cai J, Weiss ML, Rao MS. (2004). In search of "stemness." Exp Hematol. 32:585–598.

Cook JA, Gius D, Wink DA, Krishna MC, Russo A, Mitchell JB. Oxidative stress, redox, and the tumor microenvironment. (2004). Semin Radiat Oncol 14(3):259–66.

Cottler-Fox MH, Lapidot T, Petit I, et al. (2003). Stem cell mobilization. Hematology Am Soc Hematol Educ Program 419–37.

Das B. (2000). Science Behind Squalene, Toronto Medical Publishing, Toronto, Canada. ISBN

1-890412-95-3. http://www.scienceforlife.eu/tekst%20sciencebehindsqualene.htm
Das B. (2007). The role of VEGF autocrine signaling in hypoxia and oxidative stress driven "stemness switch": implications in solid tumor progression and metastasis. PhD thesis: Institute of Medical Sciences, University of Toronto.
Das B, Bayat-Mokhtari R, Tsui M, Lotfi S, Tsuchida R, Felsher DW, Herman Yeger. (2012). HIF-2alpha suppresses p53 to enhance the stemness and regenerative potential of human embryonic stem cells. Stem Cells 30 (8): 1685–95.
Das B, Tsuchida R, Baruchel S, Yeger H. (2005b). Idea and evidence of "tumor stemness switch." In: International Workshop on Cancer Stem Cell; November 10–12; Milan, Italy; p. 34.
Das B, Tsuchida R, Malkin D, Baruchel S, Yeger H. (2008). Hypoxia enhances tumor stemness by increasing the invasive and tumorigenic side-population fraction. Stem Cells 26 (7):1818–30.
Das B, Tsuchida R, et al. (2009). The idea and evidence for the tumor stemness switch. Book chapter in the Regulatory Networks in Stem Cells. Humana Press, NJ.
Das B, Yeger H, Tsuchida R, et al. (2005a). A hypoxia-driven vascular endothelial growth factor/Flt1 autocrine loop interacts with hypoxia-inducible factor-1alpha through mitogen-activated protein kinase/extracellular signal-regulated kinase 1/2 pathway in neuroblastoma. Cancer Res 65:7267–7275.
Dick JE. (2003). Breast cancer stem cells revealed. Proc Natl Acad Sci USA 100(7):3547–9.
Draper JS, Smith K, Gokhale P, et al. (2004). Recurrent gain of chromosomes 17q and 12 in cultured human embryonic stem cells. Nat Biotechnol 22:53–54.
Durand R. (2001). Intermittent blood flow in solid tumors-an under-appreciated source of "drug-resistance." Cancer and Metastasis Reviews 20:57–61.
Fowler JF. (1991). Rapid repopulation in radiotherapy: a debate on mechanism. The phantom of tumor treatment–continually rapid proliferation unmasked. Radiother Oncol 22(3):156–8.
Fuchs E, Tumbar T, Guasch G. (2004). Socializing with the neighbors: stem cells and their niche. Cell 116(6):769–78.
Furusawa C, Kaneko K. (2012). A dynamical-systems view of stem cell biology. Science 338: 215–217.
Geva-Zatorsky N, Rosenfeld N, Itzkovitz S, Milo R, Sigal A, Dekel E, Yarnitzky T, et al. (2006). Oscillations and variability in the p53 system. Mol Syst Biol 2: 2006–2033.
Hill RP. (2006). Identifying cancer stem cells in solid tumors: case not proven. Cancer Res 66(4):18911896.
Hirschmann-Jax C, Foster AE, Wulf GG, et al. (2004). A distinct "side population" of cells with high drug efflux capacity in human tumor cells. Proc Natl Acad Sci USA 101(39):14228–33.
Ito K, Hirao A, Arai F, Matsuoka S et al. (2004). Regulation of oxidative stress by ATM is required for self-renewal of hematopoietic stem cells. Nature 431:997–1002.
Jogi A, Ora I, Nilsson H, et al. (2002). Hypoxia alters gene expression in human neuroblastoma cells toward an immature and neural crest-like phenotype. Proc Natl Acad Sci USA 99(10):7021–6.
Kim CF, Jackson EL, Woolfenden AE, et al. (2005). Identification of bronchioalveolar stem cells in normal lung and lung cancer. Cell 121(6):823–35.
Kim JJ, Tannock IF. (2005). Repopulation of cancer cells during therapy: an important cause of treatment failure. Nat Rev Cancer 5(7):516–52.
Li L, Xie T. (2005). Stem cell niche: structure and function. Annu Rev Cell Dev Biol 21: 605–31.
Mackillop W, Ciampi A, Till J, Buick R. (1983). A stem cell model of human tumor growth: implications for tumor cell clonogenic assay. JNCI 70(1):9–16.
Maenhaut C, Dumont JE, Roger PP, van Staveren WCG. (2010). Cancer stem cells: a reality, a myth, a fuzzy concept or a misnomer? An analysis. Carcinogenesis 31 (2): 149–158.
Mikkers H, Frisen J. (2005). Deconstructing stemness. Embo J. 24:2715–2719.
Nowell PC. (1986). Mechanisms of tumor progression. Cancer Res 46(5):2203–7.
Pan G, Thomson JA. (2007). Nanog and transcriptional networks in embryonic stem cell pluripotency. Cell Res 17:42–49.
Pardal R, Clarke MF, Morrison SJ. (2003). Applying the principles of stem-cell biology to cancer. Nat Rev Cancer 3(12):895–902.
Patrawala L, Calhoun T, Schneider-Broussard R, Zhou J, Claypool K, Tang DG. (2005). Side population is enriched in tumorigenic, stem-like cancer cells, whereas ABCG2+ and ABCG2- cancer cells are similarly tumorigenic. Cancer Res 65(14):6207–19.

Pai R, Jones MK, Tomikawa M, Tarnawski AS. (1999). Activation of Raf-1 during experimental gastric ulcer healing is Ras-mediated and protein kinase C-independent. Am J Pathol 155(5):1759–66.

Prockop DJ. (2009). Repair of tissues by adult stem/progenitor cells (MSCs): controversies, myths, and changing paradigms. Mol Ther. 17:939–946.

Ramalho-Santos M, Yoon S, Matsuzaki Y, et al. (2002). "Stemness": transcriptional profiling of embryonic and adult stem cells. Science. 298:597–600.

Reya T, Morrison SJ, Clarke MF, Weissman IL. (2001). Stem cells, cancer, and cancer stem cells. Nature 414(6859):105–11.

Rich JN. (2007). Cancer stem cells in radiation resistance. Cancer Res 67(19):8980–4.

Rinzel J, Terman D, Wang X, Ermentrout B. (1998). Propagating activity patterns in large-scale inhibitory neuronal networks. Science. 279:1351–1355.

Silva J, Smith A. (2008). Capturing pluripotency. Cell. 132:532–536.

Singh AM, Hamazaki T, Hankowski KE, et al. (2007). A heterogeneous expression pattern for Nanog in embryonic stem cells. Stem Cells. 25:2534–2542.

Singh SK, Hawkins C, Clarke ID, et al. (2004). Identification of human brain tumour initiating cells. Nature 432(7015):396–401.

Suda T, Arai F, Shimmura S. (2005). Regulation of stem cells in the niche. Cornea 24(8 Suppl): S12–S7.

Trott KR. (1994). Tumour stem cells: the biological concept and its application in cancer treatment. Radiother Oncol 30(1):1–5.

Tscharntke M, Pofahl R, Krieg T, Haase I. (2005). Ras-induced spreading and wound closure in human epidermal keratinocytes. Faseb J 19 (13):1836–8.

Tsuchida R*, Das B*, Yeger H, et al. (2008). Cisplatin treatment increases survival and expansion of a highly tumorigenic side-population fraction by upregulating VEGF/Flt1 autocrine signaling Oncogene. Online issue, March 8. *Equal contribution 2007.

Turcotte M, Garcia-Ojalvo J, Suel GM. (2008). A genetic timer through noise-induced stabilization of an unstable state. Proc Natl Acad Sci USA. 105:15732–15737.

Tyson JJ. (1998). Another turn for p53. Molecular Systems Biology 2:2006.0032.

Wicha M, Liu S, Dontu G. (2006). Cancer stem cells: an old idea-a paradigm shift. Cancer Res 66:1883–90.

Youk H and Qudenaarden VA. (2010). Microbiology: Altruistic defense. Nature. 467:34–35.

Zhao J, Yu H, Luo JH, Cao ZW, Li YX. (2006). Hierarchical modularity of nested bow-ties in metabolic networks. BMC Bioinformatics 7 (386):1–16.

Chapter 8
The Emerging Concept of EMT-Induced Cancer Stem Cells

Jeremy Bastid
Orega Biotech, L'espace Européen, Ecully, France

Tumor as Organs: Existence of Cancer Stem Cells

Tumors are not just a mass of proliferating cancer cells. In fact, in many aspects, tumors resemble our organs and comprised multiple cell types and extracellular matrix (Egeblad et al., 2010). Tumor cells are hierarchically organized and communicate with each other and with the rest of our body. Like normal organs surrounded by their supportive stroma, tumors are surrounded by the tumor stroma, which supports cancer growth. In fact, the main difference between normal organs and tumors is that organs have positive functions for the host whereas tumors will ultimately kill the patient.

Accumulating evidence shows that both hematological and some solid tumors are hierarchically organized and that tumor growth is sustained by a subpopulation of cancer cells called "cancer stem cells" (CSC) (Visvader, 2011; Visvader and Lindeman, 2008). Like normal stem cells, CSCs endow self-renewal capacity and can generate the multiple lineages of tumorigenic and nontumorigenic cancer cells that form the tumor. Prospective identification of CSCs has often been based on the use of markers of normal stem cell from the corresponding tissue. The demonstration of the existence of CSCs in human tumors is not an easy task and has been achieved using serial transplantation of cancer cell populations in mice. To match the definition of "CSCs," the cellular subpopulation has to be able to re-establish a primary tumor with the same phenotypic heterogeneity compared to the lesion of origin and propagate the tumor over serial passages (i.e., possess self-renewal capacities). Of note, one should keep in mind that these xenograft models in immunocompromised mice have several issues that should be taken into account when interpreting data. However, three recent breakthrough papers have used lineage-tracing technologies to demonstrate the existence of CSCs in mouse models of tumors that arise, develop, and progress "naturally" in the host. They demonstrated that a subpopulation of quiescent CSCs exists in glioma, squamous cell carcinoma, and intestinal cancer. These CSCs generate the tumor and propagate tumor regrowth following therapy (Chen et al., 2012; Driessens et al., 2012; Schepers et al., 2012).

Historically, cancer stem cells were identified in chronic myeloid leukemia (Bonnet and Dick, 1997). The ability to initiate leukemia

Cancer Stem Cells, First Edition. Edited by Vinagolu K. Rajasekhar.

in mice was allotted to the $CD34^{++}$ $CD38^{-}$ cells that possess many attributes of "leukemic stem cells" including self-renewal capacity and the ability to proliferate and differentiate in order to produce a disease that is phenotypically identical to that of the donor. Then, CSCs have been identified using various markers in numerous solid tumors such as breast cancer (Al-Hajj et al., 2003), brain tumors (Singh et al., 2004), colon cancer (O'Brien et al., 2007; Ricci-Vitiani et al., 2007), head and neck cancer (Prince et al., 2007), pancreatic cancer (Li et al., 2007), prostate cancer (Collins et al., 2005), lung cancer (Eramo et al., 2008), liver cancer (Ma et al., 2007; Yang et al., 2008a), and melanoma (Schatton et al., 2008).

Origin of Cancer Stem Cells

Evidence of CSCs accumulates in numerous hematological and solid malignancies. Because this subpopulation sustains tumor growth, is resistant to therapy, and may be the source of metastatic cells (Magee et al., 2012), CSCs appear as attractive targets. However, targeting CSCs requires a good understanding of this subpopulation of (rare) cells, and the question remains of their origin. Indeed, it is debated whether CSCs arise from the oncogenic transformation of normal stem cells (which, *de facto*, already endow stem cell—including self-renewal—properties) or whether they arise from the transformation of already differentiated somatic cells or both.

Several studies have shown that CSCs can arise from the transformation of normal stem cells. Using cell-specific promoters to drive the expression of oncogenes in stem cell and nonstem cell compartments, Nick Barker and colleagues (2009) demonstrated that APC deletion $LGR5^{+}$ normal crypt stem cells led to rapid transformation and development of adenoma in mice. On the other hand, APC deletion in a nonstem cell compartment was unable to promote intestinal neoplasia. Thus, in this model, the first oncogenic hit has to occur in a stem cell. Similar conclusions were raised using CD133 as another crypt stem cell marker (Zhu et al., 2009). Along similar lines, medulloblastomas can arise from activation of the sonic hedgehog pathway in neural stem cells (Yang et al., 2008b). Alternatively, evidence is accumulating that CSCs can also arise from the transformation of restricted progenitors or differentiated cells that lack self-renewal properties. For example, it has been widely reported that leukemia can arise from the transformation of committed progenitors (Cozzio et al., 2003; Huntly et al., 2004; Krivtsov et al., 2006; Zhao et al., 2010). Similarly, medulloblastoma, but not other types of brain tumors, can arise from oncogenic insults in unipotent neuronal progenitors (Schuller et al., 2008; Yang et al., 2008b). Therefore, it is suspected that CSCs may be generated through the transformation of both stem and nonstem cells. For excellent reviews of this topic, see Magee et al. (2012) and Visvader (2011). If CSCs arise from the transformation of differentiated cells that lack stem cell properties, they will, *de facto*, have to reacquire some characteristics of stemness including self-renewal ability. One of the mechanisms by which these cells may dedifferentiate and reacquire stemness properties is the process termed epithelial to mesenchymal transition (EMT).

What is EMT?

EMT has gained enormous popularity in the field of oncology over the past decade. EMT is a transdifferentiation process that was first described during embryogenesis. It converts joined, apical-basal-polarized and poorly motile epithelial cells into individualized mesenchymal ones with spindle cell morphology and enhanced migratory properties. This process is essential for the formation of the mesoderm during gastrulation and also participates in wound healing during which epithelial cells surrounding the wound undergo EMT in order to migrate to the injury site. For an

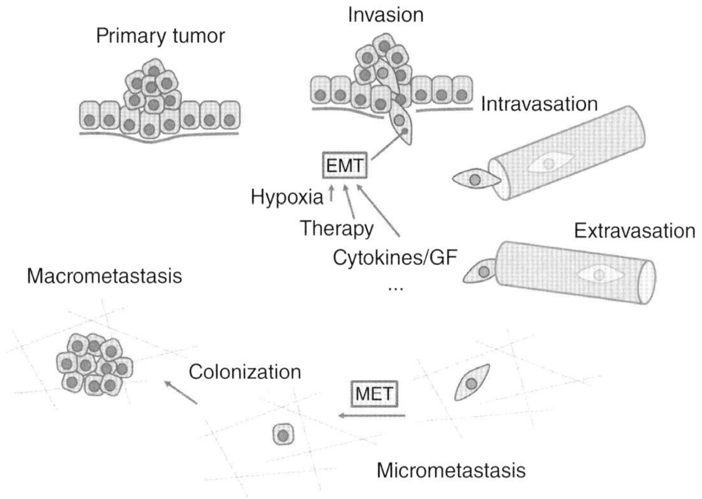

Figure 8.1. EMT and MET in carcinoma progression. At the primary tumor, EMT is induced by different stimuli such as hypoxia, several cytokines, and growth factors (GF) as well as stress induced by anti-neoplastic drugs. Cancer cell undergo EMT and acquire a mesenchymal phenotype with increased motility and ability to degrade the basement membrane and the extracellular matrix, thereby leading to local invasion. Mesenchymal cancer cells then intravasate into the blood or lymphatic vessels, disseminate and extravasate at secondary site, forming a dormant micrometastasis. Disseminated cancer cells then undergo MET (because of the lack of EMT-promoting factors), reacquire an epithelial phenotype, and reinitiate growth to generate a macrometastasis. For color detail, please see color plate section.

excellent review on the role of EMT in embryogenesis, wound healing and cancer, refer to Thiery et al. (2009). In experimental models of cancer, carcinoma cells undergo EMT upon stimulation by various stimuli including growth factors, cytokines, hypoxia, and stress induced by therapeutic agents, thereby displaying increased invasive and metastatic ability as well as broad resistance to therapy (Thiery et al., 2009). Cancer cells in the mesenchymal state can leave the primary tumor and migrate to lymphatic or blood vessels through which they will reach and seed in a distant organ (micrometastasis). There, disseminated cancer cells undergo mesenchymal to epithelial transition (most likely because of the lack of the EMT-inducing factor[s] at the secondary site), reacquire an epithelial phenotype, and can colonize the organ (macrometastasis as shown in Figure 8.1). Of important note, the relevance of EMT/MET in human cancer is still debated (Bastid, 2012). This is mainly because carcinoma cells that have undergone EMT become undistinguishable from any other tumor-associated stromal cells (e.g., cancer-associated fibroblasts). This question has been elegantly addressed by Andrew D. Rhim and colleagues (2012) in a mouse model of pancreatic carcinogenesis. In summary, EMT has been widely and unambiguously demonstrated in vitro, in xenograft and transgenic mouse models. The two most convincing demonstrations were performed by Rhim and colleagues (2012) and Trimboli and colleagues (2008). EMT has been suggested but not demonstrated so far in human tumors. Nevertheless, this situation reflects technical issues (Bastid, 2012) that will, certainly, be solved shortly. In conclusion, EMT participates in carcinoma cells metastasis and resistance to therapy, and this has been illustrated in most types of cancers.

EMT-Induced Cancer Stem Cells

In 2008, two independent groups simultaneously uncovered an unexpected property of EMT: its ability to generate cells with property of stem cells (Mani et al., 2008; Morel et al., 2008). The group of Robert A. Weinberg demonstrated that EMT could generate both normal and cancer stem cells from normal or transformed breast cells, respectively (Mani et al., 2008). Induction of EMT in human mammary epithelial cells (HMLE) following overexpression of Twist, Snail, or stimulation with TGFβ indeed promoted a $CD24^{low}$ $CD44^{high}$ stem cell phenotype with strikingly increased capacity to form mammospheres, an in vitro assay that gauges stemness. Vice versa, $CD24^{low}$ $CD44^{high}$ breast cells isolated from cultured cells and mouse or human mammary glands displayed both stem cells traits and EMT features. More importantly, induction of EMT in ERBB2 transformed HMLE promoted the generation of CSCs with drastically increased ability to form tumorospheres in vitro and to initiate tumors in vivo. However, it remains unproven that EMT by itself increased the tumorigenicity of these cells, because EMT was induced by genetically overexpressing Twist or Snail transcription factors, which both intrinsically display oncogenic properties including the ability to protect from apoptosis and senescence. For instance, it would have been interesting to test whether comparable results are obtained using TGFβ-treated cells. Similar conclusions were raised by the group of Alain Puisieux (Morel et al., 2008). Using in vitro-transformed HMEC and MCF10A mammary epithelial cells, they demonstrated that tumorigenic capacity is confined to the $CD24^{low}$ $CD44^{high}$ subpopulation, which additionally harbors mesenchymal morphology. The team postulated that the CSC population may emerge through EMT, and they indeed demonstrated that tumorigenic $CD24^{low}$ $CD44^{high}$ breast cancer cells can arise from nontumorigenic $CD44^{low}$ $CD24^{+}$ cells through TGFβ-induced EMT in vitro. However, the TGFβ-treated cells were not tested in vivo, and their identification as "CSCs" is only based on phenotypic characterization. Although the two papers laid the groundwork in the field, one important caveat of these studies is the use of spontaneously immortalized cells (MCF10A) or HMEC cells immortalized with the human telomerase, which questions the capacity of EMT to confer self-renewal ability.

Further characterization of these EMT-derived breast stem cells revealed that they are phenotypically and functionally similar to mesenchymal stem cells (MSC) (Battula et al., 2010). They express several MSC-associated markers such as CD10, CD44, CD73, CD90, CD105, CD106, and CD166. Furthermore, they have the capacity to differentiate into multiple tissue lineages including osteoblasts, adipocytes, or chondrocytes and to participate in wound healing, a hallmark of MSCs. In fact, the plasticity of the stem and nonstem cell compartments is huge, and differentiated epithelial cells can spontaneously convert into stem cells. Christine L. Chaffer and colleagues (2011) indeed demonstrated that primary and transformed differentiated cells revert to a stem-like state, potentially through EMT. This work is of particular interest because this dedifferentiation occurs without genetic modification of the cells. Therefore, the hierarchical tree in the stem cell biology field should encompass bidirectional conversion from one compartment to the other, at least in the mammary gland. The ability of EMT to generate breast CSCs has then been evidenced in response to a large number of stimuli including IL-6 (Xie et al., 2012), TNFα and TGFβ (Asiedu et al., 2011) cytokines and growth factors, LBX1 (Yu et al., 2009), Slug (Bhat-Nakshatri et al., 2010) and Twist2 (Fang et al., 2011) transcription factors, immune cells (Santisteban et al., 2009), or complex stimulatory networks (Morel et al., 2012; Scheel et al., 2011).

Later on, EMT-induced CSCs have been evidenced in other cancer types. Colon CSCs have a mesenchymal phenotype and express Snail, while Snail expression in differentiated colon cancer cells induced EMT and promoted a CSC phenotype through activation of IL-8 (Hwang et al., 2011). Similar observations were made when EMT was induced upon downregulation of the E-cadherin (Ye et al., 2012). In

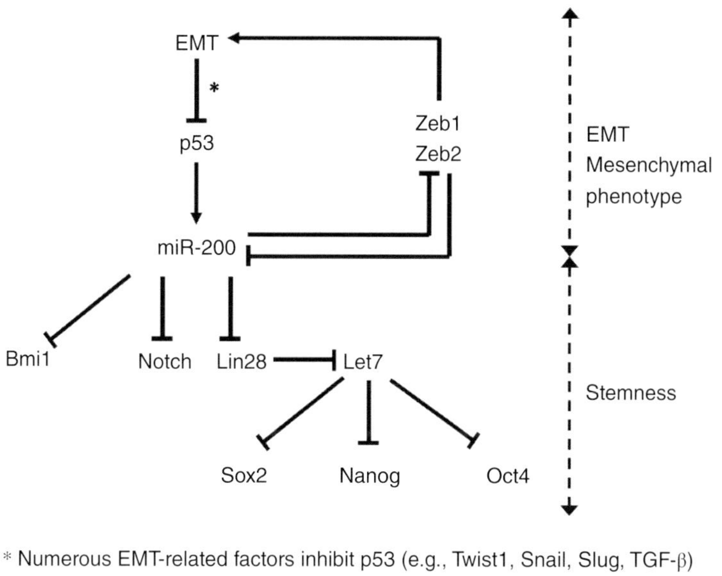

Figure 8.2. A p53-miR-200-Zeb1/2 circuit linking EMT and acquisition of stem cell properties. Induction of EMT in carcinoma cell induces decrease in miR-200b/c. This is potentially mediated by the inhibition of p53 (a direct transactivator of miR-200) by EMT-related transcription factors such as Twist, Snail, or Slug. Inhibition of miR-200 leads to the upregulation of Zeb1 and Zeb2, which maintains the mesenchymal phenotype, and to the expression of several stem cell factors such as Bim1, Lin28, Sox2, Nanog, Oct4, and the Notch pathway.

prostate cancer cell, EMT was accompanied by decreased abundance of let-7 and miR-200 miRNAs, which, in turn, mediated the acquisition of a stem cell phenotype and sustained tumor growth (Kong et al., 2010). Similarly, EMT might be associated with the acquisition of stemness properties in liver tumors (Caja et al., 2011; Oishi et al., 2012), in keratinocytes (Oh et al., 2011), in squamous cell carcinoma (Geng et al., 2012; Zhu et al., 2012), in pancreatic cancer (Bao et al., 2011), and in non-small cell lung cancer (Pirozzi et al., 2011). Is EMT always associated with the acquisition of stemness features? Is this phenomenon limited to certain cell types or certain EMT-inducing factors? Those questions remain unanswered so far.

The molecular mechanisms by which EMT promotes stem cell properties are still elusive, but clues accumulate and key signaling pathways are being uncovered. The p53-miR-200 family-Zeb1/2 axis seems to have a central role in the link between EMT and stem cell phenotype. In cancer cells, EMT is accompanied by decreased abundance of miR 200b and miR-200c (Chang et al., 2011; Kong et al., 2010; Oishi et al., 2012). This is possibly due to EMT-mediated inhibition of p53, which is a direct transactivator of miR-200 family (Chang et al., 2011; Kim et al., 2011). Downregulation of the miR-200 family sustains expression of the EMT-promoting transcription factors Zeb1 and Zeb2 (Chang et al., 2011; Kim et al., 2011; Korpal et al., 2008) (contributing to the maintenance of the mesenchymal state) and promotes stem cell phenotype by inducing, directly or indirectly through other miRNAs, activation of the Notch pathway (Brabletz et al., 2011; Kong et al., 2010), expression of the self-renewal gene Bmi1 (Chang et al., 2011) as well as upregulation of Sox2, Nanog, Oct4, and Lin28 (Kong et al., 2010; Oh et al., 2011), four factors that are sufficient for reprogramming human somatic cells to pluripotent stem cells (iPSC) (Yu et al., 2007). This pathway is illustrated in Figure 8.2 and has been the focus of several recent reviews (Brabletz and Brabletz, 2010; Hill et al., 2012). Thus, the Zeb/miR-200 feedback loop, along with p53, appears to play a central role in linking EMT with stem cell features.

Clinical Implications of EMT-Induced Cancer Stem Cells

If true, one key implication from this discovery is the necessity to rethink the way we envision therapies of advanced malignancies. Indeed, conventional therapies (e.g., chemotherapies) mainly kill the tumor bulk, leaving CSCs unaffected (Rasheed et al., 2011) (Figure 8.3A), which explains tumor recurrence. It is therefore necessary to design therapeutic molecules that eliminate CSCs ("anti-cancer stem cell" therapies), which are responsible for tumor regrowth and resistance to treatment. Of note, we do not have such tools in hands at this stage, but efforts and discoveries have already been made in this direction (Gupta et al., 2009). Using a novel high-throughput screening method based on EMT-induced CSCs, Piyush B. Gupta and colleagues (2009) identified a set of compounds that selectively target CSCs. One hit, the ionophore antibiotic called salinomycin, consistently exhibited high selectivity for CSCs. Its mechanism of action remains elusive, but it was shown that it is a potent inhibitor of multidrug resistance protein (p-glycoprotein) (Riccioni et al., 2010). However, salinomycin exerts severe neuronal and cardiac toxicity in mammals and its transferability to the clinic is at least uncertain (Huczynski, 2012). Alternatively, several key pathways such as Wnt, Notch, and Hedgehog pathways may be targeted to affect CSCs (Takebe et al., 2011).

Conceptually, elimination of the CSCs by novel selective agents would leave a bulk of differentiated, nontumorigenic cancer cells that are more sensitive to chemotherapy (Figure 8.3B). However, as discussed in this article, this is certainly a simplistic view, and the ability of differentiated tumor cells to generate, through EMT, cancer stem cells able to support the growth of the tumor certainly complexify the picture. Elimination of CSCs may be followed by the emergence of EMT-induced CSCs that would inevitably lead to tumor recurrence (Figure 8.3C). Having this notion in mind, one may think about several possible strategies. A first intuitive option would be to target both cancer stem cells and nonstem cells, for example, by combining "anti-cancer stem cell" therapies with conventional chemotherapeutic agents (Figure 8.3D). If such a therapeutic cocktail existed, this would, conceptually, leave the patient with no or "minimal" residual disease. The chances of success of such an approach rely on what is left behind, that is, what "minimal" means. If the number of surviving cancer cells is small enough, there are good chances that cancer will not recur, at least not recur in a life-threatening manner before a reasonable period of time. However, if their number is significant enough, those residual cells may undergo EMT, reacquire cancer stem cell properties, and regrow as an aggressive malignant lesion (Figure 8.3E).

In a second scenario, one may combine "anti-cancer stem cell" therapies with "anti-EMT" therapies. Once again, such therapeutic molecules do not exist, but efforts are being engaged to screen drug candidates (Chua et al., 2012). The team of Jean-Paul Thiery implemented an EMT inhibition drug-screening assay and identified several hits among the 267 preclinical or clinical compounds tested. Using such combinatory approach, we would get rid of CSCs while blocking the plasticity of differentiated cancer cells, thereby preventing the emergence of EMT-induced CSCs (Figure 8.3B and 8.3C). A more plausible scenario, in fact, relies on the combination of the three types of approaches in order to eradicate CSCs, to significantly decrease the tumor mass and to prevent EMT events in residual tumor cells (Figure 8.3D and 8.3E). Yet, those assumptions remain hypotheses that will have to be tested in relevant mouse models.

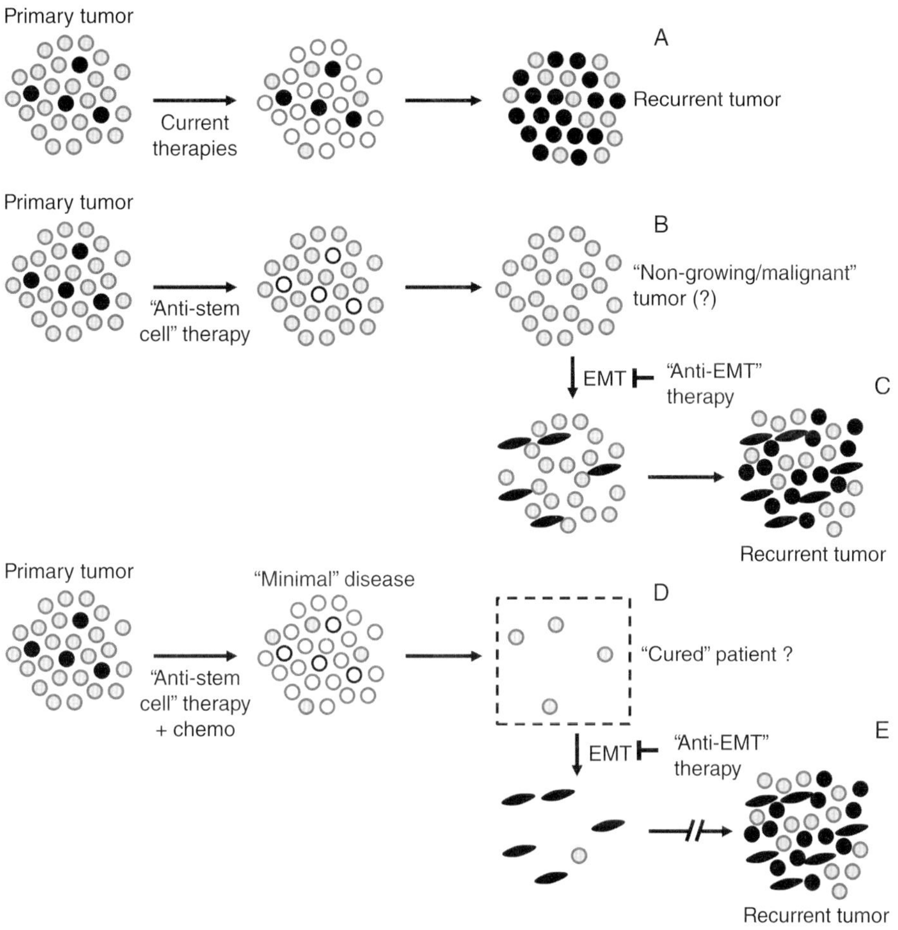

Figure 8.3. Implication of EMT-induced CSCs for therapy. **A.** In primary tumors treated with conventional chemotherapeutic agents, CSCs persist and lead to tumor recurrence. **B.** and **C.** Therapies that would selectively kill CSCs ("anti-stem cell" therapy) would eliminate cancer stem cells that sustain tumor growth, conceptually leading to a nonmalignant tumor composed of more differentiated cells **B**. However, as discussed here, differentiated cancer cells might generate de novo CSCs through EMT, thereby promoting tumor recurrence **C**. **D.** and **E.** Eradication of CSCs combined with conventional chemotherapy targeting more differentiated cancer cells should conceptually and statistically offer limited risk of recurrence **D**. However, again, residual cells may undergo EMT and may lead to tumor recurrence **E**. It is therefore expected that therapeutic agents that would block the EMT process would prevent this plasticity of tumor cells (generation of EMT-induced CSCs) and would lower the risk of recurrence (**C** and **E**).

Concluding Remarks and Perspectives

A growing body of evidence supports a role for CSCs in tumor propagation, recurrence, and resistance to therapy. CSCs may thus be the Achilles' heel of the cancer, and therefore the ones to target. Identifying CSCs, understanding their localization, their origin, their plasticity, and the genetic and epigenetic events that render them tumorigenic is key to design future cancer therapies properly. In this regard, the unexpected ability of EMT to generate CSCs from differentiated tumor cells may have enormous impact from the patient standpoint. Therefore, further work is urgently needed to better understand this process and its relevance in cancer pathophysiology.

References

Al-Hajj, M., Wicha, M.S., Benito-Hernandez, A., Morrison, S.J., and Clarke, M.F. (2003). Prospective identification of tumorigenic breast cancer cells. Proc Natl Acad Sci USA *100*, 3983–3988.

Asiedu, M.K., Ingle, J.N., Behrens, M.D., Radisky, D.C., and Knutson, K.L. (2011). TGFbeta/TNF(alpha)-mediated epithelial-mesenchymal transition generates breast cancer stem cells with a claudin-low phenotype. Cancer Res *71*, 4707–4719.

Bao, B., Wang, Z., Ali, S., Kong, D., Banerjee, S., Ahmad, A., Li, Y., Azmi, A.S., Miele, L., and Sarkar, F.H. (2011). Over-expression of FoxM1 leads to epithelial-mesenchymal transition and cancer stem cell phenotype in pancreatic cancer cells. J Cell Biochem *112*, 2296–2306.

Barker, N., Ridgway, R.A., van Es, J.H., van de Wetering, M., Begthel, H., van den Born, M., Danenberg, E., Clarke, A.R., Sansom, O.J., and Clevers, H. (2009). Crypt stem cells as the cells-of-origin of intestinal cancer. Nature *457*, 608–611.

Bastid, J. (2012). EMT in carcinoma progression and dissemination: facts, unanswered questions, and clinical considerations. Cancer Metastasis Rev *31*, 277–283.

Battula, V.L., Evans, K.W., Hollier, B.G., Shi, Y., Marini, F.C., Ayyanan, A., Wang, R.Y., Brisken, C., Guerra, R., Andreeff, M., *et al.* (2010). Epithelial-mesenchymal transition-derived cells exhibit multilineage differentiation potential similar to mesenchymal stem cells. Stem Cells *28*, 1435–1445.

Bhat-Nakshatri, P., Appaiah, H., Ballas, C., Pick-Franke, P., Goulet, R., Jr., Badve, S., Srour, E.F., and Nakshatri, H. (2010). SLUG/SNAI2 and tumor necrosis factor generate breast cells with CD44+/CD24− phenotype. BMC Cancer *10*, 411.

Bonnet, D., and Dick, J.E. (1997). Human acute myeloid leukemia is organized as a hierarchy that originates from a primitive hematopoietic cell. Nat Med *3*, 730–737.

Brabletz, S., Bajdak, K., Meidhof, S., Burk, U., Niedermann, G., Firat, E., Wellner, U., Dimmler, A., Faller, G., Schubert, J., *et al.* (2011). The ZEB1/miR-200 feedback loop controls Notch signalling in cancer cells. Embo J *30*, 770–782.

Brabletz, S., and Brabletz, T. (2010). The ZEB/miR-200 feedback loop—a motor of cellular plasticity in development and cancer? EMBO Rep *11*, 670–677.

Caja, L., Bertran, E., Campbell, J., Fausto, N., and Fabregat, I. (2011). The transforming growth factor-beta (TGF-beta) mediates acquisition of a mesenchymal stem cell-like phenotype in human liver cells. J Cell Physiol *226*, 1214–1223.

Chaffer, C.L., Brueckmann, I., Scheel, C., Kaestli, A.J., Wiggins, P.A., Rodrigues, L.O., Brooks, M., Reinhardt, F., Su, Y., Polyak, K., *et al.* (2011). Normal and neoplastic nonstem cells can spontaneously convert to a stem-like state. Proc Natl Acad Sci USA *108*, 7950–7955.

Chang, C.J., Chao, C.H., Xia, W., Yang, J.Y., Xiong, Y., Li, C.W., Yu, W.H., Rehman, S.K., Hsu, J.L., Lee, H.H., *et al.* (2011). p53 regulates epithelial-mesenchymal transition and stem cell properties through modulating miRNAs. Nat Cell Biol *13*, 317–323.

Chen, J., Li, Y., Yu, T.S., McKay, R.M., Burns, D.K., Kernie, S.G., and Parada, L.F. (2012). A restricted cell population propagates glioblastoma growth after chemotherapy. Nature *488*, 522–526.

Chua, K.N., Sim, W.J., Racine, V., Lee, S.Y., Goh, B.C., and Thiery, J.P. (2012). A cell-based small molecule screening method for identifying inhibitors of epithelial-mesenchymal transition in carcinoma. PLoS One *7*, e33183.

Collins, A.T., Berry, P.A., Hyde, C., Stower, M.J., and Maitland, N.J. (2005). Prospective identification of tumorigenic prostate cancer stem cells. Cancer Res *65*, 10946–10951.

Cozzio, A., Passegue, E., Ayton, P.M., Karsunky, H., Cleary, M.L., and Weissman, I.L. (2003). Similar MLL-associated leukemias arising from self-renewing stem cells and short-lived myeloid progenitors. Genes Dev *17*, 3029–3035.

Driessens, G., Beck, B., Caauwe, A., Simons, B.D., and Blanpain, C. (2012). Defining the mode of tumour growth by clonal analysis. Nature *488*, 527–530.

Egeblad, M., Nakasone, E.S., and Werb, Z. (2010). Tumors as organs: complex tissues that interface with the entire organism. Dev Cell *18*, 884–901.

Eramo, A., Lotti, F., Sette, G., Pilozzi, E., Biffoni, M., Di Virgilio, A., Conticello, C., Ruco, L., Peschle, C., and De Maria, R. (2008). Identification and expansion of the tumorigenic lung cancer stem cell population. Cell Death Differ *15*, 504–514.

Fang, X., Cai, Y., Liu, J., Wang, Z., Wu, Q., Zhang, Z., Yang, C.J., Yuan, L., and Ouyang, G. (2011). Twist2 contributes to breast cancer progression by promoting an epithelial-mesenchymal transition and cancer stem-like cell self-renewal. Oncogene *30*, 4707–4720.

Geng, S., Guo, Y., Wang, Q., Li, L., and Wang, J. (2012). Cancer stem-like cells enriched with CD29 and CD44 markers exhibit molecular characteristics with epithelial-mesenchymal transition in squamous cell carcinoma. Arch Dermatol Res. doi: 10.1007/s00403-012-1260-2.

Gupta, P.B., Onder, T.T., Jiang, G., Tao, K., Kuperwasser, C., Weinberg, R.A., and Lander, E.S. (2009). Identification of selective inhibitors of cancer stem cells by high-throughput screening. Cell *138*, 645–659.

Hill, L., Browne, G., and Tulchinsky, E. (2012). ZEB/miR-200 feedback loop: At the crossroads of signal transduction in cancer. Int J Cancer. doi: 10.1002/ijc.27708.

Huczynski, A. (2012). Salinomycin: a new cancer drug candidate. Chem Biol Drug Des *79*, 235–238.

Huntly, B.J., Shigematsu, H., Deguchi, K., Lee, B.H., Mizuno, S., Duclos, N., Rowan, R., Amaral, S., Curley, D., Williams, I.R., *et al.* (2004). MOZ-TIF2, but not BCR-ABL, confers properties of leukemic stem cells to committed murine hematopoietic progenitors. Cancer Cell *6*, 587–596.

Hwang, W.L., Yang, M.H., Tsai, M.L., Lan, H.Y., Su, S.H., Chang, S.C., Teng, H.W., Yang, S.H., Lan, Y.T., Chiou, S.H., *et al.* (2011). SNAIL regulates interleukin-8 expression, stem cell-like activity, and tumorigenicity of human colorectal carcinoma cells. Gastroenterology *141*, 279–291, 291 e271–275.

Kim, T., Veronese, A., Pichiorri, F., Lee, T.J., Jeon, Y.J., Volinia, S., Pineau, P., Marchio, A., Palatini, J., Suh, S.S., *et al.* (2011). p53 regulates epithelial-mesenchymal transition through microRNAs targeting ZEB1 and ZEB2. J Exp Med *208*, 875–883.

Kong, D., Banerjee, S., Ahmad, A., Li, Y., Wang, Z., Sethi, S., and Sarkar, F.H. (2010). Epithelial to mesenchymal transition is mechanistically linked with stem cell signatures in prostate cancer cells. PLoS One *5*, e12445.

Korpal, M., Lee, E.S., Hu, G., and Kang, Y. (2008). The miR-200 family inhibits epithelial-mesenchymal transition and cancer cell migration by direct targeting of E-cadherin transcriptional repressors ZEB1 and ZEB2. J Biol Chem *283*, 14910–14914.

Krivtsov, A.V., Twomey, D., Feng, Z., Stubbs, M.C., Wang, Y., Faber, J., Levine, J.E., Wang, J., Hahn, W.C., Gilliland, D.G., *et al.* (2006). Transformation from committed progenitor to leukaemia stem cell initiated by MLL-AF9. Nature *442*, 818–822.

Li, C., Heidt, D.G., Dalerba, P., Burant, C.F., Zhang, L., Adsay, V., Wicha, M., Clarke, M.F., and Simeone, D.M. (2007). Identification of pancreatic cancer stem cells. Cancer Res *67*, 1030–1037.

Ma, S., Chan, K.W., Hu, L., Lee, T.K., Wo, J.Y., Ng, I.O., Zheng, B.J., and Guan, X.Y. (2007). Identification and characterization of tumorigenic liver cancer stem/progenitor cells. Gastroenterology *132*, 2542–2556.

Magee, J.A., Piskounova, E., and Morrison, S.J. (2012). Cancer stem cells: impact, heterogeneity, and uncertainty. Cancer Cell *21*, 283–296.

Mani, S.A., Guo, W., Liao, M.J., Eaton, E.N., Ayyanan, A., Zhou, A.Y., Brooks, M., Reinhard, F., Zhang, C.C., Shipitsin, M., *et al.* (2008). The epithelial-mesenchymal transition generates cells with properties of stem cells. Cell *133*, 704–715.

Morel, A.P., Hinkal, G.W., Thomas, C., Fauvet, F., Courtois-Cox, S., Wierinckx, A., Devouassoux-Shisheboran, M., Treilleux, I., Tissier, A., Gras, B., *et al.* (2012). EMT inducers catalyze malignant transformation of mammary epithelial cells and drive tumorigenesis towards claudin-low tumors in transgenic mice. PLoS Genet *8*, e1002723.

Morel, A.P., Lievre, M., Thomas, C., Hinkal, G., Ansieau, S., and Puisieux, A. (2008). Generation of breast cancer stem cells through epithelial-mesenchymal transition. PLoS One *3*, e2888.

O'Brien, C.A., Pollett, A., Gallinger, S., and Dick, J.E. (2007). A human colon cancer cell capable of initiating tumour growth in immunodeficient mice. Nature *445*, 106–110.

Oh, J.E., Kim, R.H., Shin, K.H., Park, N.H., and Kang, M.K. (2011). DeltaNp63alpha protein triggers epithelial-mesenchymal transition and confers stem cell properties in normal human keratinocytes. J Biol Chem *286*, 38757–38767.

Oishi, N., Kumar, M.R., Roessler, S., Ji, J., Forgues, M., Budhu, A., Zhao, X., Andersen, J.B., Ye, Q.H., Jia, H.L., *et al.* (2012). Transcriptomic profiling reveals hepatic stem-like gene signatures and interplay of mir-200c and EMT in intrahepatic cholangiocarcinoma. Hepatology. doi: 10.1002/hep.25890

Pirozzi, G., Tirino, V., Camerlingo, R., Franco, R., La Rocca, A., Liguori, E., Martucci, N., Paino, F., Normanno, N., and Rocco, G. (2011). Epithelial to mesenchymal transition by TGFbeta-1 induction increases stemness characteristics in primary non small cell lung cancer cell line. PLoS One *6*, e21548.

Prince, M.E., Sivanandan, R., Kaczorowski, A., Wolf, G.T., Kaplan, M.J., Dalerba, P., Weissman, I.L., Clarke, M.F., and Ailles, L.E. (2007). Identification of a subpopulation of cells with cancer stem cell properties in head and neck squamous cell carcinoma. Proc Natl Acad Sci USA *104*, 973–978.

Rasheed, Z.A., Kowalski, J., Smith, B.D., and Matsui, W. (2011). Concise review: Emerging concepts in clinical targeting of cancer stem cells. Stem Cells *29*, 883–887.

Rhim, A.D., Mirek, E.T., Aiello, N.M., Maitra, A., Bailey, J.M., McAllister, F., Reichert, M., Beatty, G.L., Rustgi, A.K., Vonderheide, R.H., *et al.* (2012). EMT and dissemination precede pancreatic tumor formation. Cell *148*, 349–361.

Ricci-Vitiani, L., Lombardi, D.G., Pilozzi, E., Biffoni, M., Todaro, M., Peschle, C., and De Maria, R. (2007). Identification and expansion of human colon-cancer-initiating cells. Nature *445*, 111–115.

Riccioni, R., Dupuis, M.L., Bernabei, M., Petrucci, E., Pasquini, L., Mariani, G., Cianfriglia, M., and Testa, U. (2010). The cancer stem cell selective inhibitor salinomycin is a p-glycoprotein inhibitor. Blood Cells Mol Dis *45*, 86–92.

Santisteban, M., Reiman, J.M., Asiedu, M.K., Behrens, M.D., Nassar, A., Kalli, K.R., Haluska, P., Ingle, J.N., Hartmann, L.C., Manjili, M.H., *et al.* (2009). Immune-induced epithelial to mesenchymal transition in vivo generates breast cancer stem cells. Cancer Res *69*, 2887–2895.

Schatton, T., Murphy, G.F., Frank, N.Y., Yamaura, K., Waaga-Gasser, A.M., Gasser, M., Zhan, Q., Jordan, S., Duncan, L.M., Weishaupt, C., *et al.* (2008). Identification of cells initiating human melanomas. Nature *451*, 345–349.

Scheel, C., Eaton, E.N., Li, S.H., Chaffer, C.L., Reinhardt, F., Kah, K.J., Bell, G., Guo, W., Rubin, J., Richardson, A.L., *et al.* (2011). Paracrine and autocrine signals induce and maintain mesenchymal and stem cell states in the breast. Cell *145*, 926–940.

Schepers, A.G., Snippert, H.J., Stange, D.E., van den Born, M., van Es, J.H., van de Wetering, M., and Clevers, H. (2012). Lineage tracing reveals Lgr5+ stem cell activity in mouse intestinal adenomas. Science *337*, 730–735.

Schuller, U., Heine, V.M., Mao, J., Kho, A.T., Dillon, A.K., Han, Y.G., Huillard, E., Sun, T., Ligon, A.H., Qian, Y., *et al.* (2008). Acquisition of granule neuron precursor identity is a critical determinant of progenitor cell competence to form Shh-induced medulloblastoma. Cancer Cell *14*, 123–134.

Singh, S.K., Hawkins, C., Clarke, I.D., Squire, J.A., Bayani, J., Hide, T., Henkelman, R.M., Cusimano, M.D., and Dirks, P.B. (2004). Identification of human brain tumour initiating cells. Nature *432*, 396–401.

Takebe, N., Harris, P.J., Warren, R.Q., and Ivy, S.P. (2011). Targeting cancer stem cells by inhibiting Wnt, Notch, and Hedgehog pathways. Nat Rev Clin Oncol *8*, 97–106.

Thiery, J.P., Acloque, H., Huang, R.Y., and Nieto, M.A. (2009). Epithelial-mesenchymal transitions in development and disease. Cell *139*, 871–890.

Trimboli, A.J., Fukino, K., de Bruin, A., Wei, G., Shen, L., Tanner, S.M., Creasap, N., Rosol, T.J., Robinson, M.L., Eng, C., *et al.* (2008). Direct evidence for epithelial-mesenchymal transitions in breast cancer. Cancer Res *68*, 937–945.

Visvader, J.E. (2011). Cells of origin in cancer. Nature *469*, 314–322.

Visvader, J.E., and Lindeman, G.J. (2008). Cancer stem cells in solid tumours: accumulating evidence and unresolved questions. Nat Rev Cancer *8*, 755–768.

Xie, G., Yao, Q., Liu, Y., Du, S., Liu, A., Guo, Z., Sun, A., Ruan, J., Chen, L., Ye, C., *et al.* (2012). IL-6-induced epithelial-mesenchymal transition promotes the generation of breast cancer stem-like cells analogous to mammosphere cultures. Int J Oncol *40*, 1171–1179.

Yang, Z.F., Ho, D.W., Ng, M.N., Lau, C.K., Yu, W.C., Ngai, P., Chu, P.W., Lam, C.T., Poon, R.T., and Fan, S.T. (2008a). Significance of CD90+ cancer stem cells in human liver cancer. Cancer Cell *13*, 153–166.

Yang, Z.J., Ellis, T., Markant, S.L., Read, T.A., Kessler, J.D., Bourboulas, M., Schuller, U., Machold, R., Fishell, G., Rowitch, D.H., *et al.* (2008b). Medulloblastoma can be initiated by deletion of Patched in lineage-restricted progenitors or stem cells. Cancer Cell *14*, 135–145.

Ye, J., Wu, D., Shen, J., Wu, P., Ni, C., Chen, J., Zhao, J., Zhang, T., Wang, X., and Huang, J. (2012). Enrichment of colorectal cancer stem cells through epithelial-mesenchymal transition via CDH1 knockdown. Mol Med Report *6*, 507–512.

Yu, J., Vodyanik, M.A., Smuga-Otto, K., Antosiewicz-Bourget, J., Frane, J.L., Tian, S., Nie, J., Jonsdottir, G.A., Ruotti, V., Stewart, R., *et al.* (2007). Induced pluripotent stem cell lines derived from human somatic cells. Science *318*, 1917–1920.

Yu, M., Smolen, G.A., Zhang, J., Wittner, B., Schott, B.J., Brachtel, E., Ramaswamy, S., Maheswaran, S., and Haber, D.A. (2009). A developmentally regulated inducer of EMT, LBX1, contributes to breast cancer progression. Genes Dev *23*, 1737–1742.

Zhao, Z., Zuber, J., Diaz-Flores, E., Lintault, L., Kogan, S.C., Shannon, K., and Lowe, S.W. (2010). p53 loss promotes acute myeloid leukemia by enabling aberrant self-renewal. Genes Dev *24*, 1389–1402.

Zhu, L., Gibson, P., Currle, D.S., Tong, Y., Richardson, R.J., Bayazitov, I.T., Poppleton, H., Zakharenko, S., Ellison, D.W., and Gilbertson, R.J. (2009). Prominin 1 marks intestinal stem cells that are susceptible to neoplastic transformation. Nature *457*, 603–607.

Zhu, L.F., Hu, Y., Yang, C.C., Xu, X.H., Ning, T.Y., Wang, Z.L., Ye, J.H., and Liu, L.K. (2012). Snail overexpression induces an epithelial to mesenchymal transition and cancer stem cell-like properties in SCC9 cells. Lab Invest *92*, 744–752.

Chapter 9

Models to Study Chronic Myeloid Leukemia Cancer Stem Cells

Sheela A. Abraham[1,2], Lisa Hopcroft[1], Ravi Bhatia[3], Steffen Koschmieder[4], Anthony D. Whetton[2], and Tessa L. Holyoake[1]
[1] *Paul O'Gorman Leukaemia Research Centre, Institute of Cancer Sciences, College of Medical, Veterinary and Life Sciences, University of Glasgow, Scotland, UK*
[2] *Stem Cell and Leukaemia Proteomics Laboratory, School of Cancer and Enabling Sciences, Manchester Academic Health Science Centre, University of Manchester, Manchester, UK*
[3] *Division of Hematopoietic Stem Cell and Leukemia Research, City of Hope National Medical Center, Duarte, CA, USA*
[4] *Department of Medicine Oncology, Hematology, and Stem Cell Transplantation, University Medical Center of Aachen, Aachen, Germany*

Cancer

Homeostasis is facilitated by normal cells that functionally support tissue architecture thereby contributing to a healthy, functional organism. Growth factor receptors containing tyrosine kinase domains have a significant role in development and growth; their deregulation and that of their downstream signalling proteins is thought to lead to the development of malignancy. Concepts to explain carcinogenesis were initially put forward by Nordling (1953) who described cancer as resulting from an accumulation of DNA mutations. This theory was built upon and expanded by others (Ashley, 1969) (Nowell, 1976) who went further to describe that inheritable mutations, with or without environmental factors, create premalignant clones (Ghiaur et al., 2012). These premalignant clones could potentially be subject to specific genetic hits that would lead to increased survival and production of progeny, thereby conferring an increased 'fitness' or a selective advantage over normal healthy cells. Steps toward carcinogenesis may in fact subscribe to Darwinian concepts describing evolution of biological species (Greaves and Maley, 2012; Merlo et al., 2006; Nowell, 1976).

In recent years our understanding of cancer has deepened so that concepts based on cells progressing through multiple steps toward carcinogenesis have incorporated new knowledge. Capacities or general functions that a cell requires in order to acquire and maintain a malignant state have been eloquently summarised by Hanahan and Weinberg (2000, 2011) who describe characteristics or 'hallmarks' that include (1) sustaining proliferative signalling, (2) evading growth suppressors, (3) resisting cell death, (4) enabling replicative immortality, (5) inducing angiogenesis, (6) activating invasion and metastasis, (7) reprogramming of energy metabolism, and (8) evading immune destruction. Interestingly, the authors emphasize that genomic stability and inflammation foster or enable the above functions to perpetuate disease (Hanahan and Weinberg, 2011). Recent research demonstrates that cancer should be regarded as far more than single cells that have lost all restraints on

Cancer Stem Cells, First Edition. Edited by Vinagolu K. Rajasekhar.

growth, but as a complex framework or assembly comprised of multiple cells types (that include normal cells) interacting with each other to provide a milieu conducive to creating neoplastic disease.

Cancer Stem Cell (CSC) Model

Tumors as previously regarded, are not simply comprised of homogenous malignant cells but may be more realistically viewed as a complex organ comprised of multiple cells types. Importantly tumors exist within a specialised 'self'-induced environment that not just houses the malignancy but maintains an interactive relationship that contributes to tumor growth and spread. The study of hematological malignancies has enabled researchers to gain a perspective of how this observed tumor heterogeneity might arise due to a specialized population termed CSCs (Makino, 1959; Bonnet and Dick, 1997). This model holds that some cancers maintain a hierarchical organization, with the CSC at the apex, similar to a normal stem cell in the hemopoietic hierarchy (Dick, 2008). CSCs, like normal stem cells, possess the ability to modulate and balance differentiation and self-renewal according to environmental stimuli and genetic constraints. Growth of tumors are thereby fueled by CSCs, which represent the sole cell type that can recapitulate the entire tumor, comprised of rapidly proliferating cells, post-mitotic differentiated cells, and the CSCs themselves. CSCs may, it should be noted, arise from a differentiated committed progenitor that acquires self-renewal capacity or from a normal stem cell (Nguyen et al., 2012).

The CSC model can potentially explain poorly understood clinical observations. CSCs have been shown to be resistant to radiation and chemical insult. By possessing ATP binding cassette (ABC) multi-drug transporter pumps, anti-apoptotic proteins, and altered DNA damage repair mechanisms (Clevers, 2011), specific cancers can become resistant to various types of treatment. If a patient obtains an excellent response to either radiation or chemotherapy, the patient is categorized as being in remission, and not 'cured.' This is due to sound epidemiological evidence supporting that reoccurrence and growth of cancerous cells are possibilities with a specific recurrence rate and likelihood depending on the cancer type and the method of treatment. This could be explained by CSCs that can evade therapy, stay dormant due to their ability to exit the cell cycle and become quiescent, and therefore survive for years until appropriate conditions present themselves in order to re-establish the tumor.

Examples of cancers thought to follow the CSC model include malignant germ cell cancers, types of leukemia, breast cancer, brain cancer, and colon cancer (Kelly et al., 2007; Shackleton et al., 2009). The CSC model has attracted much debate in the literature because techniques have vastly improved over the time CSCs were first identified and studied. For example, a study assessing tumor initiation potential using xenograft models compared the transplantation of human primary melanoma cells into NOD/SCID mice to NOD/SCID interleukin-2 receptor gamma chain null mice, a more immune-compromised model (Quintana et al., 2008). The authors observed a 25,000-fold increase in tumor-initiating efficiency when using the more immune-compromised mice. This landmark study revealed general limitations to techniques commonly employed to identify and quantify CSCs and demonstrated that improvement of techniques will inevitably expand our understanding and challenge existing assumptions. This study illustrated that the xeno-transplantation model using NOD/SCID interleukin-2 receptor gamma chain null mice may be an excellent tool to assess the metastatic potential of human cells injected into mice, despite the different species-specific microenvironment, defined by the mouse physiology. However, this model may not be sufficiently robust to indicate tumor-initiating cell frequency for extrapolation to human disease. Normally the malignant human tumor-initiating cell would not only be surrounded by an environment that potentially played a part in the creation of the pre-malignant clone, it would also have the equivalent species-specific cytokine milieu provided by the host and most importantly a fully functional immune system normally designed to rid the body of disease.

Chronic Myeloid Leukemia

Chronic myeloid leukemia (CML) is a clonal disorder of hemopoietic stem cells and provides an elegant model to study CSCs. CML progresses through three phases: chronic, accelerated, and the blast phase, which evolves over 3 to 5 years. The disease is initiated from a genetic reciprocal translocation between the long arms of chromosomes 9 and 22 [t(9;22)(q34;q11)] leading to the fusing of the c-abl (Abelson murine leukemia viral oncogene homolog 1) oncogene 1 (ABL1) and the breakpoint cluster region (BCR) gene, generating a constitutively active protein tyrosine kinase (PTK) (Ren, 2005; Savona and Talpaz, 2008). The consequences of BCR-ABL constitutive activity include increased cell proliferation, resistance to apoptosis, deregulation of cellular adhesion and migration functions, and genomic stability resulting in a transformed neoplastic phenotype. Detection of the resultant mutation forms the basis of CML diagnosis, and both in vitro and in vivo studies confirm that the introduction of BCR-ABL to a normal hemopoietic stem cell is sufficient to establish CML (Daley et al., 1990).

Imatinib mesylate or IM (Glivec, STI571 or CGP 57148) was modeled on a compound selected from a PTK inhibitor screen. Its mechanism of action is based on competitively binding the ATP binding site of the kinase, which subsequently inhibits the activation of all pathways normally modulated by BCR-ABL through phosphorylation events. In addition to targeting BCR-ABL, IM is active against platelet-derived growth factor receptors A/B (PDGFR-A/B), stem cell factor (SCF) receptor c-kit PTK, discoidin domain receptors 1/2 (DDR-1/2), NADPH dehydrogenase quinone 2 (oxidoreductase NQO2), ABL1, and ABL-related-gene (Arg) (Druker et al., 2001).

With the introduction of IM in 1998, the treatment of CML has experienced tremendous strides forward, allowing a complete cytogenetic response in 82% of patients (Heaney and Holyoake, 2007) that translates to dramatically improved survival rates of 85% at 8 years. Because of this success, clinical trials have been undertaken in order to ascertain if patients with a complete molecular response (CMR) for at least 2 years on IM, could indefinitely terminate treatment, as current administration of IM is lifelong. The French STIM trial enrolled 100 patients and results indicated that 60% of patients relapsed within 4 months of treatment withdrawal, with Skokal scores a good indicator of relapse versus remission (Mahon et al., 2010). Relapse is thought to arise from a pool of resistant cells, and this most clinically significant malignant cell population is responsible for disease re-initiation. If CSCs indeed represent this resistant population, they should be the dominant population in minimal residual disease (Ghiaur et al., 2012) and it is critical to study them.

Models to Study CML CSCs

Previous work provides substantial evidence that BCR-ABL has oncogenic potential and promotes CML pathogenesis. Further elucidation of the molecular mechanisms underlying the development of CML and pathways propagating the disease will allow advancement of the CML field. Methods to study CML within a cellular context include human cell lines such as K562 (BCR-ABL expressing immortalized blast crisis cell line), mouse cells lines such as Ba/F3 and 32D transduced with BCR-ABL (immortalized mouse bone marrow-derived pro-B-cell line dependent on IL-3), and human primary $CD34^+$ cells derived from patients with CML. A major advantage of using cell lines is the ability to obtain significant numbers of clonal cells for biochemical, genetic, and biological analyses that require large cell numbers for specific applications such as mass spectrometry or ChIP-Chip techniques. Cell lines are also excellent for focusing on specific pathways and critical components involved in aspects of CML biology, but these in vitro cell culture models do not provide an interactive environment. In vitro cell cultures generally do not include other cell types found in the bone marrow milieu, nor do they replicate physiological concentrations of cytokines or oxygen that are thought to play vital roles in pathogenesis and disease maintenance. Other limitations include the fact that particular cell lines may skew final interpretations due to missing components. For example, the blast crisis-derived K562 cell line lacks the important transcription factor p53. Furthermore, cell lines may not be ideal models to study the hierarchy of hemopoietic CSCs, because they are theoretically clonal.

In vivo techniques, such as xenograft models, involve the transplantation of peripheral blood or bone marrow-derived mononuclear or purified $CD34^+$ cells from human CML patients into immuno-deficient mice such as NOD/SCID interleukin-2 receptor gamma chain null mice that best support engraftment of human haemopoietic cells (Eisterer et al., 2005; Shultz et al., 2005). These xenograft models offer the opportunity to assess the ability of human CML cells to engraft and to produce leukemic progeny with enhanced myelopoiesis, excessive production of erythroid and megakaryopoietic cells, and the generation of primitive ($CD34^+$) leukemic cells displaying an autocrine IL-3 and G-CSF phenotype, all characteristics of primary CML cells (Eisterer et al., 2005). These mice however do not develop lethal leukemia within their average 7-month life span. Low levels of engraftment of chronic phase CML cells combined with the lack of mice succumbing to leukemia in cancer xenograft transplant models limit its use for in vivo drug treatment studies (Peng and Li, 2010).

Efforts to overexpress BCR-ABL in mice include several alternative models. Overexpression of BCR-ABL within mouse models was initiated with retroviral-based and conventional transgenic mouse models (Daley et al., 1990; Elefanty et al., 1990; Kellliher et al., 1990). The former methodology involves the pretreatment of donor mice with 5-fluorouracil, followed by transduction of their bone marrow stem cells with retroviral constructs expressing BCR-ABL, and subsequent transplantation of these transduced bone marrow stem cells into irradiated recipient mice (Li et al., 1999). Retroviral transduction and transplantation induces a fatal CML−like disease with a dramatic increase of neutrophils and mature myeloid cells in peripheral blood, bone marrow, spleen, and liver, with mice dying of their disease around 21 days after transplantation. Due to the short latency of disease onset, this model is ideal for testing drugs.

An alternative transgenic SCLtTa/BCR-ABL mouse model utilizes an inducible transgenic system that places the tetracycline-controlled transactivator protein (tTA) under the control of the murine stem cell leukemia (SCL) gene 3′ enhancer element (Huettner et al., 2000; Huettner et al., 2003; Koschmieder et al., 2005). Reversible induction of BCR-ABL expression is achievable with this mouse model system, controlled by the presence or absence of tetracycline. Selective expression of BCR-ABL in primitive hemopoietic cells is achievable using the SCL enhancer because SCL-expressing progenitor cells are mostly multipotent. Expression of this vector in mice induces a CML-like myeloproliferative neoplasm characterized by neutrophilia and leukocytosis as well as splenomegaly, and mice become moribund within 29 to122 days (Koschmieder et al., 2005).

With the advent of multiple models to study CML CSCs, questions arise as to which model would most closely resemble human CML disease. Using human CML primary cells may appear a logical model, yet the pre-selection of $CD34^+$ cells that are mostly collected prior to IM treatment can be fraught with complications. Human CML primary cells require extensive processing that involves selection (from either blood or bone marrow) followed by cryopreservation. Once thawed, cells are cultured in media artificially supplemented with cytokine cocktails to facilitate growth, and may be maintained over the course of days to months depending on the specific in vitro techniques being used. The overexpression of BCR-ABL in murine cells using various vector constructs, offers an alternative whereby these in vivo systems provide a complex environment that includes multiple cell types, expressing various receptors, naturally derived cytokines, and limiting oxygen conditions. The mouse model, however, has a 'synthetic' component in that it is a host system that normally does not encounter CML as a disease and the forced expression of human BCR-ABL is, in this case, solely responsible for inducing the disease.

It is important that the most appropriate model be chosen and to the best of our knowledge, a direct comparison of each CML model has not been performed because parameters for their comparison are difficult to develop. With a view to investigating the elimination of residual disease in CML, the CSC hypothesis suggests that the best model to use would closely resemble primary human CML stem cells. We have considered transcriptional data to assess how closely primitive ($Lin^{-}Sca1^{+}c$-kit^{+} [LSK]) double transgenic (DTG) mouse cells (Koschmieder et al., 2005; Schemionek et al., 2010),−theoretically one of the best available murine models for human CML−correspond to primitive ($CD34^{+}HST^{lo}Py^{lo}$) primary human CML cells. This work was designed to reveal what extent a mouse model accurately reflects the functionality of human CML stem cells.

A Comparison of Human and Mouse Model Transcriptional Data to Study CML CSCs

In order to understand fundamental differences between mouse and human models of CML CSCs, we have compared transcriptional data derived from human normal and CML cells (Graham et al., 2007) with transcriptional data from mice +/−BCR-ABL. Importantly, these primitive cell populations were selected to contain putative CML CSC. The human cells were collected from patient or normal donor leukapheresis products, which were $CD34^{+}$ enriched and cryo-preserved. Cells were thawed, cultured overnight, and further selected on the basis of Hoechst and pyronin Y staining using fluorescence-activated cell sorting (FACS) (Graham et al., 2007). This gating strategy allows the selection of primitive, quiescent (G0) cells that exhibit low Hoechst and pyronin levels. This cell type was compared to LSK selected bone marrow cells (known to contain primitive stem cells) isolated from mice expressing the SCLtTA/BCR-ABL transgenes (DTG) or not (Koschmieder et al., 2005; Schemionek et al., 2010).

The series matrix file for the mouse gene expression data (Affymetrix Mouse Genome 430 2.0 GeneChips, >44,000 probe sets representing >34,000 different murine genes) were obtained from the Gene Expression Omnibus (series GSE18446). These data were normalized as described by Schemionek and colleagues (2010) and a log function is applied (base 2). The human gene expression data (Affymetrix Human Genome U133A GeneChips, >22,000 probe sets representing >14,500 different genes) were obtained directly from the raw Affymetrix files (CEL files), and normalized using robust microarray average (RMA) (with background correction) with a log function (base 2) being applied as well. Both datasets were analyzed using analysis method (LIMMA) to identify genes differentially expressed in the comparison of CML (or CML-like) versus non-CML cells. Following the methodology used by Schemionek and colleagues (2010), significantly differentially expressed genes were identified using a p-value threshold of 0.05 in each dataset.

Biomart webservices (via R 2.15.1, biomaRt 2.12.0) were used to identify human/murine orthologs in the Ensembl database, and to allow mapping of expression data and LIMMA results across these datasets. In total, 39,665 mappings were made between

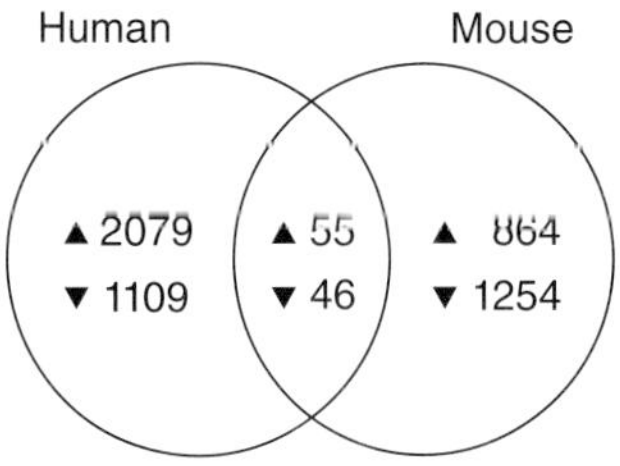

Figure 9.1. Differential expression overlap in human CML (*left*) and the mouse model of CML (*right*). Arrows indicate direction of differential expression. See text for explanation of differential expression analysis.

the 11621 mouse genes (21,455 probe sets) and 11,427 human genes (17,539 probe sets); representing adequate coverage of both genomes: ~61.94% murine genes and ~79.87% human genes.

When analyzing and comparing only significantly modulated genes between human and mouse from both data sets (Figure 9.1), of all genes detected, 7.6% (3,012 of 39,665 genes) were significantly upregulated in CML as compared to normal and 6.4% (2,543 of 39,665 genes) were significantly downregulated in CML as compared to normal within both human and mouse data sets. Surprisingly, of these genes only 1.9% were shared between the mouse and human, whether the gene was upregulated (55 genes indicated as up arrow) or downregulated (46 genes indicated as down arrow).

The lack of overlap at the individual gene level may not translate to the pathway level. A specific pathway could be affected by BCR-ABL in both species, with different genes resulting in the same eventual phenotypic change. To assess similarities at the pathway level, the two lists of differentially expressed genes (one for the human dataset and one for the mouse dataset) were uploaded to MetaCore. In both datasets, the human Ensembl ID was used to describe each gene to avoid differences in pathway coverage between human and mouse. These gene lists were analyzed using the Enrichment by Pathway Maps function in MetaCore. Given a list of n genes, this analysis generates a p-value for a GeneGo pathway map M, describing how likely a list of length n is to contain members of M by chance (using the hyper-geometric distribution). These p-values can be converted to q-values and a false discovery rate (FDR) applied.

Because we were interested in the significant differences or similarities in the human and mouse datasets, we extracted all pathways that were significantly (FDR < 0.05) enriched for in either dataset. To place these pathways in the broad context of cancer, we grouped these enriched pathways into the cancer hallmarks or enabling characteristics identified to promote carcinogenesis (Hanahan and Weinberg, 2011), and thereby planned to observe pertinent or relevant pathways (and genes) to CML pathogenesis. These results are shown in Figure 9.2 where pathways that were significantly represented (FDR < 0.05) are highlighted with asterisks (**).

It was clear that all hallmarks and enablers were represented by at least one pathway (and therefore multiple genes) in either the mouse or human transcriptional dataset. Interestingly, although 4 of the 10 hallmarks/enablers were significantly represented by *both* the mouse and human gene sets, specific hallmarks, such as activation and invasion and metastasis; avoiding immune destruction; enabling replicative immortality; inducing angiogenesis; sustaining proliferative signaling and tumor promoting inflammation, were only significantly represented in the mouse dataset. Our data suggest that one model may potentially be able to represent specific pathways more thoroughly than the other. For example, to study CD40 signaling, the mouse model may provide a better model because this pathway was significantly represented (10 of

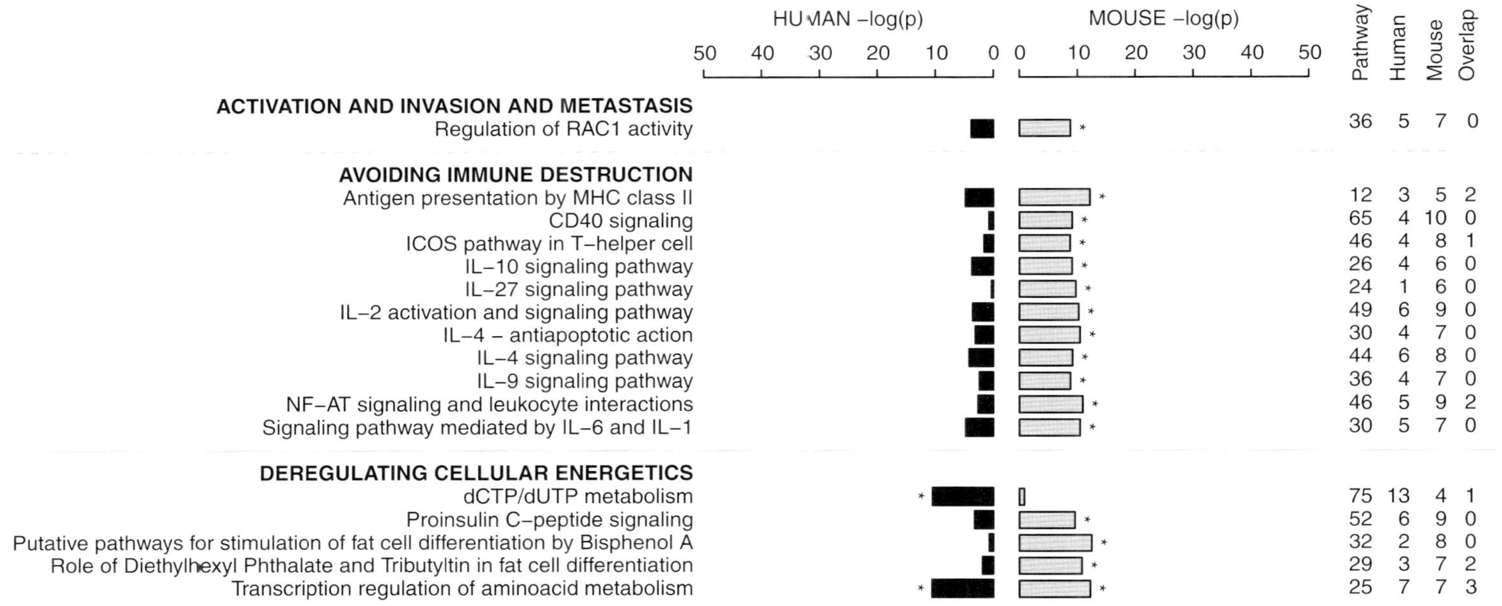

Figure 9.2. In-silico functional characterization of differentially expressed genes in human CML (*black bars, left*) and mouse model of CML (*gray bars, right*) in terms of GeneGo pathway maps. Bars represent the log(p-value) for each GeneGo pathway map. Pathway enrichment exceeding FDR thresholds are indicated by ** (FDR < 0.01), * (FDR < 0.05) and ~ (FDR < 0.10). Protein counts pertaining to each pathway map are given to the right of the bar plot and describe the number of proteins in each pathway (Pathway); the number of pathway proteins differentially expressed in the human dataset (Human); the number of proteins differentially expressed in the mouse dataset ("Mouse"); and the number of proteins differentially expressed in *both* datasets (Overlap), respectively. Pathway maps are grouped according to Weinberg's cancer hallmarks (Hanahan and Weinberg, 2011). These cancer hallmarks are emboldened on the left of the figure, above the pathway map names.

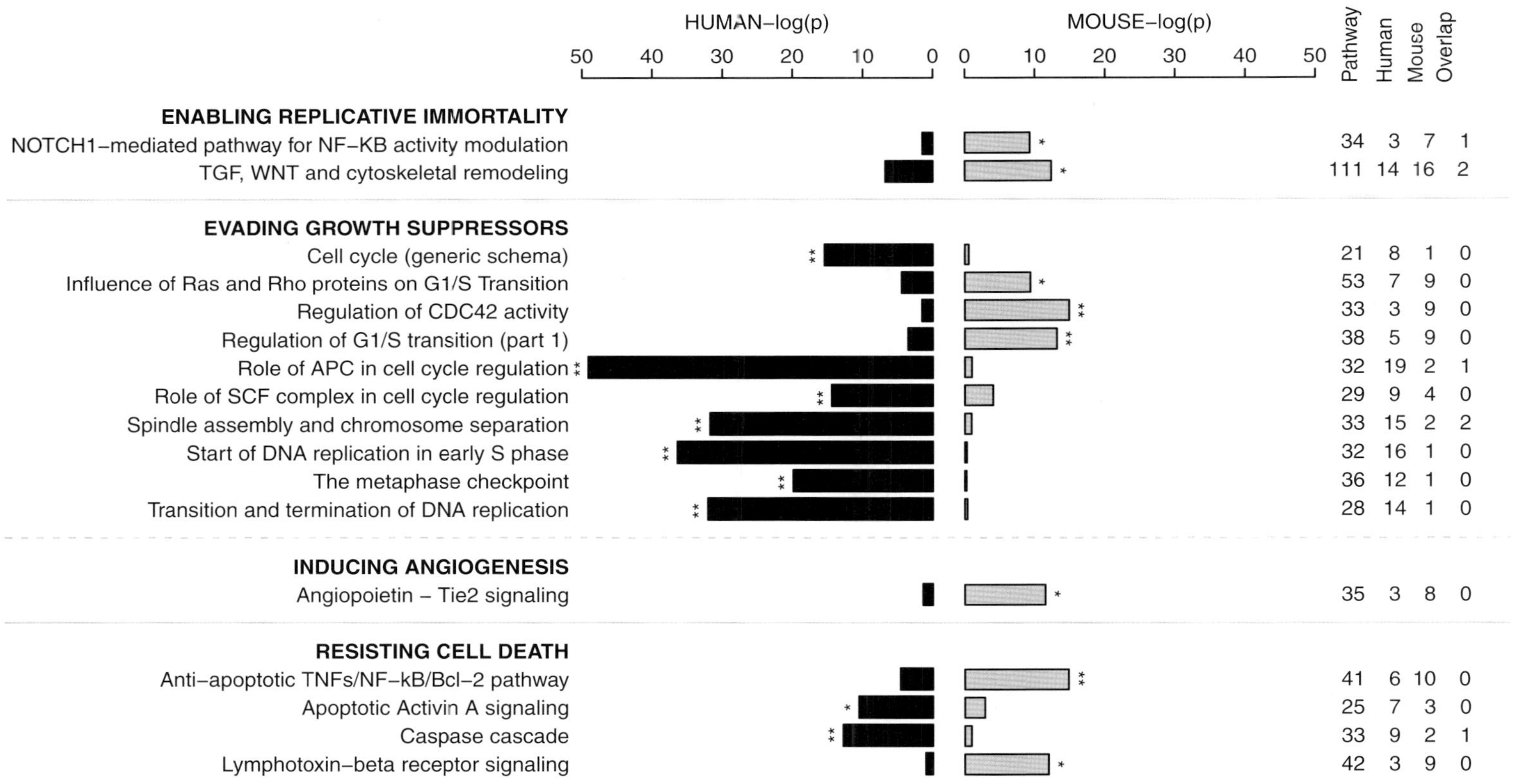

Figure 9.2 (*Continued*).

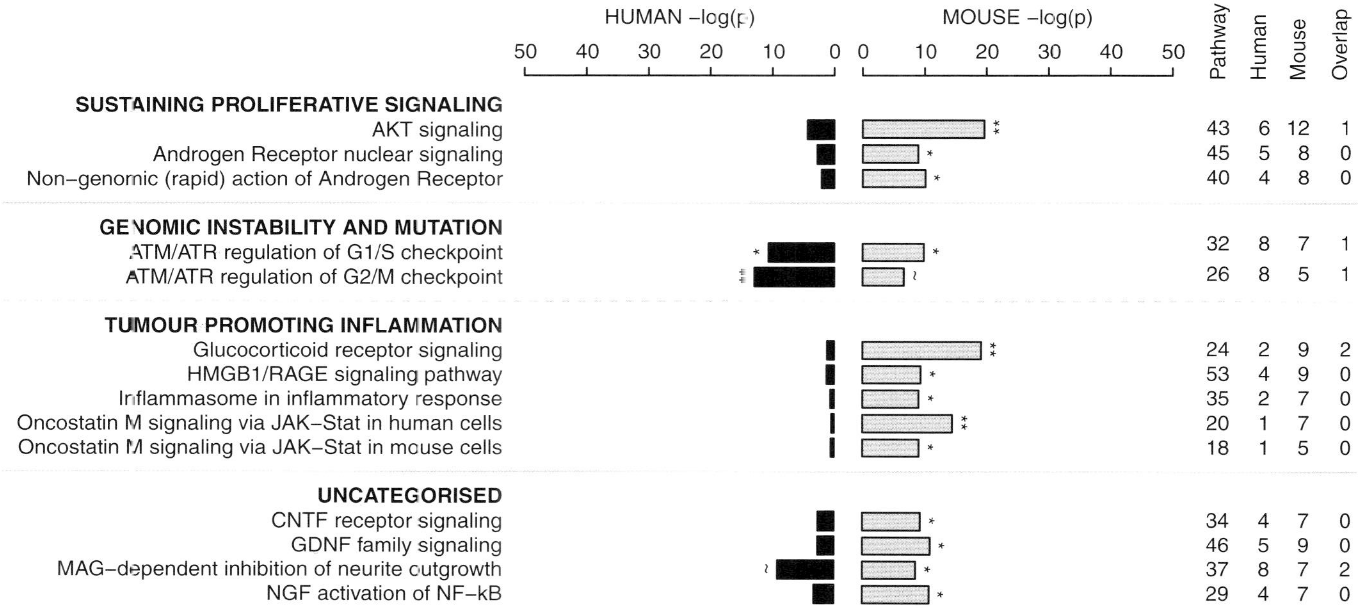

Figure 9.2 *(Continued)*.

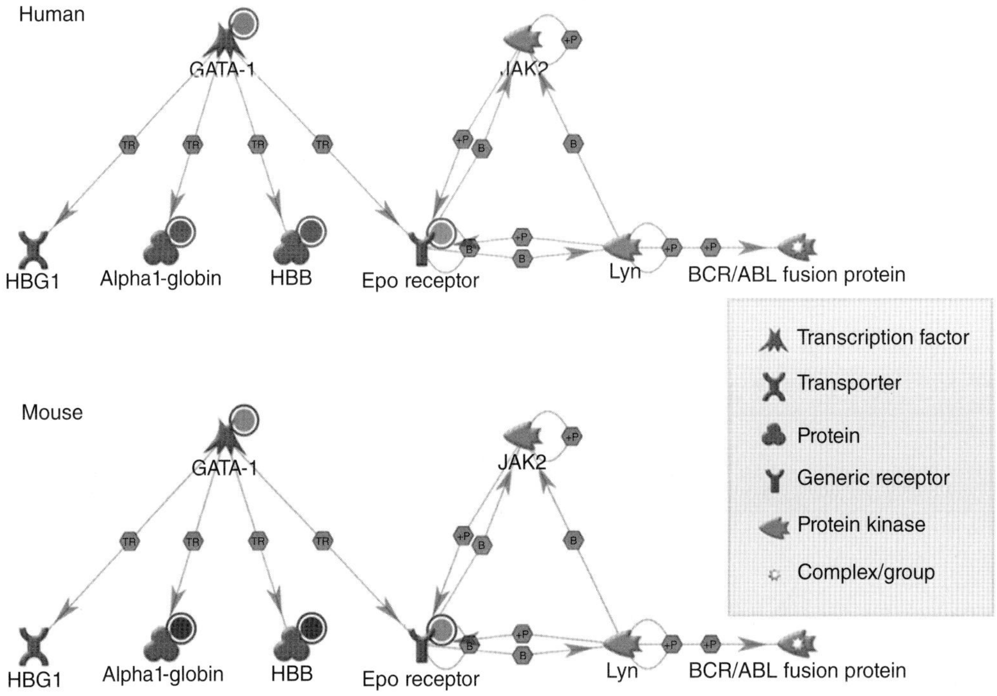

Figure 9.3. Differential expression of members of the Epo induced PI3K/AKT Ca(2+) influx pathway: Alpha1-globin (HBA1), HBB, EpoR, and GATA-1 are consistently upregulated in the human dataset and consistently downregulated in the mouse dataset. Upregulation is indicated with red circles, downregulation is indicated with blue circles, the extent of deregulation is indicated by color intensity. (See MetaCore legend for symbol meanings.) For color detail, please see color plate section.

the 65 genes in the pathway were deregulated in the mouse). On the other hand, using human primary samples may give an advantage when interrogating dCTP/dUTP metabolism (where 13 of the 75 pathway genes were deregulated, as opposed to 4 in the mouse dataset). Transcriptional regulation of amino acid metabolism was observed to be well represented by both the human and mouse models (7 of 25 genes for each). Our study therefore appears to demonstrate that existing models may offer complementary information, depending on the specific protein and pathway being investigated.

To further interrogate the two data sets, with a view to identifying what is different in the mouse model as compared to the human model, we identified those genes that were significantly regulated in the *opposite* direction in both species (i.e., genes that were significantly upregulated in the mouse but significantly downregulated in the human, or vice versa). We identify 84 genes matching such a profile. In considering interactions and functionality characterizing this set of genes, a small cluster of genes was identified. As indicated in Figure 9.3, GATA-1, haemoglobin α (HBA), haemoglobin β (HBB), and the epo receptor (EpoR) were all significantly upregulated in the human dataset. GATA-1 is a known transcription factor for HBA, HBB, and EpoR, and in these human cells may therefore be responsible for the upregulation of these three targets genes. However, in the mouse dataset GATA-1 appeared to be *down*regulated, as were the same three target genes. Epo-independent proliferation and differentiation of erythroid progenitors in vitro is a property common to all myeloproliferative disorders (Ghaffari et al., 1999) and

CML patients exhibit an increased number of erythroid progenitors in their peripheral blood. It is interesting to note that the human transcriptional data were revealing an increased modulation of the critical EpoR that was not observed in the mouse data. The differences may be attributable to the difference between peripheral blood- and bone marrow-derived hematopoietic stem cells, which is also in line with differences found between BM and spleen in regard to the erythroid potential of the BCR-ABL positive hematopoietic stem cells. Alternatively, the inducible SCLtTA/BCR-ABL transgene may result in skewing primitive progenitors toward myeloid and lymphoid rather than erythroid phenotype (Schemionek et al., 2012).

Discussion

Our comparative analyses using CML mouse and CML human transcriptional data sets has demonstrated that these BCR-ABL containing systems have significantly modulated genes. Analysis shows that these genes can modulate programs associated with the CML cancer phenotype. So, despite having very few genes shared between the mouse and human, the majority of these genes are contributing to critical pathways known to support leukemic/cancer growth. Much to our surprise, there is no evidence of a common BCR-ABL-like signature. One of our hypotheses was that all critical proteins known to play a role in CML pathogenesis would be readily identifiable, being the most well represented, and would link clearly to BCR-ABL in the models described. Not only have we analyzed (and presented) the significantly modulated genes between the two datasets, we have individually analyzed each data set and are unable to see a transcriptional BCR-ABL signature that can be identified using pathway analysis.

Our data additionally identifies that although pathways important to cancer or oncogenesis-enabling processes are well represented by our significantly regulated genes in the mouse versus human data sets, each model may be more or less enriched for each pathway. Alternatively, it is also possible that current knowledge behind each pathway component has yet to be determined and that lack of significance in our study is based on the lack of literature and representation and that further investigation is warranted.

When determining significantly regulated genes that appeared to be opposed to each other in the mouse versus human data sets, we identify 84 genes. Although we have attempted to compare the most functionally relevant cell populations available (i.e., primitive progenitor populations that will contain CML CSCs), differences in the cell types compared may play a significant role in the transcriptional differences observed. The human transcriptional data were based on $CD34^+$ cells that were, isolated, frozen and thawed, followed by culturing and primitive populations were then further purified based on quiescence. The mouse transcriptional data were based on freshly isolated cells that were $Lin^-Sca1^+c\text{-}kit^+$, a primitive population that will inherently include quiescent and nonquiescent cells. Fundamentally, species cellular context may also play a role in genes that are observed to be regulated in opposite directions.

Results from our comparison suggest that there are transcriptional differences between mouse and human CML models at the level of genes. These differences may in part be of a lesser overall significance, due to the fact that key pathways involved in regulating normal growth mechanisms are being deregulated by the one protein that has been shown central to CML: BCR-ABL. The end result of BCR-ABL expression is also the same in both models, neoplastic transformation of a hemopoietic stem cell, followed by granulocyte compartment expansion, followed by each organism succumbing to disease.

Concluding Prospective

Cancer stem cells may represent a resistant population in minimal residual disease, and it is critical to study them. We have attempted to analyze two important models in CML with respect to their transcriptional profiles in order to gain insight into the similarities and differences between the two systems. By understanding fundamental differences in the multiple models available to researchers, deeper knowledge can be obtained to true differences between healthy and malignant processes.

References

Ashley, D.J. (1969). The two "hit" and multiple "hit" theories of carcinogenesis. British journal of cancer *23*, 313–328.

Bonnet, D., and Dick, J.E. (1997). Human acute myeloid leukemia is organized as a hierarchy that originates from a primitive hematopoietic cell. Nat Med *3*, 730–737.

Clevers, H. (2011). The cancer stem cell: premises, promises and challenges. Nat Med *17*, 313–319.

Daley, G.Q., Van Etten, R.A., and Baltimore, D. (1990). Induction of chronic myelogenous leukemia in mice by the P210bcr/abl gene of the Philadelphia chromosome. Science *247*, 824–830.

Dick, J.E. (2008). Stem cell concepts renew cancer research. Blood *112*, 4793–4807.

Druker, B.J., Sawyers, C.L., Kantarjian, H., Resta, D.J., Reese, S.F., Ford, J.M., Capdeville, R., and Talpaz, M. (2001). Activity of a specific inhibitor of the BCR-ABL tyrosine kinase in the blast crisis of chronic myeloid leukemia and acute lymphoblastic leukemia with the Philadelphia chromosome. N Engl J Med *344*, 1038–1042.

Eisterer, W., Jiang, X., Christ, O., Glimm, H., Lee, K.H., Pang, E., Lambie, K., Shaw, G., Holyoake, T.L., Petzer, A.L., *et al.* (2005). Different subsets of primary chronic myeloid leukemia stem cells engraft immunodeficient mice and produce a model of the human disease. Leukemia *19*, 435–441.

Elefanty, A.G., Hariharan, I.K., and Cory, S. (1990). bcr-abl, the hallmark of chronic myeloid leukaemia in man, induces multiple haemopoietic neoplasms in mice. The EMBO journal *9*, 1069–1078.

Ghaffari, S., Wu, H., Gerlach, M., Han, Y., Lodish, H.F., and Daley, G.Q. (1999). BCR-ABL and v-SRC tyrosine kinase oncoproteins support normal erythroid development in erythropoietin receptor-deficient progenitor cells. Proc Natl Acad Sci USA *96*, 13186–13190.

Ghiaur, G., Gerber, J., and Jones, R.J. (2012). Concise review: Cancer stem cells and minimal residual disease. Stem Cells *30*, 89–93.

Graham, S.M., Vass, J.K., Holyoake, T.L., and Graham, G.J. (2007). Transcriptional analysis of quiescent and proliferating $CD34^+$ human hemopoietic cells from normal and chronic myeloid leukemia sources. Stem Cells *25*, 3111–3120.

Greaves, M., and Maley, C.C. (2012). Clonal evolution in cancer. Nature *481*, 306–313.

Hanahan, D., and Weinberg, R.A. (2000). The hallmarks of cancer. Cell *100*, 57–70.

Hanahan, D., and Weinberg, R.A. (2011). Hallmarks of cancer: the next generation. Cell *144*, 646–674.

Heaney, N.B., and Holyoake, T.L. (2007). Therapeutic targets in chronic myeloid leukaemia. Hematol Oncol *25*, 66–75.

Huettner, C.S., Zhang, P., Van Etten, R.A., and Tenen, D.G. (2000). Reversibility of acute B-cell leukaemia induced by BCR-ABL1. Nat Genet *24*, 57–60.

Huettner, C.S., Koschmieder, S., Iwasaki, H., Iwasaki-Arai, J., Radomska, H.S., Akashi, K., and Tenen, D.G. (2003). Inducible expression of BCR/ABL using human CD34 regulatory elements results in a megakaryocytic myeloproliferative syndrome. Blood *102*, 3363–3370.

Kelliher, M.A., McLaughlin, J., Witte, O.N., and Rosenberg, N. (1990). Induction of a chronic myelogenous leukemia-like syndrome in mice with v-abl and BCR/ABL. Proc Natl Acad Sci USA *87*, 6649–6653.

Kelly, P.N., Dakic, A., Adams, J.M., Nutt, S.L., and Strasser, A. (2007). Tumor growth need not be driven by rare cancer stem cells. Science *317*, 337.

Koschmieder, S., Gottgens, B., Zhang, P., Iwasaki-Arai, J., Akashi, K., Kutok, J.L., Dayaram, T., Geary, K., Green, A.R., Tenen, D.G., *et al.* (2005). Inducible chronic phase of myeloid leukemia with expansion of hematopoietic stem cells in a transgenic model of BCR-ABL leukemogenesis. Blood *105*, 324–334.

Li, S., Ilaria, R.L., Jr., Million, R.P., Daley, G.Q., and Van Etten, R.A. (1999). The P190, P210, and P230 forms of the BCR/ABL oncogene induce a similar chronic myeloid leukemia-like syndrome in mice but have different lymphoid leukemogenic activity. J Exp Med *189*, 1399–1412.

Mahon, F.X., Rea, D., Guilhot, J., Guilhot, F., Huguet, F., Nicolini, F., Legros, L., Charbonnier, A., Guerci, A., Varet, B., *et al.* (2010). Discontinuation of imatinib in patients with chronic myeloid leukaemia who have maintained complete molecular remission for at least 2 years: the prospective, multicentre Stop Imatinib (STIM) trial. The lancet oncology *11*, 1029–1035.

Makino, S. (1959). The role of tumor stem-cells in regrowth of the tumor following drastic applications. Acta – Unio Internationalis Contra Cancrum *15(Suppl 1)*, 196–198.

Merlo, L.M., Pepper, J.W., Reid, B.J., and Maley, C.C. (2006). Cancer as an evolutionary and ecological process. Nat Rev Cancer *6*, 924–935.

Nguyen, L.V., Vanner, R., Dirks, P., and Eaves, C.J. (2012). Cancer stem cells: an evolving concept. Nat Rev Cancer *12*, 133–143.

Nordling, C.O. (1953). A new theory on cancer-inducing mechanism. British journal of cancer *7*, 68–72.

Nowell, P.C. (1976). The clonal evolution of tumor cell populations. Science *194*, 23–28.

Peng, C., and Li, S. (2010). CML mouse model in translational research. Methods Mol Biol *602*, 253–266.

Quintana, E., Shackleton, M., Sabel, M.S., Fullen, D.R., Johnson, T.M., and Morrison, S.J. (2008). Efficient tumour formation by single human melanoma cells. Nature *456*, 593–598.

Ren, R. (2005). Mechanisms of BCR-ABL in the pathogenesis of chronic myelogenous leukaemia. Nat Rev Cancer *5*, 172–183.

Savona, M., and Talpaz, M. (2008). Getting to the stem of chronic myeloid leukaemia. Nat Rev Cancer *8*, 341–350.

Schemionek, M., Elling, C., Steidl, U., Baumer, N., Hamilton, A., Spieker, T., Gothert, J.R., Stehling, M., Wagers, A., Huettner, C.S., *et al.* (2010). BCR-ABL enhances differentiation of long-term repopulating hematopoietic stem cells. Blood *115*, 3185–3195.

Schemionek, M., Spieker, T., Kerstiens, L., Elling, C., Essers, M., Trumpp, A., Berdel, W.E., Muller-Tidow, C., and Koschmieder, S. (2012). Leukemic spleen cells are more potent than bone marrow-derived cells in a transgenic mouse model of CML. Leukemia *26*, 1030–1037.

Shackleton, M., Quintana, E., Fearon, E.R., and Morrison, S.J. (2009). Heterogeneity in cancer: cancer stem cells versus clonal evolution. Cell *138*, 822–829.

Shultz, L.D., Lyons, B.L., Burzenski, L.M., Gott, B., Chen, X., Chaleff, S., Kotb, M., Gillies, S.D., King, M., Mangada, J., *et al.* (2005). Human lymphoid and myeloid cell development in NOD/LtSz-scid IL2R gamma null mice engrafted with mobilized human hemopoietic stem cells. J Immunol *174*, 6477–6489.

Chapter 10

Cancer Stem Cells in Melanoma: Biomarkers and Mathematical Models

Stefano Zapperi[1] and Caterina A.M. La Porta[2]
[1] *CNR-IENI, Milano, Italy and ISI Foundation, Torino, Italy*
[2] *Laboratory of Molecular Oncology, Department of Bioscience, University of Milan, Italy*

Introduction

Melanomas represent about 5% of skin tumors and arise within any anatomic territory occupied by melanocytes. Cutaneous melanoma develops from epidermal melanocytes of the skin, representing the most common site of origination, although noncutaneous melanocytes may occasionally undergo malignant transformation. Melanocytes are neural crest-derived cells whose maturation seems to occur through well-defined stages. Cells at each maturation step may transform to either a nevus or a malignant melanoma. Nevi may further progress to melanoma or can regress by differentiating to cell types that have been histopathologically defined as (nonmalignant) Schwann cells (Clark et al., 1984). Clinical morphologies divided traditionally the cutaneous disease into several subgroups such as superficial spreading melanoma, nodular melanoma, acral lentiginous melanoma, and lentigo maligna melanoma. According to the current general accepted view of melanoma progression, melanocytes give rise to dysplastic nevi, which upon further evolution, grow radially. In the radial growth phase (RGP), melanocytes are confined to the epidermis, with occasional minor interdigitation into the dermis in a nontumorigenic neoplastic process. The vertical growth phase (VGP) is characterized by melanocyte proliferation in the dermis and invasion (Bishop et al., 2000) and it is tumorigenic. Conversion from RGP to VGP is considered the point at which the cells have acquired metastatic potential (Koh, 1991); however, this may be an oversimplification. Local invasion is necessary but not sufficient to enable metastasis. In fact, the development of a metastasis is dependent on the interplay between host factors and intrinsic characteristics of malignant tumor cells (Fidler, 1978). Neoplasms, which are predominately heterogeneous, contain a variety of cellular subpopulations with differing metastatic potential. When the tumor size is larger than 1–2 mm, tumor cells can not receive all nutrients by diffusion so that further growth depends on the development of an adequate blood supply through angiogenesis (Folkman, 1986; Folkman, 2007).

From the histological point of view, nevi are considered to be precursors of melanoma. Clinical and histologic criteria to distinguish disease grades are, however, ambiguously defined (Piepkorn et al., 1994; Helm and Marks, 1997). The prognosis in radial growth phase is excellent irrespective

Cancer Stem Cells, First Edition. Edited by Vinagolu K. Rajasekhar.

of thickness or other variables. Curable radial growth phase melanomas can be recognized by surveillance in patients identified by screening for risk markers that include dysplastic nevi, common nevi, freckles, and other indicators of chronic or acute sun exposure or sun sensitivity (Farmer et al., 1996). The prognosis in vertical growth phase depends on attributes of the tumor and of the host. The tumor mitotic rate, the presence of host tumor-infiltrating lymphocytes (TIL) within the vertical growth phase, and the tumor thickness are the most powerful predictors of survival. Cutaneous melanoma is among the most aggressive types of human cancer and, if untreated, virtually every melanoma has the potential to metastasize. Fortunately, the majority of cases are cured due to early detection and surgical excision (Koh, 1991). However, whenever metastases are developed, practically no therapy can cure them. In a recent meeting on melanoma at Queensland (Australia) (Zalaudek et al., 2011), researchers proposed a new idea for the development of human melanoma. Recent data have shown that both environmental and genetic factors are involved in the development of this disease. Nevi appear to be precursors and risk markers for melanoma. Traditionally nevi were considered to be genetically determined and also influenced by the environment. Many people have the propensity to develop many nevi, whereas others have little propensity (Zalaudek et al., 2011). This leads to the "divergent pathway hypothesis" according to which people with low propensity to develop nevi, after repeated and cumulative sun exposure, have an increased probability to develop melanoma. By contrast, in people with high nevus propensity, only a modest amount of sun exposure is required to develop melanoma, and this exposure seems to be more harmful during the early time of the life. Recent findings in genome-wide association studies of melanoma integrated with previous research from epidemiological and molecular studies, seems to support this hypothesis (Whiteman et al., 2011).

Biomarkers

To improve melanoma early detection and start a more efficient treatment, a crucial question is whether the identification of specific biomarkers could predict the clinical behavior of melanoma and thus drive the therapy. There are many recent reviews summarizing recent biomarkers described in melanoma including ingNRAS, KIT, PTEN, and NADD9 (Hensin et al., 2012). Herein, we focus on BRAF because of the recent approval by U.S. Food and Drug Administration of a small molecule BRAF inhibitor, vemarafenib.

BRAF and Melanoma

BRAF is a member of the Raf family of serine threonine kinases (ARAF, BRAF, and CRAF), which are part of the Ras/MEK/ERK mitogen-activated protein kinase (MAPK) signal transduction cascade. The MAPK pathway is a key mediator of growth signaling that links cell surface growth factor receptors to the increased transcription of genes required for cell cycle entry (Figure 10.1). More than 80% of mutated BRAF result in the substitution of valine to glutammic acid (BRAF V600E mutation) (Davies et al., 2002).

Acquisition of the V600E mutation destabilizes the inactive conformation of the BRAF kinase, shifting the equilibrium to the active state (Wan et al., 2004). In addition to melanoma, BRAF mutations are also common in many other cancers including papillary thyroid carcinoma, ovarian carcinoma, and colorectal carcinoma (Welbrock et al., 2004). BRAF mutation also occurs at high frequencies (>80%) in melanocytic nevi (Pollock et al., 2003; Yazdi et al., 2003; Kumar et al., 2004; Aldanha et al., 2004; La Porta et al., 2006), suggesting that those somatic alterations occur early in melanomagenesis. The master melanocyte regulator MITF was found to cooperate with BRAF in melanoma tumor formation in vitro (Garrawat et al., 2005) and in vivo (June-Valbuena et al., 2010). In zebrafish, BRAF activation alone resulted in benign nevus

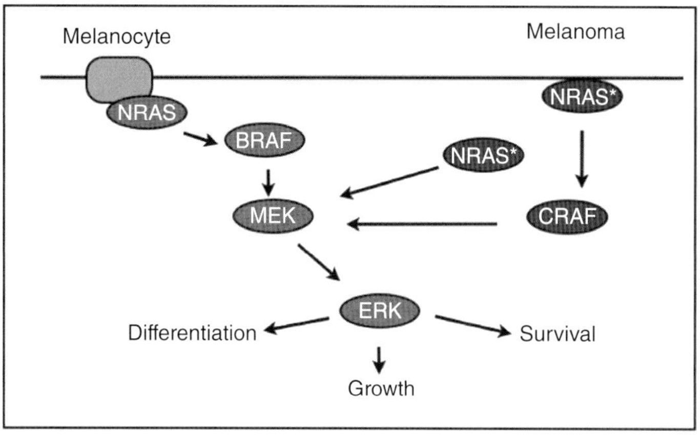

Figure 10.1. BRAF signal transduction in melanocytes and melanoma cells.

formation, while malignant transformation requires concurrent loss of p53 (Patton et al., 2005). Two similar murine models have emerged, confirming the tumorigenic potential of the mutated Braf allele (Dankort et al., 2009).

Constitutive activation of BRAF-MEK-ERK pathway in melanoma drives the invasive and motile behavior of melanoma cells through the reorganization of the cytoskeleton, the activation of the cells migration capability and upregulation of Matrix-metalloprotease expression (Woods et al., 2001; Pritchard et al., 2004). More details regarding to the mechanisms are recently reported by Arozarena and colleagues (2011), Klein and colleagues (2009), and Klein and colleagues (2008). There is also interesting evidence that BRAF mutated regulates the interaction of melanoma cells with the host microenvironment, in particular cells of the immune system (Boni et al., 2010; Kono et al., 2006; Sumimoto et al., 2006).

Vemurafenib is a drug recently approved by FDA to treat advanced-stage melanoma with a known V600E mutation, providing a specific example of how identifying a specific somatic mutation in melanoma tumors has led to direct-targeted therapy. Firstly, it is important to detect the mutation in patients, in order to decide if vemurafenib might be useful in a combined therapy. In a clinical trial, approximately 80% of patients with BRAF mutation treated at the recommended phase II dose of the drug responded to the therapy. The 1-year survival is about 50% and the 2-year survival is 38% with a median overall survival of 13.8 months (Smalley et al., 2012). However, despite the shrinkage of the tumor observed in patients with BRAF mutant melanoma and treated with vemurafenib, about 50% patients did not seem to meet the Response Evaluation Criteria in Solid Tumors, suggesting an intrinsic resistance (Chapman et al., 2011; Flaherty et al., 2010; Fedorenko et al., 2011). A possible explanation is that the target is not the correct one. It is very hard to define a biomarker, and most of the biomarkers identified are imperfect (Zapperi and La Porta, 2012). Another important aspect is the presence in melanoma of an aggressive cell subpopulation, usually defined as cancer stem cells.

Cancer Stem Cells in Melanoma and Biomarkers

In contrast with conventional views of cancer where all cells are tumorigenic, the cancer stem cell (CSC) theory proposed decades ago (Reva et al., 2001; Pardal et al., 2003) that tumors contain a subset of cells that both self-renew and give rise to differentiated progeny. Although many aspects of this theory remain speculative and are still evolving, CSCs are viewed as those cells at the apex of the tumor hierarchy, highlighting the role of aberrant differentiation in tumorigenesis. In analogy with normal adult tissue stem cells, CSCs are the driving force of the tumor. CSCs are defined at least by two minimal properties. The first one is the capability to regrow the tumor from which they were isolated or identified, which implies that the tumor-initiating cells can only be defined experimentally in vivo. The use of serial transplantation to validate a candidate cancer stem cell subpopulation allows the ability to monitor the capability to recapitulate the heterogeneity of the primary tumor. Both xeno- and syngeneic transplantations might, however, misrepresent the real network of interactions with diverse support such as fibroblasts, endothelial cells, macrophages, mesenchymal stem cells, and many of the cytokines and receptors involved in these interactions. As a consequence, contrasting results have appeared in the literature for the frequency of CSCs in xenograft tumors. In a recent paper we have suggested that the number of CSCs should not be stated in absolute terms, but only relative to the animal model used (La Porta, 2009). The second property necessary to define CSCs is the multipotency of lineage differentiation, which is likely but necessary to be frequent. In addition to the aforementioned properties, there are at least two other aspects that are not completely clear: the origin of CSCs and their frequency in vivo. It is important to highlight that "stemness" does not necessarily involve a set of genes common to all of the stem cells, since gene expression patterns of stem cells in distinct tissues differ widely (Wallenfang and Matunis, 2003). Another important aspect is the mechanism by which stem cells decide to remain in either the niche or leave. This could be a major factor in the balancing act between stem cell self-renewal and differentiation (Haas and Herlyn, 2005). The investigation of normal melanocyte homeostasis might help us to define how melanoma and, in particular, melanoma CSCs escape the microenvironment created by epidermal keratinocytes and how they develop new cellular partners in fibroblasts and endothelial cells which support their growth and invasion.

The use of multiple markers to define such a CSC subpopulation is, in our view, strongly recommended. A panel of markers defines hematopoietic stem cells as well. The literature is sometimes confusing since researchers often use potential biomarkers without a specific biological motivation, employing some markers alone in some papers and in a combination in others. In the following narrative, we discuss recent findings for two tumors: breast cancer and melanoma. The picture that comes out is still fuzzy and confused, but we will suggest some ideas to clarify it.

In melanoma seven papers came out from 2005 to 2008 showing that a CSCs subpopulation exists (Schatton et al., 2008; Dou et al., 2007; Monzani et al., 2007; Fang et al., 2005; Klein et al., 2007; Hadnagy et al., 2006). These papers used different markers like as CD20, CD133, ABCG2, and ABCB5. However, in 2008 a paper argued against the existence CSCs based on the following observations: a relatively large fraction of melanoma cells (up to ~25%) was shown to initiate tumors in severely immunocompromised NOD/SCID IL2Rγnull mice; the fraction of tumor-inducing cells depends upon assay conditions; several putative CSC markers appear to be reversibly expressed (Quintana et al., 2008). In conclusion, this paper suggested that the best model to confirm the presence of CSCs is severe immunocompromised mice. In a follow-up study, the same authors analyzed the expression of more than 50 surface markers on melanoma cells derived from several patients (A2B5, cKIT, CD44, CD49B, CD49D, CD49F, CD133, CD166), but they focused on CD133 and CD166 (Quintana et al., 2010). Using these markers, they did not find any enrichment and did find a high frequency of tumorigenic cells. In a recent paper it was shown, however, that CD133 is highly expressed in melanoma cells, and it is not a good marker to sort CSCs (Monzani et al.,

2007). Moreover, in 2010, Boiko and colleagues (2010) using the same immunocompromised mice did not confirm Quintana data. Boiko and colleagues (2010) used CD271, nerve growth factor receptor as marker to identify CSCs. In 2010 our group showed for the first time that a marker connected with a functional property of stem cells like the asymmetric division, CXCR6, is expressed by human melanoma cell lines and biopsies (Taghizadeh et al., 2010). CXCR6 subpopulation showed a stronger self-renewal capability in xenograft model (Taghizadeh et al., 2010). We also showed that CXCR6 positive cells were able to grow better when treated with the ligand (CXCL16) (Taghizadeh et al., 2010).

Mathematical Models for Cancer Growth: The Case of Melanoma

Mathematical models are slowly entering biological sciences, although their relevance is often not fully appreciated. Quantitative methods borrowed from hard sciences such as physics can sometimes help overcome the limitations of biological analysis and provide new insight on well-studied problems. Cancer research is a very promising field where mathematical modeling combined with a quantitative analysis of experimental and clinical data can help discriminate between different broad theoretical ideas, such as the cancer stem cell hypothesis (La Porta et al., 2012) or the idea that a cancer cell can dynamically switch into the cancer stem cell phenotype (Zapperi and La Porta, 2012; Gupta et al., 2011). Mathematically based approaches are not only interesting to understand the basic mechanisms of cancer progression but could also be useful to simulate complex effects of therapeutic strategies.

The kinetics of cancer cells is traditionally modeled by population dynamics approaches, using the theory of branching processes (Harris, 1989; Kimmel and Axelrod, 2002). Branching processes have been used extensively in the past to model the growth of stem cells (Vogel et al., 1968; Matioli et al., 1970; Potten and Morris, 1988; Clayton et al., 2007; Antal and Krapivsky, 2010; Itzkovitz et al., 2012) and more recently of cancer stem cells (Michor et al., 2005; Ashkenazi et al., 2008; Michor, 2008; Tomasetti and Levy, 2010). The main limitation of branching processes is due to their mean-field nature that does not take into account the geometry of the cellular arrangement inside a tumor. Despite this shortcoming, those models allow for a quantitative description of experimental growth curves of cancer cells, providing an indirect confirmation of the CSC hypothesis in melanoma La Porta et al., 2012). CSC localization in tumors is still not well understood although recent experiments tracked their spatio-temporal kinetics in vivo (Driessens et al., 2012) in analogy with tissue stem cells (Driessens et al., 2012; Sato et al., 2011; Snippert et al., 2010; Lopez-Garcia et al., 2010). Statistical mechanics models that take into account the spatial dimensions have been used to understand the stem cell results (Lopez-Garcia et al., 2010) and may in the future be generalized to the case of tumors.

Summary and Perspectives

The discovery of CSCs has opened new perspectives in cancer research. Recently, senescence has been proposed as a new possible therapeutic strategy (La Porta et al., 2012). We have investigated this possibility in light of the CSCs hypothesis, showing that a considerable fraction of melanoma cells grown in vitro turn to senescent (La Porta et al., 2012). This finding may look puzzling at first but can be understood thanks to a stochastic model of tumor growth in the presence of cancer stem cells (La Porta et al., 2012). In the model we consider a population composed by cancer stem cells, that can either self-renew or give rise to more differentiated progeny, and cancer cells that divide a finite number of times and then turn senescent

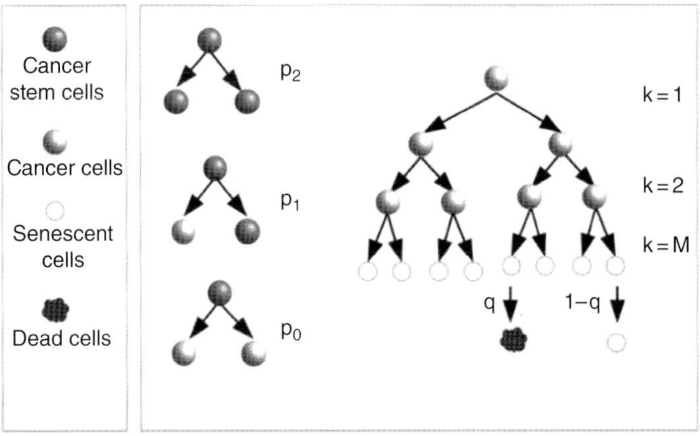

Figure 10.2. A mathematical model for cancer growth. Cells are divided into subpopulations composed by CSCs (*red*) and cancer cells (*blue*). Cancer cells divide a finite number of times M (M=3 in the figure) before turning senescent (*white*). Senescent cells die with probability q. CSCs can divide symmetrically, giving rise to two CSCs with probability p2 or to two CCs with probability p0, or asymmetrically with probability p1 = 1 − p2 − p0 giving rise to a CSC and a CC. (See La Porta et al., 2012 for more details.) For color detail, please see color plate section.

(Figure 10.2). These data show that enhancing senescence is unlikely to provide a useful therapeutic strategy to fight cancer, unless the cancer stem cells are specifically targeted. Therefore, we believe that a deep comprehension of CSCs biology is the right way to fight cancer. Mathematical models can help to reach this goal.

References

Aldanha, G., Purnell, D., Fletcher, A., Potter, L., Gillies, A., Pringle, J.H. (2004). High BRAF mutation frequency does not characterize all melanocytic tumor types, Int. J. Cancer 111, 705–710.

Antal, T., Krapivsky, P.L. (2010). Exact solution of a two-type branching process: clone size distribution in cell division kinetics. Journal of Statistical Mechanics: Theory and Experiment, P07028.

Arozarena, I., Sanchez-Laorden, B., Packer, L., Hidalgo-Carcedo, C., Hayward, R., Viros, A., Sahai, E., Marais, R. (2011). Oncogenic BRAF induces melanoma cell invasion by downregulating the cGMP specific phosphodiesterase PDE5A. Cancer Cell. 19, 45–57.

Ashkenazi, R., Gentry, S.N., Jackson, T.L. (2008). Pathways to tumorigenesis-modeling mutation acquisition in stem cells and their progeny. Neoplasia 10, 1170–1182.

Bishop, J.A., Wachsmuth, R.C., Harland, M., Bataille, V., Pinney, E., MacK, P., Baglietto, L., Cuzick, J., Bishop, D.T. (2000). Genotype/phenotype and penetrance studies in melanoma families with germline CDKN2A mutations. J Invest Dermatol. 114, 28–33.

Boiko, A.D., Razorenova, O.V., van de Rijn, M., Swetter, S.M., Johnson, D.L., Ly, D.P., Butler, P.D., Yang, G.P., Joshua, B., Kaplan, M.J., Longaker, M.T., Weissman, I.L. (2010). Human melanoma-initiating cells express neural crest nerve growth factor receptor CD271. Nature 466, 133–137.

Boni, A., Cogdill, A.P., Dang, P., Udayakumar, D., Njauw, C.J., Sloss, C.M., Ferrone, C.R., Flaherty, K.T., Lawrence, D.P., Fisher, D.E., Tsao, H., Wargo, J.A. 2010. Selective BRAFV600E inhibition enhances T-cell-recognition of melanoma without affecting lymphocyte function. Cancer Res. 70, 5213–5219.

Chapman, P.B., Hauschild; A., Robert, C., Haanen, J.B., Ascierto, P., Larkin, J., Dummer, R., Garbe, C., Testori, A., Maio, M., Hogg, D., Lorigan, P., Lebbe, C., Jouary, T., Schadendorf, D., Ribas, A., O'Day, S.J., Sosman, J.A., Kirkwood, J.M., Eggermont, A.M., Dreno, B., Nolop, K., Li, J., Nelson, B., Hou. J., Lee, R.J., Flaherty, K.T., McArthur, G.A. BRIM-3 Study Group. (2011). Improved survival with vemurafenib in melanoma with BRAF V600E mutation. New England J. Med. 364, 2507–2516.

Clark, Jr., W.H., Elder, D.E., Guerry, D., Epstein, M.N., Greene, M.H., Van Horn, M. (1984). A study of tumor progression: the precursor lesions of superficial spreading and nodular melanoma. Human Pathol. 15, 1147–1165.

Clayton, E. Doupé, D.P., Klein, A.M., Winton, D.J., Simons, B.D., Jones, P.H. (2007). A single type of progenitor cell maintains normal epidermis. Nature 446, 185–189.

Dankort, D., Curley, D.P., Cartlidge, R.A., Nelson, B., Karnezis, A.N., Damsky, W.E. Jr., You, M.J., DePinho, R.A., McMahon, M., Bosenberg, M. (2009). Braf (V600E) cooperates with Pten loss to induce metastaic melanoma. Nature Genet. 41, 344–552.

Davies, H., Bignell, G.R., Cox, C., Stephens, P., Edkins, S., Clegg, S., Teague, J., Woffendin, H., Garnett, M.J., Bottomley, W., Davis, N., Dicks, E., Ewing, R., Floyd, Y., Gray, K., Hall, S., Hawes, S., Hughes, J., Kosmidou, V., Menzies, A., Mould, C., Parker, A., Stevens, C., Watt, S., Hooper, S., Wilson, R., Jayatilake, H., Gusterson, B.A., Cooper, C., Shipley, J., Hargrave, D., Pritchard-Jones, K., Maitland, N., Chenevix-Trench, G., Riggins, G.J., Bigner, D.D., Palmieri, G., Cossu, A., Flanagan, A., Nicholson, A., Ho, J.W., Leung, S.Y., Yuen, S.T., Weber, B.L., Seigler, H.F., Darrow, T.L., Paterson, H., Marais, R., Marshall, C.J., Wooster, R., Stratton, M.R., Futreal, P.A. 2002. Mutation of the Braf gene in human cancer. Nature 417: 949–954.

Dou, J., Pan, M., Wen, P., Li. Y., Tang, Q., Chu, L., Zhao, F., Jiang, C., Hu, W., Hu, K., Gu, N. (2007). Isolation and identification of cancer stem cell-like cells from murine melanoma cell lines. Cellular and Molecular Immunology. 4, 467–472.

Driessens, G., Beck, B., Caauwe, A., Simons, B.D., Blanpain, C. (2012). Defining the mode of tumour growth by clonal analysis. Nature 488, 527–530.

Fang, D., Nguyen, T.K., Leishear, K., Finko, R., Kulp, A.N., Hotz, S., Van Belle, P.A., Elder, D.E., Herlyn, M. (2005). A tumorigenic subpopulation with stem cell properties in melanoma. Cancer Research 65, 9328–9337.

Farmer, E.R., Gonin, R., Hanna, N.P. (1996). Discordance in the histopathologic diagnosis of melanoma and melanocytic nevi between expert pathologists. Human Pathol. 27, 528–531.

Fedorenko, IV, Paraiso, K.H., Smalley, K.S. (2011). Acquired and intrisic BRAF inhibitor resistance in BRAF V600E mutant melanoma. Biochem. Pharmacol. 82: 201–209.

Fidler, I.J. (1978). Tumor heterogeneity and the biology of cancer invasion and metastasis. Cancer Res. 38:2651–2660.

Flaherty, K.T, Puzanov, I., Kim, K.B., Ribas, A., McArthur, G.A., Sosman, J.A., O'Dwyer, P.J., Lee, R.J., Grippo, J.F., Nolop, K., Chapman, P.B. (2010). Inhibition of mutated activated BRAF in metastatic melanoma. New Engl. J. Med. 363, 809–819.

Folkman, J. (1986). How is blood vessel growth regulated in normal and neoplastic tissue? G.H.A. Clowes memorial Award lecture. Cancer Res. 46: 467–473.

Folkman, J. (2007). Angiogenesis: an organizing principle for drug discovery? Nat Rev Drug Discov. 6, 273–286.

Garraway, L.A., Widlund, H.R., Rubin, M.A., Getz, G., Berger, A.J., Ramaswamy, S., Beroukhim, R., Milner, D.A., Granter, S.R., Du, J., Lee, C., Wagner, S.N., Li, C., Golub, T.R., Rimm, D.L., Meyerson, M.L., Fisher, D., Sellers W.R. (2005). Integrative genomic analyses identify MITF as a lineage survival oncogene amplified in malignant melanoma. Nature 436, 117–122.

Gupta, P.B., Fillmore, C.M., Jing, G., Shapira, S.D., Tao, K., Kuperwasser, C., Lander, E.S. (2011). Stochastic state transitions give rise to phenotypic equilibrium in populations of cancer cells. Cell. 146, 633–644.

Haas, N.K., Herlyn, M. (2005). Normal human melanocyte homeostasis as a paradigm for understanding melanoma. J. Invest. Dermatol. Symp. Proc. 10, 153–163.

Hadnagy, A., Gaboury, L., Beaulieu, R., Balicki, D. (2006). SP analysis may be used to identify cancer stem cell population. Exp. Cell Res. 312, 3701–3710.

Harris, T.E. (1989). The theory of branching processes (Dover, New York).

Helm, K.F., Marks, J.G. (1997). Differential diagnosis in dermathology. New York, Churchill Livingston.

Hensin, S., Chin, L., Garraway, L.A., Fisher, D.E. (2012). Melanoma: from mutation to medicine, genes and development. 26: 1131–1151. Wilson, M.A.; Nathanson, K.L. 2012. Molecular testing in melanoma. Cancer J. 18, 117–123.

Itzkovitz, S., Blat, I.C., Jacks, T., Clevers, H., van Oudenaarden, A. (2012). Optimality in the development of intestinal crypts. Cell. 148, 608–611.

Jane-Valbuena, J., Widlund, H.R., Perner, S., Johnson, L.A., Dibner, A.C., Lin, W.M., Baker, A.C., Nazarian, R.R.M., Vijayendran, K.G., Sellers, W.R., Hahn, W.C., Duncan, L.M., Rubin, M.A., Fisher, D.E., Garraway, L.A. (2010). An oncogenic role for ETV1 in melanoma. Cancer Res. 70, 2075–2084.

Kimmel, M., Axelrod, D.E. (2002). Branching processes in biology (Springer, New York).

Klein, W.M., Wu, B.P., Zhao, S., Wu, H., Klein-Szanto, A.J., Tahan, S.R. (2007). Increased expression of stem cell markers in malignant melanoma. Modern Pathology 20,102–107.

Klein, R.M., Spofford, L.S., Abel, E.V., Ortiz, A., Aplinet, A.E. (2008). BRAF regulation of Rnd3 participates in actin cytoskeleton and focal adhesion organisation. Mol Cell Biol. 19, 498–508.

Klein, R.M., Aplin, A.E. (2009). Rnd3 regulation of the actin cytoskeleton promotes melanoma migration and invasive outgrowth in three dimensions. Cancer Cells. 69, 2224–2233.

Koh, H.K. (1991). Cutaneous melanoma. N. Engl. J. Med. 325: 171–182.

Kono, M., Dunn, I.S., Durda, P.J., Butera, D., Rose, L.B., Haggerty, T.J., Benson, E.M., Kurnick, J.T. (2006). Role of mitogenic activated protein kinase signaling pathway in the regulation of human melanocytic antigen expression. Mol. Cancer Res. 4, 779–792.

Kumar, R., Angelini, S., Snellman, E., Hemminki, K. (2004). BRAF mutations are common somatic events in melanocytic nevi. J. Invest. Dermatol. 122, 342–348.

La Porta, C.A.M., Cardano, R., Facchetti, F., Presicce, P., Rao, S., Privitera, E., Clemente, C., Mihm, M. Jr. (2006). BRAF V599E mutation occurs in Spitz and Reed nevi. JEAVD Journal of the European Academy of Dermatology and Venerology. 20, 1164–1165.

La Porta, C.A.M. (2009). Cancer Stem cells: a lesson from melanoma. Stem Cell Reviews and Rep. 5, 61–65.

La Porta, C.A.M., Zapperi, S., Sethna, J.P. (2012). Senescent cells in growing tumors: population dynamics and cancer stem cells. PLoS Comput Biol 8, e1002316.

Lopez-Garcia, C., Klein, A.M., Simons, B.D., Winton, D. J. (2010). Intestinal stem cell replacement follows a pattern of neutral drift. Science. 330, 822–825.

Matioli, G., Niewisch, H., Vogel, H. (1970). Stochastic stem cell renewal. Rev. Eur. Etud. Clin. Biol. 15, 20–22.

Michor, F. (2008). Mathematical models of cancer stem cells. J Clin Oncol 26, 2854–61.

Michor, F., Timothy P., Hughes, I., Branford, S., Shah, N.P., Sawyers, C.L., Nowak, M.A. (2005). Dynamics of chronic myeloid leukaemia. Nature 435, 1267–1270.

Monzani, E., Facchetti, F., Galmozzi, E., Corsini, E., Benetti, A., Cavazzin, C., Gritti, A., Piccinini, A., Porro, D., Santinami, M., Invernici, G., Parati,E., Alessandri, G., La Porta, C.A.M. (2007). Melanoma contains CD133 and ABCG2 positive cells with enhanced tumorigenic potential. Eur J. Cancer 43, 935–946.

Pardal, R., Clarke, M.F., Morrison, S.J. (2003). Applying the principle of stem cell biology to cancer. Nature Rev. Cancer 3, 895–902.

Patton, E.E., Widlung, H.R., Kutok, J.L., Kopani, K.R., Amatruda, J.F., Murphey, R.D., Berghmans, S., Mayhall, E.A., Traver, D., Fletcher, C.D., Aster, J.C., Granter, S.C., Thomas Look A., Lee, C., Fisher, D.E., Zon, L.I. (2005). BRAF mutation are sufficient to promote nevi formation and cooperate with p53 in the genesis of melanoma. Curr. Biol. 15: 249–254.

Piepkorn, M.W., Barnhill, R.L., Cannon-Albright, A.L., Elden, D.E., Golgar, D.E., Lewis, C.M., Maize, J.C., Meyer, L.J., Rabkin, M.S., Sagebile, R.W., Skonick, M.H. Zone, J.J. (1994). A multiobserver, population-based analysis of histologic dysplasia in melanocytic nevi. J. Amer. Acad.Dermatol. 30, 707–714.

Pollock, P.M., Harper, U.L., Hansen, K.S., Yudt, L.M., Stark, M., Robbins, C.M., Moses, T.Y., Hostetter, G., Wagner, U., Kakareka, J., Salem, G., Pohida, T., Heenan, P., Duray, P., Kallioniemi, O., Hayward, N.K., Trent, J.M., Meltzer, P.S. (2003). High frequency of BRAF mutations in nevi. Nat. Genet 33, 19–20.

Potten, C.S., Morris, R.J. (1988). Epithelial stem cells in vivo. J. Cell Sci. Suppl. 10, 45–62.

Pritchard, C.A., Hayes, L., Wojnowski, L., Zimmer, A., Marais, R.M., Norman, J.C. (2004). B-Raf acts via the ROCKII/LIMK cofilm pathway to mantain actin stress fibers in fibroblasts. Mol. Cell Biol. 24, 5937–5952.

Quintana, E., Shackleton, M., Sabel, M.S., Fullen, D.R., Johnson, T.M., Morrison, S.J. (2008). Efficient tumor formation by single human melanoma cells. Nature 456, 593–598.

Quintana, E., Shackleton, M., Foster, H.R., Fullen, D.R., Sabel, M.S., Johnson, T.M., Morrison, S.J. (2010). Phenotypic heterogeneity among tumorigenic melanoma cells from patients that is reversible and not hierarchically organized. Cancer Cell. 201,510–523.

Sato, T., van Es, J.H., Snippert, J.H., Stange R.G., van den Born V.M., Barker N., Shroyer, N.F., van de Wetering, M., Clevers, H. (2011). Paneth cells constitute the niche for lgr5 stem cells in intestinal crypts. Nature 469, 415–418.

Schatton, T., Murphy, G.F., Frank, N.Y., Yamaura, K., Waaga-Gasser, A.M., Gasser, M., Zhan, Q., Jordan, S., Duncan, L.M., Weishaupt, C., Fuhlbrigge, R.C., Kupper, T.S., Sayegh, M.H., Frank, M.H. (2008). Identification of cells initiating human melanomas. Nature. 451, 345–349.

Smalley, K.S., Aplin, A.E., Flaherty, K.T. (2012). Meeting report from the 7th International Melanoma Congress Tampa, Florida. Pigment Cell Melanoma Res. 25, E1–E11.

Smalley, K.S., Xiao, M., Villanueva, J., Nguyen, T.K., Flaherty, K.T., Letrero, R., Van Belle, P., Elder, D.E., Wang, Y., Nathanson, K.L., Herlyn, M. (2009). CRAF inhibition induces apoptosis in melanoma cells with non-V600E BRAF mutations. Oncogene 28, 85–94.

Snippert, H.J., van der Flier, L.G., Sato, T.H., van Es, J., van den Born, M., Kroon-Veenboer, C., Barker, N., Klein, A.M., van Rheenen, J., Simons, B.D., Clevers, H. (2010). Intestinal crypt homeostasis results from neutral competition between symmetrically dividing lgr5 stem cells. Cell. 143, 134–144.

Sumimoto, H., Imabayashi, F., Iwata, T., Kawakami, Y. (2006). The BRAF-MAPK signaling pathway is essential for cancer immune evasion in human melanoma cells. J. Exp. Medicine 203, 1651–1656.

Taghizadeh, R., Noh, M., Huh, Y.H., Ciusani, E., Sigalotti, L., Maio, M., Arosio, B., Nicotra, M.R., Natali, P.G., Sherley, J.L., La Porta, C.A.M. (2010). CXCR6, a newly defined biomarker of tissue-specific stem cell asymmetric self-renewal, identified more aggressive human melanoma cancer stem cells. PLoS ONE 5: e15183.

Tomasetti, C., Levy, D. (2010). Role of symmetric and asymmetric division of stem cells in developing drug resistance. Proc Natl Acad Sci USA 107, 16766–16771.

Vogel, H., Niewisch, H., Matioli, G. (1968). The self renewal probability of hemopoietic stem cells. J Cell Physiol. 72, 221–228.

Wallenfang, M.R., Matunis, E. (2003). Developmental biology orienting stem cells. Science. 301, 1490–1491.

Wan, P.T., Garret, M.I., Roe, S.M., Lee, S., Niculescu-Duvaz, D., Good, V.M., Jones, C.M., Marshall, C.J., Springer, C.J., Barford, D., Marais, R. (2004). Mechanism of activation of the RAF-ERK signalling pathway to oncogenic mutation of BRAF. Cell 116, 855–867.

Welbrock, C., Karasarides, M., Marais, R. (2004). The RAF protein take centre stage. Nat. Rev. Mol. Cell Biol. 5, 875–885.

Whiteman, D.C., Pavan, W.J., Bastian, B.C. (2011). The melanomas: a synthesis of epidemiological, clinical, histopathological, genetic and biological aspects, supporting distinct subtypes, causal pathways and cell origin. Pig. Cell Melanoma Res, 24, 879–897.

Woods, D., Cherwinski, H., Venetsanakos, E., Bhat, A., Gysin, S., Humbert, M., Bray, P.F., Saylor, V.L., McMahon, M. (2001). Induction of beta 3 integrin gene expression by sustained activation of the Ras-regulated Raf-MEK extracellular signal-regulated kinase signaling pathway. Mol. Cell Biol, 21, 3192–3205.

Yazdi, A.S., Palmedo, G., Flaig, M.J., Puchta, U., Reckwerth, A., Rutten, A., Mentzel, T., Hugel, H., Hantscke, M., Schmid-Wendtner, M.H., Kutzner, H., Sander, C.A. (2003). Mutations of the BRAF gene in benign and malignant melanocytic lesions. J. Invest. Dermatol. 121, 1160–1162.

Zalaudek I, Whiteman D., Rosendahl C., Menzies S.W., Green A.C., Hersey P., Argenzano G. (2011). Update on melanoma and non-melanoma skin cancer. Expert Rev Anticancer Ther 11, 1829–1832.

Zapperi, S., La Porta, C.A.M. (2012). Do cancer cells undergo phenotypic switching? The case for imperfect cancer stem cell markers. Scientific Reports 2, 441, DOI: 10.1038/srep00441.

Reya, T., Morrison, S.J., Clarke, M.F., Weissman I.L. (2001). Stem cells, cancer and cancer stem cells. Nature. 414, 105–111.

Section II

Stem Cells in Liquid Tumors

Chapter 11

Acute Myeloid Leukemia Stem Cells—Updates and Controversies

Stephen S. Chung[1] and Christopher Y. Park[2]

[1] *Leukemia Service, Department of Medicine, Human Oncology and Pathogenesis Program, Memorial Sloan-Kettering Cancer Center, New York, NY, USA*

[2] *Human Oncology and Pathogenesis Program, Departments of Pathology and Laboratory Medicine, Memorial Sloan-Kettering Cancer Center, New York, NY, USA*

Background—Normal Hematopoiesis

Since acute myeloid leukemia (AML) arises from immature hematopoietic cells, efforts to purify leukemia stem cells (LSC) have been significantly aided by our understanding of normal hematopoietic stem and progenitor cell (HSPC) development and the cell surface molecules that can be utilized to prospectively isolate these populations (reviewed in Bryder et al., 2006). Technical advances, including the development of monoclonal antibodies to cell surface proteins (Kohler and Milstein, 1975) and fluorescence-activated cell sorting (FACS) (Hulett et al., 1969), allowed for the first prospective isolation of mouse HSCs in 1988 (Spangrude et al., 1988). We now have an exquisitely refined immunophenotypic definition of mouse hematopoietic stem cells (HSC), allowing for the prospective isolation of HSCs with a purity level as high as 1 in 1.3 (Kiel et al., 2005).

Our understanding of human hematopoiesis has lagged behind that of the mouse due to limitations in transplantation modeling across species barriers, but the development of immunodeficient mouse strains capable of multilineage human engraftment have allowed for the identification of increasingly refined populations of human HSPCs (reviewed in Doulatov et al., 2012). Initial long term culture initiating cell (LTC IC) assays localized human HSCs to the Lin-CD34+CD90+ and Lin−CD34+CD38−/lo fractions of BM (Baum et al., 1992; Craig et al., 1993). Subsequent xenotransplantation models utilizing severe combined immunodeficiency mice engrafted with human fetal bone and thymus (SCID-Hu) demonstrated that sustained production of donor B- and T-cells indicative of stem/progenitor activity can be initiated by Lin-CD34+CD90+ cells from the BM (McCune et al., 1988; Murray et al., 1995; Peault et al., 1993). However, myeloid engraftment was poor in this model unless mice were treated with human cytokines such as interleukin-3, stem cell factor (SCF), and granulocyte macrophage colony stimulating factor (GM-CSF) (Kamel-Reid and Dick, 1988; Lapidot et al., 1992). The development of the non-obese diabetic (NOD)-SCID mouse harboring more severe defects in innate immunity and a more supportive BM microenvironment, in part because of improved host macrophage tolerance (Takenaka et al.,

Cancer Stem Cells, First Edition. Edited by Vinagolu K. Rajasekhar.

2007), allowed for a high level of multilineage engraftment without the need for continuous cytokine administration. Using this model, investigators could assess long-term human HSC function in secondary and tertiary transplants. Studies utilizing the NOD/SCID model confirmed the presence of long-term repopulating human HSCs in the Lin-CD34+CD38− fraction of BM (Bhatia et al., 1997; Cashman et al., 1997; Hogan et al., 2002; Takenaka et al., 2007), and more recently NOD/SCID/IL-2Rγ null strains (NOG/NSG, devoid of B, T, and NK cells) have allowed for the prospective isolation of Lin-CD34+CD38−CD90+CD45RA−CD49f+Rho(lo) HSCs such that 1 in 3 cells can engraft recipients long-term (Notta et al., 2011). Within the Lin−CD34+CD38− compartment, non-HSCs include CD90−CD45RA- cells that appear to have transient but multilineage engraftment characteristics of multipotent progenitors (MPP) (Majeti et al., 2007), and CD90−CD45RA+cells that have limited myeloid potential and no self-renewal capacity, similar to what has been described in mouse as a lymphoid-primed multipotent progenitor (LMPP) (Goardon et al., 2011).

Identification of the Leukemic Stem Cell

In the 1970s, radioactive isotope labeling experiments in patients with AML demonstrated the presence of a minor population of small and rarely dividing cells with high clonogenic potential, leading to the theory that leukemic cells are hierarchically organized (Clarkson et al., 1970; Clarkson, 1969; Killmann et al., 1963). This functional heterogeneity and the observation that most leukemic cells were induced to enter the cell cycle upon administration of chemotherapy led to the prediction that slow cycling leukemic cells might represent chemo-resistant cells that mediate disease relapse (Clarkson et al., 1975; Clarkson and Fried, 1971; Cronkite, 1970). Unfortunately, methods to prospectively isolate what was estimated to be a very rare cell population were lacking, and studies examining cell cycle status and response to chemotherapy in AML patients using unpurified bone marrow samples demonstrated no positive association between quiescence and disease relapse (an inverse association was observed, proposed to be secondary to the faster growth of rapidly cycling types of leukemia) (Raza et al., 1990).

In a landmark study published in 1994, Lapidot and colleagues reported engraftment of AML cells from 17 patients in SCID mice treated with human cytokines (GM-CSF, IL-3, and SCF). Using Southern blotting to assess for human chimerism following transplantation with limiting numbers of leukemic cells, they estimated that human AMLs exhibit an "SCID leukemia-initiating cell" (SL-IC) frequency of approximately 1 in 250,000 (Lapidot et al., 1994). FACS was used to purify leukemic cells based on CD34 and CD38 expression, and using one AML sample (FAB subtype M1), it was demonstrated that only CD34+CD38− leukemic cells were capable of engrafting SCID mice, with no engraftment potential among CD34+CD38+ and CD34− cells. Importantly, transplantation of CD34+ CD38− cells resulted in engraftment of leukemia-containing cells exhibiting all three immunophenotypes, thereby recapitulating the immunophenotypic diversity of the original leukemia. This observation, along with the fact that the SL-ICs were about 1,000-fold less frequent than immature leukemic blast colony-forming units (AML-CFU), supported the idea that there is a hierarchy composed of LSCs and more differentiated, non-self-renewing leukemic progenitors in human AML.

The theory that SL-ICs represent LSCs was not fully validated until the development of the NOD/SCID mouse, which allowed for the engraftment of limiting numbers of purified human hematopoietic cells (10 to 20 times more efficiently than in SCID mice), and thus the capability to perform serial transplants. Bonnet and Dick (1997) used seven AML specimens (FAB subtypes M1, M4, and M5) to demonstrate that despite comprising only 0.02–2% of the total leukemic specimen, as few as 5,000 CD34+CD38− cells were capable of leukemic engraftment, consistent with significant enrichment for SL-ICs in this population. The engrafted cells were similar to the original AML sample with respect to differentiation status, and their self-renewal capability was demonstrated in secondary transplants that revealed an estimated 30-fold expansion of SL-ICs in the primary recipients. Thus, AML represented the first example of a human cancer fulfilling the predictions of the cancer stem cell model, with SL-IC displaying the requisite abilities to self-renew and to differentiate into non-self-renewing progeny that comprise the bulk of the malignancy. Given the shared properties between SL-ICs and normal stem cells, with time and further validation, the term SL-IC has largely given way to simply LSC.

Hematopoietic Stem Cell as Cell of Origin

The fact that LSCs share a common CD34+CD38− immunophenotype with normal HSCs initially led to the hypothesis that LSCs arise from HSCs that undergo the requisite steps for malignant transformation (Weissman, 2005). This is an attractive model given that the long life span of normal HSCs would allow for the stepwise acquisition of the full complement of genetic and epigenetic lesions required for leukemic transformation. Furthermore, the essential and cancer stem cell-defining capability of self-renewal would already exist in HSCs, potentially lowering the complement of lesions required for full transformation. Finally, this model would be consistent with other hematologic malignancies for which there is ample evidence for a HSC cell of origin, including MDS (Nilsson et al., 2002; Nilsson et al., 2007; Park et al., 2011; Tehranchi et al., 2010; Will et al., 2012) and chronic phase CML (Bhatia et al., 2003; Fialkow et al., 1977; Graham et al., 2002; Martin et al., 1980), both of which demonstrate clonal involvement of the HSC compartment with cytogenetically abnormal malignant stem cells with multilineage potential. Recent evidence supporting an HSC origin for AML is discussed below in the section on pre-leukemic HSCs.

Progenitor as Cell of Origin

Studies by Blair and colleagues (1997) utilized the NOD/SCID engraftment model to demonstrate that engrafting LSCs lack the HSC markers CD90 and c-Kit (Blair and Sutherland, 2000). In studies of *AML1/ETO* translocated (t[8;21]) AML, Miyamoto and colleagues (2000) demonstrated that only the CD34+CD38−CD90− compartment contained LSCs capable of forming leukemic blast colonies (CFU-L) *in vitro*. Together, these studies suggest that the final target of transformation is either an HSC that downregulates normal HSC cell surface markers upon transformation, or a non-HSC hematopoietic progenitor.

Beginning in 2002, mouse models of AML generated from enforced expression of MLL-associated fusion genes by retroviruses demonstrated that transformation of HSCs and myeloid progenitors resulted in leukemia types in which the bulk of leukemic blasts had a Lin-Sca-1-c-Kit+CD34+ FcγRII/III+ immunophenotype, comparable to the normal granulocyte macrophage progenitor, and these were designated "GMP-like leukemia cells" (L-GMP) (Cozzio et al., 2003; DiMartino et al., 2002; Wang et al., 2005). In 2006 Krivtsov and colleagues demonstrated that (MSCV)-MLL-AF9 transduced purified GMPs could initiate AML in mice, and that purified L-GMPs from primary recipients could serially transplant AML with a LSC frequency of one in six cells. Transcriptional profiling revealed that L-GMPs exhibited a gene expression signature most similar to normal GMPs and not HSCs. The exception was a subset of genes also highly expressed in HSCs that appeared to represent reactivation of a "self-renewal program," including known MLL target genes such as *HoxA9* and *Meis1*, as well as new genes not previously implicated in MLL rearranged leukemia types such as *Mef2c*. These findings strongly support the contention that LSCs arise in a committed myeloid progenitor that aberrantly reactivates a self-renewal program to allow for unlimited replicative potential.

Recent studies support a progenitor cell of origin for the LSC in human disease as well. In 2006, Jamieson and colleagues demonstrated that leukemic blasts from patients with CML in blast crisis are characterized by aberrant activation of the Wnt/β-catenin pathway, acquired self-renewal, and an immunophenotype similar to normal GMPs (CD34+CD38+CD123+CD45RA+). It was thus proposed that in CML, a disease in which the chronic phase cell of origin is likely the HSC, additional "hits" occur in downstream committed progenitors, leading to aberrant self-renewal and progression to blast crisis. The HSC in CML may thus be considered a "pre-leukemic" HSC, with a GMP-like progenitor representing the downstream target for full transformation into an LSC.

In 2008, Taussig and colleagues demonstrated that the specific monoclonal antibody clone routinely used to identify CD38 on AML blasts led to Fc-mediated clearance in the xenograft setting. Reduced AML engraftment was most marked in NOD/SCID mice, with improved engraftment seen in mice treated with anti-CD122 Abs, as well as NSG or NOD/SCID/$\beta_{2m}{}^{-/-}$ (β2M) mice with decreased innate immunity due to NK-cell functional defects. Interestingly, anti-CD38 Ab treatment reproducibly led to impaired engraftment of AML samples out of proportion to normal cord blood (in which HSCs are CD34+CD38−), suggesting that the CD34+CD38+ compartment of AML may also contain LSCs.

This possibility has been borne out in a recent study of a large number of primary AML specimens by Goardon and colleagues (2011), in which the CD34+ compartment of ~80% of 100 AML specimens was found to contain two predominant populations—one that is CD38+CD110−CD45RA+and one that is CD38−CD90−CD45RA+, with the former representing a GMP-like immunophenotype and the latter representing what has been proposed to be an LMPP. Using serial transplantation assays in NSG mice, LSC activity was found to be most highly enriched in the LMPP-like population (as frequent as 1 in 1,000 cells). However, LSC activity was also consistently present in the GMP-like compartment, although at a lower frequency, and the two LSC populations appeared to coexist in a hierarchy, with LMPP-like LSCs giving rise to GMP-like LSCs, but not vice versa.

Plasticity of the LSC Phenotype

Given the clinical, biologic, and molecular heterogeneity of AML, perhaps it is not surprising that LSCs may not always share the immunophenotypic characteristics discussed above. A minority (~20%) of AMLs evaluated in the study by Goardon and colleagues (2011) contained LSCs in the CD34+ compartment with alternative phenotypes (CD38−CD45RA− and CD38+CD110−CD45RA−) similar to normal MPPs and CMPs instead of normal LMPPs and GMPs. Taussig and colleagues (2010) characterized NPM1-mutated AMLs, which usually lack CD34 expression on the majority of cells. Engraftment assays in β2M and NSG mice demonstrated LSC activity exclusively in the CD34− fraction in 6 out of 15 AMLs, with the remaining samples engrafting from both the CD34+ and CD34− fractions.

Recent work by Sarry and colleagues (2011) examined engraftment of 11 cytogenetically and molecularly diverse primary AML specimens in NSG mice and determined that LSC activity is present in diverse immunophenotypic compartments. Although LSC activity was confirmed to be most highly enriched in CD34+CD38− blasts in the majority of samples, CD34 expressing AMLs were also found to have LSC activity within the more "mature" cells, including the CD34− and lineage (dim) fractions. Provocatively, "more differentiated" LSCs gave rise to the original phenotypic diversity of the primary specimen in xenografts, at odds with the theory that AML development mirrors the unidirectional hierarchical model of normal hematopoiesis. Several explanations were proposed: (1) LSCs are enriched in immature immunophenotypes, and cells largely differentiate in a manner similar to normal hematopoiesis, but the loss of stem cell activity that typically accompanies acquisition of lineage-specific markers is attenuated, and LSCs harbor some capacity to dedifferentiate; (2) LSCs do not recapitulate normal hematopoiesis and are in a "neomorphic" state with aberrant expression of lineage markers; and (3) LSCs are immunophenotypically distinct from HSCs, and they are characterized by the presence of novel cell surface markers, or "neoantigens." Discrimination of the first two theories will rely on further studies with a quantitative assessment of the engraftment potential of populations at different stages of "maturation," while evidence for the last hypothesis has been emerging over the past decade (to be discussed in the following section). In addition to these possibilities, we propose that technical limitations of cell-sorting techniques may lead to contamination of "mature" leukemic cells with a small number of cells from LSC-enriched fractions, leading to the artifactual identification of LSC activity when very large numbers of cells are transplanted.

Finally, evaluation of the *HoxA9-Meis1* driven mouse model of AML reveals that LSC activity is not confined to one, but rather to three immunophenotypic populations corresponding to stem/progenitors, committed myeloid progenitors, and lymphoid lineage cells (Gibbs et al., 2012). Transplantation of as few as 100 cells from each immunophenotype was sufficient to engraft disease in secondary recipients. Not only did LSCs with these disparate immunophenotypes share common signaling and survival pathways as determined by microarray analysis and mass spectrometry, they also clonally recapitulated the full immunophenotypic diversity of the original leukemia in secondary transplants. Together, these studies of mouse and human leukemia types demonstrate that AML can defy the lineage marker boundaries and unidirectional differentiation characteristics found in normal hematopoiesis. Figure 11.1 summarizes the conventional schema of hierarchical normal hematopoiesis, overlaid with the transformation, aberrant self-renewal, and even de-differentiation that may comprise leukemic hematopoiesis.

Leukemic Stem Cell Markers

Given the heterogeneity of LSCs with respect to expression of normal differentiation associated markers, the identification of "neoantigens" present on LSCs but not normal HSCs might allow for the identification of LSC enriched populations with higher sensitivity and specificity. The identification of novel LSC markers might also allow for the development of therapies that specifically target LSCs, as well as for the prospective isolation of residual normal HSCs in AML. This latter application might provide a powerful tool for the performance of autologous stem cell transplants, using grafts "purged" of LSCs.

The first identified LSC marker was CD123, or the IL-3 receptor, which was described by Jordan and colleagues (2000) to be expressed on nearly all CD34+CD38− cells from 18 primary AML specimens. In xenografts, although all of the mice transplanted with CD34+CD38−CD123+ cells engrafted with AML, none of the mice transplanted with CD34+CD38−CD123− cells did. However, a subsequent study demonstrated that CD34+CD38−CD123+ cord blood cells were capable of multilineage engraftment in NOD/SCID mice, suggesting that some normal HSCs may also express CD123 (Taussig et al., 2005). In 2006, Jin and colleagues (2006) reported that CD44 was highly expressed on CD34+CD38− blasts in multiple AML samples. Although the engraftment potential of CD44+ and

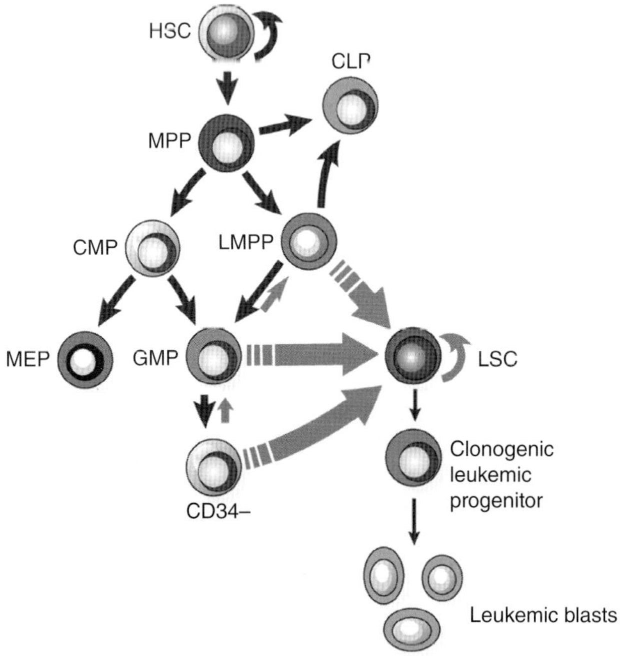

Immunophenotype	LSC frequency	References
LMPP CD34+CD38–CD90–CD45RA+	1 in 1064 to 1 in 10777 (3 samples)	(Goardon et al., 2011)
GMP CD34+CD38+CD110–CD45RA+	1 in 8879 to >1 in 10^5 (3 samples)	(Goardon et al., 2011)
CD34–	1 in 458 to 1 in 254,454 (6 samples) not quantified (10^5 to 10^6 cells transplanted)	(Taussig et al., 2010) (Sarry et al., 2011)

Figure 11.1. Hematopoietic stem and progenitor cells are putative targets for transformation into the Leukemic Stem Cell. The hierarchical model of normal hematopoiesis is illustrated with black arrows. Curved arrows indicate self-renewal capability. Dotted red lines indicate leukemic transformation. Other red arrows indicate the heretofore unresolved hypothesis that leukemic cells may "de-differentiate" with regards to lineage marker expression. Listed in the table (inset) are LSC frequencies in the putative targets for transformation as estimated by limiting dilution analyses. Abbreviations: Hematopoietic stem cell (HSC), common lymphoid progenitor (CLP), multipotent progenitor (MPP), lymphoid-primed multipotent progenitor (LMPP), common myeloid progenitor (CMP), megakaryocyte erythroid progenitor (MEP), granulocyte macrophage progenitor (GMP), leukemic stem cell (LSC). For color detail, please see color plate section.

CD44− negative fractions was not tested, incubation with an activating anti-CD44 antibody impaired the engraftment of AML CD34+CD38− cells. This inhibitory effect on engraftment was not observed in normal CB or BM CD34+CD38− cells, suggesting a unique function of CD44 on LSCs.

Several additional LSC specific markers capable of distinguishing LSCs from normal HSCs have been identified over the subsequent decade including CLL-1 (van Rhenen et al., 2007), CD96 (Hosen et al., 2007), CD47 (Majeti et al., 2009), and TIM3 (Jan et al., 2011). Some LSC markers, such as CD32 and CD25, were identified by bioinformatic analyses of genes differentially expressed in purified CD34+CD38− cells from AML compared with normal BM

(Saito et al., 2010). Development of monoclonal antibody-based therapeutics targeting CD44 (Jin et al., 2006), CD123 (Jin et al., 2009), and CD47 (Majeti et al., 2009) is ongoing, but to date no LSC-directed therapies based on these markers has been approved for clinical use. Additionally, it is important to note that while these markers are expressed on LSCs, they are also present on mature leukemia cells as well. The identification of markers truly specific for LSCs continues to be an area of active investigation. It may be useful to utilize classical stem cell assays that do not rely on cell surface markers such as ALDH, which has recently been demonstrated to detect a chemo-resistant leukemia cell population that predicts for disease relapse (Gerber et al., 2012).

The Pre-Leukemic HSC

The short life span of hematopoietic progenitor cells makes them unlikely to be able to retain the sequential genetic changes required for transformation into the LSC. To reconcile the need for multiple genetic "hits" (e.g., class I mutations promoting proliferation and class II mutations impairing differentiation (reviewed by Deguchi and Gilliland, 2002), with the mounting evidence for a progenitor cell of origin for the LSC, the concept of a "pre-leukemic" HSC has been proposed. This concept proposes that over time, a subpopulation of self-renewing HSCs acquires the complement of mutations necessary, but not sufficient, for transformation, with the final transforming lesions occurring in a downstream progenitor that acquires aberrant self-renewal. There are now several lines of evidence to support this hypothesis.

In the early to mid-1990s, several groups reported that *AML1/ETO* transcripts could be detected in the bone marrow of nearly all t(8;21) AML patients in long-term remission (Downing, 1999; Kusec et al., 1994; Miyamoto et al., 1996; Miyamoto et al., 1995; Nucifora et al., 1993; Yin and Tobal, 1999). To determine whether cells harboring these cytogenetic abnormalities represented minimal residual disease (MRD), Miyamoto and colleagues (2000) examined AML1/ETO mRNA expression in rigorously purified HSCs, progenitors, and mature hematopoietic cells. Interestingly, the *AML1/ETO* translocation was found to be present in up to 40% of non-leukemic HSCs, and these HSCs gave rise *in vitro* to *AML1/ETO* harboring mature myeloid cells and B cells. This demonstrated that *AML1/ETO* is not sufficient for leukemic transformation, and it also provided evidence for the existence of a pre-leukemic HSC. Furthermore, it suggested that genetic lesions commonly associated with AML do not necessarily represent "driver" mutations in AML.

Direct evidence for the presence of pre-leukemic genetic lesions in HSCs come from a recent analysis of six AML specimens (all characterized by a normal karyotype and the *Flt3-ITD* mutation) (Jan et al., 2012). Mutations were identified in bulk AML cells by exome sequencing, and residual normal HSCs were purified based on the lack of expression of LSC-specific cell surface markers. The residual "normal" HSCs were subjected to single cell assays targeted to the identified mutations. In five of the assayed specimens, mutations were identified in residual HSCs, with 32 of the 51 total mutations identified between the five samples being found in residual, pre-leukemic HSCs. Several known recurrent mutations in AML such as *NPM1, TET2, and SMC1A*, were found in residual HSCs from multiple specimens, suggesting a potential role of these as founder mutations. In contrast, *Flt3-ITD* was invariantly absent in residual normal HSCs, demonstrating that it is a secondary lesion that occurs closer to disease initiation. Three specimens were demonstrated to have a subclonal architecture reflective of the classic clonal evolution model of cancer, with sequential acquisition of mutations that likely confer clonal dominance (e.g., bi-allelic *TET2* mutant HSCs being more frequent than mono-allelic *TET2* mutant HSCs, which were in turn more frequent than *TET2* wild-type HSCs). The remaining specimens that did not conform to this model might reflect the role of the identified mutations as "passenger mutations" that do not contribute to disease pathogenesis, or they may reflect AMLs that do not conform to the classic clonal evolution model.

Recent genome sequencing studies of AML support the presence of a plethora of "passenger" mutations in *de novo* AML. Welch and colleagues (2012) analyzed data from whole genome sequencing of 24 leukemia cases including 12 FAB M3 leukemia types, in which the presumed disease-initiating lesion is known (*PML-RARalpha*), and 12 *de novo* FAB M1 leukemia types with normal cytogenetics, in which the disease-initiating lesion is unknown. Exome sequencing was also performed on CD34+CD38− cells (containing both HSCs and progenitors [HSPCs]) from seven normal controls of varying ages. An age-dependent accumulation of somatic mutations was observed in normal HSPCs on the order of 0.13 exonic mutations per life-year. Remarkably, the vast majority of mutations detected in the AML specimens appeared to reflect these age-related lesions, suggesting that normal self-renewing HSCs accumulate largely random and benign age-related mutations that are simply "captured" in an AML founding clone upon the occurrence of a "driver" mutation. Furthermore, a comparison of FAB M1 and FAB M3 genomes allowed the investigators to infer that lesions only present in M1 genomes were likely "driver" mutations, while mutations shared by both M1 and M3 genomes were likely either later "cooperating" mutations or background passenger mutations. *NPM1*, *DNMT3a*, and *IDH1* were predicted to be driver mutations, and *Flt3-ITD* was predicted to be a cooperating mutation, corroborating the results of Jan et al. (2012). Notably, *SMC1A*, identified by Jan et al. (2012) to be present in pre-leukemic HSCs, was predicted by this analysis to be a cooperating lesion, suggesting that despite its presence in the HSC, it may still occur secondary to an antecedent founding lesion.

To understand why some genetic lesions have a predilection to appear as secondary events, while others are more likely to be founding events, it is helpful to consider the well-described association between quiescence and self-renewal in hematopoietic stem cells (Passegue et al., 2005). The pre-leukemic HSC must balance the proliferative characteristics necessary to gain clonal dominance with the self-renewal necessary for long-term persistence and the sequential acquisition of leukemogenic mutations. Unless these two properties can be uncoupled, mutations in powerful oncogenes or tumor suppressors such as *K-ras* (Braun et al., 2004), *PTEN* (Yilmaz et al., 2006; Zhang et al., 2006), and *Flt3-ITD* (Chu et al., 2012) may not serve as suitable pre-leukemic mutations since they would likely lead to HSC depletion/exhaustion. It has been proposed that "weaker" oncogenic mutations such as *N-ras*^{G12D} might confer a more modest proliferative advantage while preserving HSC self-renewal (Li et al., 2011). A closer examination of the effects of candidate founder gene mutations on the proliferation and self-renewal of HSCs using genetic models will be invaluable in understanding the molecular mechanisms by which LSCs usurp these typically divergent cellular characteristics to generate LSCs with high rates of unlimited proliferation.

Despite having made significant inroads into understanding which cell populations and mutations might be important for disease initiation, it has not been shown that targeting such lesions can eradicate stem cells in myeloid neoplasms. For example, despite the presence of *Bcr-abl* in CML HSCs and del(5q) in HSCs from a specific subtype of myelodysplasia, therapies that eliminate virtually all mature disease cells in these diseases fail to eliminate the stem cells harboring the disease specific cytogenetic abnormalities (Chu et al., 2005; Tehranchi et al., 2010). Perhaps the only example of a successful therapy directed toward a disease-initiating lesion is retinoic acid and arsenic trioxide in acute promyelocytic leukemia, but this represents a distinct subtype of

AML in which the disease-initiating lesion, PML-RARα, is not present in immature CD34+CD38− cells (Turhan et al., 1995).

Clinical Correlations with Prognosis and Response to Treatment

Beginning with the seminal experiments by Clarkson and colleagues (1970) in the 1970s that introduced the concept of the LSC, it was postulated that this rare cell population is chemo-resistant and serves as the reservoir for disease relapse. Additional circumstantial data for this hypothesis came from studies in leukemia reporting an association of a higher percentage of "immature" CD34+ cells with increased chemo-resistance and worse clinical outcome (Chang et al., 2004; Raspadori et al., 1997), although this association has not been consistently observed (Kanda et al., 2000). Overexpression of LSC markers such as CD123 (Graf et al., 2004; Testa et al., 2004; Testa et al., 2002) and CD47 (Majeti et al., 2009) have been shown to correlate with increased tumor burden and overall survival, respectively.

More recently, two large bioinformatic analyses identified LSC-associated gene expression signatures. Gentles and colleagues (2010) compared FACS purified CD34+CD38− "LSCs" with CD34+CD38+ "leukemic progenitor cells (LPC)" from seven AML samples to derive a 52-gene "LSC" signature. This signature consisted of many genes also expressed in normal CD34+CD38− cells, and enrichment for this signature in unfractionated AML cells was associated with a worse clinical outcome. Eppert and colleagues (2011) performed a similar analysis but restricted their analyses to LSC populations functionally validated to engraft in NOD/SCID mice. Four populations were FACS purified from each of 16 AML specimens based on CD34 and CD38 expression. Gene expression profiles were derived from 25 engrafting populations and compared with that of 29 non-engrafting populations to derive an LSC-related gene signature (LSC-R). Similar to the findings of Gentles and colleagues (2010), there was significant overlap between the LSC-R gene signature and a separately derived HSC-related gene signature (HSC-R), with high expression of either LSC-R or HSC-R in unfractionated AML blasts significantly associated with a worse clinical outcome.

Despite the associations among an LSC-specific gene, cell surface marker expressions, and clinical outcome, there is little direct evidence for the widely held notion that LSCs represent the reservoir for disease relapse. Recent genomic analyses have characterized the clonal evolution that accompanies relapse in AML, demonstrating selection for either a surviving founding clone that gains mutations and evolves into the relapse clone, or a subclone of the founding clone that survives chemotherapy, acquires additional mutations, and expands at relapse (Ding et al., 2012). Whether these surviving clones reflect the mutational complement of chemo-resistant LSCs or pre-leukemic HSCs remains to be determined.

Perhaps the most direct evidence that LSCs represent a chemo-resistant subpopulation comes from a study by Ishikawa and colleagues (2007), who engrafted NSG mice with FACS purified CD34+CD38− cells from primary AMLs and demonstrated stable engraftment of CD34+CD38− cells as well as their differentiation into CD34+CD38+ and CD34− cells. Following *in vivo* administration of cytarabine, a drug commonly used to treat AML patients, CD34+CD38− cells were significantly enriched in the BM of engrafted/treated animals. These cells were shown to resist apoptosis and were capable of engrafting AML in secondary recipients upon transplantation with as few as 1,000 cells. It remains to be determined whether LSCs with other immunophenotypes can also contribute to relapse, but one might presume that this is the case given the heterogeneity of LSC immunophenotypes and the high rate of relapse in AML. When a hierarchy of LSCs is present within the same AML, as proposed by Goardon and colleagues (2011), it will be important to identify the relative contributions of each LSC population to treatment resistance and relapse.

Still, the relevance of these findings in the xenograft setting to human AML is not clear since there is no similar enrichment for CD34+CD38− cells in human AML after induction therapy (Chung, Park et al., unpublished findings). It is possible that the LSC represents a dynamic survival state in the context of chemotherapy, akin to what has been described by Sharma and colleagues (2010) in lung cancer cell lines. Such a transient state might be consistent with the phenotypic plasticity described by Sarry and colleagues (2010) (discussed above), lending further credence to the idea that LSCs do not adhere to a strict developmental hierarchy.

Future Directions

AML represents the first human malignancy shown to fulfill the predictions of the cancer stem cell hypothesis, and significant advances have been made over the past several decades to understand the molecular pathogenesis and clonal heterogeneity of this disease. However, there remains a great divide between the advances made in the laboratory and our ability to treat this disease effectively, as AML patient clinical outcomes remain dismal. With our increased understanding of LSC biology and ability to identify LSCs and study their function using in vivo and in vitro systems, it will be important to employ novel approaches to study the molecular mechanisms that regulate key LSC characteristics including self-renewal as well as therapy resistance.

LSC cell surface proteins will continue to afford ample opportunities for research. The ability to prospectively separate LSCs from normal hematopoietic cells will greatly enhance our ability to study LSC and AML pathogenesis. They may also serve as potential tools to assess MRD after treatment to help predict prognosis and select optimal therapies for patients. Thus, it will be important to determine the sensitivity and specificity of the expression of these LSC markers across different subtypes of AML. Because all described LSC markers thus far appear to be expressed on all "mature" bulk leukemic blasts as well, further enrichment of LSC-containing populations may require new functional genomics initiatives to discover new LSC surface markers and/or the combination of these markers with classical stains used to identify stem cells including ALDH (Kastan et al., 1990) or rhodamine (Kim et al., 1998). Understanding the molecular differences between LSCs and normal HSPCs may also allow for the identification of LSC-specific signaling pathways that might be targeted by existing or emerging small molecule therapeutics. The development of therapeutic monoclonal antibodies targeting LSC cell surface markers is already under way and holds great promise.

Although assessment of LSC function and drug responses in immunodeficient mice will continue to be an important research asset, it is important to recognize the limitations of existing xenograft models, which cannot recapitulate the growth and stromal factors present in human BM and lack the immune surveillance mechanisms present in humans. The development of increasingly "humanized" mice such as the NSGS mouse (Wunderlich et al., 2010), as well as engraftment of human stromal/immune system components into immunodeficient mice (akin to the SCID-Hu model), may improve our ability to model the functional heterogeneity of human AML *in vivo*. Fortunately, existing models are sufficiently robust to perform screens of small molecule therapeutics on primary AML samples *in vivo*, as has been recently reported (McDermott et al., 2012). We propose that modeling current AML therapies utilized in the clinic (e.g., cytarabine, daunorubicin, and hypomethylating agents) using xenografted AMLs will provide invaluable insight into the mechanisms underlying drug response and chemoresistance. An important adjunct to these studies will be the parallel evaluation of LSCs in mouse models of AML.

To best understand human LSC biology and the role of the LSC in therapeutic resistance, future studies should aim to prospectively isolate putative human LSC populations over the course of disease treatment. Although existing studies have focused on evaluating blasts at diagnosis and relapse, we believe that the most informative specimens will be obtained during the time period immediately following the administration of chemotherapy. Fortunately, in some institutions, it is now routine to perform BM examinations at this time point ("day +14 bone marrow"), and we look forward to assessing this valuable resource for evidence of human LSCs.

Together, efforts to develop better strategies to identify and prospectively isolate LSCs, model human AML *in vivo*, and acquire patient specimens at the most informative treatment time points will enhance our understanding of LSC function and generate novel approaches to target them therapeutically. A complete characterization of this multifaceted cell holds the key to providing curative therapies for our patients.

References

Baum, C.M., Weissman, I.L., Tsukamoto, A.S., Buckle, A.M., and Peault, B. (1992). Isolation of a candidate human hematopoietic stem-cell population. Proc Natl Acad Sci USA *89*, 2804–2808.

Bhatia, M., Wang, J.C., Kapp, U., Bonnet, D., and Dick, J.E. (1997). Purification of primitive human hematopoietic cells capable of repopulating immune-deficient mice. Proc Natl Acad Sci USA *94*, 5320–5325.

Bhatia, R., Holtz, M., Niu, N., Gray, R., Snyder, D.S., Sawyers, C.L., Arber, D.A., Slovak, M.L., and Forman, S.J. (2003). Persistence of malignant hematopoietic progenitors in chronic myelogenous leukemia patients in complete cytogenetic remission following imatinib mesylate treatment. Blood *101*, 4701–4707.

Blair, A., Hogge, D.E., Ailles, L.E., Lansdorp, P.M., and Sutherland, H.J. (1997). Lack of expression of Thy-1 (CD90) on acute myeloid leukemia cells with long-term proliferative ability in vitro and in vivo. Blood *89*, 3104–3112.

Blair, A., and Sutherland, H.J. (2000). Primitive acute myeloid leukemia cells with long-term proliferative ability in vitro and in vivo lack surface expression of c-kit (CD117). Exp Hematol *28*, 660–671.

Bonnet, D., and Dick, J.E. (1997). Human acute myeloid leukemia is organized as a hierarchy that originates from a primitive hematopoietic cell. Nat Med *3*, 730–737.

Braun, B.S., Tuveson, D.A., Kong, N., Le, D.T., Kogan, S.C., Rozmus, J., Le Beau, M.M., Jacks, T.E., and Shannon, K.M. (2004). Somatic activation of oncogenic Kras in hematopoietic cells initiates a rapidly fatal myeloproliferative disorder. Proc Natl Acad Sci USA *101*, 597–602.

Bryder, D., Rossi, D.J., and Weissman, I.L. (2006). Hematopoietic stem cells: the paradigmatic tissue-specific stem cell. Am J Pathol *169*, 338–346.

Cashman, J.D., Lapidot, T., Wang, J.C., Doedens, M., Shultz, L.D., Lansdorp, P., Dick, J.E., and Eaves, C.J. (1997). Kinetic evidence of the regeneration of multilineage hematopoiesis from primitive cells in normal human bone marrow transplanted into immunodeficient mice. Blood *89*, 4307–4316.

Chang, H., Salma, F., Yi, Q.L., Patterson, B., Brien, B., and Minden, M.D. (2004). Prognostic relevance of immunophenotyping in 379 patients with acute myeloid leukemia. Leuk Res *28*, 43–48.

Chu, S., Xu, H., Shah, N.P., Snyder, D.S., Forman, S.J., Sawyers, C.L., and Bhatia, R. (2005). Detection of BCR-ABL kinase mutations in CD34+ cells from chronic myelogenous leukemia patients in complete cytogenetic remission on imatinib mesylate treatment. Blood *105*, 2093–2098.

Chu, S.H., Heiser, D., Li, L., Kaplan, I., Collector, M., Huso, D., Sharkis, S.J., Civin, C., and Small, D. (2012). FLT3-ITD Knockin Impairs Hematopoietic Stem Cell Quiescence/ Homeostasis, Leading to Myeloproliferative Neoplasm. Cell Stem Cell *11*, 346–358.

Clarkson, B., Fried, J., Strife, A., Sakai, Y., Ota, K., and Okita, T. (1970). Studies of cellular proliferation in human leukemia. 3. Behavior of leukemic cells in three adults with acute leukemia given continuous infusions of 3H-thymidine for 8 or 10 days. Cancer *25*, 1237–1260.

Clarkson, B.D. (1969). Review of recent studies of cellular proliferation in acute leukemia. Natl Cancer Inst Monogr *30*, 81–120.

Clarkson, B.D., Dowling, M.D., Gee, T.S., Cunningham, I.B., and Burchenal, J.H. (1975). Treatment of acute leukemia in adults. Cancer *36*, 775–795.

Clarkson, B.D., and Fried, J. (1971). Changing concepts of treatment in acute leukemia. Med Clin North Am *55*, 561–600.

Cozzio, A., Passegue, E., Ayton, P.M., Karsunky, H., Cleary, M.L., and Weissman, I.L. (2003). Similar MLL-associated leukemias arising from self-renewing stem cells and short-lived myeloid progenitors. Genes & development *17*, 3029–3035.

Craig, W., Kay, R., Cutler, R.L., and Lansdorp, P.M. (1993). Expression of Thy-1 on human hematopoietic progenitor cells. J Exp Med *177*, 1331–1342.

Cronkite, E.P. (1970). Acute leukemia: Is there a relationship between cell growth kinetics and response to chemotherapy? Proc Natl Cancer Conf *6*, 113–117.

Deguchi, K., and Gilliland, D.G. (2002). Cooperativity between mutations in tyrosine kinases and in hematopoietic transcription factors in AML. Leukemia *16*, 740–744.

DiMartino, J.F., Ayton, P.M., Chen, E.H., Naftzger, C.C., Young, B.D., and Cleary, M.L. (2002). The AF10 leucine zipper is required for leukemic transformation of myeloid progenitors by MLL-AF10. Blood *99*, 3780–3785.

Ding, L., Ley, T.J., Larson, D.E., Miller, C.A., Koboldt, D.C., Welch, J.S., Ritchey, J.K., Young, M.A., Lamprecht, T., McLellan, M.D., *et al.* (2012). Clonal evolution in relapsed acute myeloid leukaemia revealed by whole-genome sequencing. Nature *481*, 506–510.

Doulatov, S., Notta, F., Laurenti, E., and Dick, J.E. (2012). Hematopoiesis: A human perspective. Cell Stem Cell *10*, 120–136.

Downing, J.R. (1999). The AML1-ETO chimaeric transcription factor in acute myeloid leukaemia: biology and clinical significance. Br J Haematol *106*, 296–308.

Eppert, K., Takenaka, K., Lechman, E.R., Waldron, L., Nilsson, B., van Galen, P., Metzeler, K.H., Poeppl, A., Ling, V., Beyene, J., *et al.* (2011). Stem cell gene expression programs influence clinical outcome in human leukemia. Nat Med *17*, 1086–1093.

Fialkow, P.J., Jacobson, R.J., and Papayannopoulou, T. (1977). Chronic myelocytic leukemia: Clonal origin in a stem cell common to the granulocyte, erythrocyte, platelet and monocyte/macrophage. Am J Med *63*, 125–130.

Gentles, A.J., Plevritis, S.K., Majeti, R., and Alizadeh, A.A. (2010). Association of a leukemic stem cell gene expression signature with clinical outcomes in acute myeloid leukemia. JAMA: the journal of the American Medical Association *304*, 2706–2715.

Gerber, J.M., Smith, B.D., Ngwang, B., Zhang, H., Vala, M.S., Morsberger, L., Galkin, S., Collector, M.I., Perkins, B., Levis, M.J., *et al.* (2012). A clinically relevant population of leukemic CD34(+) CD38(−) cells in acute myeloid leukemia. Blood *119*, 3571–3577.

Gibbs, K.D., Jr., Jager, A., Crespo, O., Goltsev, Y., Trejo, A., Richard, C.E., and Nolan, G.P. (2012). Decoupling of tumor-initiating activity from stable immunophenotype in HoxA9-Meis1-driven AML. Cell Stem Cell *10*, 210–217.

Goardon, N., Marchi, E., Atzberger, A., Quek, L., Schuh, A., Soneji, S., Woll, P., Mead, A., Alford, K.A., Rout, R., *et al.* (2011). Coexistence of LMPP-like and GMP-like leukemia stem cells in acute myeloid leukemia. Cancer Cell *19*, 138–152.

Graf, M., Hecht, K., Reif, S., Pelka-Fleischer, R., Pfister, K., and Schmetzer, H. (2004). Expression and prognostic value of hemopoietic cytokine receptors in acute myeloid leukemia (AML): implications for future therapeutical strategies. Eur J Haematol *72*, 89–106.

Graham, S.M., Jorgensen, H.G., Allan, E., Pearson, C., Alcorn, M.J., Richmond, L., and Holyoake, T.L. (2002). Primitive, quiescent, Philadelphia-positive stem cells from patients with chronic myeloid leukemia are insensitive to STI571 in vitro. Blood *99*, 319–325.

Hogan, C.J., Shpall, E.J., and Keller, G. (2002). Differential long-term and multilineage engraftment potential from subfractions of human CD34+ cord blood cells transplanted into NOD/SCID mice. Proc Natl Acad Sci USA *99*, 413–418.

Hosen, N., Park, C.Y., Tatsumi, N., Oji, Y., Sugiyama, H., Gramatzki, M., Krensky, A.M., and Weissman, I.L. (2007). CD96 is a leukemic stem cell-specific marker in human acute myeloid leukemia. Proc Natl Acad Sci USA *104*, 11008–11013.

Hulett, H.R., Bonner, W.A., Barrett, J., and Herzenberg, L.A. (1969). Cell sorting: Automated separation of mammalian cells as a function of intracellular fluorescence. Science *166*, 747–749.

Ishikawa, F., Yoshida, S., Saito, Y., Hijikata, A., Kitamura, H., Tanaka, S., Nakamura, R.,

Tanaka, T., Tomiyama, H., Saito, N., *et al.* (2007). Chemotherapy-resistant human AML stem cells home to and engraft within the bone-marrow endosteal region. Nat Biotechnol *25*, 1315–1321.

Jan, M., Chao, M.P., Cha, A.C., Alizadeh, A.A., Gentles, A.J., Weissman, I.L., and Majeti, R. (2011). Prospective separation of normal and leukemic stem cells based on differential expression of TIM3, a human acute myeloid leukemia stem cell marker. Proc Natl Acad Sci USA *108*, 5009–5014.

Jan, M., Snyder, T.M., Corces-Zimmerman, M.R., Vyas, P., Weissman, I.L., Quake, S.R., and Majeti, R. (2012). Clonal evolution of preleukemic hematopoietic stem cells precedes human acute myeloid leukemia. Sci Transl Med *4*, 149ra118.

Jin, L., Hope, K.J., Zhai, Q., Smadja-Joffe, F., and Dick, J.E. (2006). Targeting of CD44 eradicates human acute myeloid leukemic stem cells. Nat Med *12*, 1167–1174.

Jin, L., Lee, E.M., Ramshaw, H.S., Busfield, S.J., Peoppl, A.G., Wilkinson, L., Guthridge, M.A., Thomas, D., Barry, E.F., Boyd, A., *et al.* (2009). Monoclonal antibody-mediated targeting of CD123, IL-3 receptor alpha chain, eliminates human acute myeloid leukemic stem cells. Cell Stem Cell *5*, 31–42.

Jordan, C.T., Upchurch, D., Szilvassy, S.J., Guzman, M.L., Howard, D.S., Pettigrew, A.L., Meyerrose, T., Rossi, R., Grimes, B., Rizzieri, D.A., et al. (2000). The interleukin-3 receptor alpha chain is a unique marker for human acute myelogenous leukemia stem cells. Leukemia *14*, 1777–1784.

Kamel-Reid, S., and Dick, J.E. (1988). Engraftment of immune-deficient mice with human hematopoietic stem cells. Science *242*, 1706–1709.

Kanda, Y., Hamaki, T., Yamamoto, R., Chizuka, A., Suguro, M., Matsuyama, T., Takezako, N., Miwa, A., Kami, M., Hirai, H., et al. (2000). The clinical significance of CD34 expression in response to therapy of patients with acute myeloid leukemia: an overview of 2483 patients from 22 studies. Cancer *88*, 2529–2533.

Kastan, M.B., Schlaffer, E., Russo, J.E., Colvin, O.M., Civin, C.I., and Hilton, J. (1990). Direct demonstration of elevated aldehyde dehydrogenase in human hematopoietic progenitor cells. Blood *75*, 1947–1950.

Kiel, M.J., Yilmaz, O.H., Iwashita, T., Yilmaz, O.H., Terhorst, C., and Morrison, S.J. (2005). SLAM family receptors distinguish hematopoietic stem and progenitor cells and reveal endothelial niches for stem cells. Cell *121*, 1109–1121.

Killmann, S.A., Cronkite, E.P., Robertson, J.S., Fliedner, T.M., and Bond, V.P. (1963). Estimation of phases of the life cycle of leukemic cells from labeling in human beings in vivo with tritiated thymidine. Lab Invest *12*, 671–684.

Kim, M., Cooper, D.D., Hayes, S.F., and Spangrude, G.J. (1998). Rhodamine-123 staining in hematopoietic stem cells of young mice indicates mitochondrial activation rather than dye efflux. Blood *91*, 4106–4117.

Kohler, G., and Milstein, C. (1975). Continuous cultures of fused cells secreting antibody of predefined specificity. Nature *256*, 495–497.

Krivtsov, A.V., Twomey, D., Feng, Z., Stubbs, M.C., Wang, Y., Faber, J., Levine, J.E., Wang, J., Hahn, W.C., Gilliland, D.G., *et al.* (2006). Transformation from committed progenitor to leukaemia stem cell initiated by MLL-AF9. Nature *442*, 818–822.

Kusec, R., Laczika, K., Knobl, P., Friedl, J., Greinix, H., Kahls, P., Linkesch, W., Schwarzinger, I., Mitterbauer, G., Purtscher, B., et al. (1994). AML1/ETO fusion mRNA can be detected in remission blood samples of all patients with t(8;21) acute myeloid leukemia after chemotherapy or autologous bone marrow transplantation. Leukemia *8*, 735–739.

Lapidot, T., Pflumio, F., Doedens, M., Murdoch, B., Williams, D.E., and Dick, J.E. (1992). Cytokine stimulation of multilineage hematopoiesis from immature human cells engrafted in SCID mice. Science *255*, 1137–1141.

Lapidot, T., Sirard, C., Vormoor, J., Murdoch, B., Hoang, T., Caceres-Cortes, J., Minden, M., Paterson, B., Caligiuri, M.A., and Dick, J.E. (1994). A cell initiating human acute myeloid leukaemia after transplantation into SCID mice. Nature *367*, 645–648.

Li, Q., Haigis, K.M., McDaniel, A., Harding-Theobald, E., Kogan, S.C., Akagi, K., Wong, J.C., Braun, B.S., Wolff, L., Jacks, T., *et al.* (2011). Hematopoiesis and leukemogenesis in mice expressing oncogenic NrasG12D from the endogenous locus. Blood *117*, 2022–2032.

Majeti, R., Chao, M.P., Alizadeh, A.A., Pang, W.W., Jaiswal, S., Gibbs, K.D., Jr., van Rooijen, N., and Weissman, I.L. (2009). CD47 is an adverse prognostic factor and therapeutic antibody target on human acute myeloid leukemia stem cells. Cell *138*, 286–299.

Majeti, R., Park, C.Y., and Weissman, I.L. (2007). Identification of a hierarchy of multipotent hematopoietic progenitors in human cord blood. Cell Stem Cell *1*, 635–645.

Martin, P.J., Najfeld, V., Hansen, J.A., Penfold, G.K., Jacobson, R.J., and Fialkow, P.J. (1980). Involvement of the B-lymphoid system in chronic myelogenous leukaemia. Nature *287*, 49–50.

McCune, J.M., Namikawa, R., Kaneshima, H., Shultz, L.D., Lieberman, M., and Weissman, I.L. (1988). The SCID-hu mouse: murine model for the analysis of human hematolymphoid differentiation and function. Science *241*, 1632–1639.

McDermott, S.P., Eppert, K., Notta, F., Isaac, M., Datti, A., Al-Awar, R., Wrana, J., Minden, M.D., and Dick, J.E. (2012). A small molecule screening strategy with validation on human leukemia stem cells uncovers the therapeutic efficacy of kinetin riboside. Blood *119*, 1200–1207.

Miyamoto, T., Nagafuji, K., Akashi, K., Harada, M., Kyo, T., Akashi, T., Takenaka, K., Mizuno, S., Gondo, H., Okamura, T., *et al.* (1996). Persistence of multipotent progenitors expressing AML1/ETO transcripts in long-term remission patients with t(8;21) acute myelogenous leukemia. Blood *87*, 4789–4796.

Miyamoto, T., Nagafuji, K., Harada, M., Eto, T., Fujisaki, T., Kubota, A., Akashi, K., Mizuno, S., Takenaka, K., Kanaji, T., *et al.* (1995). Quantitative analysis of AML1/ETO transcripts in peripheral blood stem cell harvests from patients with t(8;21) acute myelogenous leukaemia. Br J Haematol *91*, 132–138.

Miyamoto, T., Weissman, I.L., and Akashi, K. (2000). AML1/ETO-expressing nonleukemic stem cells in acute myelogenous leukemia with 8;21 chromosomal translocation. Proc Natl Acad Sci USA *97*, 7521–7526.

Murray, L., Chen, B., Galy, A., Chen, S., Tushinski, R., Uchida, N., Negrin, R., Tricot, G., Jagannath, S., Vesole, D., et al. (1995). Enrichment of human hematopoietic stem cell activity in the CD34+Thy-1+Lin- subpopulation from mobilized peripheral blood. Blood *85*, 368–378.

Nilsson, L., Astrand-Grundstrom, I., Anderson, K., Arvidsson, I., Hokland, P., Bryder, D., Kjeldsen, L., Johansson, B., Hellstrom-Lindberg, E., Hast, R., *et al.* (2002). Involvement and functional impairment of the CD34(+)CD38(−)Thy-1(+) hematopoietic stem cell pool in myelodysplastic syndromes with trisomy 8. Blood *100*, 259–267.

Nilsson, L., Eden, P., Olsson, E., Mansson, R., Astrand-Grundstrom, I., Strombeck, B., Theilgaard-Monch, K., Anderson, K., Hast, R., Hellstrom-Lindberg, E., *et al.* (2007). The molecular signature of MDS stem cells supports a stem-cell origin of 5q myelodysplastic syndromes. Blood *110*, 3005–3014.

Notta, F., Doulatov, S., Laurenti, E., Poeppl, A., Jurisica, I., and Dick, J.E. (2011). Isolation of single human hematopoietic stem cells capable of long-term multilineage engraftment. Science *333*, 218–221.

Nucifora, G., Larson, R.A., and Rowley, J.D. (1993). Persistence of the 8;21 translocation in patients with acute myeloid leukemia type M2 in long-term remission. Blood *82*, 712–715.

Park, C.Y., Pang, W.W., Price, E., Pluvinage, J.A., Schrier, S.L., and Weissman, I.L. (2011). Hematopoietic Stem Cells Are the Disease-Initiating Cells in the Myelodysplastic Syndromes. ASH Annual Meeting Abstracts *118*, 789.

Passegue, E., Wagers, A.J., Giuriato, S., Anderson, W.C., and Weissman, I.L. (2005). Global analysis of proliferation and cell cycle gene expression in the regulation of hematopoietic stem and progenitor cell fates. J Exp Med *202*, 1599–1611.

Peault, B., Weissman, I., and Baum, C. (1993). Analysis of candidate human blood stem cells in "humanized" immune-deficiency SCID mice. Leukemia *7 Suppl 2*, S98–101.

Raspadori, D., Lauria, F., Ventura, M.A., Rondelli, D., Visani, G., de Vivo, A., and Tura, S. (1997). Incidence and prognostic relevance of CD34 expression in acute myeloblastic leukemia: analysis of 141 cases. Leuk Res *21*, 603–607.

Raza, A., Preisler, H.D., Day, R., Yasin, Z., White, M., Lykins, J., Kukla, C., Barcos, M., Bennett, J., Browman, G., *et al.* (1990). Direct relationship between remission duration in acute myeloid leukemia and cell cycle kinetics: A leukemia intergroup study. Blood *76*, 2191–2197.

Saito, Y., Kitamura, H., Hijikata, A., Tomizawa-Murasawa, M., Tanaka, S., Takagi, S., Uchida, N., Suzuki, N., Sone, A., Najima, Y., *et al.* (2010). Identification of therapeutic targets for quiescent, chemotherapy-resistant human leukemia stem cells. Sci Transl Med *2*, 17ra19.

Sarry, J.E., Murphy, K., Perry, R., Sanchez, P.V., Secreto, A., Keefer, C., Swider, C.R., Strzelecki, A.C., Cavelier, C., Recher, C., *et al.* (2011). Human acute myelogenous leukemia stem cells are rare and heterogeneous when assayed in NOD/SCID/IL2Rgammac-deficient mice. J Clin Invest *121*, 384–395.

Sharma, S.V., Lee, D.Y., Li, B., Quinlan, M.P., Takahashi, F., Maheswaran, S., McDermott, U., Azizian, N., Zou, L., Fischbach, M.A., *et al.* (2010). A chromatin-mediated reversible drug-tolerant state in cancer cell subpopulations. Cell *141*, 69–80.

Spangrude, G.J., Heimfeld, S., and Weissman, I.L. (1988). Purification and characterization of mouse hematopoietic stem cells. Science *241*, 58–62.

Takenaka, K., Prasolava, T.K., Wang, J.C., Mortin-Toth, S.M., Khalouei, S., Gan, O.I., Dick, J.E., and Danska, J.S. (2007). Polymorphism in Sirpa modulates engraftment of human hematopoietic stem cells. Nat Immunol *8*, 1313–1323.

Taussig, D.C., Miraki-Moud, F., Anjos-Afonso, F., Pearce, D.J., Allen, K., Ridler, C., Lillington, D., Oakervee, H., Cavenagh, J., Agrawal, S.G., *et al.* (2008). Anti-CD38 antibody-mediated clearance of human repopulating cells masks the heterogeneity of leukemia-initiating cells. Blood *112*, 568–575.

Taussig, D.C., Pearce, D.J., Simpson, C., Rohatiner, A.Z., Lister, T.A., Kelly, G., Luongo, J.L., Danet-Desnoyers, G.A., and Bonnet, D. (2005). Hematopoietic stem cells express multiple myeloid markers: Implications for the origin and targeted therapy of acute myeloid leukemia. Blood *106*, 4086–4092.

Taussig, D.C., Vargaftig, J., Miraki-Moud, F., Griessinger, E., Sharrock, K., Luke, T., Lillington, D., Oakervee, H., Cavenagh, J., Agrawal, S.G., *et al.* (2010). Leukemia-initiating cells from some acute myeloid leukemia patients with mutated nucleophosmin reside in the CD34(-) fraction. Blood *115*, 1976–1984.

Tehranchi, R., Woll, P.S., Anderson, K., Buza-Vidas, N., Mizukami, T., Mead, A.J., Astrand-Grundstrom, I., Strombeck, B., Horvat, A., Ferry, H., *et al.* (2010). Persistent malignant stem cells in del(5q) myelodysplasia in remission. The New England journal of medicine *363*, 1025–1037.

Testa, U., Riccioni, R., Diverio, D., Rossini, A., Lo Coco, F., and Peschle, C. (2004). Interleukin-3 receptor in acute leukemia. Leukemia *18*, 219–226.

Testa, U., Riccioni, R., Militi, S., Coccia, E., Stellacci, E., Samoggia, P., Latagliata, R., Mariani, G., Rossini, A., Battistini, A., *et al.* (2002). Elevated expression of IL-3Ralpha in acute myelogenous leukemia is associated with enhanced blast proliferation, increased cellularity, and poor prognosis. Blood *100*, 2980–2988.

Turhan, A.G., Lemoine, F.M., Debert, C., Bonnet, M.L., Baillou, C., Picard, F., Macintyre, E.A., and Varet, B. (1995). Highly purified primitive hematopoietic stem cells are PML-RARA negative and generate nonclonal progenitors in acute promyelocytic leukemia. Blood *85*, 2154–2161.

van Rhenen, A., van Dongen, G.A., Kelder, A., Rombouts, E.J., Feller, N., Moshaver, B., Stigter-van Walsum, M., Zweegman, S., Ossenkoppele, G.J., and Jan Schuurhuis, G. (2007). The novel AML stem cell associated antigen CLL-1 aids in discrimination between normal and leukemic stem cells. Blood *110*, 2659–2666.

Wang, J., Iwasaki, H., Krivtsov, A., Febbo, P.G., Thorner, A.R., Ernst, P., Anastasiadou, E., Kutok, J.L., Kogan, S.C., Zinkel, S.S., *et al.* (2005). Conditional MLL-CBP targets GMP and models therapy-related myeloproliferative disease. EMBO J *24*, 368–381.

Weissman, I. (2005). Stem cell research: paths to cancer therapies and regenerative medicine. JAMA: the journal of the American Medical Association *294*, 1359–1366.

Welch, J.S., Ley, T.J., Link, D.C., Miller, C.A., Larson, D.E., Koboldt, D.C., Wartman, L.D., Lamprecht, T.L., Liu, F., Xia, J., *et al.* (2012). The origin and evolution of mutations in acute myeloid leukemia. Cell *150*, 264–278.

Will, B., Zhou, L., Vogler, T.O., Ben-Neriah, S., Schinke, C., Tamari, R., Yu, Y., Bhagat, T.D., Bhattacharyya, S., Barreyro, L., *et al.* (2012). Stem and progenitor cells in myelodysplastic syndromes show aberrant stage-specific expansion and harbor genetic and epigenetic alterations. Blood *120*, 2076–2086.

Wunderlich, M., Chou, F.S., Link, K.A., Mizukawa, B., Perry, R.L., Carroll, M., and Mulloy, J.C. (2010). AML xenograft efficiency is significantly improved in NOD/SCID-IL2RG mice constitutively expressing human SCF, GM-CSF and IL-3. Leukemia *24*, 1785–1788.

Yilmaz, O.H., Valdez, R., Theisen, B.K., Guo, W., Ferguson, D.O., Wu, H., and Morrison, S.J. (2006). Pten dependence distinguishes haematopoietic stem cells from leukaemia-initiating cells. Nature *441*, 475–482.

Yin, J.A., and Tobal, K. (1999). Detection of minimal residual disease in acute myeloid leukaemia: Methodologies, clinical and biological significance. Br J Haematol *106*, 578–590.

Zhang, J., Grindley, J.C., Yin, T., Jayasinghe, S., He, X.C., Ross, J.T., Haug, J.S., Rupp, D., Porter-Westpfahl, K.S., Wiedemann, L.M., *et al.* (2006). PTEN maintains haematopoietic stem cells and acts in lineage choice and leukaemia prevention. Nature *441*, 518–522.

Chapter 12

Leukemia-Initiating Cells in Acute Lymphoblastic Leukemia

Thorsten Raff and Monika Brüggemann
2nd Medical Department, University Hospital Schleswig-Holstein, Campus Kiel, Kiel, Germany

Clonal Evolution and Cancer Stem Cell Hierarchy—Alternative Origins of Leukemia-Initiating Cells

Most human cancers are assumed to be clonal, i.e., to arise from a single cell that has acquired tumor-initiating potential through multiple genetic and possibly epigenetic alterations in a process called clonal evolution. This was assumed to be a process of selection of subclones from the pre-malignant clone that had gained a competitive advantage by further mutation leading to clonal expansion of the fittest clone similar to Darwinian evolution (Nowell, 1976). Recent research employing genome-wide methods has revealed a more complex pattern of clonal development in many human cancers including acute lymphoblastic leukemia (ALL) (Mullighan et al., 2008). According to this hypothesis, different leukemic clones are present at the time of diagnosis that show similar as well as different genetic lesions suggesting their origin from a common ancestral clone that has some (the common) but not all (the different) genetic lesions required for leukemogenesis. Alternatively, the leukemic clones may stem from an already leukemic but only slowly proliferating cell while the dominant clone present at diagnosis has acquired additional genetic lesions that confer a selective growth advantage. This growth advantage may be reversed by therapy leading to the emergence of a different clone resistant to therapy during treatment.

A complementary hypothesis is based on the proposition that tumors are replenished by a (relatively rare) population of cancer stem cells (CSC) that are capable of extensive proliferation, self-renewal, and differentiation (Clarke et al., 2006) whereas their progeny has irreversibly lost the potential for extensive proliferation. This CSC hierarchy hypothesis (Dick, 2008) implies that CSCs are clearly ancestral to the more differentiated cells of the tumor (i.e., a hierarchical order exists much in the same way as it does in normal tissue). As CSCs tend to be rare in tumors that clearly follow a CSC hierarchy model, epigenetic differences are thought to be responsible for the loss of tumorigenicity of the progeny as it is stochastically improbable that the genotype necessary for extensive proliferation is confined to rare cells (Shackleton et al., 2009). The mere observation that not

Cancer Stem Cells, First Edition. Edited by Vinagolu K. Rajasekhar.

all leukemic cells are able to cause relapse or initiate leukemia in appropriate models cannot be seen as indirect evidence for the existence of leukemia stem cells (LSC) in the sense of a hierarchical CSC model as it is not clear whether the cells that failed to be leukemogenic have really lost their extensive proliferation potential or simply have not encountered permissive conditions. The operational term leukemia-initiating cell (LIC) has been coined to describe these cells because of the uncertainty regarding their frequency and nature of, which is likely to differ between leukemia types. Sometimes the term leukemia-repopulating cell (LRC) is used to emphasize the potential to reconstitute the leukemia in the original host. In this chapter, the more common term LIC is used also for LRC while acknowledging the fact that the ultimate goal of research is to prevent the repopulating potential (i.e., relapse in the patient rather than leukemogenesis in mice).

The question of whether LICs in a particular leukemia are indeed equivalent to rare LSCs or are the result of clonal evolution while retaining some "stemness" has important implications for disease-related research and therapy. If only very few cells are able to function as LICs as postulated in the CSC model, research should focus on isolating and investigating LICs, and therapy should be focused on eliminating these cells while the post-mitotic majority of the leukemic cells are of less interest. If on the other hand LICs are the result of clonal evolution, retain or regain their "stemness" and abound in the leukemia, investigation of clonal heterogeneity of all leukemic cells is the key, keeping in mind that subclones may exist that are more resistant to therapy than the dominant clone at the time of diagnosis and therefore are a potential cause of relapse.

Heterogeneity of Leukemic Cells Regarding Their Initiating Potential

With respect to leukemia, the CSC hierarchy hypothesis seems to be firmly established in acute myeloid leukemia (Lapidot et al., 1994) although the relationship between the founding clone and its subclones to the LIC has not yet been clarified. Within this hierarchy, short-term, long-term, and quiescent long-term leukemic stem cells have been identified through serial transplantation experiments, which differ in their capacity to self-renew and proliferate. In chronic myeloid leukemia (CML), the hierarchy hypothesis is not only supported by transplantation studies onto SCID mice but also from the observations that (i) the blood of human CML patients contains populations that are similar to normal hematopoietic stem cells (HSC) and are able to initiate multi-lineage hematopoiesis, (ii) CML stem cells or at least quiescent CML stem cells seem not to be susceptible to treatment with tyrosine kinase inhibitors (TKI), and (iii) experimental data showing that asymmetric cell division thought to be a key property of HSCs might also be the pathophysiological basis for development of blast crisis in CML. The evidence for a hierarchical CSC model in ALL leading to heterogeneity of the tumor cells regarding their leukemia-initiating potential is less obvious.

Lessons From Disease Biology Regarding the Nature of ALL-Initiating Cells

Based to a large extend on clinical observations of different sensitivity of childhood and adult ALL to chemotherapy an inverse relation between "stemness" of the LICs and curability of the disease was already postulated in the early 1990s (Greaves, 1993). The suggestion was that pediatric ALL originated in committed lymphoid progenitor cells with finite self-renewal capacity and a low probability of exiting into the G_0-phase of the cell cycle, whereas the clonogenic population in adult ALL consists of more primitive multi-potential stem cells, which divide infrequently. Support for the first part of the hypothesis comes from studies showing that in one of the genetically most favorable subgroups in childhood ALL, those with *ETV6-RUNX1 (TEL-AML1)*-fusion, the genetical abnormalities are confined to committed, CD19⁺ B-cell precursors (Hotfilder et al., 2002).

The example of *ETV6-RUNX1*-positive pediatric ALL is particularly instructive because it constitutes up to one-third of the cases in large series, and the underlying chromosomal translocation t(12;21)(p13;q22) is considered to be necessary while not sufficient for the development of the malignant clone. Based on studies of identical twins with concordant ALL, it is current doctrine that the premalignant clone arises already in utero in one of the twins, and the clonogenic progeny spreads to the second twin prenatally through the monochorionic placenta. The clonal origin of the concordant ALL is strongly suggested by their common breakpoints and fusion sequences that are not present in the germline and thus have to be acquired during fetal life. Indeed, candidate preleukemic cells with t(12;21) can be detected in normal cord blood samples in about 1% of cases and are committed B-cell progenitors. The presence of clonotypic cells in Guthrie cards of children who later developed *ETV6-RUNX1*-positive ALL provides further evidence for the prenatal origin of the preleukemic clone. The translocation, however, by itself does not lead to overt ALL as the incidence of *ETV6-RUNX1*-positive cells in cord blood is about 100-fold higher than the cumulative frequency of the overt leukemia with the same fusion gene. Further evidence for an obligatory second genetic hit is provided by concordance rates of only about 10% between twins and the long postnatal latency of up to 14 years. One of the apparently more common second genetic hits present in this disease affects the normal *ETV6*-allele, which bears deletions that are subclonal to *ETV6-RUNX1*-fusions, distinct between twins with concordant *ETV6-RUNX1*-positive ALL, and suggests a tumor suppressor function of *ETV6*. Interestingly, *ETV6*-mutations have recently been identified as a recurrent genetic lesion in immature adult T-cell leukemia types, which, contrary to *ETV6-RUNX1*-positive pediatric ALL, have a poor prognosis. Although detailed studies on immature precursor B-ALL with t(9;22) or t(4;11) have also shown no evidence of non-lymphoid committed progenitor cells (Castor et al., 2005), there is some evidence for involvement of more primitive cells in some ALL patients based on cytogenetic data. To further clarify whether indeed non-lymphoid committed precursor cells function as LICs in ALL or the majority of blasts retain some "stemness," functional in vivo models have been developed.

Some Xenotransplantation Models Suggest a Stem Cell Origin of ALL-Initiating Cells

After demonstration of the feasibility to transplant pediatric pre-B ALL into SCID mice with striking resemblance to the human disease and the observation that cells from patients who relapsed later on were more likely to lead to overt leukemia in this model, several groups tried to identify the responsible cell population. Initial studies showing that only blasts with a CD34⁺CD38⁻ immunophenotype lead to development of Ph⁺ ALL when injected into immune-deficient NOD SCID mice seemed to provide evidence for the existence of a stem cell hierarchy (Cobaleda et al., 2000). These results were later confirmed by demonstration that only CD34⁺CD10⁻CD19⁻CD133⁺ blasts from children with precursor B-ALL were able to proliferate in immune-deficient mice for prolonged periods of time, leading the authors to conclude that "ALL-initiating cells can be enriched using the most primitive stem cell antibody routinely available for hematopoietic cell purification" (Cox et al., 2009). Similar results have been obtained in pediatric T-ALL where only CD34⁺CD7⁻ and CD34⁺CD4− but not CD34⁻, CD34⁺/CD7⁺, and CD34⁺/CD4⁺ fractions had the potential to engraft and reconstitute the leukemia in serial transplants onto NOD SCID mice (Cox et al., 2007). Similarly, the development of T-ALL in mice xenografted with human AML blasts apparently devoid of AML-LICs where the LIC population was restricted to CD34⁺ cells (Risueño et al.,

2011) might be interpreted to point toward an immature cell of origin, not committed to the lymphoid lineage.

Data From Relapsed Patients Make a Stem Cell Origin of ALL-Initiating Cells Unlikely

These xenotransplantation models, however, are challenged by clinical data available on ALL relapses. Like normal lymphoid cells, ALL blasts rearrange their immune genes by somatic recombination of the variable (V), diversity (D), and joining (J) segments. This is a process involving the RAG1- and RAG2-mediated recognition of conserved recombination signal sequences flanking all Ig and TCR gene segments. First D segments are joined to J segments and afterward V to DJ segments with the possible insertion of non-template (N−)nucleotides added at the junction between V, D, and J segments for which the enzyme terminal deoxynucleotide transferase (TdT) is required. The sequence of these events seems to be enforced by the expression of transcription factors (e.g., PAX5 for the joining of V to DJ segments during the maturation of B-cells), and occurs at defined stages of B- and T-cell maturation. These stages can be identified by the expression of immunophenotypic markers (Noordzij et al., 2002; Asnafi et al., 2003) and show distinct gene expression profiles (Dik et al., 2005; Hystad et al., 2007). Since the leukemic cells—other than their physiologic counterparts or mature lymphoid neoplasms—are not dependent on a functional rearrangement for survival, all kinds of incomplete and complete rearrangements can be observed including biallelic rearrangements utilizing nonfunctional segments and cross-lineage rearrangements (Szczepański et al., 1999). The resulting (V)(N)D(N) J-junctions are highly specific for the leukemic clone and can be used to assess minimal residual disease (MRD) in both childhood and adult ALL with a high predictive value regarding both risk of relapse by measuring initial reduction of the leukemic clone and the prediction of impending relapse. Contrary to what might be suggested by recent data on the evolution of the immunoglobulin heavy chain locus in children with B-precursor ALL, immungene rearrangements, if appropriately chosen, are indeed relatively stable between diagnosis and relapse or at least can be traced to a common ancestor clone. As these rearrangements involve excision of genomic DNA, the process is not reversible. These findings have implications regarding the stem cell hierarchy hypothesis in ALL. Since at least one immune gene rearrangement is detected in the vast majority of ALL patients and these rearrangements can be detected at the time of relapse, the LIC in the majority of cases is assumed to correspond at least to the pro B and pro T development stages of normal lymphoid cells where RAG proteins are expressed at levels appropriate for rearrangement of immune genes. Although lymphoid precursors may rearrange their immunoglobulin (Ig−) or T-cell receptor (TCR−)genes abortively early, there is strong evidence that the rearrangement pattern of both precursor B and T-ALL closely resembles that of their physiological counterparts and therefore even ALL blasts with a more mature immunophenotype seem to be able to have repopulating potential.

Even Mature Lymphoid Precursors Can Function as ALL-Initiating Cells in Appropriate Models

More recent xenotransplantation studies seem to support this hypothesis. Using more sensitive models, the presence of LICs in sorted $CD34^{+}CD19^{-}$, $CD34^{+}CD19^{+}$, and $CD34^{-}CD19^{+}$ populations was confirmed (le Viseur et al., 2008). Cells from all of the populations were able to engraft and reconstitute the leukemia when serially transplanted onto primary, secondary, tertiary, and quaternary recipient mice and reestablished the complete immunophenotype of the original leukemia. They also were able to reconstitute cell populations expected to be less mature than the graft. Moreover, transcriptional differences between the blast populations regarding the expression of genes related to B-cell development were observed mimicking the physiological maturation of B-cell precursors (e.g., the expression of *IRF4*, a transcription factor needed to make the immunoglobulin light chain loci accessible for recombination only in the mature $CD34^-CD19^+$ population). These results have been confirmed independently using cells from three children with precursor B-ALL, two of them with infant leukemia and *MLL*-translocation, generally considered to be a very immature ALL (Kong et al., 2008). Consequently, this leads to the conclusion that "the hierarchy model, which seems to describe AML so faithfully, does not seem to be suitable in B precursor ALL" (Bomken et al., 2010). The model to most accurately describe ALL-LICs currently seems to be one of clonal selection combined with a certain plasticity ("malleability") (le Viseur et al., 2008). According to this model, ALL-LICs are cells that at the same time are not positively selected to continue differentiation, resistant to apoptosis, and able to change their differentiation pattern back and forth within a certain range.

Although the situation is less clear in T-ALL, where the number of studies is more limited, recent results show that the LICs in this disease are not restricted to the $CD34^+$ compartment (Armstrong et al., 2009). A possible exception to this seems to be early T-cell precursor ALL (ETP-ALL), a subgroup of T-ALL, which is characterized by poor response to conventional therapy both in children and in adults. As the name implies, ETP-ALL is assumed to originate from early T-cell precursors (ETP) only recently having left the bone marrow and settled in the thymus. ETPs retain the potential for multilineage differentiation, and ETP-ALL shows a gene expression profile reminiscent to that of hematopoietic stem cells (HSC) (Zhang et al., 2012).

ALL-LIC-Activity Is Related to Activation Pathways That Can Be Pharmacologically Targeted

Contrary to that, typical T-ALL cells show gene expression signatures indicative of arrest at specific stages of normal T-cell development (Ferrando et al., 2002). Furthermore, activation of *NOTCH1*, which was initially described in rare cases of T-ALL with t(7;9)(q34;q34.4), was later shown to be present in over 50% of T-ALL cases. Activation of the NOTCH pathway seems to be essential for T-ALL-initiating potential (Armstrong et al., 2009) and can be inhibited by γ-secretase inhibitors. Despite development of T-ALL in murine models through overexpression of the intracellular domains of *NOTCH1*, *NOTCH2*, or *NOTCH3*, only *NOTCH1* mutations seem to play a role in human T-ALL, rendering *NOTCH1* an attractive target for pharmacological inhibition. Known prognostic molecular aberrations in T-ALL and how they relate to T-ALL-LIC activity are summarized in Table 12.1.

In precursor B-ALL, genome-wide analysis of genetic alterations identified *EBF1*, *PAX5*, and *IZKF1* as the most common targets (Mullighan et al., 2007), which is not astonishing given their prominent roles in B-cell development. These results were confirmed by sequencing of 120 candidate genes in a large set of childhood precursor B-ALL samples where aberrations involving genes related to B-cell development were found in 68% of cases, related to the TP53/RB tumor suppressor pathway in 54% of cases, to Ras signaling in 50% of cases, and to Janus kinases in 11% of cases (Zhang et al., 2011). Interestingly, the pathway related to Janus kinases, JAK-STAT, is activated in a substantial proportion of about 25% of cases in a recently identified subgroup of *BCR-ABL*-negative childhood ALL cases with a gene expression profile very similar to that of *BCR-ABL*-positive ALL and a bad prognosis (Mullighan et al., 2009). This JAK-STAT activation can be inhibited

Table 12.1. Prognostically significant genetic alterations in T-ALL and their correlation to LIC function

Molecular finding	Effect	Pathophysiological consequence	Clinical relevance	Relevance for LIC function	References
ETP-ALL (gene expression, immunophenotype, SNP)	early T-precursor cell phenotype	Immature (stem-cell like) phenotype	10-year OS only about 20%	Probably a stem cell disease	(Coustan-Smith et al., 2009)
Mutation of *NOTCH1* or *FBXW7*	Constitutive activation of NOTCH-pathway	Prevention of TCR-induced apoptosis, expansion of thymocytes	Depending on protocol	Activation of NOTCH-pathway apparently essential for LIC-activity	(Weng et al., 2004; Breit et al., 2006; Armstrong et al., 2009; Zuurbier et al., 2010; Ma et al., 2012)
TLX1-overexpression, TLX1-TCRD@, TLX1-TCRB@	Overexpression of transcription factor (TF) TLX1	T-cell maturation arrest	>90% 5-year OS	Unknown, but probably related to maturation arrest	(Ferrando et al., 2002; Dadi et al., 2012)
TLX3-overexpression	Overexpression of TF TLX3	T-cell maturation arrest	~30% 5-year OS	Unknown, but probably related to maturation arrest	(Ferrando et al., 2002; Dadi et al., 2012)
TAL1-overexpression *LMO2*-overexpression *TAL1-TCRD@* *LMO2-TCRD@*	Overexpression of TF TAL1 or transactivating factor LMO2	Probably similar to *LYL1*-overexpression, see below	~40% 5-year OS	Unknown, but probably needed for survival	(Ferrando et al., 2002; Zohren et al., 2012)
LYL1-overexpression	Overexpression of transcription factor LYL1	Anti-apoptotic in early T-lineage progenitors	~30% 5-year OS	Unknown, but probably needed for survival	(Ferrando et al., 2002; Zohren et al., 2012)
HOXA-cluster	Overexpression of HOXA-cluster	Block of differentiation, extended self-renewal	100% 6-year OS	Maintenance of LIC via extended self-renewal	(Soulier et al., 2005; Dickson et al., 2009; Starkova et al., 2010)
Methylator phenotype	CIMP + versus CIMP−	Unknown, probable gene silencing, possibly in the NOTCH-pathway	20% versus 100% 12-year DFS	Unknown, probable gene silencing, possibly in the NOTCH-pathway	(Roman-Gomez et al., 2007; Yatim et al., 2012)

in human leukemic cells and seems to be relevant in LICs as demonstrated in a murine xenograft model of *BCR-ABL*-like ALL (Maude et al., 2012). Consequently, JAK inhibition is an attractive addition to standard chemotherapy in these patients and currently evaluated clinically (Loh and Mullighan, 2012). Other pharmacological targets with proven effects on LICs in precursor B-ALL include TNF-related apoptosis-inducing ligand (TRAIL) (Castro Alves et al., 2012), inhibition of BCL6 (Duy et al., 2011) and the PI3K/AKT/mTOR pathway (Castro Alves et al., 2012; Fuka et al., 2012). Although many of the genetically driven approaches are attractive on a theoretical basis, they remain to be proven in models and ultimately in clinical studies.

Conclusion

According to current knowledge, ALL-LICs seem to originate through a process of clonal selection and retain their "stemness" by being able to activate intracellular signaling pathways that enable self-renewal and extensive proliferation. The ability to assume properties of more immature and mature cells within a certain developmental range, their plasticity, is based on their resistance to apoptosis as well as an apparent block to undergo further differentiation. ALL-LICs seem not to be rare stem cells in most of the patients but present in most immunophenotypically identifiable populations. Consequently, there is currently no easy way to distinguish ALL-LICs from the rest of the leukemic cells whatever their fraction of the total leukemic population may be. The targets identified to be involved in leukemogenesis and/or evolution of the leukemic clone and their related pathways certainly represent attractive targets for pharmacological intervention. However, the substantial genetic complexity of the disease makes predictions based on genetic data extremely difficult, underlining the necessity of appropriate models for initial evaluation of candidate targets.

Future Perspectives

It seems to be a safe prognosis that advances in ALL treatment will to a substantial part depend on the ability to abolish the leukemia-repopulating potential of ALL-LICs. This will almost certainly not be possible in a uniform manner because disease characteristics differ substantially between and within immunologic ALL subgroups. The most promising research will probably be that directed to investigate rapidly relapsing or primarily refractory disease; in these cases, the properties of the readily accessible cells will likely play a major role, and functional models are available. The problem of late relapses, however, will be far more difficult to target, as we know from MRD studies that the cells responsible for relapse can go undetected for long periods of time, and it seems probable that the interaction of the host and the latent leukemia is of major importance in these cases, which is not easily reproducible in an ex vivo model.

References

Armstrong, F., Brunet de la Grange, P., Gerby, B., Rouyez, M.-C., Calvo, J., Fontenay, M., Boissel, N., Dombret, H., Baruchel, A., Landman-Parker, J., et al. (2009). NOTCH is a key regulator of human T-cell acute leukemia initiating cell activity. Blood *113*, 1730–1740.

Asnafi, V., Beldjord, K., Boulanger, E., Comba, B., Le Tutour, P., Estienne, M.-H., Davi, F., Landman-Parker, J., Quartier, P., Buzyn, A., et al. (2003). Analysis of TCR, pT alpha, and RAG-1 in T-acute lymphoblastic leukemias improves understanding of early human T-lymphoid lineage commitment. Blood *101*, 2693–2703.

Bomken, S., Fiser, K., Heidenreich, O., and Vormoor, J. (2010). Understanding the cancer stem cell. Br. J. Cancer *103*, 439–445.

Breit, S., Stanulla, M., Flohr, T., Schrappe, M., Ludwig, W.-D., Tolle, G., Happich, M., Muckenthaler, M.U., and Kulozik, A.E. (2006). Activating NOTCH1 mutations predict favorable early treatment response and long-term outcome in childhood precursor T-cell lymphoblastic leukemia. Blood *108*, 1151–1157.

Castor, A., Nilsson, L., Astrand-Grundström, I., Buitenhuis, M., Ramirez, C., Anderson, K., Strömbeck, B., Garwicz, S., Békéssy, A.N., Schmiegelow, K., et al. (2005). Distinct patterns of hematopoietic stem cell involvement in acute lymphoblastic leukemia. Nat. Med. *11*, 630–637.

Castro Alves, C., Terziyska, N., Grunert, M., Gündisch, S., Graubner, U., Quintanilla-Martinez, L., and Jeremias, I. (2012). Leukemia-initiating cells of patient-derived acute lymphoblastic leukemia xenografts are sensitive toward TRAIL. Blood *119*, 4224–4227.

Clarke, M.F., Dick, J.E., Dirks, P.B., Eaves, C.J., Jamieson, C.H.M., Jones, D.L., Visvader, J., Weissman, I.L., and Wahl, G.M. (2006). Cancer stem cells—perspectives on current status and future directions: AACR Workshop on cancer stem cells. Cancer Res. *66*, 9339–9344.

Cobaleda, C., Gutiérrez-Cianca, N., Pérez-Losada, J., Flores, T., García-Sanz, R., González, M., and Sánchez-García, I. (2000). A primitive hematopoietic cell is the target for the leukemic transformation in human Philadelphia-positive acute lymphoblastic leukemia. Blood *95*, 1007–1013.

Coustan-Smith, E., Mullighan, C.G., Onciu, M., Behm, F.G., Raimondi, S.C., Pei, D., Cheng, C., Su, X., Rubnitz, J.E., Basso, G., et al. (2009). Early T-cell precursor leukaemia: a subtype of very high-risk acute lymphoblastic leukaemia. Lancet Oncol *10*, 147–156.

Cox, C.V., Diamanti, P., Evely, R.S., Kearns, P.R., and Blair, A. (2009). Expression of CD133 on leukemia-initiating cells in childhood ALL. Blood *113*, 3287–3296.

Cox, C.V., Martin, H.M., Kearns, P.R., Virgo, P., Evely, R.S., and Blair, A. (2007). Characterization of a progenitor cell population in childhood T-cell acute lymphoblastic leukemia. Blood *109*, 674–682.

Dadi, S., Le Noir, S., Payet-Bornet, D., Lhermitte, L., Zacarias-Cabeza, J., Bergeron, J., Villarèse, P., Vachez, E., Dik, W.A., Millien, C., et al. (2012). TLX homeodomain oncogenes mediate T cell maturation arrest in T-ALL via interaction with ETS1 and suppression of TCRα gene expression. Cancer Cell *21*, 563–576.

Dick, J.E. (2008). Stem cell concepts renew cancer research. Blood *112*, 4793–4807.

Dickson, G.J., Kwasniewska, A., Mills, K.I., Lappin, T.R.J., and Thompson, A. (2009). Hoxa6 potentiates short-term hemopoietic cell proliferation and extended self-renewal. Exp. Hematol. *37*, 322–333.e3.

Dik, W.A., Pike-Overzet, K., Weerkamp, F., de Ridder, D., de Haas, E.F.E., Baert, M.R.M., van der Spek, P., Koster, E.E.L., Reinders, M.J.T., van Dongen, J.J.M., et al. (2005). New insights on human T cell development by quantitative T cell receptor gene rearrangement studies and gene expression profiling. The Journal of Experimental Medicine *201*, 1715–1723.

Duy, C., Hurtz, C., Shojaee, S., Cerchietti, L., Geng, H., Swaminathan, S., Klemm, L., Kweon, S., Nahar, R., Braig, M., et al. (2011). BCL6 enables Ph + acute lymphoblastic leukaemia cells to survive BCR-ABL1 kinase inhibition. Nature *473*, 384–388.

Ferrando, A.A., Neuberg, D.S., Staunton, J., Loh, M.L., Huard, C., Raimondi, S.C., Behm, F.G., Pui, C.H., Downing, J.R., Gilliland, D.G., et al. (2002). Gene expression signatures define novel oncogenic pathways in T cell acute lymphoblastic leukemia. Cancer Cell *1*, 75–87.

Fuka, G., Kantner, H.-P., Grausenburger, R., Inthal, A., Bauer, E., Krapf, G., Kaindl, U., Kauer, M., Dworzak, M.N., Stoiber, D., et al. (2012). Silencing of ETV6/RUNX1 abrogates PI3K/AKT/mTOR signaling and impairs reconstitution of leukemia in xenografts. Leukemia *26*, 927–933.

Greaves, M.F. (1993). Stem cell origins of leukaemia and curability. Br. J. Cancer *67*, 413–423.

Hotfilder, M., Röttgers, S., Rosemann, A., Jürgens, H., Harbott, J., and Vormoor, J. (2002). Immature CD34+CD19− progenitor/stem cells in TEL/AML1-positive acute lymphoblastic leukemia are genetically and functionally normal. Blood *100*, 640–646.

Hystad, M.E., Myklebust, J.H., Bø, T.H., Sivertsen, E.A., Rian, E., Forfang, L., Munthe, E., Rosenwald, A., Chiorazzi, M., Jonassen, I., et al. (2007). Characterization of early stages of human B cell development by gene expression profiling. J. Immunol. *179*, 3662–3671.

Kong, Y., Yoshida, S., Saito, Y., Doi, T., Nagatoshi, Y., Fukata, M., Saito, N., Yang, S.M., Iwamoto, C., Okamura, J., et al. (2008). CD34+CD38+CD19+ as well as CD34+CD38-CD19+ cells are leukemia-initiating cells with self-renewal capacity in human B-precursor ALL. Leukemia *22*, 1207–1213.

Lapidot, T., Sirard, C., Vormoor, J., Murdoch, B., Hoang, T., Caceres-Cortes, J., Minden, M., Paterson, B., Caligiuri, M.A., and Dick, J.E. (1994). A cell initiating human acute myeloid leukaemia after transplantation into SCID mice. Nature *367*, 645–648.

Loh, M.L., and Mullighan, C.G. (2012). Advances in the genetics of high-risk childhood B-progenitor acute lymphoblastic leukemia and juvenile myelomonocytic leukemia: implications for therapy. Clin. Cancer Res. *18*, 2754–2767.

Ma, W., Gutierrez, A., Goff, D.J., Geron, I., Sadarangani, A., Jamieson, C.A.M., Court, A.C., Shih, A.Y., Jiang, Q., Wu, C.C., et al. (2012). NOTCH1 Signaling Promotes Human T-Cell Acute Lymphoblastic Leukemia Initiating Cell Regeneration in Supportive Niches. PLoS ONE *7*, e39725.

Maude, S.L., Tasian, S.K., Vincent, T., Hall, J.W., Sheen, C., Roberts, K.G., Seif, A.E., Barrett, D.M., Chen, I.-M., Collins, J.R., et al. (2012). Targeting JAK1/2 and mTOR in murine xenograft models of Ph-like acute lymphoblastic leukemia (ALL). Blood.

Mullighan, C.G., Goorha, S., Radtke, I., Miller, C.B., Coustan-Smith, E., Dalton, J.D., Girtman, K., Mathew, S., Ma, J., Pounds, S.B., et al. (2007). Genome-wide analysis of genetic alterations in acute lymphoblastic leukaemia. Nature *446*, 758–764.

Mullighan, C.G., Phillips, L.A., Su, X., Ma, J., Miller, C.B., Shurtleff, S.A., and Downing, J.R. (2008). Genomic analysis of the clonal origins of relapsed acute lymphoblastic leukemia. Science *322*, 1377–1380.

Mullighan, C.G., Su, X., Zhang, J., Radtke, I., Phillips, L.A.A., Miller, C.B., Ma, J., Liu, W., Cheng, C., Schulman, B.A., et al. (2009). Deletion of IKZF1 and prognosis in acute lymphoblastic leukemia. N. Engl. J. Med *360*, 470–480.

Noordzij, J.G., de Bruin-Versteeg, S., Comans-Bitter, W.M., Hartwig, N.G., Hendriks, R.W., de Groot, R., and van Dongen, J.J.M. (2002). Composition of precursor B-cell compartment in bone marrow from patients with X-linked agammaglobulinemia compared with healthy children. Pediatr. Res. *51*, 159–168.

Nowell, P.C. (1976). The clonal evolution of tumor cell populations. Science *194*, 23–28.

Risueño, R.M., Campbell, C.J.V., Dingwall, S., Levadoux-Martin, M., Leber, B., Xenocostas, A., and Bhatia, M. (2011). Identification of T-lymphocytic leukemia-initiating stem cells residing in a small subset of patients with acute myeloid leukemic disease. Blood *117*, 7112–7120.

Roman-Gomez, J., Jimenez-Velasco, A., Barrios, M., Prosper, F., Heiniger, A., Torres, A., and Agirre, X. (2007). Poor prognosis in acute lymphoblastic leukemia may relate to promoter hypermethylation of cancer-related genes. Leuk. Lymphoma *48*, 1269–1282.

Shackleton, M., Quintana, E., Fearon, E.R., and Morrison, S.J. (2009). Heterogeneity in cancer: cancer stem cells versus clonal evolution. Cell *138*, 822–829.

Soulier, J., Clappier, E., Cayuela, J.-M., Regnault, A., García-Peydró, M., Dombret, H., Baruchel, A., Toribio, M.-L., and Sigaux, F. (2005). HOXA genes are included in genetic and biologic networks defining human acute T-cell leukemia (T-ALL). Blood *106*, 274–286.

Starkova, J., Zamostna, B., Mejstrikova, E., Krejci, R., Drabkin, H.A., and Trka, J. (2010). HOX gene expression in phenotypic and genotypic subgroups and low HOXA gene expression as an adverse prognostic factor in pediatric ALL. Pediatr Blood Cancer *55*, 1072–1082.

Szczepański, T., Beishuizen, A., Pongers-Willemse, M.J., Hählen, K., Van Wering, E.R., Wijkhuijs, A.J., Tibbe, G.J., De Bruijn, M.A., and Van Dongen, J.J. (1999). Cross-lineage T cell receptor gene rearrangements occur in more than ninety percent of childhood precursor-B acute lymphoblastic leukemias: alternative PCR targets for detection of minimal residual disease. Leukemia *13*, 196–205.

le Viseur, C., Hotfilder, M., Bomken, S., Wilson, K., Röttgers, S., Schrauder, A., Rosemann, A., Irving, J., Stam, R.W., Shultz, L.D., et al. (2008). In childhood acute lymphoblastic leukemia, blasts at different stages of immunophenotypic maturation have stem cell properties. Cancer Cell *14*, 47–58.

Weng, A.P., Ferrando, A.A., Lee, W., Morris, J.P., 4th, Silverman, L.B., Sanchez-Irizarry, C., Blacklow, S.C., Look, A.T., and Aster, J.C. (2004). Activating mutations of NOTCH1 in human T cell acute lymphoblastic leukemia. Science *306*, 269–271.

Yatim, A., Benne, C., Sobhian, B., Laurent-Chabalier, S., Deas, O., Judde, J.-G., Lelievre, J.-D., Levy, Y., and Benkirane, M. (2012). NOTCH1 Nuclear Interactome Reveals Key Regulators of Its Transcriptional Activity and Oncogenic Function. Mol. Cell.

Zhang, J., Ding, L., Holmfeldt, L., Wu, G., Heatley, S.L., Payne-Turner, D., Easton, J., Chen, X., Wang, J., Rusch, M., et al. (2012). The genetic basis of early T-cell precursor acute lymphoblastic leukaemia. Nature *481*, 157–163.

Zhang, J., Mullighan, C.G., Harvey, R.C., Wu, G., Chen, X., Edmonson, M., Buetow, K.H., Carroll, W.L., Chen, I.-M., Devidas, M., et al. (2011). Key pathways are frequently mutated in high-risk childhood acute lymphoblastic leukemia: a report from the Children's Oncology Group. Blood *118*, 3080–3087.

Zohren, F., Souroullas, G.P., Luo, M., Gerdemann, U., Imperato, M.R., Wilson, N.K., Göttgens, B., Lukov, G.L., and Goodell, M.A. (2012). The transcription factor Lyl-1 regulates lymphoid specification and the maintenance of early T lineage progenitors. Nat. Immunol. *13*, 761–769.

Zuurbier, L., Homminga, I., Calvert, V., te Winkel, M.L., Buijs-Gladdines, J.G.C.A.M., Kooi, C., Smits, W.K., Sonneveld, E., Veerman, A.J.P., Kamps, W.A., et al. (2010). NOTCH1 and/or FBXW7 mutations predict for initial good prednisone response but not for improved outcome in pediatric T-cell acute lymphoblastic leukemia patients treated on DCOG or COALL protocols. Leukemia *24*, 2014–2022.

Section III

Stem Cells in Solid Tumors

Chapter 13

Lung Cancer Stem Cells and Resistance to Radiotherapy

Scott V. Bratman[1] and Maximilian Diehn[1,2]
[1] *Department of Radiation Oncology, Stanford University, Stanford, CA, USA*
[2] *Institute for Stem Cell Biology and Regenerative Medicine and Cancer Institute, Stanford University, Stanford, CA, USA*

Abbreviations

DSB	double-strand DNA break
IAP	inhibitor of apoptosis
IR	ionizing radiation
NSCLC	non-small cell lung cancer
ROS	reactive oxygen species
SABR	stereotactic ablative radiotherapy
SCLC	small cell lung cancer

Introduction

Cancer causes one in eight deaths worldwide, and lung cancer is the leading cause of cancer death (Siegel et al., 2012). Approximately 160,000 deaths are attributed to lung cancer annually in the United States. Standard treatments for lung cancer have limited efficacy, and survival of lung cancer patients is less than 20% at 5 years from the time of diagnosis. Although curative treatments (i.e., surgery and radiotherapy) are available for patients diagnosed with early-stage localized lung cancer, in most cases, the cancer has spread to regional lymph nodes or other organs by the time of diagnosis. However, promising results from the National Lung Screening Trial (Aberle et al., 2011) suggest that screening high-risk populations can help save lives by identifying individuals with early-stage lung cancer who may be cured by surgery or radiotherapy.

Lung cancer is comprised of small cell lung cancer (SCLC) and non-small cell lung cancer (NSCLC). This distinction is based on histologic features and clinical behavior. SCLC is a neuroendocrine malignancy that responds initially to chemotherapy and/or radiotherapy but spreads quickly. NSCLC is a more heterogeneous group of diseases and includes adenocarcinoma, squamous

Cancer Stem Cells, First Edition. Edited by Vinagolu K. Rajasekhar.

cell carcinoma, and large cell carcinoma. NSCLC, which is the focus of this chapter, accounts for ~85–90% of lung cancer cases.

The treatment of NSCLC depends on the stage at diagnosis and requires a multidisciplinary approach. For early-stage NSCLC, surgery is the standard-of-care treatment. The surgeon aims to remove the primary tumor and the entire anatomic lobe to which it is confined. Hilar and mediastinal lymph nodes are simultaneously dissected to diagnose regionally metastatic disease. Postoperative radiotherapy is reserved for cases in which positive surgical margins or mediastinal lymph nodes are detected.

Many NSCLC patients cannot tolerate surgery due to low pulmonary reserve or other medical comorbidities. Radiotherapy can be effective in this group of patients. Recent studies have shown that stereotactic ablative radiotherapy (SABR) can achieve similar rates of cure to surgery (Hadziahmetovic et al., 2010). SABR, also called stereotactic body radiotherapy, delivers high doses of ionizing radiation (IR) to the tumor in 5 or fewer fractions and implements specialized technology to target the tumor and limit exposure of the surrounding normal lung.

For patients with unresectable stage III NSCLC, external beam radiotherapy is delivered with concurrent platinum-based chemotherapy. The radiotherapy is typically delivered with conventional fractionation, meaning approximately 2 gray (Gy) per daily fraction. This fractionation is used for the treatment of many cancer types because it is generally well tolerated. Outcomes for stage III NSCLC patients are poor, however, with 5-year overall survival rates of approximately 15%. Treatment failures are due in part to resistance of NSCLC cells to conventionally fractionated radiotherapy, as local recurrence in the treated region occurs in more than half of patients. This illustrates the need for novel strategies for sensitizing NSCLC cells to IR in order to boost the survival of these patients.

Resistance to Ionizing Radiation in Lund Cancer

Radiotherapy for the treatment of solid tumors relies on the relative sensitivity of cancer cells compared to normal tissues. This therapeutic window is maximized by fractionating the total radiotherapy dose into multiple smaller doses given over the course of days or weeks. The optimal combination of dose and number of fractions continues to be tested in clinical trials. For lung cancer, conventionally fractionated radiotherapy is used for locally advanced and stage III NSCLC, while more dose intensive SABR has been found to give better results for early-stage NSCLC.

The Clinical Ramifications of Cancer Radioresistance

Resistance to radiotherapy in NSCLC is an important clinical problem. If a tumor recurs following radiotherapy, subsequent treatment options become much more limited due to (1) the difficulty with operating on a previously irradiated lung and (2) the toxicity associated with using additional radiotherapy. Furthermore, radioresistant tumor cells may also be resistant to chemotherapy and may be more prone to spreading to other organs. Therefore, cure rates and toxicity could be improved if resistance to radiotherapy in NSCLC could be overcome.

Mechanism of Cancer Cell Killing by Ionizing Radiation

Therapeutic IR leads to cell death by inflicting DNA damage. Double strand breaks (DSB) are the most lethal form of DNA damage caused by IR because these are the most difficult for the cell to accurately repair. The high-energy photons generated by clinical linear accelerators cause DSBs by producing free radical reactive oxygen species (ROS) within cells. Cells produce antioxidant enzymes that neutralize ROS (Kobayashi and Suda, 2012). Higher levels of ROS scavengers reduce DNA damage and cause resistance to IR and cytotoxic chemotherapy in human cancers (Fagerholm et al., 2008).

DSBs trigger the p53 pathway, which facilitates DNA repair or, in the case of catastrophic DNA damage, triggers apoptosis. However, in carcinoma cells, the p53 pathway and downstream signaling can go awry in a number of ways that contribute to radioresistance. First, DNA repair machinery in cancer cells may be aberrant, leading to incomplete or improper repair. Second, deranged p53 signaling and/or other pro-survival genetic lesions help cancer cells to avoid cell death. Third, genomic instability resulting from misrepair and p53 checkpoint failure can lead to higher mutation rates and rapid selection of resistant cells. As a result, in lung cancer and other carcinomas, tumor cells generally bypass apoptosis following therapeutic doses of IR. Many cells with irreparable DNA damage will instead die from mitotic catastrophe due to defective segregation of chromosomes during cell division.

Lung Stem Cells and Cancer Stem Cells

A discussion of lung normal and cancer stem cells requires an understanding of the complex cellular architecture of the adult lung (Figure 13.1). The adult lung contains large proximal airways that connect with narrow intralobar bronchioles and terminate in distal alveoli, the sites of gas exchange. The epithelium is composed of heterogeneous cell types, including Clara cells, ciliated cells, neuroendocrine cells, basal cells, and goblet cells. The trachea and bronchi are lined with pseudostratified columnar epithelium, whereas a single layer cuboidal or simple squamous epithelium lines the bronchioles and alveoli, respectively.

Somatic Stem Cells in the Adult Lung

Adult stem cells contribute to the maintenance and repair of mature tissues through (1) self-renewal and (2) the generation of more differentiated progenitor or effector cells. Unlike many epithelial tissues, where a linear hierarchy of cellular differentiation has been uncovered (Rock and Hogan,

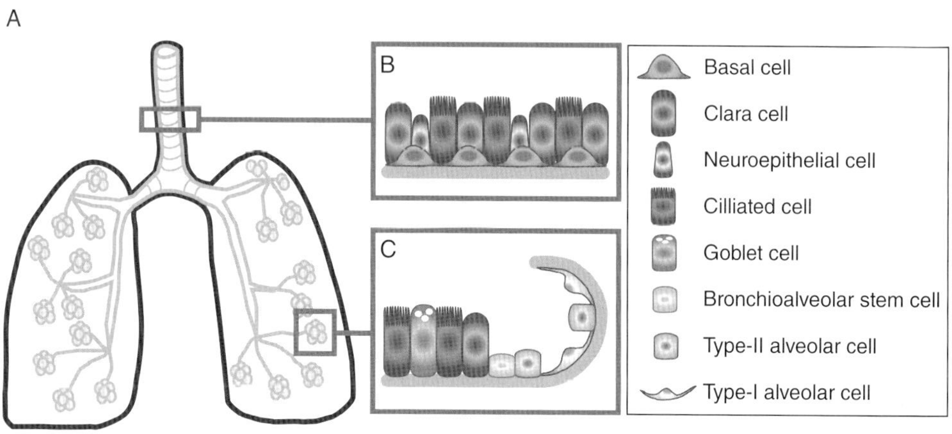

Figure 13.1. Cellular architecture in the normal adult lung. **A**. Schematic of the trachea, bronchi, bronchioles, and alveolar sacs. **B**. Graphical representation of the architecture of the airway epithelium in the conducting airways, including the trachea, bronchi, and bronchioles with diameters down to 0.5 mm. **C**. Airway epithelium in the distal airways, including the terminal bronchioles (*left, straight*) and alveoli (*right, curved*). For color detail, please see color plate section.

2011), the identification of somatic epithelial stem cells in the normal adult lung has been challenging.

There are two models for the organization of somatic stem cells in the adult lung. The most commonly held view is that distinct pools of stem or progenitor cells contribute to regeneration and maintenance of proximal and distal airways (Rock and Hogan, 2011). A competing model was recently put forth with the identification of self-renewing and pleuripotent c-kit-expressing cells that may be capable of giving rise to all cell types in the adult lung (Kajstura et al., 2011). Whether these c-kit-expressing cells might give rise to the other lung progenitor cells that have been described is not known.

Lung Stem Cells and Tumor-Initiating Cells

Tumor-initiating cells, like stem and multipotent progenitor cells, have the ability to undergo perpetual self-renewal. Normal somatic stem cells are thus primed to become tumor-initiating cells through loss of the tight regulation of self-renewal divisions (Reya et al., 2001). Conversely, more differentiated cells such as early progenitor cells could become CSCs by acquiring the capacity to self-renew via mutations or epigenetic changes. Which normal cell type is most often the pre-transformation precursor in NSCLC remains poorly understood. Some evidence points to putative bronchioalveolar stem cells situated at the bronchiolar-alveolar junction as the target of transformation in adenocarcinoma (Dovey et al., 2008; Kim et al., 2005). However, other recent data suggest that while oncogenic K-RAS expression in Clara cells and putative bronchioalveolar stem cells can lead to adenoma formation, only expression in type II alveolar cells leads to invasive adenocarcinoma (Xu et al., 2012). Squamous cell carcinoma is less well understood, but it is hypothesized that it may originate from tracheal basal cell progenitors (Sutherland and Berns, 2010).

Not all lung epithelial progenitor cells have been linked with malignant transformation. Certain cell types may be more susceptible than others for reasons that are not yet known. Because of the inherent difficulty of studying the origins of human lung cancer, animal models of NSCLC will be needed to investigate the oncogenic potential of different cell populations. For example, lineage tracing of individual cell types within the mouse adult lung (Rawlins et al., 2009) could help to shed light on this difficult problem.

Cancer Stem Cells in Lung Cancer

Cancer stem cells (CSC) are malignant cells that self-renew and give rise to more differentiated progeny within tumors. CSCs are defined experimentally by their ability to reconstitute an entire tumor upon transplantation into immunodeficient mice. The first description of CSCs in solid cancer was in breast adenocarcinoma (Al-Hajj et al., 2003), and CSCs have since been identified in many different primary tumors (Nguyen et al., 2012). Many studies have also investigated "CSC-like" cells in long-term cell lines; although these cells may share certain properties with bona fide CSCs from primary tumors such as expression of cell surface markers, they are often not significantly enriched for tumor initiation compared to the remaining cells, or in vivo tumor initiation activity has not been formally tested.

Not all CSCs within a given tumor are identical. This reflects the inherent heterogeneity within most solid tumors. Genetic instability, microenvironmental effects such as hypoxia, and/or other selective pressures due to treatment could lead to new mutations that enable previously non-self-renewing cancer cells to acquire stem-cell-like properties. This would result in tumors with multiple related but genetically distinct CSC pools. It is important to bear in mind this heterogeneity when devising methods for identifying and/or eliminating CSCs in solid cancer. Although a number of methods can enrich for cellular fractions containing CSCs within NSCLC, no one method may be sufficient to identify all CSCs in every tumor.

There are at least four flow cytometry-based methods that have been used to isolate NSCLC CSCs: (1) cell surface CD166 expression (Zhang et al., 2012); (2) cell surface CD133 expression (Eramo et al., 2008); (3) the Aldefluor assay for ALDH1 activity (Jiang et al., 2009; Sullivan et al., 2010); and (4) the Hoechst efflux assay for "side population" cells (Ho et al., 2007). Each of these methods has been reported to enrich for cancer cells with higher relative efficiency for clonogenic growth in vitro and/or in xenografts compared to the remaining cancer cells within the same tumor. It is not known whether these assays identify the same cellular populations or which one correctly identifies CSCs for the largest fraction of lung tumors. Thus, future studies are needed to test if combinations of these markers and/or novel markers could enrich further for CSCs.

Subtypes of NSCLC may be affected differently by these assays. For example, genetic subtypes of murine lung adenocarcinoma have different CSC populations requiring distinct identification strategies (Curtis et al., 2010). This makes generalizable strategies for identifying CSCs based on marker expression in all types of NSCLC particularly problematic. Instead, it is clear that for an inherently heterogeneous disease such as NSCLC, purported CSC populations require independent functional validation using in vivo transplantation assays.

Radioresistance in Cancer Stem Cells

Relative resistance of CSCs to IR has been demonstrated for a number of cancer types. Multiple factors contribute to sensitivity to IR (Figure 13.2). Both cell-extrinsic and cell-intrinsic factors have been proposed.

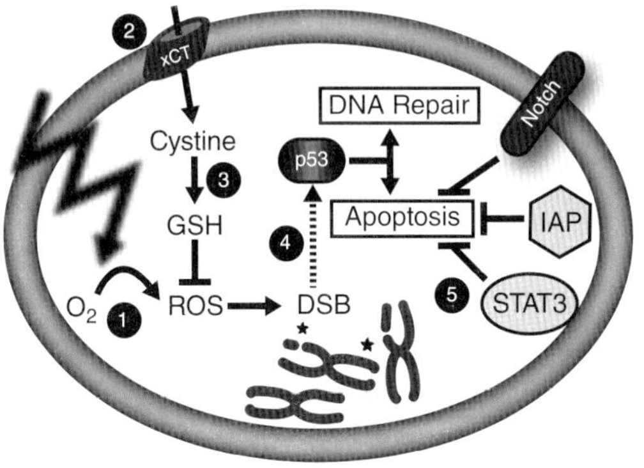

Figure 13.2. Mechanisms of resistance to IR-induced cell death in CSCs. *Hypoxia:* Low levels of molecular oxygen (1) found in some CSC niches contribute to reduced levels of ROS and oxygen-dependent fixation of free radical damage. *GSH generation for ROS scavenging.* GSH is a potent ROS scavenger in CSCs. The cystine importer, xCT (2), provides the substrate needed for intracellular GSH synthesis through a sequence of enzymatic steps (3). CSCs express higher levels of both the cystine importer and enzymes necessary for GSH synthesis. *Checkpoint signaling:* CSCs display robust p53 activation through the action of the checkpoint kinases ATM and CHK1/CHK2 in response to DNA damage leading to enhanced DNA repair (4). *Survival signaling:* Pro-survival signals from pathways such as STAT3, IAP, and Notch are activated in CSCs (5), and targeting these factors may have important therapeutic effects.

The Tumor Microenvironment and CSC Radioresistance

There are many cell types within the tumor microenvironment, including leukocytes, endothelial cells, fibroblasts, normal epithelial cells, and bone marrow-derived cells. Although all of these might contribute to radioresistance of CSCs, by far the best-studied attribute of the tumor microenvironment as it relates to CSC radioresistance is the tumor vasculature.

Dysfunctional tumor vasculature creates regions of hypoxia. Some CSCs have been proposed to occupy these hypoxic niches (Kobayashi and Suda, 2012). Furthermore, hypoxia inducible factors (HIF), which are upregulated in hypoxic cells, promote CSC self-renewal and survival (Li et al., 2009).

Hypoxia interferes with effective cell killing by IR due to reduced levels of ROS and oxygen-dependent fixation of free radical damage. Tumor hypoxia has been associated with radioresistance and treatment failure in head and neck cancer and cervical cancer (Brizel et al., 1999; Rofstad et al., 2000), and hypoxia has been documented in resectable NSCLC (Le et al., 2006). Reducing tumor hypoxia during radiotherapy has produced clinical benefits (Overgaard, 2011), but its broad clinical application remains controversial, and specific effects on CSCs have not been evaluated.

Cell-Intrinsic Factors Contributing to CSC Radioresistance

The stem cell-like properties of CSCs influence their biology in important ways that contribute to their radioresistance. These mechanisms can be broadly categorized as contributing to (1) reduced DNA damage, and (2) enhanced DNA repair in response to DNA damage.

Reduced DNA Damage in CSCs

Reduced DNA damage in CSCs has been attributed to relatively low levels of ROS. The major factors contributing to the lower levels of ROS in CSCs are thought to be (1) the hypoxic microenvironment of many CSCs, and (2) the higher expression of ROS scavengers such as glutathione (GSH) in CSCs (Figure 13.2). Supporting this view, breast cancer CSCs and CSC-like cells display lower levels of ROS and express high levels of GSH-synthesizing enzymes (Diehn et al., 2009; Phillips et al., 2006). Furthermore, in gastric and colorectal cancers, cystine import is upregulated for enhanced GSH synthesis (Ishimoto et al., 2011).

Enhanced DNA Repair and Checkpoint Signaling in CSCs

When DNA damage has occurred, CSCs may be well equipped to repair and recover to improve their chances of survival. Slow progression through the cell cycle by some CSCs provide them with additional time for DNA repair (Moore and Lyle, 2011). Quiescent CSCs are less likely to undergo mitosis—the most radiosensitive phase of the cell cycle—at the time of IR delivery.

Alterations in checkpoint signaling in response to IR also appears to favor DNA repair and survival in some CSCs. ATM activation is more efficient in breast CSC-like cells (Yin and Glass, 2011), and glioma CSCs display robust repair of IR-induced DNA DSBs in a manner dependent upon the Chk1/Chk2 checkpoint kinases (Bao et al., 2006). In a mouse model of medulloblastoma, CSC-like cells preferentially undergo IR-induced p53-dependent cell cycle arrest and recovery as opposed to apoptosis (Hambardzumyan et al., 2008). Similarly, enhanced activation of the Akt pathway appears to contribute to CSC radioresistance and enhanced DSB repair in some CSCs (Zhang et al., 2010). Taken together, CSCs appear to be protected against catastrophic DNA damage through a combination of ROS scavenging and checkpoint activation leading to enhanced repair of DSBs.

Radioresistance in Lung CSCs

Several of the mechanisms discussed above likely contribute to radioresistance of CSCs in lung cancer. NSCLC tumors are hypoxic, and the degree of hypoxia correlates with prognosis (Le et al., 2006). CD133-positive lung CSC-like cells contain higher levels of ROS scavengers (Salnikov et al., 2010). In addition, Notch signaling supports self-renewal and proliferation of ALDH-positive cells in lung adenocarcinoma (Sullivan et al., 2010). Thus, novel approaches at sensitizing CSCs to IR may need to target a number of these mechanisms simultaneously to show any clinical benefit.

Targeting Cancer Stem Cells in NSCLC

A principal challenge for translating CSC-targeted therapies into clinical practice will be to limit toxicities associated with treatment. A CSC-targeted therapy could be combined with IR by (1) sensitizing CSCs to IR, or (2) acting independently of IR to selectively kill the CSCs that would otherwise be spared. The former strategy might result in more local toxicity from damage to normal stem cells in the irradiated field, whereas the latter strategy might exhibit more systemic toxicity. As described below, a number of strategies for implementing CSC-targeted therapies into clinical practice are currently being investigated.

Inhibiting Canonical Stem Cell Signaling Pathways

Small molecule inhibitors of cell signaling pathways that are broadly involved in cellular differentiation are now in clinical development. The Wnt, Notch, and Hedgehog pathways have each been implicated in normal stem cell and CSC self-renewal. Inhibiting these pathways could have therapeutic effect by (1) direct cytotoxic effects on CSCs, or (2) driving cellular differentiation and thus inhibiting self-renewal. Although promising early results have been reported (LoRusso et al., 2011), the toxicities associated with systemic inhibition of these pathways are not yet known.

Promoting Apoptosis in CSCs

Although apoptosis is not the predominant mode of cell death caused by IR in most solid tumors, some CSCs rely heavily on blocking apoptosis to survive the effects of radiotherapy. STAT3 and the Inhibitor of Apoptosis Protein (IAP) family are key regulators of apoptosis in cancer cells. Inhibitors of these factors have shown promise as radiosensitizing agents in NSCLC cells in vitro (Cao et al., 2004; Hsu et al., 2011). Further studies are needed to evaluate the effects of IAP and STAT3 inhibition on the radiosensitivity of CSCs and normal tissues.

Inhibiting CSC Cell Surface Proteins

Selective inhibition of CSCs through cell surface markers is an area of active investigation. Targeting of c-kit and CD133 has been tested in lung cancer and glioma, respectively. The c-kit inhibitor, imatinib, has additive effects with cisplatin on growth inhibition of lung cancer CSC-like cells in vitro (Levina et al., 2010). Whether additive effects are also seen with radiotherapy is not known. In glioblastoma multiforme, a CD133 antibody-conjugated cytotoxin results in targeted killing of CD133-positive cells in vitro (Wang et al., 2011). These proof-of-concept experiments suggest that therapies targeting CSC cell surface markers might result in eradication of CSCs.

Hypofractionated Radiotherapy and Tumor Hypoxia

Conventionally fractionated radiotherapy has limited efficacy for killing some CSCs. SABR is clinically more effective and leads to improved local control, indicating enhanced elimination of CSCs. This is most likely due to the fact that SABR delivers such high doses that even the aforementioned CSC resistance mechanisms are overcome. Additionally SABR may be more effective than conventionally fractionated radiotherapy due to stimulation of acute endothelial disruption (Garcia-Barros et al., 2003) and a potent cytotoxic antitumor inflammatory response (Lee et al., 2009). Unfortunately, use of SABR is currently limited to early stage tumors or focal recurrences, since the high doses delivered would be unacceptably toxic if delivered to large regional fields. Additionally, hypoxic CSCs may still escape cell killing by SABR because of less inter-fraction tumor cell reoxygenation associated with the smaller number of fractions. The addition of a hypoxic cell radiosensitizer could help eliminate hypoxic CSCs during SABR (Brown et al., 2010). In support of this strategy, reducing tumor hypoxia during radiotherapy appears to produce greater clinical benefits when larger fraction sizes are used (Overgaard, 2011).

Conclusion

Resistance of lung cancer to IR is an important clinical problem. CSCs are likely responsible for treatment failures because of enhanced radioresistance mechanisms in these cells. CSCs that survive treatment have the potential to regrow the primary tumor or seed metastases.

Efforts to translate the CSC hypothesis into clinical practice in lung cancer would benefit from more robust methods for identifying CSC populations in vivo within primary tumors. This would increase the power of downstream analyses on CSCs and comparisons with non-CSC cancer cells and normal adult stem cells. Such comparisons will be crucial in order to exploit differences between these cell populations to achieve high therapeutic ratios with novel CSC-based therapies.

References

Aberle, D.R., Adams, A.M., Berg, C.D., Black, W.C., Clapp, J.D., Fagerstrom, R.M., Gareen, I.F., Gatsonis, C., Marcus, P.M., and Sicks, J.D. (2011). Reduced lung-cancer mortality with low-dose computed tomographic screening. The New England Journal of Medicine *365*, 395–409.

Al-Hajj, M., Wicha, M.S., Benito-Hernandez, A., Morrison, S.J., and Clarke, M.F. (2003). Prospective identification of tumorigenic breast cancer cells. Proceedings of the National Academy of Sciences of the United States of America *100*, 3983–3988.

Bao, S., Wu, Q., McLendon, R.E., Hao, Y., Shi, Q., Hjelmeland, A.B., Dewhirst, M.W., Bigner, D.D., and Rich, J.N. (2006). Glioma stem cells promote radioresistance by preferential activation of the DNA damage response. Nature *444*, 756–760.

Brizel, D.M., Dodge, R.K., Clough, R.W., and Dewhirst, M.W. (1999). Oxygenation of head and neck cancer: changes during radiotherapy and impact on treatment outcome. Radiotherapy and oncology: Journal of the European Society for Therapeutic Radiology and Oncology *53*, 113–117.

Brown, J.M., Diehn, M., and Loo, B.W., Jr. (2010). Stereotactic ablative radiotherapy should be combined with a hypoxic cell radiosensitizer. International journal of radiation oncology, biology, physics *78*, 323–327.

Cao, C., Mu, Y., Hallahan, D.E., and Lu, B. (2004). XIAP and survivin as therapeutic targets for radiation sensitization in preclinical models of lung cancer. Oncogene *23*, 7047–7052.

Curtis, S.J., Sinkevicius, K.W., Li, D., Lau, A.N., Roach, R.R., Zamponi, R., Woolfenden, A.E., Kirsch, D.G., Wong, K.K., and Kim, C.F. (2010). Primary tumor genotype is an important determinant in identification of lung cancer propagating cells. Cell stem cell *7*, 127–133.

Diehn, M., Cho, R.W., Lobo, N.A., Kalisky, T., Dorie, M.J., Kulp, A.N., Qian, D., Lam, J.S., Ailles, L.E., Wong, M., *et al.* (2009). Association of reactive oxygen species levels and radioresistance in cancer stem cells. Nature *458*, 780–783.

Dovey, J.S., Zacharek, S.J., Kim, C.F., and Lees, J.A. (2008). Bmi1 is critical for lung tumorigenesis and bronchioalveolar stem cell expansion. Proceedings of the National Academy of Sciences of the United States of America *105*, 11857–11862.

Eramo, A., Lotti, F., Sette, G., Pilozzi, E., Biffoni, M., Di Virgilio, A., Conticello, C., Ruco, L., Peschle, C., and De Maria, R. (2008). Identification and expansion of the tumorigenic lung cancer stem cell population. Cell death and differentiation *15*, 504–514.

Fagerholm, R., Hofstetter, B., Tommiska, J., Aaltonen, K., Vrtel, R., Syrjakoski, K., Kallioniemi, A., Kilpivaara, O., Mannermaa, A., Kosma, V.M., *et al.* (2008). NAD(P) H:quinone oxidoreductase 1 NQO1*2 genotype (P187S) is a strong prognostic and predictive factor in breast cancer. Nature genetics *40*, 844–853.

Garcia-Barros, M., Paris, F., Cordon-Cardo, C., Lyden, D., Rafii, S., Haimovitz-Friedman, A., Fuks, Z., and Kolesnick, R. (2003). Tumor response to radiotherapy regulated by endothelial cell apoptosis. Science *300*, 1155–1159.

Hadziahmetovic, M., Loo, B.W., Timmerman, R.D., Mayr, N.A., Wang, J.Z., Huang, Z., Grecula, J.C., and Lo, S.S. (2010). Stereotactic body radiation therapy (stereotactic ablative radiotherapy) for stage I non-small cell lung cancer—updates of radiobiology, techniques, and clinical outcomes. Discovery medicine *9*, 411–417.

Hambardzumyan, D., Becher, O.J., Rosenblum, M.K., Pandolfi, P.P., Manova-Todorova, K., and Holland, E.C. (2008). PI3K pathway regulates survival of cancer stem cells residing in the perivascular niche following radiation in medulloblastoma in vivo. Genes & development *22*, 436–448.

Ho, M.M., Ng, A.V., Lam, S., and Hung, J.Y. (2007). Side population in human lung cancer cell lines and tumors is enriched with stem-like cancer cells. Cancer research *67*, 4827–4833.

Hsu, H.S., Huang, P.I., Chang, Y.L., Tzao, C., Chen, Y.W., Shih, H.C., Hung, S.C., Chen, Y.C., Tseng, L.M., and Chiou, S.H. (2011). Cucurbitacin I inhibits tumorigenic ability and enhances radiochemosensitivity in nonsmall cell lung cancer-derived CD133-positive cells. Cancer *117*, 2970–2985.

Ishimoto, T., Nagano, O., Yae, T., Tamada, M., Motohara, T., Oshima, H., Oshima, M., Ikeda, T., Asaba, R., Yagi, H., *et al.* (2011). CD44 variant regulates redox status in cancer cells by stabilizing the xCT subunit of system xc(−) and thereby promotes tumor growth. Cancer cell *19*, 387–400.

Jiang, F., Qiu, Q., Khanna, A., Todd, N.W., Deepak, J., Xing, L., Wang, H., Liu, Z., Su, Y., Stass, S.A., *et al.* (2009). Aldehyde dehydrogenase 1 is a tumor stem cell-associated marker in lung cancer. Molecular cancer research: MCR *7*, 330–338.

Kajstura, J., Rota, M., Hall, S.R., Hosoda, T., D'Amario, D., Sanada, F., Zheng, H., Ogorek, B., Rondon-Clavo, C., Ferreira-Martins, J., et al. (2011). Evidence for human lung stem cells. The New England Journal of Medicine *364*, 1795–1806.

Kim, C.F., Jackson, E.L., Woolfenden, A.E., Lawrence, S., Babar, I., Vogel, S., Crowley, D., Bronson, R.T., and Jacks, T. (2005). Identification of bronchioalveolar stem cells in normal lung and lung cancer. Cell *121*, 823–835.

Kobayashi, C.I., and Suda, T. (2012). Regulation of reactive oxygen species in stem cells and cancer stem cells. Journal of cellular physiology *227*, 421–430.

Le, Q.T., Chen, E., Salim, A., Cao, H., Kong, C.S., Whyte, R., Donington, J., Cannon, W., Wakelee, H., Tibshirani, R., *et al.* (2006). An evaluation of tumor oxygenation and gene expression in patients with early stage non-small cell lung cancers. Clinical cancer research: an official journal of the American Association for Cancer Research *12*, 1507–1514.

Lee, Y., Auh, S.L., Wang, Y., Burnette, B., Wang, Y., Meng, Y., Beckett, M., Sharma, R., Chin, R., Tu, T., *et al.* (2009). Therapeutic effects of ablative radiation on local tumor require CD8+ T cells: changing strategies for cancer treatment. Blood *114*, 589–595.

Levina, V., Marrangoni, A., Wang, T., Parikh, S., Su, Y., Herberman, R., Lokshin, A., and Gorelik, E. (2010). Elimination of human lung cancer stem cells through targeting of the stem cell factor-c-kit autocrine signaling loop. Cancer research *70*, 338–346.

Li, Z., Bao, S., Wu, Q., Wang, H., Eyler, C., Sathornsumetee, S., Shi, Q., Cao, Y., Lathia, J., McLendon, R.E., et al. (2009). Hypoxia-inducible factors regulate tumorigenic capacity of glioma stem cells. Cancer cell *15*, 501–513.

LoRusso, P.M., Rudin, C.M., Reddy, J.C., Tibes, R., Weiss, G.J., Borad, M.J., Hann, C.L., Brahmer, J.R., Chang, I., Darbonne, W.C., *et al.* (2011). Phase I trial of hedgehog pathway inhibitor vismodegib (GDC-0449) in patients with refractory, locally advanced or metastatic solid tumors. Clinical cancer research : an official journal of the American Association for Cancer Research *17*, 2502–2511.

Moore, N., and Lyle, S. (2011). Quiescent, slow-cycling stem cell populations in cancer: a review of the evidence and discussion of significance. Journal of oncology *2011*.

Nguyen, L.V., Vanner, R., Dirks, P., and Eaves, C.J. (2012). Cancer stem cells: an evolving concept. Nature reviews Cancer *12*, 133–143.

Overgaard, J. (2011). Hypoxic modification of radiotherapy in squamous cell carcinoma of the head and neck—a systematic review and meta-analysis. Radiotherapy and oncology: journal of the European Society for Therapeutic Radiology and Oncology *100*, 22–32.

Phillips, T.M., McBride, W.H., and Pajonk, F. (2006). The response of CD24(−/low)/CD44+ breast cancer-initiating cells to radiation. Journal of the National Cancer Institute *98*, 1777–1785.

Rawlins, E.L., Okubo, T., Xue, Y., Brass, D.M., Auten, R.L., Hasegawa, H., Wang, F., and Hogan, B.L. (2009). The role of Scgb1a1+ Clara cells in the long-term maintenance and repair of lung airway, but not alveolar, epithelium. Cell stem cell *4*, 525–534.

Reya, T., Morrison, S.J., Clarke, M.F., and Weissman, I.L. (2001). Stem cells, cancer, and cancer stem cells. Nature *414*, 105–111.

Rock, J.R., and Hogan, B.L. (2011). Epithelial progenitor cells in lung development, maintenance, repair, and disease. Annual review of cell and developmental biology *27*, 493–512.

Rofstad, E.K., Sundfor, K., Lyng, H., and Trope, C.G. (2000). Hypoxia-induced treatment failure in advanced squamous cell carcinoma of the uterine cervix is primarily due to hypoxia-induced radiation resistance rather than hypoxia-induced metastasis. British journal of cancer *83*, 354–359.

Salnikov, A.V., Gladkich, J., Moldenhauer, G., Volm, M., Mattern, J., and Herr, I. (2010). CD133 is indicative for a resistance phenotype but does not represent a prognostic marker for survival of non-small cell lung cancer patients. International journal of cancer Journal international du cancer *126*, 950–958.

Siegel, R., Naishadham, D., and Jemal, A. (2012). Cancer statistics, 2012. CA: a cancer journal for clinicians *62*, 10–29.

Sullivan, J.P., Spinola, M., Dodge, M., Raso, M.G., Behrens, C., Gao, B., Schuster, K., Shao, C., Larsen, J.E., Sullivan, L.A., et al. (2010). Aldehyde dehydrogenase activity selects for lung adenocarcinoma stem cells dependent on notch signaling. Cancer research *70*, 9937–9948.

Sutherland, K.D., and Berns, A. (2010). Cell of origin of lung cancer. Molecular oncology *4*, 397–403.

Wang, C.H., Chiou, S.H., Chou, C.P., Chen, Y.C., Huang, Y.J., and Peng, C.A. (2011). Photothermolysis of glioblastoma stem-like cells targeted by carbon nanotubes conjugated with CD133 monoclonal antibody. Nanomedicine: nanotechnology, biology, and medicine *7*, 69–79.

Xu, X., Rock, J.R., Lu, Y., Futtner, C., Schwab, B., Guinney, J., Hogan, B.L., and Onaitis, M.W. (2012). Evidence for type II cells as cells of origin of K-Ras-induced distal lung adenocarcinoma. Proceedings of the National Academy of Sciences of the United States of America *109*, 4910–4915.

Yin, H., and Glass, J. (2011). The phenotypic radiation resistance of CD44+/CD24(−or low) breast cancer cells is mediated through the enhanced activation of ATM signaling. PloS one *6*, e24080.

Zhang, M., Atkinson, R.L., and Rosen, J.M. (2010). Selective targeting of radiation-resistant tumor-initiating cells. Proceedings of the National Academy of Sciences of the United States of America *107*, 3522–3527.

Zhang, W.C., Shyh-Chang, N., Yang, H., Rai, A., Umashankar, S., Ma, S., Soh, B.S., Sun, L.L., Tai, B.C., Nga, M.E., et al. (2012). Glycine decarboxylase activity drives non-small cell lung cancer tumor-initiating cells and tumorigenesis. Cell *148*, 259–272.

Chapter 14

Prostate Cancer Cell Heterogeneity and Prostate Cancer Stem Cells

Mark A. Badeaux and Dean G. Tang
Department of Molecular Carcinogenesis, University of Texas MD Anderson Cancer Center, Science Park, Smithville, TX, USA

The Idealized Prostate Cancer Stem Cell

To fulfill the obligate criteria of a cancer stem cell (CSC), a cell must be capable of both self-renewal and differentiation, of regenerating and of generating anew. In so doing, it is instructive as to the nature of the original cancerous growth. The term cancer stem cell is often replaced by or used synonymously with the phrase tumor-initiating cell (TIC), and with good reason. The functional test of a CSC is its ability to initiate tumor growth in a xenograft setting; to corroborate self-renewal, this capability must be demonstrable through several rounds of tumor growth.

The CSC, or hierarchical, model arose in part as a way to explain the omnipresent intratumoral cellular heterogeneity seen in the clinic. One logical extension of this idea is that CSCs occupy the apex of the hierarchy and give rise to subordinate lineages of relatively more mature and differentiated cells. An important test that must be applied to CSCs, or CSC-enriched populations, is that they are capable of giving rise to diverse cell types; functionally, this means that the secondary and subsequent xenograft growths must resemble the original tumor in possessing a broad spectrum of cell types, all of which are ostensibly and ultimately derived from the CSC. To truly occupy the apex of a cellular hierarchy implies that cells that occupy lower positions in the scheme cannot give rise to the "alpha" CSC. This is not to say that such cells cannot dedifferentiate to become a distinct population that fulfills the CSC criteria, but rather that such a dedifferentiation event cannot give rise to the population that originally resided upon the hierarchy's summit.

The ability of a prostate CSC to effect tumor growth in an artificial xenograft setting has direct clinical parallels. Following androgen ablation therapy, the most common frontline weapon brought to bear on an offending mass in the prostate, a small population of cells may yet remain. If these cells are incapable of re-initiating tumor growth, then for all intents and purposes, treatment may be considered successful. However, oftentimes androgen withdrawal is marked by tumor regression followed by a regrowth of the tumor as an androgen-independent entity.

Metastasis of prostate cancer, which travels primarily to the bone or lungs, relies on

Cancer Stem Cells, First Edition. Edited by Vinagolu K. Rajasekhar.

initiation of a new tumorous growth by prostate cancer cells, albeit in an environment unlike that in which they were originally able to thrive. Intuitively, it seems unlikely that cells that are incapable of initiating secondary tumors in an experimental setting would be able to do so with a high degree of efficiency in a foreign microenvironment.

In practice, owing to experimental limitations and technical hurdles, one does not study the idealized, single CSC, but rather one examines populations that are enriched for these cells. In certain cancers, such as late-stage melanomas, CSCs may be quite abundant, and therefore studies of these populations may be preferentially weighted toward the CSC fraction, but in other solid tumors where CSCs are believed to be more rare, including prostate cancer, the true molecular properties of the CSC may still be partially enshrouded by its more numerous non-tumorigenic brethren.

There is one caveat to the study of prostate CSCs that needs be mentioned up front: Primary tumor-derived prostate cancer cells are not amenable to successful tumor reconstitution following dissociation and isolation from the tumor proper. As a result, most work concerning identification of prostate CSCs has been carried out using cultured (e.g., LNCaP, DU145, PC3) and xenograft tumor lines (e.g., LAPC4, LAPC9). A major focus of several prostate cancer research laboratories, including our own, is in designing a system to overcome these technical hurdles and to allow for analysis, at the single cell level, of parental tumor reconstitution. To this end, exhaustive optimization protocols have been put forth, including but not limited to the use of extracellular matrix-based substrates to facilitate tumor engraftment, co-culture with carcinoma-derived fibroblasts (CAF), and utilization of progressively more immunodeficient host mouse models. A recent finding that combining prostate cancer epithelial cells with murine neonatal seminal vesicle mesenchyme allows for proliferative growth in the kidney capsule is cause for optimism (Toivanen et al., 2011), but it is unclear whether frank tumors can be reinitiated using this protocol and whether these tumors resemble the parental growth from which the cells were derived. It should be noted, however, that because prostate cancer cell lines and xenografts recapitulate many important clinical aspects of prostate cancer, there is reason to believe that many of the findings gleaned from their study will be applicable to primary tumor systems.

Enrichment of Prostate CSCs: Cell Surface Marker-Dependent and Independent Strategies

If one could imagine the easiest scenario by which prostate CSCs, relatively few in number and surrounded by a sea of more differentiated progeny, may be isolated, it would first be helpful to know what they "look" like. If prostate CSCs do not "appear" the same as nontumorigenic cells, then we can spot them and potentially extract them from the cellular milieu in which they dwell. In fact, we can apply this rationale to separating various cellular populations from one another, as all cells express a plethora of cellular surface antigens, some of which are specific to or preferentially expressed in a given subpopulation. Experimentally, these antigens can be visualized by first tagging them with a fluorescently labeled antibody with specificity toward the cell surface antigen of interest. This technique makes it possible to use fluorescence-activated cell sorting (FACS) to separate cells based on their emission wavelength, i.e., color profile, and allows for the simultaneous evaluation of multiple cell surface markers on a single cell. Owing in large part to the relative facility and unquestioned success of this approach in defining hierarchical relationships in the hematopoietic system, it has become the most common method of prospectively enriching CSC populations in solid tumors.

It is important to note, however, that this method has limitations. First, CSCs have doppelgangers; that is to say that the cell surface markers that CSCs express are not unique to them. In some cases, CSCs express cell surface molecules present in normal stem or progenitor cell populations, as best exemplified by the expression of the hematopoietic stem/progenitor cell marker combination $CD34^+$, $CD38^-$ in acute myelogenous leukemia cancer-initiating cells (Bonnet and Dick, 1997). In other cases, the cell surface protein may confer a specific function and thus be co-opted among many cancer cells in the bulk tumor.

Importantly, the use of cell surface marker profiles to identify CSCs may only *enrich* for their presence. In many solid tumors, such protocols produce populations in which generally no more than 1 in 100 (and oftentimes many fewer) cancer cells has the capacity to initiate a tumor, even in the most immunocompromised mouse models available. It is therefore important to combine these markers, when possible, in order to identify a cell population more concentrated in CSC activity.

CD44, a cell surface glycoprotein with known roles in cell-cell interactions, migration, and adhesion, seems to mark a population enriched for CSCs in a variety of solid tumors, including prostate cancer (Patrawala et al., 2006). Importantly, $CD44^+$ cells can readily give rise to $CD44^-$ prostate cancer, implying that the $CD44^+$ cells are hierarchically superior to the isogenic $CD44^-$ cells. Other groups have confirmed CD44 as a chief cell surface molecule capable of enriching for a population of cells with increased CSC activity, but a number of other molecules may also have CSC-enriching ability. $\alpha 2\beta 1$ integrin, for instance, can be used to select a prostate cancer cell population that is enriched in in vitro clonal and sphere-forming activity, although in vivo it marks a CSC-enriched population only when used in conjunction with $CD44^+$ selection (Patrawala et al., 2007). Using a combinatorial marker strategy may result in a degree of CSC enrichment that is even more pronounced. The Maitland lab reported prostate cancer cells capable of in vitro differentiation, proliferation, and self-renewal (Collins et al., 2005); these cells bore the phenotype $CD44^+$, $CD133^+$, and $\alpha 2\beta 1$ integrin$^+$, which can mark a population of normal prostatic epithelial stem cells. Similarly, we have found the greatest enrichment of CSC activity in $CD44^+$, $\alpha 2\beta 1^+$, and Aldefluor$^+$ LAPC9 cells (see below).

Some markers are more nebulous with regard to both their ability to enrich for prostate CSC activity as well as their potential function in mediating the properties ascribed to CSCs. CD133, for instance, a cell surface glycoprotein whose function is currently unknown, can be used to enrich for CSCs in colon cancer (O'Brien et al., 2007), although its status as a marker of stem-like cells in other tissues, including prostate cancer, is somewhat controversial. It may be that such markers are useful for subdividing other marker-demarcated hierarchies and not for enriching CSCs as the sole parameter of separation. It is critical that the functions of many such cell surface markers be illuminated, for it will help us in understanding which of this class of molecules may play an active role in effecting a CSC phenotype and which molecules represent "happy accidents" that mark but do not contribute to a cancer cell's stem-like behavior.

Although enrichment by way of cell surface marker profile is currently the most common method of enriching for CSCs, a number of marker-independent strategies can be used effectively to do the same. These methods rely on a particular activity or behavior of the cell population in question, and therefore have the potential to be more informative measures of CSC functions.

A clonal assay is a relatively simple means of assessing a cell's proliferative capacity; this technique found prominence in studies of epidermal keratinocytes in the 1980's but has since been adapted to analyze single-cell behavior in many other tissues. In this assay, cells are plated at a very low density, known eponymously as a clonal density. This ensures that the clonal growths of cells that are seen after several weeks of culture are each derived from a single original cell. The clonal assay has been used as an in vitro or ex vivo surrogate for CSC activity in some instances, as the capacity to proliferate extensively is a prerequisite for a cell to fit the definition of cancer stem cell. However, because this assay does not measure self-renewal, cells with a finite but pronounced proliferative capacity, such as progenitor cells, may form clones as well. Importantly, clones are not morphologically homogenous—holoclones are tightly packed, paraclones are sparsely populated, and meroclones are intermediate between the two, and nor do they share the same capacity for in vivo tumor growth. Holoclones derived from PC3 prostate cancer cells have an increased ability to generate tumors in immunodeficient mice, whereas meroclones and paraclones were entirely unable to give rise to tumors (Li et al., 2008). Because epithelial cancer cells tend to give rise to these colony types when plated at very low density, we believe that holoclones from many, if not all epithelial cancers are enriched in tumor initiating ability.

The clonogenic, or sphere-forming, assay is similar to the clonal assay save for the parameter of attachment: Instead of growing on a solid growth substrate, cells are cultured at clonal density in anchorage-independent conditions. In this environment, only cells that are able to proliferate extensively and resist *anoikis* (apoptosis brought about by the lack of attachment) will eventually give rise to sphere-shaped colonies composed of the founding cell and its progeny. Spheres can be harvested and passaged numerous times to assess the ability of sphere-forming cells to self-renew at the population level. This assay can be used as a tool to measure a prospectively isolated cancer cell population's ability to proliferate and self-renew through multiple generations, or it can, like the clonal assay, be used to enrich for viable CSCs (Lee et al., 2006), as has been demonstrated previously in breast cancer (mammosphere), glioblastoma (neurosphere), and now in prostate cancer (prostasphere). The assay can be modified in order to grow cancer cell-derived spheres in a semi-solid matrix, but this may remove the experimental parameter of anchorage-independence. Both the clonal and clonogenic assays, being carried out ex vivo or in vitro, have the added benefit of allowing for the study of media composition that may favor clonal growth; this, in turn, may tell us something about the conditions that favor the manifestation of the CSC phenotype.

The side population, or SP, is a peculiarity relating to a cell's ability to efflux chemicals or dyes. Cells that have this characteristic can be identified using flow cytometry, and have been shown to possess increased expression of drug transporter proteins. Although this assay has been used successfully in identifying prostate cells with tumor-initiating properties, it has fallen out of favor as it greatly skews cell viability toward the side population and away from less efflux-capable populations.

In many normal mammalian adult tissues, somatic stem cells are a relatively (though not necessarily absolutely) quiescent population, dividing asymmetrically to yield both themselves and a progenitor population that proliferates extensively and accounts for most of the specialized tissue-specific cells present at any given time. This slow-cycling behavior of adult somatic stem cells has been documented for stem cell populations in the blood, skin, brain, and intestines, among other organs. This line of thought has led many to conjecture that CSCs may also cycle relatively slowly, though it must be noted that because of the convergence of oncogenic activation and loss of tumor-suppressive braking mechanisms, even relatively "slow-cycling" cells are likely to proliferate much faster than an ordinary, untransformed cell from the same tissue of origin. The lipophilic, fluorescent dye PKH26 can be used to label live cells; it does so by associating with cell membranous structures. Because the dye is diluted out with each subsequent cell division, retention of this dye can be used to track and isolate cells that cycle relatively slowly. Such a technique has been used to successfully identify normal mammary gland stem cells and enrich for a gene signature derived there from that correlates with CSC content in breast cancer (Pece et al., 2010).

Relatively quiescent cancer cells can also be identified, but not preserved in a viable state, using 5-bromo-2-deoxyuridine (BrdU), a nucleotide analog that is incorporated into cells as they divide their DNA during mitosis, and is diluted through subsequent cell divisions. Using this method, we showed that PSA$^{-/low}$ prostate cancer cells (see below) divide more slowly than the isogenic PSA^{+} prostate cancer cell population.

PSA$^{-/low}$ Prostate Cancer Cells: A Clinically Relevant CSC-Enriched Population

The chief hope inherent in the CSC model is that it guides scientific efforts toward identifying a cancer cell population that is responsible for drug resistance and tumor recurrence, and that by targeting this population, treatment options and outcome for cancer patients are improved. Therefore, such a prospectively identified population should illuminate some fundamental truth about the course of the disease as seen in patients.

It has been well-established that the serum levels of prostate-specific antigen (PSA) directly correlate with disease state; within prostatic neoplasm, however, cells are heterogeneous for expression of this luminal cell differentiation marker. Our lab sought to answer the simple question of whether cells that express PSA are intrinsically different from those that do not. To track the two cell populations, we used a promoter-reporter system in which the expression of a green fluorescent protein (GFP) was driven by the PSA promoter; in practice, this system marks cells that express PSA by tagging them with GFP.

We found that PSA$^{-/low}$ cells and PSA^{+} cells both could regenerate tumors in androgen-proficient hosts but that tumor growth in mice whose androgen production was blocked (mimicking the clinical setting wherein androgen deprivation therapy is a frontline therapy) over several generations was the preferential province of cells that express little or no PSA. Furthermore, PSA$^{-/low}$ cells could generate both PSA$^{-/low}$ cells and PSA^{+} cells, whereas PSA^{+} cells could only divide symmetrically to produce more of themselves, i.e., could not produce cells that lack PSA expression. This finding demonstrates a clear hierarchy between the two cell populations and also helps to explain several important pathological observations regarding prostate cancer. There is an inverse correlation between tumor grade and PSA messenger RNA (mRNA) expression in prostate cancer, and this is echoed in the finding that immunohistochemical staining performed on late-stage tumors often shows very few PSA-expressing cells. Furthermore, patients who relapse following androgen-deprivation therapy often have recurrent tumors that are fairly bereft of PSA expression. It seems likely that the bulk tumor prior to androgen withdrawal is composed of cells that express PSA and those that do not (or that do so at very low levels). However, the PSA-negative population likely mediates tumor re-initiation following androgen deprivation therapy, and the bulk of the rebound growth is refractory to hormone-based therapy.

Although PSA$^{-/low}$ cells seem to fit the criteria necessary to be deemed CSCs, there is heterogeneity within this class of prostate cancer cells; for instance, not all PSA$^{-/low}$ cells express androgen receptor (AR). Therefore, although there seems to be an established hierarchy with respect to the PSA differentiation marker, it remains to be seen which subpopulations within the PSA-negative cells are important for mediating tumor regrowth and which are responsible for generation of PSA-expressing progeny. Additionally, we speculate that the PSA-positive and PSA-negative cells may cooperate in establishing the initial tumor, and there may be interactions between the two cell populations that facilitate optimal tumor growth. These avenues are now open to exploration using this clinically relevant model.

MicroRNAs as Effectors or Inhibitors of CSCs

MicroRNAs (miRNAs), small endogenous RNA species of roughly 19–25 nucleotides in length, play a role in virtually every important cellular process including metabolic processes, proliferation, and differentiation. They act by targeting a wide spectrum of messenger RNA (mRNA) species, each possessing a different affinity for the microRNA. In embryonic stem cells, microRNAs have been shown to extensively regulate these cells' ability to self-renew and to maintain the pluripotent state, and the tug-of-war between the tumor-suppressive microRNA *let-7* family and the RNA-binding protein Lin28, which sequesters these microRNAs is central to this regulation.

Although several reports have examined the microRNA species that are altered between clinical prostate adenocarcinoma samples and corresponding matched normal prostate tissue (Ambs et al., 2008), there has been no comprehensive study to determine if these same fluctuations occur between CSC and non-CSC populations *within* prostate cancers. Our lab has partially addressed this issue by profiling the microRNA species in sorted cell populations known to be enriched for CSCs (e.g., DU145 and LAPC9 CD44$^+$ cells), and then comparing these profiles to the appropriate marker-negative populations (e.g., DU145/LAPC9 CD44$^-$ cells). This produced a short list of miRNAs that are diminished in prostate CSC-enriched populations, as well as a handful that are expressed at higher levels in these cells (Liu et al., 2011; Liu et al., 2012). We validated these findings by analyzing miRNA levels from the FACS-sorted corresponding marker-positive and −negative epithelial populations purified from primary human prostate tumors.

Unsurprisingly, the *let-7* family is represented in the diminished microRNAs category, as let-7b is expressed at lower levels among prostate CSC-enriched populations than in non-CSC-enriched populations, but its effects are likely not exclusive to the CSC pool, as it targets omnipresent oncogenes such as C-Myc and H-Ras. Intriguingly, miR-34a expression is drastically dampened in the CSC-enriched populations, but at face value this miRNA does not seem to possess any predicted stem cell self-renewal-related targets. Instead, we found that miR-34a targets CD44 itself; this demonstrates that not only may certain microRNAs represent an impediment to establishment of a CSC phenotype, but that certain CSC cell surface markers such as CD44 may also play key functional roles (Liu et al., 2011).

Although our miRNA CSC screen was quite comprehensive, it stands to reason that some miRNAs that are expressed at undetectable (low) levels, or that are not expressed at all in the cell lines assayed, may have been missed. Likely candidate miRNA regulators of CSC activity include miRNAs that target genes or pathways that are important for self-renewal. Bmi-1, a protein that reads Polycomb repressive complex 2-derived epigenetic marks and modifies histones by adding ubiquitin moieties, is an important mediator of self-renewal in both normal prostate stem cells and in prostate cancer as well (Lukacs et al., 2010). MicroRNA-128a (miR-128a) is one of the most commonly diminished miRNAs in primary prostate cancer, perhaps because it represents a barrier to increased Bmi-1 expression (Godlewski et al., 2008), and therefore is an impediment to prostate cancer cell self-renewal. Interestingly, this microRNA may target two self-renewal pathways in prostate cancer, as miR-128a may also act on the mRNA of Nanog, one of the core triad of transcription factors necessary for human and mouse ES cell self-renewal that is often aberrantly re-expressed in prostate cancer. More work is required to determine if either of these two activities is preferentially exerted in the CSC compartment. Furthermore, as new markers for prostate CSCs are uncovered (miRNA profiling of PSA-negative versus PSA-positive prostate cancer cells will be an important undertaking in this regard) or as existing populations are able to be further enriched, additional microRNAs capable of acting on CSC populations are likely to be identified.

Prostate CSCs and the Metastatic Microenvironmental Mystery

The subfield of CSCs as integral components of the metastatic process is in its infancy, although intuitively the connection seems plausible. Since CSCs (or tumor-initiating cells) are judged fundamentally by their ability to reinitiate a semblance of the parental tumor in a foreign microenvironment, and since the end result of metastatic seeding *is* the formation of such a secondary growth, it is tempting to speculate that CSCs must be the population(s) that fuel this growth. However, there are a multitude of steps that occur before the seeding of a distant site by cells that have escaped from the primary tumor locale can occur.

The initial cancerous growth must invade its host environment, in the process overcoming the inhibitory signals that normally constrain migration. This process needs not solely be limited to CSC activity. For instance, in a fish model of embryonal rhabdomyosarcoma (ERMS), it has been shown that myf5$^+$ CSCs rely on the activity of non-tumorigenic, myogenin$^+$ cells to invade (Ignatius et al., 2012). This latter population migrates extensively and can enter the vasculature, whereas the myf5$^+$ cells cannot, and migrate scarcely at all unaided. Thus, myogenin$^+$ cells seem to act as trailblazers that open up avenues of invasion for the less-migratory CSCs to traverse. It is uncertain, however, whether such a paradigm may be operative in human cancers, although many human ERMS tumors do show organization of distinct myf5$^+$ and myogenin$^+$ regions. Whether this model is applicable to prostate cancer is also an open question, although it does suggest that the non-tumorigenic populations in prostate cancer should not be dismissed out of hand.

The primary tumor microenvironment itself may also condition cells toward a state of metastatic proficiency. Interestingly, our lab has found that prostate cancer cells grown in an orthotopic microenvironment (the dorsal prostate) preferentially metastasize and seem to upregulate several stem cell-related genes relative to genetically-identical prostate cancer cells grown subcutaneously in an ectopic microenvironment; this phenomenon correlates with host stroma-derived TGF-Beta production. This suggests a link between a stem-like cancer cell phenotype, one that is dependent upon microenvironmental cues and one that results in metastatic growth. Indeed, integrins, which are key intermediaries of communication between an epithelial cell and its extracellular environment, have been shown to be essential for metastasis. In PC3 prostate cancer cells, loss of αv integrin reduces the CD44$^+$, Aldefluor$^+$, and α2β1-integrin cell populations that are each enriched for CSCs, and this reduction in potential CSCs correlates with a decrease in metastasis to the bone (van den Hoogen et al., 2011).

There is evidence that shedding of cells into the circulation needs not be a late-stage event in tumor progression, as was previously thought. Instead, one early rate-limiting step in the formation of metastatic colonies may be the conditioning of the organ to be colonized. This pre-metastatic niche must be receptive to the growth of circulating cancer cells in order for frank metastasis to result. In a mouse model of breast cancer, it has been shown that the CSC fraction of cells induces upregulation of periostin in the soon-to-be-colonized lung stroma and that this extracellular matrix component in turn augments Wnt signaling in the newly-arriving CSC population, resulting in the expansion of these cells as lung metastases (Malanchi et al., 2012). This interaction is apparently necessary for metastatic growth to occur. It remains to be seen if this model may be applicable to other human cancers including those involving the prostate, which metastasizes to the bone and lung.

Prospective

There are a number of unanswered questions within the broad realm of prostate CSCs, the foremost of which is this: What do CSCs in primary human prostate cancer look like? Although the last 6 or 7 years' study of CSCs in cell lines and xenograft models have yielded some semblance of *bona fide* human prostate CSCs, these findings must be confirmed in frank human disease; this will require creation of new systems or extensive optimization of existing systems to successfully grow primary prostate-derived cells in murine hosts. Beyond this, one largely unexplored area of research centers on potential roles for non-tumorigenic, i.e., non CSC, populations. The CSC model has inspired hope that therapies that target the tumor-initiating cell population can eradicate a tumor in its entirety, but this assumption is predicated on the notion that non-tumorigenic cancer cells have no integral roles in maintaining, protecting, or serving as a niche for CSCs. Because primary prostate cancer is a largely indolent disease and prostate cancer-associated death results almost entirely from metastatic spread, it is critically important that the roles of both tumor-initiating and non-tumor-initiating cell populations in dissemination of primary prostate cancer be understood.

Acknowledgments

We thank the other Tang lab members for their support and assistance in actual experiments. Research in our lab was supported in part by grants from NIH (R01-CA155693-01A), Department of Defense (W81XWH-11-1-0331), CPRIT funding (RP120380), and the M.D. Anderson Cancer Center the Center for Cancer Epigenetics and Laura and John Arnold Foundation RNA Center pilot grant (to D.G. Tang). We apologize to the colleagues whose work could not be cited due to space constraints.

References

Ambs, S., Prueitt, R.L., Yi, M., Hudson, R.S., Howe, T.M., Petrocca, F., Wallace, T.A., Liu, C.G., Volinia, S., Calin, G.A., et al. (2008). Genomic profiling of microRNA and messenger RNA reveals deregulated microRNA expression in prostate cancer. Cancer Res. *68*, 6162–6170.

Bonnet, D., and Dick, J.E. (1997). Human acute myeloid leukemia is organized as a hierarchy that originates from a primitive hematopoietic cell. Nat. Med. *3*, 730–737.

Collins, A. T., Berry, P.A., Hyde, C., Stower, M.J., and Maitland, N.J. (2005). Prospective identification of tumorigenic prostate cancer stem cells. Cancer Res. *65*, 10946–10951.

Godlewski, J., Nowicki, M.O., Bronisz, A., Williams, S., Otsuki, A., Nuovo, G., Raychaudhury, A., Newton, H.B., Chiocca, E.A., Lawler, S. (2008). Targeting of the Bmi-1 oncogene/stem cell renewal factor by microRNA-128 inhibits glioma proliferation and self-renewal. Cancer Res. *68*, 9125–9130.

Ignatius, M. S., Chen, E., Elpek, N.M., Fuller, A.Z., Tenente, I.M., Clagg, R., Liu, S., Blackburn, J.S., Linardic, C.M., Rosenberg, A.E., et al. (2012). In Vivo imaging of tumor-propagating cells, regional tumor heterogeneity, and dynamic cell movements in embryonal rhabdomyosarcoma. Cancer Cell *21*, 680–693.

Lee, J., S. Kotliarova, Kotliarov, Y., Li, A., Su, Q., Donin, N.M., Pastorino, S., Purow, B.W., Christopher, N., Zhang, W., et al. (2006). Tumor stem cells derived from glioblastomas cultured in bFGF and EGF more closely mirror the phenotype and genotype of primary tumors than do serum-cultured cell lines. Cancer Cell *9*, 391–403.

Li, H., Chen, X., Calhoun-Davis, T., Claypool, K., and Tang, D.G. (2008). PC3 human prostate carcinoma cell holoclones contain self-renewing tumor-initiating cells. Cancer Res. *68*, 1820–1825.

Liu, C., Kelnar, K., Liu, B., Chen, X., Calhoun-Davis, T., Li, H., Patrawala, L., Yan, H., Jeter, C., Honorio, S., et al. (2011). The microRNA miR-34a inhibits prostate cancer stem cells and metastasis by directly repressing CD44. Nat Med. *17*, 211–215.

Liu, C., Kelnar, K., Vlassov, A.V., Brown, D., Wang, J., and Tang, D.G. (2012). Distinct microRNA Expression Profiles in prostate cancer stem/progenitor cells and tumor-suppressive functions of let-7. Cancer Res. *72*, 3393–3404.

Lukacs, R.U., Memarzadeh, S., Wu, H., and Witte, O.N. (2010). Bmi-1 is a crucial regulator of prostate stem cell self-renewal and malignant transformation. Cell Stem Cell *7*, 682–693.

Malanchi, I., Santamaria-Martinez, A., Susanto, E., Peng, H., Lehr, H.A., Delaloye, J.F., and Huelsken, J. (2012). Interactions between cancer stem cells and their niche govern metastatic colonization. Nature *481*, 85–89.

O'Brien, C.A., Pollett, A., Gallinger, S., and Dick, J.E. (2007). A human colon cancer cell capable of initiating tumour growth in immunodeficient mice. Nature *445*, 106–110.

Patrawala, L., Calhoun, T., Schneider-Broussard, R., Li, H., Bhatia, B., Tang, S., Reilly, J.G., Chandra, D., Zhou, J., Claypool, K., Coghlan, L., et al. (2006). Highly purified $CD44^+$ prostate cancer cells from xenograft human tumors are enriched in tumorigenic and metastatic progenitor cells. Oncogene *25*, 1696–1708.

Patrawala, L., Calhoun-Davis, Schneider-Broussard, R., and Tang, D.G. (2007). Hierarchical organization of prostate cancer cells in xenograft tumors: the CD44+ alpha2beta1+ cell population is enriched in tumor-initiating cells. Cancer Res *67*, 6796–6805.

Pece, S., Tosoni, D., Confalonieri S, Mazzarol G, Vecchi M, Ronzoni S, Bernard L, Viale G, Pelicci PG, Di Fiore PP. (2010). Biological and molecular heterogeneity of breast cancers correlates with their cancer stem cell content. Cell *140*, 62–73.

Toivanen, R., Berman, D.M., Wang, H., Pedersen, J., Frydenberg, M., Meeker, A.K., Ellem, S.J., Risbridger, G.P., Taylor, R.A. (2011). Brief report: a bioassay to identify primary human prostate cancer repopulating cells. Stem Cells *29*, 1310–1314.

van den Hoogen, C., G. van der Horst, G., Cheung H, Buijs JT, Pelger RC, van der Pluijm G. (2011). Integrin alphav expression is required for the acquisition of a metastatic stem/progenitor cell phenotype in human prostate cancer. Am J Pathol *179*, 2559–2568.

Chapter 15

Glioblastoma Stem Cells Drive Tumor Recurrence and Patient Relapse: What's the Evidence?

Aneet Mann*, Randy van Ommeren*, Branavan Manoranjan, Nicole McFarlane, Parvez Vora, Chitra Venugopal, and Sheila Singh
Faculty of Health Sciences, McMaster University Stem Cell and Cancer Research Institute, McMaster University, Hamilton, ON, Canada
* *Equal contribution*

Abbreviations

BCL	B cell lymphoma
bFGF	basic fibroblast growth factor
BTIC	brain tumor initiating cell
CDK2NA	cyclin-dependent kinase inhibitor 2A
CNS	central nervous system
CSC	cancer stem cell
DCX	double cortin
DMEM	Dulbecco's Modified Eagle's Medium
EGF	epidermal growth factor
EGFR	epidermal growth factor receptor
FBS	fetal bovine serum
FLIP	FLICE-like inhibitory protein
GABRA1	gamma-aminobutyric acid receptor subunit alpha-1
GBM	glioblastoma
GFAP	glial fibrillary acidic protein
HIF1α	hypoxia inducing factor 1α
HSC	hematopoietic stem cell
IDH1	isocitrate dehydrogenase 1
IGL	internal granular layer
LDA	limiting dilution assay
LeX	Lewis X
LIF	leukemia inhibitory factor
MB	medulloblastoma
MDM2	murine double minute 2
MGM	O6-methylguanine DNA-methyltransferase

(Continued)

Cancer Stem Cells, First Edition. Edited by Vinagolu K. Rajasekhar.

(*Continued*)

NEFL	neurofilament light polypeptide
NF-κB	nuclear factor kappa-light-chain-enhancer of activated B cells
NF1	neurofibromatosis type 1
NOD-SCID	non-obese diabetic severe combined immunodeficiency
NSC	neural stem cell
p53	(TP53) tumor protein 53
PDGFR	platelet derived growth factor receptor
PI(3)K	phosphotidylinositol-3-OH kinase
PROM1	prominin 1
Ptc	patched
PTEN	phosphatase and tensin homolog
RB1	retinoblastoma protein
SCZ	subcallosal zone
SGZ	subgranular zone
SLC12A5	solute carrier family 12 (potassium/chloride transporter), member 5
SSEA-1	stage specific embryonic antigen 1
SVZ	subventricular zone
TGF-β	transforming growth factor β
TNF	tumor necrosis factor
VEGF	vascular endothelial growth factor

Introduction to Gliomas

A vast majority of adult neurological tumors are of glial cell origin and termed as gliomas. The World Health Organization (WHO) grading system classifies tumors as gliomas based on the presence of histological, morphological, and ultrastructural indicators of glial differentiation. Gliomas are graded on a scale of I to IV corresponding to degree of malignancy as determined by various histological features, such as cellular polymorphism, microvascular proliferation, necrosis, high cellularity, and mitotic activity (Louis et al., 2007; Maher et al., 2001; Zhu and Parada, 2002). The most malignant glioma is glioblastoma multiforme (GBM), a WHO-grade IV tumor characterized histopathologically by nuclear pleomorphism, extensive microvascular proliferation, and/or necrosis. GBMs are the most common primary brain tumors in adults, and are typically categorized as primary or secondary GBMs. Primary GBMs, arising de novo in the absence of pre-existing low-grade tumors, are highly aggressive and invasive, with the majority of patients succumbing to their disease within a year following diagnosis (Maher et al., 2001). Alternatively, secondary GBMs arise usually within 5 to 10 years in patients initially diagnosed with low-grade astrocytomas. Clinically, primary and secondary GBMs are indistinguishable from each other in terms of patient survival and pathology (Maher et al., 2001). Recent advances in therapeutic interventions have only added modest improvement to survival, and patient outcome remains grim (Stupp et al., 2005). Common genetic events occurring in human GBMs include dysregulation of growth factor signaling pathways via amplification or constitutive activation of receptor tyrosine kinase genes, dysregulation of the phosphotidylinositol-3-OH kinase (PI[3]K) pathway, and mutational inactivation of tumor-suppressor pathways involving mainly p53 and retinoblastoma protein 1 (RB1). Most genes that have been found to be implicated in GBM are associated with regulation of cell cycle, proliferation, and apoptosis, which include EGFR, TP53, MDM2, PTEN, PDGFR (Huse and Holland, 2010; Verhaak et al., 2010). Integrated genomic analysis has provided a means for molecular classification of GBM into four subtypes—classical, mesenchymal, proneural, and neural—based on differential gene expression patterns (Huse and Holland, 2010; Verhaak et al., 2010). The classical subtype is characterized by amplification of EGFR, homozygous deletion of CDK2NA, which encodes for two tumor-suppressor proteins involved in cell cycle control, and a low frequency of TP53 tumor-suppressor gene mutations (Verhaak et al., 2010). The mesenchymal subtype typically has low expression of NF1 and high expression of genes associated with the tumor necrosis factor (TNF) family and NF-kB pathways, suggesting a possible association with higher overall inflammation and necrosis (Verhaak et al., 2010). The proneural subtype exhibits a higher frequency of focal amplification in PDGFRA and also displays point mutations in IDH1 and mutations in TP53. Lastly, the neural subtype is classified based on the presence of neuronal markers such as NEFL, GABRA1, SYT1, and SLC12A5 (Verhaak et al., 2010). Comparative gene expression profiles of the classical, mesenchymal, proneural, and neural subtypes were highly-enriched for distinct cell-type gene signatures associated with murine astrocytes, cultured astrocytes, oligodendrocytes, and cells showing association with both glial and neuronal gene signatures, respectively (Verhaak et al., 2010). Consequently, distinct cells of origin appear to drive each GBM subtype.

Characterizing Glioblastoma Cancer Stem Cells

CSC Hypothesis

The cancer stem cell (CSC) hypothesis has received much attention in recent years because of its ability to explain poorly understood clinical observations including therapy resistance and tumor relapse. This hypothesis proposes that a small subset of tumor cells with stem cell properties initiate and maintain tumorigenesis (Reya et al., 2001). These rare fractions of cells within the bulk tumor maintain the potential to self-renew and differentiate into more committed lineages, properties associated with stem cells. The CSC model suggests a hierarchical organization of cells with a CSC population at the top that is capable of generating the more differentiated and heterogeneous bulk of the tumor (Reya et al., 2001). Although the CSC is defined solely based on its stem-like properties and has no bearing on a tumor's cell of origin, it is plausible that more differentiated cells may acquire a stem-like phenotype through multiple mutations (Clarke et al., 2006) (see Figure 15.1).

Given the limited clinical utility of targeting entire tumor cell populations with a single treatment modality due to intra- and inter-tumoral heterogeneity (Magee et al.,

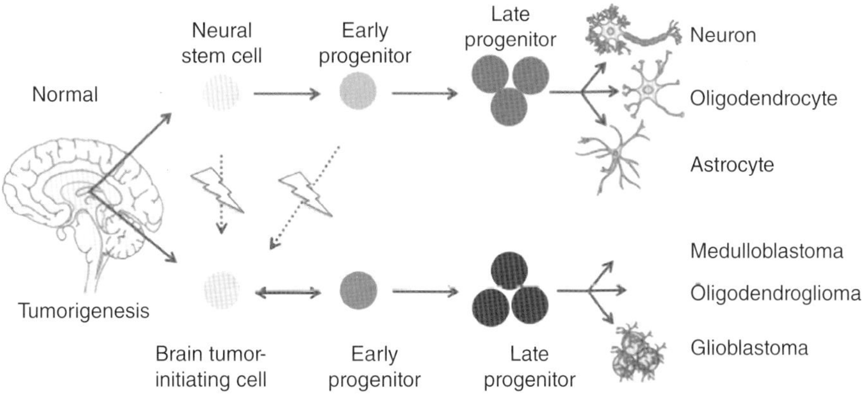

Figure 15.1. Brain tumorigenesis according to the cancer stem cell hypothesis. Brain tumors contain a cellular hierarchy that resembles neural stem cells and their lineage-restricted progeny. Lineage-specific progenitors resulting from the brain tumor-initiating cell may result in cell-type specific tumors. For color detail, please see color plate section.

2012), the possibility of targeting a small subset of cells at the apex of the hierarchy along with the tumor bulk appears more plausible. Therapeutic strategies capable of targeting this initiating and maintaining population may lead to tumor regression and/or attenuate disease progression. However, this strategy relies on the assumption that the CSC population is static, and that tumor-initiating potential is unique to these cells. Recent data suggests that the CSC phenotype may be more dynamic than initially anticipated, and that it may in fact be strongly regulated by its tumor microenvironment (Vermeulen et al., 2012). Signals from the local environment may induce more differentiated cells to revert to a stem cell-like phenotype in a phenomenon known as dedifferentiation, which may pose a significant challenge when targeting the CSC population (Vermeulen et al., 2012).

Neural Stem Cells

The importance of neural stem cells (NSC) in the adult brain is unclear. Some propose it to be a vestigial trait of evolution while others suggest the limited capacity to self-renew plays a pivotal role in learning and memory (Gage, 2000; Gage, 2002; Gross, 2000; Temple, 2001). The term "neural stem cells" refer to cells that satisfy the following criteria: (1) they are capable of generating neural tissue or originating from the nervous system; (2) they possess the capacity to self-renew; and (3) they give rise to more differentiated cells through asymmetrical cell division (Gage, 2000). The discovery of neural stem cells in the adult brain (Lois and Alvarez-Buylla, 1993; Reynolds and Weiss, 1992), long thought to be a post-mitotic organ (Gage, 2000), supported earlier evidence of adult neurogenesis in the 1960s (Gross, 2000). These cells, isolated from the adult mammalian brain, were observed to maintain themselves through asymmetrical division and differentiate into neurons, astrocytes, and oligodendrocytes (Reynolds and Weiss, 1992; Reynolds and Weiss, 1996).

There exist four distinct regions in the postnatal mammalian brain where cells with stem cell properties have shown to reside: (1) the subventricular zone (SVZ) that lines the lateral ventricles (Doetsch et al., 1999); (2) the subgranular zone (SGZ) located between the granule cell layer and hilus of the dentate gyrus (Seri et al., 2001); (3) the subcallosal zone (SCZ) located between the hippocampus and the corpus callosum (Seri et al., 2006); and (4) the border between white matter and the internal granular layer (IGL) of the cerebellum (Lee et al., 2005). The stem cell niche or the microenvironment surrounding the stem cells interacts with these cells to regulate their fate (Moore and Lemischka, 2006). Many studies support the observation that brain tumors arise at a higher frequency from these anatomical stem cell niches, and are more frequent still in developmental phases when NSC numbers are highest. Periventricular tumors were demonstrated to occur in the rat subventricular region after intraventricular inoculation with avian sarcoma viruses, with a much higher rate of tumors occurring in neonatal rats versus adult rats (Copeland and Bigner, 1977; Copeland et al., 1975; Vick et al., 1977). The incidence of brain tumors in mice inoculated with a pellet of carcinogen was much greater if the pellet was placed in the subventricular region versus the peripheral cortex. A single dose of ethyl-nitrosourea, a potent carcinogen, administered to pregnant rats also induced periventricular tumors in the their offspring (Koestner et al., 1971). Transformation, at a cellular level, refers to the collective changes, including uncontrolled cell proliferation and morphological alterations, responsible for converting a normal cell into a cancer cell (Werbowetski-Ogilvie et al., 2011). These observations regarding tumorigenesis and the NSC niche suggest that NSCs may represent the most conceivable targets for transformation into cancer cells, since their longevity permits accumulation of genetic and epigenetic alterations that could perturb their controlled self-renewal machinery (Werbowetski-Ogilvie et al., 2011).

CD133 as a NSC and a GBM Cancer Stem Cell Marker

CD133 was first identified in the hematopoietic stem cell (HSC) field as a marker for fetal/adult bone marrow stem cells. Molecular cloning of CD133 revealed the antigen to be a novel hematopoietic surface antigen consisting of 5-transmembrane domains with an extracellular N-terminus and a cytoplasmic C-terminus (Miraglia et al., 1997). Human CD133 is localized in the plasma membrane protrusions at the apical surface of the cells (Corbeil et al., 1999; Weigmann et al., 1997). Despite its association with a stemness phenotype, the function of CD133 in stem cells and their malignant counterparts remain unclear.

Subsequent to its discovery in HSCs, CD133 expression has been found to identify other primitive cell populations, including NSCs (Mizrak et al., 2008; Uchida et al., 2000). The expression of CD133 in fetal brain tissue represents a rare subset of central nervous system (CNS) cells capable of initiating neurosphere cultures, and differentiating into neurons and glial cells. *In vivo*, CD133+ fetal brain cells transplanted into the lateral ventricles of NOD-SCID newborn mice give rise to progeny that migrate and differentiate, and these rare cells are also observed to localize to sites of neurogenesis, the SVZ and dentate gyrus, where they continue to proliferate (Uchida et al., 2000). A CD133+, lineage-negative cell population capable of sustained self-renewal and sphere formation over repeated passages has also been identified in the postnatal cerebellum (Lee et al., 2005).

By applying the paradigms of stem cell biology to heterogeneous cancers such as GBM, the identification and purification of prospective CSCs has become possible. In contrast to CD133− GBM cells, a subpopulation of cells from human GBM samples purified based on CD133 positivity exhibit stem cell properties of proliferation, differentiation, and self-renewal as assayed through *in vitro* methods normally employed to study human NSCs (Bao et al., 2006a; Galli et al., 2004; Hemmati et al., 2003; Singh et al., 2003). As few as 100 CD133+ GBM cells transplanted into the frontal cortex of adult NOD-SCID mice gave rise to tumors with CD133+ and CD133− cell populations. The new tumors were serially transplanted, and they recapitulated the phenotype of the original tumor (Singh et al., 2004). While recent publications have begun to contest the exclusivity of CD133 positivity as a phenotypic requirement for self-renewing, tumor-initiating GBM CSCs (Beier et al., 2007), increased self-renewal capacity and recapitulation of the original tumor following xenotransplantation supports the idea that cells expressing CD133 form a true CSC population in GBM (Singh et al., 2004).

CSCs Sustain Tumor Growth

The CSC hypothesis postulates the presence of a stem-cell-like population that is capable of maintaining and sustaining tumor growth (Reya et al., 2001). In keeping with this philosophy, CD133+ cells isolated from human GBMs were shown to generate tumors that were highly angiogenic, proliferative, and hemorrhagic and contained widespread necrosis when implanted into immunocompromised mice. Tumors derived from CD133+ cells were strikingly more vascular than those derived from CD133− cells. Further experimentation revealed that CD133+ cells expressed elevated levels of vascular endothelial growth factor (VEGF) compared to CD133− cells, possibly explaining the difference in vascularity observed in tumors derived from the two populations. It is plausible that the CD133+ stem cell fraction in GBMs is required for maintaining tumor growth and supporting non-stem cell tumor cells by promoting angiogenesis (Bao et al., 2006b). In addition, CD133+ cells were shown to interact closely with endothelial cells. The endothelial environment may provide a vascular niche allowing for tumor growth and infiltration, and therefore may be an attractive target for cancer treatment (Calabrese et al., 2007). Studies have demonstrated a marked decrease in GBM proliferation, vascularisation, and migration with the use of anti-VEGF therapies (Bao et al., 2006b). Recent data, however, alludes to a pro-invasive phenotype of a subset of GBMs following inhibition of VEGF signaling, and demonstrates that the presence of VEGF signaling negatively regulates tumor cell invasion through a MET/VEGFR2 heterocomplex (Lu et al., 2012). Anti-VEGF therapy may still provide survival benefits if those GBMs not likely to develop a pro-invasive phenotype upon treatment can be selected for and isolated by evaluation of VEGF and MET expression, or if given concurrently with anti-MET treatment. Anti-MET therapy has been shown to attenuate tumor cell invasion and epithelial mesenchymal transition, induced by VEGF ablation in mouse models of GBMs (Lu et al., 2012).

Therapy Resistance

The CSC hypothesis proposes that the CSC population is more resistant to chemo- and radio-therapy than progenitors and the more differentiated bulk tumor. The CSC escapes therapy and then drives tumor recurrence (Reya et al., 2001). CD133+ cells isolated from primary GBM specimens have demonstrated dramatic resistance to traditional chemotherapeutic drugs compared to CD133− cells. Upon characterization, CD133+ cells were shown to have increased expression of genes associated with multi-drug resistance and DNA mismatch repair, specifically BCRPI and MGMT, respectively (Liu et al., 2006). In fact, GBMs containing a methylated MGMT promoter behaved positively in response to chemotherapy, and patients saw survival benefits (Hegi et al., 2005). Furthermore, CD133+ cells exhibited higher expression of anti-apoptotic genes, namely FLIP, BCL-2, and BCL-XL (Liu et al., 2006).

The most effective therapy for treating GBMs is ionizing radiation, but its therapeutic potential remains limited because of radioresistance of CSCs. Following treatment with ionizing radiation, short-term cultures and GBM tumor xenografts were enriched for the original CD133+ CSC population. CD133+ cells exhibited greater cell cycle checkpoint activation in response to ionizing radiation suggesting a greater ability to recover from DNA damage (Bao et al., 2006a). Alternatively, Jamal and colleagues (2012) observed no difference in response from CD133+ and CD133− cells to radiation upon quantification of radiation-induced DNA double strand breaks and DNA repair in *in vitro* cultures. However, irradiation of orthotopic xenografts derived from CD133+ cells resulted in more radioresistance observed in CD133+ cells compared to CD133− cells derived from the same tumor xenograft. These data suggest a role for the tumor microenvironment in conferring cellular radioresistance (Jamal et al., 2012), possibly mediated through the secretion of cytokines, namely Transforming Growth Factor-β (TGF-β) (Hardee et al., 2012). In addition, CD133+ cells isolated from irradiated tumors were highly tumorigenic compared to their CD133− counterparts, and demonstrated decreased latencies in tumor formation compared to CD133+ cells derived from untreated xenografts (Bao et al., 2006a). Conceivably, those cells possessing increased radioresistance may also be the most aggressive clones of the CD133+ GBM-initiating population thereby leading to reduced tumor latencies and an aggressive relapse after treatment. (See Figure 15.2.)

Controversies

CD133

The assertion that CD133 is a marker of every true CSC in GBM has become increasingly contested. CD133− cells dissociated from primary GBMs can behave as CSCs, capable of self-renewal, asymmetric division, tumorsphere formation, and tumorigenesis (Beier et al., 2007; Chen et al., 2010; Joo et al., 2008; Ogden et al., 2008; Wang et al., 2008). It has been shown that an A2B5+CD133− cell population in GBM harbors a CSC population capable of tumorigenesis (Ogden et al., 2008). Further, a significant number of

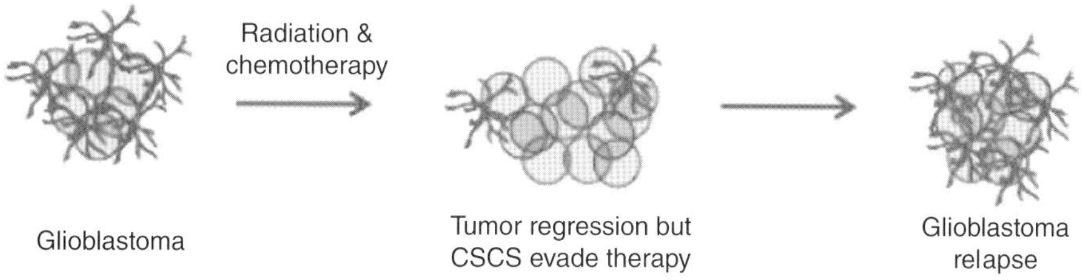

Figure 15.2. BTICs evade therapy leading to tumor relapse. Although radiation and chemotherapy may target the bulk of the tumor population, BTICs are able to evade therapy, resulting in tumor relapse.

GBM cases lack CD133+ cells completely, suggesting that the marker is only expressed by a subset of GBM CSCs found within a distinct subset of all GBM tumors. A popular interpretation of these results has been that the GBM CSC forms unique subsets along a hierarchy of differentiation where CD133 positivity may demarcate only part of the CSC population in a tumor. Consequently, variation in CD133 positivity may be reflective of both intra- and inter-tumoral heterogeneity (Chen et al., 2010).

However, despite the demonstration that CD133− CSCs do exist, CD133 still has relevance in demarcating populations of CSCs with unique properties dissimilar from those of CD133− cells. For example, in addition to roles in tumor generation, CD133+ cell populations may be implicated in specific roles associated with tumor maintenance such as promotion of angiogenesis and infiltration (Wang et al., 2008). Along with the possible interactions with the vascular niche, CD133 positivity has also correlated to other phenotypic attributes including unique gene expression and a distinctive molecular profile. A 24-gene signature can be used to classify GBM CSCs into subtypes that cluster well with either fetal NSC lines ($CD44^{low}$, $CD133^{high}$, $PDGFR\alpha^{high}$, DCX^{pos}, and $EGFR^{high}$) or adult NSC lines ($CD44^{high}$, $CD133^{low}$, $PDGFR\alpha^{low}$, DCX^{neg}, and $EGFR^{low}$), depending on their expression of CD133 (Lottaz et al., 2010; Maisel et al., 2007). Gene expression analysis of $CD133^{high}$ and $CD133^{low}$ CSC's shows that they identify closely with the proneural and mesenchymal GBM subtypes, respectively (Lottaz et al., 2010; Phillips et al., 2006). Taken together, these data suggest that CD133+ cells are phenotypically unique compared to CD133− cells and likely have clinical relevance as a targetable subpopulation in a subset of patients with GBMs.

Conflicting results on the correlation of GBM prognosis to CD133 expression have fostered further controversy over the utility of CD133 as a CSC marker. Some have shown that levels of CD133 expression by various GBM subtypes are not necessarily correlated to the prognosis associated with that subtype. For example, GBMs of the mesenchymal (poor prognosis) subtype appear to have fewer CD133+ stem cells than GBMs of subtypes with lower mortality rates (Joo et al., 2008). However, studies that forgo subtyping and directly correlate CD133 expression to patient outcome show that CD133 percentage is inversely correlated with patient outcome (Beier et al., 2008; Sato et al., 2010; Zeppernick et al., 2008). GBM patients with greater than 1% CD133 positivity die significantly sooner after diagnosis and display much faster and aggressive metastatic progression to higher grade gliomas (Zeppernick et al., 2008). High-grade oligodendroglial tumors that demonstrate poor sphere formation and lack CD133− expressing CSCs are associated with better progression-free survival (PFS) times. Although histological grading of oligodendroglial tumors is subject to pathological interpretation, this method of diagnosis has been demonstrated to be inferior to CD133 positivity assays in predicting PFS (Beier et al., 2008). While CD133 expression in patients with disseminating GBM fluctuates from very high to essentially non-detectable, tumors without dissemination are always low in CD133 expression (Sato et al., 2010). These data resonate well with reports from other malignancies such as colon and rectal cancer (Choi et al., 2009; Li et al., 2009; Saigusa et al., 2009; Wang et al., 2009).

Several important practical considerations with regard to protocol should also influence the interpretation of data involving CD133. For example, techniques for tissue dissociation such as the use of the serine protease Trypsin may cleave the CD133 marker and complicate cell sorting and assay readouts (Sakariassen et al., 2007). Additionally, hypoxic environments have been shown to increase the CD133+ GBM stem cell fraction through hypoxia-inducible factor-1α (HIF-1α), illustrating the need to accurately recapitulate oxygen exposure characteristic of the physiological conditions to avoid biasing CD133-based sorting and analysis (Soeda et al., 2009). The impact of other microenvironmental factors on CD133 expression remain enigmatic and may also influence the growth and expansion of CSC populations within *in vitro* and *in vivo* GBM models. Finally, concerns regarding the utility of current monoclonal antibodies in detecting the full spectrum of CD133 splice variants and glycosylation configurations suggest an underestimation of true CD133 ratios and incomplete isolation of CD133+ fractions (Bidlingmaier et al., 2008; Fargeas et al., 2011; Sakariassen et al., 2007). Regardless of the degree to which CD133 is expressed on GBM CSCs, it is important to remember that no explicit functional connection has been established between CD133 and the (cancer) stem cell.

CD15

As debate has increased regarding the use of CD133 to isolate CSCs from brain tumors, the use of CD15 as a CSC marker has become increasingly common. CD15, also known as stage-specific embryonic antigen 1 (SSEA1) or Lewis X (LeX), is a carbohydrate surface moiety—trisaccharide 3-fucosyl-N-acetyllactosamine (Gooi et al., 1983). In 2002, Capela and Temple published a study confirming that CD15 expression localizes to adult neurogenic zones of the mouse brain, most notably, the SVZ (Capela and Temple, 2002). In addition, CD15+ cells retained BrdU suggesting that they are slowly dividing stem cells of the adult brain. CD15+ cells readily form neurospheres in a density-dependent fashion, in contrast to CD15- cells, which lack sphere-forming ability (Capela and Temple, 2002; Capela and Temple, 2006). Further, CD15 has been shown to demarcate a subset of GFAP + NSCs within the adult mouse periventricular tissue, which co-express Nestin and form neurospheres. These CD15+ cells have the potential to differentiate into all three neural lineages (neuron, astrocyte, and oligodendrocytes), illustrating an undifferentiated and stem-like phenotype (Imura et al., 2006). In contrast to CD133, the correlation of CD15 expression to the stem-like properties of a cell is better defined. CD15 tends to co-express with the proteins FGF8 and Wnt-1, both of which have roles in stemness and development (Capela and Temple, 2006). It has been shown that fibroblast growth factor 8 (FGF8) has a role in midbrain development of the embryo and may interact with Wnt-1 (Crossley et al., 1996). Wnt-1, in turn, has also been associated with stem cell growth, especially in the hematopoietic field. Immunoprecipitation has shown a direct association of Wnt-1 and CD15 antigen, further establishing a functional relationship between CD15 positive cells and Wnt-1 (Reya et al., 2003; Willert et al., 2003). Although an understanding of the mechanics of CD15 regulation of NSC proliferation remains somewhat elusive, these data suggest a convincing relationship between the marker and NSC regulation.

Early investigation of CD15 expression in the context of GBM found that the marker is rarely expressed on high-grade gliomas (WHO grades III and IV), and more frequently on low-grade gliomas (WHO grades I and II) (Reifenberger et al., 1992). More recently however, it has been shown that CD15 can in fact mark for a GBM tumor stem cell with self-renewal and tumor-initiating properties (Mao et al., 2009; Son et al., 2009). Purification of CD15+ populations from primary GBM cells cultured in stem cell promoting conditions significantly enriches for BTICs, where 1,000 CD15+ cells will consistently generate tumors when xenografted in mice. In comparison, 300,000 CD15- GBM cells generate tumors in an unpredictable frequency.

Furthermore, lineage tracking with fluorescent markers shows most neurospheres arise from CD15+ cells. CD15+ cells, shown to be associated with stemness markers Sox2, Nestin, Integrin beta1, Integrin alpha6, and Olig2, lose their stem-like properties and marker expression in differentiation conditions, proceeding to express the markers GFAP and Tuj1 associated with astroglial cells and neurons, respectively (Mao et al., 2009; Son et al., 2009). In contrast to CD133, CD15 is expressed more frequently in GBM primary samples and GBM cell lines, and tends to co-express with CD133. Interestingly, CD15 expression trends observed in GBM have been shown to have parallels in other brain tumor types such as ependymoma and medulloblastoma (MB) (Mao et al., 2009; Ward et al., 2009). In MB, a pediatric neuroectodermal tumor, it has been demonstrated that CD15 is more effective in enriching for cells with tumorigenic potential compared to CD133 in Ptc+/− mouse models of MB (Read et al., 2009; Ward et al., 2009).

Culture Conditions

The conditions used to culture GBM CSCs *in vitro* are derived from protocols developed in the normal NSC field prior to the discovery of the CSC in brain tumors. Culturing dissociated striatal tissue in serum-free media containing epidermal growth factor (EGF) facilitates the growth of proliferating clusters of cells (neurospheres) while removal of EGF terminates further proliferation. When the spheres are grown in adherent conditions, cells migrate from the spheres and differentiate into the various CNS cell types (Reynolds et al., 1992). Culturing neurospheres in adherent conditions with media containing fetal bovine serum (FBS) allows the spheres to flatten down and differentiate into neurons, astrocytes, and oligodendrocytes (Reynolds and Weiss, 1996). Basic fibroblast growth factor (bFGF) induces proliferation of unipotent and bipotent progenitor cells in neurospheres produced in EGF positive media (directed to neuronal and neuron/astrocyte pathways, respectively) (Vescovi et al., 1993). In the context of GBM CSCs, media supplemented with EGF, bFGF, and occasionally LIF propagates GBM cells with stem-like phenotypes. To induce differentiation, cells are typically grown in media characterized primarily by the addition of FBS to cell media. Comparative analysis has shown that primary GBM cultures grown in serum-free conditions show a long-term maintenance of a CSC population that is genotypically and phenotypically similar to the orignal population found in the primary sample (Lee et al., 2006). In addition, adherent versus non-adherent growth surfaces also impact the degree to which the expansion of cancer stem cell populations are supported within a culture. Growth of cultures on adherent culture plates such as those coated with poly-L-ornithine induces a more differentiated state in the cultured cells while non-adherent plates promote the growth of free-floating neurospheres and stem-like cells. (See Figure 15.3.) To date, rigorous comparison of the adherent conditions versus non-adherent conditions has not been published, and the relative impact of each culture environment on cell phenotype is poorly described.

The Sphere Formation Assay

The proliferating clusters or neurospheres observed when culturing dissociated striatal tissues in serum-free conditions are also seen with solid cancers such as GBM (Hemmati et al., 2003; Reynolds et al., 1992; Singh et al.,

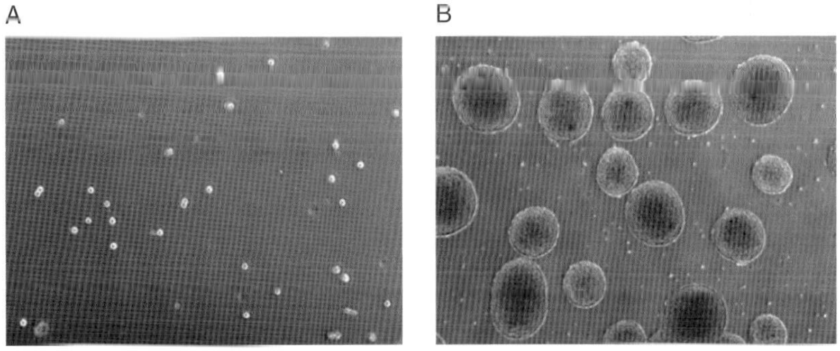

Figure 15.3. GBM BTIC populations form neurospheres in culture. **A.** Primary GBM samples are dissociated and plated as single cells in complete neural stem cell media supplemented with EGF, bFGF and LIF. **B.** Following incubation, the cells form tumorspheres. Reprinted from Stem Cell Research, Venugopal et al., 2012, Bmi1 marks intermediate precursors during differentiation of human brain tumor-initiating cells, 141–153(2012)8, with permission from Elsevier.

2003; Uchida et al., 2000). Each neurosphere is thought to arise from the self-renewal of a single stem cell or multipotent progenitor that undergoes an asymmetric division to populate a sphere with differentiated cells while maintaining a stem cell population (Reynolds et al., 1992; Reynolds and Weiss, 1996). Naturally, the realization that neurosphere generation is a property of prospective CSCs has led to the use of sphere formation as an assay for quantification of CSC frequency and/or presence. The assay readout is made more conclusive by secondary neurosphere formation, where a primary neurosphere, produced by a single stem or primitive cell, is dissociated and cultured to produce a second generation of spheres spawned by individual stem cells of the first sphere. This suggests the existence of progenitor or stem cells in the first neurosphere, discounting the possibility that a prospective neurosphere is simply a clump of cells adhering to one another (Reynolds et al., 1992; Singec et al., 2006). Tertiary and quarternary neurosphere assays may be able to distinguish between a progenitor and stem cell population. By serially diluting a cell population and culturing these progressively more dilute aliquots, a standard curve of neurosphere generation as a function of cell concentration can be produced, allowing for calculation of the number of neurosphere-generating cells in a population of interest. This procedure, commonly referred to as the LDA (limiting dilution assay), serves a fundamental role in normal and CSC biology as a functional readout of stem cell activity in a field that often lacks conclusive marker systems for cancer stem cell isolation.

Although indeed useful and a simple, elegant assay for an otherwise elusive cell, the neurosphere assay is not without its challenges. The first and most significant concern is that neurospheres arise not only from true CSCs but also from more differentiated progenitors (Bull and Bartlett, 2005; Seaberg and van der Kooy, 2002). As a result, the assay can be imprecise and the value of the readout may be heavily diluted by its low-specificity. Indeed, it has been suggested that the neurosphere assay may overestimate the true stem cell numbers by an order of magnitude (Reynolds and Rietze, 2005). Further, in order for the assay to readout stem cell frequency as a direct correlate of sphere formation, each sphere must arise from a single cell and be composed of cells of a single clone. However, time-lapse video microscopy of early spheres marked with separate reporter genes (e.g., fluorescent proteins) have revealed that large spheres are often the product of aggregations of smaller spheres, even when cultured at the low densities (termed clonal densities) previously thought to circumvent this situation (Singec et al., 2006). The otherwise simple solution of reducing cell density down to one or a few

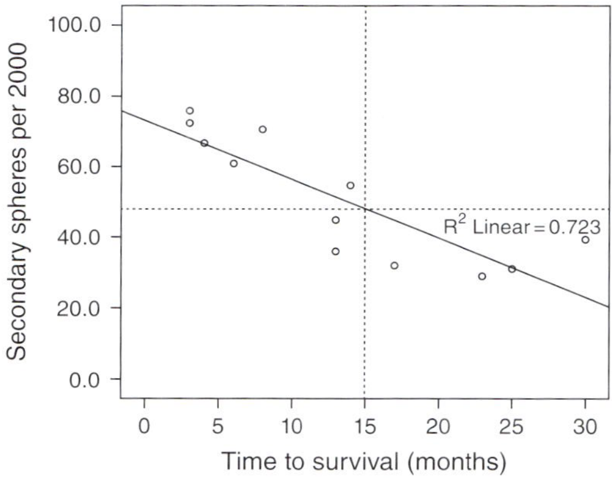

Correlations

		Secondary spheres per 2000	Time to survival
Secondary spheres per 2000	Pearson correlation	1	−.850**
	Sig. (2-tailed)		.000
	N	12	12
Time to survival	Pearson correlation	−.850**	1
	Sig. (2-tailed)	.000	
	N	12	12

** Correlation is significant at the 0.01 level (2-tailed).

Figure 15.4. BTIC stem cell assays correlate with patient outcome. We analyzed 12 GBM patients who had at least a 24-month period of documented clinical follow-up and plotted their BTIC sphere formation rates against months of survival. There is an 85% negative correlation between the secondary sphere formation rate and time to survival, indicating that as the number of secondary spheres increases, time to survival decreases ($R^2 = 0.723$, $p < 0.01$). Reprinted from *Stem Cell Research*, Venugopal et al., 2012, Bmi1 marks intermediate precursors during differentiation of human brain tumor initiating cells, 141–153, 8, with permission from Elsevier.

cells is complicated by the dependency of cells on paracrine signals from neighbouring cells, which provide a microenvironment essential for sustaining proliferation (Seaberg and van der Kooy, 2002; Taupin and Gage, 2002). These issues can be remedied by adding paracrine factors to promote proliferation of cells at low densities or culturing in media conditioned by other cells.

Conclusions (The Future Therapeutic Approach)

Regardless of the controversy surrounding the efficacy of assays, culture conditions, markers, and in vivo models used in the study of brain tumor CSCs, one theme must remain of utmost importance to all: what role does the CSC have in initiating and maintaining GBMs and how can this be prevented? Although controversies in the field provide limitations to *in vitro* and *in vivo* data supporting the CSC hypothesis, the most compelling evidence for the presence of CSCs in GBMs comes from patient correlative data. Recently, Venugopal and colleagues (2012) have shown GBM patient survival to be inversely correlated to secondary sphere formation. Conceivably, those tumors containing the most CSCs result in high secondary sphere formation and are clinically more aggressive (Venugopal et al., 2012) (Figure 15.4).

Conceptually, future therapies will likely treat GBM and many other cancers as consisting of two roughly distinguished cell populations requiring a multi-pronged chemotherapeutic regimen. If future research findings confirm the contemporary sentiment that the CSC population in GBM is maintained by signaling pathways different from those of the bulk tumor, multiple sets of therapeutics could be recruited to target the characteristics unique to each. Regardless of the way in which future therapeutic strategies are impacted, pursuit of new and better approaches to the study of the GBM CSC promises to shed further light on the nature of this rare tumor cell population.

References

Bao, S., Wu, Q., McLendon, R.E., Hao, Y., Shi, Q., Hjelmeland, A.B., Dewhirst, M.W., Bigner, D.D., and Rich, J.N. (2006a). Glioma stem cells promote radioresistance by preferential activation of the DNA damage response. Nature *444*, 756–760.

Bao, S., Wu, Q., Sathornsumetee, S., Hao, Y., Li, Z., Hjelmeland, A.B., Shi, Q., McLendon, R.E., Bigner, D.D., and Rich, J.N. (2006b). Stem cell-like glioma cells promote tumor angiogenesis through vascular endothelial growth factor. Cancer research *66*, 7843–7848.

Beier, D., Hau, P., Proescholdt, M., Lohmeier, A., Wischhusen, J., Oefner, P.J., Aigner, L., Brawanski, A., Bogdahn, U., and Beier, C.P. (2007). CD133(+) and CD133(−) glioblastoma-derived cancer stem cells show differential growth characteristics and molecular profiles. Cancer Res *67*, 4010–4015.

Beier, D., Wischhusen, J., Dietmaier, W., Hau, P., Proescholdt, M., Brawanski, A., Bogdahn, U., and Beier, C.P. (2008). CD133 expression and cancer stem cells predict prognosis in high-grade oligodendroglial tumors. Brain Pathol *18*, 370–377.

Bidlingmaier, S., Zhu, X., and Liu, B. (2008). The utility and limitations of glycosylated human CD133 epitopes in defining cancer stem cells. J Mol Med (Berl) *86*, 1025–1032.

Bull, N.D., and Bartlett, P.F. (2005). The adult mouse hippocampal progenitor is neurogenic but not a stem cell. J Neurosci *25*, 10815–10821.

Calabrese, C., Poppleton, H., Kocak, M., Hogg, T.L., Fuller, C., Hamner, B., Oh, E.Y., Gaber, M.W., Finklestein, D., Allen, M., *et al.* (2007). A perivascular niche for brain tumor stem cells. Cancer cell *11*, 69–82.

Capela, A., and Temple, S. (2002). LeX/ssea-1 is expressed by adult mouse CNS stem cells, identifying them as nonependymal. Neuron *35*, 865–875.

Capela, A., and Temple, S. (2006). LeX is expressed by principle progenitor cells in the embryonic nervous system, is secreted into their environment and binds Wnt-1. Dev Biol *291*, 300–313.

Chen, R., Nishimura, M.C., Bumbaca, S.M., Kharbanda, S., Forrest, W.F., Kasman, I.M., Greve, J.M., Soriano, R.H., Gilmour, L.L., Rivers, C.S., *et al.* (2010). A hierarchy of self-renewing tumor-initiating cell types in glioblastoma. Cancer Cell *17*, 362–375.

Choi, D., Lee, H.W., Hur, K.Y., Kim, J.J., Park, G.S., Jang, S.H., Song, Y.S., Jang, K.S., and Paik, S.S. (2009). Cancer stem cell markers CD133 and CD24 correlate with invasiveness and differentiation in colorectal adenocarcinoma. World J Gastroenterol *15*, 2258–2264.

Clarke, M.F., Dick, J.E., Dirks, P.B., Eaves, C.J., Jamieson, C.H., Jones, D.L., Visvader, J., Weissman, I.L., and Wahl, G.M. (2006). Cancer stem cells—perspectives on current status and future directions: AACR Workshop on cancer stem cells. Cancer research *66*, 9339–9344.

Copeland, D.D., and Bigner, D.D. (1977). The role of the subependymal plate in avian sarcoma virus brain tumor induction: comparison of incipient tumors in neonatal and adult rats. Acta neuropathologica *38*, 1–6.

Copeland, D.D., Vogel, F.S., and Bigner, D.D. (1975). The induction of intractranial neoplasms by the inoculation of avian sarcoma virus in perinatal and adult rats. Journal of neuropathology and experimental neurology *34*, 340–358.

Corbeil, D., Roper, K., Hannah, M.J., Hellwig, A., and Huttner, W.B. (1999). Selective localization of the polytopic membrane protein prominin in microvilli of epithelial cells—a combination of apical sorting and retention in plasma membrane protrusions. J Cell Sci *112 (Pt 7)*, 1023–1033.

Crossley, P.H., Martinez, S., and Martin, G.R. (1996). Midbrain development induced by FGF8 in the chick embryo. Nature *380*, 66–68.

Doetsch, F., Caille, I., Lim, D.A., Garcia-Verdugo, J.M., and Alvarez-Buylla, A. (1999). Subventricular zone astrocytes are neural stem cells in the adult mammalian brain. Cell *97*, 703–716.

Fargeas, C.A., Karbanova, J., Jaszai, J., and Corbeil, D. (2011). CD133 and membrane microdomains: old facets for future hypotheses. World J Gastroenterol *17*, 4149–4152.

Gage, F.H. (2000). Mammalian neural stem cells. Science (New York, NY) *287*, 1433–1438.

Gage, F.H. (2002). Neurogenesis in the adult brain. The Journal of neuroscience: the official journal of the Society for Neuroscience *22*, 612 613.

Galli, R., Binda, E., Orfanelli, U., Cipelletti, B., Gritti, A., De Vitis, S., Fiocco, R., Foroni, C., Dimeco, F., and Vescovi, A. (2004). Isolation and characterization of tumorigenic, stem-like neural precursors from human glioblastoma. Cancer Res *64*, 7011–7021.

Gooi, H.C., Thorpe, S.J., Hounsell, E.F., Rumpold, H., Kraft, D., Forster, O., and Feizi, T. (1983). Marker of peripheral blood granulocytes and monocytes of man recognized by two monoclonal antibodies VEP8 and VEP9 involves the trisaccharide 3-fucosyl-N-acetyllactosamine. Eur J Immunol *13*, 306–312.

Gross, C.G. (2000). Neurogenesis in the adult brain: death of a dogma. Nature reviews Neuroscience *1*, 67–73.

Hardee, M.E., Marciscano, A.E., Medina-Ramirez, C.M., Zagzag, D., Narayana, A., Lonning, S.M., and Barcellos-Hoff, M.H. (2012). Resistance of Glioblastoma-Initiating Cells to Radiation Mediated by the Tumor Microenvironment Can Be Abolished by Inhibiting Transforming Growth Factor-beta. Cancer research.

Hegi, M.E., Diserens, A.C., Gorlia, T., Hamou, M.F., de Tribolet, N., Weller, M., Kros, J.M., Hainfellner, J.A., Mason, W., Mariani, L., *et al.* (2005). MGMT gene silencing and benefit from temozolomide in glioblastoma. The New England journal of medicine *352*, 997–1003.

Hemmati, H.D., Nakano, I., Lazareff, J.A., Masterman-Smith, M., Geschwind, D.H., Bronner-Fraser, M., and Kornblum, H.I. (2003). Cancerous stem cells can arise from pediatric brain tumors. Proc Natl Acad Sci USA *100*, 15178–15183.

Huse, J.T., and Holland, E.C. (2010). Targeting brain cancer: advances in the molecular pathology of malignant glioma and medulloblastoma. Nat Rev Cancer *10*, 319–331.

Imura, T., Nakano, I., Kornblum, H.I., and Sofroniew, M.V. (2006). Phenotypic and functional heterogeneity of GFAP-expressing cells *in vitro*: differential expression of LeX/CD15 by GFAP-expressing multipotent neural stem cells and non-neurogenic astrocytes. Glia *53*, 277–293.

Jamal, M., Rath, B.H., Tsang, P.S., Camphausen, K., and Tofilon, P.J. (2012). The brain microenvironment preferentially enhances the radioresistance of CD133(+) glioblastoma stem-like cells. Neoplasia (New York, NY) *14*, 150–158.

Joo, K.M., Kim, S.Y., Jin, X., Song, S.Y., Kong, D.S., Lee, J.I., Jeon, J.W., Kim, M.H., Kang, B.G., Jung, Y., *et al.* (2008). Clinical and biological implications of CD133-positive and CD133-negative cells in glioblastomas. Lab Invest *88*, 808–815.

Koestner, A., Swenberg, J.A., and Wechsler, W. (1971). Transplacental production with ethylnitrosourea of neoplasms of the nervous system in Sprague-Dawley rats. Am J Pathol *63*, 37–50.

Lee, A., Kessler, J.D., Read, T.A., Kaiser, C., Corbeil, D., Huttner, W.B., Johnson, J.E., and Wechsler-Reya, R.J. (2005). Isolation of neural stem cells from the postnatal cerebellum. Nature neuroscience *8*, 723–729.

Lee, J., Kotliarova, S., Kotliarov, Y., Li, A., Su, Q., Donin, N.M., Pastorino, S., Purow, B.W., Christopher, N., Zhang, W., *et al.* (2006). Tumor stem cells derived from glioblastomas cultured in bFGF and EGF more closely mirror the phenotype and genotype of primary tumors than do serum-cultured cell lines. Cancer Cell *9*, 391–403.

Li, C.Y., Li, B.X., Liang, Y., Peng, R.Q., Ding, Y., Xu, D.Z., Zhang, X., Pan, Z.Z., Wan, D.S., Zeng, Y.X., *et al.* (2009). Higher percentage of CD133+ cells is associated with poor prognosis in colon carcinoma patients with stage IIIB. J Transl Med *7*, 56.

Liu, G., Yuan, X., Zeng, Z., Tunici, P., Ng, H., Abdulkadir, I.R., Lu, L., Irvin, D., Black, K.L., and Yu, J.S. (2006). Analysis of gene expression and chemoresistance of CD133+ cancer stem cells in glioblastoma. Molecular cancer *5*, 67.

Lois, C., and Alvarez-Buylla, A. (1993). Proliferating subventricular zone cells in the adult mammalian forebrain can differentiate into neurons and glia. Proceedings of the National Academy of Sciences of the United States of America *90*, 2074–2077.

Lottaz, C., Beier, D., Meyer, K., Kumar, P., Hermann, A., Schwarz, J., Junker, M., Oefner, P.J., Bogdahn, U., Wischhusen, J., *et al.* (2010). Transcriptional profiles of CD133+ and CD133− glioblastoma-derived cancer stem cell lines suggest different cells of origin. Cancer Res *70*, 2030–2040.

Louis, D.N., Ohgaki, H., Wiestler, O.D., Cavenee, W.K., Burger, P.C., Jouvet, A., Scheithauer, B.W., and Kleihues, P. (2007). The 2007 WHO classification of tumours of the central nervous system. Acta neuropathologica *114*, 97–109.

Lu, K.V., Chang, J.P., Parachoniak, C.A., Pandika, M.M., Aghi, M.K., Meyronet, D., Isachenko, N., Fouse, S.D., Phillips, J.J., Cheresh, D.A., et al. (2012). VEGF Inhibits Tumor Cell Invasion and Mesenchymal Transition through a MET/VEGFR2 Complex. Cancer cell *22*, 21–35.

Magee, J.A., Piskounova, E., and Morrison, S.J. (2012). Cancer stem cells: impact, heterogeneity, and uncertainty. Cancer cell *21*, 283–296.

Maher, E.A., Furnari, F.B., Bachoo, R.M., Rowitch, D.H., Louis, D.N., Cavenee, W.K., and DePinho, R.A. (2001). Malignant glioma: genetics and biology of a grave matter. Genes & development *15*, 1311–1333.

Maisel, M., Herr, A., Milosevic, J., Hermann, A., Habisch, H.J., Schwarz, S., Kirsch, M., Antoniadis, G., Brenner, R., Hallmeyer-Elgner, S., *et al.* (2007). Transcription profiling of adult and fetal human neuroprogenitors identifies divergent paths to maintain the neuroprogenitor cell state. Stem Cells *25*, 1231–1240.

Mao, X.G., Zhang, X., Xue, X.Y., Guo, G., Wang, P., Zhang, W., Fei, Z., Zhen, H.N., You, S.W., and Yang, H. (2009). Brain Tumor Stem-Like Cells Identified by Neural Stem Cell Marker CD15. Transl Oncol *2*, 247–257.

Miraglia, S., Godfrey, W., Yin, A.H., Atkins, K., Warnke, R., Holden, J.T., Bray, R.A., Waller, E.K., and Buck, D.W. (1997). A novel five-transmembrane hematopoietic stem cell antigen: isolation, characterization, and molecular cloning. Blood *90*, 5013–5021.

Mizrak, D., Brittan, M., and Alison, M.R. (2008). CD133: molecule of the moment. J Pathol *214*, 3–9.

Moore, K.A., and Lemischka, I.R. (2006). Stem cells and their niches. Science (New York, NY) *311*, 1880–1885.

Ogden, A.T., Waziri, A.E., Lochhead, R.A., Fusco, D., Lopez, K., Ellis, J.A., Kang, J., Assanah, M., McKhann, G.M., Sisti, M.B., et al. (2008). Identification of A2B5 + CD133− tumor-initiating cells in adult human gliomas. Neurosurgery *62*, 505–514; discussion 514–505.

Phillips, H.S., Kharbanda, S., Chen, R., Forrest, W.F., Soriano, R.H., Wu, T.D., Misra, A., Nigro, J.M., Colman, H., Soroceanu, L., *et al.* (2006). Molecular subclasses of high-grade glioma predict prognosis, delineate a pattern of disease progression, and resemble stages in neurogenesis. Cancer Cell *9*, 157–173.

Read, T.A., Fogarty, M.P., Markant, S.L., McLendon, R.E., Wei, Z., Ellison, D.W., Febbo, P.G., and Wechsler-Reya, R.J. (2009). Identification of CD15 as a marker for tumor-propagating cells in a mouse model of medulloblastoma. Cancer Cell *15*, 135–147.

Reifenberger, G., Sieth, P., Niederhaus, M., and Wechsler, W. (1992). Expression of CD15 in tumours of the nervous system. Histochem J *24*, 890–901.

Reya, T., Duncan, A.W., Ailles, L., Domen, J., Scherer, D.C., Willert, K., Hintz, L., Nusse, R., and Weissman, I.L. (2003). A role for Wnt signalling in self-renewal of haematopoietic stem cells. Nature *423*, 409–414.

Reya, T., Morrison, S.J., Clarke, M.F., and Weissman, I.L. (2001). Stem cells, cancer, and cancer stem cells. Nature *414*, 105–111.

Reynolds, B.A., and Rietze, R.L. (2005). Neural stem cells and neurospheres—re-evaluating the relationship. Nat Methods *2*, 333–336.

Reynolds, B.A., Tetzlaff, W., and Weiss, S. (1992). A multipotent EGF-responsive striatal embryonic progenitor cell produces neurons and astrocytes. J Neurosci *12*, 4565–4574.

Reynolds, B.A., and Weiss, S. (1992). Generation of neurons and astrocytes from isolated cells of the adult mammalian central nervous system. Science (New York, NY) *255*, 1707–1710.

Reynolds, B.A., and Weiss, S. (1996). Clonal and population analyses demonstrate that an EGF-responsive mammalian embryonic CNS precursor is a stem cell. Developmental biology *175*, 1–13.

Saigusa, S., Tanaka, K., Toiyama, Y., Yokoe, T., Okugawa, Y., Ioue, Y., Miki, C., and Kusunoki, M. (2009). Correlation of CD133, OCT4, and SOX2 in rectal cancer and their association with distant recurrence after chemoradiotherapy. Ann Surg Oncol *16*, 3488–3498.

Sakariassen, P.O., Immervoll, H., and Chekenya, M. (2007). Cancer stem cells as mediators of treatment resistance in brain tumors: status and controversies. Neoplasia *9*, 882–892.

Sato, A., Sakurada, K., Kumabe, T., Sasajima, T., Beppu, T., Asano, K., Ohkuma, H., Ogawa, A., Mizoi, K., Tominaga, T., *et al.* (2010). Association of stem cell marker CD133 expression with dissemination of glioblastomas. Neurosurg Rev *33*, 175–183; discussion 183–174.

Seaberg, R.M., and van der Kooy, D. (2002). Adult rodent neurogenic regions: the ventricular subependyma contains neural stem cells, but the dentate gyrus contains restricted progenitors. J Neurosci *22*, 1784–1793.

Seri, B., Garcia-Verdugo, J.M., McEwen, B.S., and Alvarez-Buylla, A. (2001). Astrocytes give rise to new neurons in the adult mammalian hippocampus. The Journal of neuroscience: the official journal of the Society for Neuroscience *21*, 7153–7160.

Seri, B., Herrera, D.G., Gritti, A., Ferron, S., Collado, L., Vescovi, A., Garcia-Verdugo, J.M., and Alvarez-Buylla, A. (2006). Composition and organization of the SCZ: a large germinal layer containing neural stem cells in the adult mammalian brain. Cerebral cortex (New York, NY: 1991) *16 Suppl 1*, i103–111.

Singec, I., Knoth, R., Meyer, R.P., Maciaczyk, J., Volk, B., Nikkhah, G., Frotscher, M., and Snyder, E.Y. (2006). Defining the actual sensitivity and specificity of the neurosphere assay in stem cell biology. Nat Methods *3*, 801–806.

Singh, S.K., Clarke, I.D., Terasaki, M., Bonn, V.E., Hawkins, C., Squire, J., and Dirks, P.B. (2003). Identification of a cancer stem cell in human brain tumors. Cancer Res *63*, 5821–5828.

Singh, S.K., Hawkins, C., Clarke, I.D., Squire, J.A., Bayani, J., Hide, T., Henkelman, R.M., Cusimano, M.D., and Dirks, P.B. (2004). Identification of human brain tumour initiating cells. Nature *432*, 396–401.

Soeda, A., Park, M., Lee, D., Mintz, A., Androutsellis-Theotokis, A., McKay, R.D., Engh, J., Iwama, T., Kunisada, T., Kassam, A.B., *et al.* (2009). Hypoxia promotes expansion of the CD133-positive glioma stem cells through activation of HIF-1alpha. Oncogene *28*, 3949–3959.

Son, M.J., Woolard, K., Nam, D.H., Lee, J., and Fine, H.A. (2009). SSEA-1 is an enrichment marker for tumor-initiating cells in human glioblastoma. Cell Stem Cell *4*, 440–452.

Stupp, R., Mason, W.P., van den Bent, M.J., Weller, M., Fisher, B., Taphoorn, M.J., Belanger, K., Brandes, A.A., Marosi, C., Bogdahn, U., *et al.* (2005). Radiotherapy plus concomitant and adjuvant temozolomide for glioblastoma. The New England journal of medicine *352*, 987–996.

Taupin, P., and Gage, F.H. (2002). Adult neurogenesis and neural stem cells of the central nervous system in mammals. J Neurosci Res *69*, 745–749.

Temple, S. (2001). The development of neural stem cells. Nature *414*, 112–117.

Uchida, N., Buck, D.W., He, D., Reitsma, M.J., Masek, M., Phan, T.V., Tsukamoto, A.S., Gage, F.H., and Weissman, I.L. (2000). Direct isolation of human central nervous system stem cells. Proc Natl Acad Sci U S A *97*, 14720–14725.

Venugopal, C., Li, N., Wang, X., Manoranjan, B., Hawkins, C., Gunnarsson, T., Hollenberg, R., Klurfan, P., Murty, N., Kwiecien, J., *et al.* (2012). Bmi1 marks intermediate precursors during differentiation of human brain tumor initiating cells. Stem Cell Res *8*, 141–153.

Verhaak, R.G., Hoadley, K.A., Purdom, E., Wang, V., Qi, Y., Wilkerson, M.D., Miller, C.R., Ding, L., Golub, T., Mesirov, J.P., *et al.* (2010). Integrated genomic analysis identifies clinically relevant subtypes of glioblastoma characterized by abnormalities in PDGFRA, IDH1, EGFR, and NF1. Cancer cell *17*, 98–110.

Vermeulen, L., de Sousa e Melo, F., Richel, D.J., and Medema, J.P. (2012). The developing cancer stem-cell model: clinical challenges and opportunities. The lancet oncology *13*, e83–89.

Vescovi, A.L., Reynolds, B.A., Fraser, D.D., and Weiss, S. (1993). bFGF regulates the proliferative fate of unipotent (neuronal) and bipotent (neuronal/astroglial) EGF-generated CNS progenitor cells. Neuron *11*, 951–966.

Vick, N.A., Lin, M.J., and Bigner, D.D. (1977). The role of the subependymal plate in glial tumorigenesis. Acta neuropathologica *40*, 63–71.

Wang, J., Sakariassen, P.O., Tsinkalovsky, O., Immervoll, H., Boe, S.O., Svendsen, A., Prestegarden, L., Rosland, G., Thorsen, F., Stuhr, L., *et al.* (2008). CD133 negative glioma cells form tumors in nude rats and give rise to CD133 positive cells. Int J Cancer *122*, 761–768.

Wang, Q., Chen, Z.G., Du, C.Z., Wang, H.W., Yan, L., and Gu, J. (2009). Cancer stem cell marker CD133+ tumour cells and clinical outcome in rectal cancer. Histopathology *55*, 284–293.

Ward, R.J., Lee, L., Graham, K., Satkunendran, T., Yoshikawa, K., Ling, E., Harper, L., Austin, R., Nieuwenhuis, E., Clarke, I.D., *et al.* (2009). Multipotent CD15+ cancer stem cells in patched-1-deficient mouse medulloblastoma. Cancer Res *69*, 4682–4690.

Weigmann, A., Corbeil, D., Hellwig, A., and Huttner, W.B. (1997). Prominin, a novel microvilli-specific polytopic membrane protein of the apical surface of epithelial cells, is targeted to plasmalemmal protrusions of non-epithelial cells. Proc Natl Acad Sci USA *94*, 12425–12430.

Werbowetski-Ogilvie, T.E., Schnerch, A., Rampalli, S., Mills, C.E., Lee, J.B., Hong, S.H., Levadoux-Martin, M., and Bhatia, M. (2011). Evidence for the transmission of neoplastic properties from transformed to normal human stem cells. Oncogene *30*, 4632–4644.

Willert, K., Brown, J.D., Danenberg, E., Duncan, A.W., Weissman, I.L., Reya, T., Yates, J.R., 3rd, and Nusse, R. (2003). Wnt proteins are lipid-modified and can act as stem cell growth factors. Nature *423*, 448–452.

Zeppernick, F., Ahmadi, R., Campos, B., Dictus, C., Helmke, B.M., Becker, N., Lichter, P., Unterberg, A., Radlwimmer, B., and Herold-Mende, C.C. (2008). Stem cell marker CD133 affects clinical outcome in glioma patients. Clin Cancer Res *14*, 123–129.

Zhu, Y., and Parada, L.F. (2002). The Molecular and Genetic Basis of Neurological Tumours. Nat Rev Cancer *2*, 616–626.

Chapter 16
Stem Cells and Pancreatic Cancer

Susana García-Silva and Christopher Heeschen
Molecular *Pathology Programme, Stem Cells & Cancer Group, Spanish National Cancer Research Center (CNIO), Madrid, Spain*

Abbreviations

ALDH	aldehyde dehydrogenase
CDF	difluorinate-curcumin
CK19	cytokeratin 19
CSC	cancer stem cell
EMT	epithelial mesenchymal transition
EpCAM	epithelial cell adhesion molecule
ESA	epithelial specific antigen
GTPase	guanosine triphosphate hydrolase
HDAC	histone deacetylase
MPCs	multipotent precursor cells
PanIN	pancreatic intraepithelial lesion
PDAC	pancreatic ductal adenocarcinoma
PMPs	pancreas-derived multipotent precursors
SAHA	suberoylanilide hydroxamic acid

Pancreatic Ductal Adenocarcinoma

The Organ

The pancreas is a dual organ with endocrine and exocrine functions. The exocrine pancreas constitutes the majority of the organ weight, and its main function involves the production of pancreatic juice, essential for a correct food digestion. Acinar cells secrete digestive zymogens that become activated enzymes such as trypsin, amylase, or carboxipeptidase once delivered to the small bowel during digestion. Ductal cells disposed along the ducts promote the transport of the pancreatic juice to the gut and produce mucins and bicarbonate contributing to a reduction of duodenum acidity.

Cancer Stem Cells, First Edition. Edited by Vinagolu K. Rajasekhar.

The endocrine pancreas is responsible for the production of insulin, glucagon, somatostatin, ghrelin, and P peptide and is physically concentrated in numerous compact locations spread all over the pancreas named islets of Langerhans.

The Tumor

Pancreatic cancer is a highly lethal disease, and unfortunately, the overall 5-year survival rate among patients has remained stable at around 5% over the past decades. Although the incidence of this type of tumors is lower than incidences for lung, colorectal, or prostate cancers, pancreatic cancer ranks fourth as the most frequent cause for cancer-related deaths (Hidalgo, 2010). This is due to absence of early symptoms, extensive metastasis, and high resistance to chemotherapy and radiation. In fact, while mortality is decreasing for other tumor types due to improved preventive measures and treatment regimens, pancreatic cancer has been steadily growing in incidence and prevalence consistent with the lack of improved preventive and therapeutic approaches despite increasing preclinical and clinical efforts (American Cancer Society, 2011).

More than 90% of pancreatic tumors are ductal adenocarcinomas arisen from the exocrine pancreas. The remaining 10% of the pancreatic malignancies are neuroendocrine tumors covering a heterogeneous group principally composed by insulin-producing tumors and gastrinomas. Surgical resection of primary tumors and gemcitabine treatment are only capable of delaying the mostly fatal outcome. Unfortunately, early spread of the tumor or erosion of nearby large vessels prevent surgery as a viable option in the majority of patients.

PDAC Risk Factors and Associated Genetic Alterations

Focusing on pancreatic ductal adenocarcinoma (PDAC), the causes for tumor development remain unknown although smoking, diabetes, and chronic pancreatitis are considered to be relevant risk factors. Around 5–10% of patients have a family history of the disease, and the risk of developing PDAC is 57 times higher in a family with four or more members affected by the disease than in one without affected members (Tersmette et al., 2001).

The current disease model establishes an initial preinvasive state termed p ancreatic intraepithelial neoplasia (PanIN) with three stages of increasing cellular atypia and mutations in key oncogenes and specific tumor suppressor genes. PDAC is characterized by near-universal presence of mutations in KRAS2 gene (90% of all tumors) that generate a continuously active Ras protein. This small GTPase involved in several signaling pathways related to proliferation, survival, and differentiation is considered to be a master regulator of PDAC initiation and progression. Indeed, activating mutations in the *Kras* gene are sufficient to initiate the development of pancreatic intraepithelial neoplasias (PanIN) and PDAC in mice (Morris et al., 2010).

The inactivation of CDKN2A gene results in the loss of p16 protein and is also a frequent event (>95% of cases) in PDAC. Abnormal TP53 can be found in more than 50% of cases. SMAD4 alterations are also commonly found in PDAC (Perez-Mancera et al., 2012). BRCA2 is mutated in less than 10% of sporadic cases, but around 20% of familial cases display mutations in this gene (Hidalgo, 2010).

In addition, frequent deregulation of crucial developmental signaling pathways such as Hedgehog and Notch is present in pancreatic tumors. This reactivation of embryonic pathways suggests that tumor cells or at least the cells that sustain tumor growth return to a dedifferentiated/progenitor-like state or indeed represent stem/precursor cells.

Cell of Origin of Pancreatic Ductal Adenocarcinoma

The generation of mouse models that resemble the human disease has provided a unique platform for better understanding

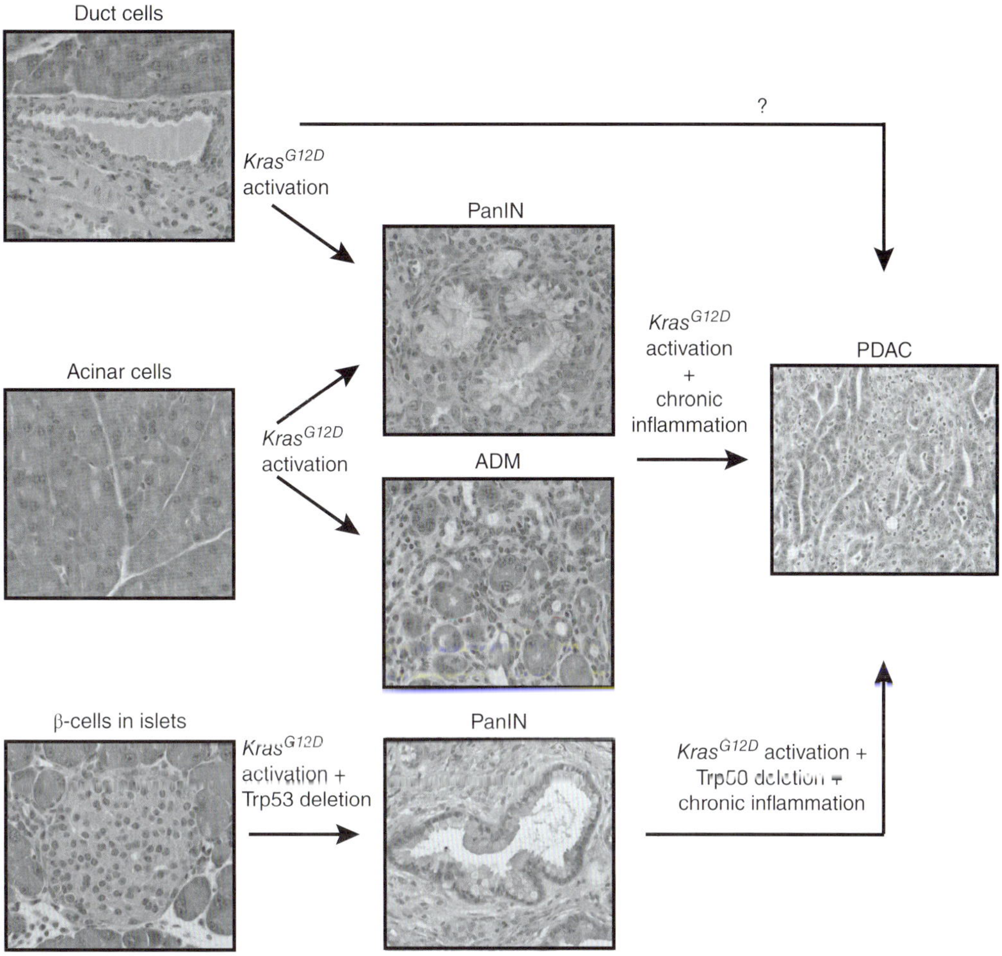

Figure 16.1. Activation of *Kras* oncogene together with chronic inflammation leads to PDAC. The activation of oncogenic *Kras* in ductal and acinar cells promotes the development of premalignant lesions (PanINs) and in the case of acinar cells, also acinar-ductal metaplasia (ADM) is observed. Formation of PanINs has also been described in Trp53-deficient, *Kras*^{G12D}-expressing β-cells. The presence of sustained inflammation in this context also promotes development of PDAC from insulin-producing cells. Photo is original to author. For color detail, please see color plate section.

which cell types are most susceptible for malignant transformation and consequently would be considered as the cell-of-origin for murine PDAC (Figure 16.1). The generation and study of numerous genetically modified mouse models has proved very useful in order to understand the initiation and development of pancreatic tumors (Perez-Mancera et al., 2012). Depending on the context, different cell types have been shown to act as the cell of origin raising the question whether plasticity of mature cells or common multipotent stem cells represent the initial cell susceptible to malignant transformation.

Pdx1-Expressing Cells

One of the most widely used mouse models for human pancreatic cancer is the KPC mouse (*Pdx1*Cre; LSL-*Kras*^{G12D}; LSL-*Trp53*^{172H}). In this model, the expression of Cre recombinase occurs in Pdx1+ pancreatic progenitor cells during embryonic development promoting the removal of a stop cassette (LSL). This allows for endogenous expression of a constitutive active *Kras* allele and an inactive mutant *Trp53* allele, a combination of mutations that triggers the development of highly metastatic and chemotherapy-resistant adenocarcinomas. These features closely resemble human disease (Hingorani et al., 2005; Olive et al., 2009). However, since *Pdx1* is an early pancreatic gene starting to be expressed from embryonic day 8.5 in the first multipotent pancreatic cells, the expression of the mutant alleles is activated in exocrine and endocrine cells causing multiple lesions in opposition to human tumors and excluding the possibility of identifying which specific cell type is the one susceptible of neoplastic transformation. Interestingly, activation of *Kras*^{G12D} alone in Pdx1+ cells starting during embryonic development causes only premalignant PanINs lesions. However, the number of those lesions is surprisingly small considering that most cells would host the active oncogene. This indicates that only a small population of Pdx1+ cells indeed represents the target for transformation and suggests that putative progenitors or stem cells could be those target cells.

Activation of *Kras*^{G12D} during adulthood by *Pdx1*CreER^{T2} also generates PanIN lesions and acinar-ductal metaplasia. Additional mutations in Ink4A/Arf or Trp53 strongly promote the formation of moderately to poorly differentiated PDAC, but again CreER^{T2} activity is detected in several cellular subsets including β cells, some ductal and acinar cells, and potential stem/progenitor cells (Gidekel Friedlander et al., 2009). Importantly, the use of more specific promoters and temporally controlled Cre recombinase have now pointed out diverse possible cell types as tumor-initiating cells that can cause PanIN lesions or invasive adenocarcinomas.

Ductal Cells

Invasive PDAC displays glandular ductal structures and common expression of cytokeratin 19 (CK19) ductal gene suggesting that one obvious candidate for the cell-of-origin in pancreatic cancer would be cells with ductal phenotype. Despite the ductal morphology of this tumor type, expression of *Kras*^{V12G} under CK19 promoter induces inflammation but not hyperplasia in the pancreas (Brembeck et al., 2003). In another study, activation of *Kras*^{G12D} by CK19-CreERT2 promoted PanIN lesions but not adenocarcinomas (Ray et al., 2011) rising doubts about a ductal origin for PDAC. The fact that histology does not reflect the cell-of-origin in PDAC is not an uncommon event. Some types of prostate tumors, breast adenocarcinomas, and basal cell carcinomas evolve from progenitors not related to the morphology and cell surface marker profiles of the respective malignancies (Visvader, 2011). For example, breast cancers developed in individuals with mutated versions of BRCA1 gene display a basal-like phenotype although they are originated from luminal progenitors.

Acinar and Centroacinar Cells

In contrast, there are well-established models demonstrating a role of acinar cells, including centroacinar cells, as putative tumor-initiating cells of PDAC. Acinar-ductal metaplasia is a phenotype associated with pre-malignant adenocarcinoma that has been proposed as the earliest pancreatic lesion and, as such, precursor of low-grade PanINs. In accordance with this hypothesis, ductal metaplasia inside acini is commonly found together with PanIN lesions when *Kras*^{G12D} is active in this cell compartment. Studies using specific promoters for these cells such as Elastase or proCPA1 show that oncogenic *Kras* and collaborative mutations in *Trp53* or *Ink4A/Arf* can induce adenocarcinomas when animals are simultaneously treated with caerulein, an inducer of chronic pancreatitis (Guerra et al., 2007; Gidekel Friedlander et al., 2009).

β Cells

Adult β cells are refractory to transformation by *Kras*^{G12D} alone or in combination with additional mutations. However, in a context of chronic pancreatitis, activation of mutant *Kras* and elimination of tumor suppressor p53 in insulin+ cells by treating *RipCreER*^{TM}; LSL-*Kras*^{G12D}; *Trp53*$^{flox/flox}$ with tamoxifen promote the development of poorly differentiated and undifferentiated adenocarcinomas (Gidekel Friedlander et al., 2009). These adenocarcinomas display a metastatic behavior similar to human tumors. Remarkably, *Kras*^{G12D} activation in adult *Pdx1*-expressing cells causes an early appearance of ductal and acinar structures inside the islets of Langerhans. These ductal lesions become elongated and produce mucin resembling PanIN lesions.

The Pro-Malignant Role of Chronic Inflammation

These studies suggest that pancreatitis promotes a degree of plasticity in different pancreatic cell types rendering them more susceptible for malignant transformation. This is in agreement with higher risk of developing PDAC in patients with chronic pancreatitis (Hidalgo, 2010). It is known that pancreatitis-driven injury promotes a proliferative and inflammatory response in the tissue associated with re-expression of *Pdx1* and activation of Notch and Hedgehog pathways previously switched off after embryonic development. Chronic pancreatitis has also been described as a favorable scenario for by-passing oncogenic-induced senescence and thus treatment of mice with non-steroidal anti-inflammatory drugs delays the progression of pancreatitis-induced PanINs (Guerra et al., 2011).

These characteristics of chronic inflammation would help to understand why activation of *Kras* oncogene together with collaborative mutations is able to progress to malignancy independently of the cell type. An alternative explanation might consider *Kras* and *Trp53* mutations to occur in tissue stem cells that become activated and proliferate in an inflammatory context due to inflammation-associated cytokines or/and signals for tissue repair.

In conclusion, there is an obvious need for genetically engineered mouse models providing more specific promoters that would allow for a tighter spatial and temporally activation of the tumor process and in particular promoters targeting tissue stem cells and progenitors.

Stem and Progenitor Cells in the Pancreas

Pancreas Development and Precursors

Pancreas development starts around embryonic day 9.5 with the formation of a dorsal bud in the foregut endoderm shortly followed by emergence of the anlage corresponding to the ventral pancreas plus common bile duct. These buds are organized in a stratified epithelium comprising early multipotent pancreatic cells and a first-wave of early-differentiated endocrine cells, mainly glycagon+ cells. These multipotent pancreatic cells display a considerable degree of plasticity and can be redirected to an intestinal lineage. However, as pancreas development progresses, multipotent pancreatic cells become more restricted to pancreatic lineages and start to express transcriptional factors involved in the initiation and commitment along the pancreatic differentiation program (Pan and Wright, 2011). *Pdx1*, *Ptf1*, and *Sox9* genes are key transcription factors in these early progenitors. Accordingly, lineage-tracing experiments have revealed that the "first wave" of either Pdx1+, Ptf1+, or Sox9+ precursors produce cells of all three pancreatic lineages (Akiyama et al., 2005; Pan and Wright, 2011).

Multipotent Stem Cells in Neonatal Pancreas

In newborn mice, cells expressing c-Met but negative for vascular and endothelial markers display colony-forming activity and are able to differentiate into acinar, ductal, and endocrine lineage after several days of culture (Suzuki et al., 2004). In recipient young mice, these cells also differentiated into different pancreatic lineages in vivo, although they were mostly scattered over the tissue and were not generating any progenies. Importantly, c-Met+ cells isolated from the adult pancreas did not show this multi-lineage potential under the same conditions. These data suggest that there are multipotent pancreatic stem cells at birth time, but multipotency is greatly reduced in adulthood.

Pancreatic islet progenitors have been isolated using CD133 and CD49f markers from mouse fetal pancreas (Sugiyama et al., 2007). Accordingly, CD133 + CD49f + cells are highly enriched for Ngn3+ cells with Ngn3 representing the consensus marker for endocrine-committed progenitors.

Multipotent Stem Cells in Adult Pancreas

The presence of stem cells in adult pancreas has been proposed in homology to other epithelial tissues. In 2004, cells isolated from the mouse pancreas were shown to be able to form spheres and differentiate not only into pancreatic exocrine and endocrine lineages but also into stellate cells and neuronal lineages (Seaberg et al., 2004). Those cells would represent only 0.02% of all pancreatic cells. In a subsequent study, the same authors identified an insulin+ cell type that displays characteristics of multipotent stem cells and was able to differentiate in vivo not only into functional β cells but also into other endocrine cell types, acinar cells, and neural cell types (Smukler et al., 2011). The so-called PMPs (pancreas-derived multipotent precursors) are derived from Pdx1+ progenitors and are mainly located in islets. Although they do express insulin, their hormone levels were lower as compared to mature β cells, and the cells do not express Glut2. This transporter is characteristic of fully functional β cells. Remarkably, this type of multipotent precursor reportedly does not only exist in the mouse, but also in the human pancreas suggesting a close homology between the species with respect to pancreas homeostasis. These results should open promising new lines of research for regenerative medicine. However, the identification of these stem cells was based on its functional properties and markers for a direct isolation of those PMPs are still missing.

Alternatively, in the adult mouse tissue, centroacinar cells (CAC) and terminal duct cells have been proposed to serve as tissue stem cells. Those cells have been isolated based on their ALDH1 activity and express low levels of pancreatic differentiation markers and Pdx-1. On the contrary, they are significantly enriched for stemness-associated genes (Rovira et al., 2010). Apart from their capacity to anchorage-independently grow in spheres, after several days of culture, those spheres contain exocrine and endocrine cells and demonstrate glucose-responsive insulin secretion indicative of functional β cells. Remarkably, ALDH1+ cells were able to contribute to endocrine and exocrine lineages when injected in microdissected dorsal pancreatic buds, whereas progenies of ALDH1 negative cells were restricted to the exocrine compartment of the pancreas.

Cancer Stem Cells in PDAC

In line with findings from other solid tumors, cancer stem cells have also been identified in human PDAC. These cells are able to self-renew and propagate the parental tumor in transplantation assays using immunodeficient mice (Figure 16.2). CSCs in PDAC have been identified by a variety of biomarkers.

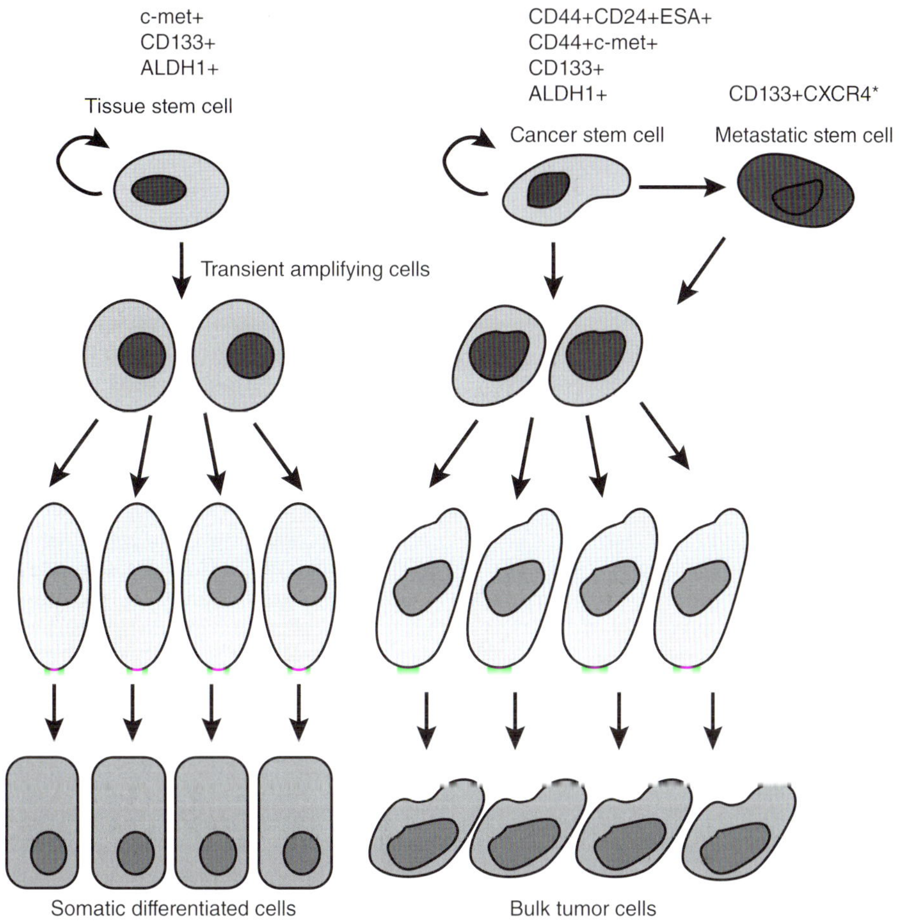

Figure 16.2. Model for cancer stem cells in PDAC. CSCs share several markers with putative tissue stem cells. Consistent with a normal stem cell phenotype, CSCs also give rise to more differentiated progenies. Importantly, the cancer stem cell pool is not homogeneous but also contains a subpopulation of metastatic cancer stem cells that express CXCR4 as a characteristic feature. A heterogeneous CSCs pool can be rationalized by the existence of different clones inside the bulk tumor. For color detail, please see color plate section.

CD44+CD24+EpCAM+ Tumor Cells

First evidence for CSCs in human pancreatic tumors was provided in 2007. In a first study, it was shown that a subpopulation of CD44+CD24+ESA+cells injected into immunodeficient mice gave rise to tumors while their triple negative counterparts did not show tumorigenic capacity (Li et al., 2007). This population represented 0.2–0.8% of the whole tumor cellularity and as few as 10^2 CD44+CD24+EpCAM+cells were sufficient to initiate tumors in 50% of transplanted mice. On the contrary, as much as 10^4 CD44-CD24-EpCAM- cells were required to detect malignant growth. Triple positive cells fulfill other cancer stem cell requirements such as self-renewal and generation of tumors with the same heterogeneity as the parental one. However, since EpCAM− cells also include stromal and vascular cells, it is possible that contamination of non-tumor cells was present in the triple negative population and affected the tumor formation capacity of this population. It is unclear if expression of different combinations of two out of those three markers could also identify cells with certain tumor-initiating potential.

Interestingly, CD44 and CD24 are cell surface markers associated with cancer stem cells in other epithelial malignancies such as the breast or prostate cancer. It could be speculated that tumor-propagating cells share some features independently of the tumor type.

CD133+ Tumor Cells

In a different study, CD133+ cells isolated from fresh human tumor samples were highly tumorigenic and represented 1–3% of the tumor cells (Hermann et al., 2007). Tumor formation in immunodeficient mice was detectable after injecting 5×10^2 CD133+ cells, but no tumor growth was detected after inoculation of 10^6 CD133− cells. The formed tumors morphologically and histologically resembled the parental tumors. In addition, CD133+ cells were able to form spheres in serum-free non-adherent conditions and generate tumors in serial transplantations demonstrating its in vitro and in vivo self-renewal ability.

In patient samples, CD133 staining was found in 80% of cases with around 1% of cells showing a cytoplasmic pattern. This distribution matched with the pattern in normal tissue where very few cells in the duct and in centroacinar position showed cytoplasmic staining (Immervoll et al., 2008). Although there was no correlation between CD133 staining and patient survival or differentiation grade, a study involving 80 patients with surgical resection of pancreatic head adenocarcinoma showed that cytoplasmic CD133 expression significantly correlated with patients' outcome (Maeda et al., 2008).

There is a certain overlap between CD133+ and CD44+CD24+EpCAM+(ranging from 10.3 to 37.4%), but since none of the proposed isolation procedures was capable of identifying single tumorigenic cells, the proposed markers select for cancer stem cells-enriched populations. Therefore, we are still missing a definitive combination of cell surface markers that is capable of identifying a pure cancer stem cell population.

Other CSC Markers

Apart from these more commonly used markers, other approaches have been proposed for the enrichment of CSCs. ALDH1, apart from being associated with pancreatic stem cells, has also been used for isolating tumorigenic cells in the human pancreatic line L3.6pl (Kim et al., 2011). In a study including 269 tumor samples, ALDH1 expression was significantly associated with poor prognosis, and ALDH1+ cells were 5 to 11 times more clonogenic as compared to their negative counterparts as well as presented an enhanced migratory and invasive potential (Rasheed et al., 2010).

Another study identified cancer stem cells via their functional properties using a fluorescent reporter system to detect proteasome activity. Cells with low activity of 26S proteasome displayed cancer stem cell properties (Adikrisna et al., 2012).

In addition, the receptor for hepatocyte growth factor c-Met has also being proposed as a CSC marker in pancreatic cancer (Li et al., 2011) and when combined with CD133 or CD44 strongly selected for cancer stem cells as demonstrated by enhanced in vivo tumorigenicity as compared to each of the single markers. In particular, the combination c-Met and CD44 was able to identify a population of cells with strongest tumorigenic potential describing that as few as 50 c-Met+CD44+ cells were able to generate subcutaneous tumors.

Metastasis-Initiating Cells

EMT and CSCs

It is well established that the switch from an epithelial phenotype to a mesenchymal one involves a series of epigenetic and gene expression changes that have been defined as epithelial to mesenchymal transition (EMT). This EMT program provides the machinery for motility and invasion required for abandoning the tumor bulk, extravasate the basal membrane, travel through the blood stream, home to new distant sites, and initiate the colonization of the new host tissue. A putative connection between metastasis and cancer stem cells has been proposed several years ago (Brabletz et al., 2005). Later on it was demonstrated that the EMT process confers many of the properties of normal and neoplastic stem cells (including sphere formation and self-renewal), and furthermore stem cells are enriched in EMT genes as compared to their epithelial progeny (Mani et al., 2008). More recently in a mouse model for PDAC, it has been shown that extravasation to bloodstream is associated not only with expression of EMT genes, but also with phenotypic and functional characteristics of pancreatic stem cells (Rhim et al., 2012). Although tumor cells that have undergone EMT display similar tumorigenic potential than other cells in the tumor bulk, the situation is different in early lesions such as PanINs where those cells manifest a strikingly enhanced tumorigenic capacity. These findings suggest that the invasive potential is acquired very early during transformation and is consistent with the hypothesis that cancer stem cells probably display certain degree of heterogeneity.

Heterogeneity of the CSC Pool

Genetic and/or epigenetic alterations as well as the microenvironment might promote the evolution of CSCs into different clones simultaneously present in the tumor, but bearing distinct invasive potential, hypoxia endurance, or susceptibility to chemotherapy. In human PDAC, CD133+ cells have enhanced tumor formation ability, but only CD133+CXCR4+ cell subset exhibits metastatic potential. CXCR4 is a chemokine receptor specific for stromal-derived factor-1 (SDF-1) involved in responses to chemotactic gradients. Depletion of CD133+ CXCR4+ cells from the CSC pool did not alter the tumor formation capacity but abolished metastatic potential (Hermann et al., 2007). In agreement with these findings, patients with lymph node metastasis demonstrated a larger fraction of CD133+CXCR4+ cells as compared to patients without detectable metastatic spread.

Importantly there seems to exist a close relationship between the pathways regulating stemness and those implicated in EMT. Hedgehog, Notch, and Wnt signaling pathways are key pathways for the biology of pancreatic CSC but also play a role in the regulation of EMT (Li et al., 2012). MicroRNAs appear also to be involved in this interconnection as the EMT-associated ZEB1 repressor has been shown to suppress the expression of microRNAs such as miR200c and miR203, which in turn have been implicated in the inhibition of stemness (Wellner et al., 2009).

Therapies Against Pancreatic Cancer Stem Cells

The most critical setback in the development of new therapeutic approaches against PDAC and many other cancers is the high percentage of relapse in resected tumors or tumors eliminated by chemotherapy and/or radiotherapy. For PDAC, very little progress has been made in the past decades since this type of tumor is particularly aggressive, invasive, and prone to metastasize.

Drugs Against CSC-Specific Properties

Resistance to chemotherapy appears to be another important hallmark of CSCs. Human CD133+ cancer stem cells display enhanced drug resistance to gemcitabine, the routine chemotherapy agent for pancreatic cancer, as compared to CD133− cells (Hermann et al., 2007). Treatment with gemcitabine also results in a relative increase in the number of c-Met+ CSCs. Interestingly, the c-Met inhibitor XL184 reversed this resistance, provoked reduced tumor sphere formation, and prevented tumor growth for a longer period than either XL184 or gemcitabine alone (Li et al., 2011).

Activin/Nodal signaling pathway has been identified as a novel and specific target for CSCs (Lonardo et al., 2011). Importantly, this pathway is not active in the adult pancreas but is reactivated in CSCs. It has been shown that Alk4/7, the TGFβ superfamily receptors for Activin/Nodal drive self-renewal of human pancreatic CSCs. Remarkably genetic targeting of this pathway reduces tumor sphere formation and virtually abrogates in vivo tumorigenicity. Combination of the Nodal/Activin inhibitor SB431542 with gemcitabine and a Hedgehog pathway inhibitor for targeting the stromal cells proved to be a successful therapy for long-term prevention of relapses in human tumor xenografts.

The use of HDAC inhibitors also appears to represent a promising strategy for targeting CSCs. Treatment of human pancreatic CSCs with 5-Aza-dC and SAHA reactivates the expression of miR34a, an effector of p53 that is down-regulated in pancreatic cancer (Nalls et al., 2011). This treatment blocks self-renewal and induces apoptosis. Interestingly, SAHA also inhibits the expression of EMT inducers such as Slug, Snail, and ZEB1.

Metformin, a common drug for type II diabetes therapy, also displays anti-tumor activity in some malignancies. In pancreatic cancer, treatment with metformin diminished tumor sphere formation (Bao et al., 2012). This reduction was associated with repression of stemness genes such as CD44, EpCAM, *Oct4,* and *Nanog*. Modulation of microRNAs such as let-7a, miR26a, miR101, and miR200b is the suggested mechanism by which metformin seem to target tumor-propagating cells.

Inhibitors of other key pathways for stem cell function such as Notch, Hegdehog, and CXCR4 have also been investigated in the context of pancreatic CSCs biology using both in vitro and in vivo models (Xia et al., 2012).

Immunotherapy Against CSCs

Immuno-based treatment strategies are a newly emerging therapeutic modality for PDAC, where immune effector mechanisms can be induced and directed toward antigens expressed by tumor cells including CSCs. These therapies could bear the putative advantage of targeting CSC irrespective of potentially very diverse genetic and epigenetic alterations in these cancer cells. A recent study by Visus and colleagues (2011) took advantage of the ALDH activity as a marker for identifying and selectively targeting CSCs using several established cell lines including pancreatic cancer cells. The authors generated in vitro ALDH1A1-specific $CD8^+$ T cells in order to eliminate $ALDH^{bright}$ CSCs in preclinical models of human tumor xenografts and observed growth inhibition and reduced metastasis. However, a major concern for this approach is the fact that ALDH1A1-specific $CD8^+$ T cells will most likely also target normal $ALDH^{bright}$ stem cells, which can for example be found in the hematopoietic system.

In a different study (Cioffi et al., 2012), the therapeutic value of the bispecific antibody MT110 targeting the T-cell receptor CD3 complex and Epithelial cell adhesion molecule (EpCAM; CD326) was investigated. EpCAM is frequently overexpressed and functionally altered in epithelial cancer cells, including CSC (Munz et al., 2009), and therefore is becoming accessible on the surface of these cells. The authors found a significant reduction in the CSC population implying that they are not spared from the cytotoxic activity of activated T cells. Moreover, sphere formation capacity and in vivo tumorigenicity of the cells was significantly reduced by MT110.

MT110 is currently tested in a dose-escalating phase I clinical trial enrolling patients with diverse epithelial cancers (lung, colon, gastric). Low toxicity and early signs of biological activity have been observed at clinically well-tolerated doses of MT110 in this first-in-human clinical trial. The promising results derived from preclinical PDAC models now suggest that this treatment regimen could also represent a new opportunity for patients with PDAC.

Other Strategies to Target CSCs

Targeted drug delivery systems have been used against the CD44+CD24+ESA+ CSC population. Taking advantage of the high level of expression of RON receptor tyrosine kinase in this cell population, the authors tested doxorubicin-liposomes coated with a RON antibody that allowed a rapid drug internalization and caused a remarkable reduction in CSC viability (Padhye et al., 2011).

Natural compounds obtained from dietary sources are emerging as alternative or complementary strategies for eliminating CSCs. The use of these natural agents or nutraceuticals would minimize noxious effects in the gastrointestinal tract or cytotoxicity for resident stem cells. For example, curcumin present in curry powders and mustard or its analogue CDF was shown to increase the sensitivity of pancreatic cancer cells to gemcitabine. More specifically, CDF reduced sphere-formation and tumor growth by increasing PTEN and miR-200 expression (Bao et al., 2011). Resveratrol, a polyphenol found in red grapes, red wine, and peanuts bears anti-tumoral properties in several malignancies. In addition to its reported

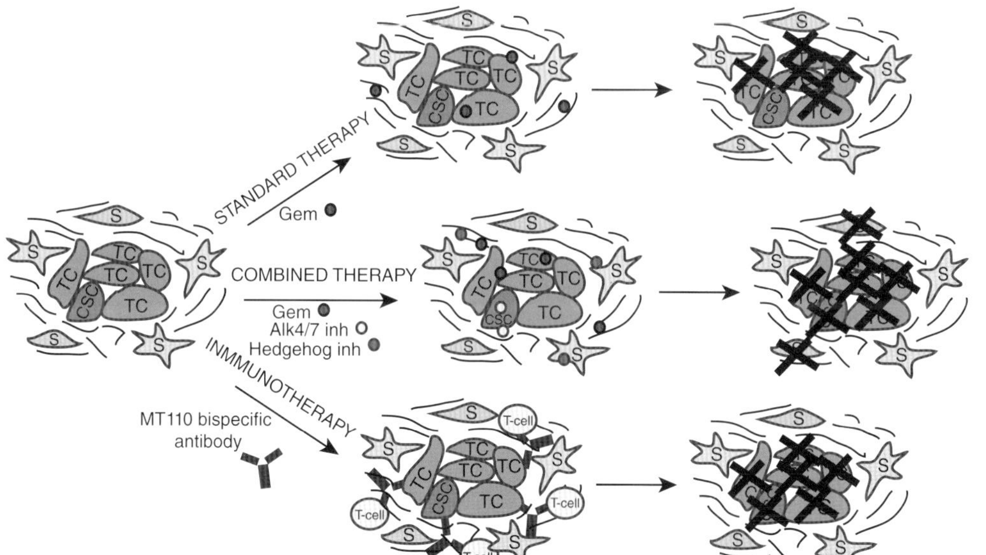

Figure 16.3. Therapeutic strategies against PDAC. The tumor is composed by cancer stem cells (CSC) and more differentiated progenies (TC) surrounded by a fibrotic matrix containing stromal cells (S). The standard therapy for pancreatic cancer represents gemcitabine treatment (Gem) in most cases, which attacks bulk tumor cells. However, cancer stem cells are more resistant to Gem causing relapse of the disease. Combined therapies include combinations of gemcitabine with drugs targeting specifically CSCs (e.g., Alk 4/7 inhibitor) and ideally stromal components of the tumor (e.g., Hedgehog pathway inhibitor) causing a total regression of the tumor and no relapse during long-term follow-up. A third approach consists of immunotherapies targeting epitopes present in both CSCs and their progenies (e.g., bispecific antibody MT110 that triggers T-cell activation against EpCAM expressing cells). For color detail, please see color plate section.

effects against CSCs in glioblastoma and breast tumors, resveratrol also inhibits self-renewal of pancreatic CSCs via activation of caspase 3/7 and inactivation of Bcl-2 (Xia et al., 2012).

In summary, targeting pancreatic cancer stem cells emerges as an essential requirement for curing this disease. However, there is an increasing awareness that a successful approach should include combined therapies attacking bulk tumor cells, tumor-initiating cells and also the stroma, an important component of the tumor mass in pancreatic tumors that modulates bulk tumor cells and cancer stem cells behavior and even hinder drug delivery due its high fibrotic content (Figure 16.3).

Conclusions and Perspectives

Tissue stem cells in the pancreas share some cell surface markers with CSCs, but it is still uncertain if the cell-of−origin for PDAC is indeed a resident stem cell that accumulates mutations over time or rather represents a progenitor or a differentiated exocrine or even endocrine cell that acquires stem cell properties during the process of malignant transformation. The use of more specific promoters driving the activation of *Kras* oncogene will be required for defining more precisely the pancreatic cell subset responsible for initiation of the carcinogenic process.

In contrast, it is now well established that pancreatic tumors contain a subset of cells with sphere-forming ability and in vivo tumorigenic capacity as assessed by limiting dilution tumorigenic transplantation assays into immunodeficient mice. These cells are considered pancreatic CSCs and are able to form tumors indistinguishable from parental ones. Furthermore, they display strong chemotherapy resistance and are implicated in tumor relapses and metastatic spread.

The next important step for advancing the field of pancreatic CSC research includes the identification and characterization of CSCs in the unperturbed in vivo setting. This has been achieved just recently for other solid tumors such as squamous cell carcinoma, glioblastomas, and intestinal cancer using clonal analysis after lineage tracing in mice (Chen et al. 2012; Driessens et al. 2012; Schepers et al. 2012). In vivo imaging of CSCs during tumor development will also help unravel the influence of fibrosis, stromal cells, and other physical constrains with respect to tumor growth, invasion, and metastasis. Finally, comprehensive studies of the tumor heterogeneity with respect to the coexistence of different clones potentially generated by distinct populations of CSCs evolving by stochastic cell fate decisions may eventually unite the CSC concept and the model of clonal evolution for pancreatic cancer.

The design of specific therapies against CSCs is opening new promising alleys to eventually improve survival of patients with PDAC. Combined therapies for target regular tumor cells, CSCs, and the protective stroma will probably be the most successful approach for the future treatment of PDAC.

References

Adikrisna, R., S. Tanaka, S. Muramatsu, et al. (2012). Identification of pancreatic cancer stem cells and selective toxicity of chemotherapeutic agents. *Gastroenterology* 143(1): 234–245 e237.

Akiyama, H., J.E. Kim, K. Nakashima, et al. (2005). Osteo-chondroprogenitor cells are derived from Sox9 expressing precursors. *Proc Natl Acad Sci USA* 102(41): 14665–14670.

American Cancer Society. (2011). Cancer Facts and Figs. www.cancer.org.

Bao, B., S. Ali, D. Kong, et al. (2011). Anti-tumor activity of a novel compound-CDF is mediated by regulating miR-21, miR-200, and PTEN in pancreatic cancer. *PLoS One* 6(3): e17850.

Bao, B., Z. Wang, S. Ali, et al. (2012). Metformin inhibits cell proliferation, migration and invasion by attenuating CSC function mediated by deregulating miRNAs in pancreatic cancer cells. *Cancer Prev Res (Phila)* 5(3): 355–364.

Brabletz, T., A. Jung, S. Spaderna, et al. (2005). Opinion: migrating cancer stem cells—an integrated concept of malignant tumour progression. *Nat Rev Cancer* 5(9): 744–749.

Brembeck, F.H., F.S. Schreiber, T.B. Deramaudt, et al. (2003). The mutant K-ras oncogene causes pancreatic periductal lymphocytic infiltration and gastric mucous neck cell hyperplasia in transgenic mice. *Cancer Res* 63(9): 2005–2009.

Chen, J., Y. Li, T.S. Yu, et al. (2012). A restricted cell population propagates glioblastoma growth after chemotherapy. *Nature*.

Cioffi, M., J. Dorado, P.A. Baeuerle, et al. (2012) EpCAM/CD3-Bispecific T-cell Engaging Antibody MT110 Eliminates Primary Human Pancreatic Cancer Stem Cells. *Clin Cancer Res*.

Driessens, G., B. Beck, A. Caauwe, et al. (2012). Defining the mode of tumour growth by clonal analysis. *Nature*.

Gidekel Friedlander, S.Y., G.C. Chu, E.L. Snyder, et al. (2009). Context-dependent transformation of adult pancreatic cells by oncogenic K-Ras. *Cancer Cell* 16(5): 379–389.

Guerra, C., M. Collado, C. Navas, et al. (2011). Pancreatitis-induced inflammation contributes to pancreatic cancer by inhibiting oncogene-induced senescence. *Cancer Cell* 19(6): 728–739.

Guerra, C., A.J. Schuhmacher, M. Canamero, et al. (2007). Chronic pancreatitis is essential for induction of pancreatic ductal adenocarcinoma by K-Ras oncogenes in adult mice. *Cancer Cell* 11(3): 291–302.

Hermann, P.C., S.L. Huber, T. Herrler, et al. (2007). Distinct populations of cancer stem cells determine tumor growth and metastatic activity in human pancreatic cancer. *Cell Stem Cell* 1(3): 313–323.

Hidalgo, M. (2010). Pancreatic cancer. *N Engl J Med* 362(17): 1605–1617.

Hingorani, S.R., L. Wang, A.S. Multani, et al. (2005). Trp53R172H and KrasG12D cooperate to promote chromosomal instability and widely metastatic pancreatic ductal adenocarcinoma in mice. *Cancer Cell* 7(5): 469–483.

Immervoll, H., D. Hoem, P.O. Sakariassen, et al. (2008). Expression of the 'stem cell marker' CD133 in pancreas and pancreatic ductal adenocarcinomas. *BMC Cancer* 8: 48.

Kim, M.P., J.B. Fleming, H. Wang, et al. (2011). ALDH activity selectively defines an enhanced tumor-initiating cell population relative to CD133 expression in human pancreatic adenocarcinoma. *PLoS One* 6(6): e20636.

Li, C., D.G. Heidt, P. Dalerba, et al. (2007). Identification of pancreatic cancer stem cells. *Cancer Res* 67(3): 1030–1037.

Li, C., J.J. Wu, M. Hynes, et al. (2011). c-Met is a marker of pancreatic cancer stem cells and therapeutic target. *Gastroenterology* 141(6): 2218–2227 e2215.

Li, Y., D. Kong, A. Ahmad, et al. (2012). Pancreatic cancer stem cells: Emerging target for designing novel therapy. *Cancer Lett*.

Lonardo, E., P.C. Hermann, M.T. Mueller, et al. (2011). Nodal/Activin signaling drives self-renewal and tumorigenicity of pancreatic cancer stem cells and provides a target for combined drug therapy. *Cell Stem Cell* 9(5): 433–446.

Maeda, S., H. Shinchi, H. Kurahara, et al. (2008). CD133 expression is correlated with lymph node metastasis and vascular endothelial growth factor-C expression in pancreatic cancer. *Br J Cancer* 98(8): 1389–1397.

Mani, S.A., W. Guo, M.J. Liao, et al. (2008). The epithelial-mesenchymal transition generates cells with properties of stem cells. *Cell* 133(4): 704–715.

Morris, J.P.T., S.C. Wang, and M. Hebrok (2010). KRAS, Hedgehog, Wnt and the twisted developmental biology of pancreatic ductal adenocarcinoma. *Nat Rev Cancer* 10(10): 683–695.

Munz, M., P. A. Baeuerle, and O. Gires (2009). The emerging role of EpCAM in cancer and stem cell signaling. *Cancer Res* 69(14): 5627–5629.

Nalls, D., S.N. Tang, M. Rodova, et al. (2011). Targeting epigenetic regulation of miR-34a for treatment of pancreatic cancer by inhibition of pancreatic cancer stem cells. *PLoS One* 6(8): e24099.

Olive, K.P., M.A. Jacobetz, C.J. Davidson, et al. (2009). Inhibition of Hedgehog signaling enhances delivery of chemotherapy in a mouse model of pancreatic cancer. *Science* 324(5933): 1457–1461.

Padhye, S.S., S. Guin, H.P. Yao, et al. (2011). Sustained expression of the RON receptor tyrosine kinase by pancreatic cancer stem cells as a potential targeting moiety for antibody-directed chemotherapeutics. *Mol Pharm* 8(6): 2310–2319.

Pan, F.C. and C. Wright (2011). Pancreas organogenesis: from bud to plexus to gland. *Dev Dyn* 240(3): 530–565.

Perez-Mancera, P.A., C. Guerra, M. Barbacid, et al. (2012). What we have learned about pancreatic cancer from mouse models. *Gastroenterology* 142(5): 1079–1092.

Rasheed, Z., Q. Wang, and W. Matsui (2010). Isolation of stem cells from human pancreatic cancer xenografts. *J Vis Exp*(43).

Ray, K.C., K. M. Bell, J. Yan, et al. (2011). Epithelial tissues have varying degrees of susceptibility to Kras(G12D)-initiated tumorigenesis in a mouse model. *PLoS One* 6(2): e16786.

Rhim, A.D., E.T. Mirek, N.M. Aiello, et al. (2012). EMT and dissemination precede pancreatic tumor formation. *Cell* 148(1–2): 349–361.

Rovira, M., S.G. Scott, A.S. Liss, et al. (2010). Isolation and characterization of centroacinar/terminal ductal progenitor cells in adult mouse pancreas. *Proc Natl Acad Sci USA* 107(1): 75–80.

Schepers, A.G., H.J. Snippert, D.E. Stange, et al. (2012). Lineage tracing reveals Lgr5+ stem cell activity in mouse intestinal adenomas. *Science* 337(6095): 730–735.

Seaberg, R. M., S. R. Smukler, T. J. Kieffer, et al. (2004). Clonal identification of multipotent precursors from adult mouse pancreas that generate neural and pancreatic lineages. *Nat Biotechnol* 22(9): 1115–1124.

Smukler, S.R., M.E. Arntfield, R. Razavi, et al. (2011). The adult mouse and human pancreas contain rare multipotent stem cells that express insulin. *Cell Stem Cell* 8(3): 281–293.

Sugiyama, T., R.T. Rodriguez, G.W. McLean, et al. (2007). Conserved markers of fetal pancreatic epithelium permit prospective isolation of islet progenitor cells by FACS. *Proc Natl Acad Sci USA* 104(1): 175–180.

Suzuki, A., H. Nakauchi, and H. Taniguchi (2004). Prospective isolation of multipotent pancreatic progenitors using flow-cytometric cell sorting. *Diabetes* 53(8): 2143–2152.

Tersmette, A.C., G.M. Petersen, G.J. Offerhaus, et al. (2001). Increased risk of incident pancreatic cancer among first-degree relatives of patients with familial pancreatic cancer. *Clin Cancer Res* 7(3): 738–744.

Visus, C., Y. Wang, A. Lozano-Leon, et al. (2011). Targeting ALDH(bright) human carcinoma-initiating cells with ALDH1A1-specific CD8 T cells. *Clin Cancer Res* 17(19): 6174–6184.

Visvader, J.E. (2011). Cells of origin in cancer. *Nature* 469(7330): 314–322.

Wellner, U., J. Schubert, U.C. Burk, et al. (2009). The EMT-activator ZEB1 promotes tumorigenicity by repressing stemness-inhibiting microRNAs. *Nat Cell Biol* 11(12): 1487–1495.

Xia, J., C. Chen, Z. Chen, et al. (2012). Targeting pancreatic cancer stem cells for cancer therapy. *Biochim Biophys Acta* 1826(2): 385–399.

Chapter 17

Melanoma Subpopulations with Cancer Stem Cell Phenotypes

Rajasekharan Somasundaram, Nicole Facompre, and Meenhard Herlyn
Tumor Microenvironment and Metastasis Program, Melanoma Research Center, The Wistar Institute, Philadelphia, PA, USA

Abbreviations

ACS	American Cancer Society
AML	acute myeloid leukemia
CAR	chimeric antigen receptor
CCND1	cyclin D1
CD	cluster definition
CDK	cyclin-dependent kinase
CTLA4	cytotoxic T-lymphocyte antigen 4
CSC	cancer stem cell
DTIC	5-(3,3-Dimethyl-1-triazenyl)imidazole-4-carboxamide (Dacarbazine)
FDA	Food and Drug Administration
IGF	insulin-dependent growth factor
IFN	interferonIL interleukin
JARID1B	Jumonji ARID1B
KDM5	lysine demethylases 5
MAA	melanoma associated antigens
MAPK	mitogen-activated protein kinase
MITF	microphthalmia-associated transcription factor
MS4A1	membrane spanning 4A1
NGFR	nerve growth factor receptor
NSG	NOD SCID IL2 receptor gamma chain knockout
PD	programmed cell death
PDGF	platelet derived growth factor
PTEN	phosphotase and tensin
RGP	radial growth phase
SCID	severe combined immunodeficiency mice

(Continued)

Cancer Stem Cells, First Edition. Edited by Vinagolu K. Rajasekhar.

(*Continued*)

SiRNA	small interfering RNA
ShRNA	short hairpin RNA
TME	tumor microenvironment
VGP	vertical growth phase

Introduction

Melanoma is an aggressive form of skin cancer whose incidence continues to rise worldwide. An estimated 76,000 new cases of melanomas (~5% of all cancers) will be diagnosed in the United States every year, and about 9,000 patients will die of metastatic disease (Siegel et al., 2012). Risk factors for the disease include increased exposure to sun or ultraviolet radiation and frequent use of tanning beds (Boniol et al., 2012; Miller and Mihm, 2006). Other contributing factors that could also influence incidence rates include family history, genetic susceptibility, environmental factors, and age-related immune-suppressions (reviewed in de Souza et al., 2012; Miller and Mihm, 2006)).

In most cases, melanoma begins with the transformation of a benign nevus that develops into a dysplastic lesion before progressing through radial- and vertical-growth phases (RGP and VGP [primary melanoma]). It can then invade into the dermis and regional lymph nodes and from there disseminate to distant organs, leading to metastatic melanoma (reviewed in Koh, 1991; Miller and Mihm, 2006). Recent advances in molecular and cellular techniques have led to better classification and understanding of melanoma. New techniques have led to the identification of a number of important genetic and biological alterations that occur during various stages of melanoma progression (reviewed in Chin et al., 2006; de Souza et al., 2012; Fecher et al., 2007; Somasundaram et al., 2012). Further, these molecular approaches have defined melanoma as a heterogeneous and rather complex neoplasm (reviewed in de Souza et al., 2012; Koh, 1991; Miller and Mihm, 2006; Somasundaram et al., 2012; Vidwans et al., 2011).

Patients with primary melanomas can be successfully treated through surgery; however, treatment of advanced stage metastatic melanoma patients remains challenging (de Souza et al., 2012; Fecher et al., 2007; Miller and Mihm, 2006). Melanoma patients undergoing chemotherapy or targeted therapy with vemurafenib or dabrafenib aimed at blocking the most frequently mutated melanoma oncogene ($BRAF^{V600E}$) are known to develop drug resistance and experience tumor recurrence (Flaherty et al., 2010a; Flaherty et al., 2010b; Long et al, 2012; Villanueva et al., 2011). Several molecular mechanisms underlying acquired drug resistance have been recently described (Johannessen et al., 2010; Nazarian et al., 2010; Villanueva et al., 2011; Villanueva et al., 2010); however, tumor recurrence can also be due, in part, to the presence and potential enrichment of tumor subpopulations that are inherently resistant to therapy (Frank et al., 2005; Monzani et al., 2007; Roesch et al., 2010). Some of the tumor subpopulations, including those described in melanoma, express cell surface markers that are often associated with stem cells in normal tissues, and because of their ability to drive tumor growth they are often referred as 'cancer stem cells' or 'tumor-initiating cells' (Fukunaga-Kalabis et al., 2011). Evidence for the presence of cancer stem cells (CSC) initially came from studies in immunodeficient mouse models, which demonstrated the ability of a small subpopulation of tumor cells to form parental tumors (see CSC overview). In the later paragraphs, we will refer to such cells as 'tumor subpopulations with CSC phenotypes.'

Here we discuss the general concept of CSC in the context of melanoma, its role in therapy resistance, and therapeutic strategies for targeting these cells to prevent tumor repopulation.

CSC Overview

The concept of CSC originated from early studies in acute myeloid leukemia (AML), where it was shown that a small subpopulation of leukemia cells that expressed high levels of CD34 and low levels of CD38 ($CD34^{high}/CD38^{low}$), when injected into severely combined immunodeficient (SCID) mice, could establish heterogeneous parental tumors. In contrast, the more differentiated $CD34^{low}$ and $CD34^{high}/CD38^{high}$ populations failed to do so. These results suggested that the AML leukemic clone has a hierarchical origin. However, some recent reports have disputed the above-mentioned studies suggesting that tumor initiating properties of AML are not necessarily confined to less differentiated CD34+ cells and that other cell types are also equally capable of contributing to tumor formation. Studies in leukemia and certain solid tumors have indicated that most tumor initiating subpopulations share surface markers with stem cells present in normal tissues. A number of these cell surface markers have been used to identify CSC in solid tumors and are summarized in Table 17.1. The main weakness of the CSC model, which mimics normal stem cell division and maturation (Figure 17.1), is the hierarchical organization of cell division and differentiation. This model does not take into account the process of dedifferentiation or reverting from a more differentiated cell to a stem cell phenotype under certain environmental conditions that is quite often observed in many tumor types. Recent studies by Schepers and colleagues (2012) have provided a significant boost to the concept of CSC. Schepers and colleagues (2012) used a 'lineage tracing technique,' in which the stem cell is marked by a fluorescence tag and traced through the subsequent divisions of daughter cells. In a mouse intestinal adenoma model, Schepers and colleagues (2012) were able to provide strong evidence for the presence of stem cell activity in precursor lesions of intestinal cancer. Using a similar lineage tracing technique, Driessens and colleagues (2012) made identical observations in mouse models of benign skin tumors and squamous skin cancer. Further support for CSC comes from studies of Chen and colleagues (2012) in which the authors used a combination of suicide-gene technology and anti-cancer drugs to target CSC and non-stem cells to prevent tumor growth in a mouse glioma model. Initial studies in melanoma supported the concept of CSC (Fang et al., 2005; Schatton et al., 2008). However, recent studies have indicated that the potential to form tumors is not restricted to a rare subpopulation, but all melanoma cells are capable of initiating tumor formation (Quintana et al., 2008; Roesch et al., 2010; Shackleton et al., 2009).

Table 17.1. Selected CSC surface markers in solid tumors

Cancer type	CSC marker	References
Breast	$CD44^{+}/CD24^{low}$ $CD44^{+}/CD24^{-}$	(Al-Hajj et al., 2003)
Colorectal	$CD133^{+}$ EPCAM/$CD44^{+}$	(Shmelkov et al., 2008)
Gliomas	$CD133^{+}$	(Singh et al., 2004)
Lung	$CD133^{+}$	(Levina et al., 2008)
Pancreas	$CD44^{+}/CD24^{+}/ESA^{+}$	(Li et al., 2007)
Squamous cell	$CD44^{+}$	(Kokko et al., 2011)

Due to conflicting views from solid tumors, there is a general consensus emerging to redefine 'CSC' as a 'tumor subpopulation with a cancer stem cell phenotype.' These cells are defined as being capable of generating heterogeneous parental tumors when injected into immunodeficient mice

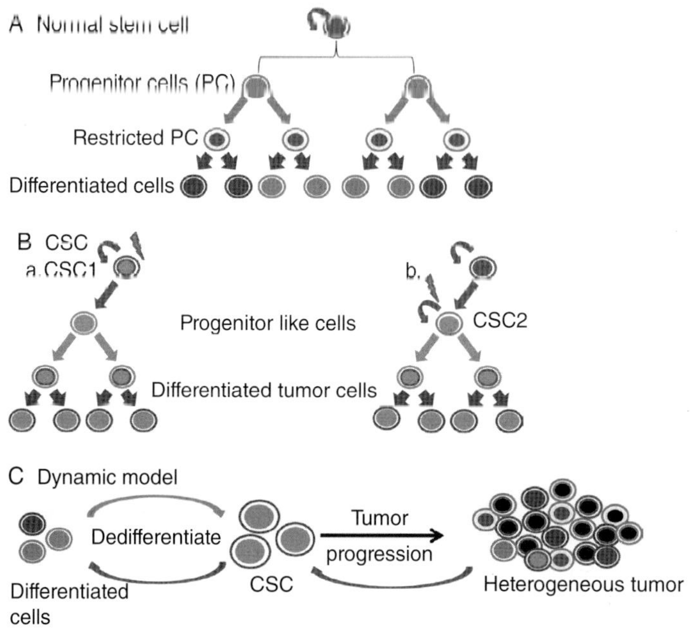

Figure 17.1. Cancer stem cell model. In the normal stem cell model; stem cells can self-renew, undergo orderly cell division and differentiation to give rise to progenitor and differentiated cells following a hierarchical pathway **A**. In the hierarchical model of CSC, CSC (originating as very early progenitor clone (CSC1) or late progenitor clone due to additional mutation or epigenetic changes CSC2) originate early with self-renewal capacity, then give rise to progenitor-like cells that can further terminally differentiate into tumor cells **B**. At this stage differentiated tumor cells have limited capacity to self-renew. In the emerging dynamic CSC model, differentiated cell can undergo de-differentiation to give rise to CSC **C**. Based on microenvironmental factors or signals, CSC now has the ability to progress further to heterogeneous tumor. For color detail, please see color plate section.

and as having the ability to maintain continuous tumor growth.

Tumor Subpopulations with CSC Phenotypes in Melanoma

Metastatic melanoma patients treated with chemo-, targeted-, or immunological-therapies show mixed responses to treatment. A small subset of melanoma patients shows dramatic responses to treatment while others show no response or initial response followed by emergence of therapy resistant tumors. Genotypic and phenotypic analyses of melanoma cells have revealed that the tumors are heterogeneous and hence exhibit a mixed response to therapy (Table 17.2). As melanoma progresses from primary to metastatic disease, the tumor acquires additional genetic and biologic properties that support tumor growth, invasion, and metastasis (reviewed in de Souza et al., 2012; Fecher et al., 2007; Somasundaram et al., 2012)). It is known that some of these acquired properties are profoundly influenced by the tumor microenvironment (TME). TME can harbor and support selective growth of tumor subpopulations that may contribute to resistance to therapies. A better understanding of the biology of tumor cell heterogeneity and their induction due to TME is needed for better targeting of cancer cells.

Table 17.2. Selected gene mutations/alterations in cell cycle regulatory genes or proteins in melanoma

Genes	Frequency (%)	Alteration type	References
BRAF	50	Mutation	(Dankort et al., 2009), (de Souza et al., 2012)
NRAS	20	Mutation	(Vidwans et al., 2011), (de Souza et al., 2012)
CKIT	28	Mutation	(Flaherty et al., 2010a), (Somasundaram et al., 2012)
PTEN	~20	Deletion/ Mutation/ Silenced	(Dankort et al., 2009), (de Souza et al., 2012), (Chin et al., 2006; Vidwans et al., 2011)
p16INK4A/ p14^{ARF} (CDKN2a)	30–70	Deletion/ Mutation	(Gallagher et al., 2008; Li et al., 2009) (Chin et al., 2006; de Souza et al., 2012)
p53	10	Mutation	(Chin et al., 2006; de Souza et al., 2012)
Cyclin D1 (CCND1)	~6	Amplification	(Chin et al., 2006; de Souza et al., 2012; Dutton-Regester et al., 2012)
Cyclin dependent kinases (CDK1-5)	3	Amplification	(Chin et al., 2006; de Souza et al., 2012)
MITF	10–20	Amplification	(Cronin et al., 2009)

In the following section, we will describe the known melanoma subpopulations with tumor-initiating properties described by us or by others.

CD20

In humans, CD20 is encoded by the membrane-spanning 4A1 (MS4A1) gene that transcribes a transmembrane protein, which is normally expressed on the surface of B cells. CD20 facilitates Ca++ channeling, B cell activation, and proliferation (Somasundaram et al., 2011; Tedder and Engel, 1994). Gene expression profiling has identified CD20 as one of the top 22 genes in melanoma that define the aggressive nature of the disease (Bittner et al., 2000). When melanoma cells are grown as 3-dimensional spheroids under in vitro culture conditions, a small subpopulation of the tumor cells express CD20 (Fang et al., 2005). This CD20+ population was previously considered to be a cancer stem-like cell or tumor-initiating cell, as it fulfilled some of the criteria of 'tumor stemness' with the ability to differentiate into multiple lineages including melanocytes, adipocytes, or chondrocytes (Fang et al., 2005). In melanomas, the concept of CSC has been challenged and remains controversial (Quintana et al., 2010; Quintana et al., 2008). Along with Morrison's group, we have demonstrated that any melanoma cell can be a tumor-initiating cell (Quintana et al., 2010; Quintana et al., 2008; Roesch et al., 2010). The significance of CD20+ melanoma cells in metastatic tumor lesions is not yet clear and is currently under investigation (Pinc et al., 2012; Schmidt et al., 2011).

In a mouse xenograft model, Schmidt and colleagues (2011) report that melanoma growth is maintained by a small population of CD20+ tumor cells. Schmidt and colleagues (2011) were able to obtain complete inhibition of tumor growth by targeting the small population of melanoma cells expressing CD20 using chimeric antigen receptor (CAR)-engineered T cells. Inhibition of tumor growth was long lasting; furthermore, no tumor relapse in mice was observed for more than 36 weeks. In a small pilot trial, we have reported that when advanced melanoma patients were treated with Rituximab (anti-CD20 antibody) in an adjuvant setting, the majority of patients remain disease free during the 3-year period of observation (Pinc et al., 2012). Similarly staged patients in historical controls showed less than 1 year of survival. In a single case study, Abken's group has confirmed the regression of metastatic melanoma lesions in a patient treated with Rituximab in a non-adjuvant setting (Schlaak et al., 2012). Overall, the above studies strongly suggest that a CD20+ melanoma subpopulation could be a major driver of tumor progression, and elimination of this subset could result in disease-free survival.

JARID1B

JARID1B is exclusively present on a small proportion of slow-cycling melanoma cells (~1–5% of all tumor cells) that have stem-like or cancer-initiating properties (Roesch et al., 2010). These cells show high expression of the histone demethylase jumonji ARID (JARID, also referred as lysine demethylase 5 [KDM5]) 1B, known to be critically involved in regulating gene expression and transcriptional activities. Our data indicate that melanoma cells grown in a hypoxic environment show up regulation of JARID1B + cells. This increased expression is transient and fully reversible when the tumor cells are returned to normoxic culture conditions (Roesch et al., 2010). In breast and prostate cancer, JARID1B up regulation is usually associated with hormone-dependent proliferation. Activation of hormone receptors is known to confer resistance to therapies (Mitra et al., 2011; Xiang et al., 2007). We have shown that isolated JARID1B + melanoma cells can give rise to a rapidly proliferating progeny that is heterogeneous (JARID1B+and JARID1B−) like the parental tumor cells (Roesch et al., 2010). Additionally, stable knockdown of JARID1B led to an initial acceleration of tumor growth followed by exhaustion, as determined by serial xeno-transplantation experiments in NOD/SCID IL2Rγ_c (NSG) mice, suggesting that JARID1B has an essential role in continuous melanoma growth (Roesch et al., 2010). Notably, Settleman's group has recently reported that JARID1A, a close homolog of JARID1B, is required for drug resistance in non small-cell lung cancer cells (Sharma et al., 2010), suggesting that slow-cycling cells can survive most conventional and targeted therapies and that this subpopulation needs to be selectively targeted.

CD271 (Nerve Growth Factor Receptor [NGFR], Also Referred as p75 Neurotrophin Receptor)

Several human neural-crest-derived tissues express CD271 or NGFR, 75 kDa transmembrane protein. NGFR is also expressed in a number of cancers including breast, colon, pancreas, prostate, ovarian, and melanoma (Civenni et al., 2011; Radfar et al., 2006; Somasundaram et al., 2012). Recently, Boiko and colleagues (2010) have shown that CD271+ melanoma subpopulations derived from patient tissues are more tumorigenic and aggressive than CD271− subpopulations when transplanted in immunodeficient *Rag2*$^{-/-}\gamma c^{-/-}$ mice (Boiko et al., 2010). Further analysis of the CD271+ melanoma cells indicated that many melanoma-associated antigens (MAA) such as MART1, MAGE, and tyrosinase were lost or down modulated in the subpopulation (Boiko et al., 2010). These antigen losses in subpopulation variants are most likely linked to the selection of immunologically resistant melanoma cells in vivo. Civenni and colleagues (2011) have also observed that CD271+ subpopulations of melanoma cells frequently show higher expression of ABCB5 transport proteins and lower expression of MAA, indicating that there may be a mechanism by which these cells have survived drug therapy and anti-melanoma reactive immune T cells. Further, expression of CD271 was correlated with higher metastatic potential and poor prognosis in patients (Civenni et al., 2011).

ABCB5/ABCG2/ABCB8

Several ABC transporters such as ABCB5, ABCB8, and ABCG2 are frequently reported to be present in various cancers including melanoma (Dean et al., 2005; Szakacs et al., 2004). In melanomas, Schatton and colleagues (2008) reported a subpopulation of cancer cells with tumor-initiating properties that have high expression of ABCB5. ABCB5+ cells were highly chemo-resistant, and their presence was higher in metastatic melanomas compared to primary or melanocytic nevi tissues. Using immunodeficient SCID mice, the authors showed that ABCB5+ cells were more tumorigenic than ABCB5− melanoma cells. Targeting ABCB5+ cells using an anti-ABCB5 antibody resulted in inhibition of tumor growth in immunodeficient nude mice. A chemo-resistant CD133+ melanoma subpopulation is known to co-express ABCG2 (Monzani et al., 2007; Taghizadeh et al., 2011). Given the selective expression of ABCG2 in a minor subpopulation of CD133+ cells, ABCG2 expression in melanoma tissue sections has not yet been confirmed. In vivo xenograft studies indicate the aggressive potential of cells that co-express CD133 and ABCG2 (Monzani et al., 2007). In an in vitro study, a subpopulation of melanoma cells resistant to drugs such as doxorubicin are known to express ABCB8 (Elliott and Al-Hajj, 2009). However, the presence of ABCB8 in melanoma tumor tissue has not been confirmed thus far.

CD133

Prominin-1 or CD133, a transmembrane glycoprotein, is normally expressed on undifferentiated cells including endothelial progenitor cells, hematopoietic stem cells, fetal brainstem cells, and prostate epithelial cells (Neuzil et al., 2007). CD133 has been identified as a CSC marker with tumor-initiating properties (Monzani et al., 2007; Shmelkov et al., 2008). Various solid tumors including brain, breast, colon, liver, lung, pancreatic, and prostate cancers show expression of CD133 (Dembinski and Krauss, 2009; Liu et al., 2006; Ricci-Vitiani et al., 2007; Salmaggi et al., 2006; Shmelkov et al., 2008). A small proportion of melanomas and primary human melanocytes are known to express CD133 (Klein et al., 2007; Rappa et al., 2008). Klein and colleagues (2007) observed a significant increase in the expression of stem-cell markers CD133, CD166, and nestin in primary and metastatic melanomas compared with benign nevi (Klein et al., 2007). Aggressive melanomas were often associated with greater expression of these markers. Studies using immunodeficient NSG mice indicate that a CD133+ subpopulation of melanoma cells is the key driver of tumor cell repopulation, whereas CD133− cells failed to form tumors (Monzani et al., 2007). However, we find that both CD133+ and CD133− melanoma cells are equally capable of forming tumors (unpublished). In colon cancer, both CD133+ and CD133− cells are capable of long-term tumor maintenance (Shmelkov et al., 2008). Drug-resistant tumor subpopulations that were obtained from breast, glioma, and lung tumors after chemotherapy frequently express CD133 (Levina et al., 2008; Liu et al., 2006; Visvader and Lindeman, 2008). Higher expression of CD133 has been associated with upregulation of anti-apoptotic proteins and increased survival mechanisms.

Aldehyde Dehydrogenase (ALDH)

ALDH is an enzyme involved in intracellular retinoic acid production and signaling (reviewed in Marcato et al., 2011a). Retinoic acid signaling is linked to cellular differentiation, apoptosis, and cell cycle arrest (Marcato et al., 2011a). ALDH activity has been used to identify tumor subpopulations with CSC phenotypes in many cancers including bladder, breast, cervical, colon, head and neck, liver, lung, ovary, pancreatic, and prostate (Marcato et al., 2011a). There are several isoforms of ALDH, and ALDH1A1 is a more prevalent isoform present in most solid tumors. The presence of ALDH1A1 is linked to chemoresistance and aggressive nature of the disease (reviewed in (Marcato et al., 2011a; Marcato et al., 2011b). ALDH has been also reported in melanomas

(Luo et al., 2012). Luo and colleagues (2012) report predominance of ALDH1A1 and ALDH1A3 isoforms in a small subset of melanoma cells with CSC phenotype. ALDH+cells were more tumorigenic in immunodeficient NSG mice when compared to ALDH− cells (Boonyaratanakornkit et al., 2010). Silencing of ALDH1A by siRNA or shRNA in tumor cells led to cell cycle arrest, apoptosis, and cell death in vitro and reduced tumorigenicity in vivo.

New Therapy Approaches to Target Tumor Subpopulations

Melanomas are heterogeneous tumors, comprised of many genotypic and phenotypic subtypes (de Souza et al., 2012; Somasundaram et al., 2012; Vidwans et al., 2011). Experimental in vitro studies and pre-clinical models have overwhelmingly supported the role of tumor subpopulations in therapy resistance. Very little is known regarding the emergence of tumor subpopulations after therapy in patients with solid tumors. Earlier therapeutic approaches to treat melanomas as a single disease using chemotherapeutic agents, such as DTIC or temozolomide, resulted in poor response rates (<15%). The majority of patients developed resistance to standard therapies very early during treatment. Lessons learned from the failures of conventional therapies have pushed the melanoma community toward designing a more personalized approach using targeted therapies. Identification of mutations in the genes involved in MAPK activation, including BRAF and NRAS, or alterations or mutations in cell-cycle regulatory genes/proteins such as CCND1/CDK4 or C-KIT in a subset of melanoma patients has further accelerated the discovery of targeted therapy drugs. Targeted therapy approaches using small molecule inhibitors that are either approved (e.g., vemurafenib) or in late stage clinical trials (e.g., the BRAF inhibitor dabrafenib, the MEK inhibitor tramatenib) have dramatically improved tumor regression and overall survival rates in melanoma patients (Flaherty et al., 2010a; Vidwans et al., 2011). Despite the dramatic regression of bulky tumor lesions in patients treated with BRAF inhibitors, many develop resistance to treatment (Flaherty et al., 2010a; Poulikakos et al., 2011; Sosman et al., 2012; Vidwans et al., 2011). Resistance to targeted agents can be mediated by diverse mechanisms, including development of secondary mutations, epigenetic changes in the target gene, and activation of compensatory signaling pathways that result in increased tumor survival (Flaherty et al., 2012; Poulikakos et al., 2011; Sosman et al., 2012; Vidwans et al., 2011; Villanueva et al., 2011; Villanueva et al., 2010). Several biological and chemical inhibitors are available to target multiple pathways that support proliferation and cell survival (Flaherty et al., 2012; Vidwans et al., 2011; Villanueva et al., 2011). A multi-modal therapy approach that combines targeting multiple pathways that promote maintenance of the bulk of the tumor with targeting melanoma subpopulations using a panel of antibodies or inhibitors may be necessary to prolong disease-free survival in melanoma patients. For this approach, each melanoma patient's tumor needs to be profiled before and after therapy to determine the best combination therapy strategy to target each individual tumor. Drug-resistant tumor subpopulations that are frequently selected after therapy need to be analyzed for epigenetic and phenotypic changes in order to design a personalized targeted approach. Efforts are under way in many cancer treatment centers to adopt a personalized therapy approach by carefully monitoring phenotypic and genotypic changes in tumor cells before and during therapy. This approach will help to design therapies with the goal of achieving long-lasting remissions.

Conclusion

Malignant melanoma, like other cancers, is a heterogeneous tumor comprised of many subpopulations with unique genotypic and phenotypic signatures. Phenotypic and molecular characterization of melanoma subpopulations will lead to the development of a number of targeted therapy agents to dramatically reduce recurrence rates in patients. Discovery of new immune checkpoint reagents such as anti-CTLA4 or anti-PD1 antibodies along with targeted therapy drugs in the treatment regimen may provide additional benefits by expansion and retention of anti-melanoma reactive T cells that have the potential to prevent emergence of drug resistant tumor cells.

References

Al-Hajj, M., Wicha. M.S., Benito-Hernandez, A., Morrison, S.J., Clarke, M.F. (2003). Prospective identification of tumorigenic breast cancer cells. *Proc Natl Acad Sci USA* 100: 3983–8

Bittner, M., Meltzer, P., Chen, Y., Jiang, Y., Seftor, E., Hendrix, M., Radmacher, M., Simon, R., Yakhini, Z., Ben-Dor, A., Sampas, N., Dougherty, E., Wang, E., Marincola, F., Gooden, C., Lueders, J., Glatfelter, A., Pollock, P., Carpten, J., Gillanders, E., Leja, D., Dietrich, K., Beaudry, C., Berens, M., Alberts, D., Sondak, V. (2000). Molecular classification of cutaneous malignant melanoma by gene expression profiling. *Nature* 406: 536–40.

Boiko, A.D., Razorenova, O.V., van de Rijn, M., Swetter, S.M., Johnson, D.L., Ly, D.P., Butler, P.D., Yang, G.P., Joshua, B., Kaplan, M.J., Longaker, M.T., Weissman, I.L. (2010). Human melanoma-initiating cells express neural crest nerve growth factor receptor CD271. *Nature* 466: 133–7.

Boniol, M., Autier, P., Boyle, P., Gandini, S. (2012). Cutaneous melanoma attributable to sunbed use: systematic review and meta-analysis. *BMJ* 345: e4757.

Boonyaratanakornkit, J.B., Yue, L., Strachan, L.R., Scalapino, K.J., LeBoit, P.E., Lu, Y., Leong, S.P., Smith, J.E., Ghadially. R. (2010). Selection of tumorigenic melanoma cells using ALDH. *J Invest Dermatol* 130: 2799–808.

Chen, J., Li, Y., Yu, T.S., McKay, R.M., Burns, D.K., Kernie, S.G., Parada, L.F. (2012). A restricted cell population propagates glioblastoma growth after chemotherapy. *Nature* 488: 522–6.

Chin, L., Garraway, L.A., Fisher, D.E. (2006). Malignant melanoma: genetics and therapeutics in the genomic era. *Genes Dev* 20: 2149–82.

Civenni, G., Walter, A., Kobert, N., Mihic-Probst, D., Zipser, M., Belloni, B., Seifert, B., Moch, H., Dummer, R., van den Broek, M., Sommer, L. (2011) Human CD271-positive melanoma stem cells associated with metastasis establish tumor heterogeneity and long-term growth. *Cancer Res* 71: 3098–109.

Cronin, J.C., Wunderlich, J., Loftus, S.K., Prickett, T.D., Wei, X., Ridd, K., Vemula, S., Burrell, A.S., Agrawal, N.S., Lin, J.C., Banister, C.E., Buckhaults, P., Rosenberg, S.A., Bastian, B.C., Pavan, W.J., Samuels, Y. (2009). Frequent mutations in the MITF pathway in melanoma. *Pigment Cell Melanoma Res* 22: 435–44.

Dankort, D., Curley, D.P., Cartlidge, R.A., Nelson, B., Karnezis, A.N., Damsky, W.E., Jr., You, M.J., DePinho, R.A., McMahon, M., Bosenberg, M. (2009). Braf(V600E) cooperates with Pten loss to induce metastatic melanoma. *Nat Genet* 41: 544–52.

de Souza, C.F., Morais, A.S., Jasiulionis, M.G. (2012). Biomarkers as key contributors in treating malignant melanoma metastases. *Dermatol Res Pract* 2012: 156068.

Dean, M., Fojo, T., Bates, S. (2005). Tumour stem cells and drug resistance. *Nat Rev Cancer* 5: 275–84.

Dembinski, J.L., Krauss, S. (2009). Characterization and functional analysis of a slow cycling stem cell-like subpopulation in pancreas adenocarcinoma. *Clin Exp Metastasis*.

Driessens, G., Beck, B., Caauwe, A., Simons, B.D., Blanpain, C. (2012). Defining the mode of tumour growth by clonal analysis. *Nature* 488: 527–30.

Dutton-Regester, K., Irwin, D., Hunt, P., Aoude, L.G., Tembe, V., Pupo, G.M., Lanagan, C., Carter, C.D., O'Connor, L., O'Rourke, M., Scolyer, R.A., Mann, G.J., Schmidt, C.W., Herington, A., Hayward, N.K. (2012). A high-throughput panel for identifying clinically relevant mutation profiles in melanoma. *Mol Cancer Ther* 11. 888–97.

Elliott, A.M., Al-Hajj, M.A. (2009). ABCB8 mediates doxorubicin resistance in melanoma cells by protecting the mitochondrial genome. *Mol Cancer Res* 7: 79–87.

Fang, D., Nguyen, T.K., Leishear, K., Finko, R., Kulp, A.N., Hotz, S., Van Belle, P.A., Xu, X., Elder, D.E., Herlyn, M. (2005). A tumorigenic subpopulation with stem cell properties in melanomas. *Cancer Res* 65: 9328–37.

Fecher, L.A., Cummings, S.D., Keefe, M.J., Alani, R.M. (2007). Toward a molecular classification of melanoma. *J Clin Oncol* 25: 1606–20.

Flaherty, K.T., Hodi, F.S., Bastian, B.C. (2010a). Mutation-driven drug development in melanoma. *Curr Opin Oncol* 22: 178–83.

Flaherty, K.T., Hodi, F.S., Fisher, D.E. (2012). From genes to drugs: targeted strategies for melanoma. *Nat Rev Cancer* 12: 349–61.

Flaherty, K.T., Puzanov, I., Kim, K.B., Ribas, A., McArthur, G.A., Sosman, J.A., O'Dwyer, P.J., Lee, R.J., Grippo, J.F., Nolop, K., Chapman, P.B. (2010b). Inhibition of mutated, activated BRAF in metastatic melanoma. *N Engl J Med* 363: 809–19.

Frank, N.Y., Margaryan, A., Huang, Y., Schatton, T., Waaga-Gasser, A.M., Gasser, M., Sayegh, M.H., Sadee, W., Frank, M.H. (2005). ABCB5-mediated doxorubicin transport and chemoresistance in human malignant melanoma. *Cancer Res* 65: 4320–33.

Fukunaga-Kalabis, M., Roesch, A., Herlyn, M. (2011). From cancer stem cells to tumor maintenance in melanoma. *J Invest Dermatol* 131: 1600–4.

Gallagher, S.J., Thompson, J.F., Indsto, J., Scurr, L.L., Lett, M., Gao, B.F., Dunleavey, R., Mann, G.J., Kefford, R.F., Rizos, H. (2008). p16INK4a expression and absence of activated B-RAF are independent predictors of chemosensitivity in melanoma tumors. *Neoplasia* 10: 1231–9.

Johannessen, C.M., Boehm, J.S., Kim, S.Y., Thomas, S.R., Wardwell, L., Johnson, L.A., Emery, C.M., Stransky, N., Cogdill, A.P., Barretina, J., Caponigro, G., Hieronymus, H., Murray, R.R., Salehi-Ashtiani, K., Hill, D.E., Vidal, M., Zhao, J.J., Yang, X., Alkan, O., Kim, S., Harris, J.L., Wilson, C.J., Myer, V.E., Finan, P.M., Root, D.E., Roberts, T.M., Golub, T., Flaherty, K.T., Dummer, R., Weber, B.L., Sellers, W.R., Schlegel, R., Wargo, J.A., Hahn, W.C., Garraway, L.A. (2010). COT drives resistance to RAF inhibition through MAP kinase pathway reactivation. *Nature* 468: 968–72.

Klein, W.M., Wu, B.P., Zhao, S., Wu, H., Klein-Szanto, A.J., Tahan, S.R. (2007). Increased expression of stem cell markers in malignant melanoma. *Mod Pathol* 20: 102–7.

Koh, H.K. (1991). Cutaneous melanoma. *N Engl J Med* 325: 171–82.

Kokko, L.L., Hurme, S., Maula, S.M., Alanen, K., Grenman, R., Kinnunen, I., Ventela, S. (2011). Significance of site-specific prognosis of cancer stem cell marker CD44 in head and neck squamous-cell carcinoma. *Oral Oncol* 47: 510–6.

Levina, V., Marrangoni, A.M., DeMarco, R., Gorelik, E., Lokshin, A.E. (2008). Drug-selected human lung cancer stem cells: cytokine network, tumorigenic and metastatic properties. *PLoS One* 3: e3077.

Li, C., Heidt, D.G., Dalerba, P., Burant, C.F., Zhang, L., Adsay, V., Wicha, M., Clarke, M.F., Simeone, D.M. (2007). Identification of pancreatic cancer stem cells. *Cancer Res* 67: 1030–7.

Li, J., Xu, M., Yang, Z., Li, A., Dong, J. (2009). Simultaneous inhibition of MEK and CDK4 leads to potent apoptosis in human melanoma cells. *Cancer Invest* 28: 350–6.

Liu, G., Yuan, X., Zeng, Z., Tunici, P., Ng, H., Abdulkadir, I.R., Lu, L., Irvin, D., Black, K.L., Yu, J.S. (2006). Analysis of gene expression and chemoresistance of CD133+ cancer stem cells in glioblastoma. *Mol Cancer* 5: 67.

Long, G.V., Trefzer, U., Davies, M.A., Kefford, R.F., Ascierto, P.A., Chapman, P.B., Puzanov, I., Hauschild, A., Robert, C., Algazi, A., Mortier, L., Tawbi, H., Wilhelm, T., Zimmer, L., Switzky, J., Swann, S., Martin, A.M., Guckert, M., Goodman, V., Streit, M., Kirkwood, J.M., Schadendorf, D. (2012). Dabrafenib in patients with Val600Glu or Val600Lys BRAF-mutant melanoma metastatic to the brain (BREAK-MB): a multicentre, open-label, phase 2 trial. *Lancet Oncol* 13: 1087–95.

Luo, Y., Dallaglio, K., Chen, Y., Robinson, W.A., Robinson, S.E., McCarter, M.D., Wang, J., Gonzalez, R., Thompson, D.C., Norris, D.A., Roop, D.R., Vasiliou, V., Fujita, M. (2012). ALDH1A Isozymes are Markers of Human Melanoma Stem Cells and Potential Therapeutic Targets. *Stem Cells* 30: 2100–13.

Marcato, P., Dean, C.A., Giacomantonio, C.A., Lee, P.W. (2011a). Aldehyde dehydrogenase: its role as a cancer stem cell marker comes down to the specific isoform. *Cell Cycle* 10: 1378–84.

Marcato, P., Dean, C.A., Pan, D., Araslanova, R., Gillis, M., Joshi, M., Helyer, L., Pan, L., Leidal, A., Gujar, S., Giacomantonio, C.A., Lee, P.W. (2011b). Aldehyde dehydrogenase activity of breast cancer stem cells is primarily due to isoform ALDH1A3 and its expression is predictive of metastasis. *Stem Cells* 29: 32–45.

Miller, A.J., Mihm, M.C., Jr. (2006). Melanoma. *N Engl J Med* 355: 51–65.

Mitra, D., Das, P.M., Huynh, F.C., Jones, F.E. (2011). Jumonji/ARID1 B (JARID1B) protein promotes breast tumor cell cycle progression through epigenetic repression of microRNA let-7e. *J Biol Chem* 286: 40531–5.

Monzani, E., Facchetti, F., Galmozzi, E., Corsini, E., Benetti, A., Cavazzin, C., Gritti, A., Piccinini, A., Porro, D., Santinami, M., Invernici, G., Parati, E., Alessandri, G., La Porta, C.A. (2007). Melanoma contains CD133 and ABCG2 positive cells with enhanced tumourigenic potential. *Eur J Cancer* 43: 935–46.

Nazarian, R., Shi, H., Wang, Q., Kong, X., Koya, R.C., Lee, H., Chen, Z., Lee, M.K., Attar, N., Sazegar, H., Chodon, T., Nelson, S.F., McArthur, G., Sosman, J.A., Ribas, A., Lo, R.S. (2010). Melanomas acquire resistance to B-RAF(V600E) inhibition by RTK or N-RAS upregulation. *Nature* 468: 973–7.

Neuzil, J., Stantic, M., Zobalova, R., Chladova, J., Wang, X., Prochazka, L., Dong, L., Andera, L., Ralph, S.J. (2007). Tumour-initiating cells vs. cancer 'stem' cells and CD133: what's in the name? *Biochem Biophys Res Commun* 355: 855–9.

Pinc, A., Somasundaram, R., Wagner, C., Hormann, M., Karanikas, G., Jalili, A., Bauer, W., Brunner, P., Grabmeier-Pfistershammer, K., Gschaider, M., Lai, C.Y., Hsu, M.Y., Herlyn, M., Stingl, G., Wagner, S.N. (2012). Targeting CD20 in Melanoma Patients at High Risk of Disease Recurrence. *Mol Ther*.

Poulikakos, P.I., Persaud, Y., Janakiraman, M., Kong, X., Ng, C., Moriceau, G., Shi, H., Atefi, M., Titz, B., Gabay, M.T., Salton, M., Dahlman, K.B., Tadi, M., Wargo, J.A., Flaherty, K.T., Kelley, M.C., Misteli, T., Chapman, P.B., Sosman, J.A., Graeber, T.G., Ribas, A., Lo, R.S., Rosen, N., Solit, D.B. (2011). RAF inhibitor resistance is mediated by dimerization of aberrantly spliced BRAF(V600E). *Nature* 480: 387–90.

Quintana, E., Shackleton, M., Foster, H.R., Fullen, D.R., Sabel, M.S., Johnson, T.M., Morrison, S.J. (2010). Phenotypic heterogeneity among tumorigenic melanoma cells from patients that is reversible and not hierarchically organized. *Cancer Cell* 18: 510–23.

Quintana, E., Shackleton, M., Sabel, M.S., Fullen, D.R., Johnson, T.M., Morrison, S.J. (2008). Efficient tumour formation by single human melanoma cells. *Nature* 456: 593–8.

Radfar, A., Stefanato, C.M., Ghosn, S., Bhawan, J. (2006). NGFR-positive desmoplastic melanomas with focal or absent S-100 staining: Further evidence supporting the use of both NGFR and S-100 as a primary immunohistochemical panel for the diagnosis of desmoplastic melanomas. *Am J Dermatopathol* 28: 162–7.

Rappa, G., Fodstad, O., Lorico, A. (2008). The stem cell-associated antigen CD133 (Prominin-1) is a molecular therapeutic target for metastatic melanoma. *Stem Cells* 26: 3008–17.

Ricci-Vitiani, L., Lombardi, D.G., Pilozzi, E., Biffoni, M., Todaro, M., Peschle, C., De Maria, R. (2007). Identification and expansion of human colon-cancer-initiating cells. *Nature* 445: 111–5.

Roesch, A., Fukunaga-Kalabis, M., Schmidt, E.C., Zabierowski, S.E., Brafford, P.A., Vultur, A., Basu, D., Gimotty, P., Vogt, T., Herlyn, M. (2010). A temporarily distinct subpopulation of slow-cycling melanoma cells is required for continuous tumor growth. *Cell* 141: 583–94.

Salmaggi, A., Boiardi, A., Gelati, M., Russo, A., Calatozzolo, C., Ciusani, E., Sciacca, F.L., Ottolina, A., Parati, E.A., La Porta, C., Alessandri, G., Marras, C., Croci, D., De Rossi, M. (2006). Glioblastoma-derived tumorospheres identify a population of tumor stem-like cells with angiogenic potential and enhanced multidrug resistance phenotype. *Glia* 54: 850–60.

Schatton, T., Murphy, G.F., Frank, N.Y., Yamaura, K., Waaga-Gasser, A.M., Gasser, M., Zhan, Q., Jordan, S., Duncan, L.M., Weishaupt, C., Fuhlbrigge, R.C., Kupper, T.S., Sayegh, M.H., Frank, M.H. (2008). Identification of cells initiating human melanomas. *Nature* 451: 345–9.

Schepers, A.G., Snippert, H.J., Stange, D.E., van den Born, M., van Es, J.H., van de Wetering, M., Clevers, H. (2012). Lineage tracing reveals Lgr5+ stem cell activity in mouse intestinal adenomas. *Science* 337: 730–5.

Schlaak, M., Schmidt, P., Bangard, C., Kurschat, P., Mauch, C., Abken, H. (2012). Regression of metastatic melanoma in a patient by antibody targeting of cancer stem cells. *Oncotarget* 3: 22–30.

Schmidt, P., Kopecky, C., Hombach, A., Zigrino, P., Mauch, C., Abken, H. (2011). Eradication of melanomas by targeted elimination of a minor subset of tumor cells. *Proc Natl Acad Sci USA* 108: 2474–9.

Shackleton, M., Quintana, E., Fearon, E.R., Morrison, S.J. (2009). Heterogeneity in cancer: cancer stem cells versus clonal evolution. *Cell* 138: 822–9.

Sharma, S.V., Lee, D.Y., Li, B., Quinlan, M.P., Takahashi, F., Maheswaran, S., McDermott, U., Azizian, N., Zou, L., Fischbach, M.A., Wong, K.K., Brandstetter, K., Wittner, B., Ramaswamy, S., Classon, M., Settleman, J. (2010). A chromatin-mediated reversible drug-tolerant state in cancer cell subpopulations. *Cell* 141: 69–80.

Shmelkov, S.V., Butler, J.M., Hooper, A.T., Hormigo, A., Kushner, J., Milde, T., St. Clair, R., Baljevic, M., White, I., Jin, D.K., Chadburn, A., Murphy, A.J., Valenzuela, D.M., Gale, N.W., Thurston, G., Yancopoulos, G.D., D'Angelica, M., Kemeny, N., Lyden, D., Rafii, S. (2008). CD133 expression is not restricted to stem cells, and both CD133+ and CD133- metastatic colon cancer cells initiate tumors. *J Clin Invest* 118: 2111–20.

Siegel, R., Naishadham, D., Jemal, A. (2012). Cancer statistics, 2012. *CA Cancer J Clin* 62: 10–29.

Singh, S.K., Clarke, I.D., Hide, T., Dirks, P.B. (2004). Cancer stem cells in nervous system tumors. *Oncogene* 23: 7267–73.

Somasundaram, R., Villanueva, J., Herlyn, M. (2011). Will engineered T cells expressing CD20 scFv eradicate Melanoma? *Mol Ther* 19: 638–40.

Somasundaram, R., Villanueva, J., Herlyn, M. (2012). Intratumoral heterogeneity as a therapy resistance mechanism: role of melanoma subpopulations. *Adv Pharmacol* 65: 335–59.

Sosman, J.A., Kim, K.B., Schuchter, L., Gonzalez, R., Pavlick, A.C., Weber, J.S., McArthur, G.A., Hutson, T.E., Moschos, S.J., Flaherty, K.T., Hersey, P., Kefford, R., Lawrence, D., Puzanov, I., Lewis, K.D., Amaravadi, R.K., Chmielowski, B., Lawrence, H.J., Shyr, Y., Ye, F., Li, J., Nolop, K.B., Lee, R.J., Joe, A.K., Ribas, A. (2012). Survival in BRAF V600-mutant advanced melanoma treated with vemurafenib. *N Engl J Med* 366: 707–14.

Szakacs, G., Annereau, J.P., Lababidi, S., Shankavaram, U., Arciello, A., Bussey, K.J., Reinhold, W., Guo, Y., Kruh, G.D., Reimers, M., Weinstein, J.N., Gottesman, M.M. (2004). Predicting drug sensitivity and resistance: profiling ABC transporter genes in cancer cells. *Cancer Cell* 6: 129–37.

Taghizadeh, R., Noh, M., Huh, Y.H., Ciusani, E., Sigalotti, L., Maio, M., Arosio, B., Nicotra, M.R., Natali, P., Sherley, J.L., La Porta, C.A. (2011). CXCR6, a newly defined biomarker of tissue-specific stem cell asymmetric self-renewal, identifies more aggressive human melanoma cancer stem cells. *PLoS One* 5: e15183.

Tedder, T.F., Engel, P. (1994). CD20: a regulator of cell-cycle progression of B lymphocytes. *Immunol Today* 15: 450–4.

Vidwans, S.J., Flaherty, K.T., Fisher, D.E., Tenenbaum, J.M., Travers, M.D., Shrager, J. (2011). A melanoma molecular disease model. *PLoS One* 6: e18257.

Villanueva, J., Vultur, A., Herlyn, M. (2011). Resistance to BRAF inhibitors: unraveling mechanisms and future treatment options. *Cancer Res* 71: 7137–40.

Villanueva, J., Vultur, A., Lee, J.T., Somasundaram, R., Fukunaga-Kalabis, M., Cipolla, A.K., Wubbenhorst, B., Xu, X., Gimotty, P.A., Kee, D., Santiago-Walker, A.E., Letrero, R., D'Andrea, K., Pushparajan, A., Hayden, J.E., Brown, K.D., Laquerre, S., McArthur, G.A., Sosman, J.A., Nathanson, K.L., Herlyn, M. (2010). Acquired resistance to BRAF inhibitors mediated by a RAF kinase switch in melanoma can be overcome by cotargeting MEK and IGF-1R/PI3K. *Cancer Cell* 18: 683–95.

Visvader, J.E., Lindeman, G.J. (2008). Cancer stem cells in solid tumours: accumulating evidence and unresolved questions. *Nat Rev Cancer* 8: 755–68.

Xiang, Y., Zhu, Z., Han, G., Ye, X., Xu, B., Peng, Z., Ma, Y., Yu, Y., Lin, H., Chen, A.P., Chen, C.D. (2007). JARID1B is a histone H3 lysine 4 demethylase up-regulated in prostate cancer. *Proc Natl Acad Sci USA* 104: 19226–31.

Chapter 18
Sarcoma Stem Cells

Filemon S. Dela Cruz and Igor Matushansky
Columbia University Medical Center, Department of Pediatrics, Division of Pediatric Oncology, New York, NY, USA

Abbreviations

ABC	adenosine triphosphate-binding cassette
ALDH	aldehyde dehydrogenase
AML	acute myeloid leukemia
CSC	cancer stem cell
ESFT	Ewing's sarcoma family of tumors
GFP	green fluorescent protein
iPSC	induced pluripotent stem cell
MFH	malignant fibrous histiocytoma
MSC	mesenchymal stem cell
NOD/SCID	nonobese diabetic/severe combined immunodeficiency
RMS	rhabdomyosarcoma
SP	side population

Introduction

Sarcomas are a biologically and clinically heterogeneous group of malignant connective tissue tumors originating from mesenchymal or ectodermal tissue, and hence, can occur at any site throughout the body. Sarcomas clinically span the entire age spectrum with an annual incidence ranging from 1 to 6 per 100,000 persons, and comprising 1.5% of all malignant tumors in adults and up to 8% in children. The categorization of different sarcoma types have classically been based upon the histological resemblance of a given sarcoma subtype to the adult tissue that it most closely resembles. There are currently over 50 sarcoma subtypes described underscoring the clinical and biologic diversity of this group of malignant cancers. Despite conventional multimodality treatment approaches (combining surgery, radiation therapy, and chemotherapy),

Cancer Stem Cells, First Edition. Edited by Vinagolu K. Rajasekhar.

patients diagnosed with sarcoma experience disproportionally higher rates of morbidity and mortality in comparison to patients with other cancers. Long-term survival rates average ~50% in adults and up to 70% in children, but a significant proportion suffer recurrence and dissemination of disease despite aggressive therapy with subsequent rates of survival no higher than 20%. Investigations into the biology of cancer resistance to therapy and cancer relapses despite initial remission of disease have led investigators to develop a hierarchical cell model of sarcoma, akin to other solid tumors. In this review, we will discuss the evidence and the methods utilized to characterize and isolate so-called sarcoma cancer stem cells (CSC) and evaluate the evidence to support the existence of sarcoma CSCs in a variety of sarcoma subtypes.

Isolation and Characterization of Sarcoma Stem Cells

Work involving the isolation and characterization of tissue stem cells has resulted in the identification of biologic markers and the development of techniques that have also been employed to isolate and characterize cancer stem cells. In this section, we will review the molecular markers and techniques that enable investigators to identify, isolate, and characterize putative sarcoma stem cells.

Serial Transplantation

The CSC is defined by its capacity to self-renew through symmetric division and recapitulate the heterogeneous lineages of cells through asymmetric cell division to yield more differentiated cells that characterize a given tumor. Given this definition of the CSC, the identification of a CSC relies on a functional assay whereby a candidate CSC generates a tumor with the same heterogeneous cellular constituents of the original tumor from which the CSC was originally isolated. The tumor-initiating and propagating properties of putative CSCs can be accomplished through serial transplantation of cells in immunodeficient mice the current gold standard used in the identification and characterization of candidate CSCs.

In general, serial transplantation experiments involve the isolation of candidate CSCs through the use of selective molecular markers, followed by the subcutaneous or intravenous injection of cancer cells into immunodeficient mice. If the inoculated cells are able to give rise to tumors reminiscent of the parental tumor phenotype, the inoculated cells are then believed to contain CSCs. In comparison to non-selected cell populations or cell populations not positive for the cell marker of interest, the CSC cell population often requires significantly lower cell inocula to initiate tumor formation in immunodeficient mice. However, there is currently no standard as to what strain of immunodeficient mouse to use for serial transplantation experiments. This point warrants mention as the immune status of recipient mice have been shown to be important for the propagation of certain tumors (Quintana et al., 2008).

Cell Surface Markers

The identification of cell surface markers that are specific to CSCs would be invaluable to the prospective identification and isolation of CSCs particularly since CSCs constitute a minute subpopulation of cells within a given tumor. Important features of a CSC cell surface marker would have to not only be specific to the CSC, but also should discriminate the CSC from non-CSC cells within the tumor and ideally not expressed by non-transformed, normal tissue cells. One of the first examples of a specific CSC cell surface marker was identified in acute myeloid leukemia (AML) cells. Subsequent work identified several other candidate CSC cell surface markers (e.g., CD44, CD90, CD117) in a variety of different solid tumors including brain, breast, prostate, melanoma, colon, liver, and sarcomas.

In sarcomas, the CD133+ cell subpopulation has been identified as a putative CSC surface marker. CD133 (also known as prominin-1) is encoded by the *PROM1* gene, which expresses a 5-transmembrane domain protein. CD133 was first recognized as a novel antigen (AC133 antigen) on CD34+ progenitor hematopoietic stem cells. Subsequent work revealed that CD133 was also expressed on other human cells and tissues including brain, kidney, prostate, pancreas, liver, colon, and stomach. However, the function of CD133 remains largely unknown. The first report of the tumorigenic properties of the CD133+ cell subpopulation and potential as a putative CSC marker came from work in brain tumors, followed thereafter by similar findings in other solid tumors. In sarcomas, the CD133+ subpopulation from Ewing's sarcoma or osteosarcoma tumors possessed tumorigenic properties when serially transplanted in immunodeficient mice.

These results suggest that CD133 may be useful for the identification of sarcoma CSCs. However, work in colon and brain cancers has revealed that CD133− cells were also tumorigenic suggesting that CD133 positivity, as a specific marker for CSCs, may not be valid for all solid tumor or sarcoma subtypes. Prospective validation of CD133 positivity as a marker for sarcoma CSCs in other sarcoma subtypes would need to be performed in the future and perhaps be combined with other assays for the identification and isolation of putative sarcoma CSCs.

Side Population and Dye Efflux Exclusion

One of the characteristic features of CSCs is its inherent or acquired resistance to cytotoxic chemotherapy agents. One explanation for the relative chemo-resistant properties of CSCs involves the overexpression of ABC (ATP-binding cassette) multidrug efflux transporters. Capitalizing on this feature of CSCs, a technique was developed involving the use of fluorescent DNA-binding dyes Hoechst 33342 or Rhodamine 123 to identify putative CSCs through flow cytometry. Expression of ABC transporters on CSCs result in the active efflux of these fluorescent dyes allowing CSCs to be identified as a "side population" (SP) of cells by flow cytometry. Similar to the isolation and characterization of CD133+ cells, SP cells from a variety of sarcomas have been isolated and shown to be capable of initiating tumor formation in immunodeficient mice as well as self-renewal properties as evidenced by serial transplantation experiments. One drawback of the dye exclusion technique to identify CSCs is the inherent toxicity associated with the dyes used. Hence, cells with the ability to efflux the dye have a survival advantage over cells that are less able to efflux the dye and that the property of dye exclusion may not necessarily be shared by all sarcoma CSCs. Indeed, non-SP cells have been evaluated and shown to possess tumor-initiating and propagating properties included in sarcomas suggesting that dye exclusion is not an essential feature of sarcoma CSCs and that use of this technique for the isolation and characterization of sarcoma CSCs be done only in conjunction with other techniques and/or CSC-specific markers (Machalinski et al., 1998).

Aldehyde Dehydrogenase (ALDH)

Chemotherapy resistance, as previously mentioned, is a characteristic feature of CSCs. In addition to the overexpression of drug efflux transporters, CSCs can also upregulate the expression of detoxification enzymes as an alternative mechanism for chemotherapy resistance. Aldehyde dehydrogenase (ALDH) catalyzes the oxidation of intracellular aldehydes and biogenic amines and has also been described to play an important role in drug detoxification in a variety of normal and malignant cells. Work in leukemia cell lines, as well as in normal hematopoietic and neural stem progenitor cells, have revealed significant expression of ALDH up to 200-fold (Chute et al., 2006).

Since progenitor stem cells can be characterized by elevated intracellular ALDH activity, investigators have examined if high

ALDH activity could identify CSC populations. High levels of ALDH activity are indeed observed in CSC subpopulations in several solid tumors including sarcomas. Utilizing a fluorescent reagent (ALDEFLUOR®), which is passively taken up by cells and converted by ALDH into a fluorescent product, several investigators have reported that candidate CSC subpopulations found in osteosarcoma, Ewing's sarcoma, and fibrosarcoma exhibit elevated ALDH activity and that these ALDH-high cell populations have the ability to form clonogenic spheres, express high levels of stem-cell related genes, exhibit chemoresistance to drugs used in the treatment of sarcomas such as doxorubicin and cisplatin, and form tumors.

The fluorescent reagent (ALDEFLUOR®) commonly utilized in identifying ALDH-high cell subpopulations functions primarily as a substrate for the ALDH 1A1 and 3A1 isoforms. It remains to be determined if other ALDH isoforms are equally present and expressed in sarcoma CSCs and if alternate ALDH isoforms can be measured utilizing this technique.

Stem Cell Gene Expression

The pluripotent and self-renewal properties of embryonic stem cells and the derived tissue stem cells are maintained by the expression of a multitude of genes that have been collectively termed "stem cell genes." Interestingly, a significant number of these stem cell-associated genes have also been implicated in tumorigenesis and has led investigators to further explore the role of stem cell-associated genes in malignant cell transformation. Re-expression of a core set of stem cell-associated genes or reprogramming genes in normal, differentiated somatic cells have resulted in the generation of pluripotent cells termed induced pluripotent stem cells (iPSCs), specifically *OCT-4*, *SOX2*, *c-MYC*, *KLF4*, *NANOG*, or *Lin28*.

In studies of CSCs, demonstration of stem cell-associated gene expression has been used as evidence to support the identification of cells with stem cell-like properties (i.e., putative CSCs) in conjunction with other cell surface markers or functional assays. However, it is important to note that expression of any or all of the stem cell-associated genes is not sufficient for the identification of sarcoma CSCs as these genes are expressed by normal stem and committed progenitor cells. It is believed that re-expression of stem cell-associated genes by cancer cells may simply reflect the malignant transformation of cells and not necessarily unique to the CSC.

Sphere-Formation Assay

The ability of cells to form spherical colonies when grown under non-adherent conditions has been used to identify tissue stem cells and consequently CSCs. The ability of tissue stem cells and CSCs to form spheres was first described from studies involving the identification of neural stem cells that formed free-floating neurospheres after dissociation of brain tissue and culture in serum-free medium (Reynolds and Weiss, 1992). Similar techniques have been employed in the propagation of CSCs involving the growth of single cell suspensions of sarcoma cells grown in serum-free conditions to ultimately yield sarcospheres. Sarcospheres isolated from human sarcoma cell lines exhibit increased tumorigenic capacity and expression of stem cell-associated genes, and were able to self-renew after dissociation of primary spheres and reculturing of secondary spheres. However, sphere formation has also been shown to arise from non-stem cell populations (Pastrana et al., 2011). Hence, the ability to form spheres in culture may not necessarily identify sarcoma CSCs but rather reflect the response of cells when removed from their normal tissue microenvironment or niche. Furthermore, sphere-formation assays may fail to identify quiescent CSCs as the success of sphere formation relies on an expanding cell population that is either poised to proliferate or are actively dividing.

Sarcoma Stem Cells

The identification and characterization of CSCs from malignant solid tumors have prompted the search for this elusive population of cells in sarcomas. Adopting the techniques just described and developed in the investigation of normal and malignant stem cells, we now examine the current evidence suggesting the existence of CSCs in sarcomas and is summarized in Table 18.1.

Osteosarcoma

Osteosarcomas represent the most common malignant tumor of the bone in children and adolescents. This group of tumors tends to occur during the second decade of life and localize to bones of the extremities (femur, tibia, and humerus) and may also consist of a soft tissue mass at initial presentation.

Table 18.1. Sarcoma Stem Cell Markers

Sarcoma Subtype	Stem Cell Marker
Osteosarcoma	CD133 CD117/Stro-1 Side population/dye efflux ALDH Sphere formation
Ewing's sarcoma family of tumors	CD133 ALDH Side population/dye efflux Sphere formation
Rhabdomyosarcoma	CD133 Sphere formation
Synovial sarcoma	CD133 Side population/dye efflux
Liposarcoma	CD133 ALDH
Fibrosarcoma	ALDH
Chondrosarcoma	CD133
Malignant fibrous histiocytoma (MFH; pleomorphic undifferentiated sarcoma)	Side population/dye efflux

Treatment incorporating multi-agent cytotoxic chemotherapy and surgery has resulted in cure rates of 60 to 70% in the localized disease setting. However, approximately 15% of patients present with disseminated or metastatic disease at initial presentation and have poor survival rates of <20% despite aggressive conventional therapy (Harris et al., 1998). Given that one of the hallmarks of CSCs is its inherent resistance to currently used cytotoxic chemotherapy drugs, the identification and direct targeting of sarcoma CSCs in osteosarcoma may improve the survival outcomes for all osteosarcoma patients, particularly those with metastatic disease.

One of the first reports of cells derived from osteosarcoma with stem cell-like properties was a study by Tsuchida and colleagues (2008) who showed that a side population (SP) of cells could be isolated by flow cytometry and expanded by exposure of parental osteosarcoma cells to the chemotherapy agent cisplatin. Analysis of SP cells treated with cisplatin revealed increased expression of stem-cell-associated genes, increased ability to form colonies on methylcellulose-based medium (an in vitro assessment of cellular transformation), and significantly larger tumors when injected into immunodeficient mice in contrast to non-SP cisplatin-treated cells. However, in contrast to other experiments evaluating the tumorigenic capacity of SP cells, untreated SP cells in this study failed to form any tumors casting doubt as to the actual presence of sarcoma CSCs in the SP population derived from this particular osteosarcoma cell line. It is conceivable that cisplatin treatment may instead initiate genetic programs that confer increased growth characteristics and tumor formation potential to cells that segregate as a side population as evaluated by flow cytometry and not actually "enrich" for a sarcoma CSC population.

A subsequent study by Tirino and colleagues (2008) evaluated the utility of CD133 expression by osteosarcoma cell lines to identify and isolate a subpopulation of cells with stem cell properties. Osteosarcoma cell lines were examined and found to express a small subpopulation of CD133+ cells that exhibited higher proliferative rates, the ability to form spherical colonies under serum-free conditions, expression of the stem cell-associated genes, the presence of a side population of cells derived from the CD133+ fraction, and growth on soft agar (indicative of in vitro transformation similar to the previously described methylcellulose-based colony assay). In addition, CD133+ cells derived from spherical colonies gave rise to both CD133+ and CD133− populations suggesting self-renewal properties reminiscent of stem cells. A follow-up study by the same group evaluated patient-derived sarcoma tissues, which showed variable expression of CD133 (Tirino et al., 2011). Although the CD133+ subpopulation isolated from tissue-derived osteosarcoma cell lines exhibited stem cell-like properties, the CD133+ fraction from the osteosarcoma cell lines were incapable of tumor formation when injected into non-obese diabetic (NOD)/severe combined immunodeficiency (SCID) mice.

Other cell surface markers have been identified that appear to more selectively enrich for cells with tumor-initiating properties. Adhikari and colleagues (2010) examined the expression of two mesenchymal stem cell markers, CD117 (also known as c-kit) and Stro-1, and discovered that mouse and human osteosarcoma cells positive for both CD117 and Stro-1 (CD117+/Stro-1+) readily formed spherical colonies, formed tumors in a nude mouse model in comparison to cells negative for both CD117 and Stro-1 (CD117-/Stro-1-), had higher metastatic potential in an orthotopic NOD/SCID mouse model, exhibited increased chemoresistance when exposed to the chemotherapy agent doxorubicin, retained multipotent differentiation ability (differentiating into bone or fat when cultured in differentiation medium), and were able to reconstitute cells doubly-positive and doubly-negative for CD117 and Stro-1 upon serial transplantation. This study has been the first to specifically identify a subpopulation of osteosarcoma cells with the hallmark features of a CSC: self-renewal capacity and the ability to reconstitute the original cellular makeup of a tumor or cell population upon serial transplantation. Validation of the stem cell-like properties of CD117+/Stro-1+ cells in a broader panel of primary, patient-derived osteosarcoma cells would further strengthen the utility of CD117 and Stro-1 marker expression as specific markers for the identification, isolation, and therapeutic targeting of this unique population in osteosarcoma.

Analysis of SP cells, a subpopulation of cells that actively efflux the Hoechst 33342 dye, for the presence of sarcoma CSCs, have also yielded cells exhibiting stem cell features. Murase and colleagues (2009) examined osteosarcoma cell lines and a malignant fibrous histiocytoma (MFH) cell line and found an identifiable SP fraction. Moreover, only the SP cells derived from the MFH cell line was able to form tumors in NOD/SCID mice, whereas SP cells from the osteosarcoma cell line could not. In contrast, Yang and colleagues (2011) identified an SP fraction in primary osteosarcoma tissues and showed that the SP cells from some tissues were capable of tumor formation in NOD/SCID mice. However, cells from both the SP and non-SP fractions were equally capable of forming tumors suggesting that the SP fraction is not specific for the enrichment of tumorigenic cells. These results underscore the importance of utilizing alternative assays and cell surface markers for the identification and isolation of putative sarcoma CSCs as SP-based isolation does not appear to be sufficient and selective for the enrichment of sarcoma CSCs. A lack of specific criteria and guidelines for the delineation of the SP fraction has likely resulted in the disparate results described. Therefore, other methods would perhaps yield improved specificity in identifying putative sarcoma CSC populations.

Lineage tracking of cells offers a novel means by which to identify and subsequently follow the fate of candidate CSCs. Levings and colleagues (2009) generated a transgenic osteosarcoma cell line which expressed a green fluorescent protein (GFP) when the stem cell-associated gene *OCT-4* was activated signifying the induction of a pluripotent state. Subsequent enrichment of GFP+cells yielded a population with increased tumor-formation and metastatic capacity compared to GFP-depleted cells, and gave rise to heterogeneous tumors. However, it is noteworthy to mention that GFP-depleted cell populations were also capable of forming tumors in NOD/SCID mice albeit at higher concentrations. Nevertheless, genetic lineage tracking of cells would be an extremely valuable tool for the prospective identification and tracking of candidate sarcoma CSCs during its evaluation for stem cell properties.

Observations of elevated aldehyde dehydrogenase (ALDH) activity in normal tissue stem cells, as well as in hematopoietic and solid tumor CSCs, have led to the use of ALDH activity as a means of identifying CSC populations in sarcomas. Wang and colleagues (2011) examined osteosarcoma cell lines and found that all cell lines contained a subpopulation of cells with high ALDH activity as evaluated by the ALDEFLUOR assay. Osteosarcoma cell lines with the highest percentage of ALDH activity (termed $ALDH^{High}$ cells) possessed increased tumorigenic capacity compared to $ALDH^{Low}$ cells, increased proliferative and clonogenic abilities, showed increased expression of the stem cell markers, and self-renewal capacity as demonstrated through serial transplantation experiments. These results support the use of $ALDH^{High}$ selection to identify and isolate a sarcoma stem cell-like population from osteosarcoma cells. However, in a similar study evaluating the $ALDH^{High}$ cells derived from an osteosarcoma cell line, $ALDH^{High}$ as well as $ALDH^{Low}$ cells equally showed the ability to reconstitute the original whole cell population putting into question the validity of ALDH activity as a specific sarcoma CSC marker (Honoki et al., 2010). Further testing in a broader panel of cell lines and primary tumors may clarify this issue. Additionally, the arbitrary cutoff for high and low ALDH expression may also contribute to differing results. Thus, similar to the issue with using side population as a means of identifying CSC populations, consensus definitions for these assays will be essential for acquiring consistent experimental outcomes and effective translation of data between investigators.

Ewing's Sarcoma Family of Tumors

The Ewing's sarcoma family of tumors (ESFT) is the second most common malignant tumor of bone and soft tissue in children and young adults. ESFTs comprise a group of tumors with a spectrum of phenotypic and clinical manifestations and are characterized by the presence of unique fusion oncoproteins that result from chromosomal translocation events. The fusion oncoproteins involve the union of the N-terminal portion of the *EWS* gene with the C-terminal portion of the *ETS* family of transcription factors, which includes the *FLI-1* gene. In over 80% of ESFTs, *EWS* joins with *FLI-1* to form the *EWS/FLI-1* fusion oncogene. Although the exact functions of *EWS/FLI-1* remain to be elucidated, *EWS/FLI-1* expression is essential to the oncogenic phenotype of ESFTs. In fact, expression of *EWS/FLI-1* in undifferentiated mesenchymal stem cells (MSC), the hypothesized cell of origin for ESFTs, result in the formation of tumors with a genetic signature reminiscent of ESFTs. Similar to osteosarcomas, a multi-modal approach to therapy involving chemotherapy and surgery, with or without radiotherapy, achieves overall survival rates of ~60–70% in the localized disease setting. In contrast, survival in the metastatic or recurrent disease setting remain abysmally low at <20%. If sarcoma CSCs do indeed exist for ESFTs and are responsible for the resistance to conventional chemotherapeutic interventions, the identification and selective targeting of the CSC subpopulation will be essential in securing durable curative responses.

Similar in approach to the identification of CSC subpopulations in osteosarcoma cells, Suvà and colleagues (2009) were the first to report on the isolation of CD133+ cells derived from primary ESFT tumors.

They showed that CD133+ cells were capable of initiating and forming tumors in NOD/SCID mice whereas CD133− cells did not. Serial transplantation of CD133+ cells successfully recapitulated the parental tumor phenotype with each successive transplant, and only the CD133+ subpopulation could renew both CD133+ and CD133− cell populations. The CD133+ could also form spherical colonies, express the stem cell markers *OCT-4* and *NANOG*, and were capable of differentiating along the mesenchymal lineage (into fat, bone, and cartilage) consistent with its purported MSC origin. In a subsequent study by the same group, the characteristic ESFT fusion oncogene *EWS/FLI-1* was expressed in undifferentiated human MSCs derived from pediatric bone marrow samples and demonstrated the generation of a CD133+ subpopulation of cells, which express stem cell markers and displayed spherical colonies lending support to the hypothesis that ESFTs are likely derived from a MSC progenitor cell. However, despite characteristic marker expression of stem cell-associated genes and sphere formation, CD133+ cells derived from human MSCs expressing *EWS/FLI-1* were unable to form tumors in NOD/SCID mice suggesting that *EWS/FLI-1* expression in human MSCs is insufficient for tumorigenesis and despite the change in phenotype reminiscent of stem-like cells.

In contrast to the previously described work, Jiang and colleagues (2010) found heterogeneous expression of CD133 in ESFT tumors examined with only a minority expressing high levels. With the exception of one cell line derived from the tumors, no differences in chemoresistance or tumorigenic potential were observed between CD133+ and CD133− subpopulations. Thus, the significance of CD133+ in identifying a tumor-initiating cell population in ESFTs remains to be seen.

The utility of side population cells in identifying stem-like cells in ESFTs has also been explored. Yang and colleagues (2010) isolated the SP fraction from an ESFT cell line and showed that the SP cells can give rise to both SP and non-SP cells, exhibit increased clonogenicity and invasive behaviour, and were more resistant to chemotherapy agents, which may be due in part to observed increases in the expression of drug efflux ABC transporters. However, the ability of the SP population to propagate tumors in vivo was not performed. SP cells have also been observed in other ESFT cell lines, but prospective validation of this cell population through serial transplantation experiments would need to be performed to determine the presence of putative sarcoma CSCs.

Isolation of cells with high ALDH activity has also shown promise in identifying cell populations in ESFTs enriched for cells with potent tumor-initiating ability. Awad and colleagues (2010) examined ESFT cell lines and patient-derived ESFT xenograft tumors and defined an ALDH^{High} and ALDH^{Low} population. Cells from the ALDH^{High} subpopulation formed spherical colonies, developed clonogenic colonies on soft agar, and were able to reconstitute a heterogeneous population of ALDH^{High} and ALDH^{Low} cells with serial cell passaging. ALDH^{High} cells also expressed stem cell marker genes and were more resistant to the chemotherapy agents doxorubicin and etoposide. Interestingly, the investigators compared the clonogenic and tumorigenic properties of ALDH^{High} cells compared to a cell population enriched for CD133+ cells and demonstrated superior colony formation and tumor induction when using cells derived from the ALDH^{High} fraction suggesting superior enrichment of tumor-initiating cells using the ALDEFLUOR-based assay. However, as previously mentioned, the delimitation of ALDH^{High} or side population fractions are arbitrary and differ among investigators, and this lack of specific consensus and guidelines will hinder the prospective analysis of data utilizing these techniques.

Rhabdomyosarcoma

Rhabdomyosarcoma (RMS) is the most common soft tissue sarcoma and the third most common extracranial solid tumor encountered in children while being exceedingly rare in adults. Similar to ESFTs, approximately 20–30% of RMS cases involve the presence of characteristic chromosomal translocations resulting in the production of an aberrant fusion oncogene. The *PAX3/FOXO1* and the less common *PAX7/FOXO1* have classically characterized a subset of RMS referred to as the alveolar variant. The more common embryonal RMS variant is not associated with any characteristic cytogenetic abnormalities and tends to have a more favourable prognosis. Also similar to both osteosarcomas and ESFTs, RMS involves multimodality treatment with generally favorable survival outcomes in the localized disease setting (~70–80% survival) and poor outcomes for metastatic disease (~20% survival).

There remains little published work regarding the determination of a CSC-like population in RMS. Walter and colleagues (2011) generated spherical colonies (termed rhabdospheres) from RMS cell lines and demonstrated that these cells possessed stem cell properties including elevated expression of stem cell markers and formed tumors at lower cell densities compared to adherent cells. Isolation of a CD133+ cell fraction from the spherical colonies showed that the CD133+ cells, in contrast to CD133− and unsorted cells, were able to generate tumors in NOD/SCID mice at lower cell densities and were more resistant to the toxic effects of chemotherapy drugs. Furthermore, analysis of CD133+ expression in patient-derived tissue samples revealed that high CD133+ expression correlated with inferior survival compared to low or intermediate expression levels in patients with the embryonal RMS variant. Hence, CD133 expression in RMS may not only be useful in enriching for candidate sarcoma CSCs, but may also have clinical utility in predicting survival outcomes. It remains to be seen if selection based on ALDH activity, perhaps in conjunction with CD133 expression, would further enrich for cells with stem cell-like properties.

Other Soft Tissue Sarcomas

Given the identification and isolation of cells with stem-cell-like properties in osteosarcoma, Ewing's sarcoma and rhabdomyosarcoma, there has been increased interest in the search for similar stem-like cells in other sarcoma subtypes. Stratford and colleagues (2011) have identified candidate sarcoma CSCs in a liposarcoma cell line after double enrichment for CD133 and ALDH high activity ($ALDH^{High}$) and showed that these cells were tumorigenic. Similarly, chondrosarcoma cells derived from primary tumors and enriched for CD133 also showed potent tumorigenic potential. Synovial sarcoma tumors and cell lines also express CD133, though it remains to be seen whether these cells possess stem-cell-like properties and are tumorigenic (Terry and Nielsen, 2010). However, a side population of cells derived from a primary synovial sarcoma tumor, as well as from other mesenchymal solid tumors (e.g., MFH), were capable of forming tumors.

Conclusion

The search for the elusive cancer stem cell in sarcomas and other solid tumors will continue to remain a challenging and controversial area of research. The incredible clinical and biological diversity of sarcomas add to the difficulties in performing any type of research within this family of malignant tumors. Additionally, if one considers that the search for the sarcoma CSC involves a search for an exceedingly rare population of cells and may even exhibit variable degrees of "stemness" depending upon microenvironmental cues as recent research suggests (Chaffer et al., 2011), the challenge of CSC research in sarcomas appears daunting and even insurmountable given current technologies. Yet it is the hope by many investigators that the unequivocal identification and validation of the existence of cancer stem cells in sarcomas, or any other cancer, will allow for a paradigm shift in how we view the evolution of cancers as well as how we develop future strategies for how we can effectively eradicate it.

Acknowledgments

FDC would like to thank the St. Baldrick's Foundation for its continuing research support through the NetApp St. Baldrick's Scholar grant. FDC would also like to thank and acknowledge research support by the Hyundai Hope on Wheels Hope Grant, the John M. Driscoll Jr. Children's Fund grant, and the Jaime Deutsch Foundation.

References

Adhikari, A.S., Agarwal, N., Wood, B.M., Porretta, C., Ruiz, B., Pochampally, R.R., and Iwakuma, T. (2010). CD117 and Stro-1 identify osteosarcoma tumor-initiating cells associated with metastasis and drug resistance. Cancer research *70*, 4602–4612.

Awad, O., Yustein, J.T., Shah, P., Gul, N., Katuri, V., O'Neill, A., Kong, Y., Brown, M.L., Toretsky, J.A., and Loeb, D.M. (2010). High ALDH activity identifies chemotherapy-resistant Ewing's sarcoma stem cells that retain sensitivity to EWS-FLI1 inhibition. PloS one *5*, e13943.

Chaffer, C.L., Brueckmann, I., Scheel, C., Kaestli, A.J., Wiggins, P.A., Rodrigues, L.O., Brooks, M., Reinhardt, F., Su, Y., Polyak, K., *et al.* (2011). Normal and neoplastic nonstem cells can spontaneously convert to a stem-like state. Proceedings of the National Academy of Sciences of the United States of America *108*, 7950–7955.

Chute, J.P., Muramoto, G.G., Whitesides, J., Colvin, M., Safi, R., Chao, N.J., and McDonnell, D.P. (2006). Inhibition of aldehyde dehydrogenase and retinoid signaling induces the expansion of human hematopoietic stem cells. Proceedings of the National Academy of Sciences of the United States of America *103*, 11707–11712.

Clarke, M.F., Dick, J.E., Dirks, P.B., Eaves, C.J., Jamieson, C.H., Jones, D.L., Visvader, J., Weissman, I.L., and Wahl, G.M. (2006). Cancer stem cells—perspectives on current status and future directions: AACR Workshop on cancer stem cells. Cancer Res *66*, 9339–9344.

Harris, M.B., Gieser, P., Goorin, A.M., Ayala, A., Shochat, S.J., Ferguson, W.S., Holbrook, T., and Link, M.P. (1998). Treatment of metastatic osteosarcoma at diagnosis: a Pediatric Oncology Group Study. Journal of clinical oncology: official journal of the American Society of Clinical Oncology *16*, 3641–3648.

Honoki, K., Fujii, H., Kubo, A., Kido, A., Mori, T., Tanaka, Y., and Tsujiuchi, T. (2010). Possible involvement of stem-like populations with elevated ALDH1 in sarcomas for chemotherapeutic drug resistance. Oncology reports *24*, 501–505.

Jiang, X., Gwye, Y., Russell, D., Cao, C., Douglas, D., Hung, L., Kovar, H., Triche, T.J., and Lawlor, E.R. (2010). CD133 expression in chemo-resistant Ewing sarcoma cells. BMC cancer *10*, 116.

Levings, P.P., McGarry, S.V., Currie, T.P., Nickerson, D.M., McClellan, S., Ghivizzani, S.C., Steindler, D.A., Gibbs, C.P. (2009). Expression of an exogenous human Oct-4 promoter identifies tumor-initiating cells in osteosarcoma, Cancer Res, *69*(14):5648–55.

Machalinski, B., Wiszniewska, B., Baskiewicz, M., Marchlewicz, M., Majka, M., Wenda-Rozewicka, L., and Ratajczak, M.Z. (1998). In vivo and in vitro studies on the toxicity of Hoechst 33342 (Ho342). Implications for employing Ho342 for the isolation of haematopoietic stem cells. Annals of transplantation: quarterly of the Polish Transplantation Society *3*, 5–13.

Murase, M., Kano, M., Tsukahara, T., Takahashi, A., Torigoe, T., Kawaguchi, S., Kimura, S., Wada, T., Uchihashi, Y., Kondo, T., *et al.* (2009). Side population cells have the characteristics of cancer stem-like cells/cancer-initiating cells in bone sarcomas. British journal of cancer *101*, 1425–1432.

Pastrana, E., Silva-Vargas, V., and Doetsch, F. (2011). Eyes wide open: a critical review of sphere-formation as an assay for stem cells. Cell stem cell *8*, 486–498.

Quintana, E., Shackleton, M., Sabel, M.S., Fullen, D.R., Johnson, T.M., and Morrison, S.J. (2008). Efficient tumour formation by single human melanoma cells. Nature *456*, 593–598.

Reynolds, B.A., and Weiss, S. (1992). Generation of neurons and astrocytes from isolated cells of the adult mammalian central nervous system. Science *255*, 1707–1710.

Stratford, E.W., Castro, R., Wennerstrom, A., Holm, R., Munthe, E., Lauvrak, S., Bjerkehagen, B., and Myklebost, O. (2011). Liposarcoma Cells with Aldefluor and CD133 Activity have a Cancer Stem Cell Potential. Clinical sarcoma research *1*, 8.

Suvà, M.L., Riggi, N., Stehle, J.C., Baumer, K., Tercier, S., Joseph, J.M., Suva, D., Clement, V., Provero, P., Cironi, L., *et al.* (2009). Identification of cancer stem cells in Ewing's sarcoma. Cancer research *69*, 1776–1781.

Terry, J., and Nielsen, T. (2010). Expression of CD133 in synovial sarcoma. Applied immunohistochemistry & molecular morphology: AIMM/official publication of the Society for Applied Immunohistochemistry *18*, 159–165.

Tirino, V., Desiderio, V., d'Aquino, R., De Francesco, F., Pirozzi, G., Graziano, A., Galderisi, U., Cavaliere, C., De Rosa, A., Papaccio, G., *et al.* (2008). Detection and characterization of CD133+ cancer stem cells in human solid tumours. PloS one *3*, e3469.

Tirino, V., Desiderio, V., Paino, F., De Rosa, A., Papaccio, F., Fazioli, F., Pirozzi, G., and Papaccio, G. (2011). Human primary bone sarcomas contain CD133+ cancer stem cells displaying high tumorigenicity in vivo. FASEB journal: official publication of the Federation of American Societies for Experimental Biology *25*, 2022–2030.

Tsuchida, R., Das, B., Yeger, H., Koren, G., Shibuya, M., Thorner, P.S., Baruchel, S., and Malkin, D. (2008). Cisplatin treatment increases survival and expansion of a highly tumorigenic side-population fraction by upregulating VEGF/Flt1 autocrine signaling. Oncogene *27*, 3923–3934.

Walter, D., Satheesha, S., Albrecht, P., Bornhauser, B.C., D'Alessandro, V., Oesch, S.M., Rehrauer, H., Leuschner, I., Koscielniak, E., Gengler, C., *et al.* (2011). CD133 positive embryonal rhabdomyosarcoma stem-like cell population is enriched in rhabdospheres. PloS one *6*, e19506.

Wang, L., Park, P., Zhang, H., La Marca, F., and Lin, C.Y. (2011). Prospective identification of tumorigenic osteosarcoma cancer stem cells in OS99-1 cells based on high aldehyde dehydrogenase activity. International journal of cancer Journal international du cancer *128*, 294–303.

Yang, M., Yan, M., Zhang, R., Li, J., and Luo, Z. (2011). Side population cells isolated from human osteosarcoma are enriched with tumor-initiating cells. Cancer science *102*, 1774–1781.

Yang, M., Zhang, R., Yan, M., Ye, Z., Liang, W., and Luo, Z. (2010). Detection and characterization of side population in Ewing's sarcoma SK-ES-1 cells in vitro. Biochemical and biophysical research communications *391*, 1062–1066.

Section IV
Cancer Stem Cells in Tumor Metastasis Perspective

Chapter 19
Cancer Stem Cells in Metastasis and Minimal Residual Disease

Joerg Huelsken and Albert Santamaria i Martínez
ISREC (Swiss Institute for Experimental Cancer Research) and National Center of Competence in Research (NCCR), Molecular Oncology, Ecole Polytechnique Fédérale de Lausanne (EPFL), Lausanne, Switzerland

Abbreviations

CSC	cancer stem cell
CTC	circulating tumor cell
DTC	disseminated tumor cell
ECM	extracellular matrix
MRD	minimal residual disease

Introduction

Many tissues are based on cellular societies with a clear hierarchical organization—similar to a pyramidal structure—that maintains tissue homeostasis and in which stem cells occupy the tip position. Stem cells possess key characteristics that distinguish them from the rest of the cells: they have long-term self-renewal potential and they can give rise to lineage restricted, sometimes transiently amplifying cells that in turn will generate terminally differentiated cells. Although the latter perform the fundamental functions of the tissue, they are often short-lived and are constantly replaced with new cells generated by the stem cell population. Stemness, which describes the combination of self-renewal capacity and suppression of lineage commitment, is achieved not only through intrinsic characteristics of stem cells, but also thanks to the presence of particular microenvironments that participate in maintaining stemness and controlling stem cell behavior—the niche.

The realization that many tumors also sustain such a hierarchical organization, with the so-called cancer stem cells at the top, has had a great influence on how we understand cancer biology. Similar to their normal tissue stem cell equivalent, this small subset of cells—also termed tumor-initiating cells based on the fundamental experimental approach—has the unique ability to initiate and perpetuate tumor growth and to give rise to all its heterogeneity in serial transplantation experiments. These cancer stem

Cancer Stem Cells, First Edition. Edited by Vinagolu K. Rajasekhar.

cells can be generated either from tissue specific normal stem cells or from more differentiated progeny that regain stem cell characteristics. Major efforts are now being made in order to understand the biological circuitry active in cancer stem cells in order to eventually target this population of cells, which evidently is essential for tumor maintenance.

Interestingly, it has recently been demonstrated that at least in some cancers, cancer stem cells are also responsible for metastatic colonization. This discovery is crucial for the treatment of cancer, given that metastases contribute notably to patient morbidity and mortality. Likewise, it also opens the door to a deeper understanding of minimal residual disease, which describes the presence of disseminated/circulating tumor cells after or during tumor remission. In this chapter we will discuss the role cancer stem cells play in the metastatic process.

Metastatic Spread

Theories for Metastatic Spread

The process of metastasis is the main cause of death by cancer due to tissue destruction and impairment of organ function; however, the exact mechanisms that enable cancer cells to grow at a distant site are still only partially understood. Furthermore, the complexity of the metastatic process is aggravated by the fact that metastases behave as new, independently developing and progressing tumors in such a way that can often generate cascades of metastases (Viadana et al., 1978). This implies parallel, genetically independent evolution of cancer cells in the primary and the secondary sites. This phenomenon has been recently proven by detailed DNA sequencing analysis to annotate genomic rearrangements in primary and secondary tumors of individual patients, which confirmed the existence of metachronous seeding to target organs (Campbell et al., 2010).

It has been known for several years that colonization of distant organs is a rather inefficient process accomplished only by few disseminated cancer cells. These founder cells have to overcome several obstacles before settling down in a new tissue: (1) they have to leave the primary tumor, (2) enter the circulation, extravasate, and (3) establish themselves as new tumors (Chambers et al., 2002). The main hindrance, which makes the overall process of metastasis rather inefficient, is the survival and outgrowth of cancer cells in varying, previously unseen environments (Psaila and Lyden, 2009). Intrinsic properties of individual tumor cells and their enforced ability to adapt to new conditions may give some cells significant advantages over others in this process, and in line with this, enhanced metastatic abilities have recently been identified in the cancer stem cell population (see below). It is known that different cancers metastasize to different target organs in a preferential manner, a process termed organ tropism (Hess et al., 2006). Whereas for example pancreatic cancer preferentially metastasizes to the liver, abdominal cavity, lymph nodes, and lungs (in order of decreasing frequencies), other primary cancers show different patterns of metastasis (e.g., breast cancer mainly seeds to the lymph nodes, bone, liver, lung, and brain). This emphasizes the fact that metastatic cells also rely on extrinsic signals, namely a proper microenvironment, in order to establish themselves as new colonies of distantly growing tumor cells. The corresponding concept of the "seed and soil" was first proposed by Stephen Paget back in 1889 (Paget, 1889). He postulated that organ-specific patterns of metastatic spread depend on the requirements of the cancer cell (the seed) for specific environments (the soil). However, in 1928, James Ewing postulated a clearly different vision of metastatic spread in his "mechanical arrest" theory, which suggests that metastatic patterns are determined by lymphatic and venous drainage and that circulating cells are mechanically arrested in the first capillary bed they encounter (Ewing, 1928). Far from being conflicting or mutually exclusive, both ideas have been supported by experimental data in different cancers. For instance, colon cancer metastasizes largely to the liver, which in principle could be explained by the circulation connecting these two organs through the portal vein. On the other hand though, hemodynamics cannot easily explain why bone is the predominant metastatic site for prostate cancer, suggesting that signals from the host organ might play a role in the colonization phase of metastasis. By taking into account not only the incidence of metastatic lesions but also hemodynamic patterns between primary and secondary sites, detailed analyses of autopsy studies provide some estimations of the relevance of the "seed and soil" and "mechanical arrest" theories (Weiss, 1992). This study concluded that about two-thirds of the secondary sites from all types of primary tumors could be explained by drainage pattern. However, a considerable number of sites that should contain metastases based on hemodynamic patterns proved to be inhospitable, an observation which supports Paget's theory. All in all, both principles seem to be valid and play a role in patterning the metastatic spread.

Cancer Stem Cells in Metastasis

As previously mentioned, metastasis is a rather unsuccessful process with several rate-limiting steps. As a matter of fact, similar to what is anticipated in human patients (Tarin et al., 1984), less than 0.1% of injected tumor cells can successfully complete the entire metastatic process and form macroscopic metastases in experimental mouse models. However, even if most tumor cells appear to be capable of survival in the circulation and to extravasate into the secondary site, the colonization phase appears to be a more restrictive step, which is only accomplished by a few cells. This implies that either these cells already have specific, inherent abilities that allow them to fulfill these different requirements, or that they acquire these abilities as a result of an adaptation to different selective pressures in a Darwinian selection manner.

In recent years, an old idea (Nowell, 1976; Virchow, 1958) has been revived to explain how tumors develop and are maintained. In this hypothesis, cancer originates from a progenitor-like cell, the so-called cancer stem cell (CSC) (Reya et al., 2001), which is responsible for tumor growth and maintenance. In the last 15 years since John E. Dick's fundamental work on leukemia stem cells (Bonnet and Dick, 1997), major efforts have been devoted to identify cancer stem cells in different tumor types. Through the use of different cell surface markers or other stem cell-specific characteristics (for example, the ability to pump out Hoechst 33342 or increased ALDH (aldehyde dehydrogenase) activity) these cells have been isolated from primary and metastatic tumors of both mouse and human cancers. But it was only until recently that primary tumor CSCs also were proven to be responsible for metastasis (Hermann et al., 2007; Malanchi et al., 2012; Pang et al., 2010). When considering the hallmarks of these cells—their increased fitness and stress-resistance, their ability to instruct and orchestrate other cancer and cancer-associated cell types, and their extended life span—these cells occur as the natural candidates to initiate the metastatic cascade. However, even if CSCs are thought to be the origin of the primary and the secondary tumor, this does not explain why metastases occur relatively late in the process of tumor progression. This is even more puzzling in the light of recent studies that have established dissemination as an early event in cancer progression (Husemann et al., 2008). It is however still not understood whether early, disseminated cells are actually metastatic or whether the ability to successfully colonize distant organs is an additional property, which is developed later. It therefore appears likely that CSCs also acquire the ability to metastasize only over a long time of evolution and progression, leading to the formation of metastatic CSC subclones.

The biological programs that determine the enhanced fitness of CSCs, which is closely related to their stemness, partially overlap with those of normal stem cells. For instance, acute myeloid leukemia (AML) stem cells can have a very similar molecular program to that of embryonic stem cells (ESC) (Somervaille et al., 2009) or normal hematopoietic stem cells (Eppert et al., 2011) depending on the origin of the tumor and the oncogenic driver mutations. Other key factors of stemness that have been involved in controlling CSC biology are for example Oct4 (Darini et al., 2012), Sox2 (Leis et al., 2012), and Nanog (Jeter et al., 2011; Lee et al., 2011). Therefore, the use of these molecular programs related to normal stem cells seems to play a role in enhancing CSC's fitness, and may provide means of targeting this population of cells.

As discussed before, cells that leave the primary tumor, extravasate, survive the circulation, and arrive to a new organ still need to establish themselves in this new tissue to grow as secondary tumors. If one takes into account the overall efficiency of the metastatic process (Chambers et al., 2002), the requirement of a supportive microenvironment—Paget's seed and soil theory discussed earlier—becomes evident. Therefore, the potential to metastasize is given not only by intrinsic properties of CSCs but also by the presence of a favorable microenvironment. This stresses the role of the stromal microenvironment at the secondary site in controlling metastatic spread, which we will discuss later.

Circulating Tumor Cells and Minimal Residual Disease (MRD)

Circulating tumor cells (CTC) and disseminated tumor cells (DTC) are rare cells that have abandoned the primary tumor and that are found in the peripheral blood or in specific target organs (in particular the bone marrow) to which they home, respectively. Their frequency is estimated to be one tumor cell per 10 million nucleated blood cells in patients with diagnosed metastatic cancer. The presence of only these cells is referred to as minimal residual disease (MRD) in patients and describes the absence of further, clinically detectable distant metastases once the tumor has been removed or is in remission. CTCs/DTCs are usually undetectable by conventional clinical imaging or laboratory tests. Although DTCs are sometimes considered to be micrometastases, this term might be misleading since it has been shown that DTCs often enter a state of latency that can last for many years, a phenomenon called dormancy. During dormancy, cells are quiescent and arrested in G0. Nevertheless, for reasons that are still not completely understood, these cells can be activated and may eventually grow as metastasis. The resumption of their proliferative activity and growth most probably involves changes in the host tissue that lead to neutralization of environmental growth restrictions. This is particularly interesting given that the presence of these DTCs has been associated with poor prognosis in some types of cancer (for example, breast, prostate, and colorectal cancers) (Braun et al., 2000b; Braun et al., 2005; Flatmark et al., 2011; Lilleby et al., 2003) indicating that DTCs may contribute to disease relapse. Moreover, for some types of cancers it has been found that neoadjuvant therapy fails to eliminate these cells (Braun et al., 2000a; Mathiesen et al., 2012). Therefore, DTC and/or CTC detection might provide estimates for the risk of relapse and overall survival. Accordingly, a numerous assays have been developed to identify and quantify DTCs, including RT-PCR and cytokeratin immunohistochemistry. However, invading tumor cells that leave the primary tumor often display characteristics of EMT and consequently downregulate epithelial markers such as cytokeratins. This poses a challenge for the correct identification and quantification of these cells in patients.

Although the factors that contribute to the eventual awakening of DTCs are hardly studied, the formation of new sprouts from preexisting blood vessels, a process known as angiogenesis, was found to play a crucial role in the outgrowth of macrometastases from micrometastasis (Holmgren et al., 1995). Therefore, the difference between a micrometastasis and a clinically detectable macrometastasis appears to be closely related to its capacity to modulate the environment in order to promote angiogenesis. Accordingly, systemic inhibition of angiogenesis may help to keep DTCs in a dormant state (Holmgren et al., 1995). In addition, the immune system has been implicated in the regulation of the dormancy state by maintaining the quiescence of DTCs in a process known as immune surveillance (Farrar et al., 1999). In a B cell lymphoma model of dormancy, the authors proved that IFN-γ and CD8+ T cells are needed to maintain tumor dormancy and promote growth arrest. Likewise, immunosurveillance was seen to control dormancy and thus restrict metastasis in a model of melanoma through cytostatic CD8+ T cells (Eyles et al., 2010). On the contrary, a bone marrow microenvironment rich in collagen-I is able to induce the metastatic outgrowth of dormant breast cancer DTCs (Barkan et al., 2010). Similarly, the frequency of breast cancer metastasis to the bone is known to be linked to osteolytic activity of osteoclast, possibly due to the release of bone-derived growth factors upon degradation of bone matrix. Interestingly, osteoclast-inhibiting bisphosphonates show disease-free survival benefit in patients, which may be due to increased clearance of DTCs from the bone marrow upon treatment (Aft et al., 2010). All these data show that signals from the microenvironment and its modulation determine whether a DTC will remain dormant or will start producing a metastasis.

Since CSCs have been shown to have superior metastatic potency, the question arises whether CTCs/DTCs are actually identical to CSCs. It is unlikely that all cells that are able to leave the primary tumor are CSCs, thus one would expect that CTCs are a heterogeneous population of cells. This seems to be the case for breast cancer, where a subpopulation of CTCs presents some features of the CSC phenotype (Aktas et al., 2009; Theodoropoulos et al., 2010). Similarly, also DTCs isolated from the bone marrow of breast cancer patients exhibit a subpopulation of CSC-like cells (Reuben, 2011). Taken together, this leads to the exciting hypothesis that similar to the thoroughly validated, preclinical experiments, also tumor cells found in the blood or bone marrow of patients contain a cancer stem cell population that may be responsible for metastasis formation. Currently, functional assays that probe the potency of such cells in experimental models are still missing. If successful, these cells could become an important clinical source of relevant tumor material, which can inform about the risk of a patient and about the genetic make-up of metastasis forming cells even before secondary tumors become clinically detectable. This may allow applying personalized treatments while the process of metastasis is just ongoing and thereby interfere at an early time point, which promises to increase the benefit of such therapy.

Extrinsic Factors Regulating Metastasis: The Niche

The local microenvironment is an essential regulator of tumor behavior. It basically comprises all of the noncancer cell components within a tumor, including cancer-associated fibroblasts, immune cells, endothelial cells and blood vessels, and the noncellular compartment (i.e., the extracellular matrix [ECM] and soluble factors). The microenvironment provides tumors not only with the nutrients and regulatory factors they require to grow and survive but also with mechanisms of defense against immune system attack. Moreover, the microenvironment regulates the mobility and invasion capacity of cancer cells, their differentiation, and their proliferation. Thus, encountering a suitable microenvironment is essential to fulfill the needs of the tumor in general and of the particular populations of cells that it comprises (Figure 19.1). Consequently, the establishment of such a tumor-permissive microenvironment has been shown to sometimes even precede the invasion of cancer cells into a secondary organ (Kaplan et al., 2005). The formation of this premetastatic niche seems to occur as a prerequisite (at least for some cancers) for metastatic cells to settle in a new organ. Tumor secreted factors such as lysyl oxidase (Erler et al., 2009), interleukins 6 and 10 (Deng et al., 2012), or VEGF-A, TFGβ and TNFα (Hiratsuka et al., 2006) induce the recruitment of myeloid cells, activation of fibroblasts, and the formation of an appropriate microenvironment that future metastasizing cells will use to colonize

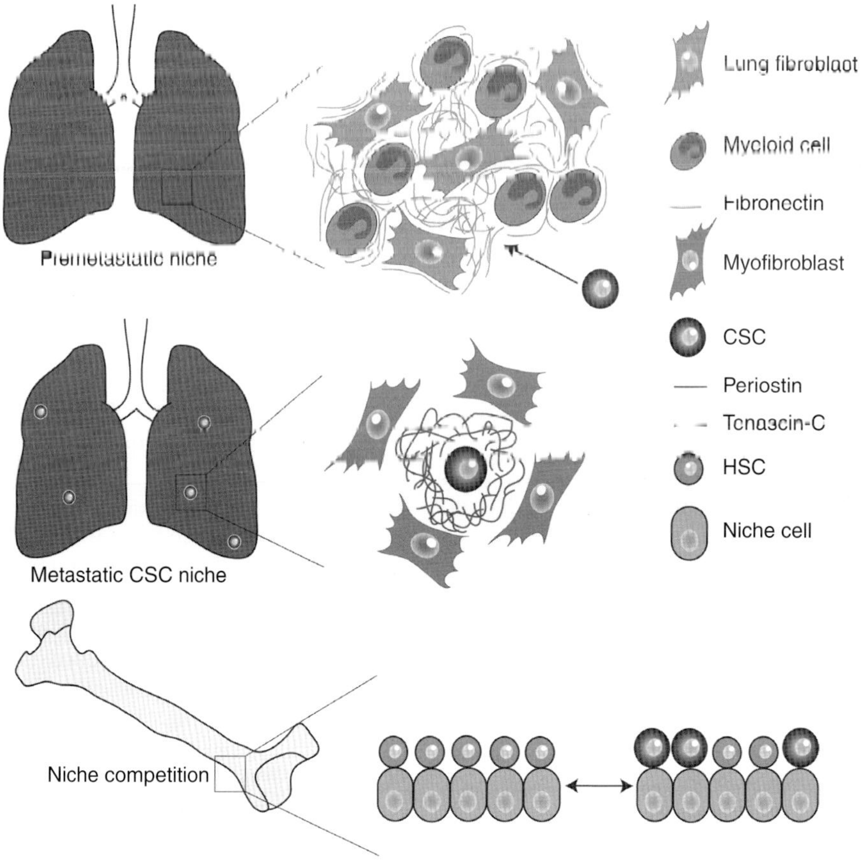

Figure 19.1. The premetastatic niche (*upper scheme*) involves the recruitment of myeloid cells to the future metastatic sites and the production, together with local fibroblasts, of ECM molecules like fibronectin. A suitable environment is created for tumor cells (and other bone marrow derived support cells) to invade these areas. The metastatic niche (*middle scheme*) is generated when seeding cancer stem cells induce the secretion of ECM molecules like POSTN and TNC by locally activated fibroblasts. These molecules enhance CSC survival and fitness in the host tissue through the activation of different signaling pathways. Niche competition (*lower scheme*) involves the parasitism of an endogenous stem cell niche by a tumor cell, i.e., the tumor cell takes the place of an HSC in the bone, displacing the latter from its natural niche. For color detail, please see color plate section.

particular organs. These factors can be delivered through exosomes, which are small secreted vesicles derived from the endocytic pathway (Hood et al., 2011; Peinado et al., 2012). In short, the development of this premetastatic niche serves the final purpose of enhancing the survival of disseminated tumor cells in an otherwise hostile environment.

Even without a premetastatic niche, furnishing their new habitat, or instructing other cells to do so, is a requirement for metastatic cancer cells to succeed in growing as secondary tumors. Recently, three independent groups provided evidence to support this concept, showing that the ECM molecules tenascin-C (TNC) and periostin (POSTN) play pivotal roles in the colonization phase. Importantly, depletion of either ECM molecule impairs metastatic colonization (Malanchi et al., 2012; O'Connell et al., 2011; Oskarsson et al., 2011). On the one hand, it has been demonstrated that TNC is initially produced by breast cancer cells when they first reach the lungs and while they establish micro-metastatic colonies. As these colonies expand, the source of TNC changes from the tumor cells themselves to the stromal compartment, and activated fibroblasts (myofibroblasts) start producing the protein. TNC is required for effective colonization, and it acts by enhancing Wnt and Notch signaling in tumor cells. Both of these pathways ultimately strengthen the fitness of cancer cells and confer pro-survival signals. On the other hand, our group demonstrated that POSTN, which is produced by lung fibroblasts only upon colonization of the tissue by invading breast tumor cells, is indispensable for successful colonization. POSTN is a niche component of the normal mammary gland and of the primary tumor and needs to get induced at the secondary site by incoming tumor cells. The way POSTN acts is through the recruitment of Wnt ligands, which POSTN enriches locally and presents to tumor cells thereby enhancing Wnt signaling in CSCs. Interestingly, periostin, which is involved in fibrosis, has been shown to regulate collagen-I fibrillogenesis (Norris et al., 2007). As mentioned before, a fibrotic, collagen-I enriched microenvironment is responsible for breaking the dormancy status of DTCs (Barkan et al., 2010). Thus, periostin could well be involved in this process, which makes it a potential candidate to target MRD.

Although deletion of these molecules affects metastasis, it does not seem to have major effects on primary tumor development. The reason for this remains to be clearly elucidated, but one could argue that the microenvironment of the autochthonous tissue (mammary gland) has an increased complexity and richness due to the fact that is better adapted to support mammary gland derived tissue-specific stem cells (or CSCs). This intricacy, which may involve a large variety of supportive factors, facilitates that several proteins may compensate for the loss of a single component. However, the secondary site is expected to be optimized to support local (lung) stem cell populations that may require different signals. In consequence, blocking a single component that is required to imitate the original niche, the disseminated cancer cells are familiarized to have dramatic effects on the ability of infiltrating cancer cells to survive. A third mechanism used by tumor cells to survive in other organs is to compete out endogenous stem cells from their niche and thereby parasitize the existing, supportive microenvironment. This has been seen in particular for prostate cancer, where the tumor cells can lodge in the hematopoietic stem cell niche of the bone marrow replacing endogenous stem cells (Shiozawa et al., 2011).

Prospectives

The cancer cell niche is the summary of the tumor's cell tolerances, and it is conceptually defined by the boundaries where the tumor can live, grow, and spread/seed. Even though most of the efforts have been put on directly targeting tumor cells, fewer researchers have focused their attempts in trying to disrupt particular elements of the supportive microenvironment. For instance, targeting de novo blood vessel formation (angiogenesis) is one of the most well-known endeavors to target the microenvironment by preventing nutrient supply to the tumor, and thus, tumor growth. In order to divide and ensure secondary tumor growth, cancer cells have certain requirements for nutrients and O_2, and as a consequence, the niche for disseminated cancer cells involves, as we have previously seen, their capacity to activate the angiogenic switch. However, it should be emphasized that a targeted therapy directed solely against such neo-angiogenesis has currently not shown the expected, broad success. This may indicate that the plasticity of cancer cells is still underestimated and that efficacious environmental targeting may require a larger, multi-modal approach. The discovery of additional targets for such anti-stromal therapies is therefore encouraging.

Thus in summary, the success of a metastatic founder is based on the support from the microenvironment in a way that mimics a familiar environment as that of the primary tumor in a distant site, provides essential factors required for the maintenance of stemness and enhanced fitness of CSCs, and can activate the angiogenic switch so that tumor cells can obtain the nutrient supply needed for continued growth until the macrometastasis state. This ensures a successful colonization and secondary tumor growth.

Acknowledgments

The authors would like to thank Dr. Ilaria Malanchi for her contribution and continuing discussions. A.S.M. would like to thank Marta Valeri, Judit, Lidia, and JM for their encouragement. J.H. and A.S.M. are supported in part by the Swiss League against Cancer, the SNF, the NCCR in Molecular Oncology, and the Anna Fuller Fund. J.H. holds the EPFL chair for Signal Transduction in Oncogenesis sponsored by Debiopharm.

References

Aft, R., Naughton, M., Trinkaus, K., Watson, M., Ylagan, L., Chavez-MacGregor, M., Zhai, J., Kuo, S., Shannon, W., Diemer, K., *et al.* (2010). Effect of zoledronic acid on disseminated tumour cells in women with locally advanced breast cancer: an open label, randomised, phase 2 trial. Lancet Oncol *11*, 421–428.

Aktas, B., Tewes, M., Fehm, T., Hauch, S., Kimmig, R., and Kasimir-Bauer, S. (2009). Stem cell and epithelial-mesenchymal transition markers are frequently overexpressed in circulating tumor cells of metastatic breast cancer patients. Breast Cancer Res *11*, R46.

Barkan, D., El Touny, L.H., Michalowski, A.M., Smith, J.A., Chu, I., Davis, A.S., Webster, J.D., Hoover, S., Simpson, R.M., Gauldie, J., *et al.* (2010). Metastatic growth from dormant cells induced by a col-I-enriched fibrotic environment. Cancer Res *70*, 5706–5716.

Bonnet, D., and Dick, J.E. (1997). Human acute myeloid leukemia is organized as a hierarchy that originates from a primitive hematopoietic cell. Nat Med *3*, 730–737.

Braun, S., Kentenich, C., Janni, W., Hepp, F., de Waal, J., Willgeroth, F., Sommer, H., and Pantel, K. (2000a). Lack of effect of adjuvant chemotherapy on the elimination of single dormant tumor cells in bone marrow of high-risk breast cancer patients. J Clin Oncol *18*, 80–86.

Braun, S., Pantel, K., Muller, P., Janni, W., Hepp, F., Kentenich, C.R., Gastroph, S., Wischnik, A., Dimpfl, T., Kindermann, G., *et al.* (2000b). Cytokeratin-positive cells in the bone marrow and survival of patients with stage I, II, or III breast cancer. N Engl J Med *342*, 525–533.

Braun, S., Vogl, F.D., Naume, B., Janni, W., Osborne, M.P., Coombes, R.C., Schlimok, G., Diel, I.J., Gerber, B., Gebauer, G., *et al.* (2005). A pooled analysis of bone marrow micrometastasis in breast cancer. N Engl J Med *353*, 793–802.

Campbell, P.J., Yachida, S., Mudie, L.J., Stephens, P.J., Pleasance, E.D., Stebbings, L.A., Morsberger, L.A., Latimer, C., McLaren, S., Lin, M.L., *et al.* (2010). The patterns and dynamics of genomic instability in metastatic pancreatic cancer. Nature *467*, 1109–1113.

Chambers, A.F., Groom, A.C., and MacDonald, I.C. (2002). Dissemination and growth of cancer cells in metastatic sites. Nat Rev Cancer *2*, 563–572.

Darini, C.Y., Pisani, D.F., Hofman, P., Pedeutour, F., Sudaka, I., Chomienne, C., Dani, C., and Ladoux, A. (2012). Self-renewal gene tracking to identify tumour-initiating cells associated with metastatic potential. Oncogene *31*, 2438–2449.

Deng, J., Liu, Y., Lee, H., Herrmann, A., Zhang, W., Zhang, C., Shen, S., Priceman, S.J., Kujawski, M., Pal, S.K., *et al.* (2012). S1PR1-STAT3 signaling is crucial for myeloid cell colonization at future metastatic sites. Cancer Cell *21*, 642–654.

Eppert, K., Takenaka, K., Lechman, E.R., Waldron, L., Nilsson, B., van Galen, P., Metzeler, K.H., Poeppl, A., Ling, V., Beyene, J., *et al.* (2011). Stem cell gene expression programs influence clinical outcome in human leukemia. Nat Med *17*, 1086–1093.

Erler, J.T., Bennewith, K.L., Cox, T.R., Lang, G., Bird, D., Koong, A., Le, Q.T., and Giaccia, A.J. (2009). Hypoxia-induced lysyl oxidase is a critical mediator of bone marrow cell recruitment to form the premetastatic niche. Cancer Cell *15*, 35–44.

Ewing J., (1928). A Treatise on Tumors, W.B. Saunders Company, Philadelphia, Pa, USA.

Eyles, J., Puaux, A.L., Wang, X., Toh, B., Prakash, C., Hong, M., Tan, T.G., Zheng, L.,

Ong, L.C., Jin, Y., *et al.* (2010). Tumor cells disseminate early, but immunosurveillance limits metastatic outgrowth, in a mouse model of melanoma. J Clin Invest *120*, 2030–2039.

Farrar, J.D., Katz, K.H., Windsor, J., Thrush, G., Scheuermann, R.H., Uhr, J.W., and Street, N.E. (1999). Cancer dormancy. VII. A regulatory role for CD8+ T cells and IFN-gamma in establishing and maintaining the tumor-dormant state. J Immunol *162*, 2842–2849.

Flatmark, K., Borgen, E., Nesland, J.M., Rasmussen, H., Johannessen, H.O., Bukholm, I., Rosales, R., Harklau, L., Jacobsen, H.J., Sandstad, B., *et al.* (2011). Disseminated tumour cells as a prognostic biomarker in colorectal cancer. Br J Cancer *104*, 1434–1439.

Hermann, P.C., Huber, S.L., Herrler, T., Aicher, A., Ellwart, J.W., Guba, M., Bruns, C.J., and Heeschen, C. (2007). Distinct populations of cancer stem cells determine tumor growth and metastatic activity in human pancreatic cancer. Cell Stem Cell *1*, 313–323.

Hess, K.R., Varadhachary, G.R., Taylor, S.H., Wei, W., Raber, M.N., Lenzi, R., and Abbruzzese, J.L. (2006). Metastatic patterns in adenocarcinoma. Cancer *106*, 1624–1633.

Hiratsuka, S., Watanabe, A., Aburatani, H., and Maru, Y. (2006). Tumour-mediated upregulation of chemoattractants and recruitment of myeloid cells predetermines lung metastasis. Nat Cell Biol *8*, 1369–1375.

Holmgren, L., O'Reilly, M.S., and Folkman, J. (1995). Dormancy of micrometastases: balanced proliferation and apoptosis in the presence of angiogenesis suppression. Nat Med *1*, 149–153.

Hood, J.L., San, R.S., and Wickline, S.A. (2011). Exosomes released by melanoma cells prepare sentinel lymph nodes for tumor metastasis. Cancer Res *71*, 3792–3801.

Husemann, Y., Geigl, J.B., Schubert, F., Musiani, P., Meyer, M., Burghart, E., Forni, G., Eils, R., Fehm, T., Riethmuller, G., *et al.* (2008). Systemic spread is an early step in breast cancer. Cancer Cell *13*, 58–68.

Jeter, C.R., Liu, B., Liu, X., Chen, X., Liu, C., Calhoun-Davis, T., Repass, J., Zaehres, H., Shen, J.J., and Tang, D.G. (2011). NANOG promotes cancer stem cell characteristics and prostate cancer resistance to androgen deprivation. Oncogene *30*, 3833–3845.

Kaplan, R.N., Riba, R.D., Zacharoulis, S., Bramley, A.H., Vincent, L., Costa, C., MacDonald, D.D., Jin, D.K., Shido, K., Kerns, S.A., *et al.* (2005). VEGFR1-positive haematopoietic bone marrow progenitors initiate the pre-metastatic niche. Nature *438*, 820–827.

Lee, T.K., Castilho, A., Cheung, V.C., Tang, K.H., Ma, S., and Ng, I.O. (2011). CD24(+) liver tumor-initiating cells drive self-renewal and tumor initiation through STAT3-mediated NANOG regulation. Cell Stem Cell *9*, 50–63.

Leis, O., Eguiara, A., Lopez-Arribillaga, E., Alberdi, M.J., Hernandez-Garcia, S., Elorriaga, K., Pandiella, A., Rezola, R., and Martin, A.G. (2012). Sox2 expression in breast tumours and activation in breast cancer stem cells. Oncogene *31*, 1354–1365.

Lilleby, W., Nesland, J.M., Fossa, S.D., Torlakovic, G., Waehre, H., and Kvalheim, G. (2003). The prognostic impact of cytokeratin-positive cells in bone marrow of patients with localized prostate cancer. Int J Cancer *103*, 91–96.

Malanchi, I., Santamaria-Martinez, A., Susanto, E., Peng, H., Lehr, H.A., Delaloye, J.F., and Huelsken, J. (2012). Interactions between cancer stem cells and their niche govern metastatic colonization. Nature *481*, 85–89.

Mathiesen, R.R., Borgen, E., Renolen, A., Lokkevik, E., Nesland, J.M., Anker, G., Ostenstad, B., Lundgren, S., Risberg, T., Mjaaland, I., *et al.* (2012). Persistence of disseminated tumor cells after neoadjuvant treatment for locally advanced breast cancer predicts poor survival. Breast Cancer Res *14*, R117.

Norris, R.A., Damon, B., Mironov, V., Kasyanov, V., Ramamurthi, A., Moreno-Rodriguez, R., Trusk, T., Potts, J.D., Goodwin, R.L., Davis, J., *et al.* (2007). Periostin regulates collagen fibrillogenesis and the biomechanical properties of connective tissues. J Cell Biochem *101*, 695–711.

Nowell, P.C. (1976). The clonal evolution of tumor cell populations. Science *194*, 23–28.

O'Connell, J.T., Sugimoto, H., Cooke, V.G., MacDonald, B.A., Mehta, A.I., LeBleu, V.S., Dewar, R., Rocha, R.M., Brentani, R.R., Resnick, M.B., *et al.* (2011). VEGF-A and Tenascin-C produced by S100A4+ stromal cells are important for metastatic colonization. Proc Natl Acad Sci USA *108*, 16002–16007.

Oskarsson, T., Acharyya, S., Zhang, X.H., Vanharanta, S., Tavazoie, S.F., Morris, P.G., Downey, R.J., Manova-Todorova, K., Brogi, E., and Massague, J. (2011). Breast cancer cells produce tenascin C as a metastatic niche component to colonize the lungs. Nat Med *17*, 867–874.

Paget, S. (1889). The distribution of secondary growths in cancer of the breast. Cancer and Metastasis Reviews *8*, 98–101.

Pang, R., Law, W.L., Chu, A.C., Poon, J.T., Lam, C.S., Chow, A.K., Ng, L., Cheung, L.W., Lan, X.R., Lan, H.Y., *et al.* (2010). A subpopulation of CD26+ cancer stem cells with metastatic capacity in human colorectal cancer. Cell Stem Cell *6*, 603–615.

Peinado, H., Aleckovic, M., Lavotshkin, S., Matei, I., Costa-Silva, B., Moreno-Bueno, G., Hergueta-Redondo, M., Williams, C., Garcia-Santos, G., Ghajar, C.M., *et al.* (2012). Melanoma exosomes educate bone marrow progenitor cells toward a pro-metastatic phenotype through MET. Nat Med *18*, 883–91.

Psaila, B., and Lyden, D. (2009). The metastatic niche: adapting the foreign soil. Nat Rev Cancer *9*, 285–293.

Reya, T., Morrison, S.J., Clarke, M.F., and Weissman, I.L. (2001). Stem cells, cancer, and cancer stem cells. Nature *414*, 105–111.

Reuben, J.M., Lee, B.N., Gao, H., Cohen, E.N., Mego, M., Giordano, A., Wang, X., Lodhi, A., Krishnamurthy, S., Hortobagyi, G.N., *et al.* (2011). Primary breast cancer patients with high risk clinicopathologic features have high percentages of bone marrow epithelial cells with ALDH activity and CD44(+) CD24(lo) cancer stem cell phenotype. Eur J Cancer *47*:1527–36.

Shiozawa, Y., Pedersen, E.A., Havens, A.M., Jung, Y., Mishra, A., Joseph, J., Kim, J.K., Patel, L.R., Ying, C., Ziegler, A.M., *et al.* (2011). Human prostate cancer metastases target the hematopoietic stem cell niche to establish footholds in mouse bone marrow. J Clin Invest *121*, 1298–1312.

Somervaille, T.C., Matheny, C.J., Spencer, G.J., Iwasaki, M., Rinn, J.L., Witten, D.M., Chang, H.Y., Shurtleff, S.A., Downing, J.R., and Cleary, M.L. (2009). Hierarchical maintenance of MLL myeloid leukemia stem cells employs a transcriptional program shared with embryonic rather than adult stem cells. Cell Stem Cell *4*, 129–140.

Tarin, D., Price, J.E., Kettlewell, M.G., Souter, R.G., Vass, A.C., and Crossley, B. (1984). Mechanisms of human tumor metastasis studied in patients with peritoneovenous shunts. Cancer Res *44*, 3584–3592.

Theodoropoulos, P.A., Polioudaki, H., Agelaki, S., Kallergi, G., Saridaki, Z., Mavroudis, D., and Georgoulias, V. (2010). Circulating tumor cells with a putative stem cell phenotype in peripheral blood of patients with breast cancer. Cancer Lett *288*, 99–106.

Viadana, E., Bross, I.D., and Pickren, J.W. (1978). The metastatic spread of cancers of the digestive system in man. Oncology *35*, 114–126.

Virchow, R. (1858). Die Cellularpathologie in ihrer Begründung auf physiologische und pathologische Gewebelehre. Verlag von August Hirschwald, Berlin, Germany.

Weiss, L. (1992). Comments on hematogenous metastatic patterns in humans as revealed by autopsy. Clin Exp Metastasis *10*, 191–199.

Chapter 20
Role of Cancer Stem Cells in Metastasis

Giovanna Merchand-Reyes[1], Rosana Pelayo[2], Lenin Pavón[3], Richard G. Pestell[4], and Marco Velasco-Velázquez[5]
[1] *Unit for Research and Development in Bioprocesses (UDIBI), National School of Biological Sciences, National Polytechnic Institute, Mexico*
[2] *Oncology Research Unit, Mexican Institute for Social Security, Mexico City, Mexico*
[3] *Department of Psychoimmunology, National Institute of Psychiatry "Ramón de la Fuente", Mexico*
[4] *Department of Cancer Biology, Kimmel Cancer Center, Thomas Jefferson University, Philadelphia, PA, USA*
[5] *Department of Pharmacology, School of Medicine, National Autonomous University of Mexico, Mexico*

Cancer Stem Cells

Cancer stem cells (CSC) are a subpopulation of tumor cells that possess several fundamental attributes similar to normal adult stem cells (SC). CSCs are capable of dividing asymmetrically to produce one stem cell, which enables self-renewal, and one progenitor cell, which allows them to produce phenotypically diverse cancer cells that constitute tumors. In addition, CSCs display enhanced capacity for tumor generation when injected into mice and increased resistance to chemo- and radiotherapy.

The concept of cancer stem cells was first documented for human leukemia types, when Dick and colleagues identified $CD34^+$/$CD38^-$ leukemia stem cells (LSC) by using fluorescence-activated cell sorting (FACS) to prospectively isolate acute myeloid leukemia (AML) cells able to initiate leukemia in transplanted mice (Bonnet and Dick, 1997; Lapidot et al., 1994). $CD34^+CD38^+$ and $CD34^-$ fractions did not contain leukemia initiating cells (LIC), indicating a hierarchical organization of the disease (Bonnet and Dick, 1997; Lapidot et al., 1994). Subsequently, CSCs have been identified in several solid tumors, such as colon (O'Brien et al., 2007; Ricci-Vitiani et al., 2007), breast (Al-Hajj et al., 2003), brain (Singh et al., 2004), pancreas (Hermann et al., 2007), and other cancers. Identification of CSCs is possible by the analysis of expression of specific markers (Table 20.1). Molecular genetic analysis in murine gene knockout or mutant allele models has identified key genes governing CSC in solid tumors including the cell-cycle control protein p21 Cip1 (Liu et al., 2009), cyclin D1, and p27 Kip1 proteins; cell signaling pathways including the canonical (Liu et al., 2010) and non-canonical NFkB pathways (Cao et al., 2007), transcription factors (i.e., c-jun) (Jiao et al., 2010) and the cell fate determination pathway factors (i.e., DACH1) (Wu et al., 2011).

Metastatic Properties of CSCs

Hierarchal theory of cancer development sustains that CSCs support the initiation, maintenance, and clinical outcome of tumors (Kakarala and Wicha, 2008). Since the discovery of CSCs, several authors have proposed that CSCs may also constitute the subpopulation of tumor cells responsible for

Cancer Stem Cells, First Edition. Edited by Vinagolu K. Rajasekhar.

Table 20.1. CSC markers described for different malignancies

Malignancy	CSC markers
Acute myeloid leukemia	$CD34^{+}/CD38^{-}$ (Bonnet and Dick, 1997)
Pancreatic cancer	$CD133^{+}$ (Hermann et al., 2007)
Breast cancer	$CD44^{+}/CD24^{-/low}/Lineage^{-}$ (Al-Hajj et al., 2003) $ALDH^{+}$ (Ginestier et al., 2007)
Colorectal cancer	$CD133^{+}/CD44^{+}$ (O'Brien et al., 2007; Ricci-Vitiani et al., 2007)
Glioma	$CD133^{+}$ (Singh et al., 2004)

local invasion and the development of metastatic tumors (Allan et al., 2006; Li et al., 2007; Wicha, 2006). Such a proposal is supported by different experimental evidence presented below.

Lessons from Hematopoietic Stem Cell (HSC) Migration

Indirect support of the participation of CSCs in metastasis comes from the study of HSC migration. Although both the most primitive HSC and differentiating hematopoietic progenitor cells (HPC) reside in the bone marrow (BM), a constant exchange of migratory HSC/HPC within BM and between BM and peripheral tissues occurs under physiological conditions (Mazo et al., 2011). Inside the marrow, primitive cells losing their quiescence migrate from an endosteal niche formed by osteoblasts lining the bone surface, toward the interstitium to proliferative reticular and vascular niches (Purizaca et al., 2012). Some of them enter the blood to migrate into multiple peripheral tissues. Parabiosis and adoptive cell transfer studies, as well as intravital and multiphoton microscopy in mice models, have been combined to dissect the mechanisms of normal HSC/HPC migration (Mazo et al., 2011). They travel from extramedullary tissues into the lymph, and from the lymph via the thoracic duct, back into the blood (Massberg et al., 2007; Mazo et al., 2011; Welner and Kincade, 2007) (Figure 20.1). Normal HSC/HPC can either reenter the BM or repeat the peripheral migration cycle. It is believed that their constant circulation contribute to the maintenance of hematopoiesis, although it remains unclear whether all HSC/HPC have the potential to patrol peripheral tissues (Welner and Kincade, 2007). N-cadherin, integrins, and CD44 are involved in cell retention within BM; the CXCL12/CXCR4 axis is crucial for BM colonization and migration to periphery, whereas exiting peripheral organs via lymphatics is dependent on gradients of the sphingolipid sphingosine 1-phosphate (S1P) (Massberg et al., 2007). HSC/HPC mobilization from the BM implies upregulation of proteolytic enzymes such as matrix metalloproteinase MMP-9, which activation promotes the cleavage of CXCL12 and VCAM-1, detaching the stromal-hematopoietic cell contacts. Many of the factors known to govern HSC migration also play an important role in metastasis of diverse cancer types.

The Metastatic Phenotype of CSCs

Current information indicates that CSCs are more invasive and metastatic than other tumor cells. In breast CSCs ($CD44^{+}/CD24^{-/low}$) the expression of genes involved in invasion (IL-1α, IL-6, IL-8, CXCR4, MMP-1, and UPA) is elevated as well as their invasive properties (Sheridan et al., 2006; Shipitsin et al., 2007). Interestingly, in clinical samples the $CD44^{+}/CD24^{-/low}$ population is preferentially found in the metastatic tissue (Kai et al., 2011), and their presence in the primary tumors is associated to a poor prognosis (Idowu et al., 2012). ALDH-positive cells isolated from breast

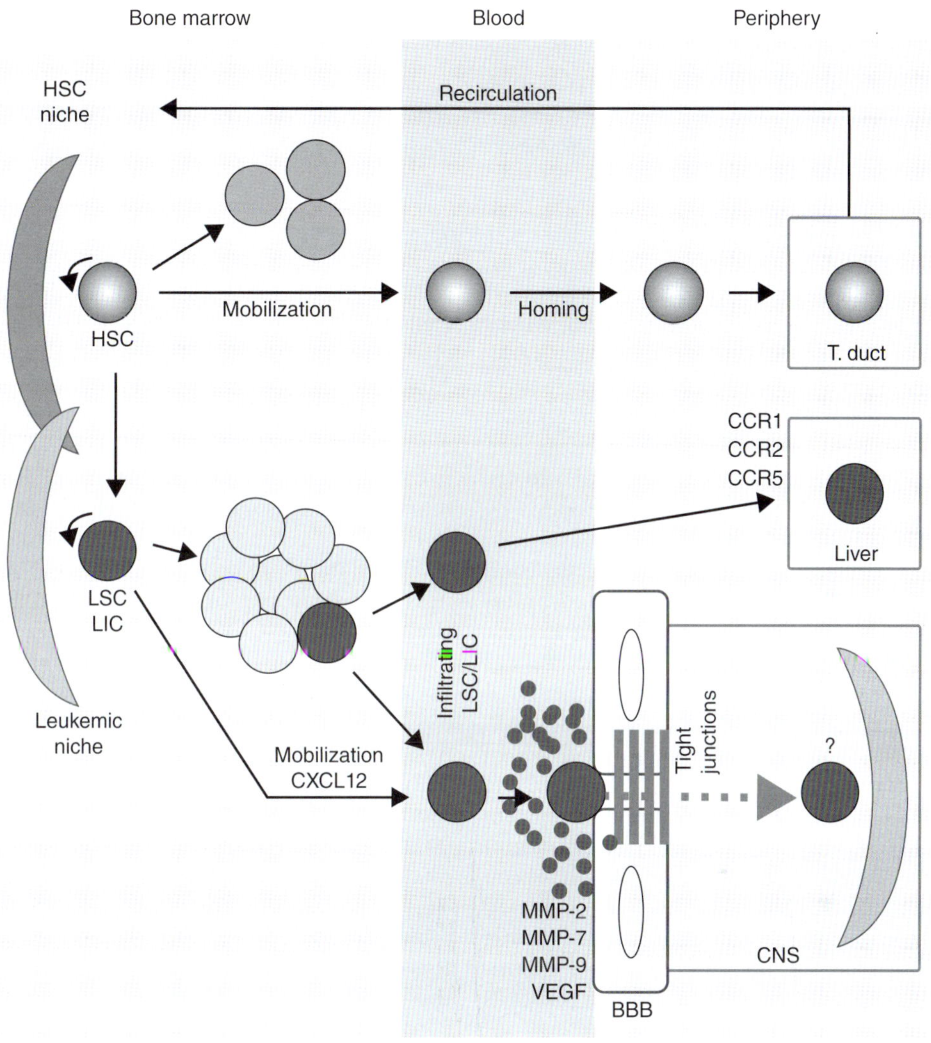

Figure 20.1. Trafficking of normal HSC and infiltrating leukemia SCs (LSC/LIC). During migration, normal hematopoietic stem cells (HSC) or infiltration of leukemic stem/initiating cells LSC/LIC leave their niche and enter circulation to migrate into multiple peripheral tissues. High expression of CXCL12 within leukemic BM contributes for mobilization of LSC/LIC expressing CXCR4. Display of a number of cytokines/chemokines by conspicuous populations of LSC/LIC is crucial for multi-organ infiltration, including liver. LSC/LIC production of MMP may break down the blood-brain barrier through disruption of epithelial tight junctions. Niche-to-niche migration during leukemia infiltration/metastasis is suggested. For color detail, please see color plate section.

cancer cell lines are more migratory and invasive than the ALDH-negative cells (Croker et al., 2009; Charafe-Jauffret et al., 2009). The ALDH-positive cells generated metastases at distinct organs when injected to into non-obese diabetic-severe combined immunodeficient (NOD/SCID) mice, whereas ALDH-negative cells produced only occasional metastases limited to lymph nodes (Charafe-Jauffret et al., 2010; Charafe-Jauffret et al., 2009). Accordingly, in MMTV-PyMT transgenic mice, breast CSCs ($CD24^+/CD90^+$) are the only cells capable of forming lung metastasis after tail vein injection (Malanchi et al., 2012). Together, these studies indicate that the metastatic ability of breast tumors is predominantly mediated by CSCs.

Leukemia SCs produce vascular endothelial growth factor (VEGF) and express functional hemangioblast-associated signaling tyrosine kinases such as VEGFR-1 and VEGFR2, resulting in the generation of an autocrine loop that may support leukemic cell survival, proliferation, MMP activation, and migration (Dias et al., 2000). Disruption of VEGFR-2 signaling with a blocking antibody inhibited tumor growth and metastasis in a murine model of human promyelocytic leukemia (Dias et al., 2000). Thus, an autocrine loop generated by VEGF/VEGFR-2 may constitute a crucial axis participating in leukemic SC invasion (Figure 20.1). Invasion of the blood-brain-barrier by leukemia requires high expression of MMP-2 and -9, which correlates with the invasive capability that is mediated by disruption of tight junctions between adjacent brain microvascular endothelial cells (Feng et al., 2011). Accordingly, the clinical signs of central nervous system (CNS) leukemia are reduced upon inhibition of MMP (Feng et al., 2011). It has been speculated that the same mechanism is used by conspicuous populations of leukemia SCs to increase the permeability of blood-brain-barrier during infiltration of the CNS (Figure 20.1).

However, at present it in not clear whether all CSCs are equally efficient at producing metastasis. Recent studies have identified within CSC-specific subpopulations with different metastatic properties. In colon CSCs ($CD133^+/CD44^+$) there are two different subpopulations defined by CD26 expression being the $CD133^+/CD44^+/CD26^+$ cells more metastatic than their $CD133^+/CD44^+/CD26^-$ counterpart (Pang et al., 2010). Clinically, the presence of $CD26^+$ cells in the primary tumors predicts future metastasis (Pang et al., 2010). Similarly, in pancreatic CSCs ($CD133^+$), the coexpression of CXCR4 favors liver metastasis, and this process is abrogated when CXCR4 is blocked by CXCR4 antagonist AMD3100 (Hermann et al., 2007).

Role of Epithelial-to-Mesenchymal Transition (EMT) in CSC-Driven Metastasis

The plasticity intrinsic to stem cells may facilitate EMT, a process required in tumor cells for local and systemic dissemination. Cells that have undergone EMT have many characteristics of SCs (Mani et al., 2008; Morel et al., 2008). Interestingly, the pro-metastatic $CD26^+$ subpopulation within colon CSCs (see above) display reduced E-cadherin expression and express EMT markers such as N-cadherin (Pang et al., 2010).

The relationship between EMT and 'stemness' has been explored in breast tumors. Breast CSCs ($CD44^+/CD24^{-/low}$) express many genes associated with cells that have undergone EMT (Mani et al., 2008; Scheel et al., 2011). Tumor growth factor β (TGF-β) signaling is responsible for inducing EMT in breast cancer cells and for maintaining their mesenchymal/CSC states (Scheel et al., 2011). TGF-β induces Pez, a tyrosine phosphatase, which in turn promotes EMT and TGF-β production, creating a positive autocrine feedback loop. Introduction of the Ras oncogene sensitize cells to EMT, and the addition of TGF-β1 accelerates this conversion, resulting in the appearance of tumorigenic $CD44^+/CD24^{-/low}$ cells (Morel et al., 2008). Another EMT inducer in breast cancer cells is CD146. High levels of CD146 control the activation of transcriptional factor Slug, downregulating E-cadherin and cytokeratin, and upregulating vimentin and fibronectin (Zeng et al., 2012). Besides promoting EMT, CD146 also contributes to the acquisition of breast CSC phenotype by increasing CD44 and reducing CD24 expressions (Zeng et al., 2012).

EMT also promotes invasiveness in CSCs. For example, the protein Caveolin-1 is a negative regulator of EMT in different CSCs. We showed that induction of Caveolin-1 expression in a pancreatic cancer cell line reduces its tumorigenicity and metastatic capability (Salem et al., 2011). Caveolin-1 expression suppressed the AKT pathway and downregulated Snail, promoting cell-to-cell contact and elevated expression of membrane E-cadherin and β-catenin (Salem et al., 2011). Suppression of caveolin-1 expression restored the migration of cancer cells (Salem et al., 2011). In breast cancer cells, we showed that the P132L mutation of caveolin-1 acts as a gain-of-function in migration and invasion (Bonuccelli et al., 2009). Caveolin-1 P132L mutation generates a specific expression signature that contains signaling pathways involved in migration and invasion, stem cell signaling, and EMT induction (Bonuccelli et al., 2009). In agreement, inhibition of EMT in tumor cells can suppress their metastatic ability. NOP14 is a nucleolar protein homolog that processes ribosomal RNA assembly. NOP14 expression correlates with an invasive phenotype in pancreatic cells, and inhibition of its expression increases E-cadherin, downregulates vimentin, MMP-9, and RhoA, and impairs liver metastasis formation in vivo (Zhou et al., 2012).

The Niche Regulates Stemness and Invasiveness

The term niche was adapted to SCs biology by Ray Shofield in 1978, and it makes reference to a microenvironment where stem cells reside. Shofield identified in animal models that transference of hematological primitive cells to the spleen resulted in the lost of their colony-formation capacity. He concluded that SCs resided in another place, probably in the BM, associated with other cells (Schofield, 1978). Further works with stem cells from *Drosophila melanogaster* and *Caenorhabditis elegans* helped to understand the composition of niches. In D*rosophila melanogaster,* the ovarian niches host have two main populations of stem cells, somatic stem cells and germ line stem cells, which are in close relationship with filament, cap, and inner sheath cells. After division, stem cells closer to cap cells keep their phenotype; this is related to the expression by cap cells of molecules like *Hedgehog*, which controls proliferation and differentiation of somatic stem cells, and *dpp*, which inhibits their differentiation and maintains the stem cell population (Jones, 2001; López-Onieva et al., 2008; Xie and Spradling, 2000). Similarly, in *Caenorhabditis elegans*, the expression of the *lag-2* gene (now recognized as a Notch ligand) at the membrane of somatic distal tip cells in the gonad environment is related to lateral signaling and stem cell division (Bertrand et al., 2011; Henderson et al., 1994).

The microenvironment at the niche also controls SC number, proliferation, apoptosis, and differentiation in mammals (Borovski et al., 2011; Lam and Adams, 2010). At present it is know that the microenvironment that constitutes the niche is composed of cells, extracellular matrix proteins, and soluble signals. In the HSC niche, osteoblasts induce the HSC population in a process dependent of N-cadherin and Jagged-1 expression. Furthermore, osteoblasts also produce G-CSF, M-CSF, GM-CSF, IL-1, IL-6, LIF, TNF-α, TGF-β, and TNF-β, giving paracrine signals (Lam and Adams, 2010). Similarly, endothelial cells regulate migration and maintenance of HSCs by expressing adhesion molecules like E-selectin, VCAM-1, ICAM-1, PECAM, and CD34 and by producing CXCL12 (Kiel and Morrison, 2006). Extracellular matrix proteins, such as fibronectin, laminin, vitronectin, thrombospondin, and collagens, are also essential components of the SC niche, affecting the homing and maturation of progenitor cells (Lam and Adams, 2010).

The Primary Niche of CSCs Promotes Metastasis

Normal homeostasis is unbalanced in the tumor microenvironment; thus, the tumor niche cannot be equal to the normal SCs niche. The main differences in the tumor niche are the presence of immune and mesenchymal cells that induce chronic inflammation and angiogenesis (Lorusso and Ruegg, 2008). This particular microenvironment not only favors tumor growth but contributes to invasiveness and therapy resistance in CSCs (Borovski et al., 2011; Jinushi et al., 2011).

Cellular Components of the Primary Niche

Immune cells can participate in the generation of microenvironments that promote tumor development by secreting paracrine factors. For example, mast cells that infiltrate tumors in the early phases of establishment release their granules in response to damage-associated molecular patterns (DAMP) via toll-like receptors (TLR) (Liu et al., 2011). The release of proteases remodels the tissue and hydrolyzes chemokines and cytokines, whereas histamine secretion activates infiltrating dendritic and T cells, producing chronic inflammation by releasing IL-6, IL-8, or IL-1β (Liu et al., 2011). Furthermore, mast cells can produce IL-10 and TGF-β, causing immune suppression (Liu et al., 2011).

Tumor-associated macrophages (TAM) contribute to an anti-inflammatory microenvironment and a poor clinical prognosis in ovarian, cervical, and lung carcinomas (Hao et al., 2012). Macrophages arrive at the tumor environment attracted by chemokines like CCL2, CCL5, CXCL12, or VEGF secreted by fibroblast in the niche or CSCs (Hao et al., 2012). Macrophages secrete COX-2-promoting angiogenesis and CSCs maintenance by stabilization of β-catenin (Lorusso and Ruegg, 2008; Porta et al., 2011). Close contact with CSCs makes F40/80$^+$ macrophages produce high quantities of milk-fat globule EGF-8 (MFG-E8), a growth factor associated with immune tolerance, phagocytosis, and angiogenesis. Reciprocally, MFG-E8 is capable of stimulating colon CSCs via activation of Stat3 and Sonic Hedgehog signaling pathways (Jinushi et al., 2011).

Cancer-associated fibroblasts (CAF) are key regulators of stemness and invasiveness in the primary tumor niche. CAFs, in response to breast cancer cell-secreted cytokines, increase CCL2 expression, which in turn activates NOTCH1 signaling and the CSC phenotype in breast tumor cells (Tsuyada et al., 2012). In prostate cancer, CAFs act on cancer cells by activating pro-inflammatory pathways involving nuclear factor kappa B (NF-κB), COX-2, and HIF-1 that lead to EMT and a stem-like phenotype (Giannoni et al., 2011). COX-2 activation in cancer cells increases reactive oxygen species (ROS) delivery, promoting dissemination of metastatic cells. Accordingly, targeting of the proinflammatory signature by interference RNA impairs metastasis (Giannoni et al., 2011). The presence of ROS in the microenvironment not only affects cancer cells, but influences CAFs (Lisanti et al., 2011). In CAFs, we have shown that ROS generate autophagy and mitophagy, turning their metabolism to produce L-lactate, glutamine, and free fatty acids that can be consumed by cancer cells in a process we called "reverse Warburg effect" (Lisanti et al., 2011; Salem et al., 2012). We showed that the presence of L-lactate and ketones is associated with metastasis in breast cancer and upregulates genes that maintain the CSC phenotype (Martinez-Outschoorn et al., 2011).

Cross Talk Between CSCs and Their Secondary Niches

According to Paget's hypothesis, metastatic processes result from cross talk between those tumor cells having special attributes (the 'seeds') and specific organ microenvironment (the 'soil') (Fidler, 2003), making metastasis determined by the biology of neoplastic cells and the microenvironment of the host tissue. Therefore, a permissive microenvironment is fundamental for the establishment of metastatic lesions at distant sites just as at the primary tumor site.

The Premetastatic Niche

Changes in the microenvironment of the target organ during metastasis can be divided in two, according to the time course:

premetastasis, where the site of invasion is altered, and metastasis, where cancer cells reside in the site of invasion and develop successfully (Hiratsuka et al., 2006).

The formation of premetastatic niches is conducted by an unique array of specific chemokines, cytokines and proteases, including VEGF, osteopontin, CXCL12, CCL2, CCL5, and MMP-1,2,3,9,11,14, among others (Kaplan et al., 2005; Wels et al., 2008). Diverse cellular components participate in establishing such a microenvironment. For example, mesenchymal stem cells (MSC) increase the metastatic potential by releasing chemotactic signals at metastatic sites. MSCs accumulate in the tumor and lung when injected intravascularly to osteosarcoma-bearing mice (Xu et al., 2009). MSCs increase the rate of lung metastasis by osteosarcoma and breast cancer cell, via secretion of CCL5 (Karnoub et al., 2007; Xu et al., 2009). The secretion of CCL5 by MSC requires tumor-derived osteopontin (Mi et al., 2011), indicating that the interaction between CSCs and their premetastatic niches is reciprocal. MSCs also stimulate breast cancer cell migration, invasion, and metastasis through the secretion of CXCL12 and tenascin C (Corcoran et al., 2008; Mi et al., 2011).

Mobilization of bone marrow-derived primitive progenitors to target organs precedes the arrival of primary cancer cells. These circulating VEGFR-1$^+$ BM cells are responsible for producing MMP-9, leading to an accelerated cell extravasation (Kaplan et al., 2005). Interestingly, BM cells prime different metastatic sites depending on the tumor type, localization of the primary tumor, and VEGF levels (Kaplan et al., 2005).

Macrophages also accumulate at the premetastatic site. TNF-α, TGF-β, and VEGF secreted by primary tumors stimulate the expression of the chemotactic proteins S100A8 and S100A9 by lung endothelial cells. S100A8 and S100A9 attract Mac$^+$ myeloid cells to the metastatic site and induce them to produce SAA3, a molecule of the heat shock protein 70 family (Hiratsuka et al., 2006). Then, SAA3 provokes macrophage accumulation and their response via TLR4; this activation generates signaling via NF-κB, promoting metastasis by activating genes like the CXCL2 chemokine (Erkel et al., 2007; Hiratsuka et al., 2008).

Infiltration of the Premetastatic Niche: Role of Receptors

In order to respond to the signals located at the metastatic site, CSCs must express appropriate receptors. Co-expression of stem cell markers and receptors that control migration/invasion has been reported in different cancer types. ALDH-positive cells isolated from breast cancer cell lines express the IL-8 receptor CXCR1, which controls chemotaxis (Charafe-Jauffret et al., 2009). IL-8 (Wu et al., 2008) or CXCR1 (Ginestier et al., 2010) blockage reduces breast CSC population and metastasis in vivo, highlighting the role of IL-8/CXCR1 axis in breast cancer metastasis.

CXCR4, the CXCL12 receptor, is expressed in cancer cells allowing them to respond to CXCL12 gradients originating in distant tissues, like lymph nodes, liver, lungs, or bones. In breast and ovarian cancer cells, CXCR4 controls their mobilization to the BM (Schuettpelz and Link, 2011). Acute myeloid leukemia (AML) SCs have shown remarkable similarities to normal HSC/HPC (Deng and Zhang, 2010; Greve et al., 2012). Accordingly, the CXCL12/CXCR4 axis controls leukemia SCs migration and homing as it does in HSC (Tavor et al., 2004).

In patients with metastatic colon cancer, some cells isolated of the peritoneum express CSC markers, such as CD133, CD44, and CD166. Genetic analysis of those cells found upregulation in stemness-related programs (i.e., Notch and Wnt), as well as in CXCL12 and CXCR4 expression (Botchkina et al., 2009). Further studies identified that the subpopulation of colon CSCs that express CXCR4$^+$ is more metastatic in vivo than their CXCR4$^-$ counterpart (Zhang et al., 2012). Increased numbers of CXCR4$^+$

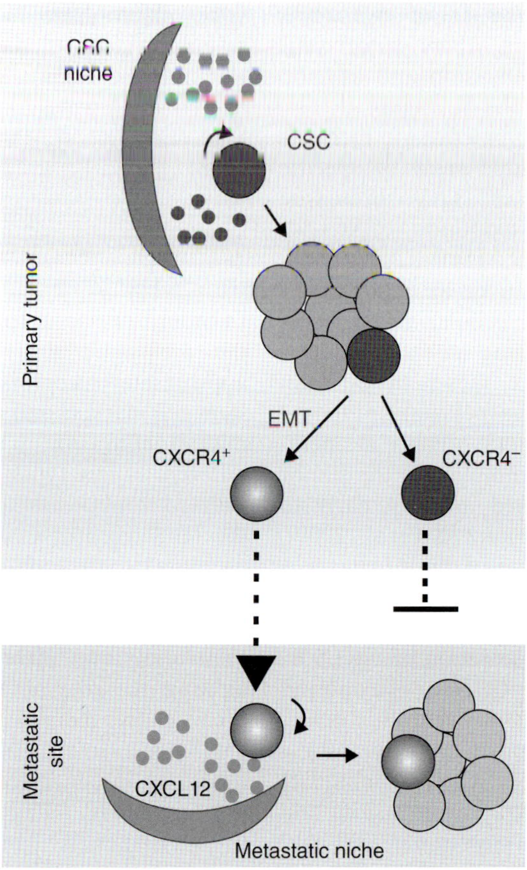

Figure 20.2. CSCs require the expression of chemotactic receptors to metastasize. Primary microenvironment supports self-renewal and differentiation of CSCs. However, CSCs require EMT-associated phenotypical changes in order to become metastatic. The chemokine receptor CXCR4 plays a key role for the invasion into secondary tissues of leukemia, colon, and pancreatic CSCs. For color detail, please see color plate section.

CSCs in colorectal tumors correlate with a reduced survival rates, since those cells favors liver metastasis (Zhang et al., 2012). Similarly, in mice transferred with pancreatic CSCs, $CD133^+/CXCR4^+$ cells localize to the bloodstream whereas $CD133^+/CXCR4^-$ cells are mainly found in the primary tumor (Hermann et al., 2007) (Figure 20.2).

CSCs Affect Their Metastatic Niche After Colonization

It has been hypothesized that when dormant CSCs are in the metastatic sites, their survival, proliferation, and/or differentiation can be activated by deregulated bidirectional signaling within the niche (Li and Neaves, 2006; Wicha et al., 2006). The recent clarification of the roles of tenascin C and periostin in breast cancer metastasis support this theory. Tenascin C, an extracellular glycoprotein, is produced by breast cancer cells at the metastatic site, where it is essential to support early metastasis outgrowth (Oskarsson et al., 2011). In addition, infiltrating breast cancer cells induce periostin expression by CAFs in the secondary target organ via TGF-β3 (Malanchi et al., 2012).

In clinical samples of breast cancer patients, periostin is expressed in 75% of lymph node metastasis (Malanchi et al., 2012). Tenascin and periostin cooperate to activate Wnt signaling at the metastatic niche, which allows CSC maintenance and expansion (Malanchi et al., 2012; Oskarsson et al., 2011). Thus, CSCs support their own metastatic niche by expressing or inducing the expression of extracellular molecules.

In particular cases, the target organ cells participate in the creation of a receptive microenvironment for CSCs. In a model of myeloproliferative disease-like myeloid leukemia (MPD-ML), it has been shown that CCR1, CCR2, and CCR5 are highly expressed by the liver cells of leukemic animals, suggesting an important role in the retention of leukemic cells in the liver (Bigildeev et al., 2011). At least 4% of the cells that invaded the liver were leukemic SCs that have activated the NF-κB pathway. Of particular interest, IL1a, which is one of the most potent activators of NF-κB, and a number of genes targeted by NF-κB such as VCAM1, ICAM1, Cyclin D1, and Myc, were all upregulated. Moreover, high expression levels of transcription factors associated with myeloid differentiation were upregulated (Bigildeev et al., 2011). Thus, the activation of NF-κB pathway in the liver leukemic SCs favors their retention at extramedullary tissues.

Conclusions

Metastasis is the principal cause of death in cancer patients. CSCs are responsible of metastasis in different cancer models since they show increased tumorigenicity and invasiveness. The microenvironment plays a key role in the establishment of CSCs in secondary organs. Within the primary and metastatic niches, there are intricate networks of bidirectional CSC–stroma signaling with multiple feedback loops. Those signals promote metastasis by inducing phenotypical changes in both the cells comprising the niches and the CSCs. The elucidation of the cellular and molecular mechanisms that regulate CSC-driven metastasis will provide new anti-metastatic targets and prognostic markers. For example, preventing de novo niche formation may be an efficient strategy for the treatment of metastatic disease. However, to develop such a strategy, it will be crucial to identify the roles of niche components in the metastatic context and distinguish them from those in normal endogenous functions.

Acknowledgments

This work was supported in part by CONACYT 252826, PIFI-IPN 20120870 (G. M-R.), CONACYT CB-2010-01-152695 (R.P.), INPRF NC092318.0, NC092318.1, PFT IC-10-002 (L.P.), PAPIIT-UNAM IN219613, a grant from "Instituto Científico Pfizer, México" (M.V-V.), the Dr. Ralph and Marian C. Falk Medical Research Trust, a grant from Pennsylvania Department of Health (R.G.P.), and the following NIH grants: R01CA086072, R01CA132115, and 1R01CA137494 (R.G.P.). The Kimmel Cancer Center was supported by the NIH Cancer Center Core grant P30CA56036 (R.G.P.).

References

Al-Hajj, M., Wicha, M.S., Benito-Hernandez, A., Morrison, S.J., and Clarke, M.F. (2003). Prospective identification of tumorigenic breast cancer cells. Proceedings of the National Academy of Sciences 100, 3983–3988.

Allan, A.L., Vantyghem, S.A., Tuck, A.B., and Chambers, A.F. (2006). Tumor dormancy and cancer stem cells: implications for the biology and treatment of breast cancer metastasis. Breast Dis 26, 87–98.

Bertrand, V., Bisso, P., Poole, Richard J., and Hobert, O. (2011). Notch-Dependent Induction of Left/Right Asymmetry in C. elegans Interneurons and Motoneurons. Current Biology 21, 1225–1231.

Bigildeev, A.E., Shipounova, I.N., Svinareva, D.A., and Drize, N.J. (2011). Leukemia cells invading the liver express liver chemokine receptors and possess characteristics of leukemia stem cells in mice with MPD-like myeloid leukemia. Experimental Hematology 39, 187–194.

Bonnet, D., and Dick, J.E. (1997). Human acute myeloid leukemia is organized as a hierarchy that originates from a primitive hematopoietic cell. Nat Med 3, 730–737.

Bonuccelli, G., Casimiro, M.C., Sotgia, F., et al. (2009). Caveolin-1 (P132L), a common breast cancer mutation, confers mammary cell invasiveness and defines a novel stem cell/metastasis-associated gene signature. Am J Pathol 174, 1650–1662.

Borovski, T., De Sousa E Melo, F., Vermeulen, L., and Medema, J.P. (2011). Cancer Stem Cell Niche: The Place to Be. Cancer Research 71, 634–639.

Botchkina, I.L., Rowehl, R.A., Rivadeneira, D.E., et al. (2009). Phenotypic Subpopulations of Metastatic Colon Cancer Stem Cells: Genomic Analysis. Cancer Genomics—Proteomics 6, 19–29.

Cao, Y., Luo, J.L., and Karin, M. (2007). IkappaB kinase alpha kinase activity is required for self-renewal of ErbB2/Her2-transformed mammary tumor-initiating cells. Proc Natl Acad Sci USA 104, 15852–15857.

Corcoran, K.E., Trzaska, K.A., Fernandes, H., et al. (2008). Mesenchymal stem cells in early entry of breast cancer into bone marrow. PLoS ONE 3, e2563.

Croker, A.K., Goodale, D., Chu, J., et al. (2009). High aldehyde dehydrogenase and expression of cancer stem cell markers selects for breast cancer cells with enhanced malignant and metastatic ability. J Cell Mol Med 13, 2236–2252.

Charafe-Jauffret, E., Ginestier, C., Iovino, F., et al. (2010). Aldehyde dehydrogenase 1-positive cancer stem cells mediate metastasis and poor clinical outcome in inflammatory breast cancer. Clin Cancer Res 16, 45–55.

Charafe-Jauffret, E., Ginestier, C., Iovino, F., et al. (2009). Breast cancer cell lines contain functional cancer stem cells with metastatic capacity and a distinct molecular signature. Cancer Res 69, 1302–1313.

Deng, C.H., and Zhang, Q.P. (2010). Leukemia stem cells in drug resistance and metastasis. Chin Med J (Engl) 123, 954–960.

Dias, S., Hattori, K., Zhu, Z., et al. (2000). Autocrine stimulation of VEGFR-2 activates human leukemic cell growth and migration. J Clin Invest 106, 511–521.

Erkel, G., Wisser, G., and Anke, T. (2007). Influence of the fungal NF-κB inhibitor panepoxydone on inflammatory gene expression in MonoMac6 cells. International Immunopharmacology 7, 612–624.

Feng, S., Cen, J., Huang, Y., et al. (2011). Matrix Metalloproteinase-2 and -9 Secreted by Leukemic Cells Increase the Permeability of Blood-Brain Barrier by Disrupting Tight Junction Proteins. PLoS ONE 6, e20599.

Fidler, I.J. (2003). The pathogenesis of cancer metastasis: the 'seed and soil' hypothesis revisited. Nat Rev Cancer 3, 453–458.

Giannoni, E., Bianchini, F., Calorini, L., and Chiarugi, P. (2011). Cancer associated fibroblasts exploit reactive oxygen species through a proinflammatory signature leading to epithelial mesenchymal transition and stemness. Antioxid Redox Signal 14, 2361–2371.

Ginestier, C., Hur, M.H., Charafe-Jauffret, E., et al. (2007). ALDH1 is a marker of normal and malignant human mammary stem cells and a predictor of poor clinical outcome. Cell Stem Cell 1, 555–567.

Ginestier, C., Liu, S., Diebel, M.E., et al. (2010). CXCR1 blockade selectively targets human breast cancer stem cells in vitro and in xenografts. J Clin Invest 120, 485–497.

Greve, B., Kelsch, R., Spaniol, K., Eich, H.T., and Götte, M. (2012). Flow cytometry in cancer stem cell analysis and separation. Cytometry Part A 81A, 284–293.

Hao, N.-B., Lü, M.-H., Fan, Y.-H., et al. (2012). Macrophages in Tumor Microenvironments and the Progression of Tumors. Clinical and Developmental Immunology 2012, 11.

Henderson, S.T., Gao, D., Lambie, E.J., and Kimble, J. (1994). lag-2 may encode a signaling ligand for the GLP-1 and LIN-12 receptors of C. elegans. Development 120, 2913–2924.

Hermann, P.C., Huber, S.L., Herrler, T., et al. (2007). Distinct Populations of Cancer Stem Cells Determine Tumor Growth and Metastatic Activity in Human Pancreatic Cancer. Cell Stem Cell 1, 313–323.

Hiratsuka, S., Watanabe, A., Aburatani, H., and Maru, Y. (2006). Tumour-mediated upregulation of chemoattractants and recruitment of myeloid cells predetermines lung metastasis. Nat Cell Biol 8, 1369–1375.

Hiratsuka, S., Watanabe, A., Sakurai, Y., et al. (2008). The S100A8-serum amyloid A3-TLR4 paracrine cascade establishes a pre-metastatic phase. Nat Cell Biol 10, 1349–1355.

Idowu, M.O., Kmieciak, M., Dumur, C., et al. (2012). CD44+/CD24−/low cancer stem/progenitor cells are more abundant in triple-negative invasive breast carcinoma phenotype and are associated with poor outcome. Human pathology 43, 364–373.

Jiao, X., Katiyar, S., Willmarth, N.E., et al. (2010). c-Jun induces mammary epithelial cellular invasion and breast cancer stem cell expansion. J Biol Chem 285, 8218–8226.

Jinushi, M., Chiba, S., Yoshiyama, H., et al. (2011). Tumor-associated macrophages regulate tumorigenicity and anticancer drug responses of cancer stem/initiating cells. Proceedings of the National Academy of Sciences 108, 12425–12430.

Jones, L. (2001). Stem cells: So what's in a niche? Current Biology 11, R484–R486.

Kai, M., Onishi, H., Souzaki, M., et al. (2011). Semi-quantitative evaluation of CD44+/CD24− tumor cell distribution in breast cancer tissue using a newly developed fluorescence immunohistochemical staining method. Cancer Science 102, 2132–2138.

Kakarala, M., and Wicha, M.S. (2008). Implications of the cancer stem-cell hypothesis for breast cancer prevention and therapy. J Clin Oncol 26, 2813–2820.

Kaplan, R.N., Riba, R.D., Zacharoulis, S., et al. (2005). VEGFR1-positive haematopoietic bone marrow progenitors initiate the pre-metastatic niche. Nature 438, 820–827.

Karnoub, A.E., Dash, A.B., Vo, A.P., et al. (2007). Mesenchymal stem cells within tumour stroma promote breast cancer metastasis. Nature 449, 557–563.

Kiel, M.J., and Morrison, S.J. (2006). Maintaining Hematopoietic Stem Cells in the Vascular Niche. Immunity 25, 862–864.

Lam, B.S., and Adams, G.B. (2010). Hematopoietic stem cell lodgment in the adult bone marrow stem cell niche. International Journal of Laboratory Hematology 32, 551–558.

Lapidot, T., Sirard, C., Vormoor, J., et al. (1994). A cell initiating human acute myeloid leukaemia after transplantation into SCID mice. Nature 367, 645–648.

Li, F., Tiede, B., Massague, J., and Kang, Y. (2007). Beyond tumorigenesis: cancer stem cells in metastasis. Cell Res 17, 3–14.

Li, L., and Neaves, W.B. (2006). Normal stem cells and cancer stem cells: the niche matters. Cancer Res 66, 4553–4557.

Lisanti, M.P., Martinez-Outschoorn, U.E., Lin, Z., et al. (2011). Hydrogen peroxide fuels aging, inflammation, cancer metabolism and metastasis: The seed and soil also needs "fertilizer." Cell Cycle 10, 2440–2449.

Liu, J., Zhang, Y., Zhao, J., et al. (2011). Mast cell: insight into remodeling a tumor microenvironment. Cancer and Metastasis Reviews 30, 177–184.

Liu, M., Casimiro, M.C., Wang, C., et al. (2009). p21CIP1 attenuates Ras- and c-Myc-dependent breast tumor epithelial mesenchymal transition and cancer stem cell-like gene expression in vivo. Proc Natl Acad Sci USA 106, 19035–19039.

Liu, M., Sakamaki, T., Casimiro, M.C., et al. (2010). The canonical NF-κB pathway governs mammary tumorigenesis in transgenic mice and tumor stem cell expansion. Cancer Res 70, 10464–10473.

López-Onieva, L., Fernández-Miñán, A., and González-Reyes, A. (2008). Jak/Stat signalling in niche support cells regulates dpp transcription to control germline stem cell maintenance in the Drosophila ovary. Development 135, 533–540.

Lorusso, G., and Ruegg, C. (2008). The tumor microenvironment and its contribution to tumor evolution toward metastasis. Histochem Cell Biol 130, 1091–1103.

Malanchi, I., Santamaria-Martinez, A., Susanto, E., et al. (2012). Interactions between cancer stem cells and their niche govern metastatic colonization. Nature 481, 85–89.

Mani, S.A., Guo, W., Liao, M.J., et al. (2008). The epithelial-mesenchymal transition generates cells with properties of stem cells. Cell 133, 704–715.

Martinez-Outschoorn, U.E., Prisco, M., Ertel, A., et al. (2011). Ketones and lactate increase cancer cell "stemness," driving recurrence, metastasis and poor clinical outcome in breast cancer: Achieving personalized medicine via Metabolo-Genomics. Cell Cycle 10, 1271–1286.

Massberg, S., Schaerli, P., Knezevic-Maramica, I., et al. (2007). Immunosurveillance by Hematopoietic Progenitor Cells Trafficking through Blood, Lymph, and Peripheral Tissues. Cell 131, 994–1008.

Mazo, I.B., Massberg, S., and von Andrian, U.H. (2011). Hematopoietic stem and progenitor cell trafficking. Trends in Immunology 32, 493–503.

Mi, Z., Bhattacharya, S.D., Kim, V.M., et al. (2011). Osteopontin promotes CCL5-mesenchymal stromal cell-mediated breast cancer metastasis. Carcinogenesis 32, 477–487.

Morel, A.P., Lievre, M., Thomas, C., et al. (2008). Generation of breast cancer stem cells through epithelial-mesenchymal transition. PLoS ONE 3, e2888.

O'Brien, C.A., Pollett, A., Gallinger, S., and Dick, J.E. (2007). A human colon cancer cell capable of initiating tumour growth in immunodeficient mice. Nature 445, 106–110.

Oskarsson, T., Acharyya, S., Zhang, X.H.F., et al. (2011). Breast cancer cells produce tenascin C as a metastatic niche component to colonize the lungs. Nat Med 17, 867–874.

Pang, R., Law, W.L., Chu, A.C.Y., et al. (2010). A Subpopulation of CD26+ Cancer Stem Cells with Metastatic Capacity in Human Colorectal Cancer. Cell Stem Cell 6, 603–615.

Porta, C., Riboldi, E., and Sica, A. (2011). Mechanisms linking pathogens-associated inflammation and cancer. Cancer Letters 305, 250–262.

Purizaca, J., Meza, I., and Pelayo, R. (2012). Early Lymphoid Development and Microenvironmental Cues in B-cell Acute Lymphoblastic Leukemia. Archives of Medical Research 43, 89–101.

Ricci-Vitiani, L., Lombardi, D.G., Pilozzi, E., et al. (2007). Identification and expansion of human colon-cancer-initiating cells. Nature 445, 111–115.

Salem, A.F., Bonuccelli, G., Bevilacqua, G., et al. (2011). Caveolin-1 promotes pancreatic cancer cell differentiation and restores membranous E-cadherin via suppression of the epithelial-mesenchymal transition. Cell Cycle 10, 3692–3700.

Salem, A.F., Whitaker-Menezes, D., Lin, Z., et al. (2012). Two-compartment tumor metabolism: Autophagy in the tumor microenvironment and oxidative mitochondrial metabolism (OXPHOS) in cancer cells. Cell Cycle 11, 2545–2556.

Scheel, C., Eaton, E.N., Li, S.H., et al. (2011). Paracrine and autocrine signals induce and maintain mesenchymal and stem cell states in the breast. Cell 145, 926–940.

Schofield, R. (1978). The relationship between the spleen colony-forming cell and the haemopoietic stem cell. Blood Cells 4, 7–25.

Schuettpelz, L.G., and Link, D.C. (2011). Niche competition and cancer metastasis to bone. The Journal of Clinical Investigation 121, 1253–1255.

Sheridan, C., Kishimoto, H., Fuchs, R.K., et al. (2006). CD44+/CD24− breast cancer cells exhibit enhanced invasive properties: an early step necessary for metastasis. Breast Cancer Res 8, R59.

Shipitsin, M., Campbell, L.L., Argani, P., et al. (2007). Molecular definition of breast tumor heterogeneity. Cancer Cell 11, 259–273.

Singh, S.K., Hawkins, C., Clarke, I.D., et al. (2004). Identification of human brain tumour initiating cells. Nature 432, 396–401.

Tavor, S., Petit, I., Porozov, S., et al. (2004). CXCR4 Regulates Migration and Development of Human Acute Myelogenous Leukemia Stem Cells in Transplanted NOD/SCID Mice. Cancer Research 64, 2817–2824.

Tsuyada, A., Chow, A., Wu, J., et al. (2012). CCL2 Mediates Cross-talk between Cancer Cells and Stromal Fibroblasts That Regulates Breast Cancer Stem Cells. Cancer Research 72, 2768–2779.

Welner, R.S., and Kincade, P.W. (2007). Stem Cells on Patrol. Cell 131, 842–844.

Wels, J., Kaplan, R.N., Rafii, S., and Lyden, D. (2008). Migratory neighbors and distant invaders: tumor-associated niche cells. Genes & Development 22, 559–574.

Wicha, M.S. (2006). Cancer stem cells and metastasis: lethal seeds. Clin Cancer Res 12, 5606–5607.

Wicha, M.S., Liu, S., and Dontu, G. (2006). Cancer stem cells: an old idea—a paradigm shift. Cancer Res 66, 1883–1890; discussion 1895–1886.

Wu, K., Jiao, X., Li, Z., et al. (2011). Cell fate determination factor Dachshund reprograms breast cancer stem cell function. J Biol Chem 286, 2132–2142.

Wu, K., Katiyar, S., Li, A., et al. (2008). Dachshund inhibits oncogene-induced breast cancer cellular migration and invasion through suppression of interleukin-8. Proc Natl Acad Sci USA 105, 6924–6929.

Xie, T., and Spradling, A.C. (2000). A Niche Maintaining Germ Line Stem Cells in the Drosophila Ovary. Science 290, 328–330.

Xu, W.T., Bian, Z.Y., Fan, Q.M., Li, G., and Tang, T.T. (2009). Human mesenchymal stem cells (hMSCs) target osteosarcoma and promote its growth and pulmonary metastasis. Cancer Lett 281, 32–41.

Zeng, Q., Li, W., Lu, D., et al. (2012). CD146, an epithelial-mesenchymal transition inducer, is associated with triple-negative breast cancer. Proceedings of the National Academy of Sciences 109, 1127–1132.

Zhang, S.-S., Han, Z.-P., Jing, Y.-Y., et al. (2012). CD133+CXCR4+ colon cancer cells exhibit metastatic potential and predict poor prognosis of patients. BMC Medicine 10, 85.

Zhou, B., Wu, Q., Chen, G., Zhang, T.-P., and Zhao, Y.-P. (2012). NOP14 promotes proliferation and metastasis of pancreatic cancer cells. Cancer Letters 322, 195–203.

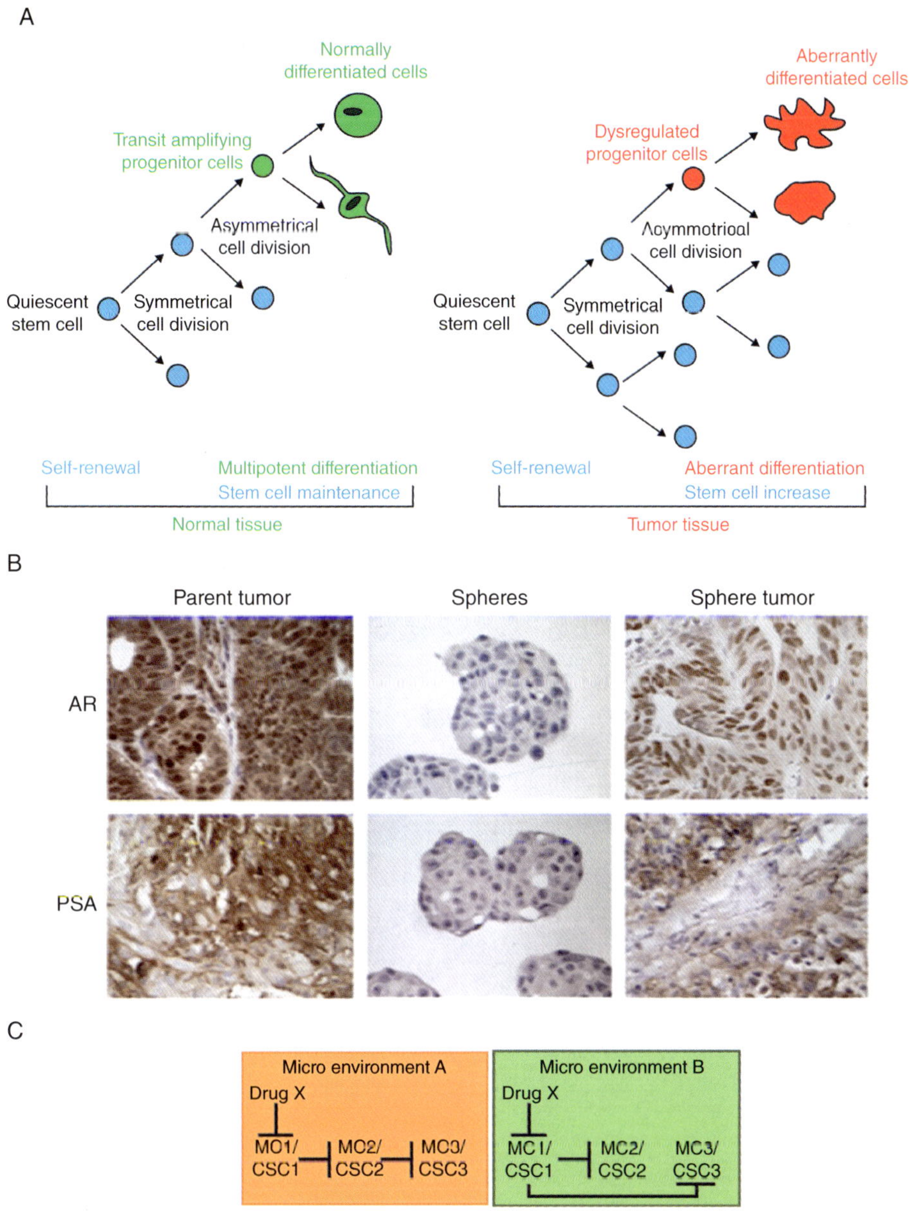

Figure 1. Cancer Stem Cells. **A**. Schematic representations of adult stem cell number maintenance and their differentiation in the normal tissues versus cancer stem cell number increase and their aberrant differentiation in the tumor tissues. **B**. Comparative immunohistochemistry of androgen receptor (AR) and prostate specific antigen (PSA) in the human primary prostate tumor (parent tumor), the tumor-derived primary spheres (spheres), and the sphere-derived tumor (sphere tumor) respectively [see http://www.nature.com/ncomms/journal/v2/n1/full/ncomms1159.html]. **C**. Hypothetical representation of hierarchical arrangement of multiple mutant clones (MC1/2/3) or progenies of different CSC pools (CSC1/2/3).

Cancer Stem Cells, First Edition. Edited by Vinagolu K. Rajasekhar.

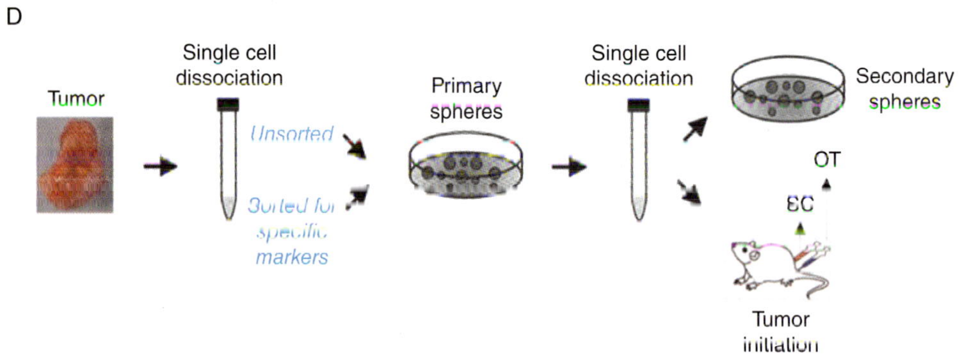

Figure 1. (*Continued*). **D**. Schematic representation of a rapid and novel mini-screen strategy for identifying the conventional therapy resistant CSCs and screening small molecule inhibitors in bulk tumor tissues [see http://www.genomeweb.com/proteomics/sloan-kettering-team-ids-non-psa-producing-cells-potentially-linked-prostate-can].

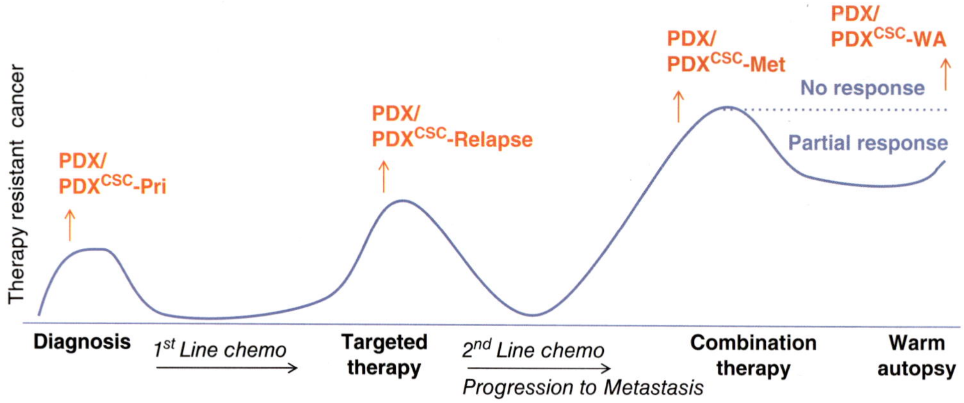

Figure 2. A schematic for strategic development of live biobanking of patient-specific and disease-stage specific tissue, including therapy-resistant PDX and PDXCSC cancer models, starting from patient diagnosis and proceeding through warm autopsy.

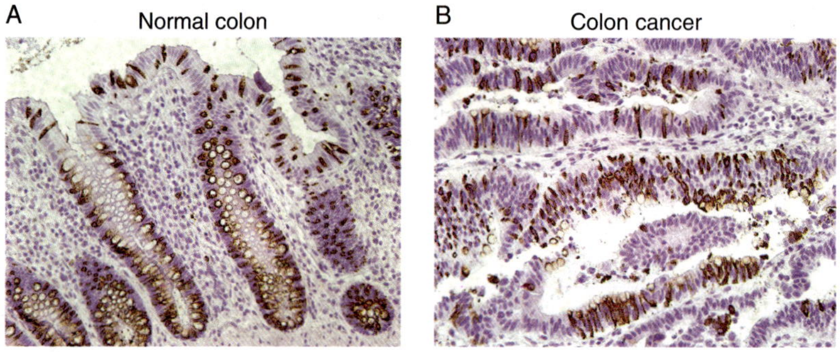

Figure 1.1. Tumors act as "*developmental caricatures*" of normal tissues. **A**. Normal colon epithelium analyzed by immunohistochemistry for expression of the MUC2 protein, a marker of mucus-producing, differentiated goblets cells (brown staining). **B**. Malignant colon cancer tissue analyzed by immunohistochemistry for expression of the MUC2 protein (brown staining). When compared to normal colon epithelia, human colon cancer tissues often reveal close similarities in terms of three-dimensional architecture (e.g., formation of crypt-like gland structures) and cell composition (e.g., presence of mucus-producing goblet-like cells).

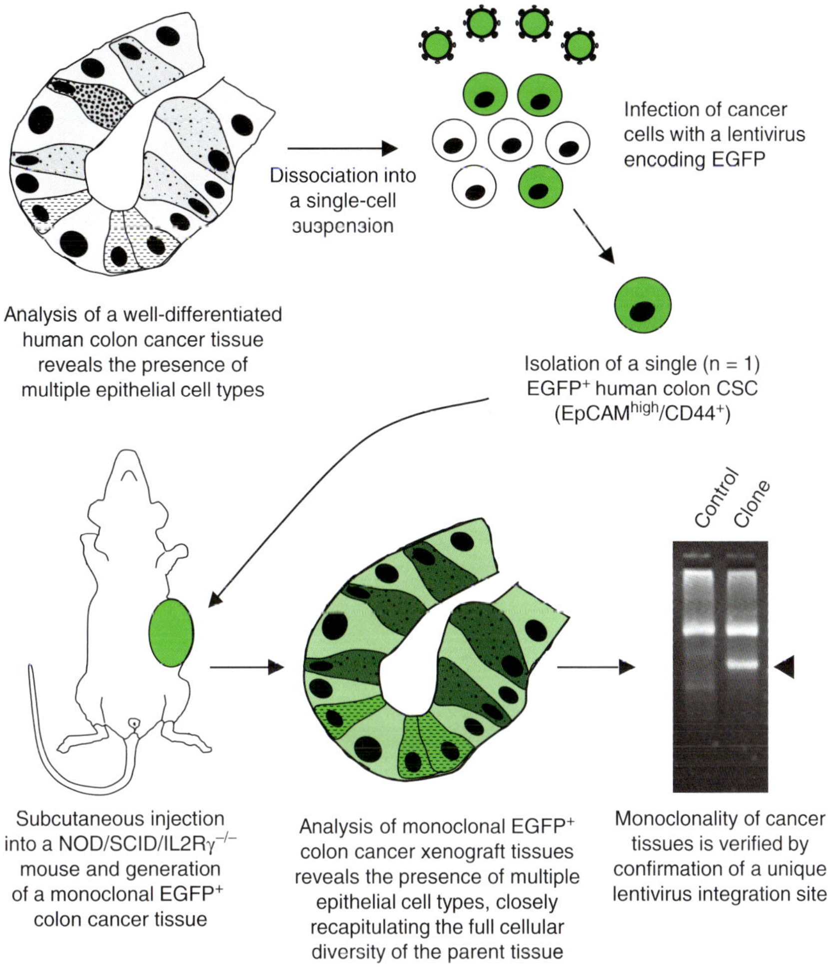

Figure 1.2. Prospective evaluation of multi-lineage differentiation capacity using transplantation experiments. Tumor tissues are frequently composed of a heterogeneous population of cancer cells, containing multiple cellular subtypes. According to the CSC model, the cellular diversity of tumor tissues can originate as the result of multi-lineage differentiation processes, akin to those sustained by normal stem cells. Formal demonstration of multi-lineage differentiation requires a prospective experiment, in which the fate of a single ($n = 1$) cancer cell is followed over time, together with careful molecular tracing of the monoclonal nature of its progeny. A possible way to perform this experiment is to inject into an immunodeficient animal host a single ($n = 1$) CSC purified from a tumor tissue known to contain a heterogeneous population of cancer cells (e.g., a single $EpCAM^{high}/CD44^{+}$ human cancer cell from a well-differentiated human colon cancer xenograft line) after infection with a lentivirus encoding for the enhanced green fluorescence protein (EGFP). The monoclonal nature of resulting tumors can be demonstrated by showing the presence of a unique lentivirus DNA integration site in the genome of cancer cells, all of which must also express EGFP (*green cells*). The observation that monoclonal tumor tissues generated by transplantation of a single cancer cell are able to recapitulate the phenotypic diversity and lineage composition of parent tumors (*various shades of green cells*) provides formal evidence of the multi-lineage differentiation capacity of the transplanted "*cancer stem cell.*"

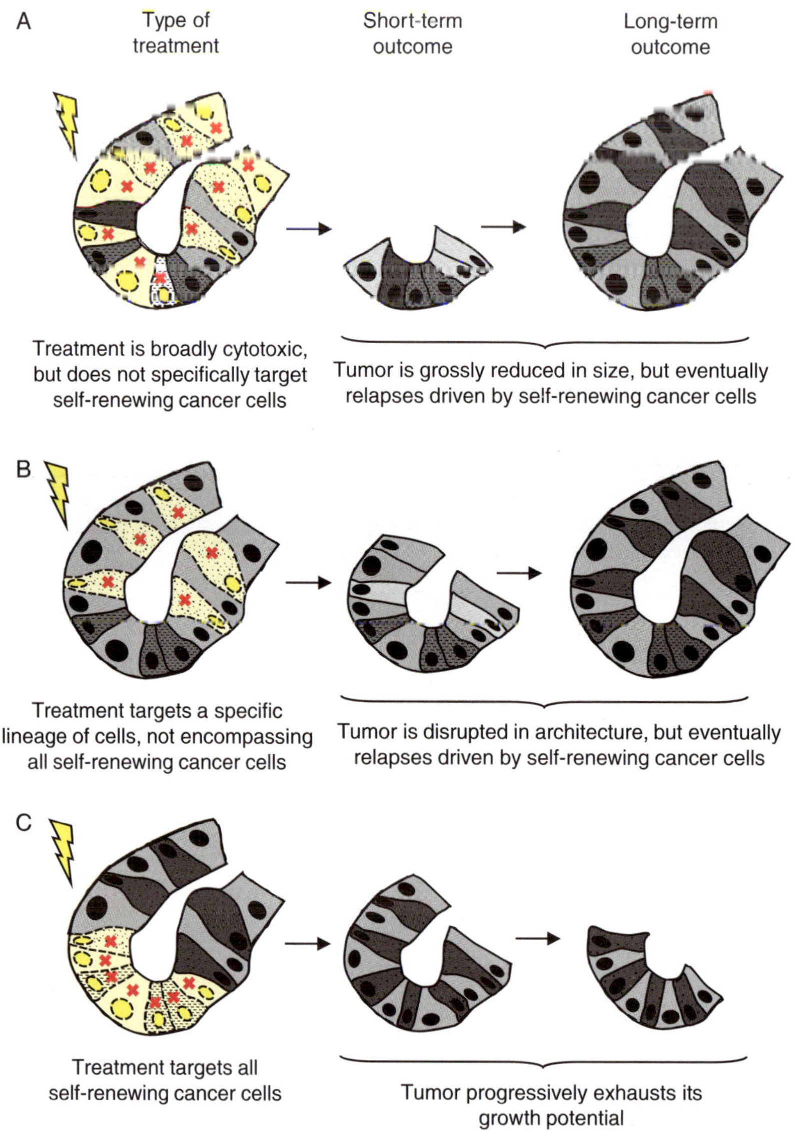

Figure 1.3. Implications of the CSC model for the design and evaluation of anti-tumor treatments. The heterogeneous cell composition of cancer tissues implies the possibility of a differential sensitivity of different cell types to various treatments. **A**. Anti-tumor treatments endowed with broad cytotoxic activity (e.g., treatments targeting rapidly proliferating cancer cells) might be able to hit multiple cell types (*yellow cells with red crosses*) and induce substantial reductions in tumor size. However, if cancer cells endowed with self-renewal capacity (*dark grey cells*) are spared, they hold the potential to cause tumor relapse. **B**. Anti-tumor treatments targeted to a specific lineage of cancer cells (*black cells*) might induce tissue disruption and temporary reduction of tumor volume. However, if the targeted lineage does not encompass all cells with self-renewal capacity, tumor tissues are likely to relapse, sustained by self-renewing cancer cells. At relapse, cells of the targeted lineage might be re-created as a result of multi-lineage differentiation processes. **C**. Anti-tumor treatments able to target and eliminate all cancer cells with self-renewal capacity, although not necessarily designed to induce rapid reductions in tumor volume, could cause long-term exhaustion of tissue growth, and hold the potential for permanent tumor eradication.

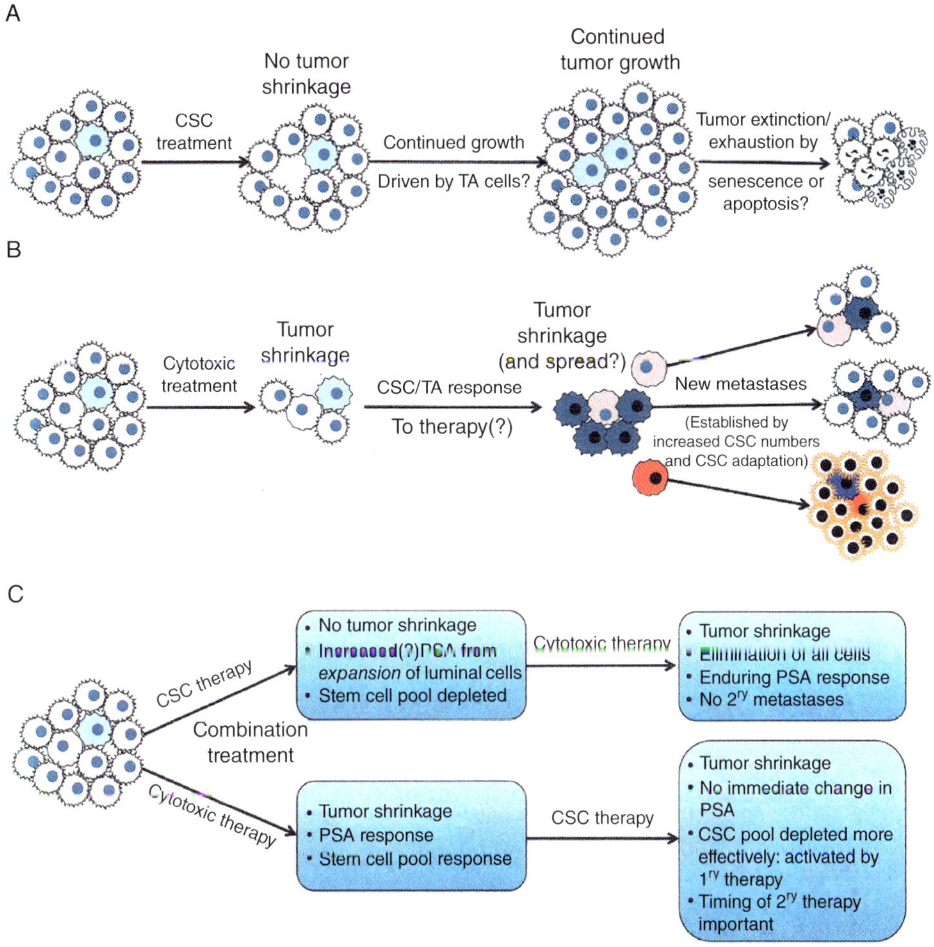

Figure 2.3. Predicted outcomes of CSC and conventional treatments on heterogeneous tumors, either as single agents or in combination. **A**. Effects of a single CSC treatment leads to unsustainable tumor growth and eventual tumor extinction. The time to tumor extinction is unknown and could vary between tumors even from the same tissue of origin. **B**. Conventional chemotherapy, which kills the replicating tumor bulk tissues, results either in enrichment of the CSC/TA fraction, or indeed to a 'wound healing' amplification of CSCs, potentially increased metastatic spread. Successful new tumors have probably undergone a new mutation/selection resulting in the emergence of new clonal variants. We do not know whether this occurs in the SC or in a TA fraction. **C**. The order and timing of a combination therapy to eliminate both the CSC and bulk (replicating tumor cells) is likely to be crucial to achieve the desired effects. The potential outcomes are illustrated.

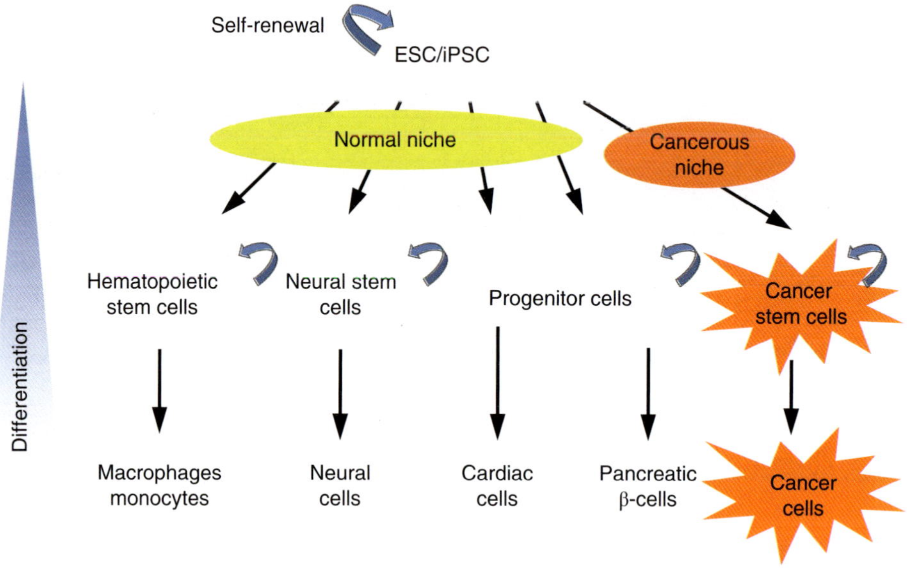

Figure 6.1. Concept of 'cancerous niche' to develop malignancy. In the appropriate niche, the stem cells should differentiate into progenitor cells and further into each mature cell. When a malignant neoplasm is considered to be a phenotype differentiated from stem cell, a cancer stem cell could be a progenitor cell giving rise to cancer cells. The niche contributing conversion of stem cells into cancer stem cells is hypothesized to be a 'cancerous niche' in this review.

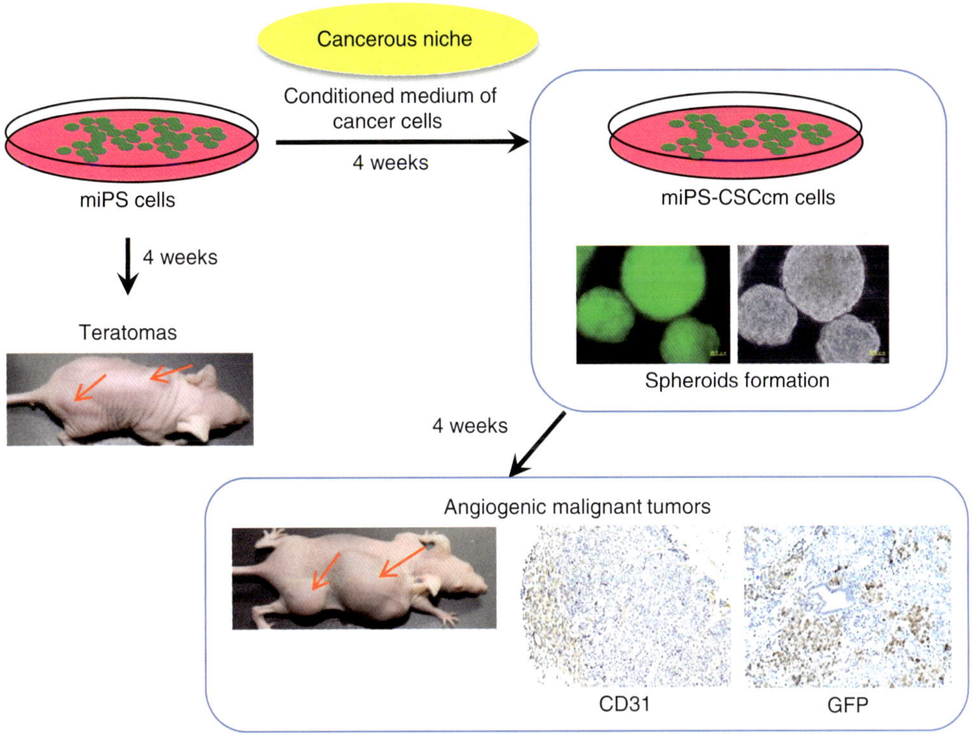

Figure 6.2. Generation of cancer stem-like cells from mouse iPS cells. miPS cells were cultured in the media containing the conditioned medium of various mouse cancer cell lines regarded as a cancerous niche. Resultant cells showed cancer stem cell properties such as sphere formation and rapid malignant tumorigenicity with extensive angiogenesis in vivo.

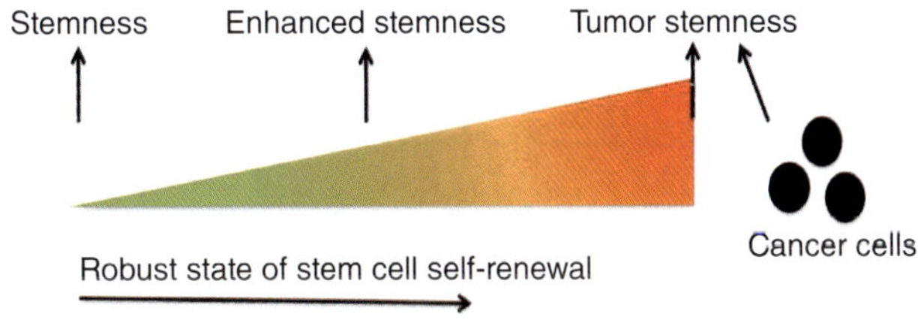

Figure 7.1. A schematic diagram of stem cell reprogramming to robust state of stemness under microenvironmental stress. The three states of stemness represent three robust states in the context of nongenetic, microenvironmental pressure. The stemness state is the resting, homeostatic state of quiescent stem cells in their niche. The enhanced stemness state is the robust state of stemness in the microenvironment of extreme oxidative stress described in Figure 7.2. The tumor stemness state is the malignant state, a robust state that tolerates genomic instability. The robustness is defined as the state of a biological system that continues to function despite genetic or nongenetic perturbations (Andreas, 2007, page 1).

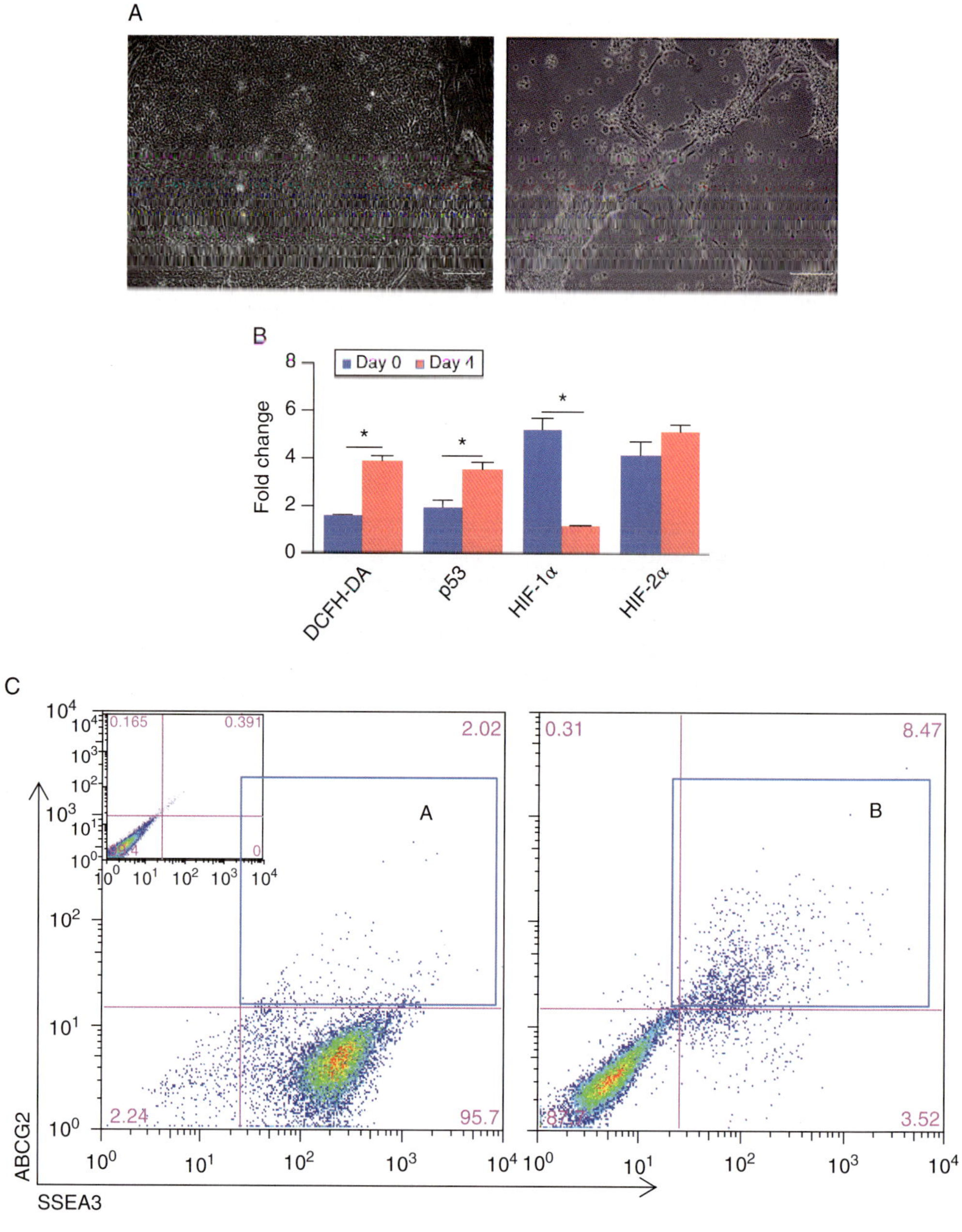

Figure 7.2. In vitro assay to study human ES cell altruism. (From Das B et al. Stem Cells, 2012). **A**. Morphological evidence of BGO1 hESC differentiation and death (image to the right) following exposure to extreme hypoxia followed by 4 days of reoxygenation. The hypoxia and re-oxygenation treated cells showed elongated neurite processes including fibroblastic-like features, as well as evidence of cell fragmentation. **B**. The fold change in ROS (DCFH-DA), p53, and HIF-2α in the BGO1 cells shown in **A**, following day 4 post-hypoxia/reoxygenation. **C**. Flow cytometry showing expansion (A versus B compartments) of the SSEA3+/ABCG2+ fraction on the day 4 post-hypoxia/reoxygenation. Insert shows the profile of isotype control. **D**. Immunofluorescence labeling of SSEA3+/ABCG2- and SSEA3+/ABCG2+ fractions on day 4 post-hypoxia/reoxygenation showing a high HIF-2α and low p53 signal in the SSEA3+/ABCG2+ fraction (Dapi, nuclear stain). **E**. The corresponding fold-change in DCFH-DA, p53, and HIFs in the SSEA3+/ABCG2+ versus SSEA3+/ABCG2− fractions on the day 4 post-hypoxia/reoxygenation. **F**. Protein levels of HIF-2α and p53 in the SSEA3+/ABCG2+ fraction were measured by ELISA. Note high HIF-2α versus low p53 until day 20. *$P<0.05$; $n=4$. Scale bar (a) 40 μm and (d) 10 μm.

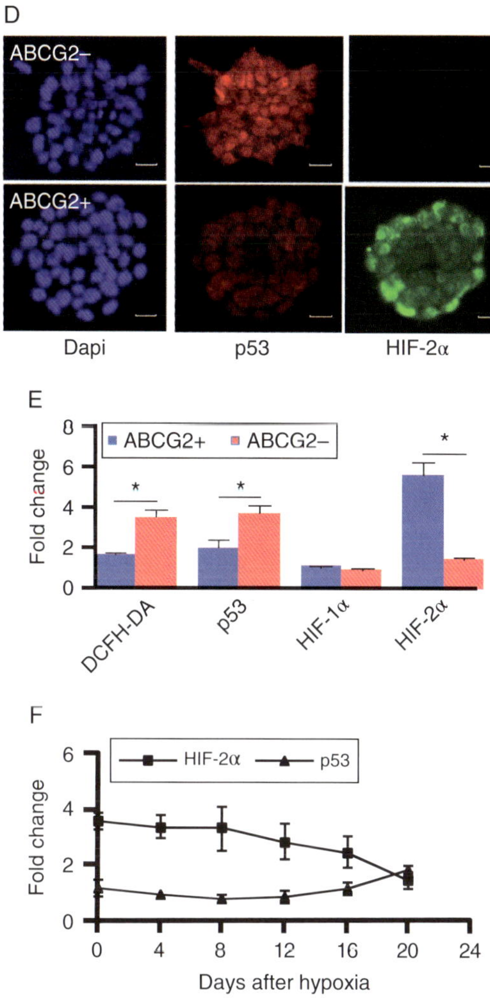

Figure 7.2. (*Continued*).

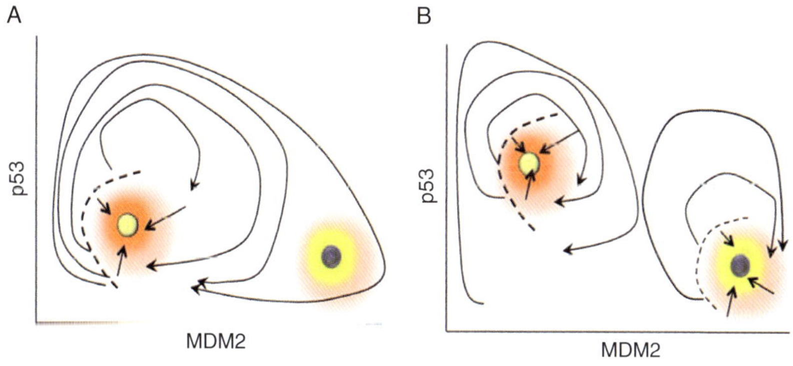

Figure 7.3. **A**. Model of p53/MDM2 oscillation during basal state. Minor fluctuation (*red zone*) away from the stable state (*yellow dot*) is immediately damped out (*short arrows*). Larger fluctuation beyond the threshold (*dashed line*) may generate unstable state of high MDM2 and low p53 (*blue dot*), which eventually bounces back to the stable state. **B**. Stabilization (Turcotte et al., 2008). Increase noise in the circuit might stochastically stabilize the unstable state. Figure **A** top panel was partially adopted from Tyson (1998).

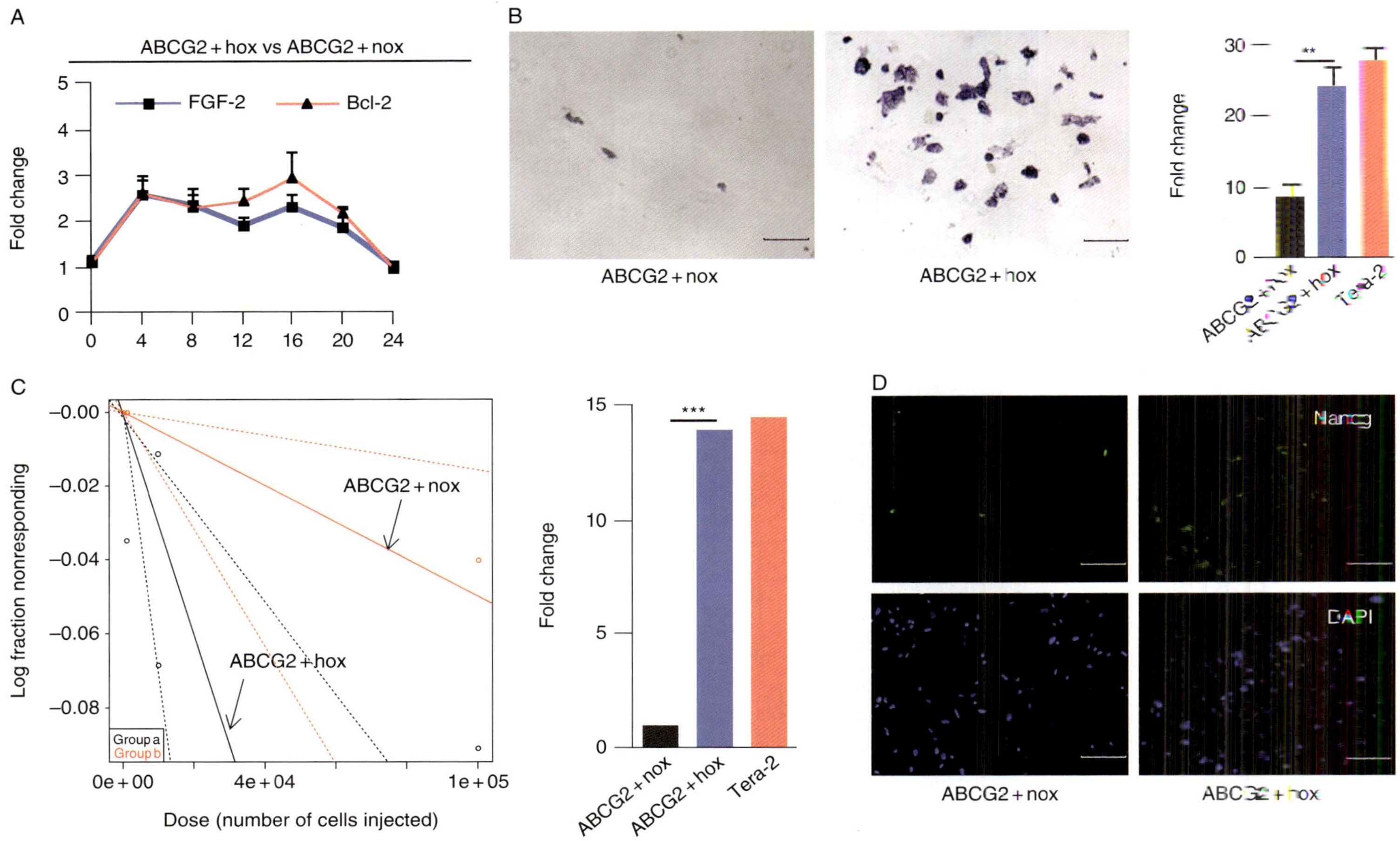

Figure 7.4. ABCG2 + hox fraction exhibits a phenotype of self-sufficiency (From Das et al., 2012). **A**. The fold change in FGF-2 and Bcl-2 in ABCG2+ hox cells relative to ABCG2 + nox cells. **B**. The single cell clonogenic assay showing the alkaline phosphatase staining of abundant colony formation in the ABCG2 + hox cell fraction. The histogram shows the significant fold increase in colonies derived from the ABCG2 + hox fraction. Tera-2 derived colonies were used as a control. **C**. The in vivo limiting dilution assay showing higher teratomagenic formation efficiency of ABCG2 + hox versus ABCG2 + nox. Tera-2 teratoma formation was used as a control. **D**. Immunofluorescence staining on a frozen section of ABCG2+ cells containing matrigel plug showing expression of Nanog positive cells. The quantification of the Nanog expressing cells is given in the Supplementary Figure 6 q1 (Das et al., 2012). *p<0.05; **p<0.001; ***p<0.00001, n = 4. Scale bar **B** and **D** 40 μm.

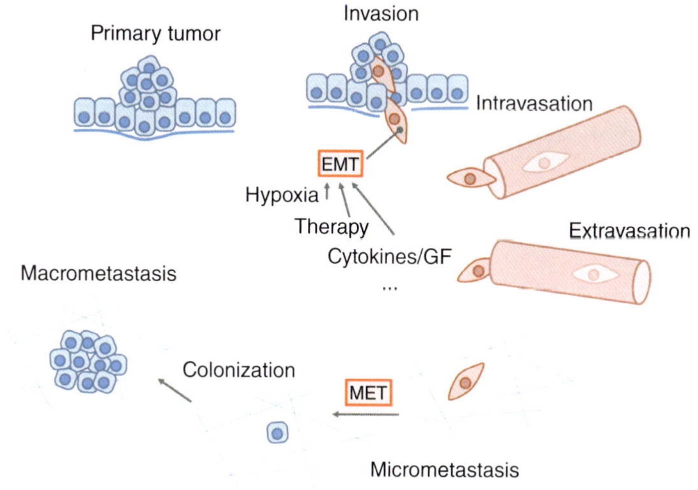

Figure 8.1. EMT and MET in carcinoma progression. At the primary tumor, EMT is induced by different stimuli such as hypoxia, several cytokines, and growth factors (GF) as well as stress induced by anti-neoplastic drugs. Cancer cell undergo EMT and acquire a mesenchymal phenotype with increased motility and ability to degrade the basement membrane and the extracellular matrix, thereby leading to local invasion. Mesenchymal cancer cells then intravasate into the blood or lymphatic vessels, disseminate and extravasate at secondary site, forming a dormant micrometastasis. Disseminated cancer cells then undergo MET (because of the lack of EMT-promoting factors), reacquire an epithelial phenotype, and reinitiate growth to generate a macrometastasis.

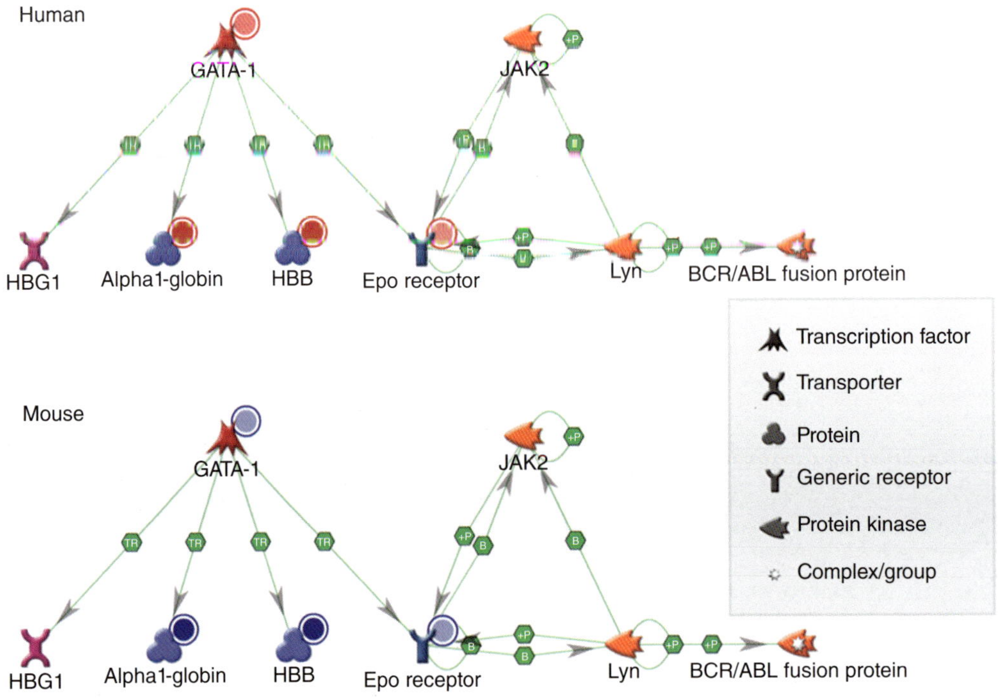

Figure 9.3. Differential expression of members of the Epo induced PI3K/AKT Ca(2+) influx pathway: Alpha1-globin (HBA1), HBB, EpoR, and GATA-1 are consistently upregulated in the human dataset and consistently downregulated in the mouse dataset. Upregulation is indicated with red circles, downregulation is indicated with blue circles, the extent of deregulation is indicated by color intensity. (See MetaCore legend for symbol meanings.)

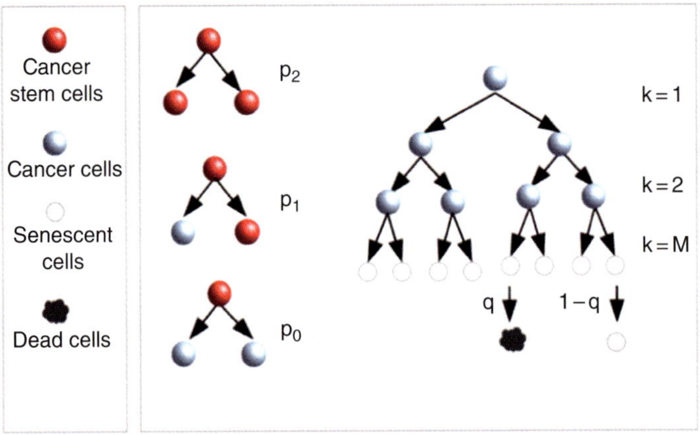

Figure 10.2. A mathematical model for cancer growth. Cells are divided into subpopulations composed by CSCs (*red*) and cancer cells (*blue*). Cancer cells divide a finite number of times M (M=3 in the figure) before turning senescent (*white*). Senescent cells die with probability q. CSCs can divide symmetrically, giving rise to two CSCs with probability p2 or to two CCs with probability p0, or asymmetrically with probability p1=1 − p2 − p0 giving rise to a CSC and a CC. (See La Porta et al., 2012 for more details.)

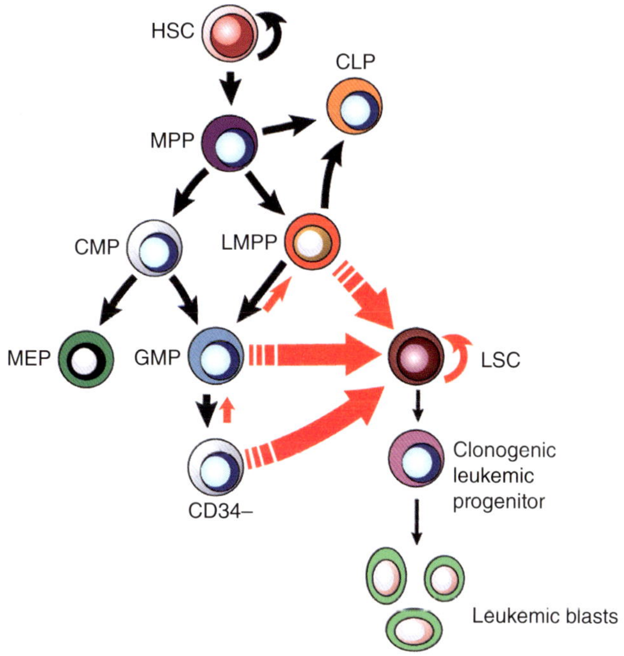

Immunophenotype	LSC frequency	References
LMPP CD34+CD38–CD90–CD45RA+	1 in 1064 to 1 in 10777 (3 samples)	(Goardon et al., 2011)
GMP CD34+CD38+CD110–CD45RA+	1 in 8879 to >1 in 10^5 (3 samples)	(Goardon et al., 2011)
CD34–	1 in 458 to 1 in 254,454 (6 samples) not quantified (10^5 to 10^6 cells transplanted)	(Taussig et al., 2010) (Sarry et al., 2011)

Figure 11.1. Hematopoietic stem and progenitor cells are putative targets for transformation into the Leukemic Stem Cell. The hierarchical model of normal hematopoiesis is illustrated with black arrows. Curved arrows indicate self-renewal capability. Dotted red lines indicate leukemic transformation. Other red arrows indicate the heretofore unresolved hypothesis that leukemic cells may "de-differentiate" with regards to lineage marker expression. Listed in the table (inset) are LSC frequencies in the putative targets for transformation as estimated by limiting dilution analyses. Abbreviations: Hematopoietic stem cell (HSC), common lymphoid progenitor (CLP),multipotent progenitor (MPP), lymphoid-primed multipotent progenitor (LMPP), common myeloid progenitor (CMP), megakaryocyte erythroid progenitor (MEP),granulocyte macrophage progenitor (GMP), leukemic stem cell (LSC).

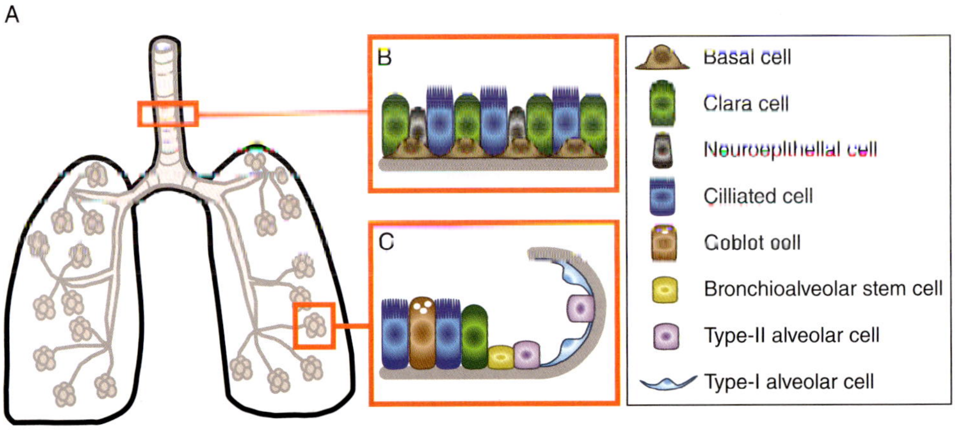

Figure 13.1. Cellular architecture in the normal adult lung. **A**. Schematic of the trachea, bronchi, bronchioles, and alveolar sacs. **B**. Graphical representation of the architecture of the airway epithelium in the conducting airways, including the trachea, bronchi, and bronchioles with diameters down to 0.5 mm. **C**. Airway epithelium in the distal airways, including the terminal bronchioles (*left, straight*) and alveoli (*right, curved*).

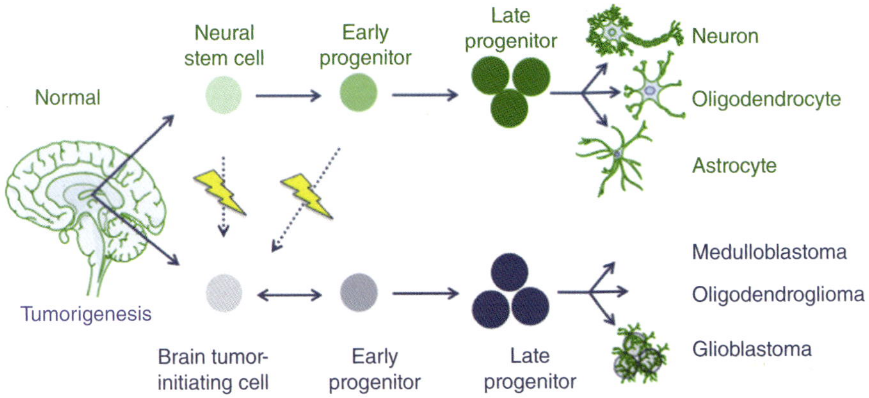

Figure 15.1. Brain tumorigenesis according to the cancer stem cell hypothesis. Brain tumors contain a cellular hierarchy that resembles neural stem cells and their lineage-restricted progeny. Lineage-specific progenitors resulting from the brain tumor-initiating cell may result in cell-type specific tumors.

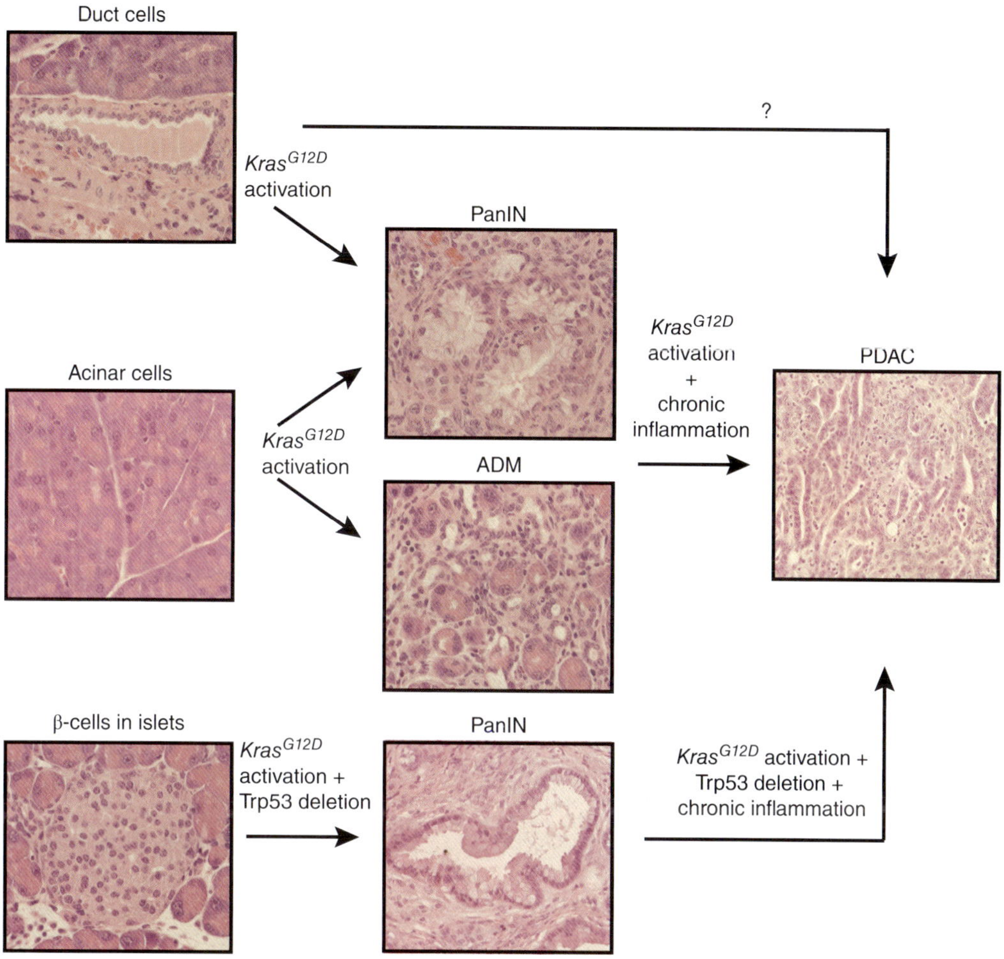

Figure 16.1. Activation of *Kras* oncogene together with chronic inflammation leads to PDAC. The activation of oncogenic *Kras* in ductal and acinar cells promotes the development of premalignant lesions (PanINs) and in the case of acinar cells, also acinar-ductal metaplasia (ADM) is observed. Formation of PanINs has also been described in Trp53-deficient, $KRas^{G12D}$-expressing β-cells. The presence of sustained inflammation in this context also promotes development of PDAC from insulin-producing cells. Photo is original to author.

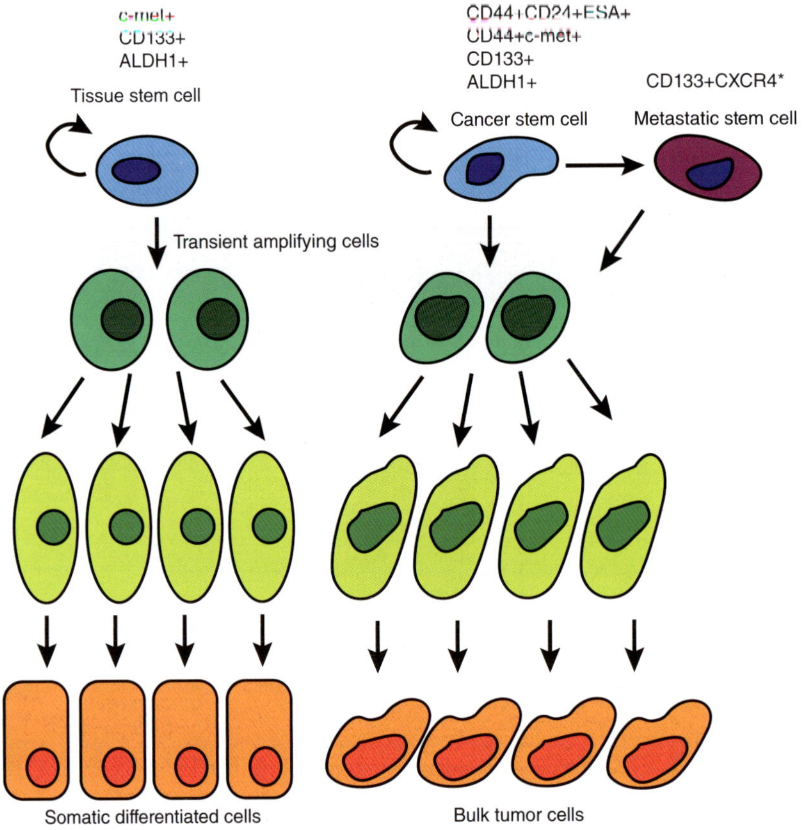

Figure 16.2. Model for cancer stem cells in PDAC. CSCs share several markers with putative tissue stem cells. Consistent with a normal stem cell phenotype, CSCs also give rise to more differentiated progenies. Importantly, the cancer stem cell pool is not homogeneous but also contains a subpopulation of metastatic cancer stem cells that express CXCR4 as a characteristic feature. A heterogeneous CSCs pool can be rationalized by the existence of different clones inside the bulk tumor.

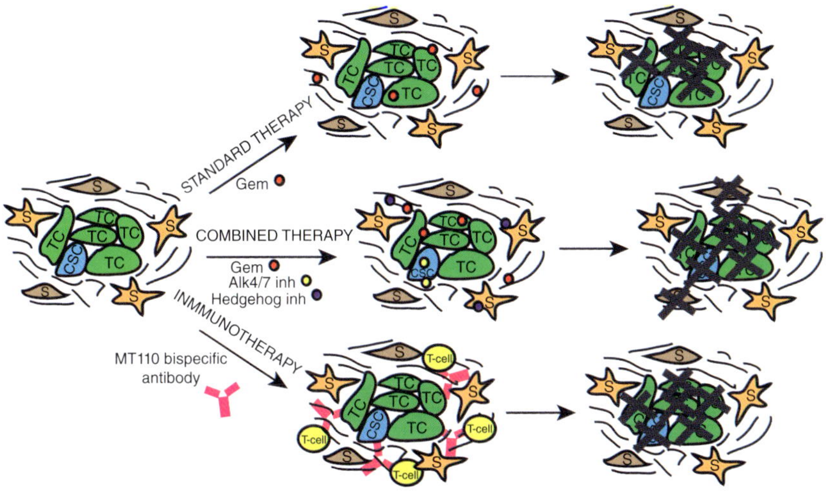

Figure 16.3. Therapeutic strategies against PDAC. The tumor is composed by cancer stem cells (CSC) and more differentiated progenies (TC) surrounded by a fibrotic matrix containing stromal cells (S). The standard therapy for pancreatic cancer represents gemcitabine treatment (Gem) in most cases, which attacks bulk tumor cells. However, cancer stem cells are more resistant to Gem causing relapse of the disease. Combined therapies include combinations of gemcitabine with drugs targeting specifically CSCs (e.g., Alk 4/7 inhibitor) and ideally stromal components of the tumor (e.g., Hedgehog pathway inhibitor) causing a total regression of the tumor and no relapse during long-term follow-up. A third approach consists of immunotherapies targeting epitopes present in both CSCs and their progenies (e.g., bispecific antibody MT110 that triggers T-cell activation against EpCAM expressing cells).

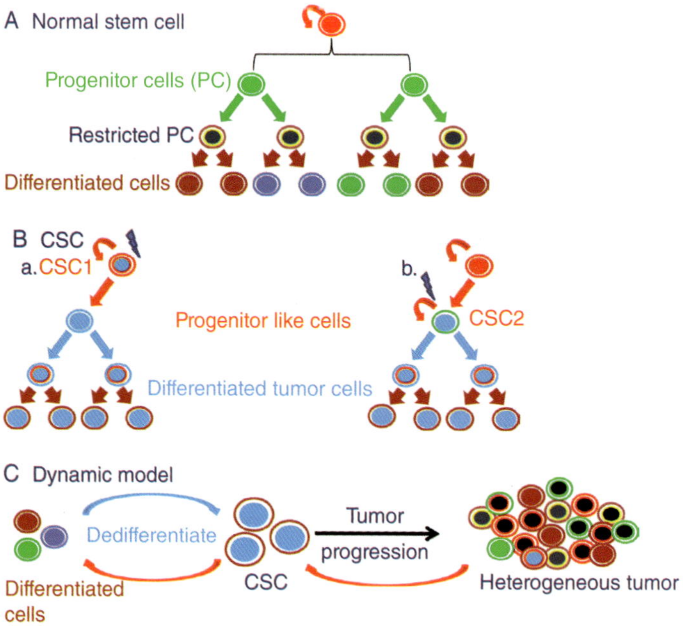

Figure 17.1. Cancer stem cell model. In the normal stem cell model; stem cells can self-renew, undergo orderly cell division and differentiation to give rise to progenitor and differentiated cells following a hierarchical pathway **A**. In the hierarchical model of CSC, CSC (originating as very early progenitor clone (CSC1) or late progenitor clone due to additional mutation or epigenetic changes CSC2) originate early with self-renewal capacity, then give rise to progenitor-like cells that can further terminally differentiate into tumor cells **B**. At this stage differentiated tumor cells have limited capacity to self-renew. In the emerging dynamic CSC model, differentiated cell can undergo de-differentiation to give rise to CSC **C**. Based on microenvironmental factors or signals, CSC now has the ability to progress further to heterogeneous tumor.

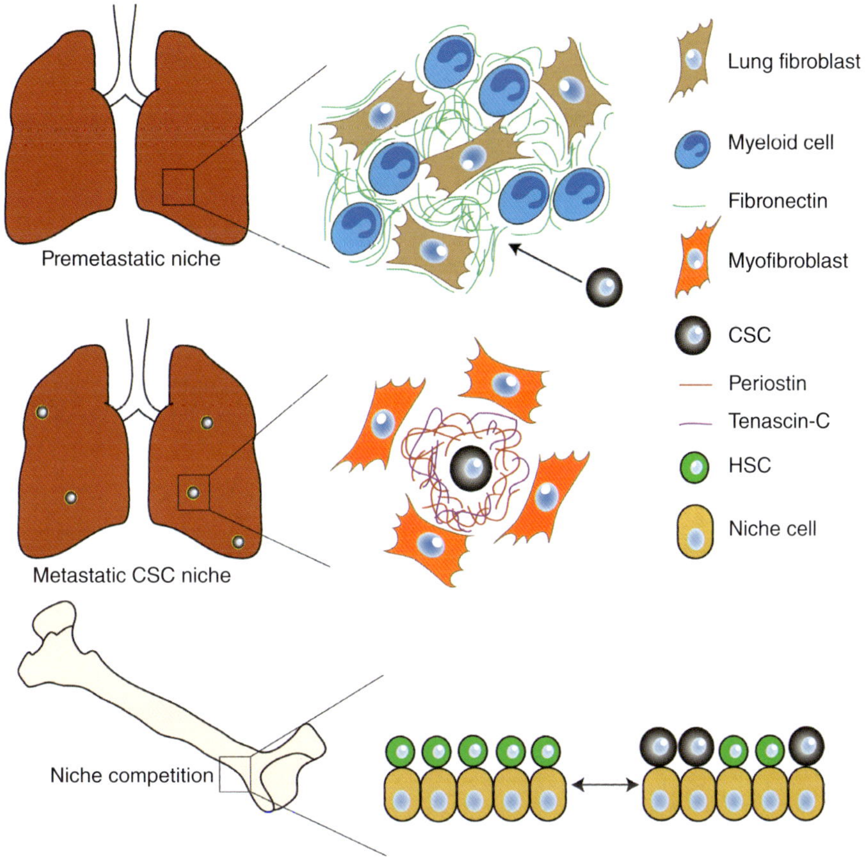

Figure 19.1. The premetastatic niche (*upper scheme*) involves the recruitment of myeloid cells to the future metastatic sites and the production, together with local fibroblasts, of ECM molecules like fibronectin. A suitable environment is created for tumor cells (and other bone marrow derived support cells) to invade these areas. The metastatic niche (*middle scheme*) is generated when seeding cancer stem cells induce the secretion of ECM molecules like POSTN and TNC by locally activated fibroblasts. These molecules enhance CSC survival and fitness in the host tissue through the activation of different signaling pathways. Niche competition (*lower scheme*) involves the parasitism of an endogenous stem cell niche by a tumor cell, i.e., the tumor cell takes the place of an HSC in the bone, displacing the latter from its natural niche.

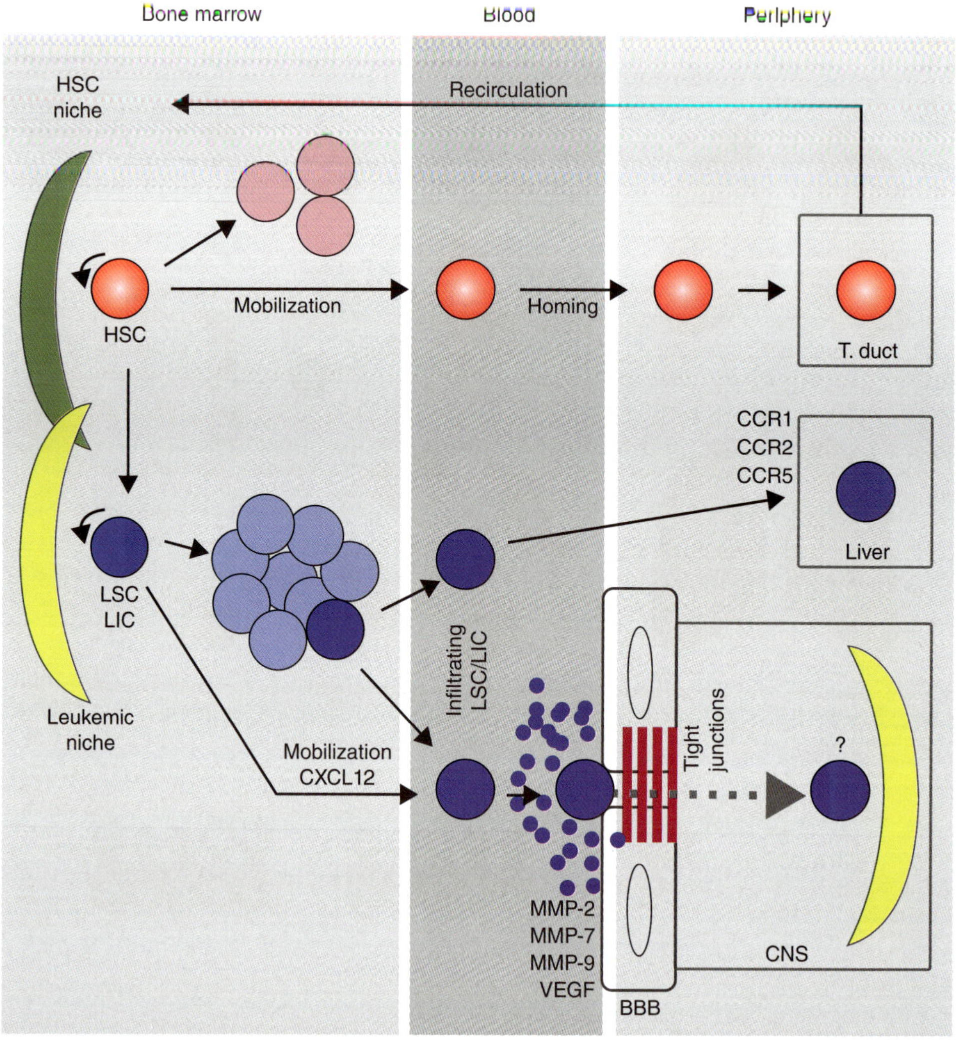

Figure 20.1. Trafficking of normal HSC and infiltrating leukemia SCs (LSC/LIC). During migration, normal hematopoietic stem cells (HSC) or infiltration of leukemic stem/initiating cells LSC/LIC leave their niche and enter circulation to migrate into multiple peripheral tissues. High expression of CXCL12 within leukemic BM contributes for mobilization of LSC/LIC expressing CXCR4. Display of a number of cytokines/chemokines by conspicuous populations of LSC/LIC is crucial for multi-organ infiltration, including liver. LSC/LIC production of MMP may break down the blood-brain barrier through disruption of epithelial tight junctions. Niche-to-niche migration during leukemia infiltration/metastasis is suggested.

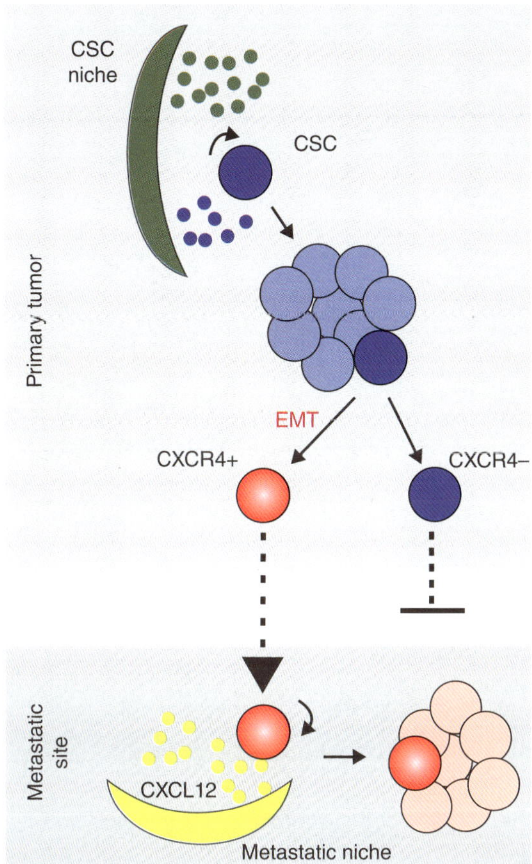

Figure 20.2. CSCs require the expression of chemotactic receptors to metastasize. Primary microenvironment supports self-renewal and differentiation of CSCs. However, CSCs require EMT-associated phenotypical changes in order to become metastatic. The chemokine receptor CXCR4 plays a key role for the invasion into secondary tissues of leukemia, colon, and pancreatic CSCs.

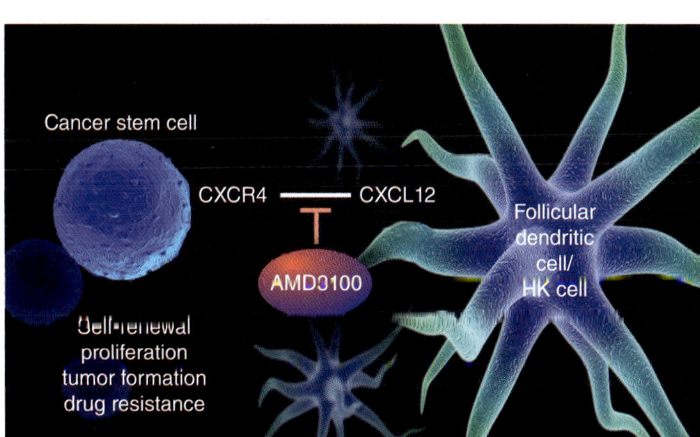

Figure 21.1. Lymph node stromal cell support CSC via paracrine CXCL12/CXCR4 signaling.

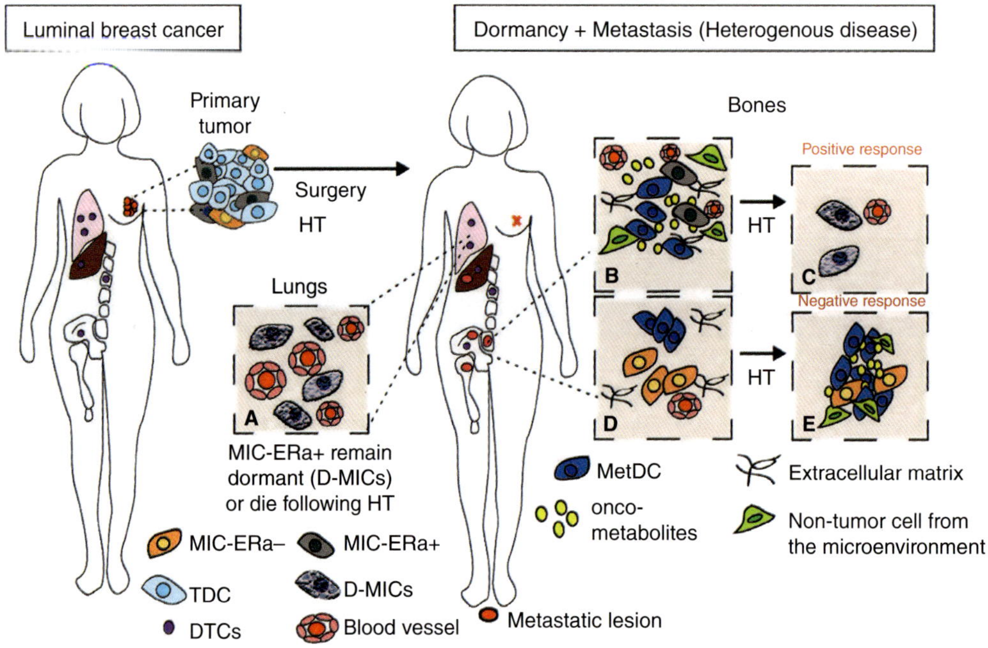

Figure 22.1. A hypothetical luminal breast cancer progression model: the ERa paradigm. Heterogeneous metastasis-initiating cells -MICs- originate in the context of the primary tumor microenvironment. These MICs can be either ERa-dependent or independent for their survival and growth (MIC-ERa + or MIC-ERa−). MICs can disseminate at distance (DTC). Following surgical removal of the primary tumor (minimal residual disease), disseminated MIC-ERa + either die or stay dormant when exposed to hormonal therapy (HT, DMICs, **A**). We also hypothesize that some MICs (especially MIC-ERa−) can self-renew giving rise to metastatic disease. Due to the heterogeneity of the secondary site microenvironments (pro-tumorigenic or not, presence of secretory cells, extracellular matrix, onco-metabolites) and the presence of distinct MICs, the kinetic of metastatic disease can vary from one site to the other. When metastases are clinically detectable, breast cancer patients will undergo HT. HT will effectively render MIC-ERa + cells/lesions dormant (**B**, **C** positive response). By contrast, lesions from MIC-ERa− will continue to self-renew exploiting an ERa-independent tumorigenesis (**D**, **E** negative response). TDC, tumor differentiated cancer cells. MetDC, metastatic differentiated cells.

A

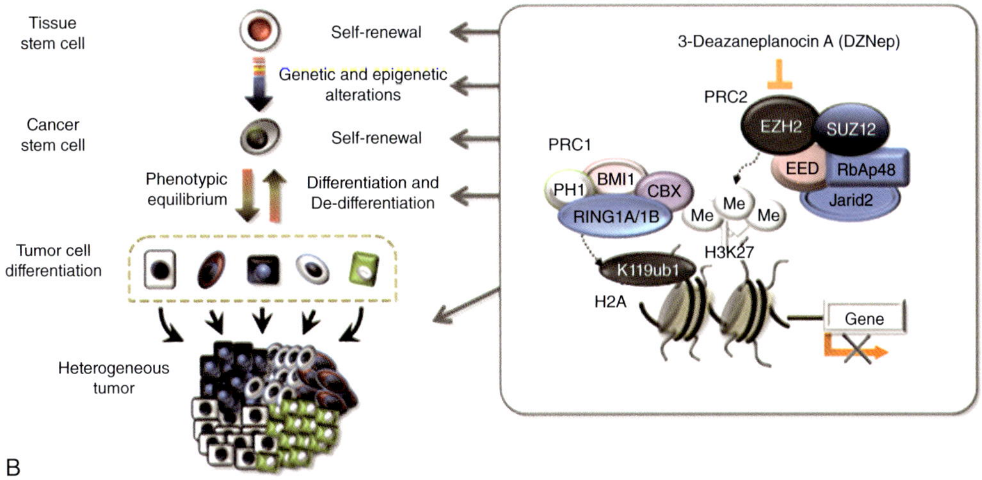

B

Overexpression/ampliciation	
EZH2	Prostate cancer, Breast cancer, Colon cancer, Liver cancer, Lung cancer, Glioblastoma, Lymphoma, Metanoma, Oral cancer, Testicular cancer
SUZ12	Colon cancer, Breast cancer, Liver cancer, Lung cancer, Lymphoma, Melanoma, Germinal cell-derived tumor, Parathyroid and pituitary adenoma
BMI1	Acute myeloid leukiemia, Breast cancer, Gastrointestinal tumor, Lymphoma, Neuroblastoma, Lung cancer, Medulloblastoma, Parathyroid and pituitary adenoma
RING1B	Colon cancer, Bladder cancer, Breast cancer, Renal cancer, Liver cancer, Lung cancer, Lymphoma, Melanoma, Pancreatic cancer, Prostate cancer, Gynecological tumor, Larynx cancer, Parathyroid and thyroid cancer, Thymus cancer, Unrinary tract cancer
Mutations	
EZH2	Lymphoma, Myelodysplastic syndrome (active and inavtive mutations)
SUZ12	Endometrial stromal sarcoma (Chromosomal translocation)

Figure 23.1. Roles of Polycomb-repressive complexes (PRC) in heterogeneous tumor formations. **A**. Cancers are mostly comprised of heterogeneous cell populations. Such multiple distinct subpopulations of cancer cells within tumors may derive from cancer stem cells (CSC). The CSCs are considered able to aberrantly differentiate into diverse cell types. Whereas CSCs can differentiate into non-CSCs, the reverse process is now also being considered. This phenotypic equilibrium in cell-state proportions is observed in CSCs (*left panel*) (Gupta et al., 2011). PRC2 has a histone methyltransferase activity with substrate specificity for H3K27 and produces H3K27me3, which is thought to recruit PRC1 via proteins of the CBX family. RING1A/1B in PRC1 induces the mono-ubiquitination of histone 2A lysine 119 (H2AK119), which is thought to affect the chromatin structure, blocks the recruitment of transcriptional activation factors (*right panel*). PRCs are required for the self-renewal of CSCs as well as CSC differentiation in response to oncogenic cues, which lead to the establishment of heterogenous tumors. **B**. PRC components were overexpressed or mutated in many types of human neoplasia (Sauvageau and Sauvageau, 2010).

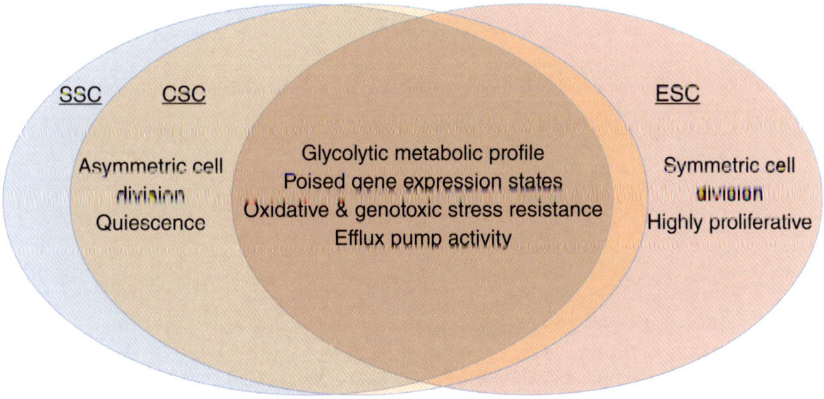

Figure 24.1. Common and distinct characteristics of SSCs, CSCs, and ESCs.

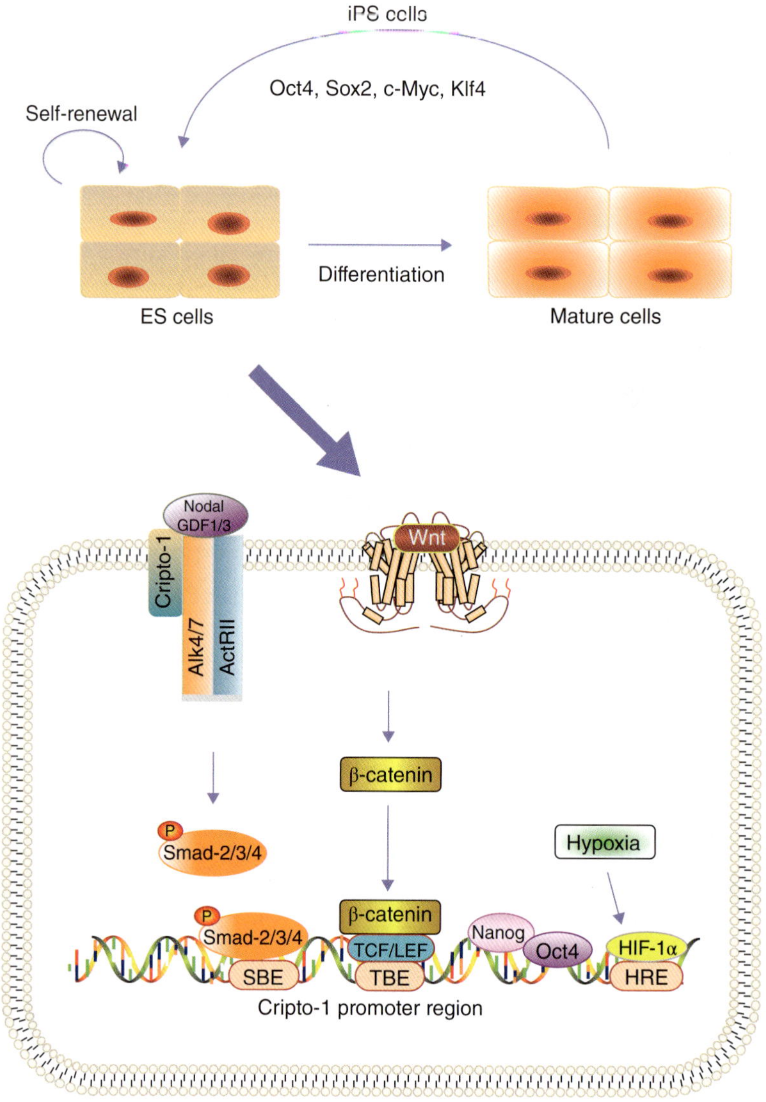

Figure 25.3. Schematic diagram of embryonic stem (ES) cells and induced pluripotent stem (iPS) cells. ES cells possess the capability to undergo numerous cell cycle divisions while maintaining their undifferentiated state (self-renewal). iPS cells are derived through reprogramming of differentiated cells into pluripotent cells by introducing the embryonic stem cells genes, Oct4, Sox2, c-Myc, and Klf4. In embryonic development, phosphorylated Smad-2/3/4 complex binds to Smad binding elements (SBE) within the Cripto-1 promoter and regulates Cripto-1 expression. HIF-1α binds to Hypoxia Reponsive Elements (HRE) within Cripto-1 promoter and directly regulates Cripto-1 expression during hypoxia. Oct-4 and Nanog can also regulate Cripto-1 expression in ES cells by binding to the Cripto-1 promoter. Finally, Cripto-1 is a direct downstream target gene that is regulated by the canonical Wnt/β-catenin signaling pathway. T-cell factor/lymphoid enhancer factor (TCF/LEF); TCF/LEF binding elements (TBE); hypoxia binding elements (HRE); Smad binding elements (SBE).

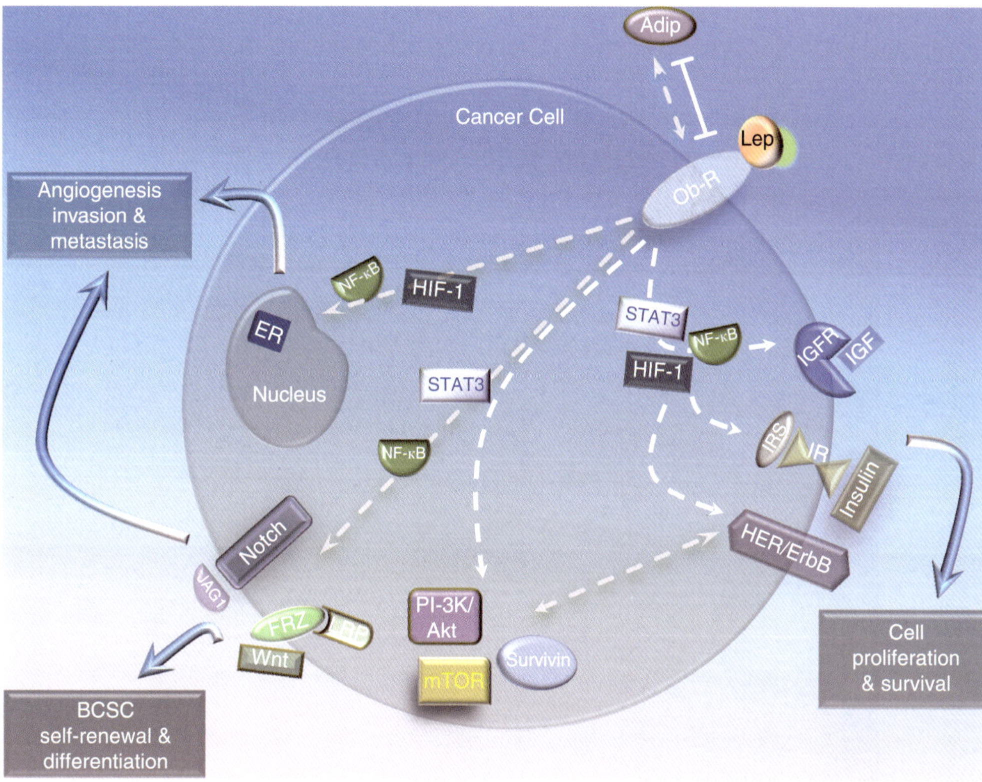

Figure 26.1. Potential cross talk of leptin signaling with other pathways in cancer stem cells (CSC). Lep: leptin; Ob-R: leptin receptor; HER/ErbB, encode for RTK-transmembrane proteins; NF-κB: a transcription factor family, nuclear factor kappa-light-chain-enhancer of activated B cells; STAT3: signal transducer and activator of transcription 3; PI-3 K: phosphatidylinositol 3-kinase; mTOR: mammalian target of rapamycin; HIF-1α: hypoxia-inducible transcription factor 1; Notch: a family of transmembrane proteins that function as receptors for membrane bound ligands; Wnt: a family of proteins linked to developmental signals with specific roles in embryogenesis and cancer; FRZ (Frizzled): G protein-coupled receptor proteins of Wnt signaling pathway; LRP: a transmembrane protein that binds Frizzled; dashed arrows indicate the main mechanisms/factors involved in leptin cross talk actions. Tubular arrows indicate how the leptin cross talk to specific oncogenic proteins/kinases/transcription factors lead to the development of key processes involved in cancer growth and metastasis.

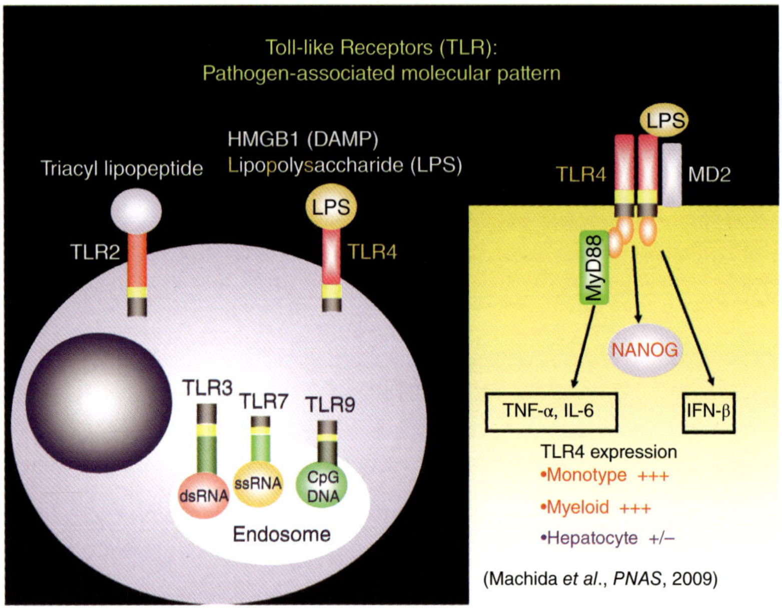

Figure 27.2. Toll-like receptor and pluripostency stem cell transcription factor NANOG.

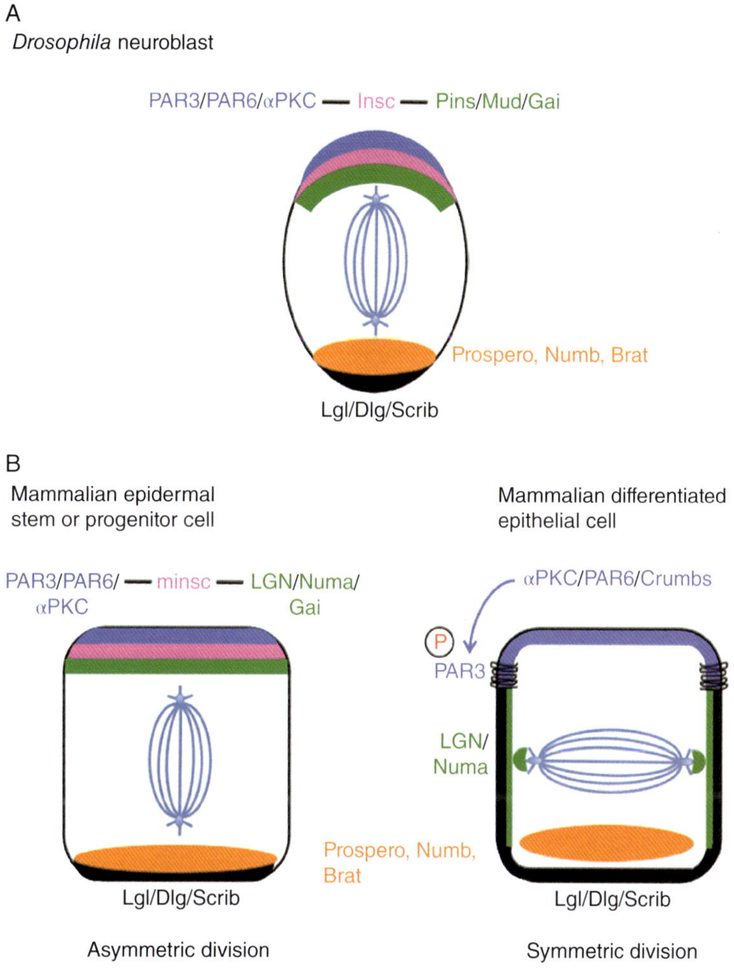

Figure 28.4. Stem cells and asymmetric division. **A**. Asymmetric division in *Drosophila* neuroblasts is controlled by apical protein complexes and basal cell fate determinants. **B**. Many of the complexes regulating *Drosophila* neuroblasts are conserved in mammalian cells. In the epidermis, the spatial distribution of these complexes is critical for the asymmetric division of stem/progenitor cells and the symmetric division of differentiated cells.

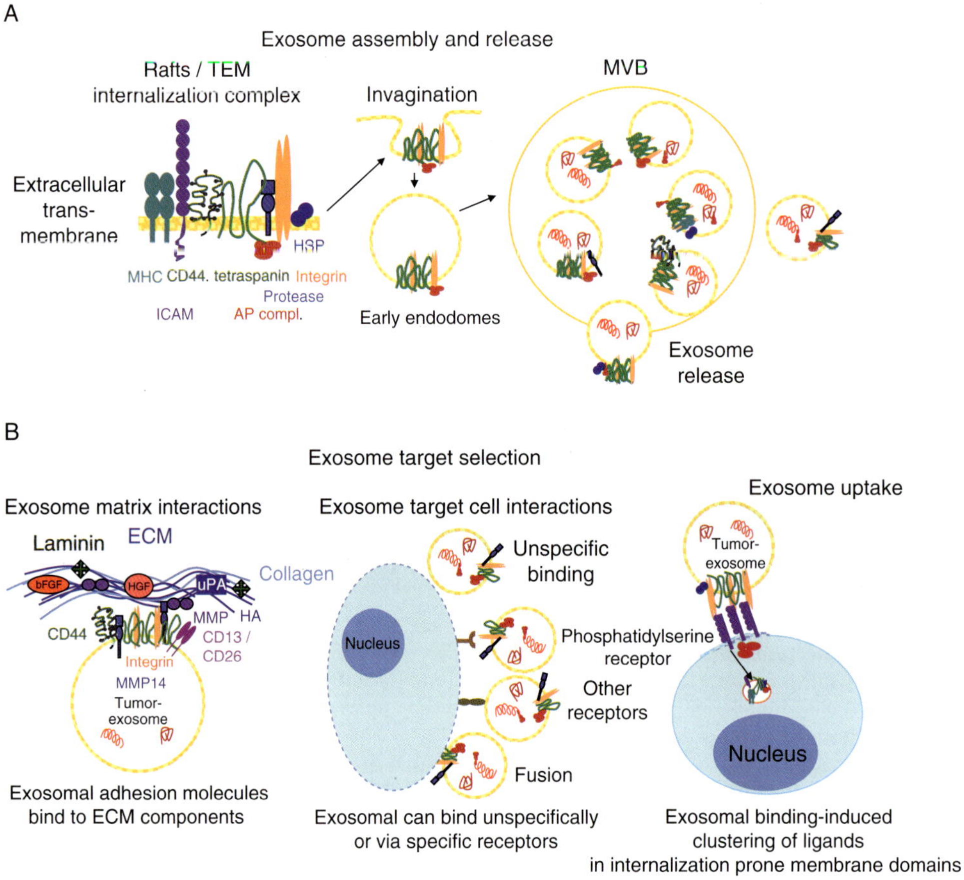

Figure 29.1. Tumor-exosomes and metastasis. **A**. Exosomes derive from early endosomes, which are generated by invagination of defined membrane domains like rafts and TEM. Accordingly, they are rich in proteins/protein complexes located in these microdomains as well as in fission, scission, and vesicular transport molecules. Upon invagination of early endosomes into MVB, cytosolic proteins, mRNA, and miRNA are recruited. **B**. Exosomes bind selective targets, which can be ECM or cellular membrane proteins. The selectivity of binding is facilitated by the engagement of protein clusters. There is strong evidence that exosome-uptake requires or is greatly facilitated by clustered ligands in internalization-prone membrane domains.

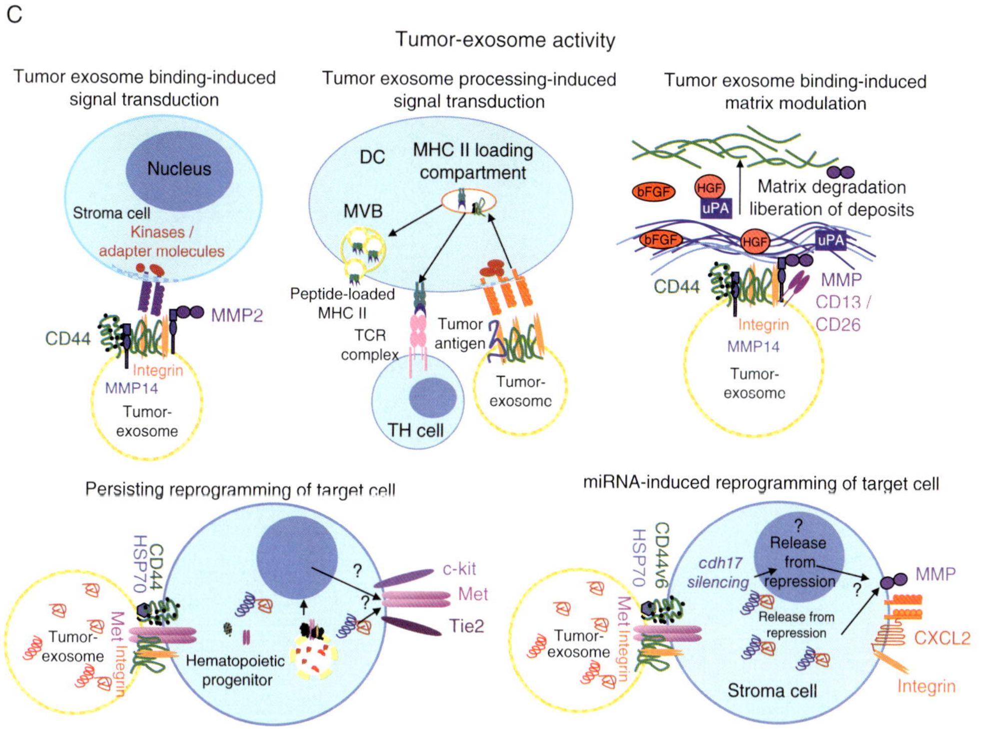

Figure 29.1. (*Continued*). **C**. Exosomal proteins, mRNA, and miRNA are function competent. Thus, binding to raft- or TEM-located ligands, these domains harboring cytosolic signal transduction molecules at the inner membrane, can initiate signal transduction in such a way that the target cell becomes activated. For DC it was described that uptaken exosomes will be transported into the MHC II assembly vesicles so that exosomal tumor antigens will be processed, and peptides are loaded on MHC II molecules, which allows for T helper activation. Alternatively, DC delivers exosomes carrying the tumor peptide-loaded MHC II as well as accessory molecules, so that DC-exosomes are fully equipped for T cell activation. Tumor-exosomes are rich in adhesion molecules like integrins, CAMs, and CD44 as well as in proteases including membrane-bound MMPs. Adhesion molecules are engaged in ECM protein binding, which become partly digested by exosomal proteases accompanied by liberation of matrix incorporated growth factors and proteases. The modulated matrix and the released factors can promote host and tumor cell motility and activation. Finally, tumor-exosomes can persistently reprogram their target cells. The mode whereby this is achieved has not been fully unraveled, but exosomal miRNA likely plays a decisive role by mRNA silencing and also by release from repression by the primary mRNA target.

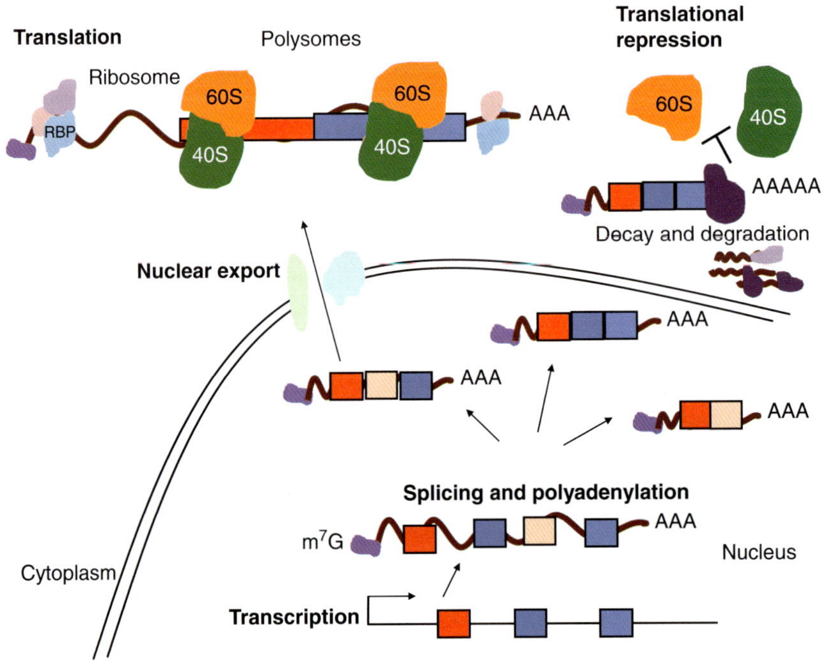

Figure 31.1. Cellular stages during mRNA biogenesis. A general pathway for mRNA processing and RBP regulation: transcription, splicing, polyadenylation, mRNA export, and translational stabilization or repression. Colored squares indicate exons and m7G; 5′7-methylguanosine cap.

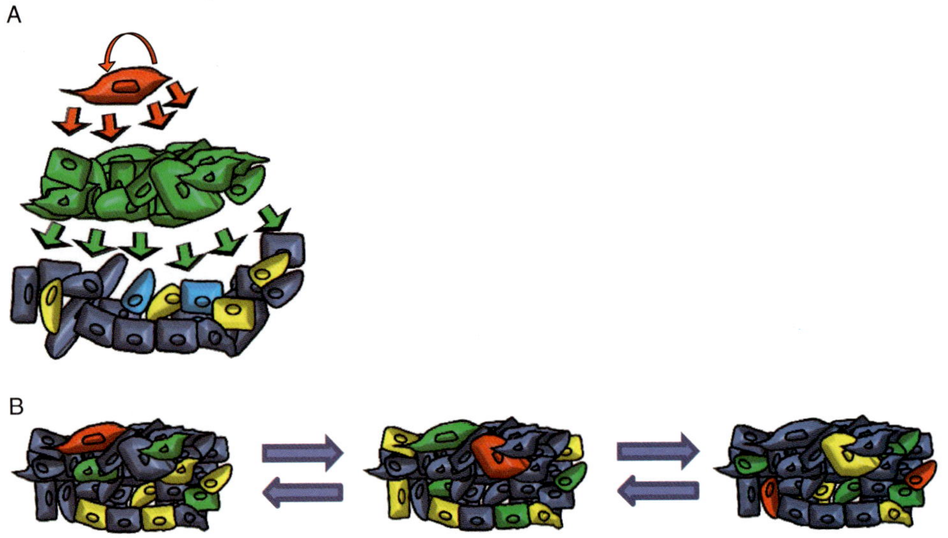

Figure 33.1. Models describing the role of putative cancer stem cells in melanoma. **A**. The hierarchic model predicts a top-down pathway in the differentiation of cancer stem cells to more differentiated cells that make up the melanoma cell mass and which lost tumor initiating capacities. **B**. The dynamic model assumes that nearly every melanoma cell has the capability of initiating cancer; the phenotype and functional capacity of each individual cell is flexible and expressed in a stochastic manner orchestrated by cell-intrinsic and/or micro-environmental conditions.

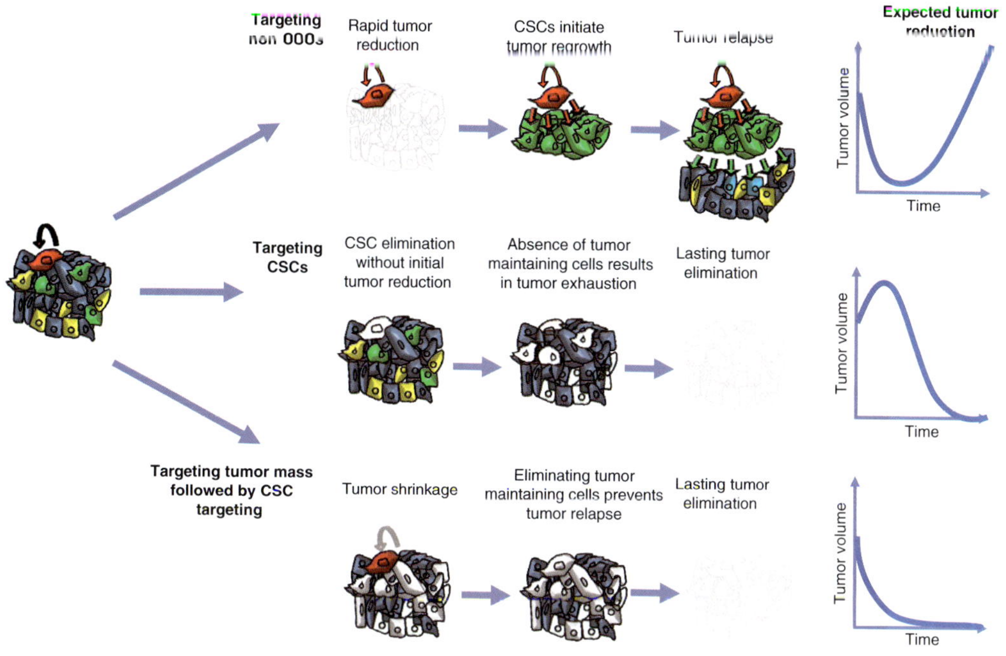

Figure 33.2. Cancer stem cell- and non-stem-cell-directed therapies will generate typical patterns of tumor responses. Targeting the tumor cell mass without targeting the melanoma sustaining cells will result in rapid tumor regression; however, the tumor will relapse fueled by drug-resistant melanoma stem cells. The tumor will initially progress when targeting CSCs due to the low number of CSCs in the tumor mass and will finally shrink due to exhaustion. This process may occur in small tumors more rapidly than in major tumor lesions. Combined strategies that "debulk" the tumor mass followed by targeting stem cells are expected to cause rapid tumor reduction and lasting tumor elimination.

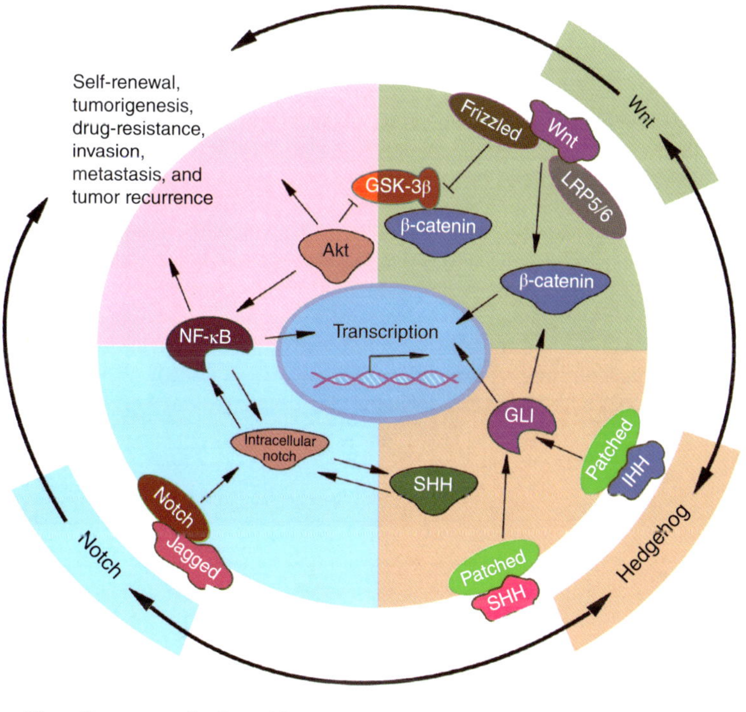

Figure 34.1. Signaling network altered in CSCs and its role in cancer development and progression.

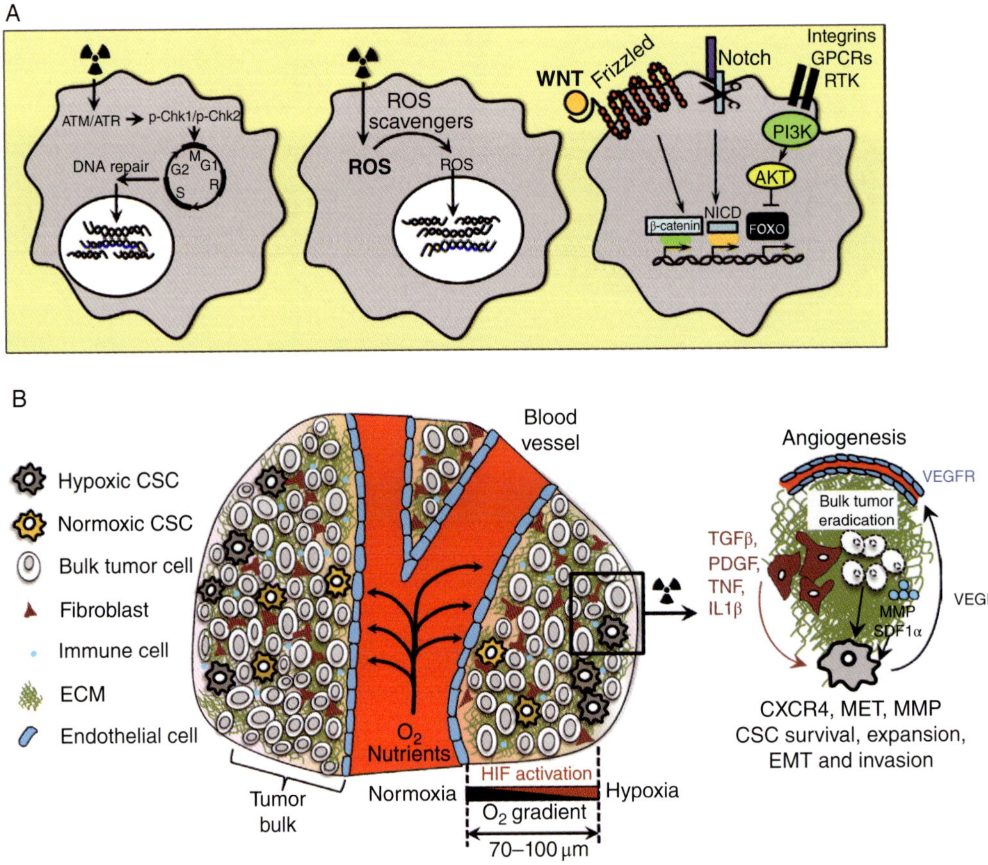

Figure 35.1. The mechanisms underlying CSC radioresistance. **A**. Intrinsic cancer stem cells radioresistance is attributed to a more efficient activation of the DNA damage response, intracellular defense against reactive oxygen species and activation of the cell survival pathways, including PI3K/AKT, WNT/β catenin, and Notch. **B**. The putative niche may protect CSCs from radiation damage. Eradication of the bulk tumor cells by conventional therapy could induce accelerated repopulation of CSCs. Hypoxic environment activates HIF signaling pathway and promotes CSC expansion and radioresistance. Ionizing radiation induces the production of the growth factors and cytokines, such as PDGF, IL1β, TNF, TGF β by irradiated fibroblasts. Irradiated immune cells, such as tumor-associated macrophages, produce CXCL12 and matrix metalloproteases (MMP) and create a microenvironment that promotes the survival of tumor and endothelial cells. In summary, an irradiated tumor bed promotes carcinogenesis and accelerates the development of highly invasive CSC phenotypes.

Chapter 21

Cancer Stem Cells and the Stromal Microenvironment

Li Li[1] and David A. Margolin[2]

[1] *Laboratory of Translational Cancer Research, Ochsner Clinic Foundation, New Orleans, LA, USA*

[2] *Department of Colon and Rectal Surgery, Ochsner Clinic Foundation, New Orleans, LA, USA and Department of Surgery, Ochsner Clinical School, the University of Queensland School of Medicine, Queensland, Australia*

Introduction

The current treatments of most solid organ tumors are surgically based, with chemotherapy and radiation either pre- or postoperative depending on tumor stage and histological grade (Siegel et al., 2012a). For example, the standard treatment of colon cancer is resectional therapy with the addition of postoperative chemotherapy for either lymph node (LN) positive cancers or those with poor pathologic features such as poor differentiation or lymphovascular invasion. Unfortunately, despite appropriate surgery and standard chemotherapeutic treatment, up to 50% of cancer will recur. Local recurrence or distant organ metastasis are the major causes of cancer mortality (Anderson et al., 2011; Vaiopoulos et al., 2012). Unfortunately, conventional treatment fails to effectively prevent extranodal recurrence, even in cases of successful eradication of all visible tumors (SEER, 2012; Siegel et al., 2012b). When such recurrences occur, the cancer cells appear to have acquired a chemoresistant phenotype. This chemoresistance can be associated with genetic alterations within the cancer cells, but recent studies proposed that recurrence is associated with the presence of cancer stem cells, or tumor-initiating cells, (jointly CSC) (Gerger et al., 2011; Merlos-Suarez et al., 2011) and their interaction with the tumor microenvironment (Grange et al., 2011; Meads et al., 2009). Current cancer therapies inadequately treat this rare, but highly significant population of CSC (Todaro et al., 2010; Vaiopoulos et al., 2012) and ignore the tumor-nurturing role that tumor microenvironment, specifically the LN microenvironment, plays in cancer recurrence. Thus, an alternative therapeutic approach is needed.

Cancer Stem Cells

Recent evidence indicates the functional heterogeneity of cancer cells is a result of cell differentiation (Dalerba et al., 2007a), and a specific cell population of CSC exists that may be identified by cell surface markers, such as CD133, CD44, aldehyde dehydrogenase 1 (ALDH1) enzyme expression, or side population (SP). However, the specific and exclusive CSC marker(s) are cancer dependent (Clevers, 2011; Visvader and Lindeman, 2008). CSCs are similar to normal stem cells in that they have the ability to

Cancer Stem Cells, First Edition. Edited by Vinagolu K. Rajasekhar.

self-renew while producing differentiated daughter cells (Reya et al., 2001). However, they are resistant to conventional chemoradiation treatment (Merlos-Suarez et al., 2011; Overdevest et al., 2011). Thus, in addition to conventional cancer treatments, targeting the CSC population may be essential to preventing recurrence or metastasis.

Colorectal Cancer Stem Cell (Co-CSC)

In the U.S., colorectal cancer (CRC) is the third most common malignancy and the second most common cause of cancer-related mortality, with an estimated incidence of 143,000 cases and 51,000 deaths per year (Siegel et al., 2012b). Co-CSC have been found to express a variety of surface markers, including CD133 (Haraguchi et al., 2008; O'Brien et al., 2007; Todaro et al., 2010), ALDH1 (Carpentino et al., 2009), epithelial-specific antigen (Ep-CAM or CD326), CD44, and CD166 (Dalerba et al., 2007b). CD133 is a transmembrane glycoprotein molecule with a molecular weight of 120 kDa on chromosome 4p15.32. $CD133^+$ cancer cells were shown to have the ability to self-renew, retain tumorigenicity, and regenerate a tumor after treatment (Dalerba et al., 2007b; O'Brien et al., 2007). In stage IIIB colon cancer, recurrence also correlates with the percentage of $CD133^+$ cells present in the original tumor (Li et al., 2009). Although $CD133^+$ cells meet some of the characteristics of CSC, its specificity as a true Co-CSC marker is still unclear (Shmelkov et al., 2008). $CD133^+$ cell might represent a heterogeneous population of cells, and CD133 might also lack a functional role in tumor initiation (Horst et al., 2009). Recently we used CRC cell lines HT-29 and HCA-7 and CRC patient specimens to show that a small proportion of Co-CSC expressing both CD133 and C-X-C chemokine receptor type 4 (CXCR4), a membrane-bound receptor for stromal cell-derived factor-1 or chemokine (C-X-C motif) ligand 12 (CXCL12), demonstrated increased tumor-initiating ability in immunodeficient mice in the presence of the LN stromal cell line, HK (Kim et al., 1994; Kim et al., 1995), and HK cell conditioned media. In addition, these double-positive Co-CSCs were enriched in chemotherapy-resistant cell populations (Margolin et al., 2011). Thus, CD133 and CXCR4 may be better markers for drug-resistant Co-CSC.

Follicular Lymphoma Cancer Stem Cell (FL-SC)

Follicular lymphoma (FL) is the second most common form of non-Hodgkin's lymphoma (NHL) in the Western Hemisphere, representing nearly 25% of all NHL (Armitage and Weisenburger, 1998). FL arises from B cells, with a clinical course that is frequently indolent and responsive to chemotherapy. However, multiple relapses following treatment are common. Over half of the patients who recur become refractory to treatment (van Oers et al., 2006).

A quiescent population of drug-resistant CSC has been identified in the SP fraction in various malignancies, including Hodgkin's lymphoma (Das et al., 2008; Ho et al., 2007; Jones et al., 2009; Wu and Alman, 2008; Zhang et al., 2009). The SP fraction expressed the drug resistant gene ATP-binding cassette subfamily G member 2 (*ABCG2*) and demonstrated higher tumorigenic capacity than the non-SP fraction (Das et al., 2008; Jakubikova et al., 2011; Kruger et al., 2006). Using a CSC enrichment technique to isolate SP fraction from FL cell line and patient specimens, we found a significantly higher percentage of cells in the SP cell fraction formed colonies, initiated tumors, and was resistant to chemo and irradiation treatment; confirming SP cells obtained from FL were highly enriched with CSC. Most importantly, FL-SCs interact with follicular dendritic cell (FDC)/HK cells in a CXCL12/CXCR4 interaction-dependent manner to maintain tumorigenicity (Lee et al., 2012).

Other Cancers

CSCs and their interaction with tumor stromal cells were also discovered in other cancers, such as bladder, brain, breast,

Table 21.1. CSC and Tumor Microenvironment Studies

Cancer Type	CSC Markers	Stromal Factors	Supporting Function	Prognostic or Therapeutic Target/Agent	References
Bladder	$CD44^{+}/CD47^{+}$	$SIRPA^{+}$ macrophage	An inhibitory signal for macrophage phagocytosis	CD47/CD47 antibody	(Chan et al., 2009)
Brain	$CD133^{+}$	MT1-MMP MMP-9	Expansion Invasion	MT1-MMP and MMP-9/	(Annabi et al., 2008)
Breast	ALDEFLUOR	CCL2 expression in activated fibroblasts	Activation of NOTCH Tumorigenesis	/NOTCH targeting agents	(Tsuyada et al., 2012)
Breast	$CD44^{+}/CD24^{-/low}$	Absence of stromal Caveolin-1	Correlated to clinic outcome	Stromal Caveolin⁻/	(Witkiewicz et al., 2009a; Witkiewicz et al., 2009b)
Colorectal	$CD133^{+}/CXCR4^{+}$	CXCL12	Migration Tumorigenesis	CXCL12 and CXCR4/ AMD3100	(Margolin et al., 2011)
Follicular lymphoma	SF	CXCL12	Migration Tumorigenesis	CXCL12 and CXCR4/ AMD3100	(Lee et al., 2012)
Melanoma	$CD133^{+}$	CXCL12/CXCR4	Metastasis	CXCL12/CXCR4	(Kim et al., 2010)
Ovarian	$CD44^{+}/CD117^{+}$	Tissue transglutaminase (TG2)	EMT Metastasis Growth	TG2 or TGF-β/	(Cao et al., 2012)
Prostate	ALDEFLUOR	α_v-Integrins	EMT Angiogenesis Bone-metastasis	α_v-Integrins/nonpeptide α_v-Integrin antagonist GLPG0187	(van der Horst et al., 2011)

ovarian, and prostate cancers (Visvader and Lindeman, 2008). CSCs exist in various cancers, and their markers differ depending on cancer type. In some cancers, CSCs interact with and are supported by tumor stromal cells for survival, proliferation, tumorgenesis, drug resistance, and possible metastasis and recurrence (Table 21.1).

Cancer Stromal Microenvironment

Studies showed that not all cancer cells are tumorigenic (Wang et al., 2007), and some cancer cells require co-injection with LN stromal cells to form tumor in immunodeficient mice (Li et al., 2004; Margolin et al., 2011), indicating that CSC are supported by microenvironmental factors produced by LN stromal cells (Vermeulen et al., 2010). In brain tumors, the interaction between brain CSC and signals from the local microenvironment is significant for region-specific tumorigenesis (Gilbertson and Gutmann, 2007). In mouse renal carcinoma cell studies, agarose macrobeads selectively support CSC in 3-dimensional culture that mimics in vivo tumor microenvironment (Smith et al., 2011). The tumor microenvironment is now recognized as a major factor influencing cancer growth and the outcome of therapy. Although the niche cells are not malignant per se, their role in supporting cancer growth is vital for the tumor survival. Thus, they have become an attractive target for chemotherapeutic agents (Basak et al., 2009). Clearly, environment-mediated drug resistance is induced by signaling events from the tumor microenvironment and is likely to be reversible (Meads et al., 2009; Roodhart et al., 2011).

Lymph Node Stromal Microenvironment—Follicular Dendritic Cell

Evidence shows that CSCs that develop in lymphoid follicles such as in FL, or CSCs that metastasize to the LN are stimulated by the LN microenvironment. The major problem with FL treatment is relapse and transformation. The transformation of FL to therapy-resistant aggressive large B cell lymphoma (de Jong and de Boer, 2009) is associated with the induction of stromal gene signatures including CXCL12 (Lenz et al., 2008). The mechanism of relapse may involve the interaction of tumorigenic FL cells with the stromal cell counterpart present in LN (Arai et al., 2000). One of the putative stromal cell types that may interact with FL cells is the FDC. FDCs are present in the germinal center (GC) of lymphoid follicles, the site of FL origin (Klein et al., 1998). FDCs were found to initiate and maintain a pro-tumorigenic microenvironment (Torlakovic et al., 2002) by producing appropriate cytokines (Li and Choi, 2002) and chemokines that promote lymphoma cell proliferation (Lee et al., 2012; Li et al., 2004; Li et al., 2000).

FDCs are the most abundant stromal cell type in the LN microenvironment. Although their tumor-promoting effects are well known, their origin was recently suggested to arise from ubiquitous perivascular precursors expressing platelet-derived growth factor receptor beta (Krautler et al., 2012). LN metastasis is one of the strongest negative prognostic factors for a number of cancers. The majority of CRC patients present with regional LN involvement, stage III disease, suggesting that the LN microenvironment plays a significant role promoting extranodal recurrence and further metastasis (SEER, 2012). For CRC, FDC are unique LN stromal cells that display both autocrine and paracrine properties, analogous to cancer-associated stromal fibroblasts, that nurture CRC cells through production of various cytokines and growth factors (Nakagawa et al., 2004). CRC cells interact with tumor-fostering stromal cells and the extracellular matrix in a protective fashion decreasing chemotherapy-induced apoptosis (Kouniavsky et al., 2002).

To evaluate the role of LN stromal cells in tumor growth, we established in vivo tumor models using CSC and FDC cell line, HK cells. Although derived from tonsilar cells, the HK cell line functionally resembles primary FDC in expressing smooth muscle antigen, von Willebrand factor, and vimentin, but not CD31, and supporting GC B cells and lymphoma growth (Li and Choi, 2002; Li et al., 2004). In our in vivo model, the addition of HK cells increased tumor formation especially in lower dosages of CSC, suggesting that FDC/HK cells play a key role in cancer cell survival, tumor initiation, and growth in vivo in both FL and CRC models (Lee et al., 2012; Margolin et al., 2011).

CXCL-12/CXCR4 Signaling

Various chemokines play important roles in stromal cell/CSC niche interaction. CXCL12 is a chemokine that regulates many essential biological processes including revascularization, cellular adhesion, and tumorigenesis (Madlambayan et al., 2009). CXCL12 is also one of many soluble microenvironmental factors produced by FDC in a paracrine fashion (Husson et al., 2001). CXCL12/CXCR4 is involved in tumor metastasis and extranodal recurrence in CRC (Kim et al., 2005; Schimanski et al., 2005; Zeelenberg et al., 2003; Kucia et al., 2005), and the metastatic activity of $CD133^{+}CXCR4^{+}$ CSCs is increased in pancreatic cancer (Hermann et al., 2007). Downregulation of CXCR4 significantly decreased cell migration and invasion only in pancreatic CSC cocultured with pancreatic stromal cells (Moriyama et al., 2010).

HK cells produce CXCL12 and CXCR4 and are active on SP cells in FL and Co-CSC. In transwell migration assays, FL-SC specifically migrated toward HK cells and CXCL12 in a dose-dependent manner, and their migration was inhibited by the presence of AMD3100, a specific small molecule inhibitor of CXCL12/CXCR4 signaling.

Similarly, the CXCL12/CXCR4 axis is also associated with in vitro CRC migration, lymphatic and distant dissemination, disease recurrence, and decreased survival rate (Kim et al., 2005; Matsusue et al., 2009; Schimanski et al., 2005). Drug-resistant CRC cells showed increased expression of CD133 and CXCR4. This is in agreement with Dessein and colleagues (2010), who demonstrated that the induction of CXCR4 acts as a major mechanism underlying invasion in drug-resistant HT-29 cells. This phenomenon is not limited to CRC, as $CD133^{+}CXCR4^{+}$ migrating CSC play a crucial role in tumor initiation, growth, and metastasis in human pancreatic cancer, prostate, and breast cancer (Darash-Yahana et al., 2004; Hermann et al., 2007).

Cell-Cell Contact

Tumor stromal microenvironment could promote cancer chemoresistance by direct cell-cell contact. Xu and colleagues (2008) proposed that TGF-β1 produced by bone marrow (BM) stromal cells promotes the survival and chemoresistance of leukemia cells via direct cell-to-cell interactions. The blockade of TGF-β signaling using LY2109761, effectively inhibits the pro-survival signaling, and enhances the efficacy of chemotherapy against myelo-monocytic leukemic cells in the BM microenvironment. Rafii and colleagues (2008) demonstrated the capacity of Hospicells, an original type of stromal cells, to confer chemoresistance to ovarian and breast cancer cells by direct cell-cell contact and the exchange of membrane patches and multiple drug-resistant proteins (Lis et al., 2011).

In FL, we found SP cells are enriched in cell populations that are adherent to HK cell. They also express ABCG2, a multidrug resistance transporter, and they are in close contact with HK cells in the initial stage of tumor formation in immunodeficient mice. The HK cell dependence of FL-SC in tumorigenesis is in both a cell-cell contact fashion and in a CXCL12/CXCR4-dependent manner, since AMD3100 effectively inhibited HK cell promoted in vivo tumor growth of the FL-SC (Lee et al., 2012).

Cytokines and Other Factors

In addition to stromal cell factors, CSCs also interact with and are regulated by other cells in the tumor microenvironment via inflammatory cytokine networks, such as interleukin (IL)-1, IL-6, and IL-8 (Korkaya et al., 2011). Recombinant human IL-1 alpha (rIL-1 alpha) enhances chemotherapy-induced growth inhibition in HCT116 colon cancer cells (Geoffroy et al., 1994). In plasma cell cancer multiple myeloma (MM), the adhesion between MM cells and BM fibroblasts leads to the increased secretion of IL-6 (Thomas et al., 1998). This pleiotropic cytokine has demonstrated the capacity to induce the resistance of MM cells to apoptotic stimuli and chemotherapeutic drugs via the Jak/STAT pathway and the expression of the anti-apoptotic protein Bcl-xL (Shain et al., 2000). Recently, IL-6-mediated STAT3 activation was found to play a specific role in maintaining an inflammatory positive feedback loop in breast CSC (Iliopoulos et al., 2009; Iliopoulos et al., 2010). In Co-CSC, it was found that STAT3 was constitutively activated and that these cells were sensitive to STAT3 or IL-6 inhibition of tumorigenesis (Lin et al., 2011). In addition, blocking CXCR1, an IL-8 receptor, using specific antibodies or small molecule inhibitors targeted breast cancer CSC (Ginestier et al., 2010).

Autocrine production of IL-4 in colon cancer cells was reported to contribute to chemoresistance (Di Stefano et al., 2010) by protecting the tumorigenic CSC through upregulation of anti-apoptotic genes (Todaro et al., 2007; Todaro et al., 2008). In contrast, overexpression of IL-12, a potent immunomodulatory cytokine, reduced the expression of IL-4 and STAT6 in Co-CSC, and decreased tumorsphere formation and tumor initiation (Yin et al., 2012). Cytokines are also used as agents for differentiation therapy. In renal cancer, IL-15 was proposed to direct the epithelial differentiation of CSC to differentiated nontumorigenic cells that are sensitive to chemotherapy (Azzi et al., 2011).

In addition to cytokins, other factors may also be involved in supporting CSC. For example, in bladder cancer, when cancer cells undergo the epithelial to mesenchymal transition (EMT), their invasiveness, drug resistance, angiogenesis, and metastatic ability are increased. These changes give rise to a more aggressive tumor type, while the tumor supportive microenvironment (tumor-associated stromal cells, extracellular matrix) plays a key role in tumorigenesis, tumor progression, and metastasis formation (van der Horst et al., 2012).

To disseminate and metastasize, the cancer cells activate the EMT pathway, switching to a migrating cancer stem cell (MCSC) phenotype. This may be induced by the tumor microenvironment that secretes EMT-inducing growth factors and the interaction with the extracellular matrix. The MCSCs can subsequently enter the blood circulation, disseminate, extravasate, and eventually colonize the target organs to form (macro) metastasis. This explains metastases at distant sites, as this distant and different microenvironment might not secrete EMT-inducing signals (Brabletz et al., 2005; van der Horst et al., 2012).

Significance

Cancer drug resistance and recurrence are major causes of mortality. The observed interaction between CSC and stromal cell provides critical insights into the mechanism of drug resistance and recurrence. The CSC models suggest that niche cell signaling plays an important role in CSC-mediated tumorigenesis and resistance to chemotherapy. In both FL and CRC studies, in vitro and in vivo drug-resistant CSCs interact with stromal cell FDC via CXCL12/CXCR4 signaling to maintain tumorigenicity, suggesting that the CXCL12 is one of the candidate cytokines that FDC secrets to modulate the tumorigenicity of these diseases (Figure 21.1). CSCs in both FL and CRC were enriched by chemotherapy in the presence of stromal cells suggesting that

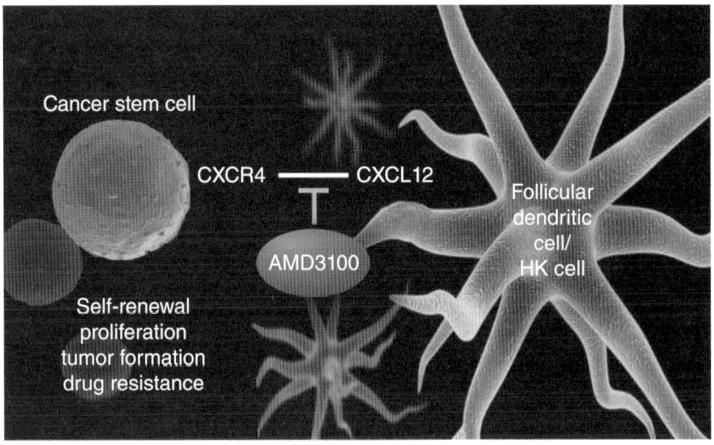

Figure 21.1. Lymph node stromal cell support CSC via paracrine CXCL12/CXCR4 signaling. For color detail, please see color plate section.

targeting CSC alone may not suffice to minimize recurrence and that future treatments must target LN microenvironmental as well.

Our findings are in agreement with a recent report that Co-CSC marker Wnt-signaling activity was regulated by microenvironmental myofibroblast-secreted factors (Vermeulen et al., 2010), implicating the microenvironment as a dominant factor in Co-CSC; as well as a recent study showing CXCL12 expression in B cell lymphoma was associated with poor prognosis (Lenz et al., 2008). FDC may promote CSC tumor growth in two ways: (1) supporting CSC survival and activation for tumor initiation by providing soluble factors, such as CXCL12, and (2) enhancing the host response to increase tumor angiogenesis.

Thus, CSC and the environment that protects them could be therapeutic targets to minimize recurrence. Studies that identify and characterize the CSC and its interactions with the tumor microenvironment are an important step in the development of biologically based, curative treatments for cancer. Understanding the essential signals for tumor chemoresistance and survival produced by tumor stromal will help develop prognostic biomarkers and novel therapeutic strategies (Burger et al., 2009). For example, a predictive gene signature was identified in bladder cancer studies that could serve as an indicator of tumor progression (Chan et al., 2009). In addition, CD47 plays a significant role in inhibiting phagocytosis, making it a prime drug target such that CD47 inhibition would enhance macrophage phagocytosis of tumor cells (Table 21.1).

Prognostic Markers

Tumor invasion and regional LN metastasis are important factors for determining cancer prognosis. The 5-year survival rate of the American Joint Committee on Cancer stage I CRC is 90% and decreases to 75% and 50% for stage II and III patients, respectively. Prognostic markers are urgently needed, especially for stage II CRC patients. Currently, two prognostic biomarkers are commercially available, the Oncotype DX Colon Cancer Assay in the United States and Coloprint in Europe (Clark-Langone et al., 2010; Tan and Tan, 2011). However, they are prognostic biomarkers and not predictive biomarkers. The development of biomarkers that are both prognostic and predictive will greatly impact the treatment of both LN-positive and negative CRC. Several studies have focused on such prognostic markers. Saiki and colleagues (2009), using RT-PCR analysis, revealed that the expression of stem cell-related genes, including

LIN28 and *SOX2*, correlated with LN metastasis indicating that the Co-CSC population is increased at metastatic sites. When putative CSC markers were tested in CRC-tumor buds, ABCG5 and EpCAM were significantly associated with a poorer prognosis (Hostettler et al., 2010). However, most of the published studies have been conducted with bulk tumor samples without differentiating the response between Co-CSC and non-CSC. Since Co-CSCs are such a small population, precise analysis of prognostic markers within Co-CSC may yield more accurate prognostic significance.

In FL, CXCL12 and CXCR4 expression may serve as prognostic markers for risk of disease transformation, since FL-SCs express higher levels of CXCR4, and inhibition of CXCL12/CXCR4 interaction abolished FLK-1 cell tumor formation in NOD/SCID mice. Recent reports showed that migrating populations enrich CSC in neuroblastoma SP cells (Das et al., 2008), and migrating CD133$^+$CXCR4$^+$ CSC are essential for pancreatic adenocarcinoma metastasis (Hermann et al., 2007).

Prognostic biomarkers will be useful in identifying those patients at greatest risk of recurrence and in need of aggressive and potentially novel therapies. Witkiewicz and colleagues (2009a; 2009b) identified the loss of stromal Caveolin-1 as a surrogate biomarker associated with an increase in cell cycle progression, the secretion of growth factors, and the angiogenic potential in the tumor microenvironment.

Novel Therapeutic Targets

Current treatments for cancers are not curative for the majority of patients and were designed and deemed successful based on the response of the bulk population of tumor cells (Reya et al., 2001). Effects on a rare CSC population or on cells in the supportive microenvironment are largely unknown, but presumed to be inadequate given the high rates of cancer recurrence. A deeper understanding of the essential signals produced by tumor stromal cells to promote CSC survival can suggest new therapeutic targets (for example, targeting the CSC niche interaction).

Antibody-mediated inhibition of CXCL12/CXCR4 signaling completely abrogates the CXCL12-mediated cell migration of lymphocytic leukemia and lymphoma, as well as the migration of lymphoma cells toward LN stromal cells (Kawaguchi et al., 2009; O'Callaghan et al., 2012; Zhang et al., 2006). Additionally, AMD300-inhibited infiltration of lymphomatous cells into liver and lung tissues by inhibiting CXCL12/CXCR4 signaling, thereby disrupting the CSC niche and making the CSC more susceptible to chemotherapy (Yang and Wechsler-Reya, 2007). The finding that inhibition of the CXCL12/CXCR4 signaling reduced the migration of lymphoma cells and Co-CSC toward stromal cells and also inhibited tumor formation is consistent with previous reports, and highlights one mechanism of action of AMD3100 in the elimination of CSC. AMD3100 has been used clinically in the setting of mobilization of normal hematopoietic stem cells prior to autologous stem cell transplantation. It is hypothesized that AMD3100 disrupts the CSC niche and makes the CSC more susceptible to chemotherapy. Clinical trials are ongoing, investigating the safety and efficacy of AMD3100 in combination with chemotherapy in a variety of hematologic malignancies, including lymphomas (ClinicalTrials.gov Identifier: NCT00694590 and NCT00512252).

Most solid tumor in vivo studies generate human tumor xenografts from immunodeficient mice (Visvader and Lindeman, 2008). We have established humanized tumor microenvironment models for FL and CRC (Lee et al., 2012; Margolin et al., 2011). In these models, co-inoculation of human stromal cells with cancer cells re-creates a humanized microenvironment similar to a LN. Results show that the humanized microenvironment is essential for CSC to form a tumor in immunodeficient mice, selectively supports CSCs survival from chemotherapeutic drugs, and enhances drug-resistant

CSC tumor formation through CXCL12/CXCR4 signaling. The CSC and stromal interaction in vivo model could serve as a humanized microenvironment model to identify and analyze key interactions and signaling molecules in CSC, and evaluate response to targeted agents.

In addition to CRC and FL, stromal/CSC interaction in various cancers requires the involvement of many molecules and agents (Table 21.1). Annabi and colleagues (2008) suggested that MT1-MMP and MMP-9, two MMPs that contribute to the blood-brain barrier opening and to the radio-resistance phenotype in brain tumor cells, may be promising new targets in the annihilation of CSC as an anticancer therapy. In the breast cancer model, Tsuyada and colleagues (2012) identified CCL2, STAT3, and NOTCH1 as potential therapeutic targets in deterring CSC-stimulating cancer-host cross talk and as a method of defeating CSC-mediated disease progression and treatment resistance. Kim and colleagues (2010) suggested that the combined targeting of the CSCL12/CXCR4 axis and the implementation of dacarbazine treatment could serve as a therapy to block chemoresistant $CD133^{+}$ melanoma CSC metastasis toward a lymphatic metastatic niche (Kim et al., 2010). In addition, the potential use of TGF-β as a therapy in ovarian cancer was revealed due to evidence of its involvement in cancer invasion and tumor progression through the regulation of tissue transglutaminase (TF2) (Cao et al., 2012). Thus, closer scrutiny of CSC and their supporting microenvironment, especially LN microenvironmental factors, will lead to a better understanding of the mechanisms of cancer recurrence and metastasis and, ultimately, to novel targeted therapies.

Prospective

In conclusion, not all cancer cells constituting a tumor are the same; a small population of CSC exists (Zeki et al., 2011). Cancer poses a problem not only within its primary tumorigenic cells but also in its involvement with a microenvironment that specifically supports CSCs. Thus, the CSC population is responsible for recurrence and metastasis with help from its stromal environment (Hermann et al., 2007; Kucia et al., 2005). Therefore, combination treatment with a chemotherapy drug in conjunction with other therapies, such as stromal/CSC signaling-targeted therapy, may effectively minimize cancer recurrence (Tysnes, 2010).

Acknowledgments

We are very grateful to Ms. Barbara Siede for artistry illustration, Ms. Christina Chen, and Dr. Theodor Borgovan for editorial assistance. This work was partially supported by NIH/NCI grant R01 CA089057 (LL) and Research Grant from Ladies Leukemia League, Metairie, Louisiana (LL).

References

Anderson, E.C., Hessman, C., Levin, T.G., Monroe, M.M., and Wong, M.H. (2011). The Role of Colorectal Cancer Stem Cells in Metastatic Disease and Therapeutic Response. Cancers *3*, 319–339.

Annabi, B., Rojas-Sutterlin, S., Laflamme, C., Lachambre, M.P., Rolland, Y., Sartelet, H., and Beliveau, R. (2008). Tumor environment dictates medulloblastoma cancer stem cell expression and invasive phenotype. Molecular cancer research: MCR *6*, 907–916.

Arai, J., Yasukawa, M., Yakushijin, Y., Miyazaki, T., and Fujita, S. (2000). Stromal cells in lymph nodes attract B-lymphoma cells via production of stromal cell-derived factor-1. Eur J Haematol *64*, 323–332.

Armitage, J.O., and Weisenburger, D.D. (1998). New approach to classifying non-Hodgkin's lymphomas: clinical features of the major histologic subtypes. Non-Hodgkin's Lymphoma Classification Project. Journal of clinical oncology: official journal of the American Society of Clinical Oncology *16*, 2780–2795.

Azzi, S., Bruno, S., Giron-Michel, J., Clay, D., Devocelle, A., Croce, M., Ferrini, S., Chouaib, S., Vazquez, A., Charpentier, B., *et al.* (2011). Differentiation therapy: targeting human renal cancer stem cells with interleukin 15. Journal of the National Cancer Institute *103*, 1884–1898.

Basak, G.W., Srivastava, A.S., Malhotra, R., and Carrier, E. (2009). Multiple myeloma bone marrow niche. Current pharmaceutical biotechnology *10*, 345–346.

Brabletz, T., Jung, A., Spaderna, S., Hlubek, F., and Kirchner, T. (2005). Opinion: migrating cancer stem cells—an integrated concept of malignant tumour progression. Nature reviews Cancer *5*, 744–749.

Burger, J.A., Ghia, P., Rosenwald, A., and Caligaris-Cappio, F. (2009). The microenvironment in mature B-cell malignancies: a target for new treatment strategies. Blood *114*, 3367–3375.

Cao, L., Shao, M., Schilder, J., Guise, T., Mohammad, K.S., and Matei, D. (2012). Tissue transglutaminase links TGF-beta, epithelial to mesenchymal transition and a stem cell phenotype in ovarian cancer. Oncogene *31*, 2521–2534.

Carpentino, J.E., Hynes, M.J., Appelman, H.D., Zheng, T., Steindler, D.A., Scott, E.W., and Huang, E.H. (2009). Aldehyde dehydrogenase-expressing colon stem cells contribute to tumorigenesis in the transition from colitis to cancer. Cancer research *69*, 8208–8215.

Chan, K.S., Espinosa, I., Chao, M., Wong, D., Ailles, L., Diehn, M., Gill, H., Presti, J., Jr., Chang, H.Y., van de Rijn, M., *et al.* (2009). Identification, molecular characterization, clinical prognosis, and therapeutic targeting of human bladder tumor-initiating cells. Proc Natl Acad Sci USA *106*, 14016–14021.

Clark-Langone, K.M., Sangli, C., Krishnakumar, J., and Watson, D. (2010). Translating tumor biology into personalized treatment planning: analytical performance characteristics of the Oncotype DX Colon Cancer Assay. BMC Cancer *10*, 691.

Clevers, H. (2011). The cancer stem cell: premises, promises and challenges. Nature medicine *17*, 313–319.

Dalerba, P., Cho, R.W., and Clarke, M.F. (2007a). Cancer stem cells: models and concepts. Annual review of medicine *58*, 267–284.

Dalerba, P., Dylla, S.J., Park, I.K., Liu, R., Wang, X., Cho, R.W., Hoey, T., Gurney, A., Huang, E.H., Simeone, D.M., *et al.* (2007b). Phenotypic characterization of human colorectal cancer stem cells. Proc Natl Acad Sci USA *104*, 10158–10163.

Darash-Yahana, M., Pikarsky, E., Abramovitch, R., Zeira, E., Pal, B., Karplus, R., Beider, K., Avniel, S., Kasem, S., Galun, E., *et al.* (2004). Role of high expression levels of CXCR4 in tumor growth, vascularization, and metastasis. FASEB journal: official publication of the Federation of American Societies for Experimental Biology *18*, 1240–1242.

Das, B., Tsuchida, R., Malkin, D., Koren, G., Baruchel, S., and Yeger, H. (2008). Hypoxia enhances tumor stemness by increasing the invasive and tumorigenic side population fraction. Stem Cells *26*, 1818–1830.

de Jong, D., and de Boer, J.P. (2009). Predicting transformation in follicular lymphoma. Leukemia & lymphoma *50*, 1406–1411.

Dessein, A.F., Stechly, L., Jonckheere, N., Dumont, P., Monte, D., Leteurtre, E., Truant, S., Pruvot, F.R., Figeac, M., Hebbar, M., *et al.* (2010). Autocrine induction of invasive and metastatic phenotypes by the MIF-CXCR4 axis in drug-resistant human colon cancer cells. Cancer research *70*, 4644–4654.

Di Stefano, A.B., Iovino, F., Lombardo, Y., Eterno, V., Hoger, T., Dieli, F., Stassi, G., and Todaro, M. (2010). Survivin is regulated by interleukin-4 in colon cancer stem cells. Journal of cellular physiology *225*, 555–561.

Geoffroy, F.J., Allegra, C.J., Sinha, B., and Grem, J.L. (1994). Enhanced cytotoxicity with interleukin-1 alpha and 5-fluorouracil in HCT116 colon cancer cells. Oncology research *6*, 581–591.

Gerger, A., Zhang, W., Yang, D., Bohanes, P., Ning, Y., Winder, T., LaBonte, M.J., Wilson, P.M., Benhaim, L., Paez, D., *et al.* (2011). Common cancer stem cell gene variants predict colon cancer recurrence. Clinical cancer research: an official journal of the American Association for Cancer Research *17*, 6934–6943.

Gilbertson, R.J., and Gutmann, D.H. (2007). Tumorigenesis in the brain: location, location, location. Cancer research *67*, 5579–5582.

Ginestier, C., Liu, S., Diebel, M.E., Korkaya, H., Luo, M., Brown, M., Wicinski, J., Cabaud, O., Charafe-Jauffret, E., Birnbaum, D., *et al.* (2010). CXCR1 blockade selectively targets human breast cancer stem cells in vitro and in xenografts. The Journal of clinical investigation *120*, 485–497.

Grange, C., Tapparo, M., Collino, F., Vitillo, L., Damasco, C., Deregibus, M.C., Tetta, C., Bussolati, B., and Camussi, G. (2011).

Microvesicles released from human renal cancer stem cells stimulate angiogenesis and formation of lung premetastatic niche. Cancer research *71*, 5346–5356.

Haraguchi, N., Ohkuma, M., Sakashita, H., Matsuzaki, S., Tanaka, F., Mimori, K., Kamohara, Y., Inoue, H., and Mori, M. (2008). CD133 + CD44+ population efficiently enriches colon cancer initiating cells. Ann Surg Oncol *15*, 2927–2933.

Hermann, P.C., Huber, S.L., Herrler, T., Aicher, A., Ellwart, J.W., Guba, M., Bruns, C.J., and Heeschen, C. (2007). Distinct populations of cancer stem cells determine tumor growth and metastatic activity in human pancreatic cancer. Cell stem cell *1*, 313–323.

Ho, M.M., Ng, A.V., Lam, S., and Hung, J.Y. (2007). Side population in human lung cancer cell lines and tumors is enriched with stem-like cancer cells. Cancer research *67*, 4827–4833.

Horst, D., Scheel, S.K., Liebmann, S., Neumann, J., Maatz, S., Kirchner, T., and Jung, A. (2009). The cancer stem cell marker CD133 has high prognostic impact but unknown functional relevance for the metastasis of human colon cancer. The Journal of pathology *219*, 427–434.

Hostettler, L., Zlobec, I., Terracciano, L., and Lugli, A. (2010). ABCG5-positivity in tumor buds is an indicator of poor prognosis in node-negative colorectal cancer patients. World journal of gastroenterology: WJG *16*, 732–739.

Husson, H., Carideo, E.G., Cardoso, A.A., Lugli, S.M., Neuberg, D., Munoz, O., de Leval, L., Schultze, J., and Freedman, A.S. (2001). MCP-1 modulates chemotaxis by follicular lymphoma cells. British journal of haematology *115*, 554–562.

Iliopoulos, D., Hirsch, H.A., and Struhl, K. (2009). An epigenetic switch involving NF-kappaB, Lin28, Let-7 MicroRNA, and IL6 links inflammation to cell transformation. Cell *139*, 693–706.

Iliopoulos, D., Jaeger, S.A., Hirsch, H.A., Bulyk, M.L., and Struhl, K. (2010). STAT3 activation of miR-21 and miR-181b-1 via PTEN and CYLD are part of the epigenetic switch linking inflammation to cancer. Molecular cell *39*, 493–506.

Jakubikova, J., Adamia, S., Kost-Alimova, M., Klippel, S., Cervi, D., Daley, J.F., Cholujova, D., Kong, S.Y., Leiba, M., Blotta, S., *et al.* (2011). Lenalidomide targets clonogenic side population in multiple myeloma: pathophysiologic and clinical implications. Blood *117*, 4409–4419.

Jones, R.J., Gocke, C.D., Kasamon, Y.L., Miller, C.B., Perkins, B., Barber, J.P., Vala, M.S., Gerber, J.M., Gellert, L.L., Siedner, M., *et al.* (2009). Circulating clonotypic B cells in classic Hodgkin lymphoma. Blood *113*, 5920–5926.

Kawaguchi, A., Orba, Y., Kimura, T., Iha, H., Ogata, M., Tsuji, T., Ainai, A., Sata, T., Okamoto, T., Hall, W.W., *et al.* (2009). Inhibition of the SDF-1alpha-CXCR4 axis by the CXCR4 antagonist AMD3100 suppresses the migration of cultured cells from ATL patients and murine lymphoblastoid cells from HTLV-I Tax transgenic mice. Blood *114*, 2961–2968.

Kim, H.S., Zhang, X., and Choi, Y.S. (1994). Activation and proliferation of follicular dendritic cell-like cells by activated T lymphocytes. J Immunol *153*, 2951–2961.

Kim, H.S., Zhang, X., Klyushnenkova, E., and Choi, Y.S. (1995). Stimulation of germinal center B lymphocyte proliferation by an FDC-like cell line, HK. J Immunol *155*, 1101–1109.

Kim, J., Takeuchi, H., Lam, S.T., Turner, R.R., Wang, H.J., Kuo, C., Foshag, L., Bilchik, A.J., and Hoon, D.S. (2005). Chemokine receptor CXCR4 expression in colorectal cancer patients increases the risk for recurrence and for poor survival. Journal of clinical oncology: official journal of the American Society of Clinical Oncology *23*, 2744–2753.

Kim, M., Koh, Y.J., Kim, K.E., Koh, B.I., Nam, D.H., Alitalo, K., Kim, I., and Koh, G.Y. (2010). CXCR4 signaling regulates metastasis of chemoresistant melanoma cells by a lymphatic metastatic niche. Cancer research *70*, 10411–10421.

Klein, U., Goossens, T., Fischer, M., Kanzler, H., Braeuninger, A., Rajewsky, K., and Kuppers, R. (1998). Somatic hypermutation in normal and transformed human B cells. Immunological reviews *162*, 261–280.

Korkaya, H., Liu, S., and Wicha, M.S. (2011). Regulation of cancer stem cells by cytokine networks: attacking cancer's inflammatory roots. Clinical cancer research: an official journal of the American Association for Cancer Research *17*, 6125–6129.

Kouniavsky, G., Khaikin, M., Zvibel, I., Zippel, D., Brill, S., Halpern, Z., and Papa, M. (2002). Stromal extracellular matrix reduces chemotherapy-induced apoptosis in colon cancer cell lines. Clin Exp Metastasis *19*, 55–60.

Krautler, N.J., Kana, V., Kranich, J., Tian, Y., Perera, D., Lemm, D., Schwarz, P., Armulik, A., Browning, J.L., Tallquist, M., *et al.* (2012). Follicular dendritic cells emerge from ubiquitous perivascular precursors. Cell *150*, 194–206.

Kruger, J.A., Kaplan, C.D., Luo, Y., Zhou, H., Markowitz, D., Xiang, R., and Reisfeld, R.A. (2006). Characterization of stem cell-like cancer cells in immune-competent mice. Blood *108*, 3906–3912.

Kucia, M., Reca, R., Miekus, K., Wanzeck, J., Wojakowski, W., Janowska-Wieczorek, A., Ratajczak, J., and Ratajczak, M.Z. (2005). Trafficking of normal stem cells and metastasis of cancer stem cells involve similar mechanisms: pivotal role of the SDF-1-CXCR4 axis. Stem Cells *23*, 879–894.

Lee, C.G., Das, B., Lin, T.L., Grimes, C., Zhang, X., Lavezzi, T., Huang, L., Cole, J., Yau, L., and Li, L. (2012). A rare fraction of drug-resistant follicular lymphoma cancer stem cells interacts with follicular dendritic cells to maintain tumourigenic potential. British journal of haematology *158*, 79–90.

Lenz, G., Wright, G., Dave, S.S., Xiao, W., Powell, J., Zhao, H., Xu, W., Tan, B., Goldschmidt, N., Iqbal, J., *et al.* (2008). Stromal gene signatures in large-B-cell lymphomas. N Engl J Med *359*, 2313–2323.

Li, C.Y., Li, B.X., Liang, Y., Peng, R.Q., Ding, Y., Xu, D.Z., Zhang, X., Pan, Z.Z., Wan, D.S., Zeng, Y.X., *et al.* (2009). Higher percentage of CD133+ cells is associated with poor prognosis in colon carcinoma patients with stage IIIB. J Transl Med *7*, 56.

Li, L., and Choi, Y.S. (2002). Follicular dendritic cell-signaling molecules required for proliferation and differentiation of GC-B cells. Semin Immunol *14*, 259–266.

Li, L., Yoon, S.O., Fu, D.D., Zhang, X., and Choi, Y.S. (2004). Novel follicular dendritic cell molecule, 8D6, collaborates with CD44 in supporting lymphomagenesis by a Burkitt lymphoma cell line, L3055. Blood *104*, 815–821.

Li, L., Zhang, X., Kovacic, S., Long, A.J., Bourque, K., Wood, C.R., and Choi, Y.S. (2000). Identification of a human follicular dendritic cell molecule that stimulates germinal center B cell growth. The Journal of experimental medicine *191*, 1077–1084.

Lin, L., Liu, A., Peng, Z., Lin, H.J., Li, P.K., Li, C., and Lin, J. (2011). STAT3 is necessary for proliferation and survival in colon cancer-initiating cells. Cancer research *71*, 7226–7237.

Lis, R., Touboul, C., Mirshahi, P., Ali, F., Mathew, S., Nolan, D.J., Maleki, M., Abdalla, S.A., Raynaud, C.M., Querleu, D., *et al.* (2011). Tumor associated mesenchymal stem cells protects ovarian cancer cells from hyperthermia through CXCL12. Int J Cancer *128*, 715–725.

Madlambayan, G.J., Butler, J.M., Hosaka, K., Jorgensen, M., Fu, D., Guthrie, S.M., Shenoy, A.K., Brank, A., Russell, K.J., Otero, J., *et al.* (2009). Bone marrow stem and progenitor cell contribution to neovasculogenesis is dependent on model system with SDF-1 as a permissive trigger. Blood *114*, 4310–4319.

Margolin, D.A., Silinsky, J., Grimes, C., Spencer, N., Aycock, M., Green, H., Cordova, J., Davis, N.K., Driscoll, T., and Li, L. (2011). Lymph node stromal cells enhance drug-resistant colon cancer cell tumor formation through SDF-1alpha/CXCR4 paracrine signaling. Neoplasia *13*, 874–886.

Matsusue, R., Kubo, H., Hisamori, S., Okoshi, K., Takagi, H., Hida, K., Nakano, K., Itami, A., Kawada, K., Nagayama, S., *et al.* (2009). Hepatic stellate cells promote liver metastasis of colon cancer cells by the action of SDF-1/CXCR4 axis. Ann Surg Oncol *16*, 2645–2653.

Meads, M.B., Gatenby, R.A., and Dalton, W.S. (2009). Environment-mediated drug resistance: a major contributor to minimal residual disease. Nature reviews Cancer *9*, 665–674.

Merlos-Suarez, A., Barriga, F.M., Jung, P., Iglesias, M., Cespedes, M.V., Rossell, D., Sevillano, M., Hernando-Momblona, X., da Silva-Diz, V., Munoz, P., *et al.* (2011). The intestinal stem cell signature identifies colorectal cancer stem cells and predicts disease relapse. Cell stem cell *8*, 511–524.

Moriyama, T., Ohuchida, K., Mizumoto, K., Cui, L., Ikenaga, N., Sato, N., and Tanaka, M. (2010). Enhanced cell migration and invasion of CD133+ pancreatic cancer cells cocultured with pancreatic stromal cells. Cancer *116*, 3357–3368.

Nakagawa, H., Liyanarachchi, S., Davuluri, R.V., Auer, H., Martin, E.W., Jr., de la Chapelle, A., and Frankel, W.L. (2004). Role of cancer-associated stromal fibroblasts in metastatic colon cancer to the liver and their expression profiles. Oncogene *23*, 7366–7377.

O'Brien, C.A., Pollett, A., Gallinger, S., and Dick, J.E. (2007). A human colon cancer cell capable of initiating tumour growth in immunodeficient mice. Nature *445*, 106–110.

O'Callaghan, K., Lee, L., Nguyen, N., Hsieh, M.Y., Kaneider, N.C., Klein, A.K., Sprague, K., Van Etten, R.A., Kuliopulos, A., and Covic, L. (2012). Targeting CXCR4 with cell-penetrating pepducins in lymphoma and lymphocytic leukemia. Blood *119*, 1717–1725.
Overdevest, J.B., Thomas, S., Kristiansen, G., Hansel, D.E., Smith, S.C., and Theodorescu, D. (2011). CD24 offers a therapeutic target for control of bladder cancer metastasis based on a requirement for lung colonization. Cancer research *71*, 3802–3811.
Rafii, A., Mirshahi, P., Poupot, M., Faussat, A.M., Simon, A., Ducros, E., Mery, E., Couderc, B., Lis, R., Capdet, J., *et al.* (2008). Oncologic trogocytosis of an original stromal cells induces chemoresistance of ovarian tumours. PloS one *3*, e3894.
Reya, T., Morrison, S.J., Clarke, M.F., and Weissman, I.L. (2001). Stem cells, cancer, and cancer stem cells. Nature *414*, 105–111.
Roodhart, J.M., Daenen, L.G., Stigter, E.C., Prins, H.J., Gerrits, J., Houthuijzen, J.M., Gerritsen, M.G., Schipper, H.S., Backer, M.J., van Amersfoort, M., *et al.* (2011). Mesenchymal stem cells induce resistance to chemotherapy through the release of platinum-induced fatty acids. Cancer cell *20*, 370–383.
Saiki, Y., Ishimaru, S., Mimori, K., Takatsuno, Y., Nagahara, M., Ishii, H., Yamada, K., and Mori, M. (2009). Comprehensive analysis of the clinical significance of inducing pluripotent stemness-related gene expression in colorectal cancer cells. Ann Surg Oncol *16*, 2638–2644.
Schimanski, C.C., Schwald, S., Simiantonaki, N., Jayasinghe, C., Gonner, U., Wilsberg, V., Junginger, T., Berger, M.R., Galle, P.R., and Moehler, M. (2005). Effect of chemokine receptors CXCR4 and CCR7 on the metastatic behavior of human colorectal cancer. Clinical cancer research: an official journal of the American Association for Cancer Research *11*, 1743–1750.
SEER (2012). SEER Cancer Statistics Review. http://seercancergov/statfacts/html/colorecthtml, accessed 9/18/2012.
Shain, K.H., Landowski, T.H., and Dalton, W.S. (2000). The tumor microenvironment as a determinant of cancer cell survival: a possible mechanism for de novo drug resistance. Current opinion in oncology *12*, 557–563.
Shmelkov, S.V., Butler, J.M., Hooper, A.T., Hormigo, A., Kushner, J., Milde, T., St Clair, R., Baljevic, M., White, I., Jin, D.K., *et al.* (2008). CD133 expression is not restricted to stem cells, and both CD133+ and CD133- metastatic colon cancer cells initiate tumors. The Journal of clinical investigation *118*, 2111–2120.
Siegel, R., DeSantis, C., Virgo, K., Stein, K., Mariotto, A., Smith, T., Cooper, D., Gansler, T., Lerro, C., Fedewa, S., *et al.* (2012a). Cancer treatment and survivorship statistics, 2012. CA: a cancer journal for clinicians *62*, 220–241.
Siegel, R., Naishadham, D., and Jemal, A. (2012b). Cancer statistics, 2012. CA: a cancer journal for clinicians *62*, 10–29.
Smith, B.H., Gazda, L.S., Conn, B.L., Jain, K., Asina, S., Levine, D.M., Parker, T.S., Laramore, M.A., Martis, P.C., Vinerean, H.V., *et al.* (2011). Three-dimensional culture of mouse renal carcinoma cells in agarose macrobeads selects for a subpopulation of cells with cancer stem cell or cancer progenitor properties. Cancer research *71*, 716–724.
Tan, I.B., and Tan, P. (2011). Genetics: an 18-gene signature (ColoPrint(R)) for colon cancer prognosis. Nat Rev Clin Oncol *8*, 131–133.
Thomas, X., Anglaret, B., Magaud, J.P., Epstein, J., and Archimbaud, E. (1998). Interdependence between cytokines and cell adhesion molecules to induce interleukin-6 production by stromal cells in myeloma. Leukemia & lymphoma *32*, 107–119.
Todaro, M., Alea, M.P., Di Stefano, A.B., Cammareri, P., Vermeulen, L., Iovino, F., Tripodo, C., Russo, A., Gulotta, G., Medema, J.P., *et al.* (2007). Colon cancer stem cells dictate tumor growth and resist cell death by production of interleukin-4. Cell stem cell *1*, 389–402.
Todaro, M., Francipane, M.G., Medema, J.P., and Stassi, G. (2010). Colon cancer stem cells: promise of targeted therapy. Gastroenterology *138*, 2151–2162.
Todaro, M., Perez Alea, M., Scopelliti, A., Medema, J.P., and Stassi, G. (2008). IL-4-mediated drug resistance in colon cancer stem cells. Cell Cycle *7*, 309–313.
Torlakovic, E., Torlakovic, G., and Brunning, R.D. (2002). Follicular pattern of bone marrow involvement by follicular lymphoma. American journal of clinical pathology *118*, 780–786.
Tsuyada, A., Chow, A., Wu, J., Somlo, G., Chu, P., Loera, S., Luu, T., Li, A.X., Wu, X., Ye, W., *et al.* (2012). CCL2 mediates cross-talk between cancer cells and stromal fibroblasts that

regulates breast cancer stem cells. Cancer research *72*, 2768–2779.

Tysnes, B.B. (2010). Tumor-initiating and -propagating cells: cells that we would like to identify and control. Neoplasia *12*, 506–515.

Vaiopoulos, A.G., Kostakis, I.D., Koutsilieris, M., and Papavassiliou, A.G. (2012). Colorectal cancer stem cells. Stem Cells *30*, 363–371.

van der Horst, G., Bos, L., and van der Pluijm, G. (2012). Epithelial plasticity, cancer stem cells, and the tumor-supportive stroma in bladder carcinoma. Molecular cancer research: MCR *10*, 995–1009.

van der Horst, G., van den Hoogen, C., Buijs, J.T., Cheung, H., Bloys, H., Pelger, R.C., Lorenzon, G., Heckmann, B., Feyen, J., Pujuguet, P., *et al.* (2011). Targeting of alpha(v)-integrins in stem/progenitor cells and supportive microenvironment impairs bone metastasis in human prostate cancer. Neoplasia *13*, 516–525.

van Oers, M.H., Klasa, R., Marcus, R.E., Wolf, M., Kimby, E., Gascoyne, R.D., Jack, A., Van't Veer, M., Vranovsky, A., Holte, H., *et al.* (2006). Rituximab maintenance improves clinical outcome of relapsed/resistant follicular non-Hodgkin lymphoma in patients both with and without rituximab during induction: results of a prospective randomized phase 3 intergroup trial. Blood *108*, 3295–3301.

Vermeulen, L., De Sousa, E.M.F., van der Heijden, M., Cameron, K., de Jong, J.H., Borovski, T., Tuynman, J.B., Todaro, M., Merz, C., Rodermond, H., *et al.* (2010). Wnt activity defines colon cancer stem cells and is regulated by the microenvironment. Nature cell biology *12*, 468–476.

Visvader, J.E., and Lindeman, G.J. (2008). Cancer stem cells in solid tumours: accumulating evidence and unresolved questions. Nature reviews Cancer *8*, 755–768.

Wang, J., Guo, L.P., Chen, L.Z., Zeng, Y.X., and Lu, S.H. (2007). Identification of cancer stem cell-like side population cells in human nasopharyngeal carcinoma cell line. Cancer research *67*, 3716–3724.

Witkiewicz, A.K., Casimiro, M.C., Dasgupta, A., Mercier, I., Wang, C., Bonuccelli, G., Jasmin, J.F., Frank, P.G., Pestell, R.G., Kleer, C.G., *et al.* (2009a). Towards a new "stromal-based" classification system for human breast cancer prognosis and therapy. Cell Cycle *8*, 1654–1658.

Witkiewicz, A.K., Dasgupta, A., Sotgia, F., Mercier, I., Pestell, R.G., Sabel, M., Kleer, C.G., Brody, J.R., and Lisanti, M.P. (2009b). An absence of stromal caveolin-1 expression predicts early tumor recurrence and poor clinical outcome in human breast cancers. The American journal of pathology *174*, 2023–2034.

Wu, C., and Alman, B.A. (2008). Side population cells in human cancers. Cancer letters *268*, 1–9.

Xu, Y., Tabe, Y., Jin, L., Watt, J., McQueen, T., Ohsaka, A., Andreeff, M., and Konopleva, M. (2008). TGF-beta receptor kinase inhibitor LY2109761 reverses the anti-apoptotic effects of TGF-beta1 in myelo-monocytic leukaemic cells co-cultured with stromal cells. British journal of haematology *142*, 192–201.

Yin, X.L., Wang, N., Wei, X., Xie, G.F., Li, J.J., and Liang, H.J. (2012). Interleukin-12 inhibits the survival of human colon cancer stem cells in vitro and their tumor initiating capacity in mice. Cancer letters *322*, 92–97.

Zeelenberg, I.S., Ruuls-Van Stalle, L., and Roos, E. (2003). The chemokine receptor CXCR4 is required for outgrowth of colon carcinoma micrometastases. Cancer research *63*, 3833–3839.

Zeki, S.S., Graham, T.A., and Wright, N.A. (2011). Stem cells and their implications for colorectal cancer. Nature reviews Gastroenterology & hepatology *8*, 90–100.

Zhang, C., Cui, G.H., Liu, F., Wu, Q.L., and Chen, Y. (2006). Inhibitory effect of triptolide on lymph node metastasis in patients with non-Hodgkin lymphoma by regulating SDF-1/CXCR4 axis in vitro. Acta Pharmacol Sin *27*, 1438–1446.

Zhang, P., Zhang, Y., Mao, L., Zhang, Z., and Chen, W. (2009). Side population in oral squamous cell carcinoma possesses tumor stem cell phenotypes. Cancer letters *277*, 227–234.

Chapter 22

A Perspective on Breast Cancer Malignant Progression: From Cancer Stem Cell Intra Tumor Heterogeneity to Metastasis-Initiating Cells

Pasquale Sansone, Vinagolu K. Rajasekhar, and Jacqueline Bromberg
Department of Medicine, Memorial Sloan-Kettering Cancer Center, New York, NY, USA

The Cancer Stem Cell Intra Tumor Heterogeneity Hypothesis

The presence of intra tumor heterogeneity has long been appreciated in breast cancer pathology (Lorico and Rappa, 2011; Polyak, 2007). Such heterogeneity depends on the number of acquired mutations, operative signaling pathways, their frequencies, and finally their combination effects on the survival, proliferation and aberrant differentiation capabilities (Perou et al., 2000). In the recent decade, CSCs have been suggested to re-create the entire original tumor heterogeneity (Hermann et al., 2007; Lorico and Rappa, 2011; Polyak, 2007). However, the mechanisms that drive the cancer stem cell intra tumor heterogeneity and whether these distinct CSCs may trigger different phenotypes are still uncovered questions in breast cancer (BCa).

Breast CSCs contain different immunophenotypes such as $CD44^{hi}/CD24^{low}$, $CD44^{hi}/CD133^{hi}$, $CD133^{hi}$, and $ALDH1^{hi}$ activity (Meyer et al., 2010; Marcato et al., 2011; Wright et al., 2008). Of note, anti-cancer therapies may foster additional mutations and epigenetic changes leading to additional mutant cell lineages and distinct CSC self-renewal potentials (Creighton et al., 2009; Lee et al., 2011). The CSC phenotype is particularly abundant in aggressive breast cancer subtypes such as triple negative and basal-like, but not in the luminal subtype (Al-Hajj et al., 2003). This is mainly due to the infrequent detection of such CSC markers and the lack of experimental in vivo models of luminal breast cancer. In this chapter we will discuss the current state of the literature on the mechanisms of the cancer stem cell intra tumor heterogeneity in breast cancer. We will also propose a model of CSC-driven metastatic disease in estrogen receptor alpha positive (ERa+, luminal) mammary tumors.

Inactivation of Tumor Suppressors

ERa negative tumors frequently carry genetic hallmarks linked to tumorigenesis such as the inactivation of tumor suppressors p53/pRb/PTEN/BRCA-1 (Wright et al., 2008; Jiang et al., 2010, Zhang et al., 2008; Mizuno et al., 2010). Experimental evidence in these investigations revealed that loss of tumor suppressors is associated with the propagation of CSCs. This is particularly true for murine mammary tumorigenesis following both p53

Cancer Stem Cells, First Edition. Edited by Vinagolu K. Rajasekhar.

and BRCA-1 loss of functions (Wright et al., 2008). Of note, Wright and colleagues (2008) first reported that BRCA-1 deficient murine cancer cell lines have distinct and non-overlapping populations of putative CSCs ($CD44^{hi}/CD24^{low}$ and $CD133^{hi}$). Moreover, Mizuno and colleagues (2010) have demonstrated a positive correlation between p53 functional inactivation and the presence of embryonic stem cell-like (ESC) transcriptional patterns in human breast tumors (Mizuno et al., 2010). In this regard, we have also confirmed that the accumulation of p53 (that experimentally mimics p53 inactivation) elicits CD133 and CD44 promoter demethylation in the p53 wild type/ERa positive/$CD133^{low}/CD44^{low}$ human luminal breast cancer model, MCF-7 (D'Anello et al., 2010). *Taken together these observations support the concept that loss of tumor suppressors may trigger the cancer stem cell intra tumor heterogeneity via a complex epigenetic mechanism in non-CSCs.*

Escape From Oncogene Addiction

Aggressive BCa subtypes frequently overexpress oncogenes and display the hallmark tumorigenic feature named oncogene addiction (Korkaya et al., 2008; Perou et al., 2000). Such a process is the dependence on one overactive gene or pathway for cancer cell survival and growth (Korkaya et al., 2008). This theory was demonstrated in transgenic models of cancer, in which even-short-term inactivation of the oncogene stops tumor growth. An example of oncogene addiction in BCa is the expression of the proto-oncogene Her-2. Interestingly, the forced overexpression of Her2 gene increases the $ALDH1^{hi}$ CSC phenotype in $Her2^{low}$ human BCa cell lines. Accordingly, Herceptin (anti-Her2 monoclonal antibody) hampers the proliferation of these de novo generated $Her2^{hi}$ cancer cell lines (Korkaya et al., 2008). Nevertheless, following the direct inhibition of oncogenes, tumor cells easily escape from such oncogene addiction by overexpressing a more aggressive CSC phenotype (Leung et al., 2012; Oliveras-Ferraros et al., 2012; Reim et al., 2009). Leung and colleagues (2012) had partly addressed this issue, observing that the withdrawal of the Myc pathway determines a de novo Myc-independent tumor growth in the context of the single oncogenic Myc mammary tumorigenesis. Tumors escaping from the oncogene addiction, rather than any further oncogene mutation, often show a remarkable increase of the $CD44^{hi}/CD24^{low}$ CSC and epithelial mesenchymal transition (EMT) phenotypes (Leung et al., 2012; Oliveras-Ferraros et al., 2012). This phenomenon was also observed in tumors recurring from herceptin resistance in Her2-dependent mammary tumorigenesis (Oliveras-Ferraros et al., 2012; Reim et al., 2009). In agreement with such observations, the promotion of an EMT-like behavior is functionally associated with the expression of CSC markers such as CD44, CD133, Nanog, OCT4, Shh, and the polycomb genes, −Bmi− (Liu et al., 2012; Liu et al., 2011a; Mani et al., 2008; Rajasekhar and Begemann, 2007; Hassiotou et al., 2013). Of note, the overexpression of OCT4/SOX2/Nanog itself induces tumorigenesis toward a lactating differentiation lineage in breast cancer (Rajasekhar and Begemann, 2007; Liu et al., 2012; Liu et al., 2011a; Hassiotou et al., 2013; Leis et al., 2012). *These data suggest that an epigenetic mechanism, rather than genetic events, may promote the escape from oncogene addiction leading to the cancer stem cell intra tumor heterogeneity and tumor progression in experimental models of breast cancer. Furthermore, these data also highlight the strict association between tumor progression and the promotion of the intra tumor cancer stem cell heterogeneity phenotype.*

The Bioenergetics of the Tumor Microenvironment

CSCs were identified by the capacity of distinct human cancer cells to survive and self-renew following serial orthotropic xenotransplantation assays in mice (Rajasekhar et al., 2011; Visvader and Lindeman, 2008). Although xenografts from CSCs are almost entirely characterized by terminally differentiated daughter cells (non-CSCs), a heterogeneous fraction of CSCs is still present in these tumors (Sansone, P., unpublished). Additionally, the ratio of CSCs/non-CSCs may vary in each experimental setting. Therefore, rather than being genetically determined, the presence and persistence of CSCs might derive from the intrinsic tumor cell plasticity of the non-CSCs component (Chaffer et al., 2011). *We can therefore hypothesize that the cancer stem cell intra tumor heterogeneity may also be the consequence of the crosstalk between cancer and non-cancer cells within the tumor microenvironment.* Perhaps the most powerful example of such hypothesis is the experimental evidence that CSCs originate following multiple xenotransplantation assays of non-CSCs (Sansone, P., unpublished). Following this reasoning, we may speculate that factors from the mammary fat pad microenvironment could promote the development, maintenance, and function of CSCs.

From a biochemical point of view, the basal metabolism of murine cells is several-fold higher (7 to 8 times) than that of human cells. As a consequence, human tumor xenografts may experience increased flux of extracellular carbon metabolites such as lactates and ketones leading to increased mitochondrial activity, intracellular oxidative stress, and tissue aging (Chiavarina et al., 2011; Sanchez et al., 2011). Such a metabolic cascade promotes higher mitochondrial respiration demand together with increased generation of reactive oxygen species (ROS). In turn, intracellular ROS may favor genetic mutation and the selection of aggressive cancer populations. Nevertheless, the increased energy demand of highly proliferative cancer cells may lead to a metabolic crisis characterized by several cycles of hypoxia/starvation/reoxygenation. This latter phenomenon further selects for cancer cells with distinct metabolic features such as high resistance to oxygen and glucose deprivation. In human breast cancer, hypoxia and IL6 are documented as key triggers of tumor stemness (Liu et al., 2011b; Conley et al., 2012; Sansone et al., 2007a; Sansone et al., 2007b). Most recently periostin, a component of extracellular matrix induced by fibrosis and oxidative stress, was shown to be required for the development of a metastatic niche in murine models of breast cancer (Malanchi et al., 2012). *Overall, we hypothesize that a bioenergetics remodeling within the tumor microenvironment can affect the cancer stem cells intra tumor heterogeneity leading to tumor progression.*

What are the metabolic features of CSCs? Can a "bioenergetics crisis" trigger the generation of CSCs? At what step of bioenergetics reshaping do CSCs originate? Before attempting to answer these questions, we need to unravel the key metabolic cascades and pathways that operate in breast cancer cells (Espina et al., 2010). Additionally, we will have to investigate whether changes in the basal metabolism can affect the differentiation/cancer stem cell status in non-CSCs. Many clinical and experimental observations lead to the conclusion that anaerobic metabolism may sustain aggressive cancer cell growth and spread. Given the role of glucose and glutamine in hypoxia, we may speculate that these two metabolites play a pivotal role in cell's survival under hypoxic culture conditions. However, glutamine and glucose are also crucial metabolites for normal cell metabolism. Indeed, from an evolutionary point of view, it is likely that CSCs exploit different metabolites (onco-metabolites) to survive and operate in the tumor microenvironment. To provide biochemical evidence in this direction, an extensive metabolome profile of CSCs and non-CSCs may help to highlight key metabolic determinants for tumoral bioenergetics.

Provocative Question: Are Metastasis-Initiating Cells (MIC) Different From Primary Tumor-Initiating Cells?

Breast cancer is a complex systemic disease showing different phenotypes such as tumor dormancy, tumor cell dissemination, and tumor cell colonization of distant organs (Polyak, 2007; Oh et al., 2006). Even if primary tumors and their matched metastases can share a similar gene expression profile, metastatic disease may exhibit a plastic biology and specific metabolic requirements. Following tumor removal, BCa cells are found in the bone marrow as disseminated tumor cells (DTC) and in the blood as circulating tumor cells (CTC), both prior to and after anti-cancer therapies (Wallwiener et al., 2013; Aktas et al., 2011).

The Cell(s) of Origin of Metastatic Disease Is Unknown in Breast Cancer

Although a great effort has been made to unravel the mechanisms beyond oncogene-derived tumorigenesis, knowledge of key aspects of the biology of tumor progression is still missing. In the pre-clinical setting, the efficacy of anti-cancer compounds is tested in cancer models, which recapitulate primary tumor growth rather than metastatic disease. This is mainly due to the lack of metastatic models and to the "*belief*" that primary tumor-initiating cells and MICs share the same biology. We also suggest that even in the case of metastatic cell line derived from experimental animal models (xenografts), most studies are based on subcutaneous transplantation assays besides the tumor cell lines themselves are passaged over many generations even by the individual investigators. Therefore, it is not uncommon that the same cancer cell line used by different investigators encounters differences in their gene expression patterns leading to the divulgation of contrasting results. In the clinical practice, chemo/targeted therapies are planned according to the histopatological phenotype of the primary tumors in metastatic breast cancer patients. *This "scientific gap" between the methods of laboratory experiences and the reasoning of clinical practice will decrease the chance to develop successful therapies for metastatic breast cancer.*

Whether CSCs Give Rise to Metastatic Disease Is Open to Debate

Besides the idea that CSCs might arise from normal stem cells, it is most likely that they originate as a consequence of tumor progression. As evidence, CSCs are found enriched in the primary tumor following current anti-cancer therapies (Creighton et al., 2009). We and others have observed that terminally differentiated cancer cells acquire a stem-cell-like phenotype to survive radiation/cytotoxic induced-cell death (Creighton et al., 2009). Accordingly, cancer cells that survive a neo-adjuvant regimen express high levels of CSC markers in luminal breast cancer (Sansone, P., unpublished). These data support the idea that the CSC phenotype is the consequence of tumor progression. The Michael Clarke's laboratory has demonstrated that xenografts procured from the inoculation of tumor-derived $CD44^{hi}$ cells are proliferative and metastatic in mice (Liu et al., 2010). However, proliferative tumor cells will metastasize to the lungs in mice bearing orthotropic xenografts in the mammary fat pad (Liu et al., 2010; unpublished observations). Accordingly, CSCs will promote metastatic disease as a consequence of a more successful tumor taken at the primary site, rather for a specific metastatic capability. This reasoning is also corroborated by the experimental observations that the primary tumor microenvironment is required for tumor cell dissemination. *Finally, further investigations need to address the role of the intra tumor cancer stem cell heterogeneity on the promotion of metastatic disease, and whether CSCs and MICs share the same immunophenotype.*

Metastasis-Initiating Cells in Luminal Breast Cancer: The ERa Paradigm

The majority of breast cancer in humans belongs to the luminal subtype (Oh et al., 2006; Powles, 2013). Luminal breast cancer cells rely on ERa pathway activity for survival and growth (Powles, 2013). Indeed, adjuvant hormonal therapy (HT) decreases the number of DTCs, leading to increased disease-free survival in luminal breast cancer

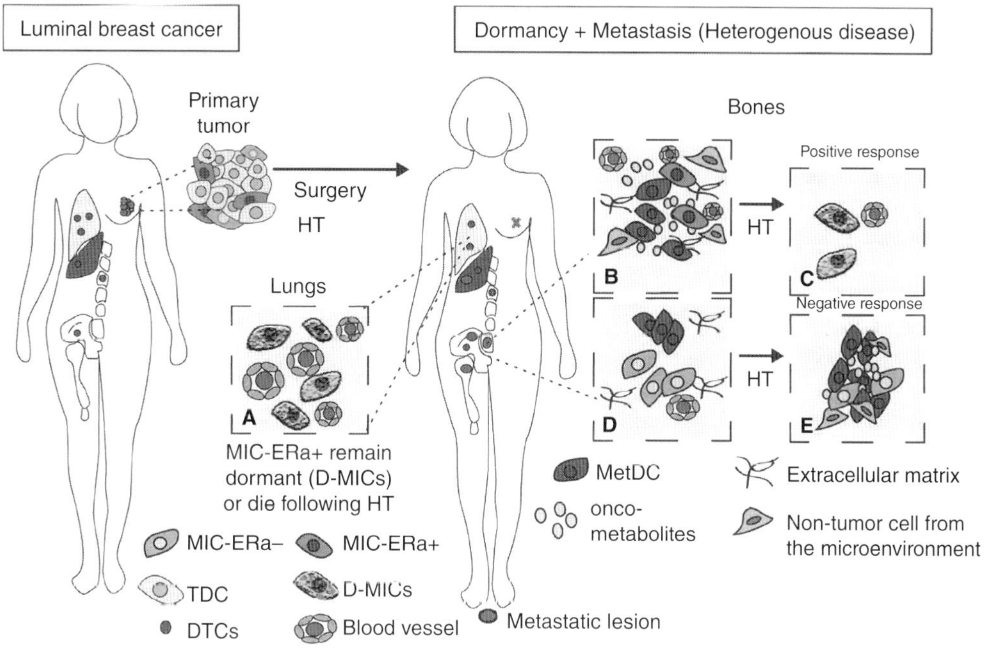

Figure 22.1. A hypothetical luminal breast cancer progression model: the ERa paradigm. Heterogeneous metastasis-initiating cells -MICs− originate in the context of the primary tumor microenvironment. These MICs can be either ERa-dependent or independent for their survival and growth (MIC-ERa+ or MIC-ERa−). MICs can disseminate at distance (DTC). Following surgical removal of the primary tumor (minimal residual disease), disseminated MIC-ERa+ either die or stay dormant when exposed to hormonal therapy (HT, DMICs, **A**). We also hypothesize that some MICs (especially MIC-ERa−) can self-renew giving rise to metastatic disease. Due to the heterogeneity of the secondary site microenvironments (pro-tumorigenic or not, presence of secretory cells, extracellular matrix, onco-metabolites) and the presence of distinct MICs, the kinetic of metastatic disease can vary from one site to the other. When metastases are clinically detectable, breast cancer patients will undergo HT. HT will effectively render MIC-ERa+ cells/lesions dormant (**B**, **C** positive response). By contrast, lesions from MIC-ERa− will continue to self-renew exploiting an ERa-independent tumorigenesis (**D**, **E** negative response). TDC, tumor differentiated cancer cells. MetDC, metastatic differentiated cells. For color detail, please see color plate section.

patients (Powles, 2013). However, when metastatic disease occurs, HT has short-lived and limited efficacy suggesting that metastatic lesions can survive without the ERa pathway. Notably, DTCs/CTCs often display an ERa^{low} phenotype and CSC phenotype in luminal breast cancer patients (Fehm et al., 2008; Hartkopf et al., 2013; Kasimir-Bauer et al., 2012). These observations led to the hypothesis that MICs may overcome ERa pathway inhibition to survive and eventually self-renew. Therefore, we can argue that heterogeneous MICs may drive metastatic disease. Due to the pivotal role of ERa in luminal breast cancer, we anticipate that some MICs will self-renew using the ERa pathway, whereas other MICs will employ unknown alternative molecular signaling pathways. As a consequence, HT therapy will effectively block the ERa-dependent MICs, leaving unaffected or even potentiating the self-renewal of ERa-independent MICs (Figure 22.1). In agreement with this idea, it is not uncommon that breast cancer metastases display a complex phenotypic heterogeneity within the same patient.

Concluding Remarks

In summary, a comprehensive investigation on the biochemistry beyond MICs-driven tumorigenesis will be crucial to unravel the complexity of metastatic disease. Furthermore, cancer stem cell biologists, oncologists, and cancer pathologists need to work together to formulate meaningful disease-driven hypotheses on the key steps of human cancer progression for the development of successful translational models and treatment strategies. *Therefore, it is pivotal to develop disease-driven hypotheses together with appropriate metastatic models from experimental and patient-derived tumor tissues.*

Acknowledgments

We thank the laboratory of Center of Applied Biomedical Research −CRBA− led by Dr. Chieco and the breast cancer unit of Bologna University, Italy (Policlinico St. Orsola-Malpighi, Drs. Taffurelli, Santini, Ceccarelli, Chieco and Bonafé) for financial support and collaborative work on breast cancer stem cell research. We are also grateful to Daniel Thorek and Laura Daly for their suggestions and technical help. We acknowledge Julia R. Vinagolu-Baur for typing the references. The Department of Defense grant W81XWH-10-1-1033 (PS) and RO1CA098234-01 (JB) support our work.

References

Aktas, B., Muller, V., Tewes, M., Zeitz, J., Kasimir-Bauer, S., Loehberg, C.R., Rack, B., Schneeweiss, A., and Fehm, T. (2011). Comparison of estrogen and progesterone receptor status of circulating tumor cells and the primary tumor in metastatic breast cancer patients. Gynecologic oncology *122*, 356–360.

Al-Hajj, M., Wicha, M.S., Benito-Hernandez, A., Morrison, S.J., and Clarke, M.F. (2003). Prospective identification of tumorigenic breast cancer cells. Proc Natl Acad Sci USA *100*, 3983–3988.

Chaffer, C.L., Brueckmann, I., Scheel, C., Kaestli, A.J., Wiggins, P.A., Rodrigues, L.O., Brooks, M., Reinhardt, F., Su, Y., Polyak, K., *et al.* (2011). Normal and neoplastic nonstem cells can spontaneously convert to a stem-like state. Proc Natl Acad Sci USA *108*, 7950–7955.

Chiavarina, B., Whitaker-Menezes, D., Martinez-Outschoorn, U.E., Witkiewicz, A.K., Birbe, R., Howell, A., Pestell, R.G., Smith, J., Daniel, R., Sotgia, F., *et al.* (2011). Pyruvate kinase expression (PKM1 and PKM2) in cancer-associated fibroblasts drives stromal nutrient production and tumor growth. Cancer Biol Ther *12*, 1101–1113.

Conley, S.J., Gheordunescu, E., Kakarala, P., Newman, B., Korkaya, H., Heath, A.N., Clouthier, S.G., and Wicha, M.S. (2012). Antiangiogenic agents increase breast cancer stem cells via the generation of tumor hypoxia. Proc Natl Acad Sci USA *109*, 2784–2789.

Creighton, C.J., Li, X., Landis, M., Dixon, J.M., Neumeister, V.M., Sjolund, A., Rimm, D.L., Wong, H., Rodriguez, A., Herschkowitz, J.I., *et al.* (2009). Residual breast cancers after conventional therapy display mesenchymal as well as tumor-initiating features. Proc Natl Acad Sci USA *106*, 13820–13825.

D'Anello, L., Sansone, P., Storci, G., Mitrugno, V., D'Uva, G., Chieco, P., and Bonafe, M. (2010). Epigenetic control of the basal-like gene expression profile via Interleukin-6 in breast cancer cells. Molecular cancer *9*, 300.

Espina, V., Mariani, B.D., Gallagher, R.I., Tran, K., Banks, S., Wiedemann, J., Huryk, H., Mueller, C., Adamo, L., Deng, J., *et al.* (2010). Malignant precursor cells pre-exist in human breast DCIS and require autophagy for survival. PLoS One *5*, e10240.

Fehm, T., Krawczyk, N., Solomayer, E.F., Becker-Pergola, G., Durr-Storzer, S., Neubauer, H., Seeger, H., Staebler, A., Wallwiener, D., and Becker, S. (2008). ERalpha-status of disseminated tumour cells in bone marrow of primary breast cancer patients. Breast Cancer Res *10*, R76.

Hartkopf, A.D., Banys, M., Meier-Stiegen, F., Hahn, M., Rohm, C., Hoffmann, J., Helms, G., Taran, F.A., Wallwiener, M., Walter, C., *et al.* (2013). The HER2 status of disseminated tumor cells in the bone marrow of early breast cancer patients is independent from primary tumor and predicts higher risk of relapse. Breast Cancer Res Treat *138*, 509–517.

Hassiotou, F., Hepworth, A.R., Beltran, A.S., Mathews, M.M., Stuebe, A.M., Hartmann, P.E., Filgueira, L., and Blancafort, P. (2013). Expression of the Pluripotency Transcription Factor OCT4 in the Normal and Aberrant Mammary Gland. Front Oncol *3*, 79.

Hermann, P.C., Huber, S.L., Herrler, T., Aicher, A., Ellwart, J.W., Guba, M., Bruns, C.J., and Heeschen, C. (2007). Distinct populations of cancer stem cells determine tumor growth and metastatic activity in human pancreatic cancer. Cell Stem Cell *1*, 313–323.

Jiang, Z., Deng, T., Jones, R., Li, H., Herschkowitz, J.I., Liu, J.C., Weigman, V.J., Tsao, M.S., Lane, T.F., Perou, C.M., *et al.* (2010). Rb deletion in mouse mammary progenitors induces luminal-B or basal-like/EMT tumor subtypes depending on p53 status. J Clin Invest *120*, 3296–3309.

Kasimir-Bauer, S., Hoffmann, O., Wallwiener, D., Kimmig, R., and Fehm, T. (2012). Expression of stem cell and epithelial-mesenchymal transition markers in primary breast cancer patients with circulating tumor cells. Breast Cancer Res *14*, R15.

Korkaya, H., Paulson, A., Iovino, F., and Wicha, M.S. (2008). HER2 regulates the mammary stem/progenitor cell population driving tumorigenesis and invasion. Oncogene *27*, 6120–6130.

Lee, H.E., Kim, J.H., Kim, Y.J., Choi, S.Y., Kim, S.W., Kang, E., Chung, I.Y., Kim, I.A., Kim, E.J., Choi, Y., *et al.* (2011). An increase in cancer stem cell population after primary systemic therapy is a poor prognostic factor in breast cancer. Br J Cancer *104*, 1730–1738.

Leis, O., Eguiara, A., Lopez-Arribillaga, E., Alberdi, M.J., Hernandez-Garcia, S., Elorriaga, K., Pandiella, A., Rezola, R., and Martin, A.G. (2012). Sox2 expression in breast tumours and activation in breast cancer stem cells. Oncogene *31*, 1354–1365.

Leung, J.Y., Andrechek, E.R., Cardiff, R.D., and Nevins, J.R. (2012). Heterogeneity in MYC-induced mammary tumors contributes to escape from oncogene dependence. Oncogene *31*, 2545–2554.

Liu, C., Cao, X., Zhang, Y., Xu, H., Zhang, R., Wu, Y., Lu, P., and Jin, F. (2012). Co-expression of Oct-4 and Nestin in human breast cancers. Molecular biology reports *39*, 5875–5881.

Liu, C.G., Lu, Y., Wang, B.B., Zhang, Y.J., Zhang, R.S., Lu, Y., Chen, B., Xu, H., Jin, F., and Lu, P. (2011a). Clinical implications of stem cell gene Oct-4 expression in breast cancer. Annals of surgery *253*, 1165–1171.

Liu, H., Patel, M.R., Prescher, J.A., Patsialou, A., Qian, D., Lin, J., Wen, S., Chang, Y.F., Bachmann, M.H., Shimono, Y., *et al.* (2010). Cancer stem cells from human breast tumors are involved in spontaneous metastases in orthotopic mouse models. Proc Natl Acad Sci USA *107*, 18115–18120.

Liu, S., Ginestier, C., Ou, S.J., Clouthier, S.G., Patel, S.H., Monville, F., Korkaya, H., Heath, A., Dutcher, J., Kleer, C.G., *et al.* (2011b). Breast cancer stem cells are regulated by mesenchymal stem cells through cytokine networks. Cancer Res *71*, 614–624.

Lorico, A., and Rappa, G. (2011). Phenotypic heterogeneity of breast cancer stem cells. Journal of oncology *2011*, 135039.

Malanchi, I., Santamaria-Martinez, A., Susanto, E., Peng, H., Lehr, H.A., Delaloye, J.F., and Huelsken, J. (2012). Interactions between cancer stem cells and their niche govern metastatic colonization. Nature *481*, 85–89.

Mani, S.A., Guo, W., Liao, M.J., Eaton, E.N., Ayyanan, A., Zhou, A.Y., Brooks, M., Reinhard, F., Zhang, C.C., Shipitsin, M., *et al.* (2008). The epithelial-mesenchymal transition generates cells with properties of stem cells. Cell *133*, 704–715.

Marcato, P., Dean, C.A., Pan, D., Araslanova, R., Gillis, M., Joshi, M., Helyer, L., Pan, L., Leidal, A., Gujar, S., *et al.* (2011). Aldehyde dehydrogenase activity of breast cancer stem cells is primarily due to isoform ALDH1A3 and its expression is predictive of metastasis. Stem Cells *29*, 32–45.

Meyer, M.J., Fleming, J.M., Lin, A.F., Hussnain, S.A., Ginsburg, E., and Vonderhaar, B.K. (2010). CD44posCD49fhiCD133/2hi defines xenograft-initiating cells in estrogen receptor-negative breast cancer. Cancer Res *70*, 4624–4633.

Mizuno, H., Spike, B.T., Wahl, G.M., and Levine, A.J. (2010). Inactivation of p53 in breast cancers correlates with stem cell transcriptional signatures. Proc Natl Acad Sci USA *107*, 22745–22750.

Oh, D.S., Troester, M.A., Usary, J., Hu, Z., He, X., Fan, C., Wu, J., Carey, L.A., and Perou, C.M. (2006). Estrogen-regulated genes predict survival in hormone receptor-positive breast cancers. J Clin Oncol *24*, 1656–1664.

Oliveras-Ferraros, C., Corominas-Faja, B., Cufi, S., Vazquez-Martin, A., Martin-Castillo, B., Iglesias, J.M., Lopez-Bonet, E., Martin, A.G., and Menendez, J.A. (2012). Epithelial-to-mesenchymal transition (EMT) confers primary resistance to trastuzumab (Herceptin). Cell Cycle *11*, 4020–4032.

Perou, C.M., Sorlie, T., Eisen, M.B., van de Rijn, M., Jeffrey, S.S., Rees, C.A., Pollack, J.R., Ross, D.T., Johnsen, H., Akslen, L.A., *et al.* (2000). Molecular portraits of human breast tumours. Nature *406*, 747–752.

Polyak, K. (2007). Breast cancer: origins and evolution. J Clin Invest *117*, 3155–3163.

Powles, T.J. (2013). Extended adjuvant tamoxifen for breast cancer—a new era? Lancet *381*, 782–783.

Rajasekhar, V.K., and Begemann, M. (2007). Concise review: roles of polycomb group proteins in development and disease: a stem cell perspective. Stem Cells *25*, 2498–2510.

Rajasekhar, V.K., Studer, L., Gerald, W., Socci, N.D., and Scher, H.I. (2011). Tumour-initiating stem-like cells in human prostate cancer exhibit increased NF-kappaB signalling. Nature communications *2*, 162.

Reim, F., Dombrowski, Y., Ritter, C., Buttmann, M., Hausler, S., Ossadnik, M., Krockenberger, M., Beier, D., Beier, C.P., Dietl, J., *et al.* (2009). Immunoselection of breast and ovarian cancer cells with trastuzumab and natural killer cells: selective escape of CD44high/CD24low/HER2low breast cancer stem cells. Cancer Res *69*, 8058–8066.

Sanchez, C.G., Penfornis, P., Oskowitz, A.Z., Boonjindasup, A.G., Cai, D.Z., Dhule, S.S., Rowan, B.G., Kelekar, A., Krause, D.S., and Pochampally, R.R. (2011). Activation of autophagy in mesenchymal stem cells provides tumor stromal support. Carcinogenesis *32*, 964–972.

Sansone, P., Storci, G., Giovannini, C., Pandolfi, S., Pianetti, S., Taffurelli, M., Santini, D., Ceccarelli, C., Chieco, P., and Bonafe, M. (2007a). p66Shc/Notch-3 interplay controls self-renewal and hypoxia survival in human stem/progenitor cells of the mammary gland expanded in vitro as mammospheres. Stem Cells *25*, 807–815.

Sansone, P., Storci, G., Tavolari, S., Guarnieri, T., Giovannini, C., Taffurelli, M., Ceccarelli, C., Santini, D., Paterini, P., Marcu, K.B., *et al.* (2007b). IL-6 triggers malignant features in mammospheres from human ductal breast carcinoma and normal mammary gland. J Clin Invest *117*, 3988–4002.

Visvader, J.E., and Lindeman, G.J. (2008). Cancer stem cells in solid tumours: accumulating evidence and unresolved questions. Nat Rev Cancer *8*, 755–768.

Wallwiener, M., Hartkopf, A.D., Baccelli, I., Riethdorf, S., Schott, S., Pantel, K., Marme, F., Sohn, C., Trumpp, A., Rack, B., *et al.* (2013). The prognostic impact of circulating tumor cells in subtypes of metastatic breast cancer. Breast Cancer Res Treat *137*, 503–510.

Wright, M.H., Calcagno, A.M., Salcido, C.D., Carlson, M.D., Ambudkar, S.V., and Varticovski, L. (2008). Brca1 breast tumors contain distinct CD44+/CD24− and CD133+ cells with cancer stem cell characteristics. Breast Cancer Res *10*, R10.

Zhang, M., Behbod, F., Atkinson, R.L., Landis, M.D., Kittrell, F., Edwards, D., Medina, D., Tsimelzon, A., Hilsenbeck, S., Green, J.E., *et al.* (2008). Identification of tumor-initiating cells in a p53-null mouse model of breast cancer. Cancer Res *68*, 4674–4682.

Section V

Novel and Potential Targets in Cancer Stem Cells

Chapter 23

Targeting Cancer Stem Cells—Modulating Embryonic Stem Cell Signaling, Epigenetics, and Tumor Metabolism

Naoko Takebe[1], Pamela Jo Harris[1], Yutaka Kondo[2], Abhilasha Nair[3], S. Percy Ivy[1], and Hideyuki Saya[4]

[1] *Cancer Therapy Evaluation Program, Division of Cancer Treatment and Diagnosis, National Cancer Institute, National Institute of Health, Rockville, MD, USA*

[2] *Division of Epigenomics, Aichi Cancer Center Research Institute, Chikusa-ku, Nagoya, Japan*

[3] *Division of Cancer Treatment and Diagnosis, National Cancer Institute, National Institute of Health, Rockville, MD, USA*

[4] *Division of Gene Regulation, Institute for Advanced Medical Research, School of Medicine, Keio University, Shinjuku-ku, Tokyo, Japan*

Introduction—Embryonic Stem Cell Targeting Therapy

One hypothesis that explains the mechanism of disease relapse argues for the presence of cancer stem cells (CSC) within the tumor (O'Brien et al., 2010) that have been described as being more resistant to conventional chemotherapy and radiation (Donnenberg and Donnenberg, 2005; Polyak and Weinberg, 2009). With the discovery that the CSCs are driven by embryonic signaling pathways (ESP) (Kelleher et al., 2006), these pathways and their components, mainly the Hedgehog (Hh), Notch, and wingless, and the mouse mammary tumor virus integration site (Wnt) have become important targets (Takebe and Ivy, 2010). Studies have shown that polycomb-repressive complexes (PRC) 1 and 2 are also involved in regulation of CSC including maintenance of self-renewal and multipotential capability (Richly et al., 2011). This chapter summarizes the discovery and current therapeutic advances in CSCs targeted agents.

Notch Signaling Pathway and Inhibitors

The first Notch gene mutation was discovered in 1913 by a young scientist, Merkel Henry Jacobs (Dexter, 1914). Notch functioning is highly context dependent (Pannuti et al., 2010) and has been implicated in a variety of biological processes (Dontu et al., 2004). The detailed Notch signaling pathway has been nicely described in the other chapter. Here, we highlight the importance of gamma

Cancer Stem Cells, First Edition. Edited by Vinagolu K. Rajasekhar.

secretase to yield Notch intracellular domain (NICD), which is then transported to the nucleus for initiation of transcription of multiple CSL-target genes (Fortini, 2009; Kopan and Ilagan, 2009).

Role in Oncogenesis and Metastasis

The role of Notch signaling in oncogenesis was described initially in T-cell ALL (Ellisen et al., 1991). Although the final outcome is context dependent, the oncogenic potential of the Notch pathway seems to arise from a variety of factors including antiapoptotic, cell growth, and angiogenesis-inducing, and EMT-facilitating effects (Harris et al., 2012). Inhibition of signaling via the Notch ligand Delta like-4 was found to have decreased tumor size including some tumors that were resistant to VEGF inhibition (Hellstrom et al., 2007). Also there appears to be extensive cross talk between Notch signaling and the major oncogenic pathways. These include the PI3K/AKT pathway (Palomero et al., 2007), Ras pathway (Fitzgerald et al., 2000), NF-κB pathway (Oswald et al., 1998), and the HIF1-alpha pathway (Gustafsson et al., 2005).

Inhibitor Discovery

Gamma Secretase Inhibitors (GSI)

The GSIs, the first Notch inhibitors to be identified, went into clinical trials as potential therapy for Alzheimer's disease. The GSIs have broad spectrum activity and target not only the Notch proteins but also the Notch ligands Delta like and Jagged, amyloid precursor protein (APP), CD44, ErbB4, LRP, syndecan-3, p75NTR, ApoER2, DCC, Nectin-1alpha, E-cadherin, and possibly N-cadherin, which are summarized in our previous review (Harris et al., 2012). In vivo and in vitro activity has been shown for this class of agents in a variety of tumor models including colorectal, lung adenocarcinoma, glioblastoma, gynecological cancer, T-cell ALL, B-cell neoplasms, and prostate cancer (Paris et al., 2005; Wang et al., 2010). Antitumor activity has been demonstrated in NSCLC xenografts for the GSI compound RO4929097 (Luistro et al., 2009) and in models of breast cancer, T-cell leukemia, lung cancer, and pancreatic cancer for the GSI MRK-003 (Konishi et al., 2007; Watters et al., 2009).

MK-0752 is a potent and selective inhibitor of gamma secretase. It was investigated first in T-cell ALL population and more recently in solid tumors. It was generally well tolerated, with GI toxicity being dose limiting (Deangelo et al., 2006; Krop et al., 2012). The potent and selective oral drug GSI R04929097 is the furthest along in clinical development with multiple doses and schedules evaluated (Strosberg et al., 2012; Lo Conte, 2012; Richter et al., 2012). The phase 2 trial of RO4929097 in metastatic melanoma reportedly failed to show target inhibition, and the investigators concluded that there was limited activity in molecularly unselected patients with melanoma (Margolin et al., 2012). A phase 2 study of RO4929097 in advanced platinum-resistant ovarian cancer showed five grade 3/4 AEs possibly related with RO: increased liver transaminases (2), diarrhea (1), headache (1), and hypophosphatemia (1) (Diaz-Padilla et al., 2012). Intracellular Notch (NIC) and JAG1 protein expression on high-grade serous ovarian cancer were correlated with response in 17 patients, and therefore, future studies may require patient enrichment based on the NIC expression for those expecting clinical activity. The results of the Phase 1b study evaluating safety and efficacy of the combination of GDC-0449 (vismodegib), a hedgehog inhibitor, with RO4929097 in patients with advanced sarcoma were also reported (Gounder et al., 2012).

Other GSIs in clinical trials are listed in Table 23.1. One of the major hindrances to the development of GSIs has been their associated GI toxicity, namely intractable diarrhea (Wei et al., 2010a). This is believed to be an on-target effect as a result of the overexpression of the gene for transcription factor kruppel-like factor4 (Klf4), which is necessary for goblet cell differentiation. In vivo models have shown that co-administration of corticosteroids with

Table 23.1. Agents Targeting the Notch Pathway

Class	Agent	Target	Clinical Phase
Gamma secretase inhibitors	MK-0752	Gamma secretase	Phase 1
	R04929097		Phase 1 & 2
	LY450139		Phase 1
	LY900009		Phase 1
	BMS-906024		Phase 1
	PF 03084014		Phase 1
Gamma secretase modifiers	MPC-7869	DLL4	Phase 1
Notch monoclonal antibodies	OMP21M18	Notch 2/3	Phase 1
Notch soluble receptor decoys	OMP-59R5		Preclinical
Genetic strategies			Preclinical

Abhilasha Nair, M.D., Yutaka Kondo, M.D., Ph.D., Pamela Jo Harris, M.D., Hideyuki Saya, M.D., Ph.D., S. Percy Ivy, M.D., and Naoko Takebe, M.D., Ph.D.

BMS-0906024 may be able to ameliorate the toxicity and improve efficacy of the GSIs (Real and Ferrando, 2009). An intermittent dosing schedule has been found with less GI toxicity while maintaining antitumor efficacy. The gamma secretase modifiers are an alternative approach to the GSI enzyme (Pannuti et al., 2010) such as MPC-7869 that inhibit gamma secretase by a different mechanism.

Monoclonal Antibodies

ANTIBODIES AGAINST DLL4 LIGAND The most extensively studied class of agents is the monoclonal antibodies against DLL4 ligand. DLL4 blockade leads to increased VEGF signaling and angiogenesis (Dufraine et al., 2008), and continuous dosing with these agents has led to new vessel formation and angiomas. One potential strategy to overcome this problem is to combine the VEGFR inhibitors with these agents, which might have the added advantage of increasing efficacy (Li and Harris, 2009). In vivo evidence shows that DLL4 antibodies do not appear to affect goblet cell differentiation or proliferation since this requires dual DLL1 and DLL4 blockade (Yan, 2011).

ANTIBODIES TO NOTCH RECEPTORS Antibodies to Notch receptors can either block or stimulate receptor signaling (Li et al., 2008), and antibodies have been developed against the negative regulatory region (anti-NRR), which can specifically block either Notch-1 or 2 activity. Twenty-four patients were enrolled to a phase 1 study of OMP-59R5, and the most frequent drug-related AEs follow: mild to moderate diarrhea, which can be reduced with dosing modification, fatigue, nausea, vomiting, decreased appetite, and constipation (Tolcher et al., 2012).

Stapled Peptide

AGENTS AGAINST MAML-CSL-NOTCH COMPLEX A novel approach to Notch inhibition has used a stapled peptide to block interaction of mastermind like with Notch intracellular domain (Moellering et al., 2009).

Hedgehog Signaling Pathway and Inhibitors

Inhibitor Discovery

The Hedgehog (Hh) pathway, one of the most critical developmental signaling pathways, was initially discovered in the fruit fly as a regulator of body segmentation and patterning (Nusslein-Volhard and Wieschaus, 1980). In 1996, three additional key findings were discovered in this pathway: (1) the mutations generated in the Hh pathway in vertebrates resulted in defective axial patterning including cyclopia or holoprocencephaly (Belloni et al., 1996; Chiang et al., 1996); (2) discovery of human sonic Hh gene mutations causing holocencephaly (Roessler et al., 1996); and(3) for the first time, the linkage between cancer development and alteration of Hh signaling pathway by mutations to patched was discovered in basal cell carcinoma (BCC) (Hahn et al., 1996; Johnson et al., 1996). Soon thereafter, the steroidal alkaloid derivative cyclopamine was discovered as an antagonist of the Hh pathway, and its mechanism of action through binding to Smoothened (Smo), a G protein coupled receptor (Cooper et al., 1998). This finding led to the discoveries in oncogenesis and pharmacologic properties against Smo or other potential target discoveries affecting the Hh signaling pathway.

Vismodegib (GDC-0449, Genentech-Roche), a first-in-class small molecule Hh pathway inhibitor was approved by the U.S. Food and Drug Administration (FDA) in January 2012 for locally advanced and metastatic basal cell carcinomas (BCC) after only 5 years since its first-in-human phase 1 clinical trial at U.S. sites. During the phase 1 study, Vismodegib was remarkably active in a patient with widely metastatic BCC at the initial dose cohort and showed further promising activity in expansion cohorts consisting of advanced BCC patients. Ultimately, 18 out of 33 metastatic BCC patients showed objective responses (Von Hoff et al., 2009). The phase 2 study confirmed the phase 1 results with objective responses of 30% in metastatic BCC and 43% in locally advanced disease including 21% who had complete responses.

In contrast to above-mentioned Hh ligand-independent or mutation driven tumors, chronic myelogenous leukemia (CML), breast cancer, and multiple myeloma (MM)-derived CSCs are also regulated by the Hh pathway in a ligand-dependent manner. Hh pathway inhibition was demonstrated in these tumor CSCs by limiting tumorigenic self-renewal or inducing terminal differentiation by cyclopamine (Liu et al., 2006; Peacock et al., 2007; Dierks et al., 2008; Zhao et al., 2009). As a result, IPI-926, a more specific and potent cyclopamine derivative was synthesized (Tremblay et al., 2008). However, the pharmaceutical company Infinity's placebo controlled randomized phase 2 study in advanced chondrosarcoma comparing placebo versus IPI-926 was stopped after interim analysis showed a lack of improved PFS in the treatment arm compared to the placebo arm. Similarly, their randomized placebo-controlled phase 2 combination trial with gemcitabine in metastatic pancreatic cancer was also terminated early after interim analysis demonstrated a higher rate of progressive disease in IPI-926 arm.

Large-scale screens of chemical libraries have identified other novel small molecules active against SMO. Novel SMO antagonists in phase 1–2 trials are LDE225 (Novartis), PF04449913 (Pfizer), BMS-833923 (XL139, BMS/Exelixis), TAK-441 (Millennium), and LEQ-506 (Novartis). Currently, vismodegib is the leading agent in this new class of therapeutics as it is the only one to have received FDA approval (for advanced or metastatic BBC as mentioned previously). A company-sponsored phase 2 randomized placebo-controlled trial of the single agent vismodegib in ovarian cancer as maintenance therapy following second or third complete remission showed marginal median PFS improvement in the treatment arm (5.8 months versus 7.5 months, HR = 0.79) (Rudin, 2012). Genentech-Roche has sponsored a second trial, a phase 2 randomized placebo-controlled trial in metastatic colorectal carcinoma comparing standard first-line chemotherapy with FOLFOX (5-fluorouracil, leucovorin, oxaliplatin) + bevacizumab or FOLFIRI (5-fluorouracil, Leucovorin, irinotecan) + bevacizumab with and without vismodegib. However, it did not meet the PFS endpoint and was a negative study (HR = 1.24) (Bendell et al., 2010).

A phase 2 randomized placebo-controlled trial in advanced metastatic pancreatic cancer sponsored the by National Cancer Institute (NCI), Cancer Therapy Evaluation Program (CTEP) studied the combination of vismodegib with gemcitabine, but it did not meet criteria for futility at interim analysis (Catenacci et al., 2012). NCI, CTEP is sponsoring a single arm biomarker driven pilot study with vismodegib followed by gemcitabine + vismodegib in metastatic pancreatic cancer to look for a predictive biomarker and reduction of tumor sphere formation. Final analysis is pending. Additional NCI, CTEP-sponsored studies under way include a 3-arm randomized study in small cell lung cancer in combination with the standard chemotherapy agent + vismodegib; a 2-arm randomized placebo-controlled study in advanced gastric cancer and lower esophageal cancer with FOLFOX with or without vismodegib; a single-arm pilot study to examine the neurosphere formation effect by a neoadjuvant therapy phase with or without vismodegib in relapsed or refractory glioblastoma patients requiring debulking surgery; and a single-arm phase 2 chondrosarcoma treated with a single agent vismodegib (Italiano et al., 2012). Among experimental Hh pathway inhibitors, there are several interesting trials. LDE225, a SMO inhibitor was tested in a phase 1/2 study with a phase 2 expansion cohort in pediatric recurrent medulloblastoma. The striking results showed 2 out of 24 patients had complete responses. Both had Hh-activated tumors by gene signature analysis. Only 14 patient samples were available for gene signature analysis, and the remaining 12 patients who did not achieve response were determined to have tumors without Hh pathway activation (Geoerger et al., 2012). Bms-833923 (XL139) has been primarily tested in chronic myelogenous leukemia patients in phase 1/2; PF-04449913, TAK-441, and LEQ506 are currently under a phase 1 stage of development.

There are other drug targets besides SMO in the Hh pathway, and some have been already tested clinically. Itraconazole, a systemic antifungal agent, has been identified as a potent Hh signaling pathway antagonist. A phase 2 study has been conducted in BCC with 4 out of 19 patients showing partial response, and 4 out of 19 had stable disease (Kim et al., 2010; Kim et al., 2012). Similarly, another old drug, arsenic trioxide, was discovered as an Hh signaling pathway inhibitor in Ewing sarcoma (Beauchamp et al., 2009; Beauchamp et al., 2011). There are several Hh signaling pathway inhibitors developed to date and tested preclinically, such as robotnikinin, which interferes with Hh ligand and receptor patched, a group of molecules named HPIs that inhibit intracellular migration of transcription factors into nucleus and primary cilia formation necessary for Gli activation, and inhibit Gli1 function (Lauth et al., 2007; Hyman et al., 2009; Stanton et al., 2009). See Table 23.2.

Table 23.2. Table of Targeted Agents of the Hedgehog Pathway

Class	Agent	Target	Clinical Phase
Smoothened inhibitors	Vismodegib	Smoothened	FDA
	IPI926	Smoothened	approved
	LDE225	Smoothened	Phase 1/2
	PF04449913	Smoothened	Phase 1/2
			Phase 1/2
	BMS-83923	Smoothened	Phase 1/2
	TAK441	Smoothened	Phase 1
	LEQ-506	Smoothened	Phase 1
	HPI1-3	Transcription factor	
HPIs		cilia	Preclinical
Hh ligand inhibitor	Robotnikinin		Preclinical
Anti-fungal	Intraconazole		Phase 2
Arsenic Trioxide	Arsenic		Phase 2
		Hh ligand	
		ND	
		ND	

Wnt Signaling Pathway and Inhibitors

Given the complexity of the Wnt pathway, it is not surprising that there are multiple agents in preclinical and clinical development to inhibit primarily canonical signaling. Although a few of these have been tested preclinically in CSC models, the preponderance of them have only been tested in mature tumor cell lines and/or in vivo models of mature tumors; a subset has been tested in human clinical trials without an emphasis on the stem cell component. We briefly review these agents, with limited reference to preclinical assessments of activity in CSCs.

Inhibitor Discovery

The most proximal way of interfering with Wnt signaling is to inhibit Wnt-ligand production and secretion. Inhibitors of Wnt production (IWP) bind to and disable the Porcupine protein. Porcupine normally adds a palmitoyl group to Wnt in the endoplasmic reticulum, which, in turn, enables secretion of Wnt. The next step in the pathway that can be targeted involves monoclonal antibodies to the Wnt proteins themselves, but antibodies against Wnt 1 and Wnt 2 have had only minimal activity in xenograft models (He et al., 2004; You et al., 2004). Because several Wnts bind to CSCs, pan-Wnt antibodies may be needed for inhibition of signaling in these early cells.

Through a complementary mechanism, proteins such as Wnt Inhibitor Factor I (WIF1) and secreted frizzled related protein (SFRP) and their analogs inhibit Wnt ligands prior to their binding to their relevant receptors (Takahashi-Yanaga and Kahn, 2010). Antibodies to the frizzled family of receptors are in preclinical and clinical development including OMP-18RF (NCT01345201). Yttrium 90-radiolabeled OTSA101, a monoclonal antibody, which targets frizzled homolog 10, is in a phase 1 trial in patients with synovial sarcomas (NCT01469975).

The soluble Fzd8-CRD-Fc has been shown to interfere with autocrine Wnt signaling (DeAlmeida et al., 2007). OMP-54F28, a proprietary fusion protein based on a truncated form of the frizzled 8 receptor, is in a dose-escalation trial in patients with solid tumors (NCT01608867). A first-in-man study investigating the safety and dose escalation of a new radiolabelled monoclonal antibody targeting frizzled homolog 10 (SYNFRIZZ) is currently under way by the French sarcoma group (NCT01469975).

Another mode of inhibition of the Wnt pathway involves disabling the LPR5/6 receptor complex by the Dickopf (DKK) protein family (Semenov et al., 2008; Niehrs and Shen, 2010; Sakane et al., 2010), inhibiting its phosphorylatation, and inducing its degradation by Salinomycin (Lu et al., 2011), and by antibodies to the Wnt-binding regions of the co-receptor complex LRP5/6 (Bourhis et al., 2010).

After Wnts are bound to frizzled and LRP5/6, interrupting transport of extracellular to cytoplasmic components represents the next way of inhibiting Wnt signaling. Three compounds, NSC66063, FJ9, and 3289–8625, in preclinical testing are directed at Dishevelled (Dvl) (Shan et al., 2005; Fujii et al., 2007; Grandy et al., 2009). Within the cytoplasm, activation of GSK-3β increases the phosphorylation and destruction of beta-catenin. Further, stabilization of axin attenuates Wnt signaling. XAV939 inhibits tankyrase 1 and tankyrase 2 and, as a consequence, has been shown to have a growth-inhibitory effect on colorectal cells (Huang et al., 2009). Agents including IWR-2, which stabilize axin, are known as inhibitors of Wnt response (IWR) (Chen et al., 2009a).

The casein kinase (CK)2 inhibitor CX-4945 has shown promise preclinically and is being investigated in two phase 1 trials in patients with advanced solid tumors (NCT00891280) and with multiple myeloma (NCT01199718). Ubiquitin-dependent proteolysis by LC-363 results in β-catenin degradation (Bol and Ebner, 2006). Additionally, hexachlorophene and isoreserpine, which upregulate "seven in absentia homolog"

(Siah), augment Siah–driven beta-catenin degradation (Park et al., 2006).

In the nucleus, unphosphorylated β-catenin forms a complex with the T cell factor/lymphoid enhancer factor (TCF/Lef) family of transcription factors. As a result, the next class of Wnt inhibitors antagonize this complex. There are several such preclinical inhibitors including PKF-115-584 and CGP049090 (Lepourcelet et al., 2004; Sukhdeo et al., 2007; Wei et al., 2010b). Both agents also inhibited the growth of human chronic lymphocytic leukemia cells (Gandhirajan et al., 2010). The synthetic compound PKF-118-310, a β-catenin/TCF inhibitor, irreversibly halted tumorsphere formation by breast tumor-initiating cells (Hallett et al., 2012). Recently, three small molecules, which are inhibitors of β-catenin-responsive-transcription (CRT), were identified; iCRT3, iCRT5, and iCRT14 specifically interfere with the protein interaction between β-catenin and TCF (Gonsalves et al., 2011). The co-activator cyclic AMP response element-binding protein (CBP) is inhibited by ICG-001, which binds to it and prevents attachment to β-catenin/TCF. ICG-001 downregulates survivin, Myc, and cyclin D1 along with increasing caspase activity in colon cancer cells (Yao et al., 2011). PRI-724, a clinical compound derived from ICG-001, is in a phase 1 trial in advanced solid tumors (NCT01302405).

β-catenin siRNA screening identified several compounds such as LC-363 and CW232291 (Yao et al., 2011; Curtin and Lorenzi, 2010). CWP232291 is currently in a clinical trial in acute myeloid leukemia (NCTCW232291).

Miscellaneous Pharmaceutical and Natural Inhibitors of the Wnt Pathway

Two classes of agents appear to work at multiple steps in Wnt signaling are nonsteroidal anti-inflammatory drugs and vitamins D. Cox-2 inhibitors, celecoxib and sulindac, have been in clinical trials for years, primarily for cancer prevention (Table 23.3). Polyphenols, quercetin, curcumin, and resveratol, abrogate Wnt/β-catenin signaling through unknown mechanisms (Takahashi-Yanaga and Kahn, 2010). These polyphenols have generated such clinical interest that there are 24 trials with just curcumin in ClinicalTrials.gov as of June 21, 2012. Pyvrinium has recently attracted attention because of its impact on Wnt signaling by activation of casein kinase 1α (CK1α) (Thorne et al., 2010). VS-507, which has been identified as a selective inhibitor of CSCs, inhibits tumorsphere formation by MDA-MB-231 breast cancer cells in a dose-dependent manner and, in a murine model of triple negative breast cancer, decreasing primary tumor growth and experimental metastasis (Pachter et al., 2012).

Polycomb-Repressive Complexes and Inhibitors

Polycomb-repressive complexes (PRC), PRC1 and PRC2, have been identified as key epigenetic regulators during development (Lund and van Lohuizen, 2004) and maintaining CSC self-renewal and multi-potential capability (Richly et al., 2011). Among those polycomb group (PcG) proteins, enhancer of zeste 2 (EZH2), a catalytic component of PRC2, is frequently overexpressed or mutated in human malignancies. EZH2 is also known to be associated with CSC biology in breast cancer and glioblastoma multiforme (Suva et al., 2009; Chang et al., 2011). Recent data indicated that inhibition of EZH2 might be a potent strategy for cancer treatment, and a number of drug development programs are ongoing now (Garber, 2012).

Roles of Two PRCs in CSCs

Aberrant trimethylation at histone H3K27 (H3K27me3), a histone modification characteristic of repressed gene expression in the process of embryonic stem cell differentiation, is frequently observed in many types of cancers (Kondo et al., 2008). EZH2/PRC2 has a histone methyltransferase activity with substrate specificity for H3K27 and produces H3K27 dimethylation (H3K27me2) or H3K27me3

Table 23.3. Agents Targeting the Wnt Pathway

Class	Agent	Target	Stage of Development
Wnt production inhibitors	*IWP-2*	*Porcupine*	*Preclinical*
	LGK974		Phase 1
Frizzled family and receptor antibodies	OMP-18RF	Frizzled homolog or receptor	Phase 1
	OTSA101		Phase 1
	SYNFRIZZ		Phase 1
Soluble frizzled receptors	Fzd-CRD-Fc	Frizzled homolog	Preclinical
	OMP-54F28		Phase 1
LRP5/6 inhibitors	salinomycin	LRP6	Preclinical
Dishevelled inhibitors	NSC66063	Dishevelled CDZ domain	Preclinical
	FJ9		Preclinical
	3289-8625		Preclinical
GSK-3β activators	Differentiating-inducing factor-3	GSK-3b	Preclinical
Axin stabilizers	XAV939	Tankyrase 1, 2; Axin	Preclinical
	IWR-2		Preclinical
Casein kinase 2 inhibitors	CX-4945	CK2	Phase 1
	CX-8184		Preclinical
	DAT		Preclinical
	TBB		Preclinical
Non-phosphorylated-catenin inhibitors	desferrioxamine	Active β-catenin	Preclinical
	deferasirox		Preclinical
	ciclopirox		Preclinical
	M-110		Preclinical
	OICR623		Preclinical
TCF/Lef/β-catenin complex inhibitors	PKF-115-584	TCF/Lef/β-catenin	Preclinical
	CGP049090		Preclinical
	PKF-118-310		Preclinical
	ICRT3, T5, T14		Preclinical
	PCNU74654		Preclinical
Co-activator cyclic AMP response element binding protein inhibitors	ICG-001	CBP	Preclinical
	PRI-724		Phase 1
Miscellaneous	Non-steroidal anti-inflammatories (celocoxib, sulindac, OXT-328)	Multiple targets	Phase 2, 3
	1α25-dihydroxyvitamin D3	β-catenin	Phase 2, 3
	Polyphenols (quercetin, curcumin, resveratol)	Unknown target	Phase 2, 3

A

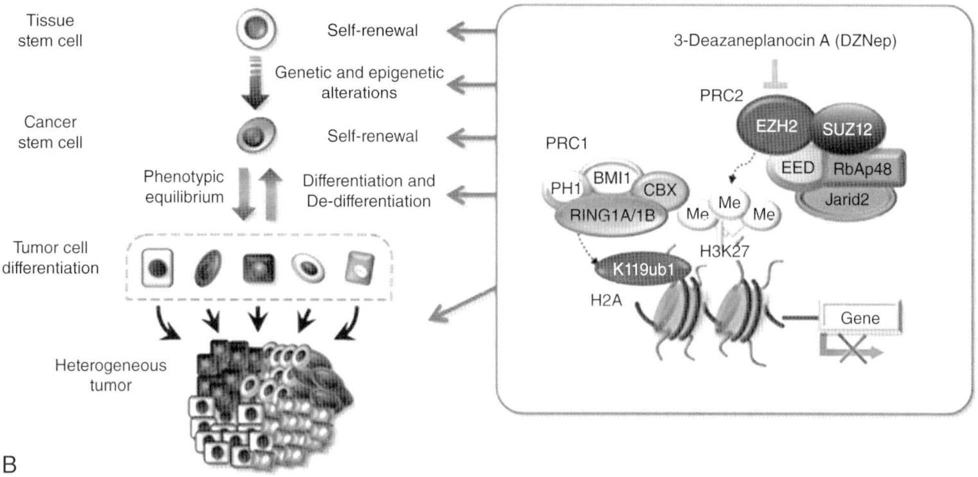

B

Overexpression/ampliciation	
EZH2	Prostate cancer, Breast cancer, Colon cancer, Liver cancer, Lung cancer, Glioblastoma, Lymphoma, Metanoma, Oral cancer, Testicular cancer
SUZ12	Colon cancer, Breast cancer, Liver cancer, Lung cancer, Lymphoma, Melanoma, Germinal cell-derived tumor, Parathyroid and pituitary adenoma
BMI1	Acute myeloid leukiemia, Breast cancer, Gastrointestinal tumor, Lymphoma, Neuroblastoma, Lung cancer, Medulloblastoma, Parathyroid and pituitary adenoma
RING1B	Colon cancer, Bladder cancer, Breast cancer, Renal cancer, Liver cancer, Lung cancer, Lymphoma, Melanoma, Pancreatic cancer, Prostate cancer, Gynecological tumor, Larynx cancer, Parathyroid and thyroid cancer, Thymus cancer, Unrinary tract cancer
Mutations	
EZH2	Lymphoma, Myelodysplastic syndrome (active and inavtive mutations)
SUZ12	Endometrial stromal sarcoma (Chromosomal translocation)

Figure 23.1. Roles of Polycomb-repressive complexes (PRC) in heterogeneous tumor formations. **A**. Cancers are mostly comprised of heterogeneous cell populations. Such multiple distinct subpopulations of cancer cells within tumors may derive from cancer stem cells (CSC). The CSCs are considered able to aberrantly differentiate into diverse cell types. Whereas CSCs can differentiate into non-CSCs, the reverse process is now also being considered. This phenotypic equilibrium in cell-state proportions is observed in CSCs (*left panel*) (Gupta et al., 2011). PRC2 has a histone methyltransferase activity with substrate specificity for H3K27 and produces H3K27me3, which is thought to recruit PRC1 via proteins of the CBX family. RING1A/1B in PRC1 induces the mono-ubiquitination of histone 2A lysine 119 (H2AK119), which is thought to affect the chromatin structure, blocks the recruitment of transcriptional activation factors (*right panel*). PRCs are required for the self-renewal of CSCs as well as CSC differentiation in response to oncogenic cues, which lead to the establishment of heterogenous tumors. **B**. PRC components were overexpressed or mutated in many types of human neoplasia (Sauvageau and Sauvageau, 2010). This figure is reprinted from Harris and colleagues with minor modifications (Harris et al., 2012). For color detail, please see color plate section.

(Margueron and Reinberg, 2011) (Figure 23.1). H3K27me3 serves as a signal for specific binding of the chromodomain of another polycomb repressor complex, PRC1 (Levine et al., 2002). In general, PRC1 and PRC2 work collaboratively to repress transcription. However, recent analyses revealed that PRC1 and PRC2 do not share all of their targets: a substantial proportion of PRC2 targets are not occupied by PRC1 and vice versa, implying independent actions of PRC1 and PRC2 (Ku et al., 2008; Leeb et al., 2010). Intriguingly, a PRC1 component, B lymphoma Moloney Murine Leukemia Virus Insertion region 1 homolog (BMI1), sometimes has opposing roles to PRC2 in the context of regulation of stem cells (Majewski et al., 2010). BMI1 maintains the self-renewal of hematopoietic and neural stem cells along with stemness function in CSCs (Lessard et al., 1999). Meanwhile PRC2 restricts hematopoietic stem cell/progenitor activities (Majewski et al., 2010). Studies also demonstrated that PRC2 not only acts to promote self-renewal, but also controls fate choices within the multipotent lineage in neural and muscle development (Palacios et al., 2010). The precise molecular basis for controlling the stemness and stem cell differentiation by PRC1 and PRC2 has though remained elusive. However, independency between PRC1 and PRC2 may enable precise and complex regulation of gene functions during development.

Targeting EZH2 as a Cancer Treatment

EZH2 has important roles in normal differentiation, maintenance of cell identity, proliferation, and stem cell plasticity. Dysregulation of this protein confers and modulates the neoplastic phenotype (Sparmann and van Lohuizen, 2006). Studies established that EZH2 is frequently overexpressed in many types of tumors, and higher level of expression of EZH2 is sometimes associated with aggressiveness of tumors (Varambally et al., 2002). Recent sequencing projects have uncovered recurrent mutations of the tyrosine 641 (Y641) residue of EZH2, which has been reported in lymphomas (Sneeringer et al., 2010). Although this mutant was initially reported to be a loss-of-function mutation (Morin et al., 2010), subsequent analysis demonstrated that the Y641 mutant showed increased activity of EZH2 (Sneeringer et al., 2010; Yap et al., 2010). These discoveries imply that increased EZH2 activities induce H3K27me3 and may act as an oncogene via repression of tumor suppressor genes. Consistently, loss of EZH2/PRC2 impaired the tumor formation in mouse models, indicating that EZH2 activity is essential for tumorigenesis (Mallen-St Clair et al., 2012; Neff et al., 2012; Tanaka et al., 2012).

Differing with those reports, homozygous and heterozygous deletions and inactive mutations of EZH2 were found in myelodysplastic syndromes, suggesting that EZH2 may have a tumor suppressor function (Ernst et al., 2010; Makishima et al., 2010; Nikoloski et al., 2010). Moreover, studies in a mouse model showed that loss of EZH2 in HSCs is sufficient to cause acute lymphoblastic leukemia (Ntziachristos et al., 2012; Simon et al., 2012).

These bifunctional roles of EZH2 as oncogene and tumor-suppressor in human malignancies have indicated context dependency of signaling outcomes downstream of dysregulated EZH2 activity. In other words, contribution of hyperactive or hypoactive EZH2 during tumor formation may reflect the complex roles of PRC-associated genes in cell fate decisions during differentiation (Sauvageau and Sauvageau, 2010). Since cancer is a disease of faulty differentiation, it makes sense that dysregulation of EZH2 are involved in different steps of carcinogenesis. Therefore, it may be rational to determine whether PRC2 is required for the differentiation and de-differentiation of CSCs (i.e., tumor cell plasticity), or for the epithelial to mesenchymal transition rather than modulating EZH2 levels to gauge its function as an oncogene versus tumor-suppressor in a defined cell context (Gupta et al., 2009; Margueron and Reinberg, 2011). However, these two aspects of EZH2 in tumorigenesis may also imply the complexity in developing agents aimed at modulating H3K27 methylation. Nonetheless, given the remarkable relationship between epigenetics and CSC biology, EZH2/PRC2 might be an attractive targeting candidate in cancer and likely in CSC treatment.

Pharmacological inhibition of EZH2/PRC2 by 3-Deazaneplanocin A (DZNep), an S-adenosylhomocysteine hydrolase inhibitor, has been shown to reduce H3K27me3 and inhibit cancer cell growth (Tan et al., 2007). By contrast, no specific inhibitor of PRC1 has been developed for cancer treatments so far (Crea et al., 2012). The experimental data support the idea that chemical inhibition of EZH2 methyltransferase activity is supposed to be a promising path for treatment of human malignancies (Chase and Cross, 2011). Since DZNep appeared to have a broad effect on histone methylation in addition to H3K27, more specific inhibitors of EZH2 are required for cancer treatment (Tan et al., 2007). Recent challenges in developing potent and selective small molecule EZH2 inhibitors with in vitro and in vivo antiproliferative activity have been reported (Garber, 2012). If such highly selective small chemicals are warranted and can be used safely and appropriately to the patients' disease, these might be novel and potent drugs for cancer treatment. Therefore, more tailored therapy approaches (e.g., treatment with EZH2 inhibitors in cancers carrying activating mutations of EZH2) are possible for safe and appropriate cancer treatment. As discussed above, experimental data raise concerns that strategies aimed at decreasing H3K27 methylation may cause or promote cancer, and careful clinical evaluation is necessary along with further clinical studies that may prove the safety and potency of these inhibitors.

Novel CSC Therapeutic Targets and Inhibitors

CD44 Variants (CD44v) Targeting Epithelial-to-Mesenchymal Transition (EMT) and CSC Therapeutics

CD44 is a major adhesion molecule for hyaluronic acid and has been implicated in a variety of biological processes including cell migration, cell proliferation as well as tumor growth and metastasis (Gunthert et al., 1991; Yu et al., 1997; Ponta et al., 2003; Nagano and Saya, 2004). CD44 has numerous isoforms generated through alternative mRNA splicing, with variability in the extracellular domain of CD44 (Nagano and Saya, 2004). Those isoforms are referred to as CD44 variants (CD44v) and are expressed in diverse cancers. Recently, CD44 has been found to be a marker for cancer stem cells (CSC) in various epithelial tumors, and CD44-expressing cancer cells have exhibited higher tumor-initiating activity than CD44-negative tumor cells (Al-Hajj et al., 2003). However, the functional link between CD44 and CSCs was not identified.

CSCs are malignant cell subsets in hierarchically organized tumors, which are selectively capable of tumor initiation and self-renewal and give rise to the nontumorigenic cancer cells through differentiation. Clinically, the most important characteristic of CSCs is the resistance to various stresses. Therefore, identification of the stress resistance mechanism of CSCs provides novel therapeutic approaches for refractory cancers.

Reactive oxygen spices (ROS) are the major environmental stressors for cancer cells. CSCs have been shown to keep their intracellular levels of ROS low and thereby enhance protection against ROS-mediated damage (Diehn et al., 2009). Indeed, it has been reported that mechanisms for protection against ROS-mediated damage are activated in subsets of CD44-positive CSCs in some tumors, which contributes to tumor resistance to chemo- and radiotherapy (Phillips et al., 2006).

Ishimoto and colleagues (2011) recently showed that CD44v attenuates ROS accumulation in cancer cells, thereby promoting tumor growth and survival. They found that CD44v interacts with the system xc^- cystine-glutamate exchange transporter complex on the cell membrane and stabilizes the expression of the transporter (Figure 23.2). The system xc^- transporter complex is composed of a light-chain subunit (xCT) and a heavy-chain subunit (CD98hc) and plays a role in regulation of the intracellular level of reduced glutathione (GSH) through the uptake of cystine. GSH, a major antioxidant acting as a free radical scavenger, protects cells from ROS. Therefore, CD44v-expressing cancer cells, which predominantly include CSC fraction, produce GSH, leading to the defense against therapeutic stresses.

Interestingly, the same research group has shown that CD44 interacts with the M2 isoform of pyruvate kinase (PKM2) (Tamada et al., 2012), which has been implicated in

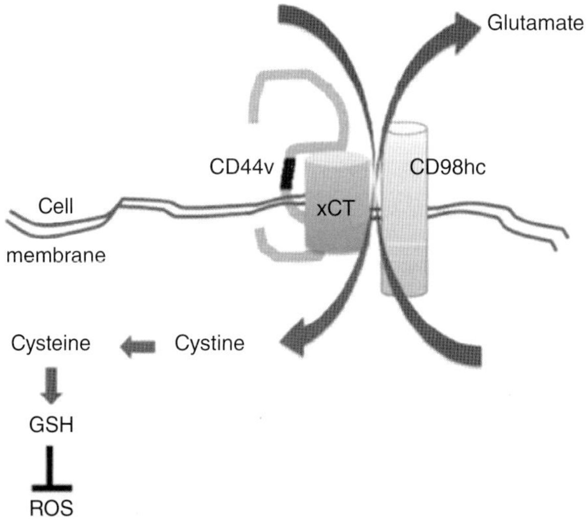

Figure 23.2. Role of CD44v in redox regulation.CD44 variant (CD44v) interacts with xCT, the light subunit of Xc- cystine transporter, and stabilizes its plasma membrane localization, leading to high glutathione (GSH) and low reactive oxygen species (ROS) levels. This figure is reprinted from Ishimoto and colleagues with minor modification (Ishimoto et al., 2011).

regulation of aerobic glycolysis in cancer cells, so called Warburg effect (Christofk et al., 2008a; Christofk et al., 2008b). Expression of CD44 enhances the glycolytic phenotype cancer cells and promotes metabolic flux to pentose phosphate pathway and thereby increased levels of NADPH, which is used for GSH synthesis. This finding indicates that CD44 plays a role in metabolic shift via regulation of PKM2 and ROS protection by GSH production in cancer cells.

Yae and colleagues (2012) have reported that orthotopic transplantation of a CD44v-expressing 4T1 mouse breast cancer cells, but not that of a CD44v-negative subpopulation, in mice results in efficient lung metastasis. They found that the metastatic ability of CD44v cells depends on the activity of the cystine transporter xCT and GSH production. Furthermore, they discovered that an isoform switch from CD44v to CD44 standard (CD44s) is regulated by epithelial splicing regulatory protein 1 (ESRP1) and that knockdown of ESRP1 in CD44v-expressing cells results in downregulation of xCT expression and suppression of lung colonization.

Genetic ablation of CD44 significantly suppressed tumorigenesis of gastric cancer in a transgenic mouse model. Additionally, RNAi-based depletion of ESRP1, which converted CD44v to CD44s, reduced lung metastasis of CD44v-expressing breast cancer cells. Based on those data, they claim that ESRP1-CD44v-xCT axis can be a potential therapeutic target for CSCs.

Sulfasalazine, which is a drug used for the treatment of ulcerative colitis, Crohn's disease, rheumatoid arthritis, or ankylosing spondylitis, was shown to have an inhibitory activity to xCT-mediated cystine transport (Gout et al., 2001; Lo et al., 2008; Chen et al., 2009b). Intraperitoneal administration of sulfasalazine inhibited the growth of tumors formed by HCT116 cells in nude mice and lung colonization of CD44v-expressing breast cancer cells. Therefore, xCT cystine transporter plays a critical role in the growth and metastasis of CD44v-expressing tumor cells in vivo.

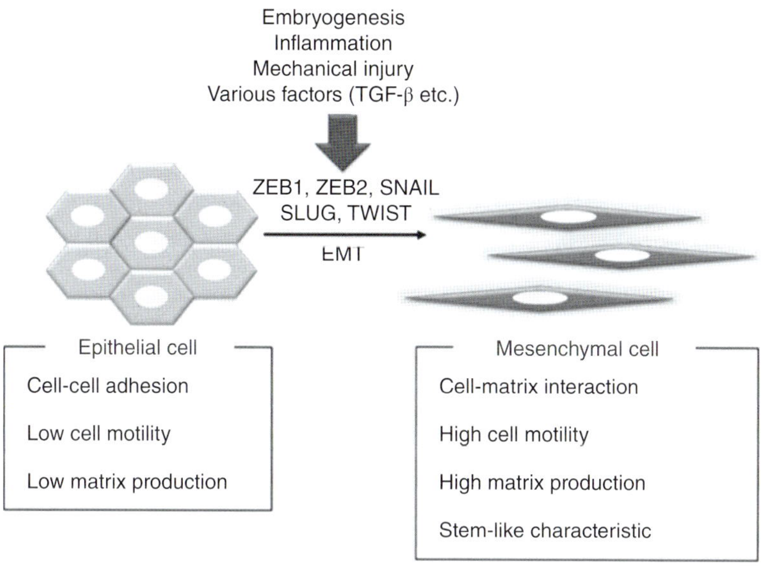

Figure 23.3. Inducible factors of epithelial mesenchymal transition (EMT) and characteristics of cells undergoing EMT.

Based on these preclinical experimental data, sulfasalazine is considered to be one of the promising agents for the management of CSCs. Expression of CD44v and GSH levels can be biomarkers to determine the effect of sulfasalazine on tumor. A clinical trial using sulfasalazine is now being planned.

Targeting Epithelial-to-Mesenchymal Transition (EMT) and CSC Therapeutics

There is a rationale for targeting the CSC population to inhibit tumor metastasis. Ongoing research has demonstrated that CSCs, similar to epithelial cells during embryogenesis, acquire mesenchymal properties and mobilize primary cells to distant locations. Thus, invasion and metastasis by CSCs occur by acquiring EMT and mesenchymal-to-epithelial transition (MET) properties to regulate cellular plasticity (Thiery, 2002). Mani and colleagues (2008) have shown that epithelial cancer cells acquiring EMT properties also acquired stem cell characteristics. Regulators of EMT have been identified; therefore, inhibitors of these regulators may be of interest to inhibit CSC population. The details of potential targets and investigational agents were reviewed previously (Takebe et al., 2011b).

Epithelial cells are connected by adherens junctions, consisting of E-cadherin, α-, β-, and γ-catenin, and are connected to the actin cytoskeleton, constructing a rigid structure. EMT, defined by the loss of epithelial characteristics and the acquisition of a mesenchymal phenotype, is triggered by various extrinsic factors (Figure 23.3). During this transition, epithelial cells downregulate cell-cell adhesions, lose their polarity, and acquire a mesenchymal phenotype associated with increased production of the extracellular matrix (ECM), promoting interaction with ECM and enhanced migratory capacity. A critical molecular feature of EMT is the downregulation of E-cadherin, a cell adhesion molecules present in the plasma membrane of most normal epithelial cells.

The EMT is a critical event in metazoan embryogenesis and physiological processes like wound healing. It is also observed in pathological settings such as fibrotic disorders as well as cancer invasion and metastasis. Furthermore, it has been suggested that breast CSCs have similar characteristics to the mesenchymal-appearing cancer cells generated by EMTs (Mani et al., 2008). Therefore, agents that block the intrinsic and extrinsic signals for induction and maintenance of EMT could be potential drugs for targeting CSC.

Arima and colleagues (2008) previously reported that inactivation of the retinoblastoma tumor suppressor protein (RB) contributes to tumor progression due to not only loss of cell cycle control, but also conversion to an invasive phenotype through EMT. Ectopic expression resulted in increased expression of E-cadherin, while reduced expression occurred with knockdown of RB. Furthermore, EMT-related transcriptional factors, including SLUG and ZEB-1, were induced by RB depletion. Given that ZEB family proteins are known to act as a transcriptional repressor of the E-cadherin gene, upregulation of ZEB-1 expression is a major mechanism for induction of EMT in RB-inactive cells. Recently, the same investigator reported that ZEB proteins are markedly upregulated by RB depletion in a manner sensitive to the miR-200 family of microRNAs (Arima et al., 2012). Therefore, RB-inactivation induces downregulation of miR-200 family members, leading to increased ZEB expression, which contributes to the EMT.

Additionally, Arima and colleagues (2012) attempted to find chemical compounds that suppress ZEB1 expression by using a SCADS (Screening Committee of Anticancer Drugs) inhibitor kit. They identified several cyclin-dependent kinase (CDK) inhibitors, including CDK1/2 inhibitor, CDK2/9 inhibitor, CDK4 inhibitor, Purvalanol A, and Alsterpaullone, which blocked ZEB1 expression and RB phosphorylation. Therefore, they tested the effect of a CDK4/6 inhibitor, PD0332991, which is currently in a clinical trial, on ZEB expression and invasive characteristics of breast cancer cells. PD0332991 significantly suppressed phosphorylation of RB, invasive activity, and in vivo tumor growth. Thus, CDK4/6 inhibitors may be effective in attenuating EMT and stem-like characteristics of cancer cells. By contrast, a recent report has shown that PD-0332991 increased Smad transcriptional activity and activated TGF-β signaling, leading to EMT induction in two pancreatic cancer cell lines, COLO-357 and PANC-1, which have a wild-type SMAD4 gene but not in AsPC-1 cells that harbor a mutated SMAD4 (Liu and Korc, 2012). Therefore, phenotypic changes induced by CDK inhibitors are potentially different depending on the cancer cell types and the signaling pathways activated. ZEB-1 can be a useful biomarker to determine the effect of CDK inhibitors on EMT properties in cancer cells.

Perspectives and Conclusions

The rationale for targeting CSCs continues to receive great interest and promise for future cancer therapy as there are increasing numbers of novel agent discoveries and efforts in discovery for new utility of old agents. However, in the context of cancer treatment, one can consider that targeting a specific pathway might not be sufficient to induce remission in patients. This is true for any molecular targeting agent therapy due to complex compensatory signaling pathways and redundancies in cancer cells. Cancer cell heterogeneity is another important factor for therapeutic resistance, which also applies to CSC therapeutics. The epigenetic regulations in cancer cells seem to be largely associated with tumor heterogeneity. We only discussed briefly the agents modulating tumor microenvironment, but this is an additional challenge in cancer therapeutics regarding the need for further understanding of the biology of how CSC proliferation can be controlled by signaling from the tumor stromal environment. This regulation includes the immunomodulation effect. Emerging findings of mutations in epigenetic regulators including EZH2 may imply the important mechanism of tumor formation and CSC biology. Therefore, targeting epigenetic regulators is also the next stratification of cancer therapeutics in selected tumor types. Targeting altered metabolism pathways in CSC is of great interest, and further studies are warranted. In summary, there can be no doubt that developing agents to control proliferation of CSCs is a very important research endeavor that will undoubtedly grow in the next few years and lead to improved treatments for patients.

Acknowledgments

We thank Dr. Richard Swerdlow for his superlative editing of the manuscript.

References

Al-Hajj, M., M.S. Wicha, A. Benito-Hernandez, *et al.* 2003. Prospective identification of tumorigenic breast cancer cells. *Proc Natl Acad Sci USA* 100(7):3983–3988.

Arima, Y., Y. Inoue, T. Shibata, *et al.* 2008. Rb depletion results in deregulation of E-cadherin and induction of cellular phenotypic changes that are characteristic of the epithelial-to-mesenchymal transition. *Cancer Res* 68(13):5104–5112.

Arima, Y., H. Hayashi, M. Sasaki, *et al.* 2012. Induction of ZEB proteins by inactivation of RB protein is key determinant of mesenchymal phenotype of breast cancer. *J Biol Chem* 287(11):7896–7906.

Beauchamp, E., G. Bulut, O. Abaan, *et al.* 2009. GLI1 is a direct transcriptional target of EWS-FLI1 oncoprotein. *J Biol Chem* 284(14): 9074–9082.

Beauchamp, E.M., L. Ringer, G. Bulut, *et al.* 2011. Arsenic trioxide inhibits human cancer cell growth and tumor development in mice by blocking Hedgehog/GLI pathway. *J Clin Invest* 121(1):148–160.

Belloni, E., M. Muenke, E. Roessler, *et al.* 1996. Identification of Sonic hedgehog as a candidate gene responsible for holoprosencephaly. *Nat Genet* 14(3):353–356.

Bendell, J.C., L.L. Hart, I. Firdaus, *et al.* 2010. Safety analysis of a randomized phase II trial of hedgehog pathway inhibitor (HPI) GDC-0449 versus placebo with FOLFOX or FOLFIRI and bevacizumab in patients with previously untreated metastatic colorectal cancer (mCRC). *J Clin Oncol* 28(15 suppl):Abstract 3530.

Bol, D., and R. Ebner. 2006. Gene expression profiling in the discovery, optimization and development of novel drugs: one universal screening platform. *Pharmacogenomics* 7(2): 227–235.

Bourhis, E., C. Tam, Y. Franke, *et al.* 2010. Reconstitution of a frizzled8.Wnt3a.LRP6 signaling complex reveals multiple Wnt and Dkk1 binding sites on LRP6. *J Biol Chem* 285(12):9172–9179.

Catenacci, D.V.T., N. Bahary, M.J. Edelman, *et al.* 2012. A phase IB/randomized phase II study of gemcitabine (G) plus placebo (P) or vismodegib (V), a hedgehog (Hh) pathway inhibitor, in patients (pts) with metastatic pancreatic cancer (PC): Interim analysis of a University of Chicago phase II consortium study. *J Clin Oncol* 30(15 suppl):Abstract 4022.

Chang, C.J., J.Y. Yang, W. Xia, *et al.* 2011. EZH2 promotes expansion of breast tumor initiating cells through activation of RAF1-beta-catenin signaling. *Cancer Cell* 19(1):86–100.

Chase, A., and N.C. Cross. 2011. Aberrations of EZH2 in cancer. *Clin Cancer Res* 17(9): 2613–2618.

Chen, B., M.E. Dodge, W. Tang, *et al.* 2009a. Small molecule-mediated disruption of Wnt-dependent signaling in tissue regeneration and cancer. *Nat Chem Biol* 5(2):100–107.

Chen, J.K., J. Taipale, M.K. Cooper, and P.A. Beachy. 2002. Inhibition of Hedgehog signaling by direct binding of cyclopamine to Smoothened. *Genes Dev* 16(21):2743–2748.

Chen, R.S., Y.M. Song, Z.Y. Zhou, *et al.* 2009b. Disruption of xCT inhibits cancer cell metastasis via the caveolin-1/beta-catenin pathway. *Oncogene* 28(4):599–609.

Chen, Z., A.M. Venkatesan, C.M. Dehnhardt, *et al.* 2009c. 2, 4-Diamino-quinazolines as inhibitors of beta-catenin/Tcf-4 pathway: Potential treatment for colorectal cancer. *Bioorg Med Chem Lett* 19(17):4980–4983.

Chiang, C., Y. Litingtung, E. Lee, *et al.* 1996. Cyclopia and defective axial patterning in mice lacking Sonic hedgehog gene function. *Nature* 383(6599):407–413.

Christofk, H.R., M.G. Vander Heiden, M.H. Harris, *et al.* 2008a. The M2 splice isoform of pyruvate kinase is important for cancer metabolism and tumour growth. *Nature* 452(7184): 230–233.

Christofk, H.R., M.G. Vander Heiden, N. Wu, *et al.* 2008b. Pyruvate kinase M2 is a phosphotyrosine-binding protein. *Nature* 452 (7184):181–186.

Cooper, M.K., J.A. Porter, K.E. Young, and P.A. Beachy. 1998. Teratogen-mediated inhibition of target tissue response to Shh signaling. *Science* 280(5369):1603–1607.

Crea, F., E. Paolicchi, V.E. Marquez, and R. Danesi. 2012. Polycomb genes and cancer: Time for clinical application? *Crit Rev Oncol Hematol* 83(2):184–193.

Curtin, J.C., and M.V. Lorenzi. 2010. Drug discovery approaches to target Wnt signaling in cancer stem cells. *Oncotarget* 1(7):563–577.

DeAlmeida, V.I., L. Miao, J.A. Ernst, *et al.* 2007. The soluble wnt receptor Frizzled8CRD-hFc inhibits the growth of teratocarcinomas *in vivo*. *Cancer Res* 67(11):5371–5379.

Deangelo, D.J., R.M. Stone, L.B. Silverman, *et al.* 2006. A phase I clinical trial of the notch inhibitor MK-0752 in patients with T-cell acute lymphoblastic leukemia/lymphoma (T-ALL) and other leukemias. *J Clin Oncol* 24(18 suppl): Abstract 6585.

Dexter, J.S. 1914. The analysis of a case of continuous variation in *Drosophila* by a study of its linkage relations. *Am Nat* 48. 712–758.

Diaz-Padilla, I., B.A. Clarke, H.W. Hirte, *et al.* 2012. A phase II study of RO4929097 (RO), a gamma-secretase inhibitor, in advanced platinum (Pt)-resistant (R) ovarian cancer (OC): A study of the PMH, Chicago, and California phase II consortia. *J Clin Oncol* 30(15 suppl):Abstract 5019.

Diehn, M., R.W. Cho, N.A. Lobo, *et al.* 2009. Association of reactive oxygen species levels and radioresistance in cancer stem cells. *Nature* 458(7239):780–783.

Dierks, C., R. Beigi, G.R. Guo, *et al.* 2008. Expansion of Bcr-Abl-positive leukemic stem cells is dependent on Hedgehog pathway activation. *Cancer Cell* 14(3):238–249.

Donnenberg, V.S., and A.D. Donnenberg. 2005. Multiple drug resistance in cancer revisited: the cancer stem cell hypothesis. *J Clin Pharmacol* 45(8):872–877.

Dontu, G., K.W. Jackson, E. McNicholas, *et al.* 2004. Role of Notch signaling in cell-fate determination of human mammary stem/progenitor cells. *Breast Cancer Res* 6(6):R605–615.

Dufraine, J., Y. Funahashi, and J. Kitajewski. 2008. Notch signaling regulates tumor angiogenesis by diverse mechanisms. *Oncogene* 27(38):5132–5137.

Duncan, J.S., L. Gyenis, J. Lenehan, *et al.* 2008. An unbiased evaluation of CK2 inhibitors by chemoproteomics: characterization of inhibitor effects on CK2 and identification of novel inhibitor targets. *Mol Cell Proteomics* 7(6):1077–1088.

Ellisen, L.W., J. Bird, D.C. West, *et al.* 1991. TAN-1, the human homolog of the Drosophila notch gene, is broken by chromosomal translocations in T lymphoblastic neoplasms. *Cell* 66(4):649–661.

Ernst, T., A.J. Chase, J. Score, *et al.* 2010. Inactivating mutations of the histone methyltransferase gene EZH2 in myeloid disorders. *Nat Genet* 42(8):722–726.

Fitzgerald, K., A. Harrington, and P. Leder. 2000. Ras pathway signals are required for notch-mediated oncogenesis. *Oncogene* 19(37): 4191–4198.

Fortini, M.E. 2009. Notch signaling: the core pathway and its posttranslational regulation. *Dev Cell* 16(5):633–647.

Fouladi, M., C.F. Stewart, J. Olson, *et al.* 2011. Phase I trial of MK-0752 in children with refractory CNS malignancies: a pediatric brain tumor consortium study. *J Clin Oncol* 29(26): 3529–3534.

Fujii, N., L. You, Z. Xu, *et al.* 2007. An antagonist of dishevelled protein-protein interaction suppresses beta-catenin-dependent tumor cell growth. *Cancer Res* 67(2):573–579.

Gandhirajan, R.K., P.A. Staib, K. Minke, *et al.* 2010. Small molecule inhibitors of Wnt/beta-catenin/lef-1 signaling induces apoptosis in chronic lymphocytic leukemia cells *in vitro* and *in vivo*. *Neoplasia* 12(4):326–335.

Garber, K. 2012. Genetic discoveries propagate new epigenetic drugs. *J Natl Cancer Inst* 104(3):174–176.

Geoerger, B., I. Aerts, M. Casanova, *et al.* 2012. A phase I/II study of LDE225, a smoothened (Smo) antagonist, in pediatric patients with recurrent medulloblastoma (MB) or other solid tumors. *J Clin Oncol* 30(15 suppl):Abstract 9519.

Gonsalves, F.C., K. Klein, B.B. Carson, *et al.* 2011. An RNAi-based chemical genetic screen identifies three small-molecule inhibitors of the Wnt/wingless signaling pathway. *Proc Natl Acad Sci USA* 108(15):5954–5963.

Gounder, M.M., M.A. Dickson, N. Wu, *et al.* 2012. A first-in-human, phase Ib combination study to assess safety, pharmacokinetics (PK), and pharmacodynamics (PD) of a hedgehog inhibitor, GDC-0449, with a Notch inhibitor, RO4929097, in patients with advanced sarcoma. *J Clin Oncol* 30(15 suppl):Abstract 10004.

Gout, P.W., A.R. Buckley, C.R. Simms, and N. Bruchovsky. 2001. Sulfasalazine, a potent suppressor of lymphoma growth by inhibition of the x(c)- cystine transporter: a new action for an old drug. *Leukemia* 15(10):1633–1640.

Grandy, D., J. Shan, X. Zhang, *et al.* 2009. Discovery and characterization of a small molecule inhibitor of the PDZ domain of dishevelled. *J Biol Chem* 284(24):16256–16263

Gunthert, U., M. Hofmann, W. Rudy, *et al.* 1991. A new variant of glycoprotein CD44 confers metastatic potential to rat carcinoma cells. *Cell* 65(1):13–24.

Gupta, P.B., C.L. Chaffer, and R.A. Weinberg. 2009. Cancer stem cells: mirage or reality? *Nat Med* 15(9):1010–1012.

Gustafsson, M.V., X. Zheng, T. Pereira, *et al.* 2005. Hypoxia requires notch signaling to maintain the undifferentiated cell state. *Dev Cell* 9(5):617–628.

Hahn, H., C. Wicking, P.G. Zaphiropoulous, *et al.* 1996. Mutations of the human homolog of Drosophila patched in the nevoid basal cell carcinoma syndrome. *Cell* 85(6):841–851.

Hallett, R.M., M.K. Kondratyev, A.O. Giacomelli, *et al.* 2012. Small molecule antagonists of the Wnt/beta-catenin signaling pathway target breast tumor-initiating cells in a Her2/Neu mouse model of breast cancer. *PLoS One* 7(3):e33976.

Harris, P.J., G. Speranza, and C. Dansky Ullmann. 2012. Targeting embryonic signaling pathways in cancer therapy. *Expert Opin Ther Targets* 16(1):131–145.

He, B., L. You, K. Uematsu, *et al.* 2004. A monoclonal antibody against Wnt-1 induces apoptosis in human cancer cells. *Neoplasia* 6(1):7–14.

Hellstrom, M., L.K. Phng, and H. Gerhardt. 2007. VEGF and Notch signaling: the yin and yang of angiogenic sprouting. *Cell Adh Migr* 1(3):133–136.

Huang, S.M., Y.M. Mishina, S. Liu, *et al.* 2009. Tankyrase inhibition stabilizes axin and antagonizes Wnt signalling. *Nature* 461(7264):614–620.

Hyman, J.M., A.J. Firestone, V.M. Heine, *et al.* 2009. Small-molecule inhibitors reveal multiple strategies for Hedgehog pathway blockade. *Proc Natl Acad Sci USA* 106(33):14132–14137.

Ishimoto, T., O. Nagano, T. Yae, *et al.* 2011. CD44 variant regulates redox status in cancer cells by stabilizing the xCT subunit of system xc(−) and thereby promotes tumor growth. *Cancer Cell* 19(3):387–400.

Italiano, A., A.L. Cesne, P.A. Cassier, *et al.* 2012. GDC-0449 in patients with advanced chondrosarcomas: A French Sarcoma Group (FSG)/French and U.S. National Cancer Institutes phase II collaborative study. *J Clin Oncol* 30(15 suppl):Abstract 10005.

Johnson, R.L., A.L. Rothman, J. Xie, *et al.* 1996. Human homolog of patched, a candidate gene for the basal cell nevus syndrome. *Science* 272(5268):1668–1671.

Kakarala, M., D.E. Brenner, H. Korkaya, *et al.* 2010. Targeting breast stem cells with the cancer preventive compounds curcumin and piperine. *Breast Cancer Res Treat* 122(3):777–785.

Kelleher, F.C., D. Fennelly, and M. Rafferty. 2006. Common critical pathways in embryogenesis and cancer. *Acta Oncol* 45(4):375–388.

Kim, D., J. Kim, K. Spaunhurst, *et al.* 2012. An open-label, exploratory phase II study of oral itraconazole for the treatment of basal cell carcinoma. *Cancer Res* 72(8 suppl 1):Abstract LB-223.

Kim, J., J.Y. Tang, R. Gong, *et al.* 2010. Itraconazole, a commonly used antifungal that inhibits Hedgehog pathway activity and cancer growth. *Cancer Cell* 17(4):388–399.

Kondo, Y., L. Shen, A.S. Cheng, *et al.* 2008. Gene silencing in cancer by histone H3 lysine 27 trimethylation independent of promoter DNA methylation. *Nat Genet* 40(6):741–750.

Konishi, J., K.S. Kawaguchi, H. Vo, *et al.* 2007. Gamma-secretase inhibitor prevents Notch3 activation and reduces proliferation in human lung cancers. *Cancer Res* 67(17):8051–8057.

Kopan, R., and M.X. Ilagan. 2009. The canonical Notch signaling pathway: unfolding the activation mechanism. *Cell* 137(2):216–233.

Krop, I., T. Demuth, T. Guthrie, *et al.* 2012. Phase I Pharmacologic and Pharmacodynamic Study of the Gamma Secretase (Notch) Inhibitor MK-0752 in Adult Patients With Advanced Solid Tumors. *J Clin Oncol* 30(19):2307–2313.

Ku, M., R.P. Koche, E. Rheinbay, *et al.* 2008. Genomewide analysis of PRC1 and PRC2 occupancy identifies two classes of bivalent domains. *PLoS Genet* 4(10):e1000242.

Lauth, M., A. Bergstrom, T. Shimokawa, and R. Toftgard. 2007. Inhibition of GLI-mediated transcription and tumor cell growth by small-molecule antagonists. *Proc Natl Acad Sci USA* 104(20):8455–8460.

Leeb, M., D. Pasini, M. Novatchkova, *et al.* 2010. Polycomb complexes act redundantly to repress genomic repeats and genes. *Genes Dev* 24(3):265–276.

Lepourcelet, M., Y.N. Chen, D.S. France, *et al.* 2004. Small-molecule antagonists of the oncogenic Tcf/beta-catenin protein complex. *Cancer Cell* 5(1):91–102.

Lessard, J., A. Schumacher, U. Thorsteinsdottir, *et al.* 1999. Functional antagonism of the Polycomb-Group genes eed and Bmi1 in hemopoietic cell proliferation. *Genes Dev* 13(20): 2691–2703.

Levine, S.S., A. Weiss, H. Erdjument-Bromage, *et al.* 2002. The core of the polycomb repressive complex is compositionally and functionally conserved in flies and humans. *Mol Cell Biol* 22(17):6070–6078.

Li, J.L., and A.L. Harris. 2009. Crosstalk of VEGF and Notch pathways in tumour angiogenesis: therapeutic implications. *Front Biosci* 14(3094–3110.

Li, K., Y. Li, W. Wu, *et al.* 2008. Modulation of Notch signaling by antibodies specific for the extracellular negative regulatory region of NOTCH3. *J Biol Chem* 283(12):8046–8054.

Li, Y., M.S. Wicha, S.J. Schwartz, and D. Sun. 2011. Implications of cancer stem cell theory for cancer chemoprevention by natural dietary compounds. *J Nutr Biochem* 22(9):799–806.

Lim, Y.C., H.J. Kang, Y.S. Kim, and E.C. Choi. 2012. All-trans-retinoic acid inhibits growth of head and neck cancer stem cells by suppression of Wnt/beta-catenin pathway. Eur J Cancer, in press electronic publication.

Liu, F., and M. Korc. 2012. Cdk4/6 Inhibition Induces Epithelial-Mesenchymal Transition and Enhances Invasiveness in Pancreatic Cancer Cells. *Mol Cancer Ther* 11(8):electronic publication.

Liu, S., G. Dontu, I.D. Mantle, *et al.* 2006. Hedgehog signaling and Bmi-1 regulate self-renewal of normal and malignant human mammary stem cells. *Cancer Res* 66(12): 6063–6071.

Lo Conte, N.K. 2012. A phase I study of the gamma-secretase inhibitor RO4929097 and capecitabine in refractory solid tumors. *J Clin Oncol* 30(suppl; abstr 3101):Abst. 3101.

Lo, M., Y.Z. Wang, and P.W. Gout. 2008. The x(c)- cystine/glutamate antiporter: a potential target for therapy of cancer and other diseases. *J Cell Physiol* 215(3):593–602.

Lu, D., M.Y. Choi, J. Yu, *et al.* 2011. Salinomycin inhibits Wnt signaling and selectively induces apoptosis in chronic lymphocytic leukemia cells. *Proc Natl Acad Sci USA* 108(32):13253–13257.

Luistro, L., W. He, M. Smith, *et al.* 2009. Preclinical profile of a potent gamma-secretase inhibitor targeting notch signaling with in vivo efficacy and pharmacodynamic properties. *Cancer Res* 69(19):7672–7680.

Lund, A.H., and M. van Lohuizen. 2004. Polycomb complexes and silencing mechanisms. *Curr Opin Cell Biol* 16(3):239–246.

Majewski, I.J., M.E. Ritchie, B. Phipson, *et al.* 2010. Opposing roles of polycomb repressive complexes in hematopoietic stem and progenitor cells. *Blood* 116(5):731–739.

Makishima, H., A.M. Jankowska, R.V. Tiu, *et al.* 2010. Novel homo- and hemizygous mutations in EZH2 in myeloid malignancies. *Leukemia* 24(10):1799–1804.

Mallen-St Clair, J., R. Soydaner-Azeloglu, K.E. Lee, *et al.* 2012. EZH2 couples pancreatic regeneration to neoplastic progression. *Genes Dev* 26(5):439–444.

Mani, S.A., W. Guo, M.J. Liao, *et al.* 2008. The epithelial-mesenchymal transition generates cells with properties of stem cells. *Cell* 133(4):704–715.

Margolin, K.A., J. Moon, B.G. Redman, *et al.* 2012. Phase II trial of RO4929097 Notch gamma-secretase inhibitor in metastatic melanoma: SWOG S0933. *J Clin Oncol* 30(15 suppl):Abstract 8525.

Messersmith, W.A., P. LoRusso, J.M. Cleary, *et al.* 2011. A phase I dose-escalation study of the novel gamma secretase inhibitor PF-03084014 in patients (pts) with advanced solid tumors. *J Clin Oncol* 29(15 suppl):Abst. 3100.

Moellering, R.E., M. Cornejo, T.N. Davis, *et al.* 2009. Direct inhibition of the NOTCH transcription factor complex. *Nature* 462(7270): 182–188.

Mori, J., F. Takahashi-Yanaga, Y. Miwa, *et al.* 2005. Differentiation-inducing factor-1 induces cyclin D1 degradation through the phosphorylation of Thr286 in squamous cell carcinoma. *Exp Cell Res* 310(2):426–433.

Morin, R.D., N.A. Johnson, T.M. Severson, *et al.* 2010. Somatic mutations altering EZH2 (Tyr641) in follicular and diffuse large B-cell lymphomas of germinal-center origin. *Nat Genet* 42(2):181–185.

Nagano, O., and H. Saya. 2004. Mechanism and biological significance of CD44 cleavage. *Cancer Sci* 95(12):930–935.

Neff, T., A.U. Sinha, M.J. Kluk, *et al.* 2012. Polycomb repressive complex 2 is required for MLL-AF9 leukemia. *Proc Natl Acad Sci USA* 109(13):5028–5033.

Nicolas, M., A. Wolfer, K. Raj, *et al.* 2003. Notch1 functions as a tumor suppressor in mouse skin. *Nat Genet* 33(3):416–421.

Niehrs, C., and J. Shen. 2010. Regulation of Lrp6 phosphorylation. *Cell Mol Life Sci* 67(15):2551–2562.

Nikoloski, G., S.M. Langemeijer, R.P. Kuiper, *et al.* 2010. Somatic mutations of the histone methyltransferase gene EZH2 in myelodysplastic syndromes. *Nat Genet* 42(8):665–667.

Ntziachristos, P., A. Tsirigos, P. Van Vlierberghe, *et al.* 2012. Genetic inactivation of the polycomb repressive complex 2 in T cell acute lymphoblastic leukemia. *Nat Med* 18(2):298–301.

Nusslein-Volhard, C., and E. Wieschaus. 1980. Mutations affecting segment number and polarity in Drosophila. *Nature* 287(5785): 795–801.

O'Brien, C.A., A. Kreso, and C.H. Jamieson. 2010. Cancer stem cells and self-renewal. *Clin Cancer Res* 16(12):3113–3120.

Oswald, F., S. Liptay, G. Adler, and R.M. Schmid. 1998. NF-kappaB2 is a putative target gene of activated Notch-1 via RBP-Jkappa. *Mol Cell Biol* 18(4):2077–2088.

Pachter, J., J. Ring, A. Neill, *et al.* 2012. The Wnt inhibitor VS-507 reduces cancer stem cell (CSC) function *in vitro* and tumorigenicity in mice. *Cancer Res* 78(8 suppl 1):Abstract LB-194.

Palacios, D., C. Mozzetta, S. Consalvi, *et al.* 2010. TNF/p38alpha/polycomb signaling to Pax7 locus in satellite cells links inflammation to the epigenetic control of muscle regeneration. *Cell Stem Cell* 7(4):455–469.

Palomero, T., M.L. Sulis, M. Cortina, *et al.* 2007. Mutational loss of PTEN induces resistance to NOTCH1 inhibition in T-cell leukemia. *Nat Med* 13(10):1203–1210.

Pannuti, A., K. Foreman, P. Rizzo, *et al.* 2010. Targeting Notch to target cancer stem cells. *Clin Cancer Res* 16(12):3141–3152.

Pant, S., S.F. Jones, C. Kurkjian, *et al.* 2012. A first-in-human phase I study of the oral Notch inhibitor LY900009 in patients with advanced cancer. *J Clin Oncol* 30(15 suppl):Abstract 3008.

Paris, D., A. Quadros, N. Patel, *et al.* 2005. Inhibition of angiogenesis and tumor growth by beta and gamma-secretase inhibitors. *Eur J Pharmacol* 514(1):1–15.

Park, S., J. Gwak, M. Cho, *et al.* 2006. Hexachlorophene inhibits Wnt/beta-catenin pathway by promoting Siah-mediated beta-catenin degradation. *Mol Pharmacol* 70(3): 960–966.

Peacock, C.D., Q. Wang, G.S. Gesell, *et al.* 2007. Hedgehog signaling maintains a tumor stem cell compartment in multiple myeloma. *Proc Natl Acad Sci USA* 104(10):4048–4053.

Phillips, T.M., W.H. McBride, and F. Pajonk. 2006. The response of CD24(−/low)/CD44+ breast cancer-initiating cells to radiation. *J Natl Cancer Inst* 98(24):1777–1785.

Polyak, K., and R.A. Weinberg. 2009. Transitions between epithelial and mesenchymal states: acquisition of malignant and stem cell traits. *Nat Rev Cancer* 9(4):265–273.

Ponta, H., L. Sherman, and P.A. Herrlich. 2003. CD44: from adhesion molecules to signalling regulators. *Nat Rev Mol Cell Biol* 4(1):33 45.

Real, P.J., V. Tosello, T. Palomero, *et al.* 2009. Gamma-secretase inhibitors reverse glucocorticoid resistance in T cell acute lymphoblastic leukemia. *Nat Med* 15(1):50–58.

Richly, H., L. Aloia, and L. Di Croce. 2011. Roles of the Polycomb group proteins in stem cells and cancer. *Cell Death Dis* 2(9):e204.

Richter, S., E. McWhirter, E.X. Chen, *et al.* 2012. A phase I study of R04929097, an oral gamma secretase inhibitor, in combination with gemcitabine in patients with advanced solid tumors (PHL-078/CTEP 8575). *J Clin Oncol* 30 (15 suppl):Abstract 3082.

Roessler, E., E. Belloni, K. Gaudenz, *et al.* 1996. Mutations in the human Sonic Hedgehog gene cause holoprosencephaly. *Nat Genet* 14(3): 357–360.

Rudin, C.M. 2012. Vismodegib. *Clin Cancer Res* 18(12):3218–3222.

Sakane, H., H. Yamamoto, and A. Kikuchi. 2010. LRP6 is internalized by Dkk1 to suppress its phosphorylation in the lipid raft and is recycled for reuse. *J Cell Sci* 123(Pt 3):360–368.

Sauvageau, M., and G. Sauvageau. 2010. Polycomb group proteins: multi-faceted regulators of somatic stem cells and cancer. *Cell Stem Cell* 7(3):299–313.

Semenov, M.V., X. Zhang, and X. He. 2008. DKK1 antagonizes Wnt signaling without promotion of LRP6 internalization and degradation. *J Biol Chem* 283(31):21427–21432.

Shan, J., D.L. Shi, J. Wang, and J. Zheng. 2005. Identification of a specific inhibitor of the dishevelled PDZ domain. *Biochemistry* 44(47): 15495–15503.

Simon, C., J. Chagraoui, J. Krosl, *et al.* 2012. A key role for EZH2 and associated genes in mouse and human adult T-cell acute leukemia. *Genes Dev* 26(7):651–656.

Smith, D., S. DC, E. P, *et al.* 2010. A first in human phase 1 trial of the DLL4 antobody (OMP-21M18) targeting cancer stem cells in patients with advance solid tumors. *22nd EORTC-NCI-AACR*

Symposium on Molecular Targets and Cancer Therapeutics.
Sneeringer, C.J., M.P. Scott, K.W. Kuntz, *et al.* 2010. Coordinated activities of wild-type plus mutant EZH2 drive tumor-associated hypertrimethylation of lysine 27 on histone H3 (H3K27) in human B-cell lymphomas. *Proc Natl Acad Sci USA* 107(49):20980–20985.
Song, S., T. Christova, S. Perusini, *et al.* 2011. Wnt inhibitor screen reveals iron dependence of beta-catenin signaling in cancers. *Cancer Res* 71(24):7628–7639.
Sparmann, A., and M. van Lohuizen. 2006. Polycomb silencers control cell fate, development and cancer. *Nat Rev Cancer* 6(11):846–856.
Stanton, B.Z., L.F. Peng, N. Maloof, *et al.* 2009. A small molecule that binds Hedgehog and blocks its signaling in human cells. *Nat Chem Biol* 5(3):154–156.
Strosberg, J.R., T. Yeatman, J. Weber, *et al.* 2012. A phase II study of RO4929097 in metastatic colorectal cancer. *Eur J Cancer* 48(7):997–1003.
Sukhdeo, K., M. Mani, Y. Zhang, *et al.* 2007. Targeting the beta-catenin/TCF transcriptional complex in the treatment of multiple myeloma. *Proc Natl Acad Sci USA* 104(18):7516–7521.
Taipale, J., J.K. Chen, M.K. Cooper, *et al.* 2000. Effects of oncogenic mutations in Smoothened and Patched can be reversed by cyclopamine. *Nature* 406(6799):1005–1009.
Takahashi-Yanaga, F., Y. Taba, Y. Miwa, *et al.* 2003. Dictyostelium differentiation-inducing factor-3 activates glycogen synthase kinase-3beta and degrades cyclin D1 in mammalian cells. *J Biol Chem* 278(11):9663–9670.
Takahashi-Yanaga, F., J. Mori, E. Matsuzaki, *et al.* 2006. Involvement of GSK-3beta and DYRK1B in differentiation-inducing factor-3-induced phosphorylation of cyclin D1 in HeLa cells. *J Biol Chem* 281(50):38489–38497.
Takahashi-Yanaga, F., and M. Kahn. 2010. Targeting Wnt signaling: can we safely eradicate cancer stem cells? *Clin Cancer Res* 16(12): 3153–3162.
Takebe, N., and S.P. Ivy. 2010. Controversies in cancer stem cells: targeting embryonic signaling pathways. *Clin Cancer Res* 16(12):3106–3112.
Takebe, N., R.Q. Warren, and S.P. Ivy. 2011b. Breast cancer growth and metastasis: interplay between cancer stem cells, embryonic signaling pathways and epithelial-to-mesenchymal transition. *Breast Cancer Res* 13(3):211–222.
Tamada, M., O. Nagano, S. Tateyama, *et al.* 2012. Modulation of glucose metabolism by CD44 contributes to antioxidant status and drug resistance in cancer cells. *Cancer Res* 72(6):1438–1448.
Tan, J., X. Yang, L. Zhuang, *et al.* 2007. Pharmacologic disruption of Polycomb-repressive complex 2-mediated gene repression selectively induces apoptosis in cancer cells. *Genes Dev* 21(9):1050–1063.
Tanaka, S., S. Miyagi, G. Sashida, *et al.* 2012. Ezh2 augments leukemogenicity by reinforcing differentiation blockage in acute myeloid leukemia. *Blood* 120(5):1107–1117.
Thiery, J.P. 2002. Epithelial-mesenchymal transitions in tumour progression. *Nat Rev Cancer* 2(6):442–454.
Thorne, C.A., A.J. Hanson, J. Schneider, *et al.* 2010. Small-molecule inhibition of Wnt signaling through activation of casein kinase 1alpha. *Nat Chem Biol* 6(11):829–836.
Tolcher, A.W., R. Chugh, G. Chambers, *et al.* 2012. A first-in-human phase I study to evaluate the fully human monoclonal antibody OMP-59R5 (anti-Notch2/3) administered intravenously to patients with advanced solid tumors. *J Clin Oncol* 30(15 suppl):Abstract 3025.
Tremblay, M.R., M. Nevalainen, S.J. Nair, *et al.* 2008. Semisynthetic cyclopamine analogues as potent and orally bioavailable hedgehog pathway antagonists. *J Med Chem* 51(21):6646–6649.
Trosset, J.Y., C. Dalvit, S. Knapp, *et al.* 2006. Inhibition of protein-protein interactions: the discovery of druglike beta-catenin inhibitors by combining virtual and biophysical screening. *Proteins* 64(1):60–67.
Varambally, S., S.M. Dhanasekaran, M. Zhou, *et al.* 2002. The polycomb group protein EZH2 is involved in progression of prostate cancer. *Nature* 419(6907):624–629.
Von Hoff, D.D., P.M. LoRusso, C.M. Rudin, *et al.* 2009. Inhibition of the hedgehog pathway in advanced basal-cell carcinoma. *N Engl J Med* 361(12):1164–1172.
Wang, M., L. Wu, L. Wang, and X. Xin. 2010. Down-regulation of Notch1 by gamma-secretase inhibition contributes to cell growth inhibition and apoptosis in ovarian cancer cells A2780. *Biochem Biophys Res Commun* 393(1):144–149.
Watters, J.W., C. Cheng, P.K. Majumder, *et al.* 2009. De novo discovery of a gamma-secretase inhibitor response signature using a novel in vivo breast tumor model. *Cancer Res* 69(23):8949–8957.
Wei, P., M. Walls, M. Qiu, *et al.* 2010a. Evaluation of selective gamma-secretase inhibitor PF-03084014 for its antitumor efficacy and

gastrointestinal safety to guide optimal clinical trial design. *Mol Cancer Ther* 9(6):1618–1628.
Wei, W., M.S. Chua, S. Grepper, and S. So. 2010b. Small molecule antagonists of Tcf4/beta-catenin complex inhibit the growth of HCC cells in vitro and in vivo. *Int J Cancer* 126(10):2426–2436.
Yae, T., K. Tsuchihashi, T. Ishimoto, *et al.* 2012. Alternative splicing of CD44 mRNA by ESRP1 enhances lung colonization of metastatic cancer cell. *Nat Commun* 3(883).
Yan, M. 2011. Therapeutic promise and challenges of targeting DLL4/NOTCH1. *Vasc Cell* 3(1):17–21.
Yao, H., E. Ashihara, and T. Maekawa. 2011. Targeting the Wnt/beta-catenin signaling pathway in human cancers. *Expert Opin Ther Targets* 15(7):873–887.
Yap, D.B., J. Chu, T. Berg, *et al.* 2010. Somatic mutations at EZH2 Y641 act dominantly through a mechanism of selectively altered PRC2 catalytic activity, to increase H3K27 trimethylation. *Blood* 117(8):2451–2459.
Yde, C.W., T. Frogne, A.E. Lykkesfeldt, *et al.* 2007. Induction of cell death in antiestrogen resistant human breast cancer cells by the protein kinase CK2 inhibitor DMAT. *Cancer Lett* 256(2):229–237.
You, L., B. He, Z. Xu, *et al.* 2004. An anti-Wnt-2 monoclonal antibody induces apoptosis in malignant melanoma cells and inhibits tumor growth. *Cancer Res* 64(15):5385–5389.
Yu, Q., B.P. Toole, and I. Stamenkovic. 1997. Induction of apoptosis of metastatic mammary carcinoma cells in vivo by disruption of tumor cell surface CD44 function. *J Exp Med* 186 (12):1985–1996.
Zhao, C., A. Chen, C.H. Jamieson, *et al.* 2009. Hedgehog signalling is essential for maintenance of cancer stem cells in myeloid leukaemia. *Nature* 458(7239):776–779.
Zhu, C., K.W. Cheng, N. Ouyang, *et al.* 2012. Phosphosulindac (OXT-328) Selectively Targets Breast Cancer Stem Cells In vitro and in Human Breast Cancer Xenografts. Stem Cells, in press.

Chapter 24
Oct4, Oct1, and Cancer Stem Cells

Jessica Maddox and Dean Tantin
Department of Pathology, University of Utah School of Medicine, Salt Lake City, UT, USA

Abbreviations

cAMP	cyclic AMP
CSC	cancer stem cell
ESC	embryonic stem cell
iPSC	induced pluripotent stem cell
SSC	somatic stem cell

Malignant Stem Cell Self-Renewal and Gene Expression

The history and complexities of the cancer stem cell (CSC) field are covered in more detail elsewhere in this work. In brief, CSCs represent a stem-like population of cells present in many developmentally heterogeneous neoplasms. They are characterized by two essential properties: the ability to self-renew and the ability to produce more differentiated progeny. These properties are shared with two other cell types: pluripotent embryonic stem cells (ESC) and somatic stem cells (SSC). CSCs have undergone intense scrutiny, providing insights into pathways governing CSC function. For example, there is evidence for activation of pluripotency gene expression programs associated with CSCs (Ben-Porath et al., 2008; Somervaille et al., 2009; Kim et al., 2010). Other properties of CSCs include relative quiescence, chemo- and radioresistance, and a glycolytic metabolic profile with simultaneous reduced reactive oxygen species (ROS) levels.

Multiple questions must be addressed in order to understand and target critical pathways underlying the CSC phenotype. For example, which master regulatory factors control gene expression to promote CSC maintenance and self-renewal? How are these factors proximally and ultimately controlled by cancer-causing mutations and activated oncogenic pathways? To what extent do the gene expression programs

Cancer Stem Cells, First Edition. Edited by Vinagolu K. Rajasekhar.

activated by these factors overlap with programs that maintain embryonic pluripotency? The goal of this chapter is to provide an introduction to the transcription factors Oct1 and Oct4, and to provide a snapshot of some current ideas regarding the roles of these factors in the maintenance and self-renewal of CSCs.

Oct4 in Pluripotency, Malignancy, and CSCs

Oct Proteins

Oct proteins comprise a large part of the POU (Pit-1, Oct1/2, UNC-86) domain transcription factor family (Kang et al., 2009b). This metazoan-specific transcription factor family is defined by the presence of a POU DNA binding domain: a bipartite structure divided into two subdomains, known as the POU-specific domain (POU_S) and POU-homeodomain (POU_H). A flexible linker connects the two subdomains, enabling Oct proteins to bind to DNA with great conformational variability. Oct proteins recognize the classical octamer DNA sequence (5′ATGCAAAT3′), but can also recognize alternative (non-canonical) DNA sequences (Remenyi et al., 2001). The ability of the DNA binding subdomains to spatially reorganize themselves relative to each other can be regulated by upstream signals (Kang et al., 2009a). Oct1 is a prototypic member of this family and is expressed in most, if not all, developing and adult mammalian tissues. Oct4 expression is much more restricted, traditionally being limited to the germline, and the blastocyst inner cell mass and epiblast. Oct4 is highly expressed in ESCs, which are derived from the inner cell mass (Nichols et al., 1998).

Oct4, Pluripotency, and Stem Cells

The most notable function of Oct transcription factors is the powerful role of Oct4 in the maintenance of ESC identity (Nichols et al., 1998). Oct4 is also a nearly ubiquitous component of transcription factor cocktails now routinely used to generate induced pluripotent stem cells (iPSC), which possess morphologic, growth, and differentiation properties of ESCs (Takahashi and Yamanaka, 2006). Hundreds of Oct4 target genes have been identified in ESCs (Chen et al., 2008). Many of the gene targets fall into just a few functional categories, three of which are itemized below.

First, in combination with other critical regulators of ESC identity, Oct4 maintains expression of the "core" pluripotency factors (including Oct4 itself, Nanog and Klf4) in a self-sustaining network (Boyer et al., 2005). The extent to which SSCs and CSCs also require the expression of core pluripotency circuitry components is not clear.

Second, Oct4 associates with an array of silent but developmentally inducible targets. Examples include *Gata2*, *Pax6*, *Hoxa5,* and *Hoxc6*. These targets are frequently characterized by a "bivalent" chromatin structure simultaneously marked by repressive and activating nucleosome modifications and a concurrent lack of DNA methylation at CpG dinucleotides (Bernstein et al., 2006; Lister et al., 2009). This chromatin state helps maintain genes in poised configurations so that in response to the correct developmental cues, they can be rapidly activated. Activation of the gene occurs simultaneously with loss of the repressive chromatin marks. In contrast, repressive marks in genes specific for other developmental trajectories are reinforced by de novo DNA methylation. Poised gene expression states are also a characteristic of stem cells, which must maintain key genes in silent but poised configurations (Figure 24.1).

Third, Oct4 regulates genes controlling metabolism (e.g., *Pcx*, *Hk1*, *Ndufa3,* and *Atp5d*). It is now appreciated that ESCs harbor a glycolytic metabolic profile (Kondoh et al., 2007; Folmes et al., 2011). As with poised gene expression states, this feature also appears to be present in SSCs and CSCs (Diehn et al., 2009). However, not all ESC properties are readily transferable to CSCs. ESCs, for example, are highly proliferative, whereas CSCs are thought to be relatively quiescent.

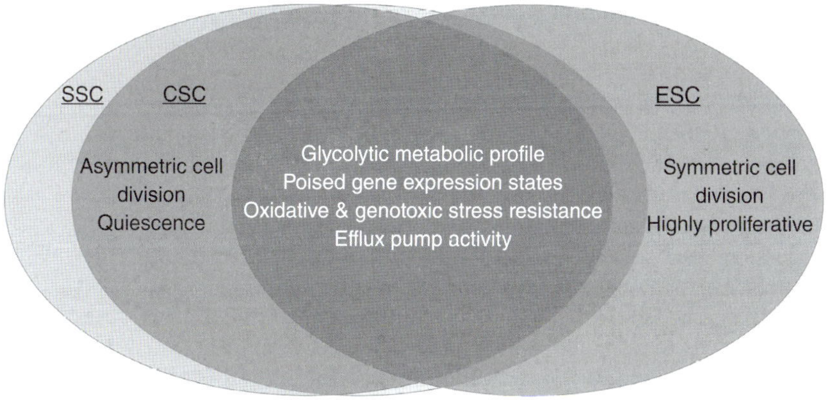

Figure 24.1. Common and distinct characteristics of SSCs, CSCs, and ESCs. For color detail, please see color plate section.

ESCs are pluripotent, and in the absence of differentiation signals, they undergo symmetric cell divisions. SSCs and CSCs are multi- or uni-potent and may undergo asymmetric cell divisions in a manner regulated by signals from the tissue-specific microenvironment (Morrison and Kimble, 2006). Therefore, only a subset of embryonic pluripotency characteristics can be consistent with "pluripotency" programs associated with CSCs (Figure 24.1).

Oct4 Expression in SSCs

As stated above, Oct4 is strongly expressed in germline and pluripotent tissues. Numerous studies have also detected Oct4 in SSCs. For example, Oct4 has been detected in mesenchymal stem cells where it works in concert with Nanog to upregulate Dnmt1, a DNA methyltransferase. Upregulated Dnmt1 represses expression of p16, p21 and genes associated with development and differentiation (Tsai et al., 2012). These data support a functional role for Oct4 in one type of SSC. On balance however, current evidence supporting a role for Oct4 in SSCs is weak. Transcriptional analysis of ESCs and neural and hematopoietic SSCs demonstrates that the gene expression profiles of these three stem cell types are considerably divergent. Also Oct4 and other transcription components of the core ESC transcription factor network are not detected in SSCs (Ivanova et al., 2002; Ramalho-Santos et al., 2002). Based on these findings and evidence from SSC regulatory networks (Miranda-Saavedra and Gottgens, 2008), it is likely that transcription regulation of SSC multipotency is largely tissue-specific. In cases where a functional role for Oct4 in SSC maintenance has been rigorously tested, the results have been negative (Lengner et al., 2007).

Oct4 Expression in Malignant Cells

A cursory literature search of Oct4 expression and cancer will return hundreds of articles supporting or refuting the claim of Oct4 expression in almost all types of cancer, including lung, bladder, and colon (Atlasi et al., 2007; Yasuda et al., 2011; Li et al., 2012). The inconsistencies in the literature are due in part to the nature of the Oct4 detection methods. Analysis of Oct4 expression in cancerous lesions typically applies immunohistochemistry, Western blotting, or qRT-PCR to formalin-fixed, paraffin-embedded tissue. Few studies relate their findings to Oct4 expression in ESCs, where it is endogenously found at meaningfully high levels (Mathieu et al., 2011). An obvious exception is germ cell tumors, which are derived from tissues that naturally express Oct4 at high levels. Oct4 functional variants and pseudogenes may also contribute to false-positives in studies that lack appropriate controls (Liedtke et al., 2008). Also, because of the similarities between Oct transcription factors, there is a possibility of antibody cross-reactivity with Oct1, Oct2, Oct6, or Oct11. Beyond these technical caveats, the functional relevance of Oct4 expression in a given malignant tissue must also be addressed. Various studies have suggested a functional role for Oct4 in cancer based on ectopic expression (Hochedlinger et al., 2005; Chang et al., 2011). Proof of a requirement for Oct4 requires loss-of-function studies. Interestingly, some poorly differentiated neoplasms demonstrate expression of downstream stem cell-related Oct4 target genes, but not Oct4 itself (Ben-Porath et al., 2008). Based on this line of thought, an attractive possibility is that limited Oct4 expression may be a *consequence* rather than a *cause* of an activated pluripotency program (e.g., a component rather than controller of a malignant "pluripotency" profile).

Oct4 Expression in CSCs

Ascertaining the role of Oct4 in CSCs confronts the same hurdles as in cancer itself. Again, many reports use ectopic Oct4 expression to support its functional role (Chiou et al., 2010; Linn et al., 2010; Beltran et al., 2011; Kumar et al., 2012). Additionally, indirect support for a role for Oct4 in CSCs comes from studies suggesting that the pathways governing the reprogramming of somatic cells to iPSCs overlap with those governing the transformation of somatic cells to CSCs (Utikal et al., 2009; Mosca et al., 2010). Recently, Bourguignon and colleagues (2012) published evidence supporting a functional role for Oct4 and other pluripotency factors (Nanog and Sox2) in CSCs of head and neck squamous cell carcinoma. These factors form a complex in the cytosol and translocate to the nucleus upon stimulation by Hyaluronan via CD44v3, a Hyaluronan receptor and known CSC marker (Bourguignon et al., 2012). Cumulatively, the evidence in support of a functional Oct4 role in CSCs is mixed.

Oct1 as an OCT4 Surrogate

Oct1: A Natural Candidate

Oct1 and Oct4 have similar DNA binding specificity and share a large number of gene targets (Kang et al., 2009b). Unlike Oct4, Oct1 is widely and abundantly expressed in adult tissues. For many years, Oct1 activity was thought to be static, consistent with its role in the regulation of housekeeping genes such as TBP-associated factor 12 (*Taf12*) and RNA polymerase II polypeptide A (*Polr2a*). Surprisingly, loss of Oct1 does not result in early embryonic lethality. Instead Oct1 deficient embryos undergo implantation and gastrulation, with the embryos failing later during development (Wang et al., 2004; Sebastiano et al., 2010). Further challenging the static activity paradigm, studies have shown that Oct1 responds to different stimuli. For example, Oct1 promotes resistance to genotoxic and oxidative stress (Tantin et al., 2005), and regulates potent stress response effectors and cell-cycle regulators such as *Gadd45a* and *Cdkn1b* (Hirose et al., 2003; Dalvai et al., 2010). Oct1 responds to genotoxic and oxidative stress by associating with alternative regulatory targets (Kang et al., 2009a), and also responds to cyclic AMP (cAMP) levels by translocating between the cytoplasm and nucleus (Wang and Jin, 2010). The last few years have witnessed the identification of additional Oct1 functions (Figure 24.2) consistent with a model in which Oct1 adopts a subset of Oct4 functions in CSCs. For example, Oct1 enforces poised transcriptional states (Shakya et al., 2011) and promotes a glycolytic metabolic profile associated with dampened mitochondrial function and ROS levels (Shakya et al., 2009). Loss of Oct1 has little impact on cell growth and viability in culture, or on immortalization by serial passage, but antagonizes oncogenic transformation in vitro and tumorigenicity in vivo (Shakya et al., 2009). These studies contribute to the mounting evidence supporting a role for Oct1 in controlling stem cell phenotypes.

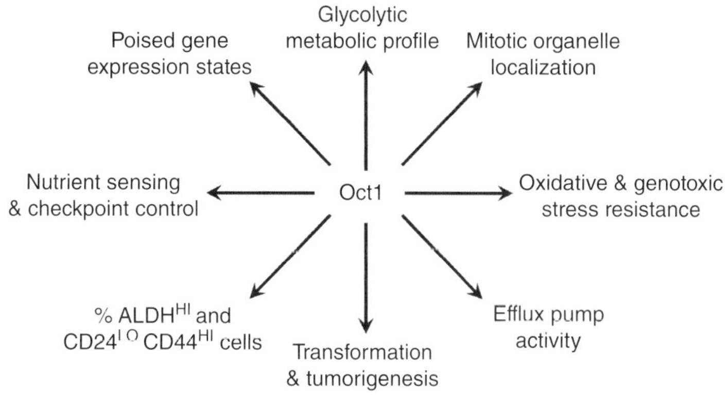

Figure 24.2. Emerging roles for Oct1.

Based on this model, we tested whether Oct1 is an upstream mediator of stem cell functions, including critical parameters of CSCs. We also tested whether Oct1 may further serve as an authentic CSC maker.

Oct1 Is a Somatic and CSC Determinant

We recently found that Oct1 controls multiple stem cell phenotypes in normal and tumor cells (Maddox et al., 2012). Oct1 deficient fetal liver hematopoietic progenitors manifest engraftment defects in competitive and serial transplantation experiments. Additionally, strong Oct1 protein expression spatially correlates with normal stem cell niches in multiple epithelial tissue types, and correlates with high levels of the known stem cell markers Lgr5, Lrig1, and Aldh1. Elevated Oct1 protein expression also correlates with elevated $ALDH^{HI}$ and $CD24^{LO}CD44^{HI}$ stem cell-like populations in tumor cell lines and primary breast cancer samples. Interestingly, the correlation with Oct1 mRNA expression is poor, suggesting that these differences reflect regulation of Oct1 at the protein level. Loss of Oct1 selectively depletes stem-like populations in multiple human tumor cell lines based on ALDH and dye efflux activity as readouts. Overexpression of Oct1 promotes the stem-like populations. These different assays measure markers associated with CSCs, but do not provide a readout of CSC function. One commonly used functional measure of CSC activity is the ability to initiate tumors in immunodeficient mice. Stable Oct1 knockdown in two different tumor cell lines reduces tumor-initiating cell frequency using limiting dilution assays in immunodeficient mice, while Oct1 ectopic expression promotes tumor initiation. Cumulatively these results indicate that Oct1 protein levels can be used as a stem cell marker and that Oct1 is an upstream SSC and CSC determinant.

Aldh1a1, *Abcg2*, *Hmgb3*, *Abcb1*, *Abcb4*, and *Foxo4* are direct Oct1 gene targets (Figure 24.3). These and other targets have been associated with stem cell identity and self-renewal. *Abcg2*, *Abcb1*, and *Abcb4* encode drug efflux pumps, which mediate an important facet of CSC chemoresistance. Another target is *Aldh1a1*, which encodes an enzyme that oxidizes toxic aldehydes and synthesizes a key intermediate in retinoic acid biosynthesis. *Hmgb3* and *Foxo4* are implicated in hematopoietic and germline stem cell function, respectively. Additional target genes remain to be identified. Delineating such targets will aid in deciphering CSC function and may provide additional opportunities for therapeutic intervention. In addition to the target genes, the underlying mechanisms that drive divergent Oct1 function in stem cell compartments are not understood. For example, Oct1 protein but not mRNA levels are

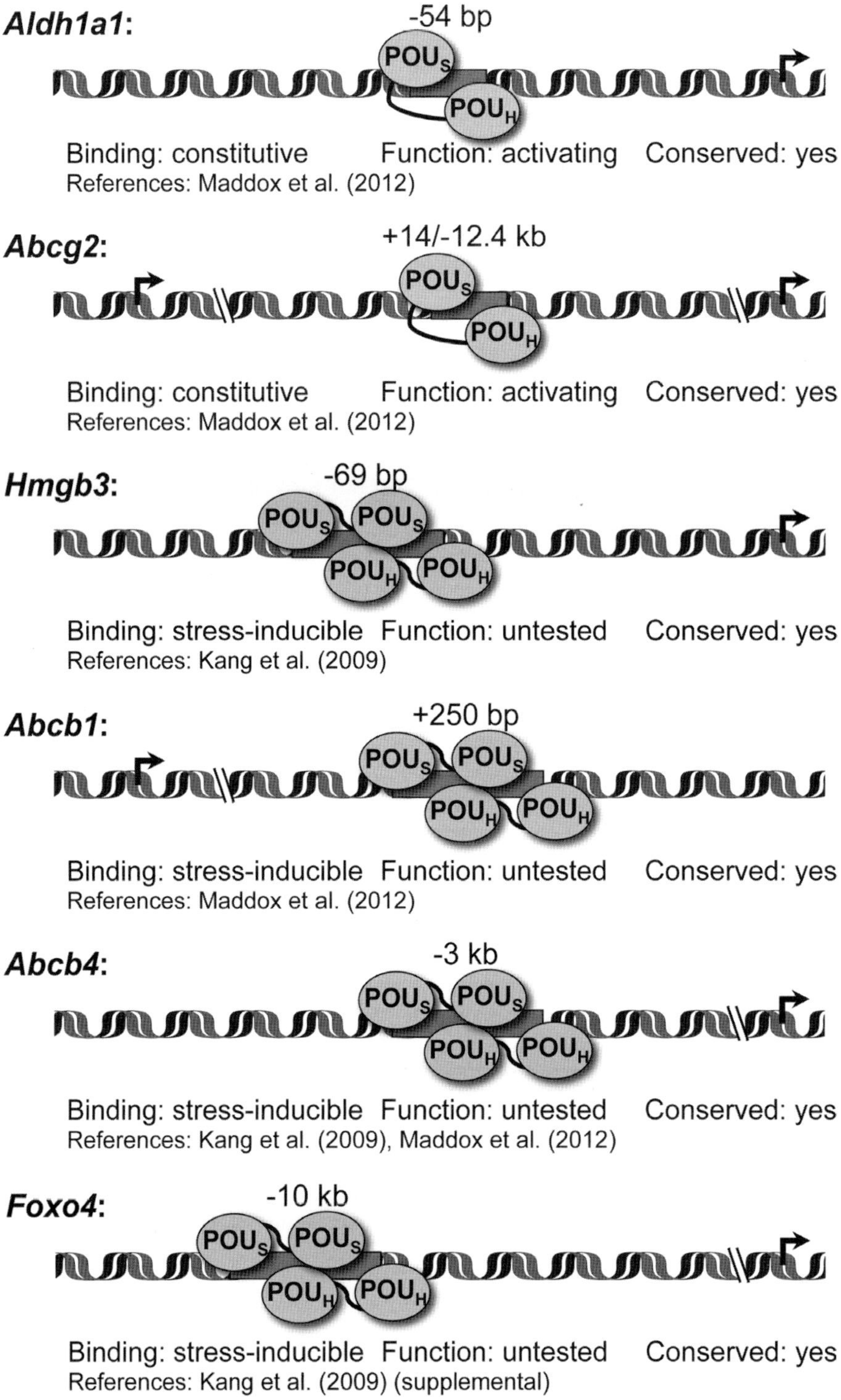

Figure 24.3. Defined Oct1 target genes with relevance for CSCs. Transcription start sites are shown with black arrows. Position of the Oct1 binding site relative to the transcription start site is shown above.

elevated in stem cell compartments. The mechanism(s) underlying this observation are unknown.

Post-Translational Modifications of Oct1 in CSCs

Oct1 protein levels and post-translational modification states in stem cell compartments and malignancy have not been studied, but they are the logical next target of investigation. Precedent exists for post-translational regulation of proteins important for self-renewal, differentiation, and proliferation in CSCs. Beta-catenin, a member of the Wnt signaling pathway, becomes stabilized and translocates to the nucleus to induce expression of genes associated with proliferation and differentiation in somatic epithelial stem cells, and is dysregulated in CSCs (Takahashi-Yanaga and Kahn, 2010). As noted above, CSC-like populations display increased Oct1 protein, but not mRNA transcript levels. Oct1 is known to be ubiquitinated and phosphorylated (Kang et al., 2009a; Kang et al., 2011). The presence of ubiquitination sites suggests that regulation of protein stability underlies the increased steady-state levels in CSCs. Potential phosphorylation sites within Oct1 that may be differentially modified in CSCs include canonical targets for extracellular signal-regulated kinase (ERK) and protein kinase A (PKA). In primary T cells, ERK is required for Oct1 coactivator switching to regulate gene expression and the establishment of poised transcriptional states (Shakya et al., 2011). In endocrine cells, PKA phosphorylates Oct1 in response to cAMP signaling (Wang et al., 2009).

Oct1 Mechanisms of Action in CSCs

Oct1 appears to act in multiple ways to regulate specific gene targets in CSCs. In many cases, the mode of regulation differs from canonical transcription factor functions (Figure 24.4). For example, a number of highly expressed ("housekeeping") Oct1 targets are unexpectedly unaffected by Oct1 loss-of-function. These targets are inducibly bound by Oct1 following exposure to genotoxic and oxidative stresses, where the function of Oct1 appears to be to insulate the gene from inappropriate repression. In the absence of Oct1, these targets become inappropriately repressed. The failure to maintain adequate expression in the absence of Oct1 may contribute to the increased stress sensitivity of Oct1 deficient cells

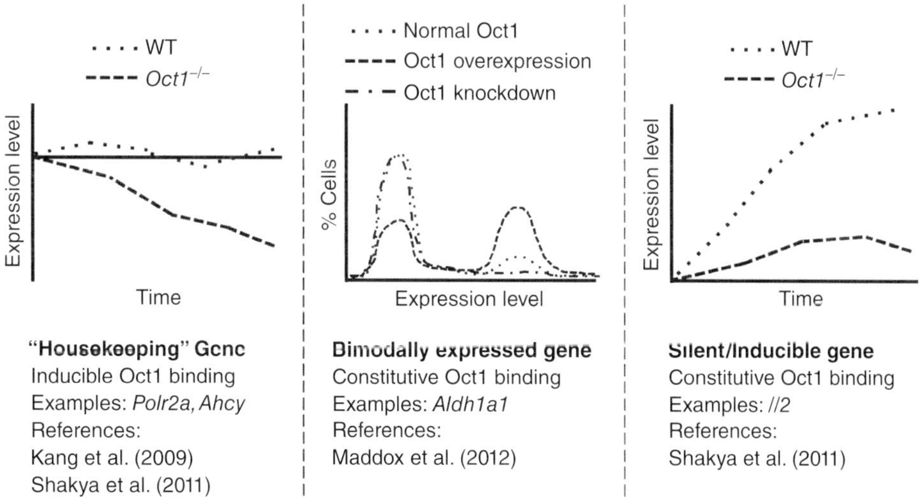

Figure 24.4. Oct1 modes of transcriptional regulation. See text for details.

(Tantin et al., 2005). Other active targets such as *Aldh1a* display an interesting property: Oct1 ablation or overexpression does not shift the degree of expression, but rather shifts the proportion of cells within a population that harbor highly elevated activity. This "digital" rather than "analog" mode of gene regulation also suggests that Oct1 need not behave as a canonical transcription factor. Finally, in the absence of Oct1 some transcriptionally silent targets remain silent but manifest poor induction following exposure to the normal induction cues. These findings indicate that Oct1 can help establish poised gene expression states. All three modes of regulation can be explained, in part, by what is known about the mechanisms that Oct1 uses to regulate gene expression. Oct1 can recruit a cofactor activity known as Jmjd1a (KDM3A), a histone lysine demethylase capable of removing an inhibitory nucleosome modification (dimethylated K9 of histone H3). This finding indicates that Oct1 need not behave as a traditional activator or repressor, but can also function as an "anti-repressor," preventing the accumulation of negative epigenetic marks and stable gene repression (Shakya et al., 2011). Oct1 could be utilizing these mechanisms to regulate CSC identity.

Possible Noncanonical Functions of Oct1

In addition to the mechanisms postulated above, Oct1 may use additional mechanisms to control CSC activity. For example, during M-phase, Oct1 is phosphorylated at Ser335 in the DNA binding domain by the kinase Nek6 (Kang et al., 2011). Phosphorylation excludes Oct1 from mitotic chromatin. Instead, Oct1^{pS335} concentrates at mitotic spindle pole bodies, kinetochores, and the midbody (Kang et al., 2011). Both Oct1 ablation and overexpression can result in mildly abnormal mitoses. Of interest, asymmetric cell division is a hallmark of some SSC and CSC populations. The asymmetric segregation of cell-fate determinants to generate mother and daughter cells relies on the correct orientation of the mitotic spindle. Although there is no direct evidence as yet, it is possible that Oct1 affects asymmetric cell division through regulation of the cell division apparatus.

Oct1 has also been associated with the nuclear envelope, particularly in close association with Lamin B1. Nuclear lamins are potent organizers of gene expression, and it is thought that Lamin B1 sequesters Oct1 and regulates target genes in response to oxidative stress (Malhas et al., 2009). Intriguingly, some nuclear lamins are associated with genetic syndromes known as laminopathies. Specific mutations in the human gene encoding Lamin A or in the Lamin A-processing enzyme, Zmpste24, cause premature aging syndromes and cancer predisposition. Interestingly Oct1 and Lamin B1 also co-localize at different mitotic structures, with both proteins being required for the proper localization of the other (Kang et al., 2011). It is unknown whether Oct1 utilizes any of these relationships in CSCs.

Elevated Oct1 protein has been detected in a number of stem cell compartments and in purified CSC-like populations from clonal cell lines (Maddox et al., 2012). Surprisingly, much of this additional Oct1 protein appears to be localized to the cytoplasm, rather than the nucleoplasm or nuclear periphery. The function of this cytoplasmic Oct1, if any, is currently unknown, but regulated Oct1 nuclear/cytoplasmic shuttling in response to cAMP and oxidative stress signals has been reported (Wang et al., 2009; Wang and Jin, 2010).

Oct1 and Oct4 Working in Concert

As discussed earlier, Oct1 and Oct4 have similar in vitro binding specificity and interact with and regulate common targets. Examples include housekeeping genes, developmental regulators, inflammatory mediators, and metabolic genes (Kang et al., 2009b). Because some Oct target sites are complex and can simultaneously interact with multiple Oct proteins, Oct1 and Oct4 can simultaneously bind the same site. Evidence for just such simultaneous interactions exists in ESCs (Ferraris et al., 2011). In addition, Oct1 and Oct4 may regulate other targets with varying degrees of selectivity. More work is required to clarify these interactions and their regulation, and their role if any in CSCs.

Conclusions

The identification of transcriptional regulators controlling CSC self-renewal, differentiation, and proliferation is key to unraveling their aberrant biology, as well as to developing potential therapeutics. Oct1 and Oct4 are natural candidates based on the commanding function of Oct4 in pluripotency, and the similar binding specificity, targets, and mechanisms shared by Oct4 and Oct1. The expression and functional role of Oct4 in CSCs is controversial. More studies are needed to fully demarcate the contribution of Oct4 to different types of CSCs. In addition, accumulating evidence indicates that Oct1 acts as a somatic and CSC determinant. In this case, the next step is to unlock the upstream signals to which Oct1 responds, and the downstream targets modulated by Oct1, to specifically control stem cell function and malignancy.

Acknowledgments

We thank J. Spangrude and C. Murtaugh for critical reading of the manuscript. This work was supported by a grant from the National Institutes of Health/National Cancer Institute (1R21CA141009) and a grant from the Concern Foundation.

References

Atlasi, Y., Mowla, S.J., Ziaee, S.A. and Bahrami, A.R. 2007. OCT-4, an embryonic stem cell marker, is highly expressed in bladder cancer, Int J Cancer 120(7): 1598–602.

Beltran, A.S., Rivenbark, A.G., Richardson, B.T., Yuan, X., Quian, H., Hunt, J.P., Zimmerman, E., Graves, L.M. and Blancafort, P. 2011. Generation of tumor-initiating cells by exogenous delivery of OCT4 transcription factor, Breast Cancer Res 13(5): R94.

Ben-Porath, I., Thomson, M.W., Carey, V.J., Ge, R., Bell, G. W., Regev, A. and Weinberg, R.A. 2008. An embryonic stem cell-like gene expression signature in poorly differentiated aggressive human tumors, Nat Genet 40(5): 499–507.

Bernstein, B.E., Mikkelsen, T.S., Xie, X., Kamal, M., Huebert, D.J., Cuff, J., Fry, B., Meissner, A., Wernig, M., Plath, K. et al. 2006. A bivalent chromatin structure marks key developmental genes in embryonic stem cells, Cell 125(2): 315–26.

Bourguignon, L.Y., Wong, G., Earle, C. and Chen, L. 2012. Hyaluronan-CD44v3 interaction with Oct4/Sox2/Nanog promotes miR-302 expression leading to self-renewal, clonal formation and cisplatin resistance in cancer stem cells from head and neck squamous cell carcinoma, J Biol Chem 287(39): 32800-24

Boyer, L.A., Lee, T.I., Cole, M.F., Johnstone, S.E., Levine, S.S., Zucker, J.P., Guenther, M.G., Kumar, R.M., Murray, H.L., Jenner, R.G. et al. 2005. Core transcriptional regulatory circuitry in human embryonic stem cells, Cell 122(6): 947–56.

Chang, C.J., Chien, Y., Lu, K.H., Chang, S.C., Chou, Y.C., Huang, C.S., Chang, C.H., Chen, K.H., Chang, Y.L., Tseng, L.M. et al. 2011. Oct4-related cytokine effects regulate tumorigenic properties of colorectal cancer cells, Biochem Biophys Res Commun 415(2): 245–51.

Chen, X., Xu, H., Yuan, P., Fang, F., Huss, M., Vega, V.B., Wong, E., Orlov, Y.L., Zhang, W., Jiang, J. et al. 2008. Integration of external signaling pathways with the core transcriptional network in embryonic stem cells, Cell 133(6): 1106–17.

Chiou, S.H., Wang, M.L., Chou, Y.T., Chen, C.J., Hong, C. F., Hsieh, W.J., Chang, H.T., Chen, Y.S., Lin, T.W., Hsu, H.S. et al. 2010. Coexpression of Oct4 and Nanog enhances malignancy in lung adenocarcinoma by inducing cancer stem cell-like properties and epithelial-mesenchymal transdifferentiation, Cancer Res 70(24): 10433–44.

Dalvai, M., Schubart, K., Besson, A. and Matthias, P. 2010 Oct1 is required for mTOR-induced G1 cell cycle arrest via the control of p27(Kip1) expression, Cell Cycle 9(19): 3933–44.

Diehn, M., Cho, R.W., Lobo, N.A., Kalisky, T., Dorie, M.J., Kulp, A.N., Qian, D., Lam, J.S., Ailles, L.E., Wong, M. et al. 2009. Association of reactive oxygen species levels and radioresistance in cancer stem cells, Nature 458(7239): 780–3.

Ferraris, L., Stewart, A.P., Kang, J., DeSimone, A.M., Gemberling, M., Tantin, D. and Fairbrother, W.G. 2011. Combinatorial binding of transcription factors in the pluripotency control regions of the genome, Genome Res 21(7): 1055–64.

Folmes, C.D., Nelson, T.J., Martinez-Fernandez, A., Arrell, D. K., Lindor, J.Z., Dzeja, P.P., Ikeda, Y., Perez-Terzic, C. and Terzic, A. 2011. Somatic oxidative bioenergetics transitions into pluripotency-dependent glycolysis to facilitate nuclear reprogramming, Cell Metab 14(2): 264–71.

Hirose, T., Sowa, Y., Takahashi, S., Saito, S., Yasuda, C., Shindo, N., Furuichi, K. and Sakai, T. 2003. p53-independent induction of Gadd45 by histone deacetylase inhibitor: coordinate regulation by transcription factors Oct-1 and NF-Y, Oncogene 22(49): 7762–73.

Hochedlinger, K., Yamada, Y., Beard, C. and Jaenisch, R. 2005. Ectopic expression of Oct-4 blocks progenitor-cell differentiation and causes dysplasia in epithelial tissues, Cell 121(3): 465–77.

Ivanova, N.B., Dimos, J.T., Schaniel, C., Hackney, J.A., Moore, K.A. and Lemischka, I.R. 2002. A stem cell molecular signature, Science 298(5593): 601–4.

Kang, J., Gemberling, M., Nakamura, M., Whitby, F.G., Handa, H., Fairbrother, W.G. and Tantin, D. 2009a. A general mechanism for transcription regulation by Oct1 and Oct4 in response to genotoxic and oxidative stress, Genes Dev 23(2): 208–22.

Kang, J., Goodman, B., Zheng, Y. and Tantin, D. 2011. 'Dynamic regulation of Oct1 during mitosis by phosphorylation and ubiquitination,' PLoS One 6(8): e23872.

Kang, J., Shakya, A. and Tantin, D. 2009b. Stem cells, stress, metabolism and cancer: a drama in two Octs, Trends Biochem Sci 34(10): 491–9.

Kim, J., Woo, A.J., Chu, J., Snow, J.W., Fujiwara, Y., Kim, C.G., Cantor, A.B. and Orkin, S.H. 2010. A Myc network accounts for similarities between embryonic stem and cancer cell transcription programs, Cell 143(2): 313–24.

Kondoh, H., Lleonart, M. E., Nakashima, Y., Yokode, M., Tanaka, M., Bernard, D., Gil, J. and Beach, D. 2007 A high glycolytic flux supports the proliferative potential of murine embryonic stem cells, Antioxid Redox Signal 9(3): 293–9.

Kumar, S.M., Liu, S., Lu, H., Zhang, H., Zhang, P.J., Gimotty, P.A., Guerra, M., Guo, W. and Xu, X. 2012. Acquired cancer stem cell phenotypes through Oct4-mediated dedifferentiation, Oncogene.

Lengner, C.J., Camargo, F.D., Hochedlinger, K., Welstead, G.G., Zaidi, S., Gokhale, S., Scholer, H.R., Tomilin, A. and Jaenisch, R. 2007. Oct4 expression is not required for mouse somatic stem cell self-renewal, Cell Stem Cell 1(4): 403–15.

Li, X., Wang, J., Xu, Z., Ahmad, A., Li, E., Wang, Y., Qin, S. and Wang, Q. 2012. Expression of sox2 and oct4 and their clinical significance in human non-small-cell lung cancer, Int J Mol Sci 13(6): 7663–75.

Liedtke, S., Stephan, M. and Kogler, G. 2008. Oct4 expression revisited: potential pitfalls for data misinterpretation in stem cell research, Biol Chem 389(7): 845–50.

Linn, D. E., Yang, X., Sun, F., Xie, Y., Chen, H., Jiang, R.,Chen, H., Chumsri, S., Burger, A. M. and Qiu, Y. 2010. A Role for OCT4 in Tumor Initiation of Drug-Resistant Prostate Cancer Cells, Genes Cancer 1(9): 908–16.

Lister, R., Pelizzola, M., Dowen, R.H., Hawkins, R.D., Hon, G., Tonti-Filippini, J., Nery, J.R., Lee, L., Ye, Z., Ngo, Q. M. et al. 2009. Human DNA methylomes at base resolution show widespread epigenomic differences, Nature 462(7271): 315–22.

Maddox, J., Shakya, A., South, S., Shelton, D., Anderson, J.N., Chidester, S., Kang, J., Jones, D.A., Spangrude, G.J., Welm, B.E. et al. 2012. Transcription Factor Oct1 is a Somatic and Cancer Stem Cell Determinant, PLoS Genetics 8(11): e1003048.

Malhas, A.N., Lee, C.F. and Vaux, D.J. 2009. Lamin B1 controls oxidative stress responses via Oct-1, J Cell Biol 184(1): 45–55.

Mathieu, J., Zhang, Z., Zhou, W., Wang, A.J., Heddleston, J.M., Pinna, C.M., Hubaud, A., Stadler, B., Choi, M., Bar, M. et al. 2011. HIF induces human embryonic stem cell markers in cancer cells, Cancer Res 71(13): 4640–52.

Miranda-Saavedra, D. and Gottgens, B. 2008. Transcriptional regulatory networks in haematopoiesis, Curr Opin Genet Dev 18(6): 530–5.

Morrison, S.J. and Kimble, J. 2006. Asymmetric and symmetric stem-cell divisions in development and cancer, Nature 441(7097): 1068–74.

Mosca, E., Cocola, C., Sabour, D., Pelucchi, P., Bertalot, G., Palumbo, O., Carella, M., Gotte, M., Scholer, H.R., Reinbold, R. et al. 2010. Overlapping Genes May Control Reprogramming of Mouse Somatic Cells into Induced Pluripotent Stem Cells (iPSCs) and Breast Cancer Stem Cells, In Silico Biol 10(5): 207–21.

Nichols, J., Zevnik, B., Anastassiadis, K., Niwa, H., Klewe-Nebenius, D., Chambers, I., Scholer, H. and Smith, A. 1998. Formation of pluripotent stem cells in the mammalian embryo depends on the POU transcription factor Oct4, Cell 95(3): 379–91.

Ramalho-Santos, M., Yoon, S., Matsuzaki, Y., Mulligan, R.C. and Melton, D.A. 2002. "Stemness": transcriptional profiling of embryonic and adult stem cells, Science 298(5593): 597–600.

Remenyi, A., Tomilin, A., Pohl, E., Lins, K., Philippsen, A., Reinbold, R., Scholer, H. R. and Wilmanns, M. 2001. Differential dimer activities of the transcription factor Oct-1 by DNA-induced interface swapping, Mol Cell 8(3): 569–80.

Sebastiano, V., Dalvai, M., Gentile, L., Schubart, K., Sutter, J., Wu, G. M., Tapia, N., Esch, D., Ju, J.Y., Hubner, K. et al. 2010. Oct1 regulates trophoblast development during early mouse embryogenesis, Development 137(21): 3551–60.

Shakya, A., Cooksey, R., Cox, J.E., Wang, V., McClain, D.A. and Tantin, D. 2009. Oct1 loss of function induces a coordinate metabolic shift that opposes tumorigenicity, Nat Cell Biol 11(3): 320–7.

Shakya, A., Kang, J., Chumley, J., Williams, M.A. and Tantin, D. 2011. Oct1 is a switchable, bipotential stabilizer of repressed and inducible transcriptional states, J Biol Chem 286(1): 450–9.

Somervaille, T.C., Matheny, C.J., Spencer, G.J., Iwasaki, M., Rinn, J.L., Witten, D.M., Chang, H.Y., Shurtleff, S.A., Downing, J.R. and Cleary, M.L. 2009. Hierarchical maintenance of MLL myeloid leukemia stem cells employs a transcriptional program shared with embryonic rather than adult stem cells, Cell Stem Cell 4(2): 129–40.

Takahashi, K. and Yamanaka, S. 2006. Induction of pluripotent stem cells from mouse embryonic and adult fibroblast cultures by defined factors, Cell 126(4): 663–76.

Takahashi-Yanaga, F. and Kahn, M. 2010. Targeting Wnt signaling: can we safely eradicate cancer stem cells?, Clin Cancer Res 16(12): 3153–62.

Tantin, D., Schild-Poulter, C., Wang, V., Hache, R.J. and Sharp, P.A. 2005. The octamer binding transcription factor Oct-1 is a stress sensor, Cancer Res 65(23): 10750–8.

Tsai, C.C., Su, P.F., Huang, Y.F., Yew, T.L. and Hung, S.C. 2012. Oct4 and Nanog Directly Regulate Dnmt1 to Maintain Self-Renewal and Undifferentiated State in Mesenchymal Stem Cells, Mol Cell 47(2): 169-82.

Utikal, J., Polo, J.M., Stadtfeld, M., Maherali, N., Kulalert, W., Walsh, R.M., Khalil, A., Rheinwald, J.G. and Hochedlinger, K. 2009. Immortalization eliminates a roadblock during cellular reprogramming into iPS cells, Nature 460(7259): 1145–8.

Wang, P. and Jin, T. 2010. Hydrogen peroxide stimulates nuclear import of the POU homeodomain protein Oct-1 and its repressive effect on the expression of Cdx-2, BMC Cell Biol 11: 56.

Wang, P., Wang, Q., Sun, J., Wu, J., Li, H., Zhang, N., Huang, Y., Su, B., Li, R.K., Liu, L. et al. 2009. POU homeodomain protein Oct-1 functions as a sensor for cyclic AMP, J Biol Chem 284(39): 26456–65.

Wang, V. E., Schmidt, T., Chen, J., Sharp, P.A. and Tantin, D. 2004. Embryonic lethality, decreased erythropoiesis, anddefective octamer-dependent promoter activation in Oct-1-deficient mice, Mol Cell Biol 24(3): 1022–32.

Yasuda, H., Tanaka, K., Okita, Y., Araki, T., Saigusa, S., Toiyama, Y., Yokoe, T., Yoshiyama, S., Kawamoto, A., Inoue, Y. et al. 2011. CD133, OCT4, and NANOG in ulcerative colitis-associated colorectal cancer, Oncol Lett 2(6): 1065–1071.

Chapter 25

The Role of Cripto-1 in Cancer and Cancer Stem Cells

Hideaki Karasawa, Nadia P. Castro, Maria Cristina Rangel, and David S. Salomon
Tumor Growth Factor Section, Laboratory of Cancer Prevention, Frederick National Laboratory for Cancer Research, Frederick, MD, USA

Abbreviations

TDGF-1	teratocarcinoma derived growth factor-1
GPI	glycophosphatidylinositol
EGF	epidermal growth factor
CFC	Cripto-FRL1-Cryptic
ES	embryonic stem
iPS	induced pluripotent stem
EMT	epithelial to mesenchymal transition
GDF	growth and differentiation factor
MAPK	mitogen-activated protein kinase
PI3K	phosphatidylinositol 3′ kinase
GRP78	78-kDa glucose-regulated protein
CSC	cancer stem cell
FAK	focal adhesion kinase
EC	embryonal carcinoma
HIF1-α	hypoxia-inducible factor-1 alpha
HSC	hematopoietic stem cell
VEGF	vascular endothelial growth factor
bFGF	basic fibroblast growth factor
WAP	whey acidic protein
MMTV	mouse mammary tumor virus
FGF	fibroblast growth factor
mAb	monoclonal antibody

Cancer Stem Cells, First Edition. Edited by Vinagolu K. Rajasekhar.

Introduction

Cripto-1, also known as teratocarcinoma derived growth factor-1 (TDGF-1), is a cell membrane glycophosphatidylinositol (GPI)-anchored protein, which is composed of an epidermal growth factor (EGF)-like domain and a cysteine-rich Cripto-FRL1-Criptic (CFC) domain (Bianco et al., 2010). Cripto-1 is highly expressed in undifferentiated pluripotential human and mouse embryonic stem cells (ESC), in early vertebrate embryos, and in induced pluripotent stem (iPS) cells. Cripto-1 expression is very low or absent in the adult mouse and human tissues (de Castro et al., 2010). Furthermore, Cripto-1 is re-expressed in a variety of human tumors, promoting cell proliferation, migration, invasion, and epithelial to mesenchymal transition (EMT). During embryogenesis, Cripto-1 functions as a co-receptor for the TGF-β family ligands, Nodal, and growth and differentiation factor-1 and -3 (GDF-1/3). Cripto-1 performs a critical role, especially in gastrulation, which is a fundamental process for the formation of the three primitive germ layers (Rangel et al., 2012). Cripto-1 directly binds to Nodal and GDF-1/3, and activates the receptor complex of the activin type I and type II receptor, leading to the phosphorylation and activation of Smad-2/3 intracellular signaling pathway (Figure 25.1). In addition

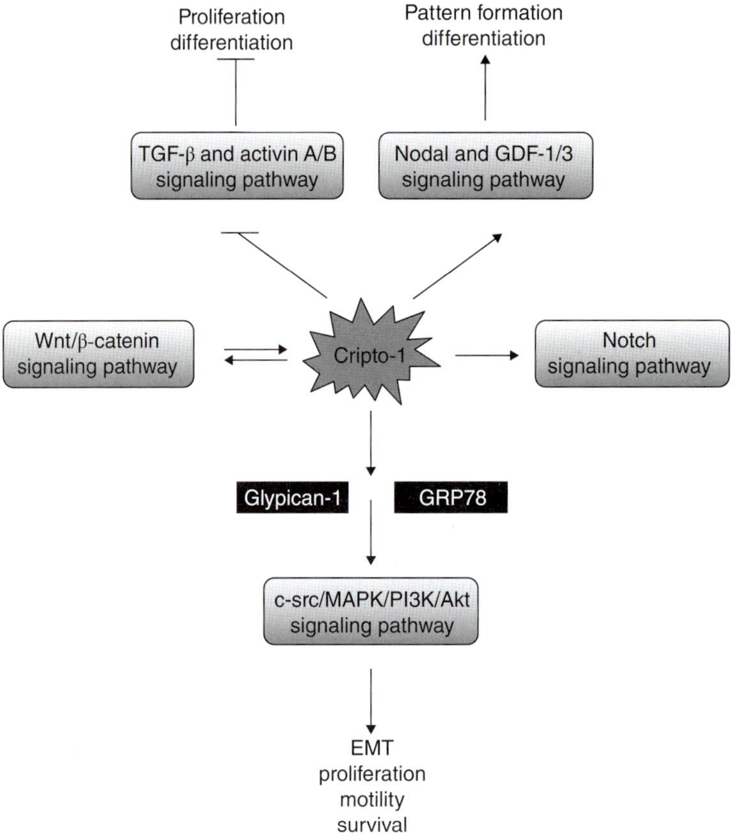

Figure 25.1. Signaling pathways regulated by Cripto-1 in development and cancer. During embryonic development, Cripto-1 functions as a co-receptor for Nodal and GDF-1/3, and contributes to pattern formation and differentiation through the activation of a Smad-2/3 signaling pathway. Cripto-1 can also inhibit TGF-β and activin A/B signaling by direct binding. In addition, c-src/MAPK/PI3K/Akt signaling pathway is activated by binding of Cripto-1 to Glypican-1 and GRP78, promoting cell proliferation, motility, survival, and EMT. Furthermore, Cripto-1 interacts with the Wnt/β-catenin and Notch signaling and contributes to their tumor-promoting activities.

to functioning as a coreceptor for Nodal/GDF-1/GDF-3 signaling, Cripto-1 can also inhibit activin A/B and TGF-β1 signaling by direct binding to these growth factors (Figure 25.1). Another major signaling pathway activated by Cripto-1 is the c-src/mitogen-activated protein kinase (MAPK)/phosphatidylinositol 3′ kinase (PI3K)/Akt signaling pathway, which is strongly involved in mediating the oncogenic activity of Cripto-1. Cripto-1 can also bind to Glypican-1 and 78-kDa glucose-regulated protein (GRP78), promoting cell proliferation, motility, and survival through the activation of c-src/MAPK/PI3K/Akt signaling pathway (Figure 25.1) (Nagaoka et al., 2012). Cripto-1 also interacts with the Wnt/β-catenin and Notch signaling pathways and contributes to their tumor-promoting activities in vitro and in vivo (Figure 25.1). Therefore, accumulating evidence clearly demonstrates that Cripto-1 plays an important role not only during early embryogenesis, but also in the malignant progression of tumors. In this review, we describe the oncogenic activities of Cripto-1 and the role that Cripto-1 performs in regulating the self-renewal and maintenance of ES cells, iPS cells, and cancer stem cells (CSC), and introduces the development of novel Cripto-1 targeted therapies in cancer treatment.

Cripto-1 Oncogenic Activities In Vitro

The initial evidence that Cripto-1 possessed oncogenic activity was demonstrated by in vitro studies, demonstrating that Cripto-1 overexpression could contribute to cellular transformation. Murine NIH-3T3 fibroblasts, and the mouse mammary epithelial cell lines, NOG-8 or CID9, when transfected with a Cripto-1 expression vector, acquired a transformed phenotype in vitro, being able to grow in anchorage-independent conditions in soft agar (Ciccodicola et al., 1989; Ciardiello et al., 1991; Niemeyer et al., 1998). Additionally, increased expression of Cripto-1 was observed in Fischer rat embryo fibroblasts cells or Fischer rat thyroid cells transformed by the activated c-Ha−ras or c-Ki−ras oncogenes, suggesting that Cripto-1 might be induced by these oncogenes, which may contribute to cellular transformation (Su et al., 1993; Mincione et al., 1998). Cripto-1 has also been shown to be involved in regulating biological processes that are involved in tumor progression, since its overexpression can enhance migration and invasion of EpH4, NOG-8, TAC-2, or NMuMG mouse mammary epithelial cells and MCF-7 human breast cancer cells (Wechselberger et al., 2001; Normanno et al., 2004). Furthermore, Cripto-1 has been shown to be directly involved in the induction of EMT, as evidenced by a significant decrease in the expression of the intercellular adhesion molecule E-cadherin and by an increase in expression of N-cadherin, vimentin, and the zinc-finger repressor transcription factor, Snail in HC-11 mouse mammary epithelial cells that are overexpressing Cripto-1 (Figure 25.2) (Strizzi et al., 2004). An increase in expression of signaling molecules that are known to be activated during EMT, such as phosphorylated c-src, focal adhesion kinase (FAK) and Akt, was also observed in these cells. The augmented cell motility and the involvement of EMT observed in HC-11 Cripto-1 overexpressing mammary epithelial cells has also been confirmed in other cell lines, such as Caski human cervical carcinoma cells that are overexpressing ectopic Cripto-1 and in human LS174-T colon cancer cells and NTERA-2/D1 embryonal carcinoma (EC) cells that express endogenous Cripto-1 (Ebert et al., 2000; Mancino et al., 2008). In fact, Caski Cripto-1 overexpressing cells acquired a more invasive phenotype, as confirmed by the upregulation of the EMT marker vimentin, suggesting that Cripto-1 may contribute to a more aggressive mesenchymal phenotype in cervical carcinoma cells (Ebert et al., 2000). Interestingly, the Cripto-1 ligand, Nodal, has been shown to promote migration and invasion of melanoma cells, and its depletion in aggressive melanoma cells was able to significantly

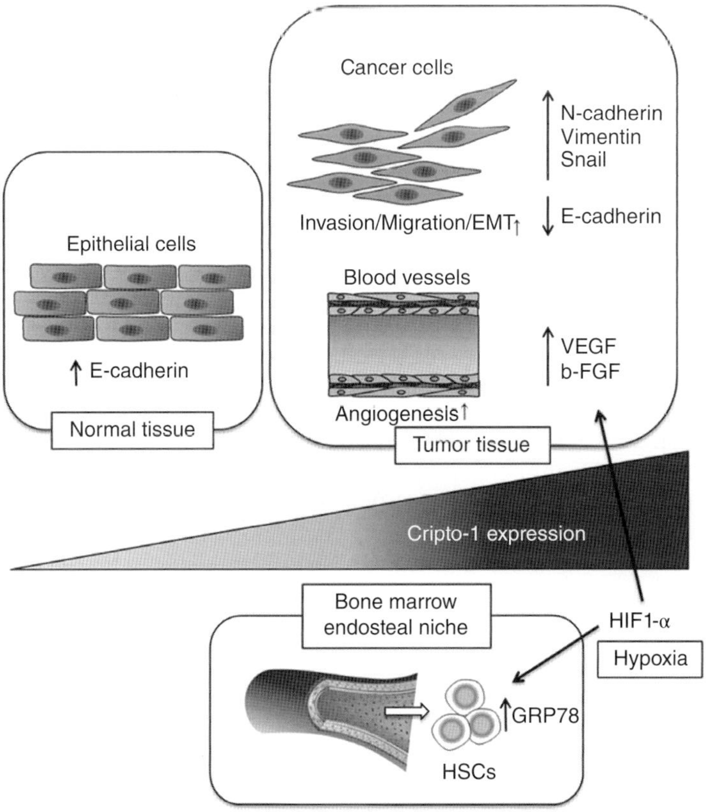

Figure 25.2. Cripto-1 expression in normal, tumor tissue, and bone marrow hematopoietic niche. Cripto-1 expression is very low in normal tissue. In tumor tissue, Cripto-1 is directly involved in the induction of EMT, through decrease in the expression of the E-cadherin and an increase in expression of N-cadherin, vimentin, and Snail during invasion, migration, and EMT. In addition, Cripto-1 stimulates angiogenesis in tumor tissue regulated by HIF1-α. HIF1-α can also activate Cripto-1/GRP78 signaling in the hypoxic bone marrow niche and can regulate hematopoietic stem cells (HSC).

decrease anchorage-independent growth and epithelial cell plasticity (Topczewska et al., 2006). Cripto-1 recombinant protein can also enhance cell motility, while the ablation of endogenous Cripto-1 expression also can result in a significant inhibition in the growth and invasiveness of melanoma cells (De Luca et al., 2011). Therefore, Cripto-1 and Nodal might be important regulators of melanoma tumorigenicity. In vitro studies with U87MG and GBM human glioma cell lines have shown that Nodal, in a Cripto-1 dependent manner, induces Smad-2 activation and promotes glioma cell proliferation, suggesting that Cripto-1 might also be involved in glioma tumorigenesis (Lee et al., 2010). In addition, Nodal together with Cripto-1 was able to enhance the growth of LNCaP prostate cancer cells in an anchorage-independent manner (Lawrence et al., 2011). Cripto-1 is also upregulated in nasopharyngeal carcinoma cell lines and inhibition of Cripto-1 expression strongly inhibited cell growth and invasiveness in these cells (Wu et al., 2009). All of this in vitro data collectively suggests that Cripto-1 overexpression might be associated with tumor progression toward a more aggressive phenotype. Interestingly, Cripto-1 also has been shown to play a role in tumor angiogenesis, since overexpression of Cripto-1 in human breast cancer cells can enhance the proliferation, migration, and invasion of human umbilical vascular endothelial cells and to stimulate their differentiation in vitro into vascular-like structures (Bianco et al., 2005). In support of a role for Cripto-1 in modulating tumor angiogenesis, recombinant Cripto-1 protein can increase microvessel formation in vivo, and can enhance tumor neovascularization in a MCF-7 human breast cancer xenograft model that is overexpressing Cripto-1 (Bianco et al., 2005). Furthermore, it has been shown that low oxygen (hypoxia) within a tumor microenvironment can induce Cripto-1 expression, which in turn can induce new vessel formation to sustain tumor growth (Bianco et al., 2009). In fact, the hypoxia-inducible factor-1 alpha (HIF1-α) enhances Cripto-1 expression in mouse ES cells and in human EC cells by directly binding to HIF1-α response element in the Cripto-1 promoter under hypoxic conditions (Bianco et al., 2009). In addition, HIF1-α has been also shown to activate Cripto-1/GRP78 signaling in the hypoxic bone marrow niche and regulates hematopoietic stem cells (HSC) (Figure 25.2) (Miharada et al., 2011; Miharada et al., 2012). Since oxygen deprivation also triggers the activation of angiogenic molecules such as vascular endothelial growth factor (VEGF) and basic fibroblast growth factor (bFGF) in growing tumors, it is probable that hypoxia can also sustain tumor growth through neovascularization by inducing Cripto-1 expression (Figure 25.2).

The Role of Cripto-1 Overexpression in Promoting Mammary Tumorigenesis in Vivo

The first clues to a potential role of Cripto-1 involvement in mammary gland tumorigenesis in vivo were reported by Kenney and colleagues (1996), who showed that a significant increase of Cripto-1 expression was observed in mammary tumors that were derived from transgenic mice overexpressing a variety of different oncogenes, including neu (erbB-2), TGF-α, int-3 (activated Notch4), polyoma middle T, or simian virus 40 large T antigens. Cripto-1 was subsequently implicated in mammary transformation in several different transgenic mouse models that were selectively overexpressing the human Cripto-1 transgene in the mouse mammary gland using mammary gland-specific promoter expression constructs that are driven by the whey acidic protein (WAP) and mouse mammary tumor virus (MMTV) promoters (Sun et al., 2005; Wechselberger et al., 2005). The WAP promoter is maximally active during mid-pregnancy and lactation, whereas the MMTV promoter is activated in the virgin pre-pubertal mammary gland. More than half of multiparous WAP-Cripto-1 female mice developed multifocal mammary tumors of mixed histological subtypes (Sun et al., 2005). Since the mammary tumors derived from WAP-Cripto-1 mice also exhibited a hyperactivation of the canonical Wnt/β-catenin pathway, the Wnt/β-catenin pathway in conjunction with Cripto-1 overexpression might play an important role during mammary gland transformation in vivo. Cripto-1 is also a direct downstream target gene that is regulated by the canonical Wnt/β-catenin pathway (Hamada et al., 2007). This suggests that a positive feedback loop between Cripto-1 and Wnt/β-catenin pathway may contribute to mammary tumor formation in this setting. Likewise, the majority of virgin MMTV-Cripto-1 transgenic mice exhibited a significant increase in lateral secondary and tertiary side branching in the mammary gland, as well as increase in the formation of intraductal hyperplasias and hyperplastic alveolar nodules. Furthermore, extensive mammary multifocal hyperplasias and papillary adenocarcinomas were observed in 30–40% of multiparous aged MMTV-Cripto-1 females (Wechselberger et al., 2005). In fact, distinct morphological changes associated with EMT were observed in the mammary tumors arising in the MMTV-Cripto-1 mice, and the expression of several EMT markers, such as N-cadherin and vimentin, were increased in MMTV-Cripto-1 mammary tumors (Figure 25.2) (Strizzi et al., 2004). These results indicated that the oncogenic role of Cripto-1 is associated with EMT. Surprisingly, after 24 months of observation, about 20% of the MMTV−Cripto-1 nulliparous or multiparous female also developed uterine leiomyosarcomas, and Cripto-1 was expressed at high levels in uterine tumors collected from these mice (Strizzi et al., 2007). This collectively suggests that Cripto-1 may be an important oncogene not only in mammary gland tumorigenesis in vivo, but also possibly in the tumorigenesis of a variety of other types of tumors.

Clinical Significance of Cripto-1 Expression

In addition to performing important oncogenic roles in vitro and in vivo, overexpression of Cripto-1 has been detected in a variety of human cancers (de Castro et al., 2010). Furthermore, Cripto-1 expression is significantly associated with poor prognosis in several types of human cancers including breast, gastric, and colorectal carcinomas. For example, Cripto-1 was found to be overexpressed in 47.5% of breast cancer patients (Gong et al., 2007), which was associated with a poor prognostic index, high histological grade, and decreased survival (Gong et al., 2007). Cripto-1 expression has also been related with poor prognosis in gastric cancer patients. Immunohistochemical studies have revealed that upregulation of Cripto-1 is positively associated with lymph node metastasis, liver metastasis, and poor prognosis (Zhong et al., 2008; Zhang et al., 2010). In addition, higher levels of Cripto-1 mRNA expression were detected in 30.9% of colorectal tumors when compared with adjacent noncancerous regions (Miyoshi et al., 2010). Patients with high Cripto-1 expression were statistically associated with early recurrence and poor disease-free survival. Therefore, Cripto-1 is considered to be associated not only with tumor formation but also with aggressive behavior in a variety of tumors. For an extensive review on Cripto-1 expression in human cancers, see de Castro and colleagues (2010).

Cripto-1 and Embryonic Stem Cells

Cripto-1 is essential for early embryonic development. During early mouse embryogenesis, Cripto-1 can only be detected prior to the gastrulation stage in the inner cell mass and in extra-embryonic trophoblast cells at day 4 of development (Johnson et al., 1994). In embryos at days 3 to 5, the blastocyst, the inner cell mass therein can differentiate into all three germ layers (ectoderm, mesoderm, and endoderm), giving rise to all the specialized tissues and organs of a mature organism (pluripotency). ES cells can be isolated from this stage and also possess the capacity to undergo numerous cell cycle divisions while maintaining their undifferentiated state (self-renewal) (Figure 25.3) (Bendall et al., 2008). One way to test for pluripotency is injecting the cells into immune-compromised mice to verify the formation of benign tumors called teratomas. Teratomas typically contain a mixture of many differentiated or partly differentiated cell types, indicating that the ES cells are able to differentiate into multiple cell types.

Interestingly, Cripto-1 was initially cloned from a human EC cell line NTERA-2/D1 (Ciccodicola et al., 1989) and has been associated with the pluripotent status of ESCs. Recently, a number of studies have been focused on the validation and characterization of ESC lines attempting to identify potential embryonic stem cell markers. In this context, Cripto-1 has been reported as stem cell marker of undifferentiated and pluripotent ESCs (Hough et al., 2009; Bianco et al., 2010). In 2007 a

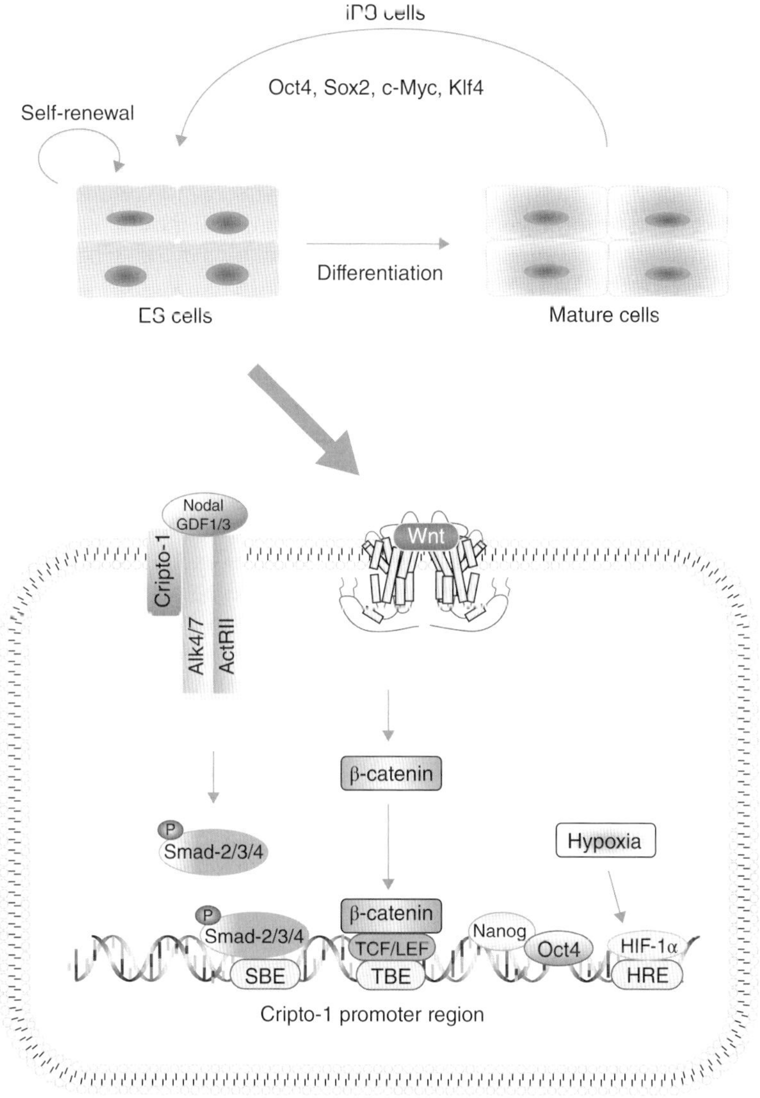

Figure 25.3. Schematic diagram of embryonic stem (ES) cells and induced pluripotent stem (iPS) cells. ES cells possess the capability to undergo numerous cell cycle divisions while maintaining their undifferentiated state (self-renewal). iPS cells are derived through reprogramming of differentiated cells into pluripotent cells by introducing the embryonic stem cells genes, Oct4, Sox2, c-Myc, and Klf4. In embryonic development, phosphorylated Smad-2/3/4 complex binds to Smad binding elements (SBE) within the Cripto-1 promoter and regulates Cripto-1 expression. HIF-1α binds to Hypoxia Reponsive Elements (HRE) within Cripto-1 promoter and directly regulates Cripto-1 expression during hypoxia. Oct-4 and Nanog can also regulate Cripto-1 expression in ES cells by binding to the Cripto-1 promoter. Finally, Cripto-1 is a direct downstream target gene that is regulated by the canonical Wnt/β-catenin signaling pathway. T-cell factor/lymphoid enhancer factor (TCF/LEF); TCF/LEF binding elements (TBE); hypoxia binding elements (HRE); Smad binding elements (SBE). For color detail, please see color plate section.

characterization study of a large and diverse set of human ESC lines showed that all cell lines exhibited a similar expression profile for several markers of human ESCs, including the glycolipid stage specific embryonic antigens, SSEA3 and SSEA4, the keratan sulfate antigens, TRA-1-60 and TRA-1-81, developmentally regulated transcription factors genes, Nanog and Oct-4, and the growth factors, Cripto-1 and GDF-3 (Bianco et al., 2010). Similarly, in five new human ESC lines, it was reported that the stemness genes such as Oct-4, Nanog, Cripto-1, and Sall4 were responsible for self-renewal and pluripotentiality (Li et al., 2006). Interestingly, the global gene expression analysis confirmed that the TGF-β superfamily was involved in early embryonic development signaling primarily through a TGF-β/activin and BMP pair of signaling pathways. The TGF-β/activin branch components such as, inhibin BC, type II receptor Acvr2A, Acvr1 (Alk2), Tgfbr1 (Alk5), and Smad-2/4 were highly expressed in undifferentiated human ESC lines and decreased upon differentiation. Conversely, the BMP members showed a significant increase in BMP7, GDF-3/15, and Nodal expression in embryoid bodies, teratomas, and placenta (Li, 2012).

A comparative study of gene expression profiles across species has shown that mouse and human ESCs share common and highly conserved signaling pathways that regulate self-renewal and pluripotency, including Cripto-1, Oct-4, Lefty, Nodal, and Sox-2 (Wei et al., 2005). A recent report regarding the establishment and characterization of several novel nonhuman primate ESC lines also has confirmed expression of the pluripotency markers Oct-4, Nanog, Sox-2, Cripto-1, LeftyA, and Rex-1 in these cells (Chang et al., 2011). Remarkably, studying the transcription factors Oct4 and Nanog, Loh and collaborators (2006) generated a list of genes that contained Nanog and Oct4 binding sites, of which Cripto-1 was found to be a direct downstream target of these key regulators of self-renewal and pluripotency in ESCs. Reciprocally, activin/Nodal and Cripto-1 are essential in maintaining the expression of Nanog and Oct-4 (Watanabe et al., 2010). It has been reported that activin/Nodal signaling and fibroblast growth factors (FGF) are essential in controlling Nanog expression in human ESCs and in mouse epiblast stem cells (Vallier et al., 2009).

Intriguingly, studies examining lineage commitment through transcript analysis of single ESCs found a gradient and a hierarchy of expression of pluripotency genes in this population (Hough et al., 2009). Although most cells were positive for ES surface markers and co-express lineage specific genes, only uncommitted ES cells with high ability to undergo self-renewal express Cripto-1 and GDF-3 (a growth factor implicated in cell renewal) (Hough et al., 2009). (For a review, see Bianco et al., 2010). Therefore, these findings suggest that Cripto-1 is a critical component of a core pathway that is used by ESCs to maintain self-renewal and pluripotency, which consists of Nanog, Oct4, and Sox2.

Cripto-1 and iPS Cells

A significant impact in ESC biology came through the reprogramming of differentiated cells into pluripotent cells and demonstrating that these iPS cells could differentiate into cell types of all three germ layers in vitro and could produce teratomas in vivo. The generation of iPS cells initially occurred by introducing the embryonic stem cell genes Oct-4, Sox-2, c-Myc, and Klf4 into mouse fibroblasts cultures (Figure 25.3) (Takahashi and Yamanaka, 2006). In late 2007, a generation of iPS cells from adult human dermal fibroblasts was reported using the same four genes (Takahashi et al., 2007). In addition, other genes, such as Oct-4, Sox-2, Nanog, and LIN28, were shown to be sufficient to reprogram human fetal fibroblasts (Yu et al., 2007). In reprogramming juvenile human primary keratinocytes by retroviral transduction using even three factors showed that all cells expressed genes and cell-surface markers characteristic of human ES cells, including Nanog, Oct-4, Sox-2, Rex-1, Cripto-1, IGF1 receptor, and SSEA3/4 (Aasen et al., 2008). Conversely, such induction by four-gene transduction conferred a high probability of stimulation of the expression of the pluripotent stem cell markers Nanog, Cripto-1, and Dnmt3b, compared to genes that are less specific for pluripotency (Masaki et al., 2007). Interestingly, activation (phosphorylation) of the downstream effectors Smad-2/3 of activin/Nodal signaling pathway, in which Cripto-1 functions primarily as a co-receptor, were found to block neuroectoderm differentiation of pluripotent cells by maintaining the expression of the key pluripotency factor Nanog in human ESCs and in mouse epiblast stem cells (Vallier et al., 2009).

Different mechanisms have been implicated in controlling the dynamic expression of embryonic genes in iPS cells and ESCs, for example, DNA methylation (Aoi et al., 2008; Chang et al., 2009; Nishino et al., 2010). Defining hypomethylated regions in the promoters of iPS/ES genes, Nishino and collaborators (2010) demonstrated that 23 genes including Sox-15, Sall4, Cripto-1, and Sox-10 as well as Oct-4 were hypomethylated in iPS/ES cells and therefore highly expressed. Likewise, the differentiated cells exhibited hypermethylation of genes such as Oct-4, Nanog, Sox-2, Lin28, Fgf4, Cripto-1, and Rex1 (De Carvalho et al., 2010).

Given their distinctive regenerative abilities, iPS cell reprogramming technology offers the potential for treating diseases such as diabetes, spinal cord injury, heart disease, and Parkinson's disease. Additionally, these findings improve our knowledge of the fundamental mechanisms that regulate cellular plasticity and epigenetic changes associated with cell fate, and consequently could lead to a better understanding of their deregulation during cancer.

Cripto-1 and Cancer Stem Cells (CSC)

Cells possessing characteristics associated with normal stem cells have been documented in hematologic malignancies as well as in solid cancers, including those arising in the breast, lung, prostate, colon, brain, head and neck, and pancreas (Lapidot et al., 1994; Bonnet and Dick, 1997; Al-Hajj et al., 2003; Singh et al., 2004), suggesting that abnormal regulation of stem cells is implicated in the generation and progression of malignant tumors. This small subpopulation of cells within the tumor is called CSCs or tumor-initiating cells and generates tumors through stem cell self-renewal and the regeneration of the original phenotype of the tumor when implanted into immunodeficient mice (Visvader and Lindeman, 2008). It has been suggested that such cells persist within the tumor as a distinct population and may contribute to relapse and metastasis by giving rise to new tumors, since these cells tend to be resistant to chemo- and radiotherapy (Morrison et al., 2011).

Recent evidence has shown that Cripto-1 is expressed at high levels in a subpopulation of human EC cells with cancer stem-like characteristics (Watanabe et al., 2010). A heterogeneous expression of Cripto-1 in EC cells was found, representing two distinct subpopulations of Cripto-1 high and Cripto-1 low expressing cells. These two subpopulations showed distinct in vitro and in vivo oncogenic capacities. Superior ability to form tumor spheres in serum-free suspension culture was observed in the Cripto-1 high EC subpopulation when compared with the Cripto-1 low expression subpopulation of cells. Moreover, increased ability to form tumors was verified in the Cripto-1 high EC subpopulation compared to Cripto-1 low expressing EC cells when EC cells were injected subcutaneously into nude mice. Nevertheless, a plasticity of EC cells was observed in the Cripto-1 low subpopulation, suggesting a reversible phenotype between Cripto-1 high and Cripto-1 low EC populations. Finally, the Cripto-1 high EC expressing cells were enriched for pluripotent ESC markers Sox-2, Nanog, and Oct-4 that could be modulated by a Smad-2/3-dependent signaling pathway involving activin, Nodal, GDF-1, or GDF-3. Similarly, activation of the Nodal/Cripto-1 signaling pathway has been shown to regulate self-renewal and in vivo tumorigenicity of pancreatic CSCs (Lonardo et al., 2011). shRNA technology or pharmacological inhibition of Nodal/activin/Cripto-1 signaling reversed the chemoresistance of orthotopically engrafted pancreatic CSCs, providing a novel therapeutic approach to target pancreatic CSCs if combined with hedgehog pathway inhibition and gemcitabine. Moreover, Cripto-1 has also been identified in a small subset of stem-like cells in human malignant melanomas and in androgen-responsive and refractory human prostate tumor cell lines (Strizzi et al., 2008; Cocciadiferro et al., 2009). Hence, development of combined therapies targeting CSCs may represent a new paradigm in the treatment of cancer.

Cripto-1 as Target for Cancer Therapy

Since Cripto-1 is highly expressed in several different types of human cancers and the expression is very low in normal adult tissues, Cripto-1 has been considered to be a potential target for cancer therapy. Furthermore, Cripto-1 is enriched in a stem-like cancer cell subpopulation in embryonal carcinoma, malignant melanoma, and prostate carcinoma cells (Strizzi et al., 2008; Cocciadiferro et al., 2009; Watanabe et al., 2010). In view of that, Cripto-1 therapy might have a significant potential in targeting CSCs. In fact, several therapeutic approaches for targeting Cripto-1 have been developed over the last 2 decades.

Antisense Oligonucleotides

Antisense oligonucleotides use RNA therapy. Antisense oligonucleotides bind to complementary mRNA and induce degradation of targeted mRNA by RNase H (Kole et al., 2012). Since 13-nucleotides-long oligonucleotides were first reported in 1978 (Zamecnik and Stephenson, 1978), several antisense oligonucleotides drugs have been developed, and several clinical trials are now ongoing. In 1994, it was reported that Cripto-1 protein expression was reduced, and cell proliferation was inhibited by antisense oligonucleotides directed against Cripto-1 in human colon cancer cells (Ciardiello et al., 1994). Furthermore, the Cripto-1 antisense oligonucleotides also inhibited the proliferation of human breast, ovarian, and lung cancer cells (De Luca et al., 1999; Casamassimi et al., 2000; Normanno et al., 2004). In a colon cancer xenograft model, oral administration of Cripto-1 antisense oligonucleotides significantly reduced tumor growth (Normanno et al., 2004). In addition, combinations of anti-Cripto-1 antisense oligonucleotides with other therapeutic agents, such as 5-FU, CDDP, and anti-EGFR monoclonal antibody, were able to induce additive growth inhibitory effects in human colon cancer cells (De Luca et al., 1997; Normanno et al., 1999).

Monoclonal Antibodies

Monoclonal antibody (mAb)-based cancer therapy has been developed over the past few decades and has become one of the most successful therapeutic strategies for cancer treatment. The mechanisms of tumor cell killing by mAbs are (1) direct action of the antibody, (2) immune-mediated cell killing mechanisms, (3) payload delivery, and (4) specific effects of an antibody on the tumor vasculature and stroma. mAbs against Cripto-1 also have been developed in the last 10 years. Adkins and colleagues (2003) have generated a mouse mAb against the CFC domain of Cripto-1. This mAb inhibits Cripto-1-dependent Nodal signaling and blocks the interaction between Cripto-1 and Alk4. In colon and breast cancer xenograft models, this mAb was able to inhibit tumor growth (Adkins et al., 2003). Another group has also produced a rat mAb against the EGF-like domain of Cripto-1 (Xing et al., 2004). This rat anti-Cripto-1 mAb reacted with several cancer tissues by immunohistochemistry, with little or no reactivity in normal tissues. The rat anti-Cripto-1 mAb inhibited cell proliferation of several cancer cell lines and enhanced the inhibitory effects of cytotoxic drugs in vitro. The mAb also inhibited tumor growth of colon cancer xenografts (Xing et al., 2004). Furthermore, the rat anti-Cripto-1 mAb inhibited multidrug resistance CEM/A7R cell growth both in vitro and in vivo (Hu et al., 2007).

Recently, Biogen Idec generated an anti-Cripto-1 N-terminal humanized mAb-conjugated to a DM4, which is a maytansinoid derivative, BIIB015. Maytansinoids are bacterial-derived anti-fungal and anti-tumor agents that are cytotoxic polyketides. The specific cytotoxity against Cripto-1 positive cancer cells was observed following BIIB015 treatment in vitro (Kelly et al., 2011). Also anticancer activity of BIIB015 was evaluated in nude mice carrying various human tumor xenografts (Kelly et al., 2011). The anti-Cripto-1-DM4 conjugated monoclonal BIIB015 antibody has recently been evaluated in phase I clinical trial (protocol ID: 207ST101/NCT00674947) for treatment of patients with refractory or relapsed Cripto-1 positive solid tumors without any serious adverse side effects.

Inhibition of Cripto-1 and GRP78 interactions is another potential aspect of anti-Cripto-1 mAb therapy. GRP78 is a member of the heat shock protein family, which are molecular chaperones that facilitate proper protein folding in the endoplasmic reticulum but can be expressed on the surface of tumor cells. Cripto-1 directly interacts with GRP78 at the cell surface of cancer cells (Shani et al., 2008). GRP78 and Cripto-1 cooperate to antagonize TGF-β signaling and to enhance cell growth in prostate cancer cells. Furthermore, the interaction between Cripto-1 and GRP78 also promotes stemness, plasticity, and tumorigenesis via a Src/MAPK/PI3K pathway (Kelber et al., 2009; Gray and Vale, 2012). Disruption of the Cripto-1/GRP78 complex by the anti-GRP78 N-20 goat polyclonal antibody (Santa Cruz Biotechnology Inc., Santa Cruz, CA, USA) inhibited activation of MAPK/PI3K pathways and modulation of activin A/B, Nodal and TGF-β signaling, and decreased cell proliferation and increased cell adhesion. In fact, the N-20 antibody inhibited Cripto-1-dependent Akt/MAPK phosphorylation and cellular proliferation of NCCIT EC cells (Kelber et al., 2009). In addition, the interaction between Cripto-1 and GRP78 is associated with the regulation of head and neck cancer tumor-initiating cells and normal adult HSCs (Wu et al., 2010; Miharada et al., 2011). Interestingly, the N-20 antibody also completely blocked the Cripto-1 effects on HSCs (Miharada et al., 2011). Taken together, blocking the GRP78/Cripto-1 interaction by mAbs may be a future candidate for CSC targeting therapy.

Conclusion

Cripto-1 plays an important role not only during early embryogenesis, but also in the malignant progression of tumors. In addition, recent findings have demonstrated that Cripto-1 is highly expressed in CSC subpopulations that tend to survive standard chemo- and radiation therapy and that are responsible for tumor progression and metastasis. Blockade of Cripto-1 oncogenic activities in human tumors could eliminate this undifferentiated subpopulation of CSCs. Therefore, Cripto-1 targeting therapy may provide a novel approach for cancer therapy.

Acknowledgments

This research was supported by the Intramural Research Program of the National Institutes of Health, National Cancer Institute.

References

Aasen, T., Raya, A., Barrero, M.J., Garreta, E., Consiglio, A., Gonzalez, F., Vassena, R., Bilic, J., Pekarik, V., Tiscornia, G., *et al.* (2008). Efficient and rapid generation of induced pluripotent stem cells from human keratinocytes. Nat Biotechnol *26*, 1276–1284.

Adkins, H.B., Bianco, C., Schiffer, S.G., Rayhorn, P., Zafari, M., Cheung, A.E., Orozco, O., Olson, D., De Luca, A., Chen, L.L., *et al.* (2003). Antibody blockade of the Cripto CFC domain suppresses tumor cell growth in vivo. J Clin Invest *112*, 575–587.

Al-Hajj, M., Wicha, M.S., Benito-Hernandez, A., Morrison, S.J., and Clarke, M.F. (2003). Prospective identification of tumorigenic breast cancer cells. Proc Natl Acad Sci USA *100*, 3983–3988.

Aoi, T., Yae, K., Nakagawa, M., Ichisaka, T., Okita, K., Takahashi, K., Chiba, T., and Yamanaka, S. (2008). Generation of pluripotent stem cells from adult mouse liver and stomach cells. Science *321*, 699–702.

Bendall, S.C., Stewart, M.H., and Bhatia, M. (2008). Human embryonic stem cells: lessons from stem cell niches in vivo. Regen Med *3*, 365–376.

Bianco, C., Cotten, C., Lonardo, E., Strizzi, L., Baraty, C., Mancino, M., Gonzales, M., Watanabe, K., Nagaoka, T., Berry, C., *et al.* (2009). Cripto-1 is required for hypoxia to induce cardiac differentiation of mouse embryonic stem cells. Am J Pathol *175*, 2146–2158.

Bianco, C., Rangel, M.C., Castro, N.P., Nagaoka, T., Rollman, K., Gonzales, M., and Salomon, D.S. (2010). Role of Cripto-1 in stem cell maintenance and malignant progression. Am J Pathol *177*, 532–540.

Bianco, C., Strizzi, L., Ebert, A., Chang, C., Rehman, A., Normanno, N., Guedez, L., Salloum, R., Ginsburg, E., Sun, Y., *et al.* (2005). Role of human cripto-1 in tumor angiogenesis. J Natl Cancer Inst *97*, 132–141.

Bonnet, D., and Dick, J.E. (1997). Human acute myeloid leukemia is organized as a hierarchy that originates from a primitive hematopoietic cell. Nat Med *3*, 730–737.

Casamassimi, A., De Luca, A., Agrawal, S., Stromberg, K., Salomon, D.S., and Normanno, N. (2000). EGF-related antisense oligonucleotides inhibit the proliferation of human ovarian carcinoma cells. Ann Oncol *11*, 319–325.

Chang, T.C., Liu, Y.G., Eddy, C.A., Jacoby, E.S., Binkley, P.A., Brzyski, R.G., and Schenken, R.S. (2011). Derivation and characterization of novel nonhuman primate embryonic stem cell lines from in vitro-fertilized baboon preimplantation embryos. Stem Cells Dev *20*, 1053–1062.

Chang, W., Song, B.W., Lim, S., Song, H., Shim, C.Y., Cha, M.J., Ahn, D.H., Jung, Y.G., Lee, D.H., Chung, J.H., *et al.* (2009). Mesenchymal stem cells pretreated with delivered Hph-1-Hsp70 protein are protected from hypoxia-mediated cell death and rescue heart functions from myocardial injury. Stem Cells *27*, 2283–2292.

Ciardiello, F., Dono, R., Kim, N., Persico, M.G., and Salomon, D.S. (1991). Expression of cripto, a novel gene of the epidermal growth factor gene family, leads to in vitro transformation of a normal mouse mammary epithelial cell line. Cancer Res *51*, 1051–1054.

Ciardiello, F., Tortora, G., Bianco, C., Selvam, M.P., Basolo, F., Fontanini, G., Pacifico, F., Normanno, N., Brandt, R., Persico, M.G., *et al.* (1994). Inhibition of CRIPTO expression and tumorigenicity in human colon cancer cells by antisense RNA and oligodeoxynucleotides. Oncogene *9*, 291–298.

Ciccodicola, A., Dono, R., Obici, S., Simeone, A., Zollo, M., and Persico, M.G. (1989). Molecular characterization of a gene of the 'EGF family' expressed in undifferentiated human NTERA2 teratocarcinoma cells. EMBO J *8*, 1987–1991.

Cocciadiferro, L., Miceli, V., Kang, K.S., Polito, L.M., Trosko, J.E., and Carruba, G. (2009). Profiling cancer stem cells in androgen-responsive and refractory human prostate tumor cell lines. Ann N Y Acad Sci *1155*, 257–262.

De Carvalho, D.D., You, J.S., and Jones, P.A. (2010). DNA methylation and cellular reprogramming. Trends Cell Biol *20*, 609–617.

de Castro, N.P., Rangel, M.C., Nagaoka, T., Salomon, D.S., and Bianco, C. (2010). Cripto-1: an embryonic gene that promotes tumorigenesis. Future Oncol *6*, 1127–1142.

De Luca, A., Casamassimi, A., Selvam, M.P., Losito, S., Ciardiello, F., Agrawal, S., Salomon, D.S., and Normanno, N. (1999). EGF-related peptides are involved in the proliferation and survival of MDA-MB-468 human breast carcinoma cells. Int J Cancer *80*, 589–594.

De Luca, A., Lamura, L., Strizzi, L., Roma, C., D'Antonio, A., Margaryan, N., Pirozzi, G.,

Hsu, M.Y., Botti, G., Mari, E., *et al.* (2011). Expression and functional role of CRIPTO-1 in cutaneous melanoma. Br J Cancer.

De Luca, A., Selvam, M.P., Sandomenico, C., Pepe, S., Bianco, A.R., Ciardiello, F., Salomon, D.S., and Normanno, N. (1997). Anti-sense oligonucleotides directed against EGF-related growth factors enhance anti-proliferative effect of conventional anti-tumor drugs in human colon-cancer cells. Int J Cancer *73*, 277–282.

Ebert, A.D., Wechselberger, C., Nees, M., Clair, T., Schaller, G., Martinez-Lacaci, I., Wallace-Jones, B., Bianco, C., Weitzel, H.K., and Salomon, D.S. (2000). Cripto-1-induced increase in vimentin expression is associated with enhanced migration of human Caski cervical carcinoma cells. Exp Cell Res *257*, 223–229.

Gong, Y.P., Yarrow, P.M., Carmalt, H.L., Kwun, S.Y., Kennedy, C.W., Lin, B.P., Xing, P.X., and Gillett, D.J. (2007). Overexpression of Cripto and its prognostic significance in breast cancer: a study with long-term survival. Eur J Surg Oncol *33*, 438–443.

Gray, P.C., and Vale, W. (2012). Cripto/GRP78 modulation of the TGF-beta pathway in development and oncogenesis. FEBS Lett *586*, 1836–1845.

Hamada, S., Watanabe, K., Hirota, M., Bianco, C., Strizzi, L., Mancino, M., Gonzales, M., and Salomon, D.S. (2007). beta-Catenin/TCF/LEF regulate expression of the short form human Cripto-1. Biochem Biophys Res Commun *355*, 240–244.

Hough, S.R., Laslett, A.L., Grimmond, S.B., Kolle, G., and Pera, M.F. (2009). A continuum of cell states spans pluripotency and lineage commitment in human embryonic stem cells. PLoS One *4*, e7708.

Hu, X.F., Li, J., Yang, E., Vandervalk, S., and Xing, P.X. (2007). Anti-Cripto Mab inhibit tumour growth and overcome MDR in a human leukaemia MDR cell line by inhibition of Akt and activation of JNK/SAPK and bad death pathways. Br J Cancer *96*, 918–927.

Johnson, S.E., Rothstein, J.L., and Knowles, B.B. (1994). Expression of epidermal growth factor family gene members in early mouse development. Dev Dyn *201*, 216–226.

Kelber, J.A., Panopoulos, A.D., Shani, G., Booker, E.C., Belmonte, J.C., Vale, W.W., and Gray, P.C. (2009). Blockade of Cripto binding to cell surface GRP78 inhibits oncogenic Cripto signaling via MAPK/PI3K and Smad2/3 pathways. Oncogene *28*, 2324–2336.

Kelly, R.K., Olson, D.L., Sun, Y., Wen, D., Wortham, K.A., Antognetti, G., Cheung, A.E., Orozco, O.E., Yang, L., Bailly, V., *et al.* (2011). An antibody-cytotoxic conjugate, BIIB015, is a new targeted therapy for Cripto positive tumours. Eur J Cancer *47*, 1736 1746.

Kenney, N.J., Smith, G.H., Maroulakou, I.G., Green, J.H., Muller, W.J., Callahan, R., Salomon, D.S., and Dickson, R.B. (1996). Detection of amphiregulin and Cripto-1 in mammary tumors from transgenic mice. Mol Carcinog *15*, 44–56.

Kole, R., Krainer, A.R., and Altman, S. (2012). RNA therapeutics: beyond RNA interference and antisense oligonucleotides. Nat Rev Drug Discov *11*, 125–140.

Lapidot, T., Sirard, C., Vormoor, J., Murdoch, B., Hoang, T., Caceres-Cortes, J., Minden, M., Paterson, B., Caligiuri, M.A., and Dick, J.E. (1994). A cell initiating human acute myeloid leukaemia after transplantation into SCID mice. Nature *367*, 645–648.

Lawrence, M.G., Margaryan, N.V., Loessner, D., Collins, A., Kerr, K.M., Turner, M., Seftor, E.A., Stephens, C.R., Lai, J., Postovit, L.M., *et al.* (2011). Reactivation of embryonic nodal signaling is associated with tumor progression and promotes the growth of prostate cancer cells. Prostate *71*, 1198–1209.

Lee, C.C., Jan, H.J., Lai, J.H., Ma, H.I., Hueng, D.Y., Lee, Y.C., Cheng, Y.Y., Liu, L.W., Wei, H.W., and Lee, H.M. (2010). Nodal promotes growth and invasion in human gliomas. Oncogene *29*, 3110–3123.

Li, S.S. (2012). Characterization and gene expression profiling of five human embryonic stem cell lines derived in Taiwan. Methods Mol Biol *873*, 127–149.

Li, S.S., Liu, Y.H., Tseng, C.N., Chung, T.L., Lee, T.Y., and Singh, S. (2006). Characterization and gene expression profiling of five new human embryonic stem cell lines derived in Taiwan. Stem Cells Dev *15*, 532–555.

Loh, Y.H., Wu, Q., Chew, J.L., Vega, V.B., Zhang, W., Chen, X., Bourque, G., George, J., Leong, B., Liu, J., *et al.* (2006). The Oct4 and Nanog transcription network regulates pluripotency in mouse embryonic stem cells. Nat Genet *38*, 431–440.

Lonardo, E., Hermann, P.C., Mueller, M.T., Huber, S., Balic, A., Miranda-Lorenzo, I.,

Zagorac, S., Alcala, S., Rodriguez-Arabaolaza, I., Ramirez, J.C., *et al.* (2011). Nodal/Activin signaling drives self-renewal and tumorigenicity of pancreatic cancer stem cells and provides a target for combined drug therapy. Cell Stem Cell *9*, 433–446.

Mancino, M., Strizzi, L., Wechselberger, C., Watanabe, K., Gonzales, M., Hamada, S., Normanno, N., Salomon, D.S., and Bianco, C. (2008). Regulation of human Cripto-1 gene expression by TGF-beta1 and BMP-4 in embryonal and colon cancer cells. J Cell Physiol *215*, 192–203.

Masaki, H., Ishikawa, T., Takahashi, S., Okumura, M., Sakai, N., Haga, M., Kominami, K., Migita, H., McDonald, F., Shimada, F., *et al.* (2007). Heterogeneity of pluripotent marker gene expression in colonies generated in human iPS cell induction culture. Stem Cell Res *1*, 105–115.

Miharada, K., Karlsson, G., Rehn, M., Rorby, E., Siva, K., Cammenga, J., and Karlsson, S. (2011). Cripto regulates hematopoietic stem cells as a hypoxic-niche-related factor through cell surface receptor GRP78. Cell Stem Cell *9*, 330–344.

Miharada, K., Karlsson, G., Rehn, M., Rorby, E., Siva, K., Cammenga, J., and Karlsson, S. (2012). Hematopoietic stem cells are regulated by Cripto, as an intermediary of HIF-1alpha in the hypoxic bone marrow niche. Ann N Y Acad Sci *1266*, 55–62.

Mincione, G., Piccirelli, A., Lazzereschi, D., Salomon, D.S., and Colletta, G. (1998). Heregulin-dependent autocrine loop regulates growth of K-ras but not erbB-2 transformed rat thyroid epithelial cells. J Cell Physiol *176*, 383–391.

Miyoshi, N., Ishii, H., Mimori, K., Sekimoto, M., Doki, Y., and Mori, M. (2010). TDGF1 is a novel predictive marker for metachronous metastasis of colorectal cancer. Int J Oncol *36*, 563–568.

Morrison, R., Schleicher, S.M., Sun, Y., Niermann, K.J., Kim, S., Spratt, D.E., Chung, C.H., and Lu, B. (2011). Targeting the mechanisms of resistance to chemotherapy and radiotherapy with the cancer stem cell hypothesis. J Oncol *2011*, 941876.

Nagaoka, T., Karasawa, H., Castro, N.P., Rangel, M.C., Salomon, D.S., and Bianco, C. (2012). An evolving web of signaling networks regulated by Cripto-1. Growth Factors *30*, 13–21.

Niemeyer, C.C., Persico, M.G., and Adamson, E.D. (1998). Cripto: roles in mammary cell growth, survival, differentiation and transformation. Cell Death Differ *5*, 440–449.

Nishino, K., Toyoda, M., Yamazaki-Inoue, M., Makino, H., Fukawatase, Y., Chikazawa, E., Takahashi, Y., Miyagawa, Y., Okita, H., Kiyokawa, N., *et al.* (2010). Defining hypomethylated regions of stem cell-specific promoters in human iPS cells derived from extra-embryonic amnions and lung fibroblasts. PLoS One *5*, e13017.

Normanno, N., De Luca, A., Maiello, M.R., Bianco, C., Mancino, M., Strizzi, L., Arra, C., Ciardiello, F., Agrawal, S., and Salomon, D.S. (2004). CRIPTO-1: a novel target for therapeutic intervention in human carcinoma. Int J Oncol *25*, 1013–1020.

Normanno, N., Tortora, G., De Luca, A., Pomatico, G., Casamassimi, A., Agrawal, S., Mendelsohn, J., Bianco, A.R., and Ciardiello, F. (1999). Synergistic growth inhibition and induction of apoptosis by a novel mixed backbone antisense oligonucleotide targeting CRIPTO in combination with C225 anti-EGFR monoclonal antibody and 8-Cl-cAMP in human GEO colon cancer cells. Oncol Rep *6*, 1105–1109.

Rangel, M.C., Karasawa, H., Castro, N.P., Nagaoka, T., Salomon, D.S., and Bianco, C. (2012). Role of Cripto-1 during epithelial-to-mesenchymal transition in development and cancer. Am J Pathol *180*, 2188–2200.

Shani, G., Fischer, W.H., Justice, N.J., Kelber, J.A., Vale, W., and Gray, P.C. (2008). GRP78 and Cripto form a complex at the cell surface and collaborate to inhibit transforming growth factor beta signaling and enhance cell growth. Mol Cell Biol *28*, 666–677.

Singh, S.K., Hawkins, C., Clarke, I.D., Squire, J.A., Bayani, J., Hide, T., Henkelman, R.M., Cusimano, M.D., and Dirks, P.B. (2004). Identification of human brain tumour initiating cells. Nature *432*, 396–401.

Strizzi, L., Abbott, D.E., Salomon, D.S., and Hendrix, M.J. (2008). Potential for cripto-1 in defining stem cell-like characteristics in human malignant melanoma. Cell Cycle *7*, 1931–1935.

Strizzi, L., Bianco, C., Hirota, M., Watanabe, K., Mancino, M., Hamada, S., Raafat, A., Lawson, S., Ebert, A.D., D'Antonio, A., *et al.* (2007). Development of leiomyosarcoma of the uterus in MMTV-CR-1 transgenic mice. J Pathol *211*, 36–44.

Strizzi, L., Bianco, C., Normanno, N., Seno, M., Wechselberger, C., Wallace-Jones, B., Khan, N.I., Hirota, M., Sun, Y., Sanicola, M., *et al.* (2004). Epithelial mesenchymal transition is a characteristic of hyperplasias and tumors in mammary gland from MMTV-Cripto-1 transgenic mice. J Cell Physiol *201*, 266–276.

Su, Z.Z., Austin, V.N., Zimmer, S.G., and Fisher, P.B. (1993). Defining the critical gene expression changes associated with expression and suppression of the tumorigenic and metastatic phenotype in Ha-ras-transformed cloned rat embryo fibroblast cells. Oncogene *8*, 1211–1219.

Sun, Y., Strizzi, L., Raafat, A., Hirota, M., Bianco, C., Feigenbaum, L., Kenney, N., Wechselberger, C., Callahan, R., and Salomon, D.S. (2005). Overexpression of human Cripto-1 in transgenic mice delays mammary gland development and differentiation and induces mammary tumorigenesis. Am J Pathol *167*, 585–597.

Takahashi, K., Tanabe, K., Ohnuki, M., Narita, M., Ichisaka, T., Tomoda, K., and Yamanaka, S. (2007). Induction of pluripotent stem cells from adult human fibroblasts by defined factors. Cell *131*, 861–872.

Takahashi, K., and Yamanaka, S. (2006). Induction of pluripotent stem cells from mouse embryonic and adult fibroblast cultures by defined factors. Cell *126*, 663–676.

Topczewska, J.M., Postovit, L.M., Margaryan, N.V., Sam, A., Hess, A.R., Wheaton, W.W., Nickoloff, B.J., Topczewski, J., and Hendrix, M.J. (2006). Embryonic and tumorigenic pathways converge via Nodal signaling: role in melanoma aggressiveness. Nat Med *12*, 925–932.

Vallier, L., Mendjan, S., Brown, S., Chng, Z., Teo, A., Smithers, L.E., Trotter, M.W., Cho, C.H., Martinez, A., Rugg-Gunn, P., *et al.* (2009). Activin/Nodal signalling maintains pluripotency by controlling Nanog expression. Development *136*, 1339–1349.

Visvader, J.E., and Lindeman, G.J. (2008). Cancer stem cells in solid tumours: accumulating evidence and unresolved questions. Nat Rev Cancer *8*, 755–768.

Watanabe, K., Meyer, M.J., Strizzi, L., Lee, J.M., Gonzales, M., Bianco, C., Nagaoka, T., Farid, S.S., Margaryan, N., Hendrix, M.J., *et al.* (2010). Cripto-1 is a cell surface marker for a tumorigenic, undifferentiated subpopulation in human embryonal carcinoma cells. Stem Cells *28*, 1303–1314.

Wechselberger, C., Ebert, A.D., Bianco, C., Khan, N.I., Sun, Y., Wallace-Jones, B., Montesano, R., and Salomon, D.S. (2001). Cripto-1 enhances migration and branching morphogenesis of mouse mammary epithelial cells. Exp Cell Res *266*, 95–105.

Wechselberger, C., Strizzi, L., Kenney, N., Hirota, M., Sun, Y., Ebert, A., Orozco, O., Bianco, C., Khan, N.I., Wallace-Jones, B., *et al.* (2005). Human Cripto-1 overexpression in the mouse mammary gland results in the development of hyperplasia and adenocarcinoma. Oncogene *24*, 4094–4105.

Wei, C.L., Miura, T., Robson, P., Lim, S.K., Xu, X.Q., Lee, M.Y., Gupta, S., Stanton, L., Luo, Y., Schmitt, J., *et al.* (2005). Transcriptome profiling of human and murine ESCs identifies divergent paths required to maintain the stem cell state. Stem Cells *23*, 166–185.

Wu, M.J., Jan, C.I., Tsay, Y.G., Yu, Y.H., Huang, C.Y., Lin, S.C., Liu, C.J., Chen, Y.S., Lo, J.F., and Yu, C.C. (2010). Elimination of head and neck cancer initiating cells through targeting glucose regulated protein78 signaling. Mol Cancer *9*, 283.

Wu, Z., Li, G., Wu, L., Weng, D., Li, X., and Yao, K. (2009). Cripto-1 overexpression is involved in the tumorigenesis of nasopharyngeal carcinoma. BMC Cancer *9*, 315.

Xing, P.X., Hu, X.F., Pietersz, G.A., Hosick, H.L., and McKenzie, I.F. (2004). Cripto: a novel target for antibody-based cancer immunotherapy. Cancer Res *64*, 4018–4023.

Yu, J., Vodyanik, M.A., Smuga-Otto, K., Antosiewicz-Bourget, J., Frane, J.L., Tian, S., Nie, J., Jonsdottir, G.A., Ruotti, V., Stewart, R., *et al.* (2007). Induced pluripotent stem cell lines derived from human somatic cells. Science *318*, 1917–1920.

Zamecnik, P.C., and Stephenson, M.L. (1978). Inhibition of Rous sarcoma virus replication and cell transformation by a specific oligodeoxynucleotide. Proc Natl Acad Sci USA *75*, 280–284.

Zhang, J.G., Zhao, J., and Xin, Y. (2010). Significance and relationship between Cripto-1 and p-STAT3 expression in gastric cancer and precancerous lesions. World J Gastroenterol *16*, 571–577.

Zhong, X.Y., Zhang, L.H., Jia, S.Q., Shi, T., Niu, Z.J., Du, H., Zhang, G.G., Hu, Y., Lu, A.P., Li, J.Y., *et al.* (2008). Positive association of up-regulated Cripto-1 and down-regulated E-cadherin with tumour progression and poor prognosis in gastric cancer. Histopathology *52*, 560–568.

Chapter 26

Leptin Signaling in the Regulation of Stem and Cancer Stem Cells

Shanchun Guo[1], Keshav K. Singh[2], James W. Lillard[1], and Lily Yang[3]
[1] *Microbiology, Biochemistry and Immunology, Morehouse School of Medicine, Atlanta, GA, USA*
[2] *Departments of Genetics, Pathology, Environmental Health, Center for Free Radical Biology, Center for Aging and University of Alabama at Birmingham Comprehensive Cancer Center, University of Alabama at Birmingham, Birmingham, Alabama, AL, USA*
[3] *Department of Surgery, Emory University School of Medicine, Atlanta, GA, USA*

Introduction

The identification of spontaneous mutations in the leptin (ob or LEP) and Ob-R (db or LEPR) genes in mice opened up a new field in obesity research. High leptin levels in obese and overweight individuals or populations are clearly correlated with body fat and adipocyte size. Calle and colleagues (2003) showed that significant trends of increasing risk with higher body-mass-index values were observed for death from breast cancer in obese/overweight women. Other studies have also found that women with higher percentage of body adiposity/leptin levels have higher incidence of breast cancer (Basen-Engquist and Chang, 2011; Han et al., 2005; Wu et al., 2009).

In this chapter we will focus on the role of leptin in the regulation of stem cells and cancer stem cells (CSC), highlighting the leptin-mediated signaling pathways and its potential as a drug target. Additionally, we will review and discuss recently identified molecular mechanisms of leptin in CSC, tumor angiogenesis, as well as its cross talk with other oncogenic signaling pathways.

Structure and Function of Leptin and Leptin Receptors

Leptin is a small 167-amino acid non-glycosilated protein with a molecular weight of 16 kDa, coded by the LEP gene, whose name is derived from the Greek word "leptos," which means thin. The LEP gene is preserved in mammals providing a high sequence identity for leptin. The biological role of leptin in energy homeostasis was demonstrated by normalization of hyperphagy and obese phenotypes with recombinant leptin administration in rodents and humans (Farooqi et al., 1999; Halaas et al., 1995). Leptin also plays critical roles in the regulation of glucose homeostasis, reproduction, growth, and the immune response (Leinninger, 2009; Myers et al., 2009; Robertson et al., 2008). The binding of Leptin to Ob-R is able to induce the extracellular domains of Ob-R to form a homodimers, which constitutes the functional unit responsible for leptin-mediated signals. The N-terminal region (94 amino acids) of leptin is essential for both the biological and the receptor-binding activities (Imagawa et al., 1998).

Cancer Stem Cells, First Edition. Edited by Vinagolu K. Rajasekhar.

The leptin receptor (Ob-R) belongs to a member of the class I cytokine receptor superfamily (Fruhbeck, 2006). This superfamily of receptors lacks autophosphorylation capabilities and needs auxiliary kinases for activation. Ob-R binding to leptin induces conformational changes that recruit JAKs (Janus kinases), which in turn phosphorylates Ob-R and activates STATs (signal transducers and activators of transcription). Six leptin receptor isoforms have been discovered so far, and are generated by mRNA alternative splicing (Friedman, 1997): the long isoform (OB-RL or OB-Rb) with full intracellular signaling capabilities and shorter isoforms with less biological activity (OB-RS) (Fruhbeck, 2006). The large extracellular domain of Ob-R (816 amino acids) is common to all Ob-R forms, and the variable length cytoplasmatic tail (300 amino acid residues) distinguishes the several isoforms (Bazan, 1990; Tartaglia et al., 1995). Leptin binding to Ob-R induces canonical (JAK2/STATs; MAPK/ERK 1/2, PI-3 K/AKT) and non-canonical signaling pathways (PKC, JNK, p38 MAPK and AMPK) in diverse cell types. Among the leptin receptor isoforms, only Ob-Rb that contains an intact intracellular domain, has the ability to activate the intracellular JAK-STAT pathway on ligand binding (Tartaglia, 1997). Tyr 1138 of Ob-Rb is required for leptin-mediated STAT3 signaling (Banks et al., 2000; Bjorbaek et al., 1997; Stahl et al., 1995). However, Ob-Ra, which contains a truncated intracellular domain, and is unable to activate the STAT pathway (Banks et al., 2000) may still transduce signals by way of activation of JAK2, IRS-1 or extracellular factor-regulated kinases (ERKS) including MAPKs (mitogen-activated protein kinases) (Bjorbaek et al., 1997). Although, leptin-induced signals occur in normal peripheral tissues, high levels of leptin in obesity could lead to increased leptin signaling, which is a hallmark for the development of obesity-associated cancers.

Leptin Signaling in the Regulation of Stem Cells

Stem cells (SC) are defined by their ability to undergo self-renewal, as well as multilineage differentiation. Human placenta was earlier identified as an important source of leptin for fetal and neonatal growth (Masuzaki et al., 1997). Further, the role of mammary glands in providing leptin to the newborn was reported (Smith-Kirwin et al., 1998). It has been known that leptin plays a critical role in regulating bone mass through osteogenic differentiation of mesenchymal stem cells (Han et al., 2010; Scheller et al., 2010). Leptin was able to promote 3T3-L1 mouse cell line differentiating into adipocytes, even at a low level (Nakamura et al., 2011; Rentsch and Chiesi, 1996). Ob-R is also detected on endothelial cells (Bouloumie et al., 1998; Sierra-Honigmann et al., 1998), hematopoietic cells (Gainsford et al., 1996), and peripheral blood-derived early and late outgrowth endothelial progenitor cells (Schroeter et al., 2008; Wolk et al., 2005). Thus, leptin can induce proliferation, differentiation, and functional activation of hemopoietic cells (Gainsford et al., 1996), as well as proliferation of endothelial progenitor cells (Amjadi et al., 2011).

Leptin produced by human mammary epithelial cells was revealed by RT/PCR analysis and immunohistochemical staining of breast tissue, cultured mammary epithelial cells, and secretory epithelial cells present in human milk (Smith-Kirwin et al., 1998). Analyses of micrographs of whole-mount preparations of mammary tissues for $Lep^{ob}Lep^{ob}$ and $Lepr^{db}Lepr^{db}$ leptin-deficient and leptin receptor-deficient mice, respectively, demonstrated a notable lack of mammary duct formation in both strains of genetically obese mice, compared with their respective lean counterparts, whose ductal branching was prominent (Hu et al., 2002). These results indicate that both leptin and Ob-R are necessary for normal mammary gland development.

Leptin Signaling in Cancer Stem Cells

Cancer stem cells (CSC) initiate and drive carcinogenesis and differentiation contributing to tumor cellular heterogeneity through deregulation of the self-renewal process (Pang and Argyle, 2009). Thus, the population of CSC may be a risk factor for carcinogenesis (Kakarala and Wicha, 2008). Many human cancers possess an enhanced tumor-initiating capacity, can undergo self-renewal, and partially re-create the cellular heterogeneity of the parental tumor. Those cancer cells seem to be generally more resistant than other cancer cells to conventional anticancer therapeutics. Previous studies have identified CSC in solid tumors that are responsible for tumor initiation, growth, and metastasis (Clarke, 2005; Clevers, 2011). Several factors are linked to self-renewal of BCSC, including Wnt, FGF (fibroblast growth factor), Notch, Hedgehog, and TGF-β/bone morphogenetic proteins (BMP). TGF-β/BMP signaling cascades constitute the stem-cell signaling network, which plays a key role in the maintenance or homeostasis of pluripotent stem cells and CSC. Recently, several reports have revealed that leptin regulates or activates certain oncogenes HER2 (Korkaya et al., 2008; Korkaya and Wicha, 2009) and Akt (Korkaya et al., 2009) as well as transcriptional factors, such as STAT3 (Zhou et al., 2007), NF-κB (Pratt et al., 2009) and Notch (Guo and Gonzalez-Perez, 2011; Guo et al., 2011), which are cellular factors critically associated with breast cancer stem cells (BCSC). In addition, leptin also activates the Notch signaling pathway (Guo and Gonzalez-Perez, 2011; Knight et al., 2011) that is important for the maintenance of SC or progenitors through the inhibition of differentiation (Artavanis-Tsakonas et al., 1999; Leong and Karsan, 2006; Radtke and Raj, 2003). Leptin deficiency in murine mammary tumor virus (MMTV)-Wnt-1 transgenic mice results in functional depletion of BCSC, leading to reduced tumor growth (Zheng et al., 2011). Feldman and colleagues (2011) also demonstrated that robust and selective expression of Ob-R is a characteristic feature of CSC and of a broad array of embryonic and induced pluripotent stem cells. CSC exhibit sensitized responses to leptin, including the phosphorylation and activation of STAT3 and induction of Oct4 and Sox2, thereby establishing a self-reinforcing signaling module (Feldman et al., 2011).

Cross Talk Between Leptin Signaling and Oncogenic Pathways in Cancer Stem Cells

Leptin is a pleitropic adipocytokine that exerts mitogenic, pro-inflammatory, and pro-angiogenic effects that promote anchorage, proliferation of cancer cells, growth of microvessels, and hematopoiesis and increases the levels of cell cycle regulators (Gonzalez et al., 2006; Pischon et al., 2008; Rene Gonzalez et al., 2009). Many of the effects of leptin are attributable to effects in the central nervous system, particularly in the hypothalaus, a site of high Ob-Rb mRNA expression. The failure of high levels of leptin to suppress feeding and decrease body weight/adiposity to prevent or mitigate obesity suggests a relative resistance to the catabolic effects of leptin action in obesity. Proposed mechanisms of leptin resistance include alterations in the transport of leptin across the blood-brain-barrier, dysregulation of Ob-Rb signaling, perturbations in developmental programming, and others (Myers et al., 2008). However, high levels of circulating leptin could regulate signaling and expression of active Ob-R in cancer and CSC through its crosstalk with multiple oncogenic pathways (Figure 26.1).

Developmental Signaling Pathways

Wnt/β-Catenin Signaling Pathway

Wnt/β-catenin signaling pathway plays a pivotal role in the regulation of cell growth, cell development, and the differentiation of

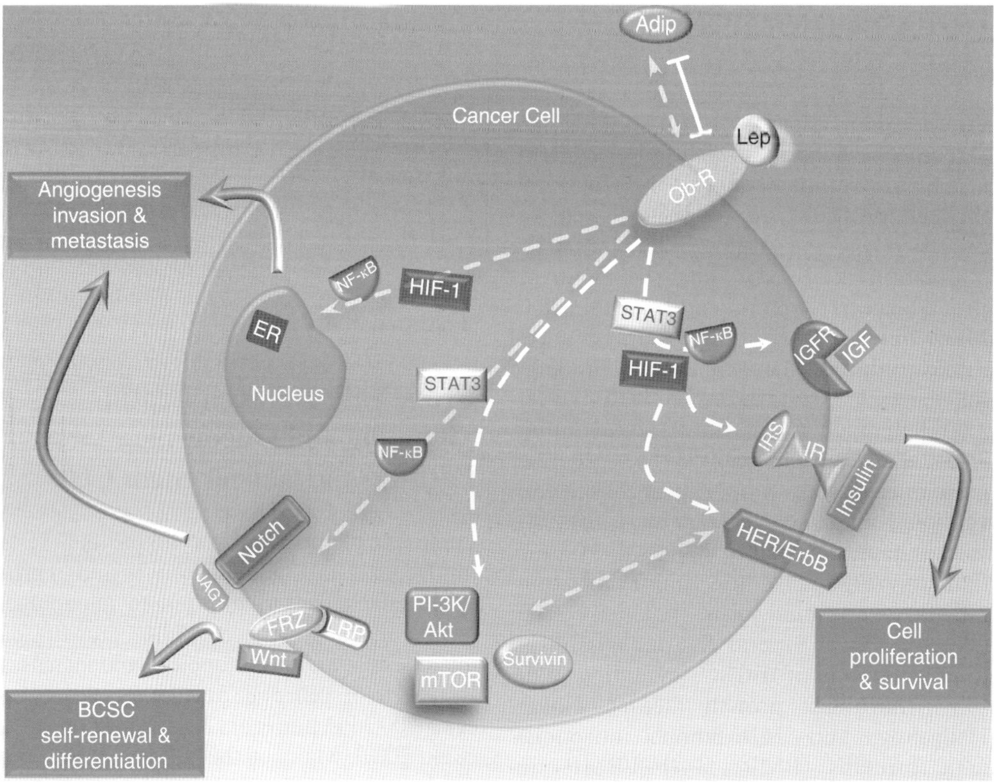

Figure 26.1. Potential cross talk of leptin signaling with other pathways in cancer stem cells (CSC). Lep: leptin; Ob-R: leptin receptor; HER/ErbB, encode for RTK-transmembrane proteins; NF-κB: a transcription factor family, nuclear factor kappa-light-chain-enhancer of activated B cells; STAT3: signal transducer and activator of transcription 3; PI-3K: phosphatidylinositol 3-kinase; mTOR: mammalian target of rapamycin; HIF-1α: hypoxia-inducible transcription factor 1; Notch: a family of transmembrane proteins that function as receptors for membrane bound ligands; Wnt: a family of proteins linked to developmental signals with specific roles in embryogenesis and cancer; FRZ (Frizzled): G protein-coupled receptor proteins of Wnt signaling pathway; LRP: a transmembrane protein that binds Frizzled; dashed arrows indicate the main mechanisms/factors involved in leptin cross talk actions. Tubular arrows indicate how the leptin cross talk to specific oncogenic proteins/kinases/transcription factors lead to the development of key processes involved in cancer growth and metastasis. For color detail, please see color plate section.

normal stem cells (Boras-Granic and Wysolmerski, 2008; Prosperi and Goss, 2010). Wnt/β-catenin signaling is also important for the proliferation of CSC (Curtin and Lorenzi, 2010; de Sousa et al., 2011; Yao et al., 2011). Genes regulating the Wnt/β-catenin-mediated pathway including Mdm2, Pik3r1, and Rb1 were upregulated by leptin inducing the survival and proliferation of colon epithelial cells possessing an Apc mutation but not normal cells (Fenton et al., 2008). These leptin effects were linked to IGF-I metabolism as in colon epithelial APC-mutated cells IGFBP-6 and IGF-1 expression was upregulated, while IGFBP-2, IGFBP-3, IGFBP-4, IGFBP-5, and nephroblastoma overexpressed gene-protein (NOV) expression was downregulated by leptin treatment (Fenton et al., 2008). Interestingly, *HES1* and *HES5*, which are the downstream transducers of Notch signaling, could be activated by NOV, thus reducing cell differentiation (Gonzalez et al., 2001). More recent data suggest that leptin could also regulate the Wnt pathway contributing to breast cancer development. Indeed, leptin deficiency suppressed MMTV-Wnt-1 mammary tumor growth in obese mice and abrogated tumor-initiating cell survival (Zheng et al., 2011).

Notch Signaling Pathway

Notch is a family of mammalian transmembrane proteins that function as receptors for membrane-bound ligands. There are four mammalian Notch genes, Notch1–Notch4, and five ligands, Jagged1, Jagged2 (homologs of Drosophila Serrate-like proteins), Delta-like 1 (DLL1), DLL3, and DLL4. The activation of a Notch receptor is triggered by ligands expressed on adjacent Jagged and Delta cells. The correlation of Notch signaling, stem cell, and CSC was well known (Guo et al., 2011; Liu et al., 2010; Wang et al., 2012). Leptin can regulate Jagged1 and Notch4 in human cord blood $CD34^+$ cells and early differentiated human endothelial cells (HUVEC) possibly promoting cell differentiation (Polus, 2003). Leptin is also able to upregulate Notch (ligands, receptors, and targets) in HUVEC cells (Lanier V et al., 2012). Moreover, leptin activated Notch signaling pathway (Notch1,3,4 NICD and CSL) in mouse (4T1, EMT6, and MMT) and human breast cancer cell lines (MCF-7 and MDA-MB-231) under normoxic conditions (Guo and Gonzalez-Perez, 2011; Guo et al., 2011). Leptin has been shown to increase the expression of both Notch receptors and ligands (Guo and Gonzalez-Perez, 2011; Guo et al., 2011). In these cells, leptin further upregulated Notch-target genes, Hey2, and Survivin (Guo et al., 2010a; Guo et al., 2010b).

Growth Factors

HER/ErbB

HER/ErbB (HER1/EGFR, HER3, and HER4) receptor tyrosine kinases bind several ligands. HER2/*neu* is an orphan receptor without a known ligand. The HER2/*neu* gene has been implicated in cancer with special emphasis in breast cancer (Tai et al., 2010; Tzahar and Yarden, 1998). HER2 can induce breast cancer by increasing the frequency of CSC (Wang et al., 2010). The clinical importance of HER2 in the renewal of CSC is associated with aromatase inhibitor-resistant breast cancer (Gilani et al., 2012).

HER2/*neu* could be transactivated by leptin/Ob-R signaling (Gonzalez and Leavis, 2001). Leptin-induced proliferation of breast cancer cell lines was HER2/*neu* dependant (Ray et al., 2007). Leptin also induced rapid-tyrosine phosphorylation of HER2 in SK-BR-3 cells, which can be partially by an EGFR inhibitor, AG1478, or a JAK inhibitor, AG490(Soma et al., 2008). In these cells, leptin also induced pERK 1/2, and this effect was mostly abrogated by a HER2 tyrosine kinase inhibitor, AG825. Therefore, leptin transactivates HER2 through activation of both EGFR and JAK 2. In line with these findings, Ob-R and HER2 were also observed to physically interact and be coexpressed in several breast cancer cell lines, further suggesting leptin may contribute to an enhanced HER2 activity (Fiorio et al., 2008). Therefore, above emerging evidence suggests the existence of cross talk between HER2/*neu* and Ob-R signaling pathways in breast cancer.

Oncogenic Kinases and Transcription Factors

PI-3K/Akt Signaling Pathway

PI-3 K/Akt pathway plays a central role in a variety of cellular processes including cell growth, proliferation, motility, survival, and angiogenesis in tumor cells (Steelman et al., 2008; Wickenden and Watson, 2010). The PI-3 K/Akt pathway is also instrumental in EMT during carcinogenesis (Sabbah et al., 2008). Strong evidence indicates that PI-3 K/Akt pathway has a key role in CSC biology (Hambardzumyan et al., 2008; Lacerda et al., 2010; Martelli et al., 2011). CSCs are more sensitive to pathway inhibition with small molecules when compared to healthy stem cells (Gallia et al., 2009). This observation provides the proof-of-principle that functional differences in signaling transduction pathways between CSC and healthy stem cells can be identified. A number of studies have already established the central role of leptin-induced regulation of the PI-3 K/Akt signaling pathway in several type of cancer (Cheng et al., 2011; Gao et al., 2009; Miyazaki et al., 2008; Riolfi et al., 2010; Saxena et al., 2007a; Shen et al., 2009; Uddin et al., 2010a; Uddin et al., 2009a; Uddin et al., 2009b; Uddin et al., 2010b).

mTOR Signaling

The mammalian Target of Rapamycin (mTOR), a key protein kinase, controls signal transduction from various growth factors and upstream proteins through the level of mRNA translation and ribosome biogenesis (Menon and Manning, 2008; Russell et al., 2011). Studies of conditional knockouts of mTOR pathway components in mice have demonstrated a role for mTOR in stem cell homeostasis and life span determination (Russell et al., 2011). mTOR also has a role in CSC from a number of cancer types (Ibanez et al., 2012; Singh et al., 2012; Zhu et al., 2012).

mTOR signaling has been reported to cross talk with the leptin signaling pathway in several types of malignant cells (Attoub et al., 2000; Carino et al., 2008; Guo and Gonzalez-Perez, 2011; Zhou et al., 2011). IGF-I induced phosporylation of Ob-R leucine could activate leptin expression via mTOR in adipocytes (Heida et al., 2010). mTOR is a ser/threo kinase that is often a downstream effector of PI-3 K/Akt signaling pathway in breast and other types of cancer cells. Leptin-induced IL-1/IL-1Ra and VEGF in endometrial cancer cells involved PI-3 K/Akt phosphorylation. Activated PI-3 K/Akt was in turn strongly linked to mTOR-mediated activation of its downstream targets, S6K1 and 4BEP-1 (Zhou et al., 2011). However, MAPK pathway was identified as the preferential upstream regulator of mTOR in the induction of inflammatory/pro-angiogenic molecules in endometrial cancer cells (Carino et al., 2008). mTOR can also phosphorylate Akt (Sparks and Guertin, 2010).

NF-κB Signaling Pathway

The family of NF-κB transcription factors is involved in the expression of key genes for innate and adaptive immunity, cell proliferation and survival, and lymphoid organ development. NF-κB family, RelA (p65), RelB, c-Rel, p105/p50, and p100/p52 are evolutionarily conserved molecules that form hetero- or homodimers. The p65/p50 heterodimer, the most abundant form of NF-κB, is regulated by the so-called canonical pathway (Hayden and Ghosh, 2004; Hayden and Ghosh, 2008). Recent studies indicate that NF-κB regulates the self-renewal of breast CSCs in a model of Her2-dependent tumourigenesis (Shostak and Chariot, 2011).

NF-κB has been found to be activated by leptin in many tumors, such as glioma (Mattace Raso et al., 2006; Yeh et al., 2009), chondrosarcoma (Yang et al., 2009), colon cancer (Ogunwobi and Beales, 2007b; Rouet-Benzineb et al., 2004), prostate cancer (Huang et al., 2011), as well as breast cancer (Gonzalez-Perez et al., 2010; Zhou et al., 2011). Leptin enhances migration, invasion, and proliferation of malignant cells, possibly through activation of NF-κB (Ogunwobi and Beales, 2007b; Yang et al., 2009; Yeh et al., 2009). Leptin also regulates pro-angiogenic, pro-inflammatory, and anti-apoptotic actions in breast cancer cells through activation of NF-κB (Gonzalez-Perez et al., 2010; Gonzalez et al., 2006; Rene Gonzalez et al., 2009; Zhou et al., 2011).

STAT3 Signaling Pathway

Signal transducer and activator of transcription (STAT3) has been placed at a central node in the development, progression, and maintenance of many human tumors (Fletcher et al., 2009; Masciocchi et al., 2011). The active form of STAT3 dimerizes and then quickly translocates to the nucleus. For many cancers, elevated levels of activated STAT3 have been associated with poor prognosis, suggesting that STAT3 could be an attractive molecular target for the development of novel cancer therapeutics (Mankan and Greten, 2011; Zhao et al., 2011). Activation of STAT3 was also necessary for proliferation and survival of CSC from several cancer types, such as bladder cancer (Ho et al., 2012), breast cancer (Marotta et al., 2011), colon cancer (Lin et al., 2011a; Lin et al., 2011b), glioblastoma (Guryanova et al., 2011; Li et al., 2010), liver cancer (Lee et al., 2011), as well as thyroid cancer (Tseng et al., 2012).

A number of studies have also demonstrated that the STAT3 pathway is involved in leptin actions in cell migration (Ratke et al., 2010; Saxena et al., 2007a), proliferation (Catalano et al., 2009; Dieudonne et al., 2002; Li et al., 2008; Pai et al., 2005; Saxena et al., 2007b; Yin et al., 2004), and anti-apoptosis (Ogunwobi and Beales, 2007a; Russo et al., 2004; Shen et al., 2009) of malignant cells. STAT3 as a transcriptional factor regulates genes including cyclin D1(Catalano et al., 2009; Saxena et al., 2007b), cyclooxygenase (COX)-2 (Gao et al., 2009), VEGF (Cascio et al., 2009; Gonzalez-Perez et al., 2010), human telomerase reverse transcriptase (hTERT) (Ren et al., 2010), Survivin (Jiang et al., 2008) as well as leptin (Cascio et al., 2009).

HIF Signaling Pathway

Hypoxia-inducible factors (HIF), key molecules overexpressed in response to oxygen deficiency, are master regulators of genes involved in tissue reoxygenation (Ke and Costa, 2006). Several studies have demonstrated that HIF-1 plays important roles in the development and progression of cancer or CSC through activation of various genes involved in angiogenesis, energy metabolism, vasomotor function, erythropoiesis, and cell survival (Li and Ye, 2010; Li and Rich, 2010).

Correlation between leptin and HIFs is bi-directional. HIFs can upregulate leptin signaling in several cell types. In trophoblast-derived BeWo cells, exogenous HIF-1α markedly increased leptin-promoter luciferase-reporter activity (Grosfeld et al., 2002). Moreover, HIF-1α induced aberrant expression of leptin in ectopic endometriotic stromal cells, which may have a role in the etiology of endometriosis (Wu et al., 2007). In endometrial cancer, the expression of HIF-1α was positively correlated with leptin (Koda et al., 2007b). Furthermore, in human colorectal cancer HIF-1α and leptin signaling were also positively correlated (Koda et al., 2007a). On the other hand, leptin could also upregulate HIF-1. Leptin activates HIF-1α DNA binding activity to VEGF promoter (Gonzalez-Perez et al., 2010). These results suggest that leptin-induced VEGF transcription involves the enhanced ability of HIF-1α to bind the HRE in the VEGF promoter in cancer cells.

Conclusion

Leptin and Ob-R have a critical role in stem cell proliferation and differentiation. Leptin signaling is also important for self-renewal and differentiation of CSC through cross talk with multiple oncogenic pathways. Better understanding of regulation of gene expression, functions, and intracellular signaling pathways of targets, as well as the complex crosstalk of leptin/Ob-R with other oncogenic signals in CSCs will be essential to ensure rational use of treatment and development of new combinatory therapeutic possibilities. Novel opportunities could emerge from the discovery of leptin cross talk with inflammatory, angiogenic cytokines and other oncogenic factors and their links to obesity-related cancers and CSC.

References

Amjadi, F., Javanmard, S.H., Zarkesh-Esfahani, H., Khazaei, M., and Narimani, M. (2011). Leptin promotes melanoma tumor growth in mice related to increasing circulating endothelial progenitor cells numbers and plasma NO production. J Exp Clin Cancer Res *30*, 21.

Artavanis-Tsakonas, S., Rand, M.D., and Lake, R.J. (1999). Notch signaling: cell fate control and signal integration in development. Science *284*, 770–776.

Attoub, S., Noe, V., Pirola, L., Bruyneel, E., Chastre, E., Mareel, M., Wymann, M.P., and Gespach, C. (2000). Leptin promotes invasiveness of kidney and colonic epithelial cells via phosphoinositide 3-kinase-, rho-, and rac-dependent signaling pathways. FASEB J *14*, 2329–2338.

Banks, A.S., Davis, S.M., Bates, S.H., and Myers, M.G., Jr. (2000). Activation of downstream signals by the long form of the leptin receptor. J Biol Chem *275*, 14563–14572.

Basen-Engquist, K., and Chang, M. (2011). Obesity and cancer risk: recent review and evidence. Curr Oncol Rep *13*, 71–76.

Bazan, J.F. (1990). Structural design and molecular evolution of a cytokine receptor superfamily. Proc Natl Acad Sci USA *87*, 6934–6938.

Bjorbaek, C., Uotani, S., da Silva, B., and Flier, J.S. (1997). Divergent signaling capacities of the long and short isoforms of the leptin receptor. J Biol Chem *272*, 32686–32695.

Boras-Granic, K., and Wysolmerski, J.J. (2008). Wnt signaling in breast organogenesis. Organogenesis *4*, 116–122.

Bouloumie, A., Drexler, H.C., Lafontan, M., and Busse, R. (1998). Leptin, the product of Ob gene, promotes angiogenesis. Circ Res *83*, 1059–1066.

Calle, E.E., Rodriguez, C., Walker-Thurmond, K., and Thun, M.J. (2003). Overweight, obesity, and mortality from cancer in a prospectively studied cohort of U.S. adults. N Engl J Med *348*, 1625–1638.

Carino, C., Olawaiye, A.B., Cherfils, S., Serikawa, T., Lynch, M.P., Rueda, B.R., and Gonzalez, R.R. (2008). Leptin regulation of proangiogenic molecules in benign and cancerous endometrial cells. Int J Cancer *123*, 2782–2790.

Cascio, S., Ferla, R., D'Andrea, A., Gerbino, A., Bazan, V., Surmacz, E., and Russo, A. (2009). Expression of angiogenic regulators, VEGF and leptin, is regulated by the EGF/PI3K/STAT3 pathway in colorectal cancer cells. J Cell Physiol *221*, 189–194.

Catalano, S., Giordano, C., Rizza, P., Gu, G., Barone, I., Bonofiglio, D., Giordano, F., Malivindi, R., Gaccione, D., Lanzino, M., *et al.* (2009). Evidence that leptin through STAT and CREB signaling enhances cyclin D1 expression and promotes human endometrial cancer proliferation. J Cell Physiol *218*, 490–500.

Cheng, S.P., Yin, P.H., Hsu, Y.C., Chang, Y.C., Huang, S.Y., Lee, J.J., and Chi, C.W. (2011). Leptin enhances migration of human papillary thyroid cancer cells through the PI3K/AKT and MEK/ERK signaling pathways. Oncol Rep *26*, 1265–1271.

Clarke, M.F. (2005). Self-renewal and solid-tumor stem cells. Biol Blood Marrow Transplant *11*, 14–16.

Clevers, H. (2011). The cancer stem cell: premises, promises and challenges. Nat Med *17*, 313–319.

Curtin, J.C., and Lorenzi, M.V. (2010). Drug discovery approaches to target Wnt signaling in cancer stem cells. Oncotarget *1*, 563–577.

de Sousa, E.M., Vermeulen, L., Richel, D., and Medema, J.P. (2011). Targeting Wnt signaling in colon cancer stem cells. Clin Cancer Res *17*, 647–653.

Dieudonne, M.N., Machinal-Quelin, F., Serazin-Leroy, V., Leneveu, M.C., Pecquery, R., and Giudicelli, Y. (2002). Leptin mediates a proliferative response in human MCF7 breast cancer cells. Biochem Biophys Res Commun *293*, 622–628.

Farooqi, I.S., Jebb, S.A., Langmack, G., Lawrence, E., Cheetham, C.H., Prentice, A.M., Hughes, I.A., McCamish, M.A., and O'Rahilly, S. (1999). Effects of recombinant leptin therapy in a child with congenital leptin deficiency. N Engl J Med *341*, 879–884.

Feldman, D.E., Chen, C., Punj, V., Tsukamoto, H., and Machida, K. (2011). Pluripotency factor-mediated expression of the leptin receptor (OB-R) links obesity to oncogenesis through tumor-initiating stem cells. Proc Natl Acad Sci USA.

Fenton, J.I., Lavigne, J.A., Perkins, S.N., Liu, H., Chandramouli, G.V., Shih, J.H., Hord, N.G., and Hursting, S.D. (2008). Microarray analysis reveals that leptin induces autocrine/paracrine cascades to promote survival and proliferation of colon epithelial cells in an Apc genotype-dependent fashion. Mol Carcinog *47*, 9–21.

Fiorio, E., Mercanti, A., Terrasi, M., Micciolo, R., Remo, A., Auriemma, A., Molino, A., Parolin, V., Di Stefano, B., Bonetti, F., *et al.* (2008). Leptin/HER2 crosstalk in breast cancer: in vitro study and preliminary in vivo analysis. BMC Cancer *8*, 305.

Fletcher, S., Drewry, J.A., Shahani, V.M., Page, B.D., and Gunning, P.T. (2009). Molecular disruption of oncogenic signal transducer and activator of transcription 3 (STAT3) protein. Biochem Cell Biol *87*, 825–833.

Friedman, J. (1997). Leptin, leptin receptors, and the control of body weight, in Blum WF, Kiess W, and Rascher W (Eds): Leptin—The Voice of Adipose Tissue. Heidelberg, JA Barth Verlag, pp 3–22.

Fruhbeck, G. (2006). Intracellular signalling pathways activated by leptin. Biochem J *393*, 7–20.

Gainsford, T., Willson, T.A., Metcalf, D., Handman, E., McFarlane, C., Ng, A., Nicola, N.A., Alexander, W.S., and Hilton, D.J. (1996). Leptin can induce proliferation, differentiation, and functional activation of hemopoietic cells. Proc Natl Acad Sci USA *93*, 14564–14568.

Gallia, G.L., Tyler, B.M., Hann, C.L., Siu, I.M., Giranda, V.L., Vescovi, A.L., Brem, H., and Riggins, G.J. (2009). Inhibition of Akt inhibits growth of glioblastoma and glioblastoma stem-like cells. Mol Cancer Ther *8*, 386–393.

Gao, J., Tian, J., Lv, Y., Shi, F., Kong, F., Shi, H., and Zhao, L. (2009). Leptin induces functional activation of cyclooxygenase-2 through JAK2/STAT3, MAPK/ERK, and PI3K/AKT pathways in human endometrial cancer cells. Cancer Sci *100*, 389–395.

Gilani, R.A., Kazi, A.A., Shah, P., Schech, A.J., Chumsri, S., Sabnis, G., Jaiswal, A.K., and Brodie, A.H. (2012). The importance of HER2 signaling in the tumor-initiating cell population in aromatase inhibitor-resistant breast cancer. Breast Cancer Res Treat.

Gonzalez, R.R., Cherfils, S., Escobar, M., Yoo, J.H., Carino, C., Styer, A.K., Sullivan, B.T., Sakamoto, H., Olawaiye, A., Serikawa, T., *et al.* (2006). Leptin signaling promotes the growth of mammary tumors and increases the expression of vascular endothelial growth factor (VEGF) and its receptor type two (VEGF-R2). J Biol Chem *281*, 26320–26328.

Gonzalez, R.R., Devoto, L., Campana, A., and Bischof, P. (2001). Effects of leptin, interleukin-1alpha, interleukin-6, and transforming growth factor-beta on markers of trophoblast invasive phenotype: integrins and metalloproteinases. Endocrine *15*, 157–164.

Gonzalez, R.R., and Leavis, P. (2001). Leptin upregulates beta3-integrin expression and interleukin-1beta, upregulates leptin and leptin receptor expression in human endometrial epithelial cell cultures. Endocrine *16*, 21–28.

Gonzalez-Perez, R.R., Xu, Y., Guo, S., Watters, A., Zhou, W., and Leibovich, S.J. (2010). Leptin upregulates VEGF in breast cancer via canonic and non-canonical signalling pathways and NFkappaB/HIF-1alpha activation. Cell Signal *22*, 1350–1362.

Grosfeld, A., Andre, J., Hauguel-De Mouzon, S., Berra, E., Pouyssegur, J., and Guerre-Millo, M. (2002). Hypoxia-inducible factor 1 transactivates the human leptin gene promoter. J Biol Chem *277*, 42953–42957.

Guo, S., Colbert, L.S., Fuller, M., Zhang, Y., and Gonzalez-Perez, R.R. (2010a). Vascular endothelial growth factor receptor-2 in breast cancer. Biochim Biophys Acta *1806*, 108–121.

Guo, S., Fuller, M., and Gonzalez-Perez, R. (2010b). Regulation of VEGFR-2 by leptin-Notch signaling crosstalk in mammary cancer cellsIn: Proceedings of the 101st Annual Meeting of the American Association for Cancer Research *Apr 17–21*, Abstract #LB-365.

Guo, S., and Gonzalez-Perez, R.R. (2011). Notch, IL-1 and leptin crosstalk outcome (NILCO) is critical for leptin-induced proliferation, migration and VEGF/VEGFR-2 expression in breast cancer. PLoS One *6(6)*: e21467.

Guo, S., Liu, M., and Gonzalez-Perez, R.R. (2011). Role of Notch and its oncogenic signaling crosstalk in breast cancer. Biochim Biophys Acta *1815*, 197–213.

Guryanova, O.A., Wu, Q., Cheng, L., Lathia, J.D., Huang, Z., Yang, J., MacSwords, J., Eyler, C.E., McLendon, R.E., Heddleston, J.M., *et al.* (2011). Nonreceptor tyrosine kinase BMX maintains self-renewal and tumorigenic potential of glioblastoma stem cells by activating STAT3. Cancer Cell *19*, 498–511.

Halaas, J.L., Gajiwala, K.S., Maffei, M., Cohen, S.L., Chait, B.T., Rabinowitz, D., Lallone, R.L., Burley, S.K., and Friedman, J.M. (1995). Weight-reducing effects of the plasma protein encoded by the obese gene. Science *269*, 543–546.

Hambardzumyan, D., Squatrito, M., Carbajal, E., and Holland, E.C. (2008). Glioma formation, cancer stem cells, and akt signaling. Stem Cell Rev *4*, 203–210.

Han, C., Zhang, H.T., Du, L., Liu, X., Jing, J., Zhao, X., Yang, X., and Tian, B. (2005). Serum levels of leptin, insulin, and lipids in relation to breast cancer in china. Endocrine *26*, 19–24.

Han, G., Jing, Y., Zhang, Y., Yue, Z., Hu, X., Wang, L., Liang, J., and Liu, J. (2010). Osteogenic differentiation of bone marrow mesenchymal stem cells by adenovirus-mediated expression of leptin. Regul Pept *163*, 107–112.

Hayden, M.S., and Ghosh, S. (2004). Signaling to NF-kappaB. Genes Dev *18*, 2195–2224.

Hayden, M.S., and Ghosh, S. (2008). Shared principles in NF-kappaB signaling. Cell *132*, 344–362.

Heida, N.M., Leifheit-Nestler, M., Schroeter, M.R., Muller, J.P., Cheng, I.F., Henkel, S., Limbourg, A., Limbourg, F.P., Alves, F., Quigley, J.P., *et al.* (2010). Leptin enhances the potency of circulating angiogenic cells via src kinase and integrin (alpha)vbeta5: implications for angiogenesis in human obesity. Arterioscler Thromb Vasc Biol *30*, 200–206.

Ho, P.L., Lay, E.J., Jian, W., Parra, D., and Chan, K.S. (2012). Stat3 activation in urothelial stem cells leads to direct progression to invasive bladder cancer. Cancer Res *72*, 3135–3142.

Hu, X., Juneja, S.C., Maihle, N.J., and Cleary, M.P. (2002). Leptin—a growth factor in normal and malignant breast cells and for normal mammary gland development. J Natl Cancer Inst *94*, 1704–1711.

Huang, C.Y., Yu, H.S., Lai, T.Y., Yeh, Y.L., Su, C.C., Hsu, H.H., Tsai, F.J., Tsai, C.H., Wu, H.C., and Tang, C.H. (2011). Leptin increases motility and integrin up-regulation in human prostate cancer cells. J Cell Physiol *226*, 1274–1282.

Ibanez, E., Agliano, A., Prior, C., Nguewa, P., Redrado, M., Gonzalez-Zubeldia, I., Plano, D., Palop, J.A., Sanmartin, C., and Calvo, A. (2012). The quinoline imidoselenocarbamate EI201 blocks the AKT/mTOR pathway and targets cancer stem cells leading to a strong antitumor activity. Curr Med Chem *19*, 3031–3043.

Imagawa, K., Numata, Y., Katsuura, G., Sakaguchi, I., Morita, A., Kikuoka, S., Matumoto, Y., Tsuji, T., Tamaki, M., Sasakura, K., *et al.* (1998). Structure-function studies of human leptin. J Biol Chem *273*, 35245–35249.

Jiang, H., Yu, J., Guo, H., Song, H., and Chen, S. (2008). Upregulation of survivin by leptin/STAT3 signaling in MCF-7 cells. Biochem Biophys Res Commun *368*, 1–5.

Kakarala, M., and Wicha, M.S. (2008). Implications of the cancer stem-cell hypothesis for breast cancer prevention and therapy. J Clin Oncol *26*, 2813–2820.

Ke, Q., and Costa, M. (2006). Hypoxia-inducible factor-1 (HIF-1). Mol Pharmacol *70*, 1469–1480.

Knight, B.B., Oprea-Ilies, G.M., Nagalingam, A., Yang, L., Cohen, C., Saxena, N.K., and Sharma, D. (2011). Survivin upregulation, dependent on leptin-EGFR-Notch1 axis, is essential for leptin induced migration of breast carcinoma cells. Endocr Relat Cancer.

Koda, M., Sulkowska, M., Kanczuga-Koda, L., Cascio, S., Colucci, G., Russo, A., Surmacz, E., and Sulkowski, S. (2007a). Expression of the obesity hormone leptin and its receptor correlates with hypoxia-inducible factor-1 alpha in human colorectal cancer. Ann Oncol *18* Suppl 6, vi116–119.

Koda, M., Sulkowska, M., Wincewicz, A., Kanczuga-Koda, L., Musiatowicz, B., Szymanska, M., and Sulkowski, S. (2007b). Expression of leptin, leptin receptor, and hypoxia-inducible factor 1 alpha in human endometrial cancer. Ann N Y Acad Sci *1095*, 90–98.

Korkaya, H., Paulson, A., Charafe-Jauffret, E., Ginestier, C., Brown, M., Dutcher, J., Clouthier, S.G., and Wicha, M.S. (2009). Regulation of mammary stem/progenitor cells by PTEN/Akt/beta-catenin signaling. PLoS Biol *7*, e1000121.

Korkaya, H., Paulson, A., Iovino, F., and Wicha, M.S. (2008). HER2 regulates the mammary stem/progenitor cell population driving tumorigenesis and invasion. Oncogene *27*, 6120–6130.

Korkaya, H., and Wicha, M.S. (2009). HER-2, notch, and breast cancer stem cells: targeting an axis of evil. Clin Cancer Res *15*, 1845–1847.

Lacerda, L., Pusztai, L., and Woodward, W.A. (2010). The role of tumor initiating cells in drug resistance of breast cancer: Implications for future therapeutic approaches. Drug Resist Updat *13*, 99–108.

Lanier, V., Gillespie, C., McGlothen, T., Dickson, T., Guo, S., and Gonzalez-Perez, R.R. (2012). Leptin induces Notch in endothelial cells: role of VEGFR-2 transactivation. AACR 102 Annual Congress.

Lee, T.K., Castilho, A., Cheung, V.C., Tang, K.H., Ma, S., and Ng, I.O. (2011). CD24(+)

liver tumor-initiating cells drive self-renewal and tumor initiation through STAT3-mediated NANOG regulation. Cell Stem Cell *9*, 50–63.
Leinninger, G.M. (2009). Location, location, location: the CNS sites of leptin action dictate its regulation of homeostatic and hedonic pathways. Int J Obes (Lond) *33 Suppl* 2, S14–17.
Leong, K.G., and Karsan, A. (2006). Recent insights into the role of Notch signaling in tumorigenesis. Blood *107*, 2223–2233.
Li, G.H., Wei, H., Lv, S.Q., Ji, H., and Wang, D.L. (2010). Knockdown of STAT3 expression by RNAi suppresses growth and induces apoptosis and differentiation in glioblastoma stem cells. Int J Oncol *37*, 103–110.
Li, L., Gao, Y., Zhang, L.L., and He, D.L. (2008). Concomitant activation of the JAK/STAT3 and ERK1/2 signaling is involved in leptin-mediated proliferation of renal cell carcinoma Caki-2 cells. Cancer Biol Ther *7*, 1787–1792.
Li, Y., and Ye, D. (2010). Cancer Therapy By Targeting Hypoxia-Inducible Factor-1. Curr Cancer Drug Targets.
Li, Z., and Rich, J.N. (2010). Hypoxia and hypoxia inducible factors in cancer stem cell maintenance. Curr Top Microbiol Immunol *345*, 21–30.
Lin, L., Fuchs, J., Li, C., Olson, V., Bekaii-Saab, T., and Lin, J. (2011a). STAT3 signaling pathway is necessary for cell survival and tumorsphere forming capacity in LDH(+)/CD133(+) stem cell-like human colon cancer cells. Biochem Biophys Res Commun *416*, 246–251.
Lin, L., Liu, A., Peng, Z., Lin, H.J., Li, P.K., Li, C., and Lin, J. (2011b). STAT3 is necessary for proliferation and survival in colon cancer-initiating cells. Cancer Res *71*, 7226–7237.
Liu, J., Sato, C., Cerletti, M., and Wagers, A. (2010). Notch signaling in the regulation of stem cell self-renewal and differentiation. Curr Top Dev Biol *92*, 367–409.
Mankan, A.K., and Greten, F.R. (2011). Inhibiting signal transducer and activator of transcription 3: rationality and rationale design of inhibitors. Expert Opin Investig Drugs *20*, 1263–1275.
Marotta, L.L., Almendro, V., Marusyk, A., Shipitsin, M., Schemme, J., Walker, S.R., Bloushtain-Qimron, N., Kim, J.J., Choudhury, S.A., Maruyama, R., *et al.* (2011). The JAK2/STAT3 signaling pathway is required for growth of CD44(+)CD24(−) stem cell-like breast cancer cells in human tumors. J Clin Invest *121*, 2723–2735.
Martelli, A.M., Evangelisti, C., Follo, M.Y., Ramazzotti, G., Fini, M., Giardino, R., Manzoli, L., McCubrey, J.A., and Cocco, L. (2011). Targeting the phosphatidylinositol 3-kinase/Akt/mammalian target of rapamycin signaling network in cancer stem cells. Curr Med Chem *18*, 2715–2726.
Masciocchi, D., Gelain, A., Villa, S., Meneghetti, F., and Barlocco, D. (2011). Signal transducer and activator of transcription 3 (STAT3): a promising target for anticancer therapy. Future Med Chem *3*, 567–597.
Masuzaki, H., Ogawa, Y., Sagawa, N., Hosoda, K., Matsumoto, T., Mise, H., Nishimura, H., Yoshimasa, Y., Tanaka, I., Mori, T., *et al.* (1997). Nonadipose tissue production of leptin: leptin as a novel placenta-derived hormone in humans. Nat Med *3*, 1029–1033.
Mattace Raso, G., Esposito, E., Iacono, A., Pacilio, M., Coppola, A., Bianco, G., Diano, S., Di Carlo, R., and Meli, R. (2006). Leptin induces nitric oxide synthase type II in C6 glioma cells. Role for nuclear factor-kappaB in hormone effect. Neurosci Lett *396*, 121–126.
Menon, S., and Manning, B.D. (2008). Common corruption of the mTOR signaling network in human tumors. Oncogene *27* Suppl 2, S43–51.
Miyazaki, T., Bub, J.D., and Iwamoto, Y. (2008). c-Jun NH(2)-terminal kinase mediates leptin-stimulated androgen-independent prostate cancer cell proliferation via signal transducer and activator of transcription 3 and Akt. Biochim Biophys Acta *1782*, 593–604.
Myers, M.G., Cowley, M.A., and Munzberg, H. (2008). Mechanisms of leptin action and leptin resistance. Annu Rev Physiol *70*, 537–556.
Myers, M.G., Jr., Munzberg, H., Leinninger, G.M., and Leshan, R.L. (2009). The geometry of leptin action in the brain: more complicated than a simple ARC. Cell Metab *9*, 117–123.
Nakamura, Y., Hinoi, E., Takarada, T., Takahata, Y., Yamamoto, T., Fujita, H., Takada, S., Hashizume, S., and Yoneda, Y. (2011). Positive regulation by GABA(B)R1 subunit of leptin expression through gene transactivation in adipocytes. Plos one *6*, e20167.
Ogunwobi, O.O., and Beales, I.L. (2007a). The anti-apoptotic and growth stimulatory actions of leptin in human colon cancer cells involves activation of JNK mitogen activated protein kinase, JAK2 and PI3 kinase/Akt. Int J Colorectal Dis *22*, 401–409.

Ogunwobi, O.O., and Beales, I.L. (2007b). Cyclo-oxygenase-independent inhibition of apoptosis and stimulation of proliferation by leptin in human colon cancer cells. Dig Dis Sci *52*, 1934–1945.

Pai, R., Lin, C., Tran, T., and Tarnawski, A. (2005). Leptin activates STAT and ERK2 pathways and induces gastric cancer cell proliferation. Biochem Biophys Res Commun *331*, 984–992.

Pang, L.Y., and Argyle, D.J. (2009). Using naturally occurring tumours in dogs and cats to study telomerase and cancer stem cell biology. Biochim Biophys Acta *1792*, 380–391.

Pischon, T., Nothlings, U., and Boeing, H. (2008). Obesity and cancer. Proc Nutr Soc *67*, 128–145.

Polus A., G.J., Piatkowska E., Dembinska-Kiec A. (2003). Differences in leptin, VEGF, and bFGF-induced angiogenic differentiation of HUVEC and human umbilical blood CD34+ progenitor cells. European Journal of Biochemistry, Abstract number: P*4*.1–67.

Pratt, M.A., Tibbo, E., Robertson, S.J., Jansson, D., Hurst, K., Perez-Iratxeta, C., Lau, R., and Niu, M.Y. (2009). The canonical NF-kappaB pathway is required for formation of luminal mammary neoplasias and is activated in the mammary progenitor population. Oncogene *28*, 2710–2722.

Prosperi, J.R., and Goss, K.H. (2010). A Wnt-ow of Opportunity: Targeting the Wnt/beta-Catenin Pathway in Breast Cancer. Curr Drug Targets *11*, 1074–1088.

Radtke, F., and Raj, K. (2003). The role of Notch in tumorigenesis: oncogene or tumour suppressor? Nat Rev Cancer *3*, 756–767.

Ratke, J., Entschladen, F., Niggemann, B., Zanker, K.S., and Lang, K. (2010). Leptin stimulates the migration of colon carcinoma cells by multiple signaling pathways. Endocr Relat Cancer *17*, 179–189.

Ray, A., Nkhata, K.J., and Cleary, M.P. (2007). Effects of leptin on human breast cancer cell lines in relationship to estrogen receptor and HER2 status. Int J Oncol *30*, 1499–1509.

Ren, H., Zhao, T., Wang, X., Gao, C., Wang, J., Yu, M., and Hao, J. (2010). Leptin upregulates telomerase activity and transcription of human telomerase reverse transcriptase in MCF-7 breast cancer cells. Biochem Biophys Res Commun *394*, 59–63.

Rene Gonzalez, R., Watters, A., Xu, Y., Singh, U.P., Mann, D.R., Rueda, B.R., and Penichet, M.L. (2009). Leptin-signaling inhibition results in efficient anti-tumor activity in estrogen receptor positive or negative breast cancer. Breast Cancer Res *11*, R36.

Rentsch, J., and Chiesi, M. (1996). Regulation of ob gene mRNA levels in cultured adipocytes. FEBS Lett *379*, 55–59.

Riolfi, M., Ferla, R., Del Valle, L., Pina-Oviedo, S., Scolaro, L., Micciolo, R., Guidi, M., Terrasi, M., Cetto, G.L., and Surmacz, E. (2010). Leptin and its receptor are overexpressed in brain tumors and correlate with the degree of malignancy. Brain Pathol *20*, 481–489.

Robertson, S.A., Leinninger, G.M., and Myers, M.G., Jr. (2008). Molecular and neural mediators of leptin action. Physiol Behav *94*, 637–642.

Rouet-Benzineb, P., Aparicio, T., Guilmeau, S., Pouzet, C., Descatoire, V., Buyse, M., and Bado, A. (2004). Leptin counteracts sodium butyrate-induced apoptosis in human colon cancer HT-29 cells via NF-kappaB signaling. J Biol Chem *279*, 16495–16502.

Russell, R.C., Fang, C., and Guan, K.L. (2011). An emerging role for TOR signaling in mammalian tissue and stem cell physiology. Development *138*, 3343–3356.

Russo, V.C., Metaxas, S., Kobayashi, K., Harris, M., and Werther, G.A. (2004). Antiapoptotic effects of leptin in human neuroblastoma cells. Endocrinology *145*, 4103–4112.

Sabbah, M., Emami, S., Redeuilh, G., Julien, S., Prevost, G., Zimber, A., Ouelaa, R., Bracke, M., De Wever, O., and Gespach, C. (2008). Molecular signature and therapeutic perspective of the epithelial-to-mesenchymal transitions in epithelial cancers. Drug Resist Updat *11*, 123–151.

Saxena, N.K., Sharma, D., Ding, X., Lin, S., Marra, F., Merlin, D., and Anania, F.A. (2007a). Concomitant activation of the JAK/STAT, PI3K/AKT, and ERK signaling is involved in leptin-mediated promotion of invasion and migration of hepatocellular carcinoma cells. Cancer Res *67*, 2497–2507.

Saxena, N.K., Vertino, P.M., Anania, F.A., and Sharma, D. (2007b). leptin-induced growth stimulation of breast cancer cells involves recruitment of histone acetyltransferases and mediator complex to CYCLIN D1 promoter via activation of Stat3. J Biol Chem *282*, 13316–13325.

Scheller, E.L., Song, J., Dishowitz, M.I., Soki, F.N., Hankenson, K.D., and Krebsbach, P.H.

(2010) Leptin functions peripherally to regulate differentiation of mesenchymal progenitor cells. Stem Cells *28*, 1071–1080.

Schroeter, M.R., Leifheit, M., Sudholt, P., Heida, N.M., Dellas, C., Rohm, I., Alves, F., Zientkowska, M., Rafail, S., Puls, M., *et al.* (2008). Leptin enhances the recruitment of endothelial progenitor cells into neointimal lesions after vascular injury by promoting integrin-mediated adhesion. Circ Res *103*, 536–544.

Shen, Y., Wang, Q., Zhao, Q., and Zhou, J. (2009). Leptin promotes the immune escape of lung cancer by inducing proinflammatory cytokines and resistance to apoptosis. Mol Med Report *2*, 295–299.

Shostak, K., and Chariot, A. (2011). NF-kappaB, stem cells and breast cancer: the links get stronger. Breast Cancer Res *13*, 214.

Sierra-Honigmann, M.R., Nath, A.K., Murakami, C., Garcia-Cardena, G., Papapetropoulos, A., Sessa, W.C., Madge, L.A., Schechner, J.S., Schwabb, M.B., Polverini, P.J., *et al.* (1998). Biological action of leptin as an angiogenic factor. Science *281*, 1683–1686.

Singh, B.N., Kumar, D., Shankar, S., and Srivastava, R.K. (2012). Rottlerin induces autophagy which leads to apoptotic cell death through inhibition of PI3K/Akt/mTOR pathway in human pancreatic cancer stem cells. Biochem Pharmacol.

Smith-Kirwin, S.M., O'Connor, D.M., De Johnston, J., Lanccy, E.D., Hassink, S.G., and Funanage, V.L. (1998). Leptin expression in human mammary epithelial cells and breast milk. J Clin Endocrinol Metab *83*, 1810–1813.

Soma, D., Kitayama, J., Yamashita, H., Miyato, H., Ishikawa, M., and Nagawa, H. (2008). Leptin augments proliferation of breast cancer cells via transactivation of HER2. J Surg Res *149*, 9–14.

Sparks, C.A., and Guertin, D.A. (2010). Targeting mTOR: prospects for mTOR complex 2 inhibitors in cancer therapy. Oncogene *29*, 3733–3744.

Stahl, N., Farruggella, T.J., Boulton, T.G., Zhong, Z., Darnell, J.E., Jr., and Yancopoulos, G.D. (1995). Choice of STATs and other substrates specified by modular tyrosine-based motifs in cytokine receptors. Science *267*, 1349–1353.

Steelman, L.S., Stadelman, K.M., Chappell, W.H., Horn, S., Basecke, J., Cervello, M., Nicoletti, F., Libra, M., Stivala, F., Martelli, A.M., *et al.* (2008). Akt as a therapeutic target in cancer. Expert Opin Ther Targets *12*, 1139–1165.

Tai, W., Mahato, R., and Cheng, K. (2010). The role of HER2 in cancer therapy and targeted drug delivery. J Control Release.

Tartaglia, L.A. (1997). The leptin receptor. J Biol Chem *272*, 6093–6096.

Tartaglia, L.A., Dembski, M., Weng, X., Deng, N., Culpepper, J., Devos, R., Richards, G.J., Campfield, L.A., Clark, F.T., Deeds, J., *et al.* (1995). Identification and expression cloning of a leptin receptor, OB-R. Cell *83*, 1263–1271.

Tseng, L.M., Huang, P.I., Chen, Y.R., Chen, Y.C., Chou, Y.C., Chen, Y.W., Chang, Y.L., Hsu, H.S., Lan, Y.T., Chen, K.H., *et al.* (2012). Targeting signal transducer and activator of transcription 3 pathway by cucurbitacin I diminishes self-renewing and radiochemoresistant abilities in thyroid cancer-derived CD133+ cells. J Pharmacol Exp Ther *341*, 410–423.

Tzahar, E., and Yarden, Y. (1998). The ErbB-2/HER2 oncogenic receptor of adenocarcinomas: from orphanhood to multiple stromal ligands. Biochim Biophys Acta *1377*, M25–37.

Uddin, S., Bavi, P., Siraj, A.K., Ahmed, M., Al-Rasheed, M., Hussain, A.R., Amin, T., Alzahrani, A., Al-Dayel, F., Abubaker, J., *et al.* (2010a). Leptin-R and its association with PI3K/AKT signaling pathway in papillary thyroid carcinoma. Endocr Relat Cancer *17*, 191–202.

Uddin, S., Bavi, P.P., Hussain, A.R., Alsbeih, G., Al-Sanea, N., Abduljabbar, A., Ashari, L.H., Alhomoud, S., Al-Dayel, F., Ahmed, M., *et al.* (2009a). Leptin receptor expression in Middle Eastern colorectal cancer and its potential clinical implication. Carcinogenesis *30*, 1832–1840.

Uddin, S., Bu, R., Ahmed, M., Abubaker, J., Al-Dayel, F., Bavi, P., and Al-Kuraya, K.S. (2009b). Overexpression of leptin receptor predicts an unfavorable outcome in Middle Eastern ovarian cancer. Mol Cancer *8*, 74.

Uddin, S., Bu, R., Ahmed, M., Hussain, A.R., Ajarim, D., Al-Dayel, F., Bavi, P., and Al-Kuraya, K.S. (2010b). Leptin receptor expression and its association with PI3K/AKT signaling pathway in diffuse large B-cell lymphoma. Leuk Lymphoma *51*, 1305–1314.

Wang, J., Sullenger, B.A., and Rich, J.N. (2012). Notch signaling in cancer stem cells. Adv Exp Med Biol *727*, 174–185.

Wang, K.H., Kao, A.P., Chang, C.C., Lee, J.N., Hou, M.F., Long, C.Y., Chen, H.S., and Tsai, E.M. (2010). Increasing CD44+/CD24(-) tumor stem cells, and upregulation of COX-2 and HDAC6, as major functions of HER2 in breast tumorigenesis. Mol Cancer *9*, 288.

Wickenden, J.A., and Watson, C.J. (2010). Key signalling nodes in mammary gland development and cancer. Signalling downstream of PI3 kinase in mammary epithelium: a play in 3 Akts. Breast Cancer Res *12*, 202.

Wolk, R., Deb, A., Caplice, N.M., and Somers, V.K. (2005). Leptin receptor and functional effects of leptin in human endothelial progenitor cells. Atherosclerosis *183*, 131–139.

Wu, M.H., Chen, K.F., Lin, S.C., Lgu, C.W., and Tsai, S.J. (2007). Aberrant expression of leptin in human endometriotic stromal cells is induced by elevated levels of hypoxia inducible factor-1alpha. Am J Pathol *170*, 590–598.

Wu, M.H., Chou, Y.C., Chou, W.Y., Hsu, G.C., Chu, C.H., Yu, C.P., Yu, J.C., and Sun, C.A. (2009). Circulating levels of leptin, adiposity and breast cancer risk. Br J Cancer *100*, 578–582.

Yang, S.N., Chen, H.T., Tsou, H.K., Huang, C.Y., Yang, W.H., Su, C.M., Fong, Y.C., Tseng, W.P., and Tang, C.H. (2009). Leptin enhances cell migration in human chondrosarcoma cells through OBRl leptin receptor. Carcinogenesis *30*, 566–574.

Yao, H., Ashihara, E., and Maekawa, T. (2011). Targeting the Wnt/beta-catenin signaling pathway in human cancers. Expert Opin Ther Targets *15*, 873–887.

Yeh, W.L., Lu, D.Y., Lee, M.J., and Fu, W.M. (2009). Leptin induces migration and invasion of glioma cells through MMP-13 production. Glia *57*, 454–464.

Yin, N., Wang, D., Zhang, H., Yi, X., Sun, X., Shi, B., Wu, H., Wu, G., Wang, X., and Shang, Y. (2004). Molecular mechanisms involved in the growth stimulation of breast cancer cells by leptin. Cancer Res *64*, 5870–5875.

Zhao, M., Jiang, B., and Gao, F.H. (2011). Small Molecule Inhibitors of STAT3 for Cancer Therapy. Curr Med Chem.

Zheng, Q., Dunlap, S.M., Zhu, J., Downs-Kelly, E., Rich, J.N., Hursting, S.D., Berger, N.A., and Reizes, O. (2011). Leptin deficiency suppresses MMTV-Wnt-1 mammary tumor growth and abrogates tumor initiating cell survival. Endocr Relat Cancer.

Zhou, J., Wulfkuhle, J., Zhang, H., Gu, P., Yang, Y., Deng, J., Margolick, J.B., Liotta, L.A., Petricoin, E., 3rd, and Zhang, Y. (2007). Activation of the PTEN/mTOR/STAT3 pathway in breast cancer stem-like cells is required for viability and maintenance. Proc Natl Acad Sci USA *104*, 16158–16163.

Zhou, W., Guo, S., and Gonzalez-Perez, R.R. (2011). Leptin pro-angiogenic signature in breast cancer is linked to IL-1 signalling. Br J Cancer *104*, 128–137.

Zhu, Y., Zhang, X., Liu, Y., Zhang, S., Liu, J., Ma, Y., and Zhang, J. (2012). Antitumor effect of the mTOR inhibitor everolimus in combination with trastuzumab on human breast cancer stem cells in vitro and in vivo. Tumour Biol.

Chapter 27

Tumor-Initiating Stem-Like Cells: Carcinogenesis through Toll-Like Receptors, Environmental Factors, and Virus

Keigo Machida
Department of Molecular Microbiology and Immunology, University of Southern California School of Medicine, Los Angeles, CA, USA and Southern California Research Center for ALPD and Cirrhosis, Los Angeles, CA, USA

Abbreviations

HBV	Hepatitis B virus
HCV	Hepatitis C virus
TIC	tumor-initiating stem-like cells
TLR	toll-like receptor

Introduction

Tumor-initiating stem-like cells (TIC) share key features with embryonic stem cells (ESC) present within pre-implantation blastocyst stage embryos, including the expression of a core pluripotency-associated transcription factor (TF) network (Kim et al., 2010; Ikushima et al., 2011). However, in contrast to ESCs, TICs fail to properly control the self-renewing mode of cell division that is fundamental property of stem cells. In untransformed stem cells, self-renewal occurs through asymmetric cell division, in which one daughter cell retains the multipotent progenitor status of its parent while the other cell commits to a specialized cell fate. TICs exhibit a striking loss of this intrinsic asymmetry, leading to the ectopic implementation of stem cell gene expression programs in both progeny cells and subsequently to unchecked expansion of the progenitor cell pool (Cicalese et al., 2009; Knoblick, 2010; Martin-Belmonte and Perez-Moreno, 2012).

Hepatocellular carcinoma (HCC) is the third leading cause of cancer deaths in the world, with more than 600,000 deaths in 2007. HCC has a very low 5-year survival rate due to the lack of therapeutic options, even in advanced countries such as the U.S. (Liang and Heller, 2004; Reid and Liang, 1995; Okuda et al., 2002; Okuda, 2000; Yao and Terrault, 2001). The high mortality rate of HCC is due in part to its unresponsiveness to treatment, and in part to the lack of sensory fibers in the noncapsular part of the liver. The latter often leads to its detection in the late stage when surgical removal of the

Cancer Stem Cells, First Edition. Edited by Vinagolu K. Rajasekhar.

tumor is no longer an option. The 5-year survival rate of HCC is less than 5% with or without therapeutic intervention (El-Serag and Mason, 1999). The major cause of HCC is hepatitis B virus (HBV) and hepatitis C virus (HCV) (Crippin et al., 2002). Other nonviral factors account for about 20% of the HCC cases (El-Serag and Mason, 1999). The lifetime risk of HCC in chronically infected HCV patients is 2–7% (Di Bisceglie et al., 2003), and it may take 30 to 40 years for HCC to develop in these patients. HCC caused by HCV may be an indirect result of chronic liver inflammation, which causes liver injury and compensatory hepatocellular proliferations. This chronic liver inflammation can lead to liver cirrhosis and HCC. Various factors such as mutations in tumor-suppressor genes or proto-oncogenes or the activity of specific growth factors during the course of chronic HCV infection may enhance hepatocellular transformation (Simonetti et al., 1992). The prolonged liver inflammation coupled with the repeated liver regeneration is the most important factor in the multi-step process of hepatocarcinogenesis (Simonetti et al., 1992).

Compelling evidence identifies obesity as a co-morbidity factor that significantly increases HCV patients' risk for HCC, especially in alcoholics or obese patients who are African Americans, Hispanic/Latinos, and low income/rural poor. HCV can induce the expression of pattern recognition molecular pattern Toll-like receptor 4 (TLR4) in hepatocytes in transgenic mouse models, in cell cultures, and during natural infection. This induction of TLR4 sensitizes hepatocytes to endotoxemia induced by obesity and alcohol. TLR4 signaling activated by endotoxins, derived from gram-negative bacteria, induces the expression of the stem cell marker NANOG. The risk for HCC, as assessed by odds ratio, increases from 8–12 to 48–54 if HCV patients have concomitant obesity, especially in alcoholics or obese patients of African American or Hispanic/Latino heritage, and low income/rural poor (Yuan et al., 2004; Hassan et al., 2002; Artinyan et al., 2010). Obesity increases the gut permeability and leads to increased blood levels of endotoxins such as lipopolysaccharides (LPS), which in turn activates toll-like receptors (TLR) (Hritz et al., 2008) and induces the production of cytokines and the inflammatory response, leading to liver injury and the development of obesity/alcohol-related liver disease (Figure 27.1).

Common chemotherapeutic agents, Rapamycin (inhibitor of PI3K/Akt/mTOR pathway) and Sorafenib (inhibitor of tyrosine kinase in the MAP Kinase, Raf/Mek/Erk, and VEGF pathways) (Llovet and Bruix, 2008; Newell et al., 2009) are used for cancer treatment, but tumors are developed after withdrawal of treatment in some cases. TICs are believed to be the source of tumor recurrence in some of cases.

Drug Resistance and Tumor-Initiating Cells (TIC)

Normal cells and TICs express high levels of ATP-binding cassette (ABC) transporters, such as *ABCB1*, and the half-transporter *ABCG2* identified in mitoxantrone-resistant cells. Cancer cells in culture become resistant to cytotoxic anticancer drugs through multiple pathways, such as the efflux pumps at the plasma membrane (MDR1, MRP family members, and MXR), reduced uptake of drugs, expression of one or more energy-dependent transporters that detect and eject anticancer drugs from cells, insensitivity to drug-induced apoptosis and induction of drug-detoxifying mechanisms (Gottesman, 2002). The drug-transporting property of TICs conferred by ABC transporters is the basis for the 'side-population' phenotype that arises from the exclusion of the fluorescent dye Hoechst 33342. Therefore, tumors might have a built-in population of drug-resistant pluripotent cells that can survive chemotherapy and repopulate the tumor (Gottesman, 2002).

To overcome drug resistance due to expansion of TICs, new therapeutic opportunities are proposed, including targeting ABCG2 inhibitors (GF120918 and tariquidar) that inhibit both ABCG2 and ABCB1, ABCG2 antibodies killing TICs to deliver toxins or radioisotopes, stem-cell inhibitors such as cyclopamine, a compound that inhibits the Hedgehog-Patched receptor signaling protein

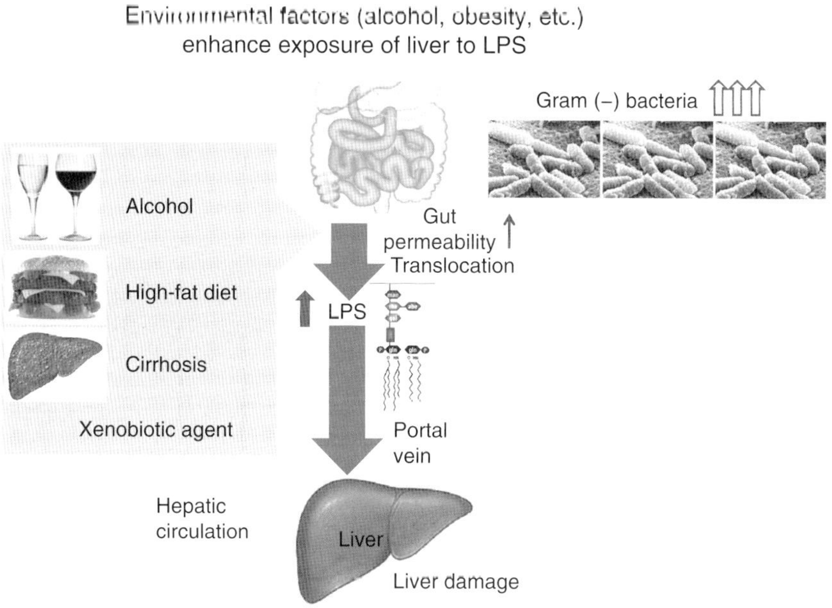

Figure 27.1. Environmental factors (alcohol, obesity, etc.) enhances exposure to livers to LPS. Original to author (Keigo Machida).

Smoothened and immunotherapy by the transplant of bone-marrow stem cells from a donor to kill the tumor cells in the recipient (Dean et al., 2005). Purified TICs from a patient could be lethally irradiated and used to 'immunize' the patient or to activate the donor's immune cells against the TICs (Dean et al., 2005). Nevertheless, these treatments also have limitation, leading to cancer metastasis.

Tumor-Initiating Stem-Like Cells (TIC) and Hepatocellular Carcinoma (HCC)

Stem cells have three major characteristics: self-renewal, asymmetric and multiple cell division (clonality), and plasticity. The liver has a high regenerative potential, and hepatic small oval progenitor cells around the peripheral branches of the bile ducts, the canals of Hering, can differentiate into biliary epithelial cells and hepatocytes (Roskams et al., 2004). These oval liver progenitor cells share molecular markers with adult hepatocytes (albumin, cytokeratin 7 [CK7], CK19, oval cell markers [OV-6, A6, and OV-1], chromogranin-A, NCAM [neural cell adhesion molecule]), and fetal hepatocytes (α-fetoprotein) (Table 27.1) (Roskams et al., 2004; Roskams, 2006). They are also positive for more common stem cell markers such as CD34+, Thy-1+, c-Kit+, and Flt-3+ (FMS-like tyrosine kinase 3) (Burke et al., 2007). Thus, it currently remains unclear whether these stem cells are derived from the bone marrow and just migrate to this niche or whether they represent true resident liver stem/progenitor cells. Binding of stroma-derived factor-1α (SDF-1α) to its surface receptor CXCR4 activates oval hepatic cells (Hatch et al., 2002). Forty percent of HCC have clonality, and thus are considered to originate from progenitor/stem cells (Roskams, 2006; Alison, 2005; Zender, 2006; Tang et al., 2008). Recent studies of HCC have centered on TICs, including detection of TICs in cancer, identification of TIC markers, and isolation of TICs from human HCC cell lines. TICs were identified as CD117+/CD133+ hepatic precursors in regenerating liver tissue (Craig et al., 2004) and a CD45−/CD90+ subpopulation of

Table 27.1. Markers for liver tumor-initiating cells

Gene name	Other name	Function	Species	Organ	References
CD133	Prominin 1 (PROM1)	Glycoprotein, membrane protrusions	Human, Mouse	Liver, Brain	(Ma et al., 2007; Rountree et al., 2008; Ma et al., 2008; Ho et al., 2006; Singh et al., 2004; Shmelkov et al., 2005)
CD49f	Integrinα chain α6 (ITGA6)	Cell adhesion, cell signaling	Mouse	Liver	(Ma et al., 2007; Rountree et al., 2008)
CK19	Cytokeratin 19	Biliary lineage marker	Mouse	Liver	(Tirnitz-Parker et al., 2007; Libbrecht et al., 2001)
OV-6	Oval cell marker	Early progenitor cells	Human	Liver	(Libbrecht et al., 2001)
CD34	Glycoprotein	Cell-cell adhesion factor	Mouse	Liver, Leukemia	(Lapidot et al., 1994)
AFP	α-fetoprotein	Fetal counterpart of serum albumin	Mouse	Liver	(Chiba et al., 2006)
CD90	Thy-1	Glycophosphatidylinositol (GPI) anchor	Mouse	Liver	(Ma et al., 2007)
CD44	Hyaluronic acid receptor	Cell adhesion and migration, metastasis	Mouse	Liver, Breast	(Ma et al., 2007; Al-Hajj et al., 2003)
CD117	KIT	C-kit receptor Cytokine receptor	Mouse	Liver	(Ma et al., 2007)

tumor cells in HCC (Yang et al., 2008). The CD90+ cells are not present in the normal liver and, when injected into immunodeficient mice, create tumors repeatedly. In human HCC and HCC cell lines, specifically CD133+ cells, not CD133- cells, had the ability to self-renew, create differentiated progenies, and form tumors (Ma et al., 2007). This coincided with the expression of genes associated with stem/progenitor status, such as β-catenin, NOTCH, BMI, and OCT3/4. When compared to CD133- cells, the CD133+ cells isolated from the HCC cell lines showed higher expression of CD44 (cell adhesion molecule) and CD34, but both CD133 subpopulations displayed similar expression for CD29 (integrin β1), CD49f (integrin α6), CD90, and CD117 (c-kit: gastrointestinal stroma tumor), indicating these makers are still not definitive TIC markers (Ma et al., 2007). Nevertheless, these makers are differentially expressed in other cancers, such as $CD44^{+}/CD24^{-/low}$ in breast cancer (Al-Hajj et al., 2003), $CD34^{+}/CD38^{-}$ in acute myeloid leukemia (Bonnet and Dick, 1997), and $CD44^{+}/\alpha2\beta1^{hi}/CD133^{+}$ in prostate cancer (Collins et al., 2005), and, for this reason, identification of other surface markers, expressed along with CD133, as a goal to better characterize liver TICs is worthwhile. Furthermore, CD133+/CD49f + cells in liver tumors correlate with tumorigenesity and "stemness" genes, such as Wnt/β-catenin, Nanog (Feldman et al., 2012), Notch, Hedgehog/SMO, Bmi, and Oct3/4 (Valk-Lingbeek et al., 2004; Chambers and Smith, 2004; Beachy et al., 2004). CD133+/CD49f + HCC cancer stem cells confer resistance to chemotherapy, and this presents a major obstacle for the treatment of HCC (Rountree et al., 2008). One potential reason for this chemoresistance may lie in the plasticity of cancer stem cells with dysregulated signaling and gene expression. Several oncogenic signaling pathways in cancer stem cells of HCC, have been described including activated PI3K/AKT (Ma et al., 2008), signal transducer, and activator of transcription 3 (STAT3) (Wurmbach et al., 2007; Yeoh et al., 2007), Notch (Dando et al., 2005), hedgehog (Sicklick et al., 2006a; Sicklick et al., 2006b), and transforming growth factor-beta (TGF-β) (Kitisin et al., 2007; Nguyen et al., 2007).

Nanog-Positive TICs Induced by Virus and Environmental Factors

Nanog is one of the core transcription factors found in pluripotent embryonic stem cells (ESC) (Martin, 1981). It is essential for maintaining self-renewal and pluripotency of both human and mouse embryonic stem cells (Loh et al., 2006; Wang et al., 2006; Rao and Orkin, 2006; Pan and Thomson, 2007). Overexpression of Nanog induces and maintains the pluripotency and self-renewing characteristics of ESCs under what normally would be differentiation-inducing culture conditions (Chambers et al., 2003). Recently, Nanog expression has been reported in human neoplasms, including germ cell tumors (Hoei-Hanson et al., 2005; Hart et al., 2005; Santagata et al., 2007; Ezeh et al., 2005), breast carcinomas (Ezeh et al., 2005), osteosarcoma (Gibbs et al., 2005), and HCC (Ma et al., 2008). Ectopic expression of Nanog induces an oncogenic potential in NIH3T3 (Zhang et al., 2005).

Nanog transduction alone is not as effective as TLR4 activation in liver tumorigenesis, as shown by our cell transplantation experiment (Machida et al., 2009). In our previous work using a cell line, we demonstrated that TLR4 promoter upregulation by HCV NS5A is mediated by PU.1, Oct-1, and AP-1 elements (Machinda et al., 2006a). We believe that TLR4 activation induces other tumor-driver genes that cooperatively work with Nanog to cause liver oncogenesis. Thus, Nanog is essential for TLR4-dependent oncogenesis, but it alone is poorly oncogenic. Therefore, alcohol and HCV NS5A synergistically induce liver tumor development via induction and activation of TLR4 in mice. The importance of Nanog as a direct downstream gene of TLR4 in this oncogenesis has also been identified (Figure 27.2) (Feldman et al.,

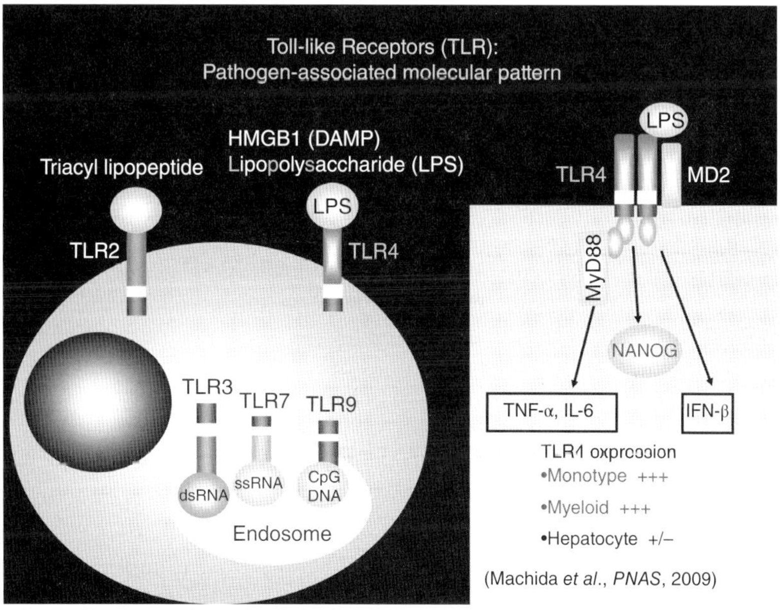

Figure 27.2. Toll-like receptor and pluripostency stem cell transcription factor NANOG. For color detail, please see color plate section.

2012). Pharmacologic inhibition of TLR4 signaling may become a novel therapeutic strategy for HCV-associated liver tumors.

TLRs Signaling in Hepatocellular Carcinoma

The TLR signaling pathway is upregulated in chronic liver diseases. Many different cell types in the liver express TLRs (Testro, 2009). Hepatocytes express TLR1 through TLR9. Stellate cells express TLR2, 3, and 4. Bile duct epithelium expresses TLR2, 3, 4 and 5. Kupffer cells express TLR2, 3, and 4. Chronic alcohol consumption activates other TLRs, such as TLR1, 2, and 6–9, which further increases the TNF-α response to LPS in mice (Testro, 2009). Human monocytes exposed to ethanol for a week develop hypersensitivity to LPS through decreased IRAK-M expression, which activates mitogen-activated protein kinase (MAPK) and NF-κB through TLR4 signaling, leading to activation of NF-κB, AP-1, and ERK (Mandrekar et al., 2009).

Virus, Environmental Factors, and Hepatocellular Carcinoma (HCC)

Compelling evidence identifies obesity and hepatitis C virus (HCV) as co-morbidity risk factors for hepatocellular carcinoma (HCC), which contains TICs. TLR4 signaling, activated by alcohol/obesity-associated endotoxemia, induces the stem marker NANOG and liver tumors in the HCV nonstructural protein Ns5a transgenic (Tg) mice but not in wild type or Ns5a Tg mice deficient in TLR4.

Chronic liver damage caused by viral infection and environmental factors (such as alcohol or metabolic syndrome) can result in increased risk for HCC. Clearly, understanding the molecular mechanisms of HCV-induced hepatocarcinogenesis is required for the eventual development of improved therapeutic modalities for this disease (Crippin et al., 2002). In particular, chronic infection with HBV or HCV represents a major risk factor for HCC (Okuda, 2000). HCV affects more than 170 million people worldwide (Okuda et al., 2002; Okuda, 2000; Yao and Terrault, 2001).

HCV may induce cellular transformation by the induction of chronic liver inflammation mediated by the immune cells. This chronic liver inflammation leads to cell deaths, hepatocellular regeneration, and the emergence of mutated cells that may be oncogenic. In addition, HCV may also induce cellular transformation directly via the induction of oxidative stress and gene mutations as well as the alteration of cellular physiology, which may also lead to malignant cellular transformation.

Alcohol

Several possible mechanisms may explain the high prevalence rate of HCV among alcoholics and the increased severity of liver diseases in these patients. First, alcohol may enhance the replication of HCV and thus increase the expression of viral RNA and proteins, resulting in more severe HCV-induced liver injury, independent of the damage induced by alcohol alone. Indeed, HCV titer has been shown to exhibit a positive correlation with the amount of alcohol consumption (Oshita et al., 1994). This enhanced effect on HCV replication could be caused directly by the metabolites of ethanol, such as acetaldehyde and free radicals, which may stimulate HCV replication and gene expression. It could also be caused indirectly through alcohol-induced inhibition of the antiviral immune response. Indeed, HCV replication is more active in immuno-deficient patients, such as HIV-infected patients (Sherman et al., 1993), and ethanol consumption can cause immunosuppression (Paronetto, 1993).

Obesity

Misregulation of a pluripotency-associated transcription factor network in adult tissues is associated with the expansion of rare, highly malignant TIC through poorly understood mechanisms. Leptin is the most important adipose-derived hormone, and its circulating concentration is proportional to the total amount of body fat. We demonstrate that robust and selective expression of the receptor for the adipocyte-derived peptide hormone leptin (OB-R) is a characteristic feature of TICs and of a broad array of embryonic and induced pluripotent stem cells and is mediated directly by the core pluripotency-associated transcription factors OCT4 and SOX2 (Feldman et al., 2012). The induction of OCT4 and SOX2 was observed in leptin-stimulated cells. TICs exhibit sensitized responses to leptin, including the phosphorylation and activation of the pluripotency-associated oncogene STAT3 and induction of *Oct4* and *Sox2*, thereby establishing a self-reinforcing signaling module. The induction of OB-R was also observed in human HCC. We also found that TICs were highly sensitized to leptin exposure in vitro, as judged by the phosphorylation of Stat3-Y705 (Feldman et al., 2012). Further, leptin promotes the growth of tumors from TICs implanted subcutaneously into immune-compromised *NOG* mice (NOD/Shi-scid/IL-2Rγnull). Exposure of cultured mouse embryonic stem cells to leptin sustains pluripotency in the absence of leukemia inhibitory factor. By implanting TICs into leptin-deficient *ob/ob* mice or into comparably overweight *Lepr*$^{db/db}$ mice that produce leptin, we provide evidence of a central role for the leptin-TIC-signaling axis in promoting obesity-induced tumor growth (Feldman et al., 2012). Differential responses to extrinsic, adipocyte-derived cues may promote the expansion of tumor cell subpopulations and contribute to oncogenesis (Feldman et al., 2012). These findings provide a direct link between the adipose-derived hormone leptin (obesity) and TICs.

Molecular Mechanism of HCV Carcinogenesis

HCV can also cause B-lymphocyte proliferative disorders, including mixed cryoglobulinemia, a disorder characterized by oligoclonal proliferation of B cells, and B-cell lymphoma (Silvestri et al., 1997; Saito et al., 1990; Ferri et al., 1994). In addition, HCV patients may also develop other diseases, including Sjögren's syndrome, lichen planus, disturbance in lipid metabolism, and diabetes or insulin resistance.

Induction of Mutator Phenotype

By infecting B-cell lines and peripheral blood mononuclear cells (PBMC) with HCV in vitro, we demonstrated that HCV infection induced a mutator phenotype, which involves enhanced mutations in many somatic genes, including immunoglobulin (Ig) genes, proto-oncogenes, and tumor suppressor genes (Machida et al., 2004). By comparing the nucleotide substitution patterns of these mutations with those of the predominant proto-oncogene mutations observed in HCV-associated B-cell lymphomas and HCC, we found that the mutations observed in HCV-infected B-cell line or PBMC had nucleotide substitution patterns similar to those found in HCV-associated lymphoma, but only the latter contained amplified mutations (Machinda et al., 2004). Furthermore, the mutations detected in the HCV-associated lymphoma, but not in the unselected HCV-infected cells, had a replacement/silent mutation ratio higher than the expected ratio in the absence of selection. Similarly, both the tumor and neighboring non-tumor tissues of HCV-associated HCC have similar nucleotide substitution patterns, but, interestingly, the tumor tissues have three to five times higher mutation rates than the non-tumor tissues. Furthermore, the ratio of replacement/silent mutations in the tumors, but not the neighboring non-tumor tissues, was significantly higher than the ratio expected in the absence of selection (Machida et al., 2004). These additional hits may provide the selection pressures for the amplification of the proto-oncogene mutations. It is striking that other types of tumors, including lymphomas not associated with HCV, HBV-associated HCC, and

HCC of nonviral origin, did not show such an amplification of proto-oncogene mutations. Thus, the ability of HCV to induce the mutator phenotype, with subsequently amplified mutations, is unique and likely plays a critical role in HCV oncogenesis. It is conceivable that HCV-induced mutations contribute to the eventual selection and amplification of certain deleterious mutations of proto-oncogenes or tumor suppressor genes in tumors, and the long latency period of HCV-associated malignancies is due to the need of multiple hits for HCV-associated oncogenesis.

Chromosome Translocation

Defects in DNA repair genes cause genetic instability, gross chromosomal rearrangements, and accumulation of mutations, leading ultimately to neoplastic transformation. Both homologous recombination and nonhomologous end joining (NHEJ) play a role in the repair of DSBs in mammalian cells (Hiom, 1999). The interaction of broken DNA with members of the Rad52 epistasis group, including Rad51, a mammalian homologue of bacterial RecA, initiates homologous recombination repair (Hiom, 1999). Following DNA damage, Rad51 is redistributed within the nucleus (Haaf et al., 1995; Baumann and West, 1998) and induces the ATP-dependent homologous strand pairing reaction that initiates recombination. In contrast, NHEJ works by non-homology-dependent ligation of broken DNA ends, DNA-dependent protein kinase (DNA-PK), and its associated proteins Ku70, Ku80, and Xrcc4 mediate NHEJ (Song et al., 2003).

Cellular DNA repair proteins prevent potential DNA mutations caused by oxidative damage, but are themselves vulnerable to nitric oxide (NO)-induced oxidative damage because of sulphenyl, tyrosyl, and/or phenolic side chains in their active sites (Starke et al., 1997; Jaiswal et al., 2000; Jaiswal et al., 2001). Suppression of DNA repair, coupled with the induction of DNA breaks by viral proteins, may increase the mutation frequency and chromosome rearrangements in virus-infected cells.

Reactive Oxygen Species

Reactive oxygen species (ROS) are generated as a consequence of mitochondrial electron transport. They are caused by a small proportion of the electron flow that interacts with oxygen molecules before reaching the cytochrome oxidase complex. HCV infection induces ROS (Choi and Ou, 2006), which leads to oxidative DNA damages and lipid peroxidation in HCV-infected cells (Machida et al., 2006b). Treatment of HCV-infected cells with ROS inhibitors effectively prevented mitochondrial damages and the production of ROS (Machida et al., 2006b).

In summary, ROS, together with the previously reported NO (Machida et al., 2004), directly or indirectly play important roles in HCV pathogenesis. ROS can induce steatosis (through lipid peroxides) and oncogenesis (through DNA mutations and STAT3 activation), and may also contribute to acute hepatocyte damages (through the production of STAT3) (Machida et al., 2006b). ROS and/or iNOS inhibitors may provide therapeutic measures for HCV patients for the protection of their liver tissues from oxidative stress.

Inhibition of DNA Damage Repair

As discussed above, we have previously demonstrated that HCV infection induces a mutator phenotype by causing double-strand DNA breaks (DSB) (Machida et al., 2004). We have further reported that HCV induces iNOS mRNA expression and enhances NO production through the action of the viral core protein and the NS3 protein, and that NO is responsible for DSBs in most of cellular genes (Machida et al., 2004). Accumulation of DSBs in HCV-infected cells suggests that an HCV-induced oxidative environment may overwhelm cellular antioxidant and DNA-repair mechanisms, leading to chromosomal abnormalities.

Conclusions

TIC population is the origin of cancer recurrence in drug-treated patients with HCCs. Chronic HCV infection results in a high frequency of HCC that displays non-metastatic and multicentric characteristics. The molecular pathways of HCV-induced carcinogenesis may involve the indirect, non-virological factors such as the induction of chronic liver inflammation and regeneration that lead to the emergence of mutated cells with high proliferation rates. In addition, it may also involve virological factors such as viral gene products that stimulate the production of ROS and the expression of error-prone DNA polymerases. These different pathways highlight the complicated interplays between the virus and its host in HCV carcinogenesis.

Conflict of Interests

The authors declare no conflict of interests.

Acknowledgments

This research was supported by NIH grants 1R01AA018857-01, pilot project funding (5P30DK048522-13), P50AA11999 (Animal Core, Morphology Core, and Pilot Project Program), R24AA012885 (Non-Parenchymal Liver Cell Core), Zumberge Foundation, AI83025U19, RC2AA019392-01, CA123328, and CA108302. This research is also supported by a Research Scholar Grant, RSG-12-177-01-MPC, and pilot funding (IRG-58-007-48) from American Cancer Society.

References

Al-Hajj, M., Wicha, M.S., Benito-Hernandez, A., Morrison, S.J., and Clarke, M.F. Prospective identification of tumorigenic breast cancer cells. Proc Natl Acad Sci USA 100, 3983–8 (2003).

Alison, M.R. Liver stem cells: implications for hepatocarcinogenesis. Stem Cell Rev 1, 253–60 (2005).

Artinyan, A. et al. Race, ethnicity, and socioeconomic status influence the survival of patients with hepatocellular carcinoma in the United States. Cancer 116, 1367–77 (2010).

Baumann, P. and West, S.C. DNA end-joining catalyzed by human cell-free extracts. Proc Natl Acad Sci USA 95, 14066–70 (1998).

Beachy, P.A., Karhadkar, S.S. and Berman, D.M. Tissue repair and stem cell renewal in carcinogenesis. Nature 432, 324–31 (2004).

Bonnet, D. and Dick, J.E. Human acute myeloid leukemia is organized as a hierarchy that originates from a primitive hematopoietic cell. Nat Med 3, 730–7 (1997).

Burke, Z.D., Thowfeequ, S., Peran, M., and Tosh, D. Stem cells in the adult pancreas and liver. Biochem J 404, 169–78 (2007).

Chambers, I. and Smith, A. Self-renewal of teratocarcinoma and embryonic stem cells. Oncogene 23, 7150–60 (2004).

Chambers, I. et al. Functional expression cloning of Nanog, a pluripotency sustaining factor in embryonic stem cells. Cell 113, 643–55 (2003).

Chiba, T. et al. Side population purified from hepatocellular carcinoma cells harbors cancer stem cell-like properties. Hepatology 44, 240–51 (2006).

Cicalese, A. et al. The tumor suppressor p53 regulates polarity of self-renewing divisions in mammary stem cells. Cell 138, 1083–95 (2009).

Collins, A.T., Berry, P.A., Hyde, C., Stower, M.J. and Maitland, N.J. Prospective identification of tumorigenic prostate cancer stem cells. Cancer Res 65, 10946–51 (2005).

Craig, C.E. et al. The histopathology of regeneration in massive hepatic necrosis. Semin Liver Dis 24, 49–64 (2004).

Crippin, J.S., McCashland, T., Terrault, N., Sheiner, P., and Charlton, M.R. A pilot study of the tolerability and efficacy of antiviral therapy in hepatitis C virus-infected patients awaiting liver transplantation. Liver Transpl 8, 350–5 (2002).

Choi, J. and Ou, J.H. Mechanisms of liver injury. III. Oxidative stress in the pathogenesis of hepatitis C virus. Am J Physiol Gastrointest Liver Physiol 290, G847–51 (2006).

Dando, J.S. et al. Notch/Delta4 interaction in human embryonic liver CD34+ CD38- cells: positive influence on BFU-E production and

LTC-IC potential maintenance. Stem Cells 23, 550–60 (2005).

Dean, M., Fojo, T., and Bates, S. Tumour stem cells and drug resistance. Nat Rev Cancer 5, 275–84 (2005).

Di Bisceglie, A.M. et al. Hepatitis C-related hepatocellular carcinoma in the United States: influence of ethnic status. Am J Gastroenterol 98, 2060–3 (2003).

El-Serag, H.B. and Mason, A.C. Rising incidence of hepatocellular carcinoma in the United States. N Engl J Med 340, 745–50 (1999).

Ezeh, U.I., Turek, P.J., Reijo, R.A., and Clark, A.T. Human embryonic stem cell genes OCT4, NANOG, STELLAR, and GDF3 are expressed in both seminoma and breast carcinoma. Cancer 104, 2255–65 (2005).

Feldman, D.E., Chen, C., Punj, V., Tsukamoto, H., and Machida, K. Pluripotency factor-mediated expression of the leptin receptor (OB-R) links obesity to oncogenesis through tumor-initiating stem cells. Proc Natl Acad Sci USA 109, 829–34 (2012).

Ferri, C. et al. Hepatitis C virus infection in patients with non-Hodgkin's lymphoma. Br J Haematol 88, 392–4 (1994).

Gibbs, C.P. et al. Stem-like cells in bone sarcomas: implications for tumorigenesis. Neoplasia 7, 967–76 (2005).

Gottesman, M.M. Mechanisms of cancer drug resistance. Annu Rev Med 53, 615–27 (2002).

Haaf, T., Golub, E.I., Reddy, G., Radding, C.M., and Ward, D.C. Nuclear foci of mammalian Rad51 recombination protein in somatic cells after DNA damage and its localization in synaptonemal complexes. Proc Natl Acad Sci USA 92, 2298–302 (1995).

Hart, A.H. et al. The pluripotency homeobox gene NANOG is expressed in human germ cell tumors. Cancer 104, 2092–8 (2005).

Hassan, M.M. et al. Risk factors for hepatocellular carcinoma: synergism of alcohol with viral hepatitis and diabetes mellitus. Hepatology 36, 1206–13 (2002).

Hatch, H.M., Zheng, D., Jorgensen, M.L., and Petersen, B.E. SDF-1alpha/CXCR4: a mechanism for hepatic oval cell activation and bone marrow stem cell recruitment to the injured liver of rats. Cloning Stem Cells 4, 339–51 (2002).

Hiom, K. Dna repair: Rad52—the means to an end. Curr Biol 9, R446–8 (1999).

Ho, J.W. et al. Significance of circulating endothelial progenitor cells in hepatocellular carcinoma. Hepatology 44, 836–43 (2006).

Hoei-Hansen, C.E. et al. Stem cell pluripotency factor NANOG is expressed in human fetal gonocytes, testicular carcinoma in situ and germ cell tumours. Histopathology 47, 48–56 (2005).

Hritz, I. et al. The critical role of toll-like receptor (TLR) 4 in alcoholic liver disease is independent of the common TLR adapter MyD88. Hepatology 48, 1224–31 (2008).

Ikushima, H. et al. Glioma-initiating cells retain their tumorigenicity through integration of the Sox axis and Oct4 protein. J Biol Chem 286, 41434–41 (2011).

Jaiswal, M., LaRusso, N.F., Burgart, L.J., and Gores, G.J. Inflammatory cytokines induce DNA damage and inhibit DNA repair in cholangiocarcinoma cells by a nitric oxide-dependent mechanism. Cancer Res 60, 184–90 (2000).

Jaiswal, M., LaRusso, N.F., Shapiro, R.A., Billiar, T.R., and Gores, G.J. Nitric oxide-mediated inhibition of DNA repair potentiates oxidative DNA damage in cholangiocytes. Gastroenterology 120, 190–9 (2001).

Kim, J. et al. A Myc network accounts for similarities between embryonic stem and cancer cell transcription programs. Cell 143, 313–24 (2010).

Kitisin, K. et al. Disruption of transforming growth factor-beta signaling through beta-spectrin ELF leads to hepatocellular cancer through cyclin D1 activation. Oncogene 26, 7103–10 (2007).

Knoblich, J.A. Asymmetric cell division: recent developments and their implications for tumour biology. Nat Rev Mol Cell Biol 11, 849–60 (2010).

Lapidot, T. et al. A cell initiating human acute myeloid leukaemia after transplantation into SCID mice. Nature 367, 645–8 (1994).

Liang, T.J. and Heller, T. Pathogenesis of hepatitis C-associated hepatocellular carcinoma. Gastroenterology 127, S62–71 (2004).

Libbrecht, L. et al. Hepatic progenitor cells in hepatocellular adenomas. Am J Surg Pathol 25, 1388–96 (2001).

Llovet, J.M. and Bruix, J. Molecular targeted therapies in hepatocellular carcinoma. Hepatology 48, 1312–27 (2008).

Loh, Y.H. et al. The Oct4 and Nanog transcription network regulates pluripotency in mouse embryonic stem cells. Nat Genet 38, 431–40 (2006).

Ma, S. et al. Identification and characterization of tumorigenic liver cancer stem/progenitor cells. Gastroenterology 132, 2542–56 (2007).

Ma, S., Lee, T.K., Zheng, B.J., Chan, K.W., and Guan, X.Y. CD133+ HCC cancer stem cells confer chemoresistance by preferential expression of the Akt/PKB survival pathway. Oncogene 27, 1749–58 (2008).

Machida, K. et al. Toll-like receptor 4 mediates synergism between alcohol and HCV in hepatic oncogenesis involving stem cell marker Nanog. Proc Natl Acad Sci USA 106, 1548–53 (2009).

Machida, K. et al. Hepatitis C virus induces toll-like receptor 4 expression, leading to enhanced production of beta interferon and interleukin-6. J Virol 80, 866–74 (2006).

Machida, K. et al. Hepatitis C virus triggers mitochondrial permeability transition with production of reactive oxygen species, leading to DNA damage and STAT3 activation. J Virol 80, 7199–207 (2006).

Machida, K. et al. Hepatitis C virus infection activates the immunologic (type II) isoform of nitric oxide synthase and thereby enhances DNA damage and mutations of cellular genes. J Virol 78, 8835–43 (2004).

Machida, K. et al. Hepatitis C virus induces a mutator phenotype: enhanced mutations of immunoglobulin and protooncogenes. Proc Natl Acad Sci USA 101, 4262–7 (2004).

Mandrekar, P., Bala, S., Catalano, D., Kodys, K., and Szabo, G. The opposite effects of acute and chronic alcohol on lipopolysaccharide-induced inflammation are linked to IRAK-M in human monocytes. J Immunol 183, 1320–7 (2009).

Martin, G.R. Isolation of a pluripotent cell line from early mouse embryos cultured in medium conditioned by teratocarcinoma stem cells. Proc Natl Acad Sci USA 78, 7634–8 (1981).

Martin-Belmonte, F. and Perez-Moreno, M. Epithelial cell polarity, stem cells and cancer. Nat Rev Cancer 12, 23–38 (2012).

Newell, P. et al. Ras pathway activation in hepatocellular carcinoma and anti-tumoral effect of combined sorafenib and rapamycin in vivo. J Hepatol 51, 725–33 (2009).

Nguyen, L.N. et al. Transforming growth factor-beta differentially regulates oval cell and hepatocyte proliferation. Hepatology 45, 31–41 (2007).

Okuda, K. Hepatocellular carcinoma. J Hepatol 32, 225–37 (2000).

Okuda, M. et al. Mitochondrial injury, oxidative stress, and antioxidant gene expression are induced by hepatitis C virus core protein. Gastroenterology 122, 366–75 (2002).

Oshita, M. et al. Increased serum hepatitis C virus RNA levels among alcoholic patients with chronic hepatitis C. Hepatology 20, 1115–20 (1994).

Pan, G. and Thomson, J.A. Nanog and transcriptional networks in embryonic stem cell pluripotency. Cell Res 17, 42–9 (2007).

Paronetto, F. Immunologic reactions in alcoholic liver disease. Semin Liver Dis 13, 183–95 (1993).

Rao, S. and Orkin, S.H. Unraveling the transcriptional network controlling ES cell pluripotency. Genome Biol 7, 230 (2006).

Reid, A. E. and Liang, T. J. Association of hepatitis C virus and hepatocellular carcinoma in the United States. Princess Takamatsu Symp 25, 41–9 (1995).

Roskams, T.A. et al. Nomenclature of the finer branches of the biliary tree: canals, ductules, and ductular reactions in human livers. Hepatology 39, 1739–45 (2004).

Roskams, T. Liver stem cells and their implication in hepatocellular and cholangiocarcinoma. Oncogene 25, 3818–22 (2006).

Rountree, C.B., Senadheera, S., Mato, J.M., Crooks, G.M., and Lu, S.C. Expansion of liver cancer stem cells during aging in methionine adenosyltransferase 1A-deficient mice. Hepatology 47, 1288–97 (2008).

Saito, I. et al. Hepatitis C virus infection is associated with the development of hepatocellular carcinoma. Proc Natl Acad Sci USA 87, 6547–9 (1990).

Santagata, S., Ligon, K.L., and Hornick, J.L. Embryonic stem cell transcription factor signatures in the diagnosis of primary and metastatic germ cell tumors. Am J Surg Pathol 31, 836–45 (2007).

Sherman, K.E. et al. Quantitative evaluation of hepatitis C virus RNA in patients with concurrent human immunodeficiency virus infections. J Clin Microbiol 31, 2679–82 (1993).

Shmelkov, S.V., St Clair, R., Lyden, D., and Rafii, S. AC133/CD133/Prominin-1. Int J Biochem Cell Biol 37, 715–9 (2005).

Sicklick, J.K. et al. Dysregulation of the Hedgehog pathway in human hepatocarcinogenesis. Carcinogenesis 27, 748–57 (2006a).

Sicklick, J.K. et al. Hedgehog signaling maintains resident hepatic progenitors throughout life. Am J Physiol Gastrointest Liver Physiol 290, G859–70 (2006b).

Silvestri, F. et al. The genotype of the hepatitis C virus in patients with HCV-related B cell

non-Hodgkin's lymphoma. Leukemia 11, 2157–61 (1997).
Simonetti, R.G. et al. Hepatitis C virus infection as a risk factor for hepatocellular carcinoma in patients with cirrhosis. A case-control study. Ann Intern Med 116, 97–102 (1992).
Singh, S.K. et al. Identification of human brain tumour initiating cells. Nature 432, 396–401 (2004).
Song, E. et al. Sustained small interfering RNA-mediated human immunodeficiency virus type 1 inhibition in primary macrophages. J Virol 77, 7174–81 (2003).
Starke, D.W., Chen, Y., Bapna, C.P., Lesnefsky, E.J. and Mieyal, J.J. Sensitivity of protein sulfhydryl repair enzymes to oxidative stress. Free Radic Biol Med 23, 373–84 (1997).
Tang, Y. et al. Progenitor/stem cells give rise to liver cancer due to aberrant TGF-beta and IL-6 signaling. Proc Natl Acad Sci USA 105, 2445–50 (2008).
Testro, A.G. and Visvanathan, K. Toll-like receptors and their role in gastrointestinal disease. J Gastroenterol Hepatol 24, 943–54 (2009).
Tirnitz-Parker, J.E., Tonkin, J.N., Knight, B., Olynyk, J.K., and Yeoh, G.C. Isolation, culture and immortalisation of hepatic oval cells from adult mice fed a choline-deficient, ethionine-supplemented diet. Int J Biochem Cell Biol 39, 2226–39 (2007).
Valk-Lingbeek, M.E., Bruggeman, S.W., and van Lohuizen, M. Stem cells and cancer; the polycomb connection. Cell 118, 409–18 (2004).
Wang, J. et al. A protein interaction network for pluripotency of embryonic stem cells. Nature 444, 364–8 (2006).
Wurmbach, E. et al. Genome-wide molecular profiles of HCV-induced dysplasia and hepatocellular carcinoma. Hepatology 45, 938–47 (2007).
Yang, Z.F. et al. Significance of CD90+ cancer stem cells in human liver cancer. Cancer Cell 13, 153–66 (2008).
Yao, F. and Terrault, N. Hepatitis C and hepatocellular carcinoma. Curr Treat Options Oncol 2, 473–83 (2001).
Yeoh, G.C. et al. Opposing roles of gp130-mediated STAT-3 and ERK-1/2 signaling in liver progenitor cell migration and proliferation. Hepatology 45, 486–94 (2007).
Yuan, J.M., Govindarajan, S., Arakawa, K., and Yu, M.C. Synergism of alcohol, diabetes, and viral hepatitis on the risk of hepatocellular carcinoma in blacks and whites in the U.S. Cancer 101, 1009–17 (2004).
Zender, L. et al. Identification and validation of oncogenes in liver cancer using an integrative oncogenomic approach. Cell 125, 1253–67 (2006).
Zhang, J. et al. Expression of Nanog gene promotes NIH3T3 cell proliferation. Biochem Biophys Res Commun 338, 1098–102 (2005).

Chapter 28

The Role of Epithelial Cell Polarity Pathways on Cancer Stem Cells

Inmaculada Bañón-Rodríguez, Ilenia Bernascone, and Fernando Martín-Belmonte
Centro de Biología Molecular Severo-Ochoa, Consejo Superior de Investigaciones Científicas (CSIC), Madrid, Spain

Abbreviations

AJ	adherens junctions
AMPK	AMP-activated protein kinases
AMOT	angiomotin
Aur-A	Aurora-A
CAMMKβ	calmodulin-dependent kinase kinase β
Dlg	discs large
EMT	epithelial mesenchymal transition
FAP	familial adenomatous polyposis
FDGC	familial diffuse gastric cancer
GMC	ganglion mother cell
Lgl	lethal giant larvae
MDCK	Madin Darby canine kidney cells
mTOR	mammalian target of rapamycin
Mud	mushroom body defect
Pins	partner of inscuteable
RAPTOR	regulatory associated protein of mTOR
Scrib	Scribble
TJ	tight junctions
TSC2	tuberin

Cancer Stem Cells, First Edition. Edited by Vinagolu K. Rajasekhar.

Introduction

Genes that control epithelial cell polarity are known tumor suppressors or proto-oncogenes, and they include proteins that form the intercellular junctions, those that regulate apical-basal polarity, and proteins that control the orientation of the mitotic spindle during cell division.

Although epithelial sheets expand during development, they maintain their cell junctions and epithelial apical-basal polarity throughout this process. Thus, growth and regeneration do not affect the structure and organization of such tissues. However, epithelial sheets can proliferate without control, thereby generating tumors. To prevent this, epithelial cells have acquired different mechanisms to control growth through the cross talk between polarity proteins and the signaling pathways that regulate cell proliferation and apoptosis, as well as energy metabolism.

Here, we highlight recent advances in our understanding of how epithelial cell polarity is established and maintained, and how these processes are implicated in cancer. We first describe the mechanisms that underlie the establishment of epithelial cell polarity and their connections with the signaling pathways involved in growth control. Subsequently, we discuss the importance of spindle orientation and asymmetric division for epithelial integrity in stem cells.

Polarity in Cancer Stem Cells

Apical-basolateral polarity allows cells to respond in the correct spatiotemporal manner to multiple cues from neighboring cells and the surrounding microenvironment, and it is fundamental for the final acquisition of cell shape and directional transport. When polarity is disrupted, cells may become unresponsive to growth inhibitory signaling and circumvent differentiation, senescence, and apoptosis, promoting cancer initiation. In addition, the disruption of epithelial cell polarity is associated with the loss of key components of intercellular junctions, which precedes cells undergoing the epithelial-mesenchymal-transition (EMT), a crucial step in metastasis (Royer and Lu, 2011).

Role of Cell Junctions in Cell Polarity

Epithelial cell polarity is sustained by the formation and maintenance of cell junction complexes. Two intercellular junctions promote adhesion between epithelial cells: adherens junctions (AJ) and tight junctions (TJ). These junctions control the signaling pathways responsible for epithelial cell proliferation. Disruption of these junctions induces depolarization and the acquisition of invasive potential.

AJs are comprised of two adhesive complexes, cadherin-catenin and nectin-afadin. The *cadherin-catenin* complex physically connects neighboring cells (Figure 28.1). The E-cadherin expression is closely linked with epithelial differentiation (Hulpiau and van Roy, 2009). In several tumors, cadherin switches appear to occur, whereby E-cadherin levels decrease to be compensated by an increase in the expression of N (neural) or P (placental) cadherin, both of which are markers of stem cells or progenitors (Mohamet et al., 2011; Wheelock et al., 2008), a further demonstration of the importance of polarity in maintaining the differentiated phenotype. The extracellular domain of cadherin establishes homophilic calcium-dependent interactions between cells (Gumbiner, 1988). Via their intercellular cytoplasmic tails, cadherins bind to p120-catenin (p120) and β-catenin (Fig 28.1), enabling the cadherin-catenin complex to bind to microtubules and the microtubule motor dynein, as well as monomeric α-catenin and different actin-binding proteins (Yonemura et al., 2010). As described for cadherins, modifications in catenins are associated with tumorigenesis, with the loss of p120 inducing E-cadherin instability and resulting in the disruption of cell junctions (Xiao et al., 2007). Moreover, loss of α-catenin increases cell proliferation, decreases apoptosis, and suppresses actin

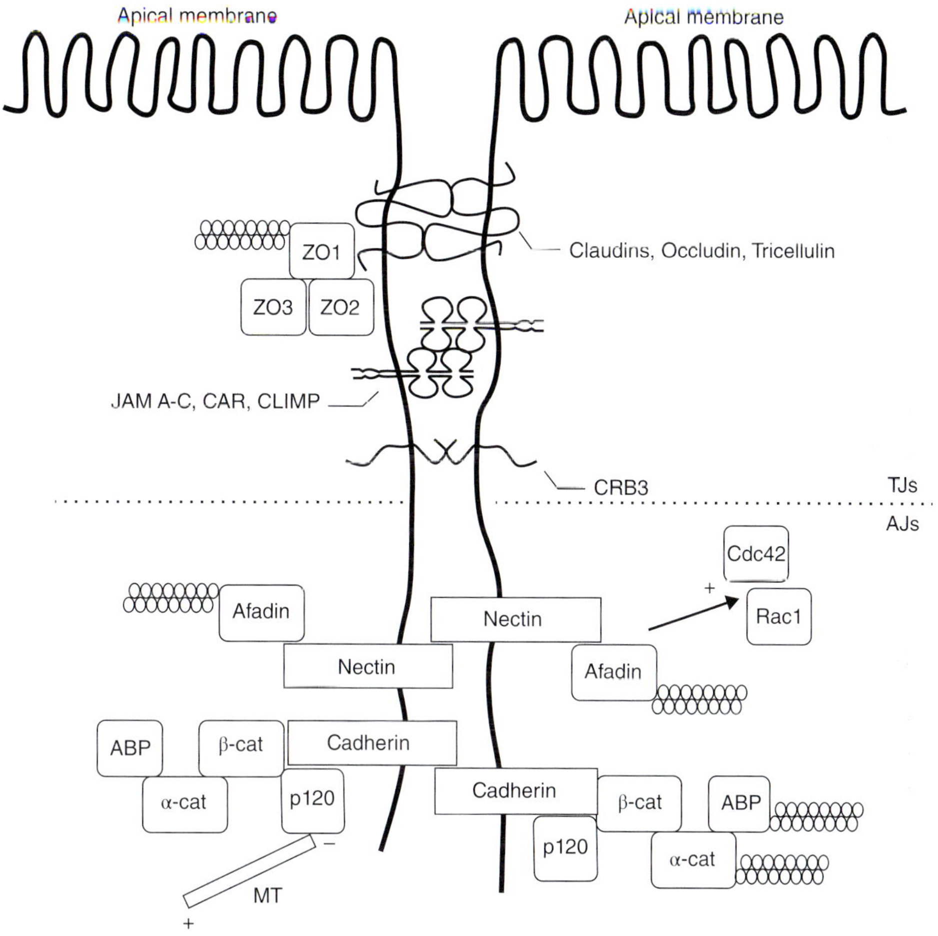

Figure 28.1. Cell junctions. The cadherin-catenin complex localizes at the AJ together with the nectin-afadin complex. TJs are situated apical to the AJs.

polymerization, which in turn decreases cell-cell adhesion (Benjamin and Nelson, 2008). In summary, the interaction of cadherins with their cytoplasmic partners links the adhesive function of this complex to the dynamic forces of the cytoskeleton, as well as to several signaling pathways. The regulation of the cadherin-catenin complex is therefore essential for tumor suppression (Table 28.1).

The *nectin-afadin* complex represents the second major adhesive structure at AJs (Figure 28.1) (Takai and Nakanishi, 2003). Nectins are a family of immunoglobulin-like proteins. The cytoplasmic domain of nectins interacts directly with the scaffolding protein afadin (AF6) (Takai and Nakanishi, 2003), which plays a key role in organizing TJs and AJs in embryonic stem cells. Indeed, AF6 interacts with F-actin, binding the nectin-based adhesion sites to the actin cytoskeleton. In addition, nectin can activate the small GTPases Cdc42 and Rac1, suggesting that this activation is required to facilitate junction formation (Figure 28.1).

It is believed that nectins and cadherins cooperate to regulate the actin cytoskeleton at cell adhesion sites (Ebnet, 2008), facilitating the formation of AJs and strengthening their adhesive activity. Disruption of

Table 28.1. Alterations of polarity proteins in human cancer. Partially extracted from Martin-Belmonte and Pérez-Moreno, 2012

PROTEIN	ALTERATION	CANCER TYPE
	AJs	
E-cadherin	Loss of function	Epithelial adenomas and carcinomas, ovarian and familial gastric cancer.
	Decreased expression	Oral squamous cell carcinoma
β-catenin	Constitutively active, frequent gene mutations	Pilomatricoma, endometrial cancer, ovarian cancer, medulloblastoma, gastrointestinal cancer.
	Loss of expression	Lobular breast carcinomas, oral squamous cell carcinoma
α-catenin	Decreased expression	Breast cancer, gastrointestinal cancers, oral squamous cell carcinoma.
	Loss of expression	Skin cancer
p120-catenin	Abnormal expression	Breast cancer
	Overexpression	Colon adenocarcinoma
	Loss of function	Skin cancer
	TJs	
Claudins	Misregulated	Different carcinomas, pancreatic cancer-ductal, prostate cancer, skin papilloma.
Occludin	Downregulated	Gastric adenocarcinomas
	Loss of function	Breast cancer
Tricellulin	Downregulated	Pancreatic adenocarcinoma, hepatocellular carcinomas, tonsillar squamous cell carcinoma.
	Null expression	Endocrine tumours
JAM A	Downregulation	Pancreatic adenocarcinoma
JAM B	Downregulation	Breast carcinomas
JAM C	Upregulated	Melanoma
CRB3	Downregulation	Human tumour epithelial-derived cell lines
	POLARITY COMPLEXES ASSOCIATED TO CELL JUNCTIONS	
PAR3	Downregulated	Esophageal cell carcinoma
PAR6	Overexpressed	Breast cancer, stromal cells in non-small-cell lung.
	Overexpressed and phosphorylated	Breast BRCA-1 associated tumour
aPKC	Overexpressed	Hepatocellular carcinoma, Bladder tumour, pancreatic cancer, precancerous breast lesion.
	Overexpressed and phosphorylated	Squamous cell carcinoma of the head and neck and dysplastic oral epithelial tissues.
aPKC	Overexpressed	Breast cancer, pancreatic cancer, non-small-cell lung cancer.
	Overexpressed and phosphorylated	Hepatocellular carcinoma
	Overexpressed and mislocalized	Ovarian cancer
ERBB2	Overexpressed	Breast cancer, pancreatic ductal adenocarcinoma, urothelial carcinoma, poor prognosis gastric cancer
	Gene mutation	Lung adenocarcinoma
Merlin	Loss of expression	Central nervous system and skin tumours, epithelial and Schwann cells cancers
	Gene mutation	Meningiomas and mesotheliomas

(*Continued*)

Table 28.1. (*Continued*)

PROTEIN	ALTERATION	CANCER TYPE
PATJ	Mediate interaction of virus like HPV and HTLV	Cervical cancer and adult T-cell leukemia
ZO2	Mediate interaction of virus like HPV and HTLV	Cervical cancer and adult T-cell leukemia
SCRIB	Overexpressed and mislocalized	Colon carcinoma and ovarian cancer
	Downregulated	Breast carcinoma
Lgl1	Loss of expression Aberrant splicing	Glioblastoma, colorectal carcinoma, melanoma, prostate, breast and lung cancer
Lgl2	Overexpressed	Colon and breast tumors
DLG1	Overexpressed and mislocalized	Well- and moderately differentiated cervical, breast and colon cancer
	Loss of expression	Advanced poorly differentiated lesions cervical, breast and colon cancer
	Downregulated	Gastric carcinoma
	Point mutations	Breast cancer
DLG3	Overexpressed	Ovarian adenocarcinoma
	Downregulated	Gastric carcinoma, papillary thyroid carcinoma
DLG4	Downregulated	Invasive gastric carcinoma
	HIPPO PATHWAY	
RASSF1A	Gene hypermethylation	Bladder cancer, melanoma, brain tumours, breast, colon, cervical, esophageal, gastric, head and neck, prostate, pancreatic, kidney and lung cancer.
LATS1/2	Gene hypermethylation	Breast cancer and astrocytomas.
MST1/2	Gene hypermethylation	Soft tissue sarcomas
YAP/TAZ	Overexpression	Intracranial ependymomas, epithelial carcinomas, breast cancer, liver, lung, pancreatic, esophageal, oral squamous and breast cancer.
	LKB1 PATHWAY	
LKB1 (STK11)	Germ line mutations	Peutz-Jeghers syndrome. Gastrointestinal, pancreatic, ovarian and breast cancer
	Mutation or loss of function	Sporadic cancers, like lung, uterine cervix cancer and cervical carcinomas
TSC2	Hyperphoshorylation	Epithelial cancers
mTOR	Overexpression	Gliomas, renal cell carcinomas

either of these AJ complexes may affect other complexes and provoke a loss of junction integrity and tissue stability, a process common to different tumor types.

The TJs in polarized cells arise apical to the AJs (Figure 28.1), and they form a belt-like structure with two main functions. Firstly, this structure acts as a paracellular barrier that prevents ions, small solutes, and water from passing between cells. Secondly, it forms a physical barrier restricting the movement of proteins and lipids between the apical and basolateral plasma membrane domains, which is necessary to develop membrane polarity. Mutations in TJ components prevent the establishment of cell polarity and favor tumorigenesis (Table 28.1).

The first protein described as a TJ component was zonula occludens-1 (ZO-1), a protein that interacts with the cytoplasmic domain of nectin to connect AJs and TJs, which is necessary to maintain cell polarity and for tumor suppression. ZO-1 interacts with the integral membrane proteins and with other cytoplasmic proteins, including its homologues ZO-2 and ZO-3, which in turn interact with F-actin (Fanning et al., 2002). The ZO protein complex provides the main link between TJs and the actin cytoskeleton.

At the molecular level, TJs are formed by three classes of integral membrane proteins that bind to different protein complexes (Figure 28.1):

- The first class is comprised of claudins, occluding, and tricellulin, characterized by the presence of two extracellular loops, four transmembrane domains and two cytoplasmic tails that correspond to the N- and C-terminal portions of the protein. This group has been strongly implicated in cancer (Table 28.1).
- The second class is comprised of Ig-SF proteins that contain two Ig-like domains, and it includes the JAM A-C, CAR, and CLMP proteins, among others (Ebnet et al., 2004).
- The third class consists exclusively of the CRB3 protein, which has a very short extracellular domain and cytoplasmic tail.

In summary, AJs and TJs control epithelial tissue formation and maintenance. In addition, they function as platforms to regulate incoming and outgoing signals, thereby providing means of controlling epithelial proliferation and differentiation. A classic example of this dual role is seen by the activity of β-catenin at AJs, where it is involved in anchoring cadherin-catenin complexes to the actin cytoskeleton, and it also acts as a component of the WNT signaling pathway that can receive and deliver proliferation signals to the cell's interior (Amin and Vincan, 2012).

Polarity Complexes and Their Role in Cancer

Polarity Complexes Associated With Cell Junctions

As indicated above, cell junctions act as platforms that anchor and stabilize polarity complexes that function as scaffolds, connecting these proteins to specific signaling pathways. Thus, cell junctions and polarity complexes can be considered biosensors, providing information about the immediate external cellular microenvironment and the spatial localization of their immediate neighbors. These processes are necessary to orient intracellular structures and to maintain cell polarity and the three-dimensional organization of the tissue.

Three multiprotein complexes are located at the cellular junctions:

- The Crumbs complex. This is a multiprotein complex found at TJs that is formed by Pals1 and PATJ, which bind to the integral-membrane protein CRB3. Knockdown of Pals1 or PATJ results in the loss of expression of the other member of the complex, inducing CRB3 internalization and redistribution of other components of the TJ (Michel et al., 2005). This redistribution impairs the TJ from fulfilling its barrier function, leading to the loss of apical-basal polarity.
- The PAR complex (PAR3/PAR6/aPKC). Studies using dominant-negative mutants of PAR3, aPKC, or PAR6 have demonstrated that these proteins are essential for TJ formation but not for TJs maintenance once the epithelial tissue is fully polarized (Chen and Macara, 2006). ERBB2 is a member of the EGFR family that localizes to the basolateral membrane and that is involved in cell proliferation during normal development. This EGFR is overexpressed in several epithelial cancers, provoking its localization to apical membranes where it binds PAR6-aPKC and sequesters it away from PAR3, thereby

impeding PAR complexes from forming at TJs (Table 28.1).

- The Scribble complex (Scrib/Lgl/DLG). Scrib is a tumor suppressor that binds to E-cadherin, promoting its localization at the basolateral membrane. This complex is antagonistic to the Crumbs and Par complexes, as it is critical to define the basolateral plasma membrane domain and to maintain apical-basal polarity. In addition, Lgl binds to myosin II, thereby regulating F actin filament contractility and cell shape. Finally, Dlg binds to different members of the Wnt pathway, implicating the Scribble complex in the regulation of the activity of this pathway. This complex has also been implicated in vesicle trafficking, which may represent the link between the loss of function of the Scribble complex and alterations in signaling (Humbert et al., 2008).

The protein Merlin (also known as NF2) also plays an important role in promoting the establishment of AJs, and it is associated with TJs through its interaction with angiomotin (AMOT)-PATJ and Pals1. Merlin provides junctional stability and mediates contact-dependent inhibition of proliferation, thereby constituting an important tumor inhibitor. Loss of Merlin has been associated with defective morphogenesis and tumorigenesis in multiple tissues (Table 28.1) (Okada et al., 2007).

Epithelial Growth Control: The Hippo Pathway

The activity of the Salvador-Warts-Hippo (Hippo) pathway is regulated by the interaction of its components with AJ proteins and other proteins involved in cell polarity (Boggiano and Fehon, 2012). The Hippo signaling pathway is a kinase cascade that is activated by cell adhesion, polarity, and stress, and it is responsible for regulating tissue growth, limiting organ size by promoting cell cycle exit and apoptosis. In fact, Hippo may be involved in processes like cell-contact inhibition and tissue repair in epithelial cells. When a tissue is damaged and some cells are removed mechanically, or cells undergo apoptosis in conditions of stress, cell adhesion is relaxed in neighboring cells and the Hippo pathway is turned off. Cells then proliferate until mature cell adhesion can be reestablished and tissues are repaired. At this point, the Hippo pathway must be reactivated to prevent tissue overgrowth. These characteristics indicate that the components of the Hippo pathway are important tumor suppressors (Yin and Zhang, 2011), and in fact, disruption of the Hippo pathway is frequently detected in human cancers (Table 28.1) (Bao et al., 2011). Recent studies have demonstrated how the Hippo pathway interacts with other growth regulatory signaling pathways, which increases the tumor suppressant capacity of Hippo, while maintaining homeostasis and stem cell pluripotency. Accordingly, the Hippo pathway is an important target for the development of new therapeutic drugs.

The molecular elements of the Hippo pathway can be divided into three interconnected parts (Figure 28.2):

- The cell surface upstream regulators that include the cell-adhesion molecules and cell-polarity complexes.
- The Hippo core kinase cascade that includes two serine-threonine kinases (Hippo and Warts) with their regulators and adaptors.
- The downstream nuclear protein (YAP/TAZ), which is a transcriptional co-activator that is negatively regulated by the Hippo pathway, which mediates the transcription of pro-proliferative and anti-apoptotic genes.

Here we shall focus on the part of the pathway in which polarity proteins are implicated.

Recent studies of imaginal disc epithelia and the developing eye of *D. melanogaster* have shown that the polarity proteins Lgl, aPKC, and CRB play an indirect role in the control of cell proliferation and survival by regulating the activity of the Hippo pathway

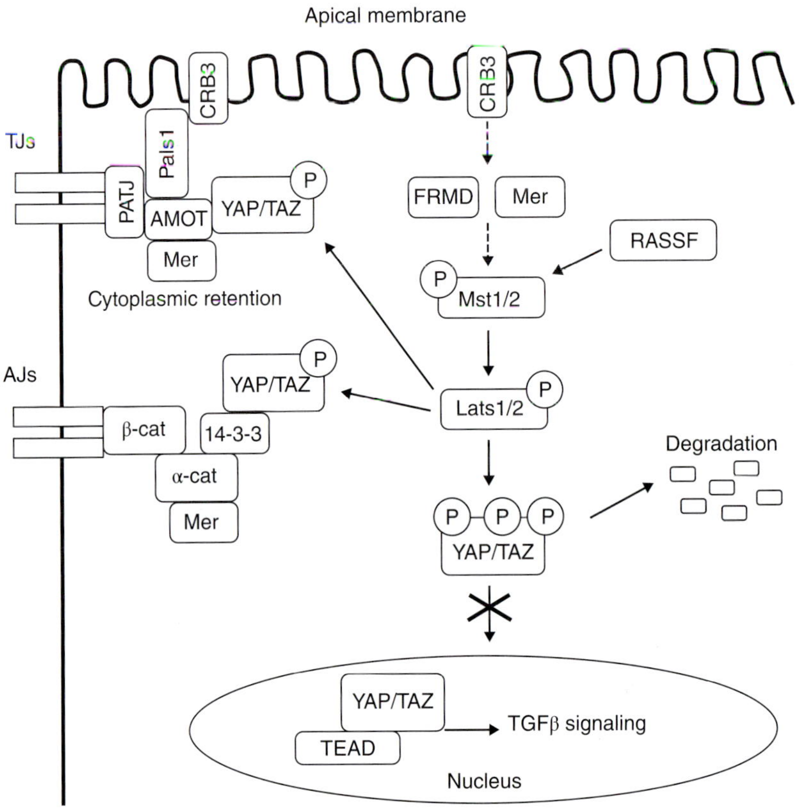

Figure 28.2. Hippo pathway. CRB3 localizes at the apical membrane. The Hippo pathway activates Mst1/2 kinases via proteins such as Merlin and FRMD, which phosphorylate Lats1/2 kinases. Mst1/2 can also be activated by RASSF. Lats1/2 phosphorylates Ser127 of YAP/TAZ, promoting YAP/TAZ binding to 14-3-3 and subsequent cytoplasmic retention. Lats1/2 can also phosphorylate YAP/TAZ at Ser381, activating a signaling cascade that leads to protein degradation. TJs can also sequester YAP/TAZ at cell junctions.

(Grzeschik et al., 2010). This connection may explain why alterations in polarity complexes induce cell proliferation and overgrowth, both in *Drosophil*a and vertebrates. When cells acquire polarity, their junctions and membrane domains become established, and CRB localizes at the apical domain and transmits information regarding cell density via the Pals1-PATJ complex. This complex binds AMOT, which sequesters YAP/TAZ and induces its phosphorylation by Lats1/2. The phosphorylation of YAP/TAZ induces its cytoplasmic retention and suppresses TGFβ signaling (Varelas et al., 2010), a signaling pathway known to promote EMT, which is a process enhanced by the disruption of CRBs. By contrast, depletion of Lgl and Dlg may trigger the expression of downstream Hippo pathway targets and likewise, activation of aPKC upregulates Hippo target genes (Sun and Irvine, 2011).

Several recent studies have described other mechanisms of YAP/TAZ inactivation. For example, the α-Catenin at AJs regulates the Hippo pathway by controlling YAP/TAZ activity and localization (Silvis et al., 2011). This control of cell growth by sequestering YAP at AJs appears to be mediated by Merlin/NF2, and interestingly, Merlin may also regulate the targeting of YAP to TJs

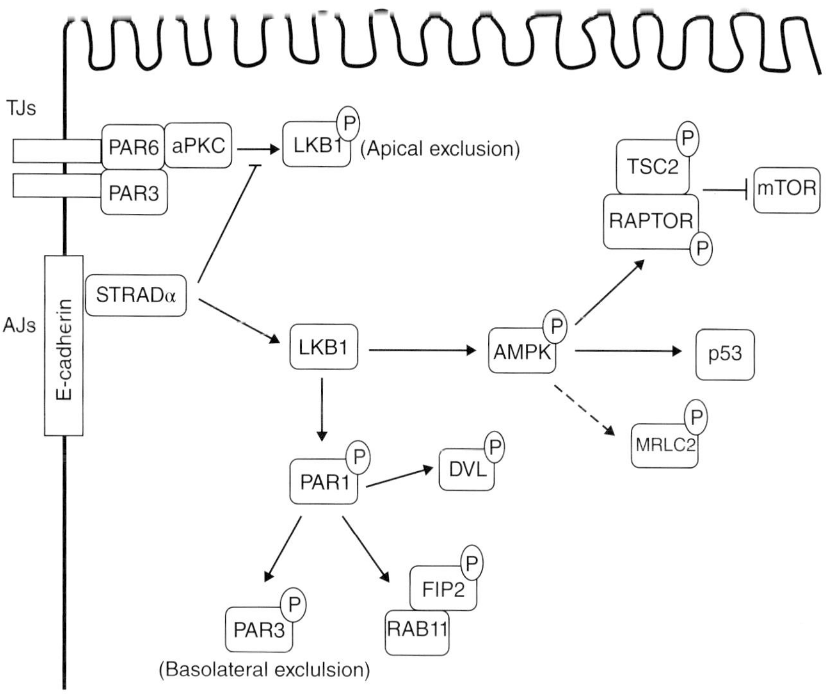

Figure 28.3. LKB1 pathway. The polarity complex PAR3/PAR6/aPKC induces LKB1 phosphorylation and its apical exclusion, while the E-cadherin binding protein, STRADα, binds LKB1 to AJs. LKB1 phosphorylates PAR1, which in turn phosphorylates PAR3, a modification that is essential for its exclusion from the basolateral region. Dishevelled controls the non-canonical Wnt pathway and assembly of E-cadherin adhesion complexes, as well as RAB11/FIP2, proteins that are implicated in cellular junction formation. LKB1 can also phosphorylate members of the AMPK family. Phosphorylated AMPK in turn controls actomyosin contractility by indirectly activating MRLC2, autophagy and cell growth via p53 activation, and it controls energy metabolism via mTOR inhibition.

by controlling its interaction with AMOT (Okada et al., 2007).

In summary, the Hippo pathway provides the cell with a central mechanism to coordinate cell proliferation through external cell contact cues. Disrupting this pathway results in loss of tissue architecture, which in turn induces tumorigenesis. However, further studies are required to unravel the key processes mediated by the Hippo pathway in oncogenesis.

Metabolic Status of the Cell and Its Polarity: The LKB1-mTOR Pathway

During both development and tumorigenesis, signaling pathways that regulate energy production are fundamental, particularly since cancer cells have limited access to metabolites for energy production. LKB1-AMPK links the metabolic status of the cell with polarity at the molecular level (Figure 28.3). LKB1, also known as STK11, has been identified as a tumor suppressor gene because the loss of LKB1 disrupts cell polarity, thereby contributing to tumor invasion and metastasis (Table 28.1) (Herrmann et al., 2011). Mammalian LKB1 encodes a master kinase that phosphorylates and activates at least 14 downstream proteins in response to modifications in the intracellular energy levels. Its major protein targets are the family of AMP-activated protein kinases (AMPK), which act as both an energy checkpoint and an important regulator of cell architecture and polarity. At the level of polarity, AMPK controls the assembly of epithelial TJs in MDCK cells. In this cell line, TJ assembly can be promoted by increasing extracellular calcium concentrations, which enhances AMPK catalysis. However, AMPK can also be activated by calmodulin-dependent kinase kinase β (CAMKKβ) that promotes TJ assembly. Thus, it seems likely that CAMKKβ and LKB1 act redundantly in certain tissues and/or physiological contexts.

LKB1-AMPK may also be involved in E-cadherin-mediated AJ formation because LKB1 acts downstream of E-cadherin to suppress tumor formation, and by binding to STRADα, E-cadherin controls LKB1 localization and consequently AMPK activation (Sebbagh et al., 2009). The main downstream target of AMPK is the mammalian target of rapamycin (mTOR), which controls cell growth in response to energy stress and is dysregulated in most human cancers (Table 28.1). Activated AMPK phosphorylates tuberin (TSC2) and the regulatory-associated protein of mTOR (RAPTOR) in order to inhibit mTOR, which in turn induces cell cycle arrest, upregulates genes involved in catabolism, and downregulates those involved in gluconeogenesis and lipogenesis, as well as halting protein synthesis. Loss of LKB1 decreases AMPK activity, which activates mTOR and relieves these growth constraints, thereby favoring tumorigenesis (Table 28.1) (Herrmann et al., 2011).

Finally, the LKB1-AMPK pathway activates myosin II, and consequently, it influences the contractile properties of the actomyosin cytoskeleton. Significantly, reorganization of the actomyosin cytoskeleton is essential for the establishment of cell polarity and cell division (Mirouse and Billaud, 2011).

In summary, LKB1-AMPK links metabolic control, cell polarity, and epithelial architecture by regulating several targets downstream of AMPK. Disruption of any of these components may affect cell polarity and favor tumorigenesis (Table 28.1).

Stem Cells and the Importance of Being Asymmetrically Divided

Asymmetric cell division is the predominant form of stem or progenitor cell division during development, adult tissue homeostasis, and regeneration. Cell division is considered asymmetric when the two daughter cells produced are of different sizes, they are differently positioned in the tissue plane, or they inherit different cellular components. Stem cells divide asymmetrically to generate two unequal daughter cells, one that retains a stem cell fate (self-renewal capacity) and another that will acquire a more differentiated fate (differentiation). Through asymmetric cell division, the size of the stem cell pool is maintained, while generating the differentiated cells required for tissue development, maintenance, and self-regeneration, such as in the intestine and in the epithelial skin layer.

Our understanding of the molecular mechanisms underlying asymmetric cell division is derived mainly from elegant studies carried out on *Drosophila melanogaster* stem cells (Neumuller and Knoblich, 2009). Here we will first discuss asymmetric division in *Drosophila* neuroblasts, the precursors of the central nervous system. The intensive research carried out in this model defines asymmetric cell division as a process dependent on three key components: (1) polarity protein complexes; (2) cell fate determinants; and (3) mitotic spindle orientation.

The apically localized PAR3/PAR6/aPKC protein complex forms a polarity axis and controls the asymmetric segregation of cell fate determinants (Prospero, Numb, Brat) that accumulate in the basal cortex of the dividing neuroblast. The mitotic spindle is aligned along the apical-basal axis to ensure that the furrow for cytokinesis cleaves the cell such that the basal cell fate determinants are segregated exclusively into a single daughter cell (Figure 28.4). The daughter cell inheriting these cell fate determinants differentiates and becomes a ganglion mother cell (GMC), while the other daughter

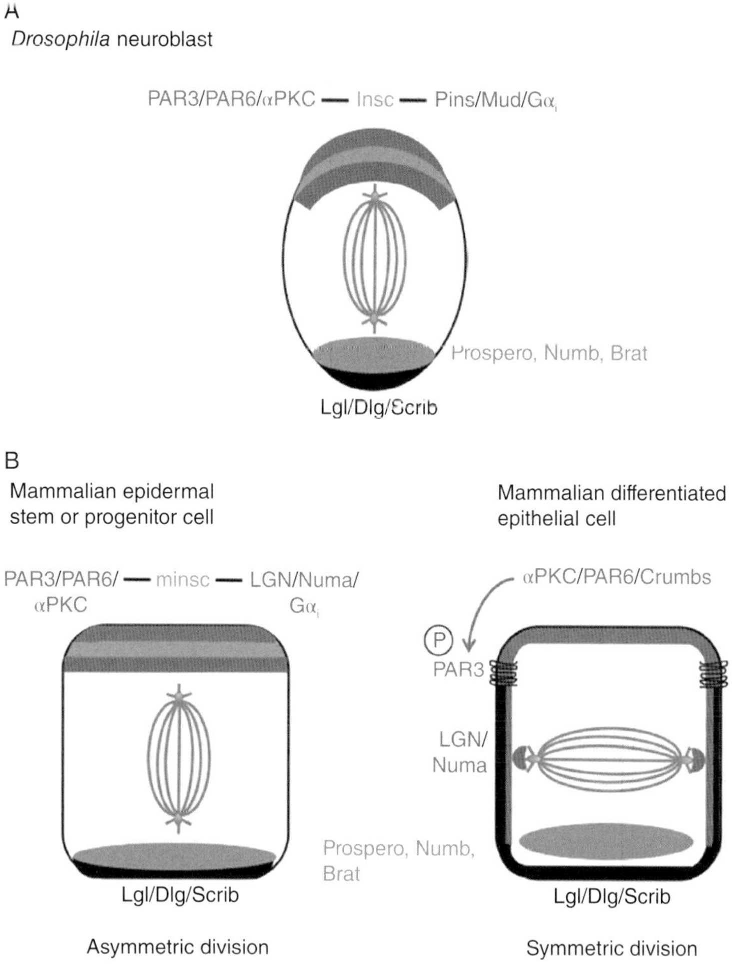

Figure 28.4. Stem cells and asymmetric division. **A**. Asymmetric division in *Drosophila* neuroblasts is controlled by apical protein complexes and basal cell fate determinants. **B**. Many of the complexes regulating *Drosophila* neuroblasts are conserved in mammalian cells. In the epidermis, the spatial distribution of these complexes is critical for the asymmetric division of stem/progenitor cells and the symmetric division of differentiated cells. For color detail, please see color plate section.

cell remains a neuroblast and continues to divide asymmetrically and to self-renew. Exciting studies using *Drosophila* neuroblasts have demonstrated that when one of the key components of the asymmetric division machinery is mutated or deleted, neuroblasts divide symmetrically, leading to the accumulation of cells with proliferative potential and accordingly, tissue overgrowth and tumor formation. These studies indicate that asymmetric stem cell division serves to suppress tumor formation and that a critical balance between asymmetric and symmetric cell division is constantly required to maintain tissue homeostasis.

To what extent are these findings in *Drosophila* relevant to mammalian tumor biology and mammalian cancer stem cells? In the following section, we will discuss how the different components involved in asymmetric division act as tumor suppressors in *Drosophila* stem cells, and we will summarize our current understanding of the putative link between dysregulated asymmetric cell division and tumorigenesis in mammalian epithelial stem cells.

Cancer Induction Due to Failed Asymmetric Division in Drosophila *Stem Cells*

Cancer Induction By the Alteration of Core Polarity Proteins

The *Drosophila* polarity proteins Lgl, Dlg, and Scrib are evidently potent tumor suppressors (see discussion above). Neuroblasts with mutant *lgl*, *dlg*, and *scrib* fail to localize cell fate determinants correctly, and they generate daughter cells that divide and accumulate rapidly, leading to tumor development in larval imaginal discs and brains (Albertson and Doe, 2003). The connection between the loss of cell polarity, defective asymmetric cell division, and tumor initiation is one of the most surprising and important findings in the field of cancer biology in the past 10 years.

Cancer Induction By Altering Cell Fate Determinants

The three main *Drosophila* cell fate determinants are Prospero, Numb, and Brat, and all three promote neuroblast differentiation via distinct pathways. Prospero is a transcription factor that regulates genes crucial for cell-cycle repression, whereas Numb acts as a Notch inhibitor. Brat is a translational repressor with dozens of possible targets, including the oncogene *myc*. When cell fate determinants are deleted or mutated, *Drosphila* neuroblasts divide symmetrically and overproliferate (at the expense of GMCs), and that is associated with in situ tumor development (Betschinger et al., 2006). Furthermore, the resulting tumors can survive for years after implantation into new hosts, indicating that the transformed cells have become immortal.

Cancer Induction By Altering Proteins That Control Spindle Orientation

In *Drosophila* neuroblasts, the mitotic spindle is positioned through dynamic interactions between astral microtubules and the cell cortex. The apical tripartite complex that is composed of Mushroom body defect (Mud), Partner of Inscuteable (Pins) and the heterotrimeric G protein α-subunit ($G\alpha_i$), links astral microtubules to the apical PAR3/PAR6/aPKC polarity complex, thereby regulating spindle orientation. Indeed, *mud* mutant neuroblasts correctly localize the apical and basal polarity complexes and cell fate determinants prior to cell division, although these neuroblasts exhibit randomly oriented spindles that are misaligned with the polarity axis of the cell. This defect provokes a mis-segregation of cell fate determinants and it gives rise to the excessive accumulation of neuroblasts (Bowman et al., 2006).

Abnormal Asymmetric Division as a Potential Source of Mammalian Cancer Stem Cells

Taken together, studies in *Drosophila* convincingly demonstrate that a failure in asymmetric stem cell division results in an abnormal accumulation of dividing cells and promotes cancer formation. In light of such intriguing findings, current cancer research aims to determine whether these abnormalities are causally involved in the emergence of cancer stem cells in mammals.

To date, asymmetric division of stem cells has been described in muscle, skin, gut, mammary glands, the hematopoietic system, and the developing brain. Furthermore, many apical-basal polarity mechanisms that have been discovered in *Drosophila* are conserved in mammalian cells (Figure 28.4). Thus, one of the most pressing questions in modern cancer biology is whether core polarity pathways that control asymmetric stem cell division are disrupted in human cancers. Although this question has been addressed using mouse cancer models and human cancer tissue samples, the data available are still not conclusive, and there is currently no definitive evidence of a direct relationship between loss of asymmetric stem cell division and tumor initiation in mammals.

Why is such evidence still not forthcoming? A number of genes that participate in the correct positioning of the mitotic spindle or that act as cell fate determinants during asymmetric division in *Drosophila* stem cells have been implicated in human cancer. These genes are either mutated or overexpressed/lost in certain human cancers. However, the direct role of these genes in mammalian cancer stem cells remains unknown. Furthermore, due to the evolution of redundancy, it has been more difficult than expected to elucidate the function of cell polarity proteins in mammals. Mammalian genomes encode several isoforms of each polarity protein, which may explain why transgenic mouse models carrying mutations in only one isoform do not exhibit any defects in asymmetric stem cell division.

Core Polarity Proteins and Cancer

To date, Lgl is the only core polarity protein shown to fulfill a role in asymmetric division. Mammals possess two Lgl homologues, Lgl1 and Lgl2, and the loss of Lgl1 in mouse neural progenitors provokes a dysregulation of mitotic spindle orientation. As a consequence, a large proportion of neural progenitor cells do not differentiate but rather they overproliferate, giving rise to a brain tumor phenotype (Klezovitch et al., 2004). In humans, Lgl1 is lost in many solid tumors, and the loss of this gene is strongly correlated with advanced stages of malignant melanoma, colorectal cancer, and endometrial cancer (Schimanski et al., 2005). Furthermore, aberrant splicing of Lgl1 transcripts is observed in human hepatocellular carcinoma (Table 28.1). Like *Lgl*, Lgl1, and Lgl2 interact with the human aPKC-PAR6 protein complex. Interestingly, knockdown of endogenous Lgl1/Lgl2 in epithelial cells induces disorganization of the mitotic spindle, resulting in the formation of multiple micronuclei (Yasumi et al., 2005).

These findings demonstrate a direct role of Lgl in regulating mitotic spindle orientation both in mice and humans. However, the existence of a direct link between Lgl loss in stem cells and tumor initiation has only been demonstrated in mice to date. Thus, further studies are required to determine whether Lgl loss contributes to the conversion of human epithelial stem cells into cells that initiate tumor formation.

Cell Fate Determinants and Cancer

To date, Brat is the only *Drosophila* cell fate determinant implicated in mammalian cancer. In the mouse neocortex, TRIM32, a Brat homologue, is enriched in one of the two daughter cells, and it is upregulated during neuronal differentiation. TRIM32 suppresses self-renewal, and it induces neuronal differentiation via two distinct pathways: by degrading the transcription factor *c-myc*, like *brat* and by activating certain microRNAs important for stem cell self-renewal. Increased levels of TRIM32 mRNA and protein have been described in human head and neck squamous-cell-carcinoma samples (Horn et al., 2004). Furthermore, in vitro studies have shown that TRIM32 overexpression inhibits proliferation and induction of cell differentiation (Schwamborn et al., 2009).

These results suggest that TRIM32 acts as a tumor suppressor gene in mammals, although more comprehensive studies are required to determine whether it plays a direct role in asymmetric cell division and if it participates in the formation of cancer stem cell populations.

Spindle Orientation and Cancer

In mammals, LGN (the mammalian homologue of Pins), Numa (the mammalian homologue of Mud), and $G\alpha_i$ all form a protein complex that directs spindle orientation through the cortical capture of astral microtubules. In the developing mammalian epidermis, the LGN/Numa/$G\alpha_i$ complex is apically localized in mitotic basal cells that undergo asymmetric division (Figure 28.4) (Lechler et al., 2005). In these cells, LGN or Numa silencing induces a shift from asymmetric to symmetric cell division, which results in defects in epidermal stratification and differentiation, due to the impairment of Notch signaling (Williams et al., 2011). Interestingly, aberrant Notch signaling is associated with skin tumors both in mice and humans (Lefort et al., 2007). Further studies are necessary to fully elucidate the role of LGN and Numa proteins in tumor initiation.

Three proteins have recently emerged as good candidates to control spindle orientation in cultured human cells: APC, E-cadherin, and VHL. In vitro knockdown by RNA interference or expression of a dominant-negative version of these proteins impairs spindle orientation (Green et al., 2005; den Elzen et al., 2009; Thoma et al., 2009), although the molecular mechanisms underlying this effect remain unclear. Intriguingly, these proteins are mutated in familial cancers. APC is the disease gene for familial adenomatous polyposis (FAP) syndrome, while patients with germline E-cadherin mutations develop familial diffuse gastric cancer (FDGC). Similarly, patients who inherit a germline mutation of the VHL gene develop von Hippel-Lindau syndrome, characterized by malignant tumors in several organs. However, it is still not clear whether these proteins contribute to tumor initiation by altering spindle orientation, although such a role has been proposed for APC.

In a recent report it was demonstrated that stem cells localized in the intestinal crypts exhibit an apical-basal spindle orientation in wild-type mice that is associated with asymmetric division. By contrast, in APC mutant mice and human patients, intestinal stem cells exhibit a planar spindle orientation (Quyn et al., 2010). When considered along with an independent study demonstrating that crypt stem cells are "cells-of-origin" of intestinal cells (Barker et al., 2009), these findings suggest that the switch from apico-basal to planar spindle orientation mediates the increase in the pool of premalignant stem cells. However, formal proof that stem cells with mutant APC are converted into tumor cells and become dominant in the crypt during tumorigenesis is necessary to unequivocally demonstrate that the APC protein influences the generation and/or behavior of cancer stem cells.

In summary, there is a significant body of evidence supporting a causative link between impaired stem cell asymmetric division and tumor growth in flies. The high degree of evolutionary conservation of some of the molecular complexes that govern stem cell division strongly suggests that such a link also applies to human cancer stem cells, although evidence remains scant. Thus, the role of polarity proteins, cell fate determinants, and proteins controlling spindle orientation in asymmetric stem cell division remains unclear, as does their role in converting mammalian stem cells into tumor-initiating cells.

Conclusions and Future Challenges

The evolution of redundant mechanisms in mammals significantly complicates the analysis of epithelial cell polarity pathways in cancer stem cells. New studies using multiple and/or conditional knockout mouse models will be essential to determine whether core polarity proteins regulate asymmetric stem cell division in mammalian epithelia, and whether their dysregulation drives tumor initiation. If proven, this will provide a novel perspective on the origin of human cancer stem cells and facilitate the development of therapies that correct cell polarity and spindle-orientation in these cells.

Acknowledgments

We thank Carmen M. Ruiz-Jarabo for comments on the manuscript and members of Martin-Belmonte lab for helpful discussions. This work was supported by grants from the Human Frontiers Science Program (HFSP-CDA 00011/2009), Marie Curie Program (IRG-209382), MICINN (BFU2008-01916 and BFU2011-22622), and CONSOLIDER Program (CSD2009-00016) to FM-B. I.B.-R. was a recipient of a postdoctoral grant from the AECC, and I.B was a recipient of a Juan de la Cierva postdoctoral grant from MINECO. An institutional grant from the Fundación Ramón Areces to the CBMSO is also acknowledged.

The authors declare that they have no competing financial interests.

References

Albertson, R. and Doe, C.Q. 2003. Dlg, Scrib and Lgl regulate neuroblast cell size and mitotic spindle asymmetry. Nat. Cell. Biol. 5:166–170.

Amin, N. and Vincan, E. 2012. The Wnt signaling pathways and cell adhesion. Front Biosci. 17:784–804.

Bao, Y., Hata, Y., Ikeda, M., and Withanage, K. 2011. Mammalian Hippo pathway: from development to cancer and beyond. J Biochem. 149(4):361–379.

Barker, N., Ridgway, R., van Es, J., van de Wetering, M., Begthel, H., van den Born, M., Danenberg, E., Clarke, A.R., Sansom, O.J. and Clevers, H. 2009. Crypt stem cells as the cells-of-origin of intestinal cancer. Nature 457(7229):608–611.

Benjamin, J.M. and Nelson, W.J. 2008. Bench to bedside and back again: molecular mechanisms of α-catenin function and roles in tumorigenesis. Semin Cancer Biol. 18(1):53–64.

Betschinger, J., Mechtler, K., and Knoblich, J.A. 2006. Asymmetric segregation of the tumor suppressor brat regulates self-renewal in Drosophila neural stem cells. Cell 124: 1241–1253.

Boggiano, J.C. and Fehon, R.G. 2012. Growth control by committee: intercellular junctions, cell polarity, and the cytoskeleton regulate Hippo signaling. Dev Cell. 22(4):695–702.

Bowman, S.K., Neumuller, R.A., Novatchkova, M., Du, Q. and Knoblich, J.A. 2006. The Drosophila NuMA homolog Mud regulates spindle orientation in asymmetric cell division. Dev. Cell. 10:731–742.

Chen, X. and Macara, I.G. 2006. RNA interference techniques to study epithelial cell adhesion and polarity. Methods Enzymol. 406:362–374.

Clevers, G., Dotto, P., and Radtke, F. 2003. Notch1 functions as a tumor suppressor in mouse skin. Nat Genet 33:416–421.

den Elzen, N., Buttery, C.V., Maddugoda, M.P., Reg, G., and Yap, A.S. 2009. Cadherin adhesion receptors orient the mitotic spindle during symmetric cell division in mammalian epithelia. Mol. Biol. Cell. 20:3740–3750.

Ebnet, K. 2008. Organization of multiprotein complexes at cell-cell junctions. Histochem Cell Biol. 130(1):1–20.

Ebnet, K., Suzuki, A., Ohno, S., and Vestweber, D. 2004. Junctional adhesion molecules (JAMs): more molecules with dual functions? J Cel Sci.117(Pt1):19–29.

Fanning, A.S., Ma, T.Y., and Anderson, J.M. 2002. Isolation and functional characterization of the binding region in the tight junction protein ZO1. FASEB J. 16(13):1835–1837.

Green, R.A., Wollman, R., and Kaplan, K.B. 2005. APC and EB1 function together in mitosis to regulate spindle dynamics and chromosome alignment. Mol.Biol.Cell 16:4609–4622.

Grzeschik, N.A., Parsons, L.M, Allott, M.L., Harvey K.F., and Richardson, H.E. 2010. Lgl, aPKC, and Crumbs regulate Salvador/Warts/ Hippo pathway through two distinct mechanisms. Curr Biol. 20(7):573–581.

Gumbiner, G. 1988. Cadherins: a family of Ca2+−dependent adhesion molecules. Trends Biochem Sci. 13(3):75–76.

Herrmann, J.L., Byekova, Y., Elmets, C.A., and Athar, M. 2011. The role of LKB1 in the pathogenesis of skin and other epithelial cancers. Cancer Lett. 306(1):1–9.

Horn, E.J., Albor, A., Liu, Y., El-Hizawi, S., Vanderbeek, G.F., Babcock, M., Bowden, G.T., Hennings, H., Lozano, G., Weinberg, W.C., and Kulesz-Martin, M. 2004. RING protein Trim32 associated with skin carcinogenesis has anti-apoptotic and E3-ubiquitin ligase properties. Carcinogenesis 25:157–167.

Hulpiau, P. and van Roy, F. 2009. Molecular evolution of the cadherin superfamily. Int J Biochem Cell Biol. 41(2):349–369.

Humbert, P.O., Grzeschik, N.A., Brumby, A.M., Galea, R., Elsum, I., and Richardson, H.E. 2008. Control of tumourigenesis by the Scribble/ Dlg/Lgl polarity module. Oncogene. 27(55): 6888–6907.

Klezovitch, O., Fernandez, T.E., Tapscott, S.J., and Vasioukhin, V. 2004. Loss of cell polarity causes severe brain dysplasia in Lgl1 knockout mice. Genes Dev. 18(5):559–571.

Lechler, T. and Fuchs, E. 2005. Asymmetric cell divisions promote stratification and differentiation of mammalian skin. Nature 437: 275–280.

Lefort, K., Mandinova, A., Ostano, P., Kolev, V., Calpini, V., Kolfschoten, I., Devgan, V., Lieb, J., Raffoul, W., Hohl, D., Neel, D., Garlick, J.,

Chiorino, G., and G. Dotto, P. 2007. Notch1 is a p53 target gene involved in human keratinocyte tumor suppression through negative regulation of ROCK1/2 and MRCKα kinases. Genes Dev 21:562–577

Martin-Belmonte, F. and Perez-Moreno, M. 2012. Epithelial cell polarity, stem cells and cancer. Nat Rev Cancer. 12(1):23–38.

Michel, D., Arsanto, J.P., Massey-Harroche,D., Béclin, C., Wijnholds, J., and Le Bivic, A. 2005. PATJ connects and stabilizes apical and lateral components of tight junctions in human intestinal cells. J Cell Sci. 118(Pt17):4049–4057.

Mirouse, V. and Billaud, M. 2011. The LKB1/AMPK polarity pathway. FEBS Lett. 585(7): 981–985.

Mohamet, L., Hawkins K., and Ward, C.M. 2011. Loss of function of E-cadherin in embryonic stem cells and the relevance to models of tumorigenesis. J. Oncol. 2011:352–616.

Neumuller, R.A. and Knoblich, J.A. 2009. Dividing cellular asymmetry: asymmetric cell division and its implications for stem cells and cancer. Genes Dev. 23:2675–2699.

Nicolas, M., Wolfer, A., Raj, K., Kummer, J.A., Mill, J., van Noort, M., Hui, CH., Clevers, H., Dotto, G.P., and Radtke, F. 2003. Notch1 functions as a tumor suppressor in mouse skin. Nat Genet 33:416–421.

Okada, T., You, L., and Giancotti, F. 2007. Shedding light on Merlin′s wizardry. Trends Cell Biol 17(5):222–9.

Quyn, A.J., Appleton, P.L., Carey, F.A., Steele, R.J., Barker, N., Clevers, H., Ridgway, R.A., Sansom, O.J., and Nathke, I.S. 2010. Spindle orientation bias in gut epithelial stem cell compartments is lost in precancerous tissue. Cell Stem Cell 6(2):175–181.

Royer, C. and Lu, X. 2011. Epithelial cell polarity: a major gatekeeper against cancer? Cell Death Differ.18(9):1470–7.

Schimanski, C.C., Schmitz, G., Kashyap, A., Bosserhoff, A.K., Bataille, F., Schafer, S.C., Lehr, H.A., Berger, M.R., Galle, P.R., Strand, S., and Strand, D. 2005. Reduced expression of Hugl-1, the human homologue of Drosophila tumor suppressor gene lgl, contributes to progression of colorectal cancer. Onocogene 24:3100–3109.

Schwamborn, J.C., Berezikov, E., and Knoblich, J.A. 2009. The TRIM-NHL protein TRIM32 activates microRNAs and prevents self-renewal in mouse neural progenitors. Cell 136:913–925.

Sebbagh, M., Santoni, M.J., Hall, B., Borg, J.P., and Schwartz, M.A. 2009. Regulation of LKB1/STRAD localization and function of E-cadherin. 19(1):37–42.

Silvis, M.R., Kroger, B.T., Lien, W.H., Klezovitch, O., Rudakova, G.M., Camargo, F.D., Lantz, D.M., Seykora, J.T., and Vasioukhin, V. 2011. α-catenin is a tumor suppressor that controls cell accumulation by regulating the localization and activity of the transcriptional coactivator Yap1. Sci Signal. 24;4(174):ra33.

Sun, G. and Irvine, K.D. 2011. Regulation of Hippo signaling by Jun kinase signaling during compensatory cell proliferation and regeneration, and in neoplastic tumors. Dev Biol. 350(1):139–51.

Takai, Y. and Nakanishi, H. 2003. Nectin and afadin: novel organizers of intercellular junctions. J Cell Sci. 116(Pt1)17–27.

Thoma, C.R., Toso, A., Gutbrodt, K.L., Reggi, S.P., Frew, I.J., Schraml, P., Hergovich, A., Moch, H., Meraldi, P., and Krek, W. 2009. VHL loss causes spindle misorientation and chromosome instability. Nat. Cell. Biol. 11:994–1001.

Varelas, X., Samavarchi-Tehrani, P., Narimatsu, M., Weiss, A., Cockburn, K., Larsen, B.G., Rossant, J., and Wrana, J.L. 2010. The Crumbs complex couples cell density sensing to Hippo-dependent control of the TGFβ-SMAD pathway. Dev Cell. 19(6):831–844.

Wheelock, M.J., Shintani, Y., Maeda, M., Fukumoto Y., and Johnson, K.R. 2008. Cadherin switching. J. Cell Sci. 121(Pt6): 727–735.

Williams, S.E., Beronja, S., Pasolli H.A., and Fuchs, E. 2011. Asymmetric cell divisions promote Notch-dependent epidermal differentiation. Nature 470:353–358.

Xiao, K., Oas, R.G., Chiasson C.M., and Kowalczyk, A.P. 2007. Role of p120-catenin in cadherin trafficking. Biochim. Biophys. Acta. 1773(1):8–16.

Yasumi, M., Sakisaka, T., Hoshino, T., Kimura, T., Sakamoto, Y., Yamanaka, T., Ohno, S., and Takai, Y. 2005. Direct binding of Lgl2 to LGN during mitosis and its requirements for normal cell division. J. Biol. Chem. 280(8):6761–6765.

Yin, M. and Zhang, L. 2011. Hippo signaling: a hub of growth control, tumor suppression and pluripotency maintenance. J Genet Genomics. 30(10):471–481.

Yonemura, S., Wada, Y., Watanabe, T., Nagafuchi, A., and Shibata, M. 2010. α-catenin as a tension transducer that induces adherens junction development. Nat Cell Biol. 12(6):533–542.

Chapter 29

Cancer-Initiating Cells, Exosomes, and the Premetastatic Niche

Margot Zöller
Department of Tumor Cell Biology, University Hospital of Surgery, Heidelberg, Germany

Abbreviations

AML	acute myeloid leukemia
ASC	adult stem cells
BMP	bone morphogenetic protein
CAF	cancer-associated fibroblasts
CIC	cancer-initiating cells
CTL	cytotoxic T cells
DC	dendritic cells
EC	endothelial cells
ECM	extracellular matrix
EMT	epithelial to mesenchymal transition
ESC	embryonic stem cells
HSC	hematopoietic SC
HSP	heat shock proteins
LIC	leukemia-initiating cells
MET	mesenchymal to epithelial transition
MSC	mesenchymal SC
MVB	multivesicular bodies
M□	macrophages
SC	stem cells
TEM	tetraspanin-enriched membrane domain

Cancer Stem Cells, First Edition. Edited by Vinagolu K. Rajasekhar.

Introduction

Cancer remains one of the leading causes of death. This mostly is due to the capacity of malignant cells to metastasize, the unpredictable spread of tumor cells frequently setting the cornerstone for curative therapy (Nicholson et al., 2011).

Recently strong evidence has been provided that tumors, a heterogeneous population of cells, contain a small population of cancer-initiating cells (CIC) (Sales et al., 2007), which accounts for primary tumor growth as well as metastatic spread (Al-Hajj and Clarke, 2004). CICs resemble embryonic stem cells (ESC) and adult stem cells (ASC), which during development, account for tissue organization and in the adult organism for tissue repair (Collas et al., 2007). CICs are slowly progressing through the cell cycle (Foijer and te Riele, 2006), are highly radiation and drug resistant (Huang et al., 2007) and use similar signaling pathways that guide the fate of ESC and ASC (Bigas and Epinosa, 2012). Finally, at least, metastasizing CICs also require a niche to support homing and survival (Bissell and Labarge, 2005). Great efforts are undertaken to unravel the pathways whereby CICs account for tumor progression and therapy resistance, because there is hope that by attacking CICs a more efficient approach than is currently available can be designed (Horch et al., 2012).

This review focuses on CICs and metastasis, discussing the current state of knowledge on the CIC niche and introducing exosomes, which are supposed to be the main actors in the cross talk between CICs and the host, thereby promoting metastatic spread.

Cancer-Initiating Cells and Metastasis

More than a century ago, Virchow (1858) proposed that cancer arises from embryonic-like cells, the hypothesis receiving strong support by the discovery of CICs (Allan et al., 2007). One of the features qualifying CICs for metastatic growth is their particular intense cross talk with the host. The fate of ESC and ASC is determined by their position and is minutely regulated by few adjacent cells creating a defined environment, the niche (Fuchs et al., 2004). CICs, too, appear to require a niche, possibly during oncogenesis as well as during settlement and growth in distant organs (Li and Neaves, 2006). The latter is called the premetastatic niche, because it is initiated by the primary tumor before metastasizing cell-arrival (Kaplan et al., 2005).

A Niche for CICs Within the Primary Tumor

Stem cell (SC) niches are composed of epithelial and mesenchymal cells and extracellular substrates; they are important for stemness maintenance (Kuhn and Tuan, 2010). Besides intrinsic regulatory systems (Groszer et al., 2001), the niche functions as an extrinsic regulatory system, which maintains and governs location, adhesiveness, retention, homing, mobilization, quiescence/activation, and symmetric/asymmetric division and differentiation (Li and Neaves, 2006). Important regulators of cell fate are Wnt, Hedgehog, Notch, TGFβ, several tyrosine kinase receptors that all use similar intracellular signaling pathways like the Ras-Raf-MAPK and PI3K-Akt pathway (Fossett, 2012).

Whether CICs require a niche has not been unequivocally answered. First, the niche for ASC provides a regulatory system, which could promote CIC proliferation, but also could prevent tumorigenesis by controlling ASC (Li and Neaves, 2006). In fact, malignant melanoma CIC could be reprogrammed by exposure to an embryonic microenvironment (Hendrix et al., 2007). On the other hand, a preformed niche can support CIC survival and homing (Bissell and Labarge, 2005). Thus, leukemia-(L)IC can survive stroma-independent but home in vivo into the niche, which provides a growth advantage (Mulloy et al., 2003). A specialized niche also was demonstrated for several CIC (Lathia et al., 2011).

Important contributors to the CIC niche are cancer-associated fibroblasts (CAF) that express αSMA (Cirri and Chiarugi, 2012). CAF contribute to extracellular matrix (ECM) remodeling (Borovski et al., 2011) by provision of HGF, IL6, PDGFβ, prostaglandins, and proteases (Giannoni et al., 2010) as well as by miR-31, downregulating the chromatin remodeling homeobox gene SATB2 (Aprelikova et al., 2010) and additional miRNA implicated in cell proliferation, which are transferred from stroma cells into tumor cells (Grange et al., 2011). Other important players are mesenchymal stem cells (MSC) (Ehninger and Trumpp, 2011) where nestin1+ MSC together with CAF (Crisan et al., 2008) and macrophages (Mϕ) (Winkler et al., 2010) support hematopoietic-(H)SC and LIC. Mϕ recruitment may also be promoted by tissue factor-secreting tumor cells (Gil-Bernabé et al., 2012). Furthermore, tumor cell-derived IL1 induces PGE2 secretion by MSC, which operates in an autocrine manner to induce secretion of cytokines that induce β-catenin signaling and CIC formation in adjacent tumor cells (Li et al., 2012).

The finding that tumor stroma promotes epithelial to mesenchymal transition (EMT) strengthens CIC requiring a niche (Biddle and Mackenzie, 2012). EMT is a reversible process allowing epithelial cells to detach and acquire a motile mesenchymal phenotype (Strauss et al., 2012). EMT is controlled on multiple levels, one of the crucial steps being the transcriptional repression of E-cadherin, where Snail1, Snail2, Zeb1, Zeb2, E47, and Klf8 are direct repressors and Twist is an important indirect repressor. Repression of E-cadherin and, in addition, of tight junction components is accompanied by activation of the mesenchymal genes N-cadherin, vimentin, fibronectin, and cyclinD1 (Yang and Weinberg, 2008). Besides these intrinsic regulators, CIC-derived TGFβ converts fibroblasts into myofibroblasts, which secrete TGFβ and by back-signaling force CIC into EMT (Biddle and Mackenzie, 2012). Additional inflammatory mediators delivered by the stroma sustain TGFβ production and attract MSC to produce CIC supportive CXCL7 (Wang et al., 2010). Furthermore, CIC may even differentiate into niche cells (Korkaya et al., 2011).

Thus, CICs shape their own environment by recruiting and activating specific cell types that support maintenance and the switch toward a motile phenotype, a prerequisite to metastasize.

CIC and the Premetastatic Niche

Of particular interest for CICs is the formation of a premetastatic niche (Kaplan et al., 2005). The authors noted that in the organs where tumor cells are likely to settle and grow, niches can be detected that later on will house the metastatic cells. How do CICs prepare a niche in advance of arrival?

It has been described that hematopoietic progenitors expressing VEGFR1 and VLA-4 are recruited into the niche. VLA-4 binds to fibronectin, which is enriched in the premetastatic niche due to upregulated expression by local fibroblasts. The local fibroblasts become stimulated by not yet defined tumor-derived factors, where the stimulation of tissue resident fibroblasts follows the metastatic profile of different tumor types. In addition, stromal fibroblasts frequently express high amounts of CXCL12 and, thus, may first attract CXCR4-expressing hematopoietic progenitors and later on CXCR4-expressing CIC (Kucia et al., 2007). CXCR4 is expressed on ESC, ASC, and several CIC (Suárez-Álvarez et al., 2012); CXCL12 is secreted by bone marrow stromal and endothelial cells (EC) and attracts hematopoietic progenitor cells to the niche (Jung et al., 2006). Conversely, G-CSF that mobilizes HSC induces CXCL12 degradation via activation of neutrophil elastase (Petit et al., 2002). Thus, organs with high CXCL12 expression, like lymph nodes, lung, liver, and bones recruit CXCR4+ CIC (Domanska et al., 2012).

Another important program for invasiveness of SCs and CICs proceeds via Met

(Boccaccio and Comoglio, 2006): HGF and Met drive mobilization and emigration of ESC and ASC (Kucia et al., 2007). Via MSCs, which express HGF, the HGF-Met axis becomes involved in angiogenesis (Forte et al., 2006). Many carcinoma cells expressing Met, the same pathway accounts for tumor cell migration and metastasis (Gherardi et al., 2012), where the CIC marker CD44v6 contributes to activation of the Met-signaling cascade (Zöller, 2011).

Finally, β-catenin is a key player in stem cell homing/mobilization and metastasis (Wray and Hartmann, 2012). Heterodimeric β-catenin interacts with membrane-bound cadherin, but upon activation translocates to the nucleus and interacts with the Tcf/Lef complex, which turns on cell cycle-related genes like *cyclin D1* and *c-Myc* (Clevers and Nusse, 2012). Costimulatory signals from the environment like HGF, EGF, TGFβ support nuclear localization (Jamieson et al., 2012).

Though all outlined pathways likely contribute to premetastatic niche preparation, the organ selectivity of metastasis is difficult to explain, as all components are rather abundant in tumor-free tissues. Thus, we suggested that exosomes, rather than individual factors, are the driving force (Jung et al., 2009), which was confirmed by several groups, the pathways, whereby exosomes fulfill this task currently being explored. In advance of lining the current state of knowledge, I will introduce exosomes.

Exosomes, the SC/CIC Communicators

Exosomes, the most important intercellular communicators (György et al., 2011), are small 30–100 nm vesicles, which derive from the fusion of intraluminal vesicles of multivesicular bodies (MVB) with the plasma membrane. Exosomes, delivered by many cells and abundantly by tumor cells, are found in all body fluids (Simons and Raposo, 2009).

Exosome Assembly and Secretion

Exosomes contain a restricted repertoire of proteins, mRNA, and miRNA, which points toward active sorting. Protein sorting depends on mono ubiquitination and the endosomal sorting complex required for transport. Protein recruitment into tetraspanin networks and other internalization prone detergent-resistant membrane domains can also be decisive (Trajkovic et al., 2008). Recruitment via raft or tetraspanin-enriched microdomains (TEM) is accompanied by recovery of exosomal protein complexes that affects exosome targeting (Rana et al., 2011; Rana et al., 2012). mRNA recruitment can be guided by a zip code in the 3′-UTR (Pant et al., 2012). miRNA recruitment is facilitated by coupling of RISCs (RNA-induced silencing complexes) to components of the sorting complex (Lee et al., 2009), the release of miRNA being controlled through a ceramide-dependent machinery associated with exosome secretion (Kosaka et al., 2010).

According to their origin, exosomes are rich in molecules located in membrane domains prone for internalization as well as molecules engaged in fission, scission, and vesicular transport (Rak, 2010). Thus, tetraspanins, abundantly recovered in intracellular vesicles (Mathivanan et al., 2010), are constitutive exosome components, CD63 serving as a marker to differentiate exosomes from other extracellular vesicles (Pols and Klumperman, 2009). Other abundantly recovered molecules are HSP, enzymes, cytoskeletal proteins, signal transduction, and MHC molecules (Henderson and Azorsa, 2012). Exosomal mRNA is less abundant than exosomal miRNA. Exosomal mRNA that differs from that of the originating cells are involved in cell cycle progression, angiogenesis, migration, or histone modification (Skog et al., 2008; Nazarenko et al., 2010). Exosomes also contain selected miRNA, which in tumor-exosomes can be tumor-specific (Henderson and Azorsa, 2012) (Figure 29.1A). Being

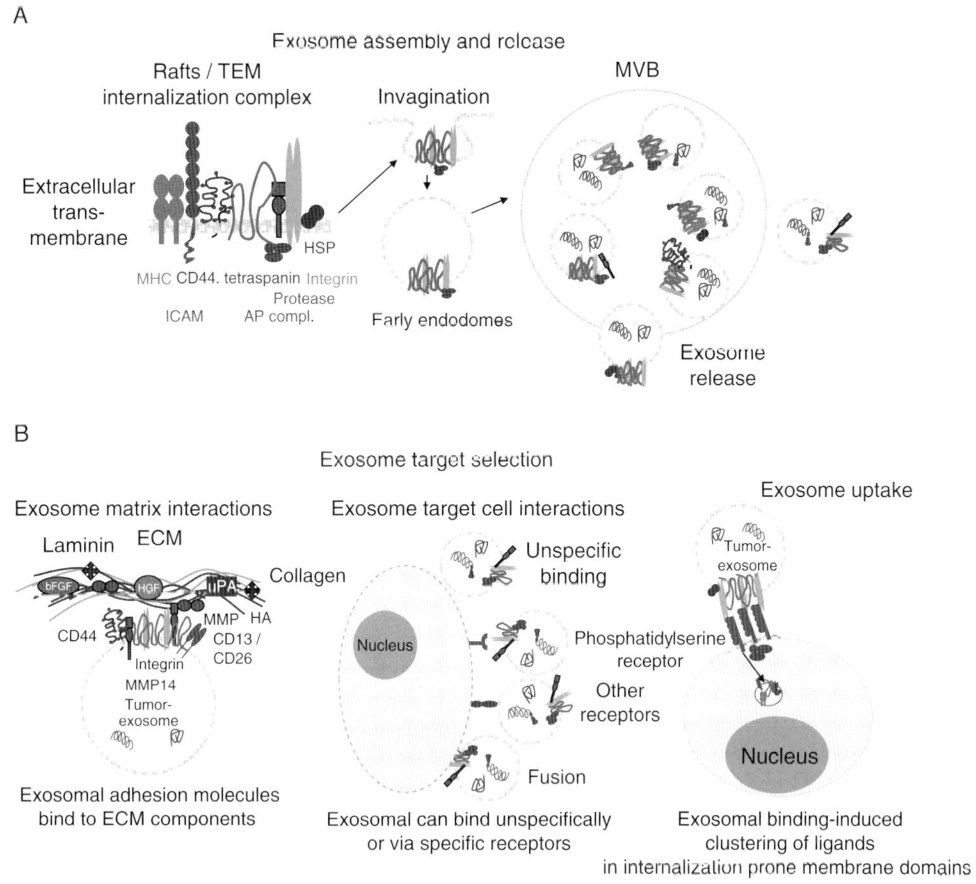

Figure 29.1. Tumor-exosomes and metastasis. **A**. Exosomes derive from early endosomes, which are generated by invagination of defined membrane domains like rafts and TEM. Accordingly, they are rich in proteins/protein complexes located in these microdomains as well as in fission, scission, and vesicular transport molecules. Upon invagination of early endosomes into MVB, cytosolic proteins, mRNA, and miRNA are recruited. **B**. Exosomes bind selective targets, which can be ECM or cellular membrane proteins. The selectivity of binding is facilitated by the engagement of protein clusters. There is strong evidence that exosome-uptake requires or is greatly facilitated by clustered ligands in internalization-prone membrane domains. For color detail, please see color plate section.

detected in all body fluids, the selective enrichment of "marker" proteins, mRNA, and miRNA makes exosomes an attractive means for noninvasive diagnosis (Wittmann and Jäck, 2010).

For exosome secretion, the intracellular Ca++ level (Merendino et al., 2010), the intracellular and extracellular pH (Parolini et al., 2009) and Rab GTPases (Hsu et al., 2010) play key roles.

Exosomes and Target Selection

Exosome target cell selection is still disputed. Exosomes can interact with recipient cells via receptor ligand pairs, attachment or fusion with the target cell membrane, or internalization (Vickers and Remaley, 2012). Irrespective of the mode of uptake, exosomes bind selected targets (Hood et al., 2011; Jung et al., 2009; Nazarenko et al., 2010; Vallhov et al., 2011).

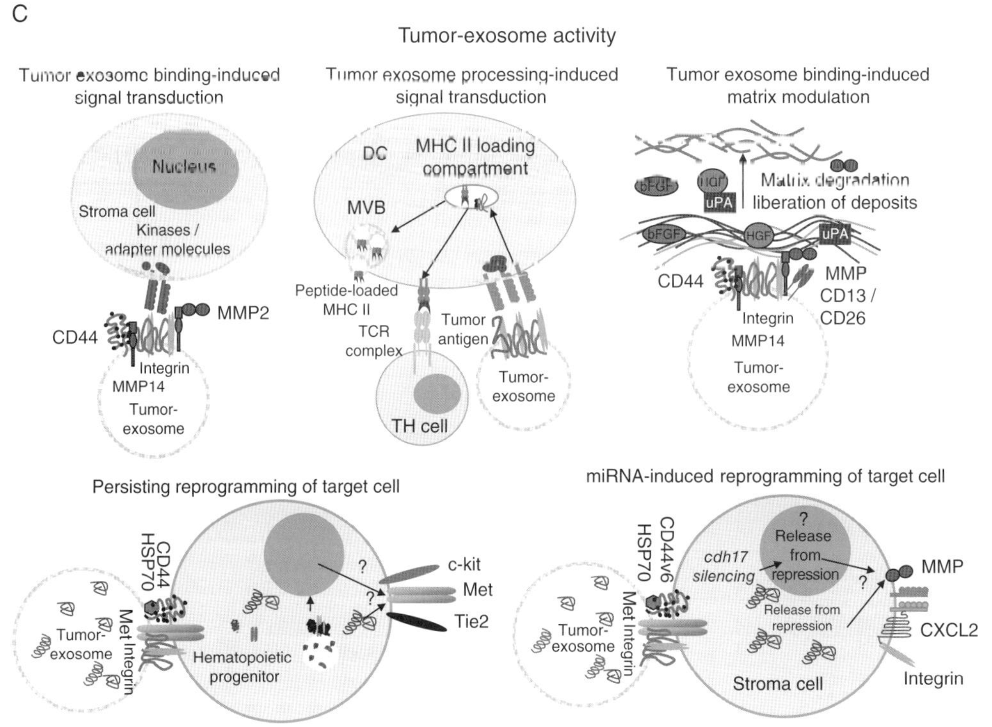

Figure 29.1. (*Continued*) **C.** Exosomal proteins, mRNA, and miRNA are function competent. Thus, binding to raft- or TEM-located ligands, these domains harboring cytosolic signal transduction molecules at the inner membrane, can initiate signal transduction in such a way that the target cell becomes activated. For DC it was described that uptaken exosomes will be transported into the MHC II assembly vesicles so that exosomal tumor antigens will be processed, and peptides are loaded on MHC II molecules, which allows for T helper activation. Alternatively, DC delivers exosomes carrying the tumor peptide-loaded MHC II as well as accessory molecules, so that DC-exosomes are fully equipped for T cell activation. Tumor-exosomes are rich in adhesion molecules like integrins, CAMs, and CD44 as well as in proteases including membrane-bound MMPs. Adhesion molecules are engaged in ECM protein binding, which become partly digested by exosomal proteases accompanied by liberation of matrix incorporated growth factors and proteases. The modulated matrix and the released factors can promote host and tumor cell motility and activation. Finally, tumor-exosomes can persistently reprogram their target cells. The mode whereby this is achieved has not been fully unraveled, but exosomal miRNA likely plays a decisive role by mRNA silencing and also by release from repression by the primary mRNA target. For color detail, please see color plate section.

Already in 2004 it was described that allogeneic exosomes are taken up by dendritic cells (DC), Kupffer cells, and some macrophages (Mϕ), exosome targeting being mediated by milk fat globulin-E8, CD11a, CD54, phosphatidyl serine, CD9, and CD81 on exosomes and requiring αvβ3, CD11a, and CD54 as ligands (Morelli et al., 2004) suggesting that exosome binding and uptake vary depending on the pathway of endocytosis and the protein pattern on exosomes and targets (Atay et al., 2011). Also, phosphatidyl serine exposure on the outer exosomal membrane triggers binding to scavenger receptors, integrins, and complement receptors (Zakharova et al., 2007). More recent evidence points toward a contribution of tetraspanins (Pols and Klumperman, 2009; Zöller, 2009). Exploring this question in detail by pulldown and blocking studies revealed that exosomes bind via TEM-localized, tetraspanin-associated adhesion molecules, where target cell ligands largely overlap with exosomal binding structures (Rana et al., 2012), suggesting similar fusion/fission machineries to account for the internalization process in the exosome donor and target cells (Lakkaraju and Rodriguez-Boulan, 2008; Tian et al., 2010) (Figure 29.1B).

Exosome Activities

Exosomal proteins are function-competent (Simons and Raposo, 2009). In fact, exosome research became highly stimulated, when it was noted that DCs release exosomes, which suffice to stimulate T cells (Bobrie et al., 2011). Exosomes also avidly bind matrix proteins. The ECM being a structural element and storage of bioactive compounds, essential in tissue repair and the cross talk between tumor cells and stroma (Sangaletti and Colombo, 2008), abundant exosomal proteases likely contribute to ECM modulation (Simpson et al., 2009).

Exosome research was strikingly boosted with the discovery of horizontal transfer of genetic material. Exosomal mRNA and miRNA, so called shuttle RNA, are transferred and translated or mediate RNA silencing (Bussolati et al., 2011; Chen et al., 2012; Valadi et al., 2007).

Taken together, exosomes display exquisite target cell selectivity, and exosomal proteins, mRNA, and miRNA are function-competent. Accordingly, therapeutic exploitation of exosomes appears promising, exosomes being already in clinical use as a vaccine (Tan et al., 2010) and are discussed as most potent gene delivery system (Seow and Wood, 2009; Lässer, 2012).

CIC-Exosomes and Metastasis

There are several modes, whereby exosomes can facilitate metastatic progress, immune suppression, angiogenesis, and activation of the premetastatic environment, which includes the impact of exosomes on hematopoietic progenitors and their recruitment. In advance of discussing these points, I would like to remember that LIC/CIC rather than non-LIC/CIC account for metastatic spread (Yu et al., 2012). Thus, even most studies being performed with exosomes from unseparated tumor cells, it appears justified to ascribe the observed effects to CIC-exosomes.

Tumor-Exosomes and the Immune System

DC-exosomes replacing DC is already therapeutically exploited in cancer immunotherapy (Tan et al., 2010), where DC also can use tumor-exosomes as an antigen (Tumne et al., 2009). However, hope for exosomes as a promise for immunotherapy was dampened by tumor-exosomes interfering with immune response induction (Taylor and Gercel-Taylor, 2011). Tumor-exosomes can induce FAS-mediated apoptosis (Abusamra et al., 2005) and inhibit $CD4^+$ T cell proliferation accompanied by stronger suppressive activity of regulatory T cells (Clayton et al., 2007). Impaired NK activity may rely on tumor-exosomes inhibiting Stat5, Jak3 activation, cyclinD3 expression, and perforin release (Zhang et al., 2007) or on blocking NK cells via NKG2D binding (Ashiru et al., 2010). By inducing TGFβ1 and IL4 secretion, tumor-exosomes also suppress antigen-specific responses (Yang et al., 2011). By suppressing inflammatory and anti-tumor immune responses, tumor-exosomes can promote tumor growth even in allogeneic mice (Lima et al., 2009).

But tumor-exosomes also can support immune response induction. Exosomes express HSP, which functions as an endogenous danger signal promoting NK activation and tumor cell lysis (Khalil et al., 2011). Exosomes from heat-stressed tumor cells also induce a tumor-antigen-specific cytotoxic T cell (CTL) response, support NK activity, IL6 and TNFα secretion, and costimulatory molecule expression in DC (Hurwitz et al., 2010). Vaccination with staphyloccocus enterotoxin A expressing tumor-exosomes inhibits tumor growth by promoting IL2 and IFNγ secretion, T helper, and CTL and NK activation (Xiu et al., 2007). Exosomal chemokines from heat-stressed tumor cells also attract DC and T cells (Chen et al., 2011). Importantly, tumor antigens presented by tumor-exosomes are a stronger immunogen than when presented by tumor cells (Zeelenberg et al., 2011) (Figure 29.1C).

Taken together, tumor-exosomes can be immunosuppressive, but also can strengthen tumor-specificimmuneresponses.Thus,tumor-exosomes in combination with DC-derived exosomes may well contribute to cancer immunotherapy.

Tumor-Exosomes and Angiogenesis

There is ample evidence that tumor-exosomes induce angiogenesis. Melanoma exosome-induced angiogenesis was accompanied by endothelial tube formation and proangiogenic cytokine secretion by EC (Hood et al., 2009). Colorectal cancer and glioblastoma-exosomes, too, promote EC proliferation by the transfer of exosomal proteins and mRNA (Hong et al., 2009; Skog et al., 2008). Also, EGFR-positive tumor-exosomes are taken-up by EC and elicit EGFR-dependent responses including VEGFR2 expression (Al-Nedawi et al., 2009). Angiogenesis also can be induced by the transfer of Notch ligand delta like 4, which inhibits Notch signaling (Sheldon et al., 2010). Tumor-exosomes expressing a Tspan8-CD49d complex induce EC/EC progenitor maturation and activation, accompanied by VEGFR2 transcription (Nazarenko et al., 20010). Finally, one of the few reports on CIC-exosomes describes CD105+ renal-CIC-exosomes to carry proangiogenic mRNA and miRNA, which trigger the angiogenic switch (Grange et al., 2011).

Taken together, tumor-exosomes induce angiogenesis, which progresses via EC/EC progenitor stimulation and/or the transfer of mRNA and miRNA.

Tumor-Exosomes, Bone Marrow Cell Progenitors and Stroma Cells in Premetastatic Organs

Organ selectivity of metastasis has remained enigmatic until the discovery of the premetastatic niche (Kaplan et al., 2005), where defining exosomes as the initiating agent has become a major breakthrough (Ge et al., 2012; Peinado et al., 2011).

An involvement of exosomes in metastasis was first described for platelet-derived exosomes, which transferred the αIIb integrin chain to lung cancer cells, stimulated the MAPK pathway, increased expression of MT1-MMP, cyclin D2, and angiogenic factors, and enhanced adhesion to fibrinogen and EC (Janowska-Wieczorek et al., 2005). We explored that exosomes together with a tumor matrix support recruitment of hematopoietic progenitors from the bone marrow and activation of stroma cells and leukocytes in premetastatic nodes allowing settlement and growth of a nonmetastatic tumor (Jung et al., 2009), subcutaneously applied exosomes reaching the draining node, the lung, the bone marrow, and spleen. Melanoma-exosomes enhance migration and homing of melanoma cells in sentinel nodes by changes in angiogenic factor secretion and matrix modulation (Hood et al., 2011). A complex of exosomal HSP90-MMP2-tissue plasminogen activator-annexin II promotes plasmin activation and thereby tumor cell motility (McCready et al., 2010). Ongoing work on the mode whereby exosomes modulate the local "premetastatic environment" confirmed recovery of mRNA and miRNA in host stroma cells, transferred miRNA significantly affecting mRNA translation, so that metastasis-suppressor downregulation is accompanied by metastasis-associated genes being released from repression (Rana et al., 2013).

The group of Lyden focused on the impact of tumor-exosomes on bone marrow progenitors. In an elegant study they first describe that the protein profile of melanomaexosomes is predictive for metastatic progression suggesting diagnostic validity of serum-exosomes. Further they noted that exosomes enhance endothelial cell permeability and, most impressively, permanently educate bone marrow progenitors through upregulation of c-Met toward a pro-vasculogenic/pro-metastatic phenotype with upregulated c-Kit, Tie2, and Met expression, speculating on an important contribution of exosomal miRNA (Peinado et al., 2012).

Last but not least, the collagenolytic and laminin-degrading activity of exosomes facilitates angiogenesis and metastasis (Hakulinen et al., 2008; Park et al., 2010), whether liberated cytokines/chemokines, proteases, and cleavage products additionally promote motility and/or angiogenesis, remains to be explored (Figure 29.1C).

Taken together, tumor/CIC-exosomes are engaged in premetastatic niche preparation. Tumor-exosomes directly affect the stroma of the premetastatic organ as well as hematopoietic progenitors that become recruited toward the niche. The link between reprogrammed hematopoietic progenitors and their recruitment into the premetastatic organ remains to be unraveled. A second question, how exosomes modulate their targets, also has not been sufficiently answered, although there is evidence that exosome binding initiates signaling cascades and transferred miRNA reprograms target cells. Since exosomes are the most powerful means of intercellular communication that function across long distance, much effort should be given to unravel these open questions.

Outlook

Exosomes are easy to manipulate and provide a powerful means of protein and gene transfer. This suggests the therapeutic use of exosomes in cancer therapy as outlined in several excellent reviews. Thus, I will only summarize my personal point of view (Figure 29.1):

- Tumor-exosomes can be immunosuppressive but also be a superior antigen source for DC. Though not fully explored, it can be hypothesized that exosomes present the whole panel of tumor antigens, which being processed by DC allows for T helper activation. A tumor-exosome vaccine may be particularly advantageous, when nominal antigens are unknown.
- Exosome-initiated tumor angiogenesis has fostered the search for therapeutic interference. According to current knowledge, transferred miRNA likely initiates the angiogenic switch. Undue angiogenesis might become balanced by an excess of exosomes displaying identical targeting features but carrying regulatory miRNA. The same therapeutic principle may be used to prevent bone marrow progenitor cell and metastatic organ stroma cell priming for niche formation. For both approaches, tailoring the therapeutic exosomal membrane proteins to interact with selected target cells will be mandatory.
- Of central importance is the elimination of CICs themselves. This again requires a most scrutinized elaboration of exosome targeting structures on CICs, which will be facilitated by the definition of CIC markers that are mostly located in internalization-prone membrane domains and will facilitate exosome uptake. Loading exosomes with miRNA to reprogram CICs toward differentiation and/or apoptosis will become feasible with further characterization of miRNA and their targets.

At the end, here is an old story. Exosomes first crossed my way when rats with a small tumor developed deathly systemic consumption coagulopathy due to exosomes that initiated overshooting angiogenesis (Claas et al., 1998; Gesierich et al., 2006). Thus, exosomes are extremely powerful, justifying all the hope set in "therapeutic" exosomes to fight CICs. However, utmost care is required to tailor exosomes for "drug" delivery only to selected targets.

References

Abusamra, A.J., Zhong, Z., Zheng, X., Li, M., Ichim, T.E., Chin, J.L., Min, W.P. (2005). Tumor-exosomes expressing Fas ligand mediate CD8+ T-cell apoptosis. Blood Cells Mol Dis 35:169–173.

Al-Hajj, M., Clarke. M.F. (2004). Self-renewal and solid tumor stem cells. Oncogene 23: 7274–7282.

Allan, A.L., Vantyghem, S.A., Tuck, A.B., Chambers, A.F. (2006–2007). Tumor dormancy and cancer stem cells: implications for the biology and treatment of breast cancer metastasis. Breast Dis 26:87–98.

Al-Nedawi, K., Meehan, B., Kerbel, R.S., Allison, A.C., Rak, J. (2009). Endothelial expression of autocrine VEGF upon the uptake of tumor-derived microvesicles containing oncogenic EGFR. Proc Natl Acad Sci USA 106: 3794–3799.

Aprelikova, O., Yu, X., Palla, J., Wei, B.R., John, S., Yi, M., Stephens, R., Simpson, R.M., Risinger, J.I., Jazaeri, A., Niederhuber, J. (2010). The role of miR-31 and its target gene SATB2 in cancer-associated fibroblasts. Cell Cycle 9:4387–4398.

Ashiru, O., Boutet, P., Fernández-Messina, L., Agüera-González, S., Skepper, J.N., Valés-Gómez, M., Reyburn, H.T. (2010). Natural killer cell cytotoxicity is suppressed by exposure to the human NKG2D ligand MICA*008 that is shed by tumor cells in exosomes. Cancer Res 70:481–489.

Atay, S., Gercel-Taylor, C., Taylor, D.D. (2011). Human Trophoblast-Derived Exosomal Fibronectin Induces Pro-Inflammatory Il-1β Production by Macrophages. Am J Reprod Immunol 66:259–269.

Biddle, A., Mackenzie, I.C. (2012). Cancer stem cells and EMT in carcinoma. Cancer Metastasis Rev Feb 3. [Epub ahead of print]

Bigas, A., Espinosa, L. (2012). Hematopoietic stem cells: to be or Notch to be. Blood 119: 3226–3235.

Bissell, M.J., Labarge, M.A. (2005). Context, tissue plasticity, and cancer: are tumor stem cells also regulated by the microenvironment? Cancer Cell 7:17–23.

Bobrie, A., Colombo, M., Raposo, G., Théry, C. (2011). Exosome secretion: molecular mechanisms and roles in immune responses. Traffic 12:1659–1668.

Boccaccio, C., Comoglio, P.M. (2006). Invasive growth: a MET-driven genetic programme for cancer and stem cells. Nat Rev Cancer 6:637–645.

Borovski, T., De Sousa, E., Melo, F., Vermeulen, L., Medema, J.P. (2011). Cancer stem cell niche: the place to be. Cancer Res 71:634–639.

Bussolati, B., Grange, C., Camussi, G. (2011). Tumor exploits alternative strategies to achieve vascularization. FASEB J 25:2874–2882.

Chen, X., Liang, H., Zhang, J., Zen, K., Zhang, C.Y. (2012). Horizontal transfer of microRNAs: molecular mechanisms and clinical applications. Protein Cell 3:28–37.

Chen, T., Guo, J., Yang, M., Zhu, X., Cao, X. (2011). Chemokine-containing exosomes are released from heat-stressed tumor cells via lipid raft-dependent pathway and act as efficient tumor vaccine. J Immunol 186:2219–2228.

Cirri, P., Chiarugi, P. (2012). Cancer associated-fibroblasts and tumour cells: a diabolic liaison driving cancer progression. Cancer Metastasis Rev 31:195–208.

Claas, C., Seiter, S., Claas, A., Savelyeva, L., Schwab, M., Zöller, M. (1998). Identification of the rat homologue of CO-029, a metastasis-associated tetraspan molecule. J Cell Biol 141:267–280.

Clayton, A., Mitchell, J.P., Court, J., Mason, M.D., Tabi, Z. (2007). Human tumor-derived exosomes selectively impair lymphocyte responses to interleukin-2. Cancer Res 67:7458–7466.

Clevers, H., Nusse, R. (2012) Wnt/β-Catenin Signaling and Disease. Cell 149:1192–1205.

Collas, P., Noer, A., Timoskainen, S. (2007). Programming the genome in embryonic and somatic stem cells. J Cell Mol Med 11:602–620.

Crisan, M., Yap, S., Casteilla, L., Chen, C.W., Corselli, M., Park, T.S., Andriolo, G., Sun, B., Zheng, B., Zhang, L., Norotte, C., Teng, P.N., Traas, J., Schugar, R., Deasy, B.M., Badylak, S., Buhring, H.J., Giacobino, J.P., Lazzari, L., Huard, J., Péault, B. (2008). A perivascular origin for mesenchymal stem cells in multiple human organs. Cell Stem Cell 3:301–313.

Domanska, U.M., Kruizinga, R.C., Nagengast, W.B., Timmer-Bosscha, H., Huls, G., de Vries, E.G., Walenkamp, A.M. (2013). A eview on CXCR4/CXCL12 axis in oncology: No place to hide. Eur J Cancer Jun 49:219–230.

Ehninger, A., Trumpp, A. (2011). The bone marrow stem cell niche grows up: mesenchymal stem cells and macrophages move in. J Exp Med 208:421–428.

Foijer, F., te Riele, H. (2006). Check, double check: the G2 barrier to cancer. Cell Cycle 5:831–836.

Forte, G., Minieri, M., Cossa, P., Antenucci, D., Sala, M., Gnocchi, V., Fiaccavento, R., Carotenuto, F., De Vito, P., Baldini, P.M., Prat, M., Di Nardo, P. (2006). Hepatocyte growth factor effects on mesenchymal stem cells: proliferation, migration, and differentiation. Stem Cells 24:23–33.

Fossett, N. (2013) Signal transduction pathways, intrinsic regulators, and the control of cell fate choice. Biochim Biophys Acta 1830:2375–2384.

Fuchs, E., Tumbar, T., Guasch, G. (2004) Socializing with the neighbors: stem cells and their niche. Cell 116:769–778.
Ge, R., Tan, E., Sharghi-Namini, S., Asada, H.H. (2012). Exosomes in Cancer Microenvironment and Beyond: Have We Overlooked These Extracellular Messengers? Cancer Microenviron 5:323–332.
Gesierich, S., Berezovkiy, I., Ryschich, E., Zöller, M. (2006). Angiogenesis induction by the tetraspanin D6.1A. Cancer Res 66:7083–7094.
Gherardi, E., Birchmeier, W., Birchmeier, C., Vande Woude, G. (2012). Targeting MET in cancer: rationale and progress. Nat Rev Cancer 12:89–103.
Giannoni, E., Bianchini, F., Masieri, L., Serni, S., Torre, E., Calorini, L., Chiarugi, P. (2010). Reciprocal activation of prostate cancer cells and cancer-associated fibroblasts stimulates epithelial-mesenchymal transition and cancer stemness. Cancer Res 70:6945–6956.
Gil-Bernabé, A.M., Ferjancic, S., Tlalka, M., Zhao, L., Allen, P.D., Im, J.H., Watson, K., Hill, S.A., Amirkhosravi, A., Francis, J.L., Pollard, J.W., Ruf, W., Muschel, R.J. (2012). Recruitment of monocytes/macrophages by tissue factor-mediated coagulation is essential for metastatic cell survival and premetastatic niche establishment in mice. Blood 119:3164–3175.
Grange, C., Tapparo, M., Collino, F., Vitillo, L., Damasco, C., Deregibus, M.C., Tetta, C., Bussolati, B., Camussi, G. (2011). Microvesicles released from human renal cancer stem cells stimulate angiogenesis and formation of lung premetastatic niche. Cancer Res 71:5346–556.
Groszer, M., Erickson, R., Scripture-Adams, D.D., Lesche, R., Trumpp, A., Zack, J.A., Kornblum, H.I., Liu, X., Wu, H. (2001). Negative regulation of neural stem/progenitor cell proliferation by the Pten tumor suppressor gene in vivo. Science 294:2186–2189.
György, B., Szabó, T.G., Pásztói, M., Pál, Z., Misják, P., Aradi, B., László, V., Pállinger, E., Pap, E., Kittel, A., Nagy, G., Falus, A., Buzás, E.I. (2011). Membrane vesicles, current state-of-the-art: emerging role of extracellular vesicles. Cell Mol Life Sci 68:2667–288.
Hakulinen, J., Sankkila, L., Sugiyama, N., Lehti, K., Keski-Oja, J. (2008). Secretion of active membrane type 1 matrix metalloproteinase (MMP-14) into extracellular space in microvesicular exosomes. J Cell Biochem 105:1211–1218.
Henderson, M.C., Azorsa, D.O. (2012). The genomic and proteomic content of cancer cell-derived exosomes. Front Oncol 2012;2:38.
Hendrix, A., Westbroek, W., Bracke, M., De Wever, O. (2010). An ex(o)citing machinery for invasive tumor growth. Cancer Res 70:9533–9537.
Hendrix, M.J., Seftor, E.A., Seftor, R.E., Kasemeier-Kulesa, J., Kulesa, P.M., Postovit, L.M. (2007). Reprogramming metastatic tumour cells with embryonic microenvironments. Nat Rev Cancer 7:246–255.
Hong, B.S., Cho, J.H., Kim, H., Choi, E.J., Rho, S., Kim, J., Kim, J.H., Choi, D.S., Kim, Y.K., Hwang, D., Gho, Y.S. (2009). Colorectal cancer cell-derived microvesicles are enriched in cell cycle-related mRNAs that promote proliferation of endothelial cells. BMC Genomics 10:556.
Hood, J.L, Pan, H., Lanza, G.M., Wickline, S.A. (2009). Paracrine induction of endothelium by tumor exosomes. Consortium for Translational Research in Advanced Imaging and Nanomedicine (C-TRAIN). Lab Invest 89: 1317–1328.
Hood, J.L., San Roman, S., Wickline, S.A. (2011). Exosomes released by melanoma cells prepare sentinel lymph nodes for tumor metastasis. Cancer Res 71:3792–801.
Horch, R.E., Kneser, U., Polykandriotis, E., Schmidt, V.J., Sun, J., Arkudas, A. (2012). Tissue engineering and regenerative medicine—where do we stand? J Cell Mol Med 16:1157–1165.
Hsu, C., Morohashi, Y., Yoshimura, S., Manrique-Hoyos, N., Jung, S., Lauterbach, M.A., Bakhti, M., Grønborg, M., Möbius, W., Rhee, J., Barr, F.A., Simons, M. (2010) Regulation of exosome secretion by Rab35 and its GTPase-activating proteins TBC1D10A-C. J Cell Biol 189:223–232.
Huang, E.H., Heidt, D.G., Li, C.W., Simeone, D.M. (2007). Cancer stem cells: a new paradigm for understanding tumor progression and therapeutic resistance. Surgery 141:415–419.
Hurwitz, M.D., Kaur, P., Nagaraja, G.M., Bausero, M.A., Manola, J., Asea, A. (2010). Radiation therapy induces circulating serum Hsp72 in patients with prostate cancer. Radiother Oncol 95:350–358.
Jamieson, C., Sharma, M., Henderson, B.R. (2012). Wnt signaling from membrane to nucleus: β-catenin caught in a loop. Int J Biochem Cell Biol 44:847–850.
Janowska-Wieczorek, A., Wysoczynski, M., Kijowski, J., Marquez-Curtis, L., Machalinski, B., Ratajczak, J., Ratajczak, M.Z. (2005) Microvesicles derived from activated platelets

induce metastasis and angiogenesis in lung cancer. Int J Cancer 113:752–760.

Jung, T., Castellana, D., Klingbeil, P., Cuesta Hernández, I., Vitacolonna, M., Orlicky, D.J., Roffler, S.R., Brodt, P., Zöller, M. (2009). CD44v6 dependence of premetastatic niche preparation by exosomes. Neoplasia 11: 1093–1105.

Jung, Y., Wang, J., Schneider, A., Sun, Y.X., Koh-Paige, A.J., Osman, N.I., McCauley, L.K., Taichman, R.S. (2006). Regulation of SDF-1 (CXCL12) production by osteoblasts; a possible mechanism for stem cell homing. Bone 38:497–508.

Kaplan, R.N., Riba, R.D., Zacharoulis, S., Bramley, A.H., Vincent, L., Costa, C., MacDonald, D.D., Jin, D.K., Shido, K., Kerns, S.A., Zhu, Z., Hicklin, D., Wu, Y., Port, J.L., Altorki, N., Port, E.R., Ruggero, D., Shmelkov, S.V., Jensen, K.K., Rafii, S., Lyden, D. (2005). VEGFR1-positive haematopoietic bone marrow progenitors initiate the pre-metastatic niche. Nature 438, 820–827.

Khalil, A.A., Kabapy, N.F., Deraz, S.F., Smith, C. (2011). Heat shock proteins in oncology: Diagnostic biomarkers or therapeutic targets? Biochim Biophys Acta 1816:89–104.

Korkaya, H., Liu, S., Wicha, M.S. (2011). Breast cancer stem cells, cytokine networks, and the tumor microenvironment. J Clin Invest 121: 3804–3809.

Kosaka, N., Iguchi, H., Yoshioka, Y., Takeshita, F., Matsuki, Y., Ochiya, T. (2010). Secretory mechanisms and intercellular transfer of microRNAs in living cells. J Biol Chem 285:17442–17452.

Kucia, M., Halasa, M., Wysoczynski, M., Baskiewicz-Masiuk, M., Moldenhawer, S., Zuba-Surma, E., Czajka, R., Wojakowski, W., Machalinski, B., Ratajczak, M.Z. (2007). Morphological and molecular characterization of novel population of CXCR4+ SSEA-4+ Oct-4+ very small embryonic-like cells purified from human cord blood: preliminary report. Leukemia 21:297–303.

Kuhn, N.Z., Tuan, R.S. (2010). Regulation of stemness and stem cell niche of mesenchymal stem cells: implications in tumorigenesis and metastasis. J Cell Physiol 222:268–277.

Lakkaraju, A., Rodriguez-Boulan, E. (2008). Itinerant exosomes: emerging roles in cell and tissue polarity. Trends Cell Biol 18:199–209.

Lässer, C. (2012). Exosomal RNA as biomarkers and the therapeutic potential of exosome vectors. Expert Opin Biol Ther 12 Suppl 1:S189–197.

Lathia, J.D., Heddleston, J.M., Venere, M., Rich, J.N. (2011). Deadly teamwork: neural cancer stem cells and the tumor microenvironment. Cell Stem Cell 8:482–485.

Lee, Y.S., Pressman, S., Andress, A.P., Kim, K., White, J.L., Cassidy, J.J., Li, X., Lubell, K., Lim do, H., Cho, I.S., Nakahara, K., Preall, J.B., Bellare, P., Sontheimer, E.J., Carthew, R.W. (2009). Silencing by small RNAs is linked to endosomal trafficking. Nat Cell Biol 11:1150–1156.

Li, H.J., Reinhardt, F., Herschman, H.R., Weinberg, R.A. (2012). Cancer-stimulated mesenchymal stem cells create a carcinoma stem-cell niche via Prostaglandin E2 signaling. Cancer Discov 2:840–855.

Li, L., Neaves, W.B. (2006). Normal stem cells and cancer stem cells: the niche matters. Cancer Res 66:4553–4557.

Lima, L.G., Chammas, R., Monteiro, R.Q., Moreira, M.E., Barcinski, M.A. (2009). Tumor-derived microvesicles modulate the establishment of metastatic melanoma in a phosphatidylserine-dependent manner. Cancer Lett 283:168–175.

Mathivanan, S., Lim, J.W., Tauro, B.J., Ji, H., Moritz, R.L., Simpson, R.J. (2010). Proteomic analysis of A33-immunoaffinity-purified exosomes released from the human colon tumor cell line LIM1215 reveals a tissue-specific protein signature. Mol Cell Proteomics 9:197–208.

McCready, J., Sims, J.D., Chan, D., Jay, D.G. (2010). Secretion of extracellular hsp90alpha via exosomes increases cancer cell motility: a role for plasminogen activation. BMC Cancer 10:294.

Merendino, A.M., Bucchieri, F., Campanella, C., Marcianò, V., Ribbene, A., David, S., Zummo, G., Burgio, G., Corona, D.F., Conway de Macario, E., Macario, A.J., Cappello, F. (2010). Hsp60 is actively secreted by human tumor cells. PLoS One 5:e9247.

Morelli, A.E., Larregina, A.T., Shufesky, W.J., Sullivan, M.L., Stolz D.B., Papworth, G.D., Zahorchak, A.F., Logar, A.J., Wang, Z., Watkins, S.C., Falo, L.D. Jr., Thomson, A.W. (2004). Endocytosis, intracellular sorting, and processing of exosomes by dendritic cells. Blood 104:3257–3266.

Mulloy, J.C., Cammenga, J., Berguido, F.J., Wu, K., Zhou, P., Comenzo, R.L., Jhanwar, S., Moore, M.A., Nimer, S.D. (2003). Maintaining the self-renewal and differentiation potential of human CD34+ hematopoietic cells using a single genetic element. Blood 102:4369–4376.

Nazarenko, I., Rana, S., Baumann, A., McAlear, J., Hellwig, A., Trendelenburg, M., Lochnit, G., Preissner, K.T., Zöller, M. (2010). Cell surface tetraspanin Tspan8 contributes to molecular pathways of exosome-induced endothelial cell activation. Cancer Res 70:1668–1678.
Nicholson, S., Whitehouse, H., Naidoo, K., Byers, R.J. (2011). Yin Yang 1 in human cancer. Crit Rev Oncog 16:245–260.
Pant, S., Hilton, H., Burczynski, M.E. (2012). The multifaceted exosome: Biogenesis, role in normal and aberrant cellular function, and frontiers for pharmacological and biomarker opportunities. Biochem. Pharmacol 83:1484–1494.
Park, J.E., Tan, H.S., Datta, A., Lai, R.C., Zhang, H., Meng, W., Lim, S.K., Sze, S.K. (2010). Hypoxia modulates tumor microenvironment to enhance angiogenic and metastastic potential by secretion of proteins and exosomes. Mol Cell Proteomics 9:1985–1099.
Parolini, I., Federici, C., Raggi, C., Lugini, L., Palleschi, S., De Milito, A., Coscia, C., Iessi, E., Logozzi, M., Molinari, A., Colone, M., Tatti, M., Sargiacomo, M., Fais, S. (2009). Microenvironmental pH is a key factor for exosome traffic in tumor cells. J Biol Chem 284:34211–34222.
Peinado, H., Alečković, M., Lavotshkin, S., Matei, I., Costa-Silva, B., Moreno-Bueno, G., Hergueta-Redondo, M., Williams, C., García-Santos, G., Ghajar, C.M., Nitadori-Hoshino, A., Hoffman, C., Badal, K., Garcia, B.A., Callahan, M.K., Yuan, J., Martins, V.R., Skog, J., Kaplan, R.N., Brady, M.S., Wolchok, J.D., Chapman, P.B., Kang, Y., Bromberg, J., Lyden, D. (2012). Melanoma exosomes educate bone marrow progenitor cells toward a pro-metastatic phenotype through MET. Nat Med 18:883–891.
Peinado, H., Lavotshkin, S., Lyden, D. (2011). The secreted factors responsible for pre-metastatic niche formation: old sayings and new thoughts. Semin Cancer Biol 21:139–146.
Petit, I., Szyper-Kravitz, M., Nagler, A., Lahav, M., Peled, A., Habler, L., Ponomaryov, T., Taichman, R.S., Arenzana-Seisdedos, F., Fujii, N., Sandbank, J., Zipori, D., Lapidot, T. (2002). G-CSF induces stem cell mobilization by decreasing bone marrow SDF-1 and up-regulating CXCR4. Nat Immunol 3:687–694.
Pols, M.S., Klumperman, J. (2009). Trafficking and function of the tetraspanin CD63. Exp Cell Res 315:1584–1592.
Rak, J. (2010). Microparticles in cancer. Semin Thromb Hemost 36:888–906.
Rana, S., Claas, C., Kretz, C.C., Nazarenko, I., Zöller, M. (2011). Activation-induced internalization differs for the tetraspanins CD9 and Tspan8: Impact on tumor cell motility. Int J Biochem Cell Biol 43:106–119.
Rana, S., Yue, S., Stadel, D., Zöller, M. (2012). Toward tailored exosomes: The exosomal tetraspanin web contributes to target cell selection. Int J Biochem Cell Biol 44:1574–1584.
Rana, S., Malinowska, K., Zöller, M. (2013). Exosomal tumor microRNA modulates premetastatic organ cells. Neoplasia 15:281–295.
Sales, K.M., Winslet, M.C., Seifalian, A.M. (2007). Stem cells and cancer: an overview. Stem Cell Rev 3:249–255.
Sangaletti, S., Colombo, M.P. (2008). Matricellular proteins at the crossroad of inflammation and cancer. Cancer Lett 267:245–253.
Seow, Y., Wood, M.J. (2009). Biological gene delivery vehicles: beyond viral vectors. Mol Ther 17:767–777.
Sheldon, H., Heikamp, E., Turley, H., Dragovic, R., Thomas, P., Oon, C.E., Leek, R., Edelmann, M., Kessler, B., Sainson, R.C., Sargent, I., Li, J.L., Harris, A.L. (2010). New mechanism for Notch signaling to endothelium at a distance by Delta-like 4 incorporation into exosomes. Blood 116:2385–2394.
Simons, M., Raposo, G. (2009). Exosomes—vesicular carriers for intercellular communication. Curr Opin Cell Biol 21:575–581.
Simpson, R.J., Lim, J.W., Moritz, R.L., Mathivanan, S. (2009). Exosomes: proteomic insights and diagnostic potential. Expert Rev Proteomics 6:267–283.
Skog, J., Würdinger, T., van Rijn, S., Meijer, D.H., Gainche, L., Sena-Esteves, M., Curry, W.T. Jr., Carter, B.S., Krichevsky, A.M., Breakefield, X.O. (2008). Glioblastoma microvesicles transport RNA and proteins that promote tumour growth and provide diagnostic biomarkers. Nat Cell Biol 10:1470–1476.
Strauss, R., Hamerlik, P., Lieber, A., Bartek, J. (2012). Regulation of stem cell plasticity: mechanisms and relevance to tissue biology and cancer. Mol Ther 20:887–897.
Suárez-Álvarez, B., López-Vázquez, A., López-Larrea, C. (2012). Mobilization and homing of hematopoietic stem cells. Adv Exp Med Biol 741:152–170.
Tan, A., De La Peña, H., Seifalian, A.M. (2010). The application of exosomes as a nanoscale cancer vaccine. Int J Nanomedicine 5:889–900.
Taylor, D.D., Gercel-Taylor, C. (2011). Exosomes/microvesicles: mediators of cancer-associated

immunosuppressive microenvironments. Semin Immunopathol 33:441–454.

Tian, T., Wang, Y., Wang, H., Zhu, Z., Xiao, Z. (2010). Visualizing of the cellular uptake and intracellular trafficking of exosomes by live-cell microscopy. J Cell Biochem 111:488–496.

Trajkovic, K., Hsu, C., Chiantia, S., Rajendran, L., Wenzel, D., Wieland, F., Schwille, P., Brügger, B., Simons, M. (2008). Ceramide triggers budding of exosome vesicles into multivesicular endosomes. Science 319:1244–1247.

Tumne, A., Prasad, V.S., Chen, Y., Stolz, D.B., Saha, K., Ratner, D.M., Ding, M., Watkins, S.C., Gupta, P. (2009). Noncytotoxic suppression of human immunodeficiency virus type 1 transcription by exosomes secreted from CD8+ T cells. J Virol 83:4354–4364.

Valadi, H., Ekström, K., Bossios, A., Sjöstrand, M., Lee, J.J., Lötvall, J.O. (2007) Exosome mediated transfer of mRNAs and microRNAs is a novel mechanism of genetic exchange between cells. Nat Cell Biol 9:654–659.

Vallhov, H., Gutzeit, C., Johansson, S.M., Nagy, N., Paul, M., Li, Q., Friend, S., George, T.C., Klein, E., Scheynius, A., Gabrielsson, S. (2011). Exosomes containing glycoprotein 350 released by EBV-transformed B cells selectively target B cells through CD21 and block EBV infection in vitro. J Immunol 186:73–82.

Vickers, K.C., Remaley, A.T. (2012) Lipid-based carriers of microRNAs and intercellular communication. Curr Opin Lipidol 23:91–97.

Virchow, R.L.K. (1858). In Cellular Pathology (Ed. Hirschwald A.), Berlin.

Wang, R., Chadalavada, K., Wilshire, J., Kowalik, U., Hovinga, K.E., Geber, A., Fligelman, B., Leversha, M., Brennan, C., Tabar, V. (2010). Glioblastoma stem-like cells give rise to tumour endothelium. Nature 468:829–833.

Winkler, I.G., Sims, N.A., Pettit, A.R., Barbier, V., Nowlan, B., Helwani, F., Poulton, I.J., van Rooijen, N., Alexander, K.A., Raggatt, L.J., Lévesque, J.P. (2010) Bone marrow macrophages maintain hematopoietic stem cell (HSC) niches and their depletion mobilizes HSCs. Blood 116:4815–4828.

Wittmann, J., Jäck, H.M. (2010). Serum microRNAs as powerful cancer biomarkers. Biochim Biophys Acta 1806:200–207.

Wray, J., Hartmann, C. (2012). WNTing embryonic stem cells. Trends Cell Biol 22:159–168.

Xiu, F., Cai, Z., Yang, Y., Wang, X., Wang, J., Cao, X. (2007). Surface anchorage of superantigen SEA promotes induction of specific antitumor immune response by tumor-derived exosomes. J Mol Med 85:511–521.

Yang, J., Weinberg, R.A. (2008). Epithelial-mesenchymal transition: at the crossroads of development and tumor metastasis. Dev Cell 14:818–829.

Yang, C., Kim, S.H., Bianco, N.R., Robbins, P.D. (2011). Tumor-derived exosomes confer antigen-specific immunosuppression in a murine delayed-type hypersensitivity model. PLoS One 6:e22517.

Yu, Y., Ramena, G., Elble, R.C. (2012). The role of cancer stem cells in relapse of solid tumors. Front Biosci (Elite Ed) 4:1528–1541.

Zakharova, L., Svetlova, M., Fomina, A.F. (2007). T cell exosomes induce cholesterol accumulation in human monocytes via phosphatidylserine receptor. J Cell Physiol 212:174–181.

Zeelenberg, I.S., van Maren, W.W., Boissonnas, A., Van Hout-Kuijer, M.A., Den Brok, M.H., Wagenaars, J.A., van der Schaaf, A., Jansen, E.J., Amigorena, S., Théry, C., Figdor, C.G., Adema, G.J. (2011) Antigen localization controls T cell-mediated tumor immunity. J Immunol 187:1281–1288.

Zhang, H.G., Kim, H., Liu, C., Yu, S., Wang, J., Grizzle, W.E., Kimberly, R.P., Barnes, S. (2007). Curcumin reverses breast tumor exosomes mediated immune suppression of NK cell tumor cytotoxicity. Biochim Biophys Acta 1773:1116–1123.

Zöller, M. (2009). Tetraspanins: push and pull in suppressing and promoting metastasis. Nat Rev Cancer 9:40–55.

Zöller, M. (2011). CD44: can a cancer-initiating cell profit from an abundantly expressed molecule? Nat Rev Cancer 11:254–267.

Chapter 30
MicroRNA Therapeutics to Target Brain Tumor Stem Cells

Derryn Xin Hui Chan[1], Srikanth Nama[1], Gopinath Sundaram[1], and Prabha Sampath[1,2]
[1] *Institute of Medical Biology, Agency for Science, Technology and Research (A*STAR), Singapore*
[2] *Department of Biochemistry, Yong Loo Lin School of Medicine, National University of Singapore, Singapore*

Cancer Stem Cells

The term "cancer stem cells" is used frequently to describe a subpopulation of cells found within the tumor that has the capability to maintain and propagate tumor growth. This phenomenon otherwise known as the "cancer stem cell hypothesis", suggests a hierarchical model of carcinogenesis. This posed a challenge to the old paradigm of cancer development based on the clonal evolution model, whereby mutant tumor cells with growth advantage are selected and have a similar potential for tumor regeneration. After its first identification in acute myeloid leukemia (Bonnet and Dick, 1997; Lapidot et al., 1994), cancer stem cells (CSC) were subsequently identified and isolated in a variety of solid neoplasm such as breast cancer (Al-Hajj et al., 2003), colon cancer (O'Brien et al., 2007), and gliomas (Galli et al., 2004; Singh et al., 2004). Studies by various groups have shown that CSCs share many characteristics as normal stem cells such as the expression of stem cells markers, similar signal transduction pathways, and the ability to undergo prolonged proliferation and self-renewal being the most important trait (Li et al., 2009b). Contemporary evidence in a variety of tumors has lent strong support to the cancer stem cell hypothesis. The hypothesis has implications for risk assessment, early detection, and prediction. The identification of cancer stem cells have certainly provided new perspectives on the current cancer therapies and ushered much interest into therapies that could effectively target these CSCs. Cancer therapeutics based on tumor regression may have yielded several agents that kill differentiated tumor cells but have failed to eliminate the stem cell population. Targeting cancer stem cells can lead to very effective cancer therapies, and elimination of these cells can prevent recurrence or relapse of cancer.

Cancer Stem Cell Hypothesis and Carcinogenesis

Several mouse models of human cancer, in many cases, have failed to recapitulate the human disease. Transgenic models use tissue-specific promoters to drive oncogene expression, but tissue-specific genes may be expressed only in differentiated cells. If cancer stem cells are real targets of transforming

Cancer Stem Cells, First Edition. Edited by Vinagolu K. Rajasekhar.

events, then the expression of oncogenes in more differentiated cells may fail to recapitulate actual carcinogenic processes. It has been demonstrated that expression of c-Met and c-Myc driven by the mammary-specific promoter from mouse mammary tumor virus fails to produce carcinomas (Welm et al., 2005). The normal stem cell niche has direct relevance to stromal epithelial interactions that occur during tumorigenesis, and the clinical course of micrometastases also reflect stem cell characteristics of disseminated cells. Cancer stem cells adopt multiple molecular mechanisms to develop resistance to apoptosis including cell cycle kinetics. Cancer stem cells are in G0 phase and resistant to cell cycle–specific chemotherapy agents (Venezia et al., 2004). Further, stem cells display enhanced DNA repair (Cairns, 2002; Potten et al., 2002) and are resistant to DNA-damaging agents by the virtue of being able to undergo asynchronous DNA synthesis. Stem cells express higher levels of antiapoptotic proteins, such as members of the Bcl-2 family and inhibitors of apoptosis (Wang et al., 2003). Cancer risk may be directly related to the number of cancer stem cells and elucidation of molecular signatures characteristic of these stem cells can provide prognostic information. Interventions that induce apoptotic death of these cancer stem cells can effectively treat carcinogenesis.

Identification of Cancer Stem Cells

Leukemia cancer stem cells capable of recapitulating the disease in an immunodeficient mouse were first identified based on known markers of hematopoietic development by John Dick (Lapidot et al., 1994). CSCs were first identified in acute myeloid leukemia in which a subpopulation of cells that expresses the surface marker $CD34^{+}CD38^{-}$ could induce leukemia upon serial transplantation in immunodeficient mice. Following this, CSCs were subsequently demonstrated in various malignancies. Generally, CSCs are isolated from the heterogeneous tumor mass using two major techniques. The first technique made use of cellular surface markers expression to separate the CSCs from the non-stem cells population such as $CD44^{+}CD24^{-/low}$ in breast CSCs, CD133 in colorectal cancer as well as brain tumor. Another technique that has proven useful in the isolation of subsets enriched for CSCs is the exclusion of $Hoechst_{33342}$ dye by the side population cells, which highly expressed drug efflux transporter gene such as ABCG2 and ABCA3 (Hirschmann-Jax et al., 2004; Zhou et al., 2001). In this chapter, we will focus on the identification and isolation of CSCs isolated from malignant brain tumors such as medulloblastoma and glioblastoma, glioblastoma mulitforme (GBM), hereafter referred to as brain tumor stem cells (BTSC).

Brain Tumor Stem Cells

Gliomas are primary brain tumors, a collective term for neoplasms that originate from glial cells or their precursors cells in the central nervous system. According to the World Health Organization (WHO), gliomas are classified into four different grades (I, II, III, and IV) depending on their rate of growth, invasiveness, and histological appearance of the tumor cells. Grade I and grade II brain tumors are referred to as low-grade gliomas, and grade III and grade IV tumors are termed as malignant gliomas. An example of grade IV brain tumor is glioblastoma multiforme (GBM), which is the most common and aggressive primary gliomas.

Despite a multimodality of treatment approaches for the disease, this lethal intracranial malignancy has a high degree of mortality. This is due to the diffuse and invasive nature of these tumor cells but also can be explained by the hierarchical model of cancer development, which suggests that only a subset of cells within the tumor possess the ability to undergo prolonged proliferation for tumor regeneration.

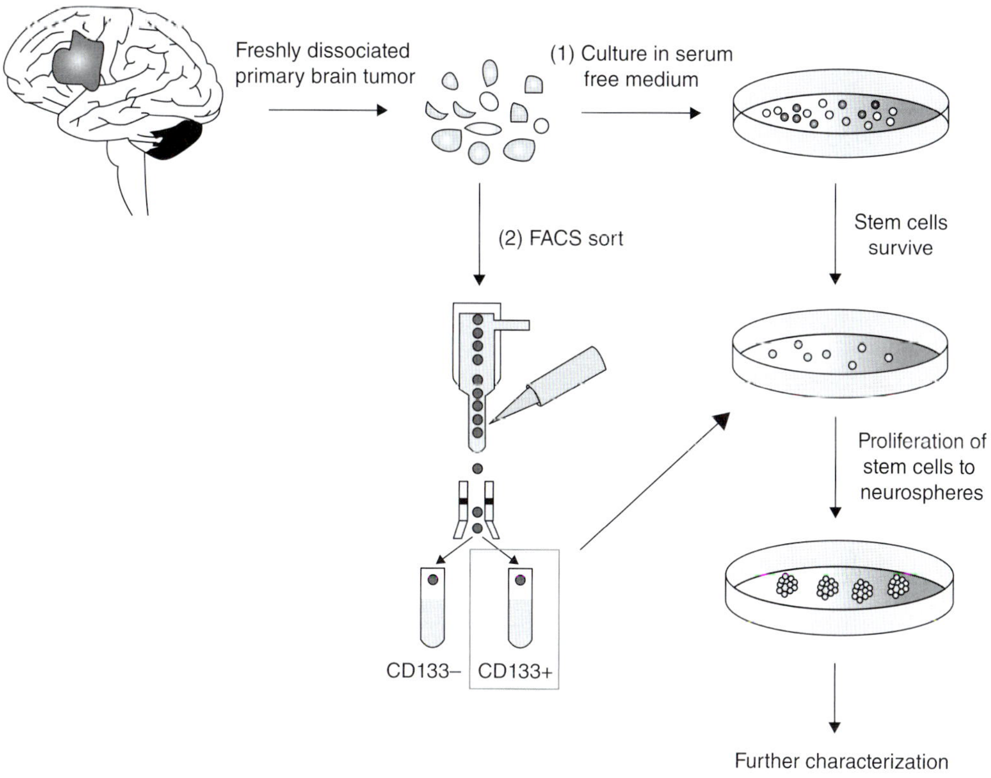

Figure 30.1. Schematic representation of isolation of brain tumor stem cells.

The observation that cancer cells and normal neural stem cells share the common defining features of incompletely differentiated state and unlimited self-renewal capacity, have led to the identification of brain tumor stem cells. Understanding normal neural stem cell (NSC) biology has allowed us to gain better understanding of the cells that shared similar characteristics. There are two main approaches to the identification of BTSCs. Using the same cultural conditions in vitro, tumor stem cells are enriched and isolated based on their ability to grow as floating neurospheres in serum-free media. Subsequently, the neurosphere assay is used to assess for self-renewal through the serial passage of the spheres formed from single cells and for proliferation, through the size of the generated spheres (Figure 30.1). However, care has to be exercised when interpreting the data of the neurosphere assays as spheres could form or increase in size due to the fusion of smaller spheres if cells are seeded at a higher density. This could possibly limit the sensitivity of the neurosphere assay (Rich and Eyler, 2008; Singec et al., 2006).

The second approach to the identification and enrichment of BTSCs would be based on stem cell markers expression. Currently, the identification and isolation of BTSCs from brain tumor such as GBM frequently and largely relies on the expression of the cell surface marker Prominin-1 (cluster of differentiation133 [CD133]), which is also a putative marker for NSCs. CD133 is a penta-span-transmembrane glycoprotein located on cellular protrusions. Its function is still unknown, however, mutations in CD133 are detected in patients with familiar macular degeneration and disrupt photoreceptor disk morphogenesis in a genetic mouse model (Yang et al., 2008). Tumors are disaggregated and sorted based on the expression of CD133 (Figure 30.1). The isolated CD133+ cells are then functionally defined based on self-renewal assay, chemo resistance, and radio resistance (Bao et al., 2006). Moreover, these cells are also capable of undergoing multipotency differentiation to various neural lineages such as neurons, astrocytes, and oligodendrocytes. In contrast to CD133+ cells, the majority of the glioma cells were CD133− and were incapable of forming self-sustaining neurospheres. In the presence of differentiating conditions, BTSCs lost the expression of primitive markers like CD133 and nestin. When serially transplanted in the brain of NOD-SCID (non-obese diabetic, severe combined immunodeficient) mice, these GSCs were able to produce an exact phenocopy of the patient's original tumor with all the histopathological features and cell surface markers of the original tumor.

Though the use of CD133 has some success in the isolation of BTSCs as well as other cancer stem cells (Beier et al., 2007), questions were raised against the variations in its expression in GBM and evidence highlighting the tumorigenic property of CD133− cell population. The use of CD133 as a marker for BTSCs was further confounded with a recent report that the expression of CD133 is cell-cycle dependent, which might potentially result in the omission of CD133+ cells during fluorescence activated cell sorting (FACS) (Sun et al., 2009). Subsequently, other markers such as L1CAM (Bao et al., 2008; Cheng et al.), CD15 (Ward et al., 2009), and integrin α6 (Lathia et al. 2011) have been used for the identification and isolation of BTSCs. As mentioned earlier, tumors consist of heterogeneous population of cells and isolation based on the expression of a single marker is highly inadequate. As such, the integration of two markers to negatively or positively select for BTSCs might prove to be more useful and informative.

Cellular Origin of BTSCs

Although there is compelling evidence attributing the refractoriness of highly invasive brain malignancies toward conventional therapies to the existence of BTSCs within these tumors, the cellular origin of these cells remains unknown and highly controversial. It has been proposed that BTSCs can arise from differentiated glial cells, restricted neural progenitors, or NSCs. In the first case, the committed glial cells can possibly acquire mutations that results in aberrant activation of the stem cell self-renewal regulatory mechanism, thus allowing them to serve as precursors for BTSCs. Alternatively, restricted neural progenitor cells with limited self-renewal potential may acquire mutations leading to the formation of BTSCs. In both of the above cases, the cells with stem-cell-like properties and genetic aberrations could lead to progression to a fully neoplastic state.

Many studies have shown that BTSCs display strikingly similar characteristics as NSCs such as sharing common growth regulatory pathways, expressing similar cell surface markers, differentiation potential, and self-renewal capacity (Figure 30.2). It is hypothesized that BTSCs are derived from NSCs that have undergone transformation allowing them to overcome constraints placed on normal stem cell expansion. It is conceivable that NSCs can be a likely source of BTSCs since these cells are perennial cells with extensive proliferation and self-renewal potential (Figure 30.2), providing them with more opportunities to accumulate genetic lesions to become attractive targets for tumorigenesis (Figure 30.2). In addition, it has previously been reported that gliomas tend to develop in germinal region of the brain such as the subventricular zone (SVZ) where neural stem cells reside (Sanai et al., 2005). Moreover, using mouse model harboring mutations

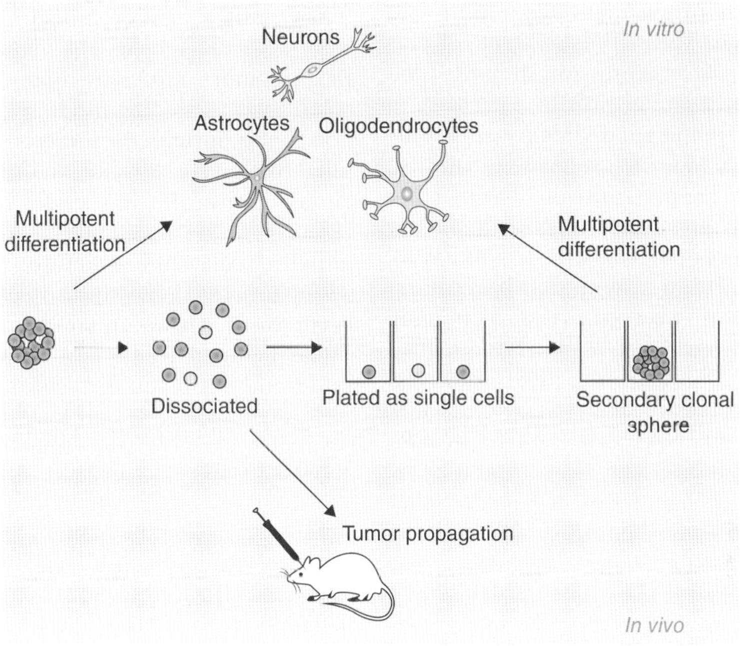

Figure 30.2. Schematic representation of multiple steps in brain tumor stem cells characterization.

in *TP53*, *NF1*, and *PTEN* in neural stem/progenitors cells and astrocytes, Alcantara Llaguno and colleagues (2009) have shown that the transformed NSCs residing in SVZ are capable of inducing astrocytomas.

BTSCs and MicroRNAs

Recent reports have shown that microRNAs (miRNAs) can affect the expression of numerous genes involved in cancer pathway either functioning as oncomir and/or tumor suppressor. miRNAs are endogenously made small non-coding RNAs of about 20–22 nucleotides in size and regulate a wide array of target mRNAs at post-transcriptional level (Kim, 2005; Kim et al., 2009) by binding to complementary regions in target mRNAs. It has been proposed that a complete complementarity between miRNA and target mRNA leads to the degradation of target mRNA (Tang et al., 2003) whereas a partial complementarity may lead to translational repression and/or mRNA degradation (He and Hannon, 2004). Djuranovic and colleagues (2012) further dissected the kinetics of miRNA-mediated repression and showed elegantly that miRNA inhibits translation initiation at an early step, followed by mRNA deadenylation and decay. A notable feature of miRNA-mediated post-transcriptional regulation is the pleotrophic effect of a single miRNA in targeting multiple mRNAs and multiple miRNAs regulating a single mRNA in a spatio-temporal manner (Segura et al., 2009; Zhang et al., 2010). Understanding the miRNA-mRNA interaction in a global scale has led to the identification of miRNA-regulatory networks regulating cohorts of mRNAs involved in a physiological and pathological process such as cancer (Gennarino et al., 2012; Plaisier et al., 2012). Kefas and colleagues (2008) were the first to discover that miR-7 was markedly reduced in BTSCs. miR-7 is weakly expressed due to reduced processing of primary miR-7 transcript to the precursor miR-7 transcript (Kefas et al., 2008). Overexpression of miR-7 via transient transfection leads to a decrease in BTSCs proliferation and viability in vitro by targeting the EGFR signaling pathway, resulting in suppression of downstream Akt activity. Subsequently, other miRNAs studied such as miR-34a, miR-128, miR-138, and miR-125b can also inhibit the glioblastoma growth in vivo and in vitro by targeting various oncogenes such as Bmi1 and CDK6 (Chan et al., 2012; Cui et al., 2009; Godlewski et al., 2008; Li et al., 2009a; Shi et al., 2010). In addition, miRNAs such as miR-124/miR-137 can also inhibit glioblastoma growth by affecting the differentiation of BTSCs (Silber et al., 2008). Similar to miR-7, miR-124 and miR-137 are both weakly expressed in the BTSCs. While miR-124 is known to play a role in neuronal differentiation, the role of miR-137 remains unknown (Silber et al., 2008). Overexpression of both miR-124 and miR-137 can induce neuronal differentiation of BTSCs and thus inhibit proliferation of BTSCs. The cancer stem cells from malignant gliomas display aberrant elevated expression of miR-138. In these cells, miR-138 functions as an oncogene whereby knocking down miR-138 leads to a decrease in proliferation and increase in apoptosis by targeting a network of genes such as BLCAP, MXD1, and CASP3. In addition to its critical role in promoting growth and survival of stem cells, the higher expression of miR-138 in BTSCs compared to non-neoplastic tissue and association with tumor recurrence and survival highlights the clinical significance of miR-138 as a prognostic biomarker and a therapeutic target for treatment of malignant gliomas (Chan et al., 2012). The identification of growth and differentiation pathways responsible for BTSCs proliferation and survival is of paramount importance for the discovery of novel therapeutic targets. Acting as novel and pivotal regulators of protein-encoding genes, miRNAs will have great potential in regulating BTSCs' biological functions by targeting BTSCs-related signal pathway molecules.

Targeting Cancer Stem Cells

Brain tumors represent a unique medical condition mainly because of a specific location that is difficult for therapeutic intervention. Similar to normal neural stem cells, glioma-derived BTSCs reside within a perivascular niche and suggest that targeting tumor-associated vasculature can interfere with glioma BTSCs growth (Gilbertson and Rich, 2007). However, development of targeted therapies to eradicate these cells can lead to more effective treatment options.

Virotherapy

Oncolytic virotherapy manipulates their lytic property to achieve tumor cell lysis by intraneoplastic virus replication. A viral vector that selectively targets the tumor and has minimal brain and systemic toxicities can achieve effective oncolytic activity. Further, this should reach all neoplastic foci beyond tumor resection border. A phase I trial by Forsyth and colleagues (2008) showed that administration of live, replication competent and genetically unmodified reovirus that selectively replicates in cells with activated ras signaling, directly into the tumors of patients with malignant glioma is safe and well tolerated with no evidence of clinical encephalitis. However, a multicenter randomized controlled phase III trial with recombinant replication deficient retrovirus (RV) vector with HSV thymidine kinase transgene insert, showed no significant benefit in the treatment group when compared to untreated group (Rainov, 2000). Although adenovirus is a very attractive candidate for virotherapy because of its non-pathogenicity and integration-deficiency in host cell genome, the biggest limitation of this strategy is limited biodistribution of these vectors. Due to very low infectivity and poor therapeutic gene transduction limited to very short distance from the site of injection, replication-deficient adenoviral vectors are no longer a vector of choice.

Optimal strategy for specifically targeting brain tumor stem cells with a successful viral vector will involve transduction of viral vectors in only CD133+ BTSCs and transcription of viral gene driven by glioma specific promoter. Taking into account the heterogeneous and evolving nature of malignant gliomas, the optimal treatment strategy will encompass a planned interplay of several distinct treatment modalities. Therefore, the synergistic effect between virotherapy and gene/microRNA-therapy may lead to development of novel effective therapy to target BTSCs.

MicroRNA Therapeutics

Identification of differential gene expression in BTSCs has resulted in the discovery of several promising therapeutic targets. OncomiRs show differential expression levels in cancer and affect cellular transformation, carcinogenesis, and metastasis, acting either as oncogenes or tumor suppressors. The hypothesis of 'oncogene addiction' suggests that despite the multimode nature of tumorigenesis, targeting specific single oncogenes may have therapeutic value. In certain cases, tumor growth strictly depends on the single genes such as c-Myc, Ras, Her2, EGFR, and Bcr-Abl, and disruption of these genes leads to tumor regression. This sort of single gene dependency is called "Gene addiction." MicroRNA-21 (miR-21) is overexpressed in most tumor types, and overexpression of miR-21 leads to a pre-B malignant lymphoid-like phenotype, demonstrating that miR-21 is truly an oncomiR. When miR-21 was inactivated, the tumors regress completely suggesting that tumors can become addicted to oncomiRs, and pharmacological inactivation of miRNAs such as miR-21 can be effective to treat malignant lymphoid-like phenotype (Medina et al., 2010). Malignant glioma biomarker research is still in its early stages because of the tumor's complex pathophysiology. Although miRNA-targeted therapy is still in its infancy, and clinical trials with glioma/brain tumor patients are under recruitment or currently running, several miRNAs have been identified as promising tumor biomarkers, with increased potential to reduce disease progression in combination to conventional first-line therapy for gliomas.

Several microRNAs have been shown to function either as an oncomiR or as a tumor suppressor. Aberrant miRNA expression can be rectified by two simple ways, either by overexpression of tumor suppressor miRNAs (Moscat and Diaz-Meco, 2009) or by downregulating the oncomiRs (Medina et al., 2010). Target miRNA overexpression or knockdown can be achieved by systemic injection of modified nucleic acid in the form of mimics (same sequence) or inhibitors (reverse complementary sequences). Systemic injection of modified small-nucleic acids can be targeted to various tissues (Krutzfeldt et al., 2005). The modified microRNAs can reach various organs (miRNA levels in liver, lung, kidney, heart, intestine, fat, skin, bone marrow, muscle, ovaries, and adrenals) in the body and shows effective downregulation target mRNAs. In colon cancer, miRNAs miR-33a and miR145 play key role in tumor regression by downregulating oncogenes. miR-33a downregulates oncogene PIM-1, and miR-145 downregulates c-Myc and ERK5. Systemic injection of low molecular weight PEI/miRNA145/33a complexes decreased the tumor size in vivo (Ibrahim et al., 2011). AntimiR therapeutics inhibit specific microRNA targets. Animal models showed that modulating microRNAs through antimiRs effectively regulates biological processes and provides therapeutic benefits. Administration of antimiR oligonucleotides through local or parenteral injection is shown to be very effective. AntimiRs are also known to be more stable and less toxic than other known cancer treatments leading us to suggest that administration of tiny LNAs as antimiRs against specific miRNAs might effectively eliminate the BTSCs.

Tiny LNAs are thermodynamically stable and effectively bind the miRNA seeds in competition with cellular target mRNAs leading to sequestration of the miRNA. LNA is a modification of RNA that contains an oxy-methylene bridge between the 2′ and 4′ carbons in the ribose ring. This bridge creates a bi-cyclic structure that locks the conformation of the ribose and is the key to the high stability and

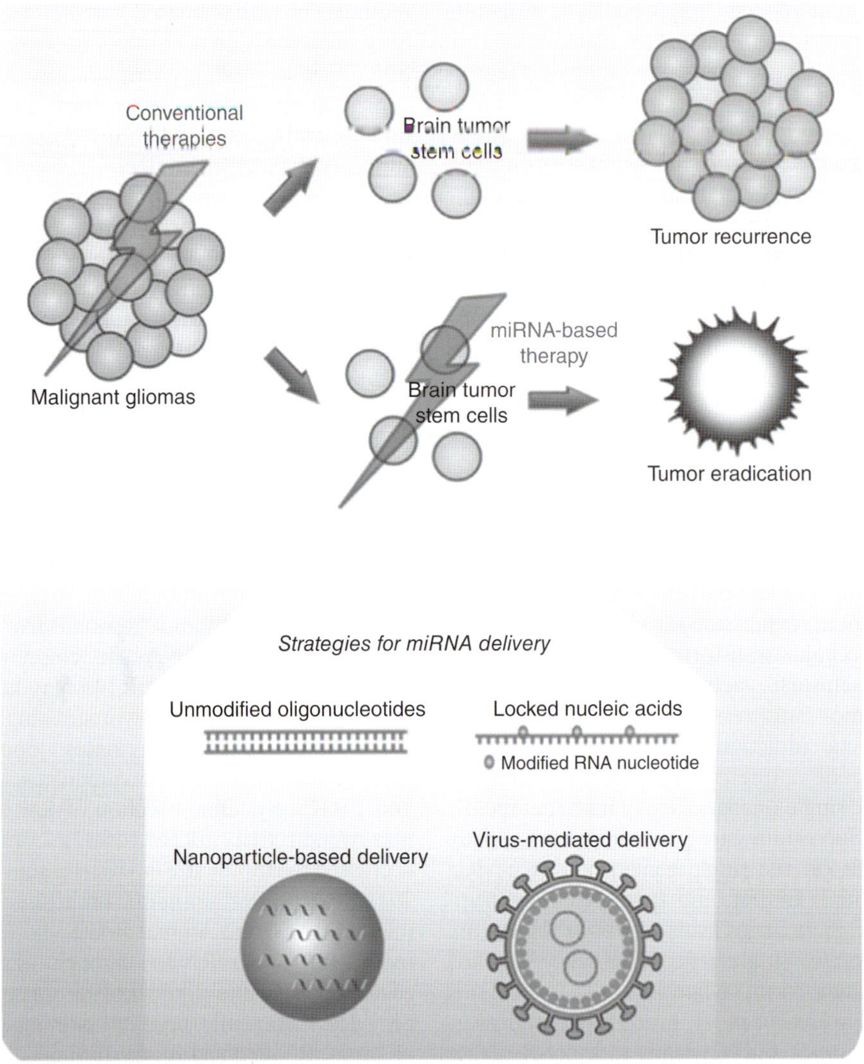

Figure 30.3. Schematic representation of miRNA therapeutics and delivery strategies.

affinity of LNA to its complementary nucleotide sequence. Phosphorothioate backbone modifications facilitate delivery of tiny LNAs in vivo. The phosphorothioate (PS) bond substitutes a sulfur atom for nonbridging oxygen in the phosphate backbone of an oligo. This modification renders the internucleotide linkage resistant to nuclease degradation. Phosphorothioates can be introduced at either the 5′- or 3′-end of the oligo to inhibit exonuclease degradation. In antisense oligonucleotides, phosphorothioates are also introduced internally to limit attack by endonucleases. The phophorothioate linkages and LNA stability allow these to be systemically administered in the absence of a specific carrier. These chemically modified and cholesterol-conjugated antimirs can be used for rapid and specific degradation of miRNAs. Cholesterol conjugation has been shown to facilitate uptake by virtually all cells (Figure 30.3).

LNA-based knockdown allows for high-potency knockdown without compromising specificity. These LNA-containing oligonucleotides have low cytotoxicity and are easy to transfect. The inclusion of LNAs in oligonucleotides dramatically increases the affinity for the target. It has been shown previously that the addition of a single LNA nucleotide typically increases the melting temperature (Tm) of the oligonucleotide-target duplex by 2–6 °C, in sequences below 30 nt in length, and additionally, LNA-containing probes are known to have excellent mismatch discrimination capabilities. These properties of LNAs make them excellent tools for studying miRNAs. The locked nucleic acid (LNA) RNA-targeted therapies overcome the limitations of antisense and siRNA technologies to deliver potent single-stranded LNA-based drugs. The chief problem with siRNA has been a relatively large molecule size, leading to poor cellular uptake and the need for complex and often toxic delivery vehicles. A main limitation with early generations of antisense has been low affinity for its RNA target, leading to insufficient potency and narrow therapeutic index in animals and humans. The unique combination of small size and excellent specificity provides optimal targeting and increased affinity to targets. This provides improved potency and favorable pharmacokinetic and tissue-penetrating properties.

However, safe and efficient systemic delivery of antimiR to brain tumors is a prerequisite to realize the therapeutic effects. Recently rabies virus glycoprotein (RVG) peptide, which targets neuronal cells expressing acetylcholine receptor, were applied for gene delivery to brain via forming complexes with siRNA. However, the stability of the complexes needed to be improved. Therefore, suitable delivery systems of antimiR are required. In comparison with other types of gene delivery systems including viral vectors, cationic polymers based gene nano-vehicles (Figure 30.3) have become more and more attractive due to many advantages including non-immunogenicity, no integration of exogeneous genes into host chromosomes, simple and less expensive good manufacturing products, convenience of handling, and modifiable multiple functionality for efficient gene delivery. But cationic polymer adopted for gene delivery to the brain is nonbiodegradable such as PAMAM or has a poorly defined structure as disulfide linked PEI (SSPEI), which will hinder the clinical applications.

Virus-Mediated miRNA Therapy

Viral-mediated gene transfer has benefits over chemically modified oligonucleotides due to their prolonged expression for months (Figure 30.3). Especially Lentiviral or Retroviral vectors have ability to integrate into the genome of target tissue and ensures long-term expression of either miRNA or inhibitors (Brown and Naldini, 2009, Xie et al., 2006). Specificity of the miRNA expression can be achieved by mode of delivery. Lentiviral pseudo typing helps to achieve tissue specificity. In recent years, the Adeno Associated Virus (AAV) system has been shown to offer tissue-specific infection with different serotypes showing specificity for a particular tissue. AAV8 serotype is specific to liver, and AAV8-mediated gene transfer has been shown to achieve 80–90% gene delivery in the liver (Kota et al., 2009). AAV is a versatile system to develop tissue or tumor-specific serotype, which increases specificity and robustness of gene delivery (Muzyczka and Warrington, 2005).

Similarly for functional knockdown of an miRNA, antagomiRs can be used. AntagomiRs are reverse complementary sequences produced upon processing of short hairpins. Due to their complementary double-stranded structure formation, this technology is also known as ZIP technology (commercially available from System Bio Sciences). Viral-mediated Antagomir delivery is efficient. Viral-mediated delivery has advantages over synthetic oligos. These advantages include higher transduction efficiency, and the virus can integrate in the genome thereby resulting in persistent long-term expression of the antagomiRs

(Gentner et al., 2009). Initially, Antagomirs were developed in Lentiviral backbone to downregulate miR-20a (Scherr et al., 2007). An added advantage of using lentiviral backbone for antagomiR technology is the ability to downregulate multiple microRNAs at the same time using principle cluster gene expression with miR-30a context sequences (Scherr et al., 2007).

Translational Relevance of BTSCs in Glioma Therapies

From the clinical perspective, the "cancer stem cell" hypothesis presented a paradigm shift in the development of anti-cancer therapies as well as evaluation of the efficacy of these therapies. Current cancer therapies assumed that all cancer cells have unlimited proliferative potential and predominantly target the bulk tumor population. Often, these targeting strategies have limited efficacies since CSCs are resistant against these therapies through increased repair of DNA damage from drug or ionizing radiation, increased expression of ABC transporter protein that allows drug efflux and preventing their accumulation. Failing to eradicate this subset of cells allow the CSCs to re-initiate and perpetuates tumorigenesis. If tumors are propagated by a rare population of cancer stem cells, then anti-cancer therapies should be developed to targeting this subset of cells. Taken together, combined treatment modalities targeting both the CSCs and bulk tumor population might increase the efficacies of current treatment regime and prove more effective for tumor eradication as well as preventing tumor recurrence (Diehn et al., 2009; Frank et al., 2010). Identification of cancer stem cells has altered the current paradigm in drug development. Eradication of cancers requires targeting and elimination of cancer stem cells. To selectively kill these cells is still a major challenge, and elimination of these cells may result in improved therapeutic outcomes for patients with even advanced cancers.

Conclusion/Perspectives

In this chapter, we have summarized the identification of CSCs in tumors with emphasis on malignant gliomas and various miRNAs that have emerged as important regulators of CSCs. Multiple roles of miRNA as pro-survival oncomiRs or as tumor-suppressors either promoting the survival and proliferation or promoting differentiation and apoptosis respectively is clearly illustrated. The most profound effect of miRNAs is the ability of a single miRNA to regulate multiple genes and pathways simultaneously. The use of miRNAs-based therapeutics as a powerful tool for targeting the BTSCs is promising. However, there are inevitable hurdles that need to be overcome. The key challenges include stability and delivery of the 8 to 11 nucleotides that act either as antimirs. We have outlined the multiple modification strategies and various modes of delivery that can be applied for efficient administration of these oligonucleotides. Though there is some success in administering the drug either via systemic administration or local delivery, effective delivery of these drugs to the brain still pose a challenge. Targeting BTSCs within the malignant gliomas using miRNA-based therapeutics is further confounded by the absence of definite marker for BTSCs. Although research in this field is still at its infancy, miRNA-based therapeutics may be the most promising, highly effective novel drugs in terms of high specificity and low toxicity. Identification of brain tumor stem cell population has important implications for the management of brain cancer patients. Mainly this includes diagnostic and predictive assays, as well as novel therapeutic strategies specifically targeting the brain tumor stem cells. Combination of this therapeutic strategy with current treatments that target the rest of the tumor population should eradicate tumors more efficiently, reduce the risk of relapse, and have a major impact on metastatic clones and tumor resistance to conventional therapies.

References

Al-Hajj, M., Wicha, M.S., Benito-Hernandez, A., Morrison, S.J., and Clarke, M.F. (2003). Prospective identification of tumorigenic breast cancer cells. Proc Natl Acad Sci USA *100*, 3983–3988.

Alcantara Llaguno, S., Chen, J., Kwon, C.H., Jackson, E.L., Li, Y., Burns, D.K., Alvarez-Buylla, A., and Parada, L.F. (2009). Malignant astrocytomas originate from neural stem/progenitor cells in a somatic tumor suppressor mouse model. Cancer Cell *15*, 45–56.

Bao, S., Wu, Q., Li, Z., Sathornsumetee, S., Wang, H., McLendon, R.E., Hjelmeland, A.B., and Rich, J.N. (2008). Targeting cancer stem cells through L1CAM suppresses glioma growth. Cancer Res *68*, 6043–6048.

Bao, S., Wu, Q., McLendon, R.E., Hao, Y., Shi, Q., Hjelmeland, A.B., Dewhirst, M.W., Bigner, D.D., and Rich, J.N. (2006). Glioma stem cells promote radioresistance by preferential activation of the DNA damage response. Nature *444*, 756–760.

Beier, D., Hau, P., Proescholdt, M., Lohmeier, A., Wischhusen, J., Oefner, P.J., Aigner, L., Brawanski, A., Bogdahn, U., and Beier, C.P. (2007). CD133(+) and CD133(−) glioblastoma-derived cancer stem cells show differential growth characteristics and molecular profiles. Cancer Res *67*, 4010–4015.

Bonnet, D., and Dick, J.E. (1997). Human acute myeloid leukemia is organized as a hierarchy that originates from a primitive hematopoietic cell. Nat Med *3*, 730–737.

Brown, B.D., and Naldini, L. (2009). Exploiting and antagonizing microRNA regulation for therapeutic and experimental applications. Nat Rev Genet *10*, 578–585.

Cairns, J. (2002). Somatic stem cells and the kinetics of mutagenesis and carcinogenesis. Proc Natl Acad Sci USA *99*, 10567–10570.

Chan, X.H., Nama, S., Gopal, F., Rizk, P., Ramasamy, S., Sundaram, G., Ow, G.S., Vladimirovna, I.A., Tanavde, V., Haybaeck, J., *et al.* (2012). Targeting Glioma Stem Cells by Functional Inhibition of a Prosurvival OncomiR-138 in Malignant Gliomas. Cell Rep *2*, 591–602.

Cheng, L., Wu, Q., Huang, Z., Guryanova, O.A., Huang, Q., Shou, W., Rich, J.N., and Bao, S. (2012). L1CAM regulates DNA damage checkpoint response of glioblastoma stem cells through NBS1. EMBO J *30*, 800–813.

Cui, J.G., Zhao, Y., Sethi, P., Li, Y.Y., Mahta, A., Culicchia, F., and Lukiw, W.J. (2009). MicroRNA-128 (miRNA-128) down-regulation in glioblastoma targets ARP5 (ANGPTL6), Bmi-1 and E2F-3a, key regulators of brain cell proliferation. J Neurooncol *98*, 297–304.

Diehn, M., Cho, R.W., and Clarke, M.F. (2009). Therapeutic implications of the cancer stem cell hypothesis. Semin Radiat Oncol *19*, 78–86.

Djuranovic, S., Nahvi, A., and Green, R. (2012). miRNA-mediated gene silencing by translational repression followed by mRNA deadenylation and decay. Science *336*, 237–240.

Forsyth, P., Roldan, G., George, D., Wallace, C., Palmer, C.A., Morris, D., Cairncross, G., Matthews, M.V., Markert, J., Gillespie, Y., *et al.* (2008). A phase I trial of intratumoral administration of reovirus in patients with histologically confirmed recurrent malignant gliomas. Mol Ther *16*, 627–632.

Frank, N.Y., Schatton, T., and Frank, M.H. (2010). The therapeutic promise of the cancer stem cell concept. J Clin Invest *120*, 41–50.

Galli, R., Binda, E., Orfanelli, U., Cipelletti, B., Gritti, A., De Vitis, S., Fiocco, R., Foroni, C., Dimeco, F., and Vescovi, A. (2004). Isolation and characterization of tumorigenic, stem-like neural precursors from human glioblastoma. Cancer Res *64*, 7011–7021.

Gennarino, V.A., D'Angelo, G., Dharmalingam, G., Fernandez, S., Russolillo, G., Sanges, R., Mutarelli, M., Belcastro, V., Ballabio, A., Verde, P., *et al.* (2012). Identification of microRNA-regulated gene networks by expression analysis of target genes. Genome Res *22*, 1163–1172.

Gentner, B., Schira, G., Giustacchini, A., Amendola, M., Brown, B.D., Ponzoni, M., and Naldini, L. (2009). Stable knockdown of microRNA in vivo by lentiviral vectors. Nat Methods *6*, 63–66.

Godlewski, J., Nowicki, M.O., Bronisz, A., Williams, S., Otsuki, A., Nuovo, G., Raychaudhury, A., Newton, H.B., Chiocca, E.A., and Lawler, S. (2008). Targeting of the Bmi-1 oncogene/stem cell renewal factor by microRNA-128 inhibits glioma proliferation and self-renewal. Cancer Res *68*, 9125–9130.

He, L., and Hannon, G.J. (2004). MicroRNAs: small RNAs with a big role in gene regulation. Nat Rev Genet *5*, 522–531.

Hirschmann-Jax, C., Foster, A.E., Wulf, G.G., Nuchtern, J.G., Jax, T.W., Gobel, U., Goodell, M.A., and Brenner, M.K. (2004). A distinct "side population" of cells with high drug efflux capacity in human tumor cells. Proc Natl Acad Sci USA *101*, 14228–14233.

Ibrahim, A.F., Weirauch, U., Thomas, M., Grunweller, A., Hartmann, R.K., and Aigner, A. (2011). MicroRNA replacement therapy for miR-145 and miR-33a is efficacious in a model of colon carcinoma. Cancer Res *71*, 5214–5224.

Kefas, B., Godlewski, J., Comeau, L., Li, Y., Abounader, R., Hawkinson, M., Lee, J., Fine, H., Chiocca, E.A., Lawler, S., *et al.* (2008). microRNA-7 inhibits the epidermal growth factor receptor and the Akt pathway and is down-regulated in glioblastoma. Cancer Res *68*, 3566–3572.

Kim, V.N. (2005). MicroRNA biogenesis: coordinated cropping and dicing. Nat Rev Mol Cell Biol *6*, 376–385.

Kim, V.N., Han, J., and Siomi, M.C. (2009). Biogenesis of small RNAs in animals. Nat Rev Mol Cell Biol *10*, 126–139.

Kota, J., Chivukula, R.R., O'Donnell, K.A., Wentzel, E.A., Montgomery, C.L., Hwang, H.W., Chang, T.C., Vivekanandan, P., Torbenson, M., Clark, K.R., *et al.* (2009). Therapeutic microRNA delivery suppresses tumorigenesis in a murine liver cancer model. Cell *137*, 1005–1017.

Krutzfeldt, J., Rajewsky, N., Braich, R., Rajeev, K.G., Tuschl, T., Manoharan, M., and Stoffel, M. (2005). Silencing of microRNAs in vivo with 'antagomirs.' Nature *438*, 685–689.

Lapidot, T., Sirard, C., Vormoor, J., Murdoch, B., Hoang, T., Caceres-Cortes, J., Minden, M., Paterson, B., Caligiuri, M.A., and Dick, J.E. (1994). A cell initiating human acute myeloid leukaemia after transplantation into SCID mice. Nature *367*, 645–648.

Lathia, J.D., Gallagher, J., Heddleston, J.M., Wang, J., Eyler, C.E., Macswords, J., Wu, Q., Vasanji, A., McLendon, R.E., Hjelmeland, A.B., *et al.* (2011). Integrin alpha 6 regulates glioblastoma stem cells. Cell Stem Cell *6*, 421–432.

Li, Y., Guessous, F., Zhang, Y., Dipierro, C., Kefas, B., Johnson, E., Marcinkiewicz, L., Jiang, J., Yang, Y., Schmittgen, T.D., *et al.* (2009a). MicroRNA-34a inhibits glioblastoma growth by targeting multiple oncogenes. Cancer Res *69*, 7569–7576.

Li, Z., Wang, H., Eyler, C.E., Hjelmeland, A.B., and Rich, J.N. (2009b). Turning cancer stem cells inside out: an exploration of glioma stem cell signaling pathways. J Biol Chem *284*, 16705–16709.

Medina, P.P., Nolde, M., and Slack, F.J. (2010). OncomiR addiction in an in vivo model of microRNA-21-induced pre-B-cell lymphoma. Nature *467*, 86–90.

Moscat, J., and Diaz-Meco, M.T. (2009). p62 at the crossroads of autophagy, apoptosis, and cancer. Cell *137*, 1001–1004.

Muzyczka, N., and Warrington, K.H., Jr. (2005). Custom adeno-associated virus capsids: the next generation of recombinant vectors with novel tropism. Hum Gene Ther *16*, 408–416.

O'Brien, C.A., Pollett, A., Gallinger, S., and Dick, J.E. (2007). A human colon cancer cell capable of initiating tumour growth in immunodeficient mice. Nature *445*, 106–110.

Plaisier, C.L., Pan, M., and Baliga, N.S. (2012). A miRNA-regulatory network explains how dysregulated miRNAs perturb oncogenic processes across diverse cancers. Genome Res *22*, 2302–2314.

Potten, C.S., Owen, G., and Booth, D. (2002). Intestinal stem cells protect their genome by selective segregation of template DNA strands. J Cell Sci *115*, 2381–2388.

Rainov, N.G. (2000). A phase III clinical evaluation of herpes simplex virus type 1 thymidine kinase and ganciclovir gene therapy as an adjuvant to surgical resection and radiation in adults with previously untreated glioblastoma multiforme. Hum Gene Ther *11*, 2389–2401.

Rich, J.N., and Eyler, C.E. (2008). Cancer stem cells in brain tumor biology. Cold Spring Harb Symp Quant Biol *73*, 411–420.

Sanai, N., Alvarez-Buylla, A., and Berger, M.S. (2005). Neural stem cells and the origin of gliomas. N Engl J Med *353*, 811–822.

Scherr, M., Venturini, L., Battmer, K., Schaller-Schoenitz, M., Schaefer, D., Dallmann, I., Ganser, A., and Eder, M. (2007). Lentivirus-mediated antagomir expression for specific inhibition of miRNA function. Nucleic Acids Res *35*, e149.

Segura, M.F., Hanniford, D., Menendez, S., Reavie, L., Zou, X., Alvarez-Diaz, S., Zakrzewski, J., Blochin, E., Rose, A., Bogunovic, D., *et al.* (2009). Aberrant miR-182 expression promotes melanoma metastasis by repressing FOXO3 and microphthalmia-associated transcription factor. Proc Natl Acad Sci USA *106*, 1814–1819.

Shi, L., Zhang, J., Pan, T., Zhou, J., Gong, W., Liu, N., Fu, Z., and You, Y. (2010). MiR-125b is critical for the suppression of human U251 glioma stem cell proliferation. Brain Res *1312*, 120–126.

Silber, J., Lim, D.A., Petritsch, C., Persson, A.I., Maunakea, A.K., Yu, M., Vandenberg, S.R., Ginzinger, D.G., James, C.D., Costello, J.F., *et al.* (2008). miR-124 and miR-137 inhibit proliferation of glioblastoma multiforme cells and induce differentiation of brain tumor stem cells. BMC Med *6*, 14.

Singec, I., Knoth, R., Meyer, R.P., Maciaczyk, J., Volk, B., Nikkhah, G., Frotscher, M., and Snyder, E.Y. (2006). Defining the actual sensitivity and specificity of the neurosphere assay in stem cell biology. Nat Methods *3*, 801–806.

Singh, S.K., Hawkins, C., Clarke, I.D., Squire, J.A., Bayani, J., Hide, T., Henkelman, R.M., Cusimano, M.D., and Dirks, P.B. (2004). Identification of human brain tumour initiating cells. Nature *432*, 396–401.

Sun, Y., Kong, W., Falk, A., Hu, J., Zhou, L., Pollard, S., and Smith, A. (2009). CD133 (Prominin) negative human neural stem cells are clonogenic and tripotent. PLoS One *4*, e5498.

Tang, G., Reinhart, B.J., Bartel, D.P., and Zamore, P.D. (2003). A biochemical framework for RNA silencing in plants. Genes Dev *17*, 49–63.

Venezia, T.A., Merchant, A.A., Ramos, C.A., Whitehouse, N.L., Young, A.S., Shaw, C.A., and Goodell, M.A. (2004). Molecular signatures of proliferation and quiescence in hematopoietic stem cells. PLoS Biol *2*, e301.

Wang, S., Yang, D., and Lippman, M.E. (2003). Targeting Bcl-2 and Bcl-XL with nonpeptidic small-molecule antagonists. Semin Oncol *30*, 133–142.

Ward, R.J., Lee, L., Graham, K., Satkunendran, T., Yoshikawa, K., Ling, E., Harper, L., Austin, R., Nieuwenhuis, E., Clarke, I.D., *et al.* (2009). Multipotent CD15+ cancer stem cells in patched-1-deficient mouse medulloblastoma. Cancer Res *69*, 4682–4690.

Welm, A.L., Kim, S., Welm, B.E., and Bishop, J.M. (2005). MET and MYC cooperate in mammary tumorigenesis. Proc Natl Acad Sci USA *102*, 4324–4329.

Xie, J., Ameres, S.L., Friedline, R., Hung, J.H., Zhang, Y., Xie, Q., Zhong, L., Su, Q., He, R., Li, M., *et al.* (2006). Long-term, efficient inhibition of microRNA function in mice using rAAV vectors. Nat Methods *9*, 403–409.

Yang, Z., Chen, Y., Lillo, C., Chien, J., Yu, Z., Michaelides, M., Klein, M., Howes, K.A., Li, Y., Kaminoh, Y., *et al.* (2008). Mutant prominin 1 found in patients with macular degeneration disrupts photoreceptor disk morphogenesis in mice. J Clin Invest *118*, 2908–2916.

Zhang, C.Z., Zhang, J.X., Zhang, A.L., Shi, Z.D., Han, L., Jia, Z.F., Yang, W.D., Wang, G.X., Jiang, T., You, Y.P., *et al.* (2010). MiR-221 and miR-222 target PUMA to induce cell survival in glioblastoma. Mol Cancer *9*, 229.

Zhou, S., Schuetz, J.D., Bunting, K.D., Colapietro, A.M., Sampath, J., Morris, J.J., Lagutina, I., Grosveld, G.C., Osawa, M., Nakauchi, H., *et al.* (2001). The ABC transporter Bcrp1/ABCG2 is expressed in a wide variety of stem cells and is a molecular determinant of the side-population phenotype. Nat Med *7*, 1028–1034.

Chapter 31

The Riboproteome Orchestrates Self-Renewal and Cell Fate in Leukemia

Elianna M. Amin and Michael G. Kharas
Molecular Pharmacology and Chemistry Program and Center for Cell Engineering, Memorial Sloan-Kettering Cancer Center, New York, NY, USA

Introduction

Leukemia stem cells (LSC) are characterized by a gain in self-renewal properties and are functionally able to recapitulate the phenotypic heterogeneity of the leukemia (Bonnet and Dick, 1997; Hope et al., 2004). Deregulation and activation of aberrant signaling pathways are considered the hallmark of cancer development, but the emerging role of post-transcriptional regulation is currently being intensively investigated (Cooper et al., 2009; David and Manley, 2010; Venables, 2004). In this chapter we will focus on the hematopoietic system and highlight the role of RNA binding proteins (RBP) in leukemia and within the LSCs.

Apart from providing diversity to the final protein product, post-transcriptional regulation is also a rapid and efficient way to modulate protein abundance. The post-transcription process involves the nascent RNA to undergo splicing, capping, 3′ end formation, and nucleocytoplasmic export to the ribosome before being translated and then degraded (Glisovic et al., 2008) (Figure 31.1).

The mRNAs bound by RBPs form dynamic complexes, which make up the riboproteome in a cell. Biogenesis of the mRNA is determined by the different RBPs. RBPs are able to recruit additional factors and enzymes to the mRNA, which determine the fate of the target. Thus, RBPs are master regulators and modulators of RNA metabolism (Glisovic et al., 2008; St Johnston, 2005). Apart from binding to mRNA, RBPs are also able to bind to microRNA (miRNA) to form microRNPs, which are then able to regulate translation and RNA stability by binding to sequences in the 3′ untranslated region (UTR) of target mRNAs.

Each RBP is able to bind to specific sequences and secondary structures that are found in mRNA. The sequences in the open reading frame (ORF) are involved in splicing, while the 5′ UTR is important in ribose methylation of the cap structure and the 3′ UTR regulates the stability of the mRNA, translation efficiency, and localization.

Several highly conserved domains known as RNA binding domains (RBD) determine the specificity for the target RNA, and this provides a classification system for the diverse set of RBPs. There are more than 40 RBDs identified to date, and the most studied are RNA recognition motif (RRM), K-homology domain (KH), double stranded

Cancer Stem Cells, First Edition. Edited by Vinagolu K. Rajasekhar.

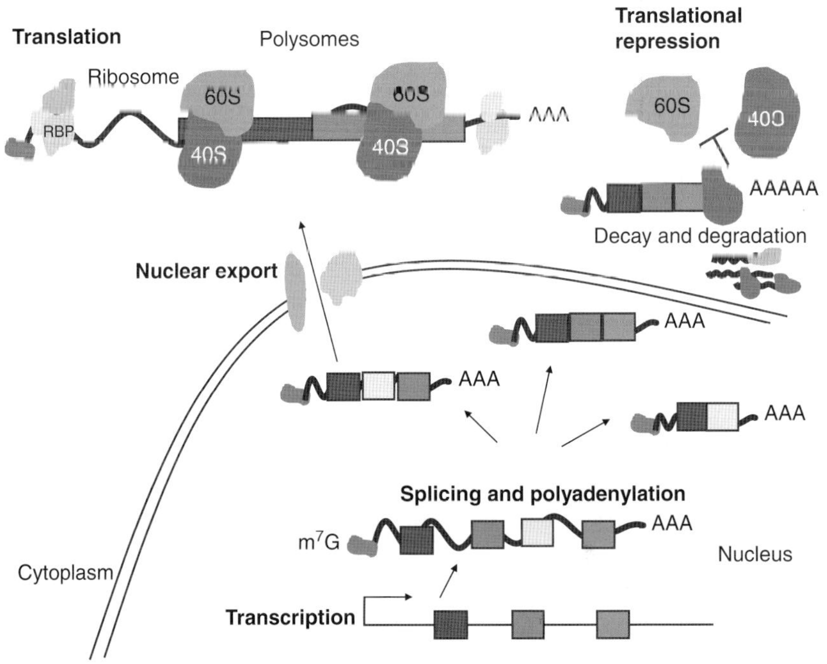

Figure 31.1. Cellular stages during mRNA biogenesis. A general pathway for mRNA processing and RBP regulation: transcription, splicing, polyadenylation, mRNA export, and translational stabilization or repression. Colored squares indicate exons and m7G; 5′7-methylguanosine cap. For color detail, please see color plate section.

RNA-binding domain (dsRBD), zinc finger domain, Arginine-rich domain, cold-shock domain (CSD), and the PAZ and PIWI domains. RBPs can contain different combinations of RBDs, which allow flexibility for interaction with different targets (Lunde et al., 2007). Additional domains in RBPs provide further specificity for protein-protein interactions and cellular localization. The specific sequences bound by all the RBPs are not yet fully elucidated. New technologies that cross-link proteins to RNA followed by high throughput sequencing (CLIP-seq, HITS-CLIP, par-CLIP, etc.) are beginning to provide mechanistic insights for the specific RBP targets (Darnell, 2010). These studies will contribute to our understanding of the multifunctional nature of RBPs by regulating mRNA processing in different cellular compartments.

As transcription progresses, the RNA processing occurs simultaneously by forming dynamic ribonucleic protein complexes (RNP). Splicing occurs in the nucleus whereby introns from pre-mRNA are excised and exons join together to form mature mRNA (De Conti et al., 2012). The splicing reaction is performed by the spliceosome, a macromolecule complex containing five small nuclear riboproteins (snRNP) and accessory proteins, in a sequential and highly coordinated process that determines and joins the exons together (Hoskins and Moore, 2012). Alterations in splicing can lead to the pathogenesis of diseases and tumorigenesis which are of great interest (David and Manley, 2010; Venables, 2004).

After splicing, the mRNA is assembled, and additional RBPs and associated factors form the mature ribonucleoprotein (mRNP) (Hocine et al., 2010). The mRNP is then exported out of the nucleus via the nuclear pore complex to the cytoplasm where additional cytoplasmic factors bind and dictate the fate of the mRNA. The exported mRNA undergoes either nonsense-mediated decay or stabilization. RBPs can protect the mRNA from rapid turnover and allow for multiple ribosomes to bind and drive efficient translation. These multiribosomal complexes are called polysomes (Warner et al., 1963).

RBPs are further controlled by post-translational modifications (Blackwell and Ceman, 2012; Lee, 2012). Phosphorylation or other modifications can result in changes in the affinity to the RNA targets, protein interactions, and/or localization within the cell.

In this chapter we will explore the RBPs that are commonly dysregulated in hematopoietic stem or progenitor cells, resulting in leukemic progression. We will then examine how RBP function can be regulated and understand the clinical and translational potential for this new class of therapeutic targets.

RNA Binding Proteins and Leukemia

SF3B1 and Splicing Factors

Genomics studies have mapped the landscape of somatic mutations and genetic alterations that are associated with hematological malignancies. Exome sequencing has revealed multiple somatic mutations in the splicing machinery in myelodysplastic syndromes (MDS), chronic lymphocytic leukemia (CLL), chronic myelomonocytic leukemia (CMML) including SF3B1, SRSF2, U2AF35, ZRSR2, SF3A1, PRP40B, SF1, U2AF1, and U2AF65 (Graubert et al., 2012; Malcovati et al., 2011; Meggendorfer et al., 2012; Papaemmanuil et al., 2011; Quesada et al., 2012; Visconte et al., 2012a; Wang et al., 2011; Yoshida et al., 2011) and is reviewed here (Visconte et al., 2012b). To date, the relevance of these mutations and the mechanism by which it contributes to leukemogenesis and LSCs are poorly understood. In this section we briefly summarize the recent studies of a select set of RBPs, their role in leukemia, and the genetic alterations that have implicated them in hematopoietic diseases (Table 31.1).

MDS patients can harbor mutations in SRSF2, which is correlated with a poor clinical outcome and most likely result in the loss of function (Makishima et al., 2012; Meggendorfer et al., 2012; Yoshida et al., 2011). Consistent with this finding, SRSF2 had a lower reconstituting ability and reduced cellular proliferation in a bone marrow transplant

Table 31.1. RNA binding proteins and self-renewal

Gene	RNA activity	Self-renewal	Leukemia
RBM15(OTT1)	Involved in splicing and mRNA export	Maintains the HSC during quiescence	MKL fusion partner
ZFP36 (ZFP36L1,ZFP36L2)	Deadenylates mRNA resulting in degradation	Essential for fetal hematopoiesis	Mutations in AML and T-ALL in ZFP36 null mice
Msi2	Translational repressor	Controls self-renewal in HSC and HSPC	Fusion partner, elevated expression determines worse prognosis in CML-BC and AML
HuR	Binds 3′UTR, RNA translation and	Maintains the HPC	Overexpressed in M4 AML and CML-BC
Lin28	RNA maturation of miRNA and mRNA translation	Expression decreases during ES stem cell differentiation	Overexpressed in CML-BC and mediates MLL leukemia in mice
Npm	Ribosome biogenesis, RNAse activity, RNA processing, and export	Increaesed HSPC in NPM heterozygous mice and decreased self-renewal	Mutated in AML, deleted in -5q therapy related MDS, fusion partner
SF3B1, U2AF1, U2AF2, SRSF2, SF1, ZRSR2	Splicing and processing	To be studied	Mutated in leukemia

model (Matsunawa et al., 2012). Moreover, conditional deletion of SRSF2 in T-cells results in a decrease in the number of thymocytes and defects in T cell maturation (Wang et al., 2001).

Through whole genome sequencing, recurrent mutations in SF3B1 were identified in 20% of patients with MDS (Cooper et al., 2009; Papaemmanuil et al., 2011) and CLL (Quesada et al., 2012; Wang et al., 2011). Although not completely consistent, MDS patients with SF3B1 mutations were less likely to have cytopenia and also had longer event-free survival than patients without the SF3B1 mutation (Malcovati et al., 2011; Papaemmanuil et al., 2011). In contrast, in CLL, SF3B1 mutations are associated with a poor prognosis (Quesada et al., 2012).

SF3B1, a core component of one of the snRNPs, is involved in the 3′ splice site at the intron-exon junction. Exome sequencing uncovered an accumulation of mutations resulting in a substitution in K700E and in other hotspot regions including within the non-identical 22-tandem carboxy-terminal HEAT (huntingtin, elongation 3, protein phosphatase 2A, target of rapamycin1) repeat region (Papaemmanuil et al., 2011). The subtype of MDS patients with the SF3B1 mutations had ring sideoroblasts (RS), refractory anemia with ring sideroblasts (RARS), and refractory cytopenia with multi-lineage dysplasia (Papaemmanuil et al., 2011; Visconte et al., 2012a). This observation is recapitulated in the mouse model whereby haploinsufficiency in SF3B1 results in ring sideroblasts (Visconte et al., 2012c). Transcriptome profiling of MDS patients with SF3B1 mutations showed that there is a down regulation of mitochondrial gene networks (Papaemmanuil et al., 2011).

The common mutations in splicing factors are heterozygous, and it remains unclear whether the pathogenesis is due to a loss of expression of the wild type allele or caused by dominant negative effects. It also remains unknown whether these mutations are themselves drivers of the disease or if there is a role in the LSC. Moreover, comprehensive RNA profiling studies will be needed to evaluate the changes in mRNA isoforms associated with mutations found in the spliceosome, and each class of mutation needs to be modeled in both normal hematopoiesis and leukemia settings.

RBM15(OTT1)

The RNA binding motif 15 gene (*RBM15*) was initially identified as a partner in the chromosomal translocation t(1;22)(p13;q13) creating a fusion with the Megakaryocytic Acute Leukemia-1 (*MKL1*) in patients with infant acute megakaryocytic leukemia (AMKL) (Ma et al., 2001). Infants with this chromosomal translocation are associated with a poor prognosis and a median survival of only 8 months.

The conditional knock in mouse model of the RBM15-MKL fusion resulted in abnormal fetal and adult hematopoiesis (Mercher et al., 2009). There is a significant increase in colony-forming efficiency in the hematopoietic stem and progenitor cells, multipotent myeloid progenitors, and progenitors in the splenocytes of adult mice, which result in a low penetrant AMKL mirroring the phenotype found in the patients (Mercher et al., 2009). Using leukemic cells derived from the knock in RBM15-AMKL mice, RBM15-MKL is able to form a complex with the Setd1b methyltransferase (Lee and Skalnik, 2012). This interaction causes alterations in epigenetic regulation. Altogether these studies indicate that the fusion protein RBM15-AMKL requires additional cooperating events to initiate leukemia.

RBM15, a member of the split end (Spen) family of proteins, contains three N-terminal RRMs and a Spen paralogue and orthologue domain at its c-terminus (Spoc). RBM15 is implicated in regulating mRNA export and splicing (Hiriart et al., 2005; Lindtner et al., 2006; Zolotukhin et al., 2009). The loss of RBM15 in mice results in embryonic lethality at E9.5 to E10.5 (Raffel et al., 2007). Conditional deletion of RBM15 in the hematopoietic system results in a block in B cell differentiation, an increase in the hematopoietic stem and progenitors, and decreased repopulation ability in vitro and in vivo (Raffel et al., 2007; Xiao et al., 2012). Additionally, HSCs deleted for RBM15 fail to maintain quiescence during replicative stress (Xiao et al., 2012).

RBM15 inhibits myeloid differentiation by stimulating Notch signaling via interactions with RBPJκ (Ma et al., 2007). Consistent with a role for RBM15 in Notch signaling, the fusion protein abrogates its function resulting in abnormal Notch signaling and a block in fetal megakaryopoiesis (Mercher et al., 2009). This further suggests a potential role for Notch in maintaining the self-renewal activity of HSCs and in megakaryocyte differentiation (Duncan et al., 2005). These studies support an important role for RBM15 in both normal and leukemia hematopoietic cells.

ZFP36

The ZFP36 (TIS11) family consists of three members—ZFP36, ZFP36LI, and ZFP36L2—all of which contain the conserved CCCH zinc finger RBD. The ZFP36 family members bind to adenine-uridine rich elements (ARE) located at the 3′UTR of the mRNA. Binding of the ZFP36 family member to the mRNA leads to deadenylation and targets the mRNA for degradation via the exosome (Baou et al., 2009).

Heterozygous mutations were identified in ZFP36L2 in 9/172 patients with de novo AML, 2/21 patients with T-lineage acute lymphoblastic leukemia (T-ALL), and 1/64 B lineage acute lymphoblastic leukemia (B-ALL) patients (Iwanaga et al., 2011). When one of the mutations was overexpressed in Hela cells (eight base pair insertion), it failed to inhibit proliferation compared to the wild-type protein. However, the clinical relevance of these mutants remains unclear as the ZFP36 family controls a wide variety of cellular functions including differentiation and apoptosis (Ciais et al., 2012).

ZFP36L2 knockout mice die shortly after birth due to a reduced number of hematopoietic stem cells and progenitors in the fetal liver, suggesting that ZFP36L2 plays a crucial role in developmental hematopoiesis (Stumpo et al., 2009). Deletion of *Zfp36l1* and *Zfp36l2* within the T-cell compartment results in perturbed thymic development with aberrant β-selection that quickly progresses into T-ALL (Hodson et al., 2010). Loss of ZFP36L1 and ZFP36L2 results in the activation of Notch1 by directly binding to the ARE. The role for Notch in T-cell development is well characterized. Furthermore, Notch is commonly mutated and constitutively activated in T-ALL (Demarest et al., 2008). These studies suggest that Zfp36 may have additional functions in regulating LSCs.

MSI2

The first hint that *MSI2* has a role in leukemia was in the identification of several patients who progressed to CML blast crisis (BC-CML) and harbored the *MSI2-HOXA9* translocation (Barbouti et al., 2003). *MSI2* is found to be upregulated in a variety of myeloid leukemia cell lines and in patients that have a worse clinical prognosis. Protein levels for MSI2 measured in AML patients, similarly showed decreased survival and correlated with increased expression of *MSI2*. Of note, the expression and survival correlation of *MSI2* was found in patients with both cytoplasmic and nuclear localization (Byers et al., 2011). Additionally, *MSI2* expression is reduced in patients with low and high-risk MDS compared to normal CD34 cells (Pereira et al., 2012).

The *MSI* gene family is evolutionarily conserved and contains two closely related proteins, MSI1 and MSI2, each of which is characterized by the presence of two tandem RRMs and a carboxy-terminal polyA binding protein association domain. They share high homology and similar binding specificities to target RNAs - GU_{3-5} (G/AG) in the 3′UTR (Imai et al., 2001). Once bound, MSI interacts with the polyA binding protein and competes for eIF4G. Thereby, interfering with translation initiation and leads to the degradation of target mRNAs (Kawahara et al., 2008).

In addition to its role in regulating translation, the Msi family has been implicated in controlling polyadenylation of specific mRNAs in *Xenopus* oocytes (Charlesworth et al., 2006). Even though the RRM motifs are 90% conserved between the homologs, mechanistic studies have focused mainly on Msi1.

In contrast, Msi2 has been functionally studied in the hematopoietic system due to the relatively low expression of Msi1 (Ito et al., 2010; Kharas et al., 2010a). Depletion or loss of *Msi2* in a CML-BC murine leukemia model led to increased survival and differentiation of the leukemic cells (Ito et al., 2010; Kharas et al., 2010a). Additionally, loss of MSI2 in human myeloid leukemia cells also result in reduced proliferation, differentiation, and apoptosis.

In addition to its role in leukemia, Msi2 is also identified to be important in hematopoietic stem cells by gene expression analysis, a retroviral insertional screen, and an shRNA screen for asymmetric regulators (de Andres-Aguayo et al., 2011; Ito et al., 2010; Kharas et al., 2010a). Hematopoietic stem and progenitor cells express Msi2 higher levels compared to more committed progenitors and differentiated myeloid cells. Moreover, mice deficient in Msi2 had reduced HSC numbers as well as a decrease in their self-renewal and a failure to maintain quiescence (Park et al., 2014) (de Andres-Aguayo et al., 2011). Interestingly, mature myeloid differentiation is normal in mice deleted for Msi2 (de Andres-Aguayo et al., 2011; Park et al., 2014). However, MSI2 overexpression in a conditional murine system results in a transient increase in HSC numbers followed by a decrease in the overall competitiveness of the HSCs and a partial block in mature myeloid differentiation. Several studies with retroviral overexpression or insertion resulted in improved engraftment with higher levels of MSI2 (Hope et al., 2010; de Andres-Aguayo et al., 2011). These studies are most likely reconciled by the system and the difference in Msi2 overexpression.

The relatively limited mechanistic studies so far have suggested that Numb might play a role in altering the differentiation and stem cell fate by affecting the Notch signaling pathway. However, deletion of Numb or Numb-like has no effect on HSC self-renewal (Wilson et al., 2007). It remains unclear what targets Msi2 binds and how it alters self-renewal and increases aggressiveness in leukemia.

Overall, these data indicate that Msi2 may maintain the leukemic stem cell by imparting stem cell characteristics to the immature myeloid cells thus providing leukemia with a more aggressive phenotype. However, forced MSI2 expression in a BCR-ABL model does not lead to increased frequency of CML leukemia initiating cells (Kharas et al., 2010a). It has not been formally tested if Msi2 maintains the leukemic stem cell but may just impart a more undifferentiated state in the mature myeloid blast cells.

HuR

HuR is a ubiquitously expressed member of the embryonic lethal abnormal vision (ELAV) family of RBPs. HuR contains three RRMs of which the first two are used to bind ARE elements, while the third RRM binds the polyA tail (Ma et al., 1997). Binding of HuR to the ARE elements of its target transcripts allows it to stabilize and modulate translation efficiency (Brennan and Steitz, 2001). HuR is able to shuttle between the nucleus and cytoplasm via a small sequence, which is located between its second and third RRM. The shuttling ability of HuR may protect the mRNA as it exits from the nucleus into the cytoplasm from the mRNA degradation machinery. Also, the presence of HuR binding sites in intronic regions suggests that HuR might play a role in mRNA processing(Lebedeva et al., 2011).

HuR is required for embryonic development as postnatal deletion of HuR results in atrophic spleens, lymph nodes, and thymus with reduced numbers of lymphoid, myeloid, and erythroid progenitors in the bone marrow (Ghosh et al., 2009). This defect was not observed at the level of the HSC, but at the multipotent progenitor. Moreover, HuR is overexpressed in the M4 subtype of AML and this correlated with high levels of eukaryotic initiation factor 4E (eIF4E), a component of the translation initiation complex that will be discussed later in this chapter (Topisirovic et al., 2009).

Apart from stabilizing mRNA, HuR is also involved in mRNA processing(Mukherjee et al., 2011). Therefore, a more complete understanding of HuR's multifunctional roles will provide answers to its contribution in both normal and malignant hematopoiesis.

Lin28

Discovered initially in C. elegans as a heterochronic gene, Lin28 family is inversely correlated with let7 miRNAs (Ambros and Horvitz, 1984). The mammalian homologs of Lin28 consisting of Lin28a and Lin28b have two CCHC-type zinc fingers and a cold shock domain. The Lin28 acts as a repressor of miRNA processing by binding to the terminal loops of the miRNA, which prevents the maturation of the miRNA precursors. Additionally, Lin28 binds to mRNA transcripts with the "GGAGA" motif, and these transcripts are mainly involved in the RNA processing and splicing machinery (Wilbert et al., 2012). Initial data revealed that Lin28a cosediments in the polysomes (Balzer and Moss, 2007), and additional data suggest that Lin28a is localized to the endoplasmic reticulum (ER) and is able to inhibit the translation of mRNA destined for the ER in mouse embryonic stem cells (Cho et al., 2012). The recognition motif for Lin28 in this context is "AAGNNG, AAGNG" (Cho et al., 2012). The Lin28 family has a wide variety of functions involving pluripotency, reprogramming, and oncogenesis (Thornton and Gregory, 2012), (Newman et al., 2011; Yu et al., 2007).

In normal development, Lin28 is highly expressed in the embryonic stem cells (ESC) and progenitor cells, however, the expression decreases as the stem cells undergo differentiation. The decrease in expression of Lin28 leads to increased maturation of let 7 miRNA. The let-7 family of miRNA, targets genes that are involved in the regulation of stem cell differentiation, cell proliferation, and apoptosis. Let-7 miRNA represses genes involved in self-renewal, which leads to lineage commitment and differentiation of stem cells. Conversely, Lin28 is also a target of let-7, thus establishing a negative feedback loop that can drive the stem cells to differentiation. Interestingly, Lin28 can also contribute to cellular reprogramming from somatic cells to pluripotent embryonic stem cells when added with Oct4, Sox 2, and Nanog (Yu et al., 2007). Lin28 accelerates reprogramming to pluripotent stem cells by promoting cell proliferation (Melton et al., 2010). Lin28b is expressed in abundance in the fetal HSC but not in terminally differentiated blood cells and HSC (Yuan et al., 2012). Interestingly, overexpression of Lin28 in adult HSC leads to the development of a fetal-specific lymphoid lineage (Yuan et al., 2012).

The inverse correlation of low abundance of let-7 miRNA overexpression of Lin28 has been described in a variety of malignancies (Thornton and Gregory, 2012; Viswanathan et al., 2009). Lin28b expression was more frequently detected in patients with BC-CML (6/14) and accelerated phase CML (4/10) compared to patients with CP-CML (1/11), and this correlated with reduced let-7 transcripts (Viswanathan et al., 2009). These results suggest the LSC population has acquired mutations or mechanisms that will allow for the increase of expression of Lin28 to resulting in a block in the differentiation of the stem cell. Myc is able to upregulate the expression of Lin28 by binding to the promoter of Lin28. In a Myc-mediated mouse model of lymphoma, Lin28 is upregulated, leading to a loss in expression of let-7 (Chang et al., 2009). Additionally, MLL fusion proteins drive Myc expression resulting in the upregulation of Lin28. Lin28 then regulates miR-150, which is itself repressed in AML. Lastly, MLL fusion proteins are also able to negatively regulate miR-150 through Lin28 (Jiang et al., 2012). Altogether, these data suggest a role for the Lin28 axis in the maintenance of the LSC in AML. Consistent with this hypothesis is the increased embryonic gene expression signature that was found in LSC (Somervaille et al., 2009).

NPM

Nucleophosmin (NPM) is an abundant phosphoprotein that is primarily localized in the nucleolus, but it is also able to shuttle to the cytoplasm. NPM plays a variety of roles in cellular processes and has been reviewed elsewhere (Colombo et al., 2011; Grisendi et al., 2006). In brief, one of the key roles of NPM is that it is a molecular chaperone that inhibits protein aggregation and provides protection against thermal and chemical denaturation by preserving the enzyme activity of its direct targets (Szebeni and Olson, 1999).

In addition to this role, NPM participates in ribosome biogenesis and is responsible for the nuclear export of the ribosomal protein L5/5 s rRNA subunit complex that contains intrinsic RNAse activity and processes pre-RNA.

Most importantly, the *NPM* gene is a frequent target of chromosomal translocations in hematopoietic malignancies. In anaplastic large-cell lymphoma, NPM is translocated to t(2;5) and forms the fusion gene NPM-ALK. In AML, NPM is translocated to t(3;5) and forms the fusion gene NPM-MLF1, while in rare cases of APL, NPM is translocated to t(5;17) forming the NPM-RARa. The contributions of NPM translocations in leukemogenesis are poorly understood. Furthermore, NPM is mapped to chromosome 5, and this is frequently deleted in therapy-related MDS. The exon 12 of NPM is frequently mutated in patients with AML (Falini et al., 2005). Mutation of exon 12 results in NPM to be localized to the cytoplasm and hence named NPMc+. Several lines of evidence indicate that the cytoplasmic localization of NPM is critical for leukemic transformation (Bolli et al., 2007; Falini et al., 2006; Mariano et al., 2006). However, the role of NPMc+ is still not well understood. NPMc+ transgenic mice had increased myeloproliferation in the bone marrow and the spleen (Cheng et al., 2010). Loss of a single copy of *NPM* results in the expansion of the HSC compartment with reduced competitive engraftment (Raval et al., 2012). A different report suggests that animals with one copy of NPM develop dysplasia in the myeloid and erythroid compartment resembling an MDS-like syndrome (Grisendi et al., 2005). Thus, this cellular regulator controls hematopoietic stem and progenitor function. However, it remains unknown whether these phenotypes are related to its RBP activity or to its other functions in the cell.

Signaling and RBPs

The PI3K/AKT/mTOR signaling pathway plays an important role in the transmission of proliferative and growth signals from membrane bound growth factor receptors to transcription factors. Dysregulation of this pathway, through mutations in upstream components, is commonly associated with resistance to therapy in many types of cancers including acute leukemia (Martelli et al., 2011). The downstream components made up of two RBPs, eIF4E and eIF4G, mediate the major changes in cellular growth, maintenance, and self-renewal (Aoki et al., 2001; Wendel et al., 2004).

A set of proteins in the PI3K signaling pathway carefully control the activity of these RBPs. More specifically, the mammalian target of rapamycin complex 1 (mTORC1) phosphorylates eukaryotic initiation factor 4E-binding protein 1 (4EBP1), which leads to the dissociation of eIF4E (Gingras et al., 1998). 4EBP1 acts as a repressor of translation by preventing eIF4e from binding to the 5′ cap and assembly of the translation initiation complex (Rousseau et al., 1996). Binding of eIF4E to the 5′ 7-methyl guanosine cap of the mRNA is critical for the translation of mRNA that have long 5′ UTRs containing highly structured secondary structures. These mRNAs termed 'weak mRNA' because they are difficult to translate are usually transcribed for genes that are important in cell growth, transformation, and prevent apoptosis. The eIF4E protein provides the critical interface between mRNA, recruitment of eIF4A and eIF4G, and the 40S ribosomal subunit in the cytoplasm. In the nucleus, eIF4E promotes the export of a subset of mRNAs that contains a 50-nucleotide element known as the eIF4E sensitive element in the 3′ UTR. eIF4E also binds to the 5′ cap of the mRNA; these transcripts also encode for genes associated with enhanced proliferation (Sonenberg and Gingras, 1998). In M4/M5 FAB subtypes of AML, eIF4E is overexpressed and elevated levels usually correlate with poor prognosis (Topisirovic et al., 2003).

Alterations in mTORC1 activity, promotes leukemogenesis and depletes HSCs (Chen et al., 2008; Lee et al., 2010; Nardella et al., 2008). Tuberous Sclerosis Complex 1 (TSC1) and phosphatase and tensin homolog (PTEN) are negative regulators of the PI3K pathway and thereby modulate mTOR activity (Laplante and Sabatini, 2012). *Tsc1* deletion in the hematopoietic system results in reduced self-renewal in HSC (Chen et al., 2008). The deletion of Tsc1 drives the HSCs from quiescence into rapid cycling with increased mitochondrial biogenesis and elevated levels of reactive oxygen species (ROS), which can be rescued by the mTOR pharmacological inhibitor rapamycin (Chen et al., 2008). However, deletion of Tsc1 did not result in any hematological malignancy in the mice, suggesting that the Tsc1/mTOR axis is probably not the key axis in initiating the LSC population.

Activation of the pathway by overexpressing a constitutively activated AKT or deletion of Pten in the hematopoiesis system drives the HSCs into rapid cycling followed by a decline in the long-term reconstituting ability (Kharas et al., 2010b; Yilmaz et al., 2006; Zhang et al., 2006). Mice typically present with a myeloproliferative disorder (MPD) that can progress into an AML or succumb to the concurrent T-ALL depending on the level of PI3K pathway activation (Yilmaz et al., 2006; Zhang et al., 2006). Pharmacological inhibition with rapamycin results in the restoration of the HSC activity and depletion of the LSC. These results indicate that apart from maintaining the normal HSC, Pten is also is able to contribute to the development of LSCs.

Studies involving the deletion of components in the mTORC1 complex provide further evidence for mTOR's role in LSC development. The mTORC1 complex is made up of mTOR, Raptor, and mLST8. Deletion of Raptor results in differentiation defects in both the myeloid and B-lymphoid compartments (Hoshii et al., 2012) and accumulation of monocytoid cells (Kalaitzidis et al., 2012). Ablation of Raptor in the context of Pten deletion or MLL-AF9 results in improved survival and reduced progression of leukemogenesis (Hoshii et al., 2012; Kalaitzidis et al., 2012).

In CML, the mTOR pathway is highly dysregulated. BCR-ABL forms a complex with GRB2-associated binding protein 2 (GAB2) to activate the pathway (Sattler et al., 2002) and also phosphorylates 4EBP1 (Prabhu et al., 2007). Primary patient CML blasts displayed aberrant expression of hnRNPK, FUS, hnRNPA1, hnRNPE2, and, La/SSB (Perrotti et al., 1998; Perrotti and Neviani, 2007). Inhibitors that block upstream signaling pathways or transduction of dominant negative RBPs reversed the upregulated RBP expression and resulted in the inhibition of leukemogenesis in vitro and in vivo (Chang et al., 2007; Iervolino et al., 2002; Perrotti et al., 1998). Complex interactions with hnRNPs mRNAs resulted in upregulation of E2F4 and contribute to CML disease progression (Eiring et al., 2008). Overall, these RBPs modulate a wide variety of targets, which are required to maintain leukemic cells.

Translating RBPs to the Clinic

Drugs That Target the Spliceosome

SF3B1, a component of the U2 snRNP, assembles on the branch point recognition sequence and crosslinks both 5′ and 3′ of the pre-mRNA intron to the branch point to facilitate splicing. Several pharmacological compounds drugs are available to specifically prevent this interaction. Spliceostatin A (SSA) binds to the SF3B complex, resulting in the destabilization of the U2 snRNP pre mRNA and the prevention of proceeding events in spliceosome assembly (Corrionero et al., 2011). In the presence of SSA, SF3B1 is unable to cross-link to the pre-mRNA suggesting that SSA induces a conformational change in SF3B1 that prevents it from binding to the RNA (Corrionero et al., 2011). E7107, another pharmacological inhibitor shown to target the splicesome, blocks the ATP-dependent remodeling of the branch point binding region essential for splicing (Folco et al., 2011).

Targeting the post-translational modifications of the RBP could be used as possible therapeutic strategy in leukemia. SR protein kinases (SRPK1/2) can phosphorylate SR proteins and modulate their function. Pharmacological inhibition of SRPK1 blocked the phosphorylation of SR proteins, ASF/SF2, and resulted in the reduction of angiogenesis and tumor regression (Amin et al., 2011). Other components of the spliceosome can also be targeted as reviewed in (Bonnal et al., 2012; Zaharieva et al., 2012).

Targeting PI3K/AKT/mTOR Pathway

Given the aberrant activation of the PI3K/AKT/mTOR signaling in cancer, therapeutic approaches were designed to inhibit the components of this pathway. Targeting mTOR was an attractive therapeutic option as inhibition of mTOR would prevent the phosphorylation of eIF4E and p70S6K, the regulators of translation efficiency of mRNA. Allosteric inhibitors of mTOR such as rapamycin, RAD001 and CC1-779, bind to FKBP12 and interfere with the FRB domain of mTOR. Pharmacological inhibition of mTOR can restore HSC activity and deplete LSC function (Hoshii et al., 2012; Kalaitzidis et al., 2012), and these inhibitors are currently undergoing clinical trials. However, the effects of these inhibitors are only partial because only mTOR kinase activity that phosphorylates p70S6K is inhibited (Choo et al., 2008; Kuo et al., 2011). Second generation inhibitors are currently in development; they work by binding to the catalytic cleft of mTOR and competing against ATP to bind the catalytic domain (e.g., PP242). These inhibitors would block the phosphorylation of all mTOR targets. The pharmacological inhibition by PP242 is more potent than rapamycin as PP242 is able to cause cell death in vitro and delay leukemic progression in the mouse model of BCR/ABL induced B-ALL leukemia (Janes et al., 2010).

Phosphorylation of 4EBP1 by mTOR results in the release of eIF4E, which is subsequently able to initiate the translation initiation complex with e1F4G. The binding of eIF4E to eIF4G increases the cap-dependent translation through recruitment of eIF4A and the e1F3-40S ribosomal subunit. The elevated expression of the RBP, e1F4E, in M4/M5 AML and the BC-CML provides a clinical opportunity. Most importantly, a small molecule pharmacological inhibitor called 4EGI-1 is able to disrupt the interaction between e1F4E and e1F4G (Moerke et al., 2007). Interestingly, binding of 4EGI-1 to e1F4E does not disrupt the interaction of e1F4E to its translational repressor 4EBP1 and appears to enhance the activity of 4EBP1 (Moerke et al., 2007). Another inhibitor called Ribavirin (1-b-D-ribofuranosyl-1,2,4-triazole-3-carboxamide) has been previously used for the treatment of viral infections. Ribavirin binds to eIF4E around the 5′ 7-methyl guanosine cap of the mRNA causing a reduction in the association of the mRNA that is required for nuclear export. Ribavirin has already been used in a phase II clinical trial, which involved 11 patients with the M4/M5 AML subtype (Assouline et al., 2009). The clinical trial had encouraging results with one patient going into complete remission, two partial remissions, and four with stable disease (Assouline et al., 2009). The resistance to Ribavirin was associated with re-entry of eIF4E into the nucleus. It remains unknown how the changes in localization leads to resistance. Altogether, these data suggest the potential for developing small molecule inhibitors that block the RBP function (Table 31.2).

Table 31.2. Drugging RNA binding proteins

Compound	Targets
Isoginkterin	Prevents the formation of different RNP complexes in splicing
Indole derivatives	Binds to SR proteins and inhibits splicing of enhancer dependent introns
PP1 and PP2A inhibitors	Inhibits phosphorylation of splicing factors
TG003 benzothioate	SR protein kinase Sty1 and CLK4 inhibitor
1,4 naphtoquinones	Prevents spliceosomal rearrangements required for second catalytic step
NB-506 glycocylated indocarbozale	Inhibits topoisomerase 1, which prevents SR protein phosphorylation
SRPIN340, nicotinomide	Inhibits SR protein kinase (SRPK1/2)
Rapamycin and its rapalogs	Targets mTORc kinase activity by interfering with FRB domain of mTOR. Second generation inhibitors bind to catalytic cleft of mTOR
4EG1	Inhibits interaction of eIF4E and eIF4G. Enhances eIF4E binding to its translational repressor 4EBP1
Ribavirin	Binds to the 5′cap of mRNA and prevents eIF4E from exporting mRNA out from the nucleus
PRMT inhibitors eg. AMI-1	Prevents methylation of RBP and thus affecting its function
Spliceostatin A (SSA)	Binds to the SF3B complex and prevents spliceosome assembly
E7107	Blocks the ATP-dependent remodeling of the branch point binding region in splicing

Summary and Future Directions

Post-transcriptional regulation provides a powerful way for cells to control their protein expression resulting in changes in cell fate. The expanding role of miRNAs, identification of genomic alterations, and dysregulation of RBPs in hematopoietic malignancies have refocused the interest toward this emerging field. Moreover, technological advances in sequencing and novel strategies to assess the riboproteome are already providing insight into the complex interplay that affects cellular function.

For example, global analysis of targets for Lin28 suggested that besides its role as a regulator of miRNA processing, Lin28 auto regulates itself and alters the abundance of splicing factors (Wilbert et al., 2012). These data suggest that RBPs control other RBPs, and through these feedback loops, the proper stoichiometry is maintained. These networks are further complicated by cell type and cell context-specific expression patterns. However, a distinction between the normal and leukemia cells and their effect on the riboproteome will need to be further examined. Current technological hurdles may limit these analyses to cells that can be grown easily instead of rare subsets such as HSCs or LSCs. Global RNA target profiling of MSI2 in a cancer cell line can provide insight into the pathways that are altered in normal hematopoietic stem cells (Park et al., 2014).

In addition to the direct RNA targets, an understanding of how RBPs are regulated needs to be further explored. Insights from mTOR signaling and the translational elongation complexes suggest that phosphorylation on RBPs can dramatically alter the activity of these complexes, and rapalogs provide a potential avenue for therapeutic intervention. In some cases phosphorylation of RBPs can alter their activity by changing its localization. For example, ASF/SF2 modulates translation in the cytoplasm and shuttles to the nucleus to modulate splicing and mRNA trafficking only after phosphorylation (Ma et al., 2010; Ma et al., 2009). This nuclear localization might result in a more oncogenic protein and suggests that its specific role as a splicing regulator is also clinically relevant.

Besides phosphorylation, additional post-translational modifications can augment the function of RBPs. For example, Arginine Protein methyltransferases (PRMT) can methylate RBPs and changes their localization and activity (Cote et al., 2003). This has been elegantly reviewed by Blackwell and Ceman (2012). PRMT1 and Sam68, a RBP, were found in the MLL-EEN transcriptional complex, and direct fusion of MLL with either Sam68 or PRMT1 enhances the self-renewal of HSCs (Cheung et al., 2007). The complex integration of post-translational modifications might provide an exciting avenue for clinical intervention.

In summary, we propose that an understanding of the riboproteome in the regulation of self-renewal and cell fate will uncover novel mechanisms that drive the most aggressive cancers. The biggest challenges for this exciting and emerging field are untangling the complexity of targets and protein interactions, and most importantly, developing small molecule inhibitors that can target RBPs specifically.

Acknowledgments

We would like to acknowledge Cynthia Almonte, Sun Mi Park, Gerard Minuesa, and Omar Abdel-Wahab for their critical reading of the manuscript and invaluable insights. We apologize in advance for any possible omissions due to space constraints or a failure to cite published work. The U.S. National Institutes of Health, National Institute of Diabetes and Digestive and Kidney Diseases Career Development Award, Louis V. Gerstner Young Investigator Award, Leukemia Research Foundation, and the American Society of Hematology Junior Scholar Award supported the research in the Kharas laboratory.

References

Ambros, V., and Horvitz, H.R. (1984). Heterochronic mutants of the nematode Caenorhabditis elegans. Science (New York, NY) *226*, 409–416.

Amin, E.M., Oltean, S., Hua, J., Gammons, M.V., Hamdollah-Zadeh, M., Welsh, G.I., Cheung, M.K., Ni, L., Kase, S., Rennel, E.S., *et al.* (2011). WT1 mutants reveal SRPK1 to be a downstream angiogenesis target by altering VEGF splicing. Cancer cell *20*, 768–780.

Aoki, M., Blazek, E., and Vogt, P.K. (2001). A role of the kinase mTOR in cellular transformation induced by the oncoproteins P3k and Akt. Proceedings of the National Academy of Sciences of the United States of America *98*, 136–141.

Assouline, S., Culjkovic, B., Cocolakis, E., Rousseau, C., Beslu, N., Amri, A., Caplan, S., Leber, B., Roy, D.C., Miller, W.H., Jr., and Borden, K.L. (2009). Molecular targeting of the oncogene eIF4E in acute myeloid leukemia (AML): a proof-of-principle clinical trial with ribavirin. Blood *114*, 257–260.

Balzer, E., and Moss, E.G. (2007). Localization of the developmental timing regulator Lin28 to mRNP complexes, P-bodies and stress granules. RNA biology *4*, 16–25.

Baou, M., Jewell, A., and Murphy, J.J. (2009). TIS11 family proteins and their roles in post-transcriptional gene regulation. J Biomed Biotechnol *2009*, 634520.

Barbouti, A., Hoglund, M., Johansson, B., Lassen, C., Nilsson, P.G., Hagemeijer, A., Mitelman, F., and Fioretos, T. (2003). A novel gene, MSI2, encoding a putative RNA-binding protein is recurrently rearranged at disease progression of chronic myeloid leukemia and forms a fusion gene with HOXA9 as a result of the cryptic t(7;17)(p15;q23). Cancer research *63*, 1202–1206.

Blackwell, E., and Ceman, S. (2012). Arginine methylation of RNA-binding proteins regulates cell function and differentiation. Molecular reproduction and development *79*, 163–175.

Bolli, N., Nicoletti, I., De Marco, M.F., Bigerna, B., Pucciarini, A., Mannucci, R., Martelli, M.P., Liso, A., Mecucci, C., Fabbiano, F., *et al.* (2007). Born to be exported: COOH-terminal nuclear export signals of different strength ensure cytoplasmic accumulation of nucleophosmin leukemic mutants. Cancer research *67*, 6230–6237.

Bonnal, S., Vigevani, L., and Valcarcel, J. (2012). The spliceosome as a target of novel antitumour drugs. Nature reviews Drug discovery *11*, 847–859.

Bonnet, D., and Dick, J.E. (1997). Human acute myeloid leukemia is organized as a hierarchy that originates from a primitive hematopoietic cell. Nature medicine *3*, 730–737.

Brennan, C.M., and Steitz, J.A. (2001). HuR and mRNA stability. Cellular and molecular life sciences: CMLS *58*, 266–277.

Byers, R.J., Currie, T., Tholouli, E., Rodig, S.J., and Kutok, J.L. (2011). MSI2 protein expression predicts unfavorable outcome in acute myeloid leukemia. Blood *118*, 2857–2867.

Chang, J.S., Santhanam, R., Trotta, R., Neviani, P., Eiring, A.M., Briercheck, E., Ronchetti, M., Roy, D.C., Calabretta, B., Caligiuri, M.A., and Perrotti, D. (2007). High levels of the BCR/ABL oncoprotein are required for the MAPK-hnRNP-E2 dependent suppression of C/EBPalpha-driven myeloid differentiation. Blood *110*, 994–1003.

Chang, T.C., Zeitels, L.R., Hwang, H.W., Chivukula, R.R., Wentzel, E.A., Dews, M., Jung, J., Gao, P., Dang, C.V., Beer, M.A., *et al.* (2009). Lin-28B transactivation is necessary for Myc-mediated let-7 repression and proliferation. Proceedings of the National Academy of Sciences of the United States of America *106*, 3384–3389.

Charlesworth, A., Wilczynska, A., Thampi, P., Cox, L.L., and MacNicol, A.M. (2006). Musashi regulates the temporal order of mRNA translation during Xenopus oocyte maturation. The EMBO journal *25*, 2792–2801.

Chen, C., Liu, Y., Liu, R., Ikenoue, T., Guan, K.L., Liu, Y., and Zheng, P. (2008). TSC-mTOR maintains quiescence and function of hematopoietic stem cells by repressing mitochondrial biogenesis and reactive oxygen species. The Journal of experimental medicine *205*, 2397–2408.

Cheng, K., Sportoletti, P., Ito, K., Clohessy, J.G., Teruya-Feldstein, J., Kutok, J.L., and Pandolfi, P.P. (2010). The cytoplasmic NPM mutant induces myeloproliferation in a transgenic mouse model. Blood *115*, 3341–3345.

Cheung, N., Chan, L.C., Thompson, A., Cleary, M.L., and So, C.W. (2007). Protein arginine-methyltransferase-dependent oncogenesis. Nature cell biology *9*, 1208–1215.

Cho, J., Chang, H., Kwon, S.C., Kim, B., Kim, Y., Choe, J., Ha, M., Kim, Y.K., and Kim, V.N. (2012). LIN28A Is a Suppressor of ER-Associated Translation in Embryonic Stem Cells. Cell *151*, 765–777.

Choo, A.Y., Yoon, S.O., Kim, S.G., Roux, P.P., and Blenis, J. (2008). Rapamycin differentially inhibits S6Ks and 4E-BP1 to mediate cell-type-specific repression of mRNA translation. Proceedings of the National Academy of Sciences of the United States of America *105*, 17414–17419.

Ciais, D., Cherradi, N., and Feige, J.J. (2012). Multiple functions of tristetraprolin/TIS11 RNA-binding proteins in the regulation of mRNA biogenesis and degradation. Cellular and molecular life sciences: CMLS.

Colombo, E., Alcalay, M., and Pelicci, P.G. (2011). Nucleophosmin and its complex network: a possible therapeutic target in hematological diseases. Oncogene *30*, 2595–2609.

Cooper, T.A., Wan, L., and Dreyfuss, G. (2009). RNA and disease. Cell *136*, 777–793.

Corrionero, A., Minana, B., and Valcarcel, J. (2011). Reduced fidelity of branch point recognition and alternative splicing induced by the anti-tumor drug spliceostatin A. Genes & development *25*, 445–459.

Cote, J., Boisvert, F.M., Boulanger, M.C., Bedford, M.T., and Richard, S. (2003). Sam68 RNA binding protein is an in vivo substrate for protein arginine N-methyltransferase 1. Molecular biology of the cell *14*, 274–287.

Darnell, R.B. (2010). HITS-CLIP: panoramic views of protein-RNA regulation in living cells. Wiley interdisciplinary reviews RNA *1*, 266–286.

David, C.J., and Manley, J.L. (2010). Alternative pre-mRNA splicing regulation in cancer: pathways and programs unhinged. Genes & development *24*, 2343–2364.

de Andres-Aguayo, L., Varas, F., Kallin, E.M., Infante, J.F., Wurst, W., Floss, T., and Graf, T. (2011). Musashi 2 is a regulator of the HSC compartment identified by a retroviral insertion screen and knockout mice. Blood *118*, 554–564.

De Conti, L., Baralle, M., and Buratti, E. (2012). Exon and intron definition in pre-mRNA splicing. Wiley interdisciplinary reviews RNA.

Demarest, R.M., Ratti, F., and Capobianco, A.J. (2008). It's T-ALL about Notch. Oncogene *27*, 5082–5091.

Duncan, A.W., Rattis, F.M., DiMascio, L.N., Congdon, K.L., Pazianos, G., Zhao, C., Yoon, K., Cook, J.M., Willert, K., Gaiano, N., and Reya, T. (2005). Integration of Notch and Wnt signaling in hematopoietic stem cell maintenance. Nature immunology *6*, 314–322.

Eiring, A.M., Neviani, P., Santhanam, R., Oaks, J.J., Chang, J.S., Notari, M., Willis, W., Gambacorti-Passerini, C., Volinia, S., Marcucci, G., *et al.* (2008). Identification of novel posttranscriptional targets of the BCR/ABL oncoprotein by ribonomics: requirement of E2F3 for BCR/ABL leukemogenesis. Blood *111*, 816–828.

Falini, B., Martelli, M.P., Bolli, N., Bonasso, R., Ghia, E., Pallotta, M.T., Diverio, D., Nicoletti, I., Pacini, R., Tabarrini, A., *et al.* (2006). Immunohistochemistry predicts nucleophosmin (NPM) mutations in acute myeloid leukemia. Blood *108*, 1999–2005.

Falini, B., Mecucci, C., Tiacci, E., Alcalay, M., Rosati, R., Pasqualucci, L., La Starza, R., Diverio, D., Colombo, E., Santucci, A., *et al.* (2005). Cytoplasmic nucleophosmin in acute myelogenous leukemia with a normal karyotype. The New England journal of medicine *352*, 254–266.

Folco, E.G., Coil, K.E., and Reed, R. (2011). The anti-tumor drug E7107 reveals an essential role for SF3b in remodeling U2 snRNP to expose the branch point-binding region. Genes & development *25*, 440–444.

Ghosh, M., Aguila, H.L., Michaud, J., Ai, Y., Wu, M.T., Hemmes, A., Ristimaki, A., Guo, C., Furneaux, H., and Hla, T. (2009). Essential role of the RNA-binding protein HuR in progenitor cell survival in mice. The Journal of clinical investigation *119*, 3530–3543.

Gingras, A.C., Kennedy, S.G., O'Leary, M.A., Sonenberg, N., and Hay, N. (1998). 4E-BP1, a repressor of mRNA translation, is phosphorylated and inactivated by the Akt(PKB) signaling pathway. Genes & development *12*, 502–513.

Glisovic, T., Bachorik, J.L., Yong, J., and Dreyfuss, G. (2008). RNA-binding proteins and post-transcriptional gene regulation. FEBS Lett *582*, 1977–1986.

Graubert, T.A., Shen, D., Ding, L., Okeyo-Owuor, T., Lunn, C.L., Shao, J., Krysiak, K., Harris, C.C., Koboldt, D.C., Larson, D.E., *et al.* (2012). Recurrent mutations in the U2AF1 splicing factor in myelodysplastic syndromes. Nature genetics *44*, 53–57.

Grisendi, S., Bernardi, R., Rossi, M., Cheng, K., Khandker, L., Manova, K., and Pandolfi, P.P. (2005). Role of nucleophosmin in embryonic development and tumorigenesis. Nature *437*, 147–153.

Grisendi, S., Mecucci, C., Falini, B., and Pandolfi, P.P. (2006). Nucleophosmin and cancer. Nature reviews Cancer *6*, 493–505.

Hiriart, E., Gruffat, H., Buisson, M., Mikaelian, I., Keppler, S., Meresse, P., Mercher, T., Bernard, O.A., Sergeant, A., and Manet, E. (2005). Interaction of the Epstein-Barr virus mRNA export factor EB2 with human Spen proteins SHARP, OTT1, and a novel member of the family, OTT3, links Spen proteins with splicing regulation and mRNA export. The Journal of biological chemistry *280*, 36935–36945.

Hocine, S., Singer, R.H., and Grunwald, D. (2010). RNA processing and export. Cold Spring Harbor perspectives in biology *2*, a000752.

Hodson, D.J., Janas, M.L., Galloway, A., Bell, S.E., Andrews, S., Li, C.M., Pannell, R., Siebel, C.W., MacDonald, H.R., De Keersmaecker, K., *et al.* (2010). Deletion of the RNA-binding proteins ZFP36L1 and ZFP36L2 leads to perturbed thymic development and T lymphoblastic leukemia. Nature immunology *11*, 717–724.

Hope, K.J., Jin, L., and Dick, J.E. (2004). Acute myeloid leukemia originates from a hierarchy of leukemic stem cell classes that differ in self-renewal capacity. Nature immunology *5*, 738–743.

Hope, K.J., Cellot, S., Ting, S.B., MacRae, T., Mayotte, N., Isocove, N.N. and Sauvageau, G. (2010). An RNAi screen identifies Msi2 and Prox1 as having opposite roles in the regulation of hematopoietic stem cell activity. Cell stem cell *1*, 101–113.

Hoshii, T., Tadokoro, Y., Naka, K., Ooshio, T., Muraguchi, T., Sugiyama, N., Soga, T., Araki, K., Yamamura, K., and Hirao, A. (2012). mTORC1 is essential for leukemia propagation but not stem cell self-renewal. The Journal of clinical investigation *122*, 2114–2129.

Hoskins, A.A., and Moore, M.J. (2012). The spliceosome: a flexible, reversible macromolecular machine. Trends in biochemical sciences *37*, 179–188.

Iervolino, A., Santilli, G., Trotta, R., Guerzoni, C., Cesi, V., Bergamaschi, A., Gambacorti-Passerini, C., Calabretta, B., and Perrotti, D. (2002). hnRNP A1 nucleocytoplasmic shuttling activity is required for normal myelopoiesis and BCR/ABL leukemogenesis. Molecular and cellular biology *22*, 2255–2266.

Imai, T., Tokunaga, A., Yoshida, T., Hashimoto, M., Mikoshiba, K., Weinmaster, G., Nakafuku, M., and Okano, H. (2001). The neural RNA-binding protein Musashi1 translationally regulates mammalian numb gene expression by interacting with its mRNA. Molecular and cellular biology *21*, 3888–3900.

Ito, T., Kwon, H. Y., Zimdahl, B., Congdon, K.L., Blum, J., Lento, W.E., Zhao, C., Lagoo, A., Gerrard, G., Foroni, L., *et al.* (2010). Regulation of myeloid leukaemia by the cell-fate determinant Musashi. Nature *466*, 765–768.

Iwanaga, E., Nanri, T., Mitsuya, H., and Asou, N. (2011). Mutation in the RNA binding protein TIS11D/ZFP36L2 is associated with the pathogenesis of acute leukemia. Int J Oncol *38*, 25–31.

Janes, M.R., Limon, J.J., So, L., Chen, J., Lim, R.J., Chavez, M.A., Vu, C., Lilly, M.B., Mallya, S., Ong, S.T., *et al.* (2010). Effective and selective targeting of leukemia cells using a TORC1/2 kinase inhibitor. Nature medicine *16*, 205–213.

Jiang, X., Huang, H., Li, Z., Li, Y., Wang, X., Gurbuxani, S., Chen, P., He, C., You, D., Zhang, S., *et al.* (2012). Blockade of miR-150 Maturation by MLL-Fusion/MYC/LIN-28 Is Required for MLL-Associated Leukemia. Cancer cell *22*, 524–535.

Kalaitzidis, D., Sykes, S.M., Wang, Z., Punt, N., Tang, Y., Ragu, C., Sinha, A.U., Lane, S.W., Souza, A.L., Clish, C.B., *et al.* (2012). mTOR complex 1 plays critical roles in hematopoiesis and Pten-loss-evoked leukemogenesis. Cell stem cell *11*, 429–439.

Kawahara, H., Imai, T., Imataka, H., Tsujimoto, M., Matsumoto, K., and Okano, H. (2008). Neural RNA-binding protein Musashi1 inhibits translation initiation by competing with eIF4G for PABP. The Journal of cell biology *181*, 639–653.

Kharas, M.G., Lengner, C.J., Al-Shahrour, F., Bullinger, L., Ball, B., Zaidi, S., Morgan, K., Tam, W., Paktinat, M., Okabe, R., *et al.* (2010a). Musashi-2 regulates normal hematopoiesis and promotes aggressive myeloid leukemia. Nature medicine *16*, 903–908.

Kharas, M.G., Okabe, R., Ganis, J.J., Gozo, M., Khandan, T., Paktinat, M., Gilliland, D.G., and Gritsman, K. (2010b). Constitutively active AKT depletes hematopoietic stem cells and induces leukemia in mice. Blood *115*, 1406–1415.

Kuo, S.H., Hsu, C.H., Chen, L.T., Lu, Y.S., Lin, C.H., Yeh, P.Y., Jeng, H.J., Gao, M., Yeh, K.H., and Cheng, A.L. (2011). Lack of compensatory pAKT activation and eIF4E phosphorylation of lymphoma cells towards mTOR inhibitor, RAD001. European journal of cancer (Oxford, England: 1990) *47*, 1244–1257.

Laplante, M., and Sabatini, D.M. (2012). mTOR signaling in growth control and disease. Cell *149*, 274–293.

Lebedeva, S., Jens, M., Theil, K., Schwanhausser, B., Selbach, M., Landthaler, M., and Rajewsky, N. (2011). Transcriptome-wide analysis of regulatory interactions of the RNA-binding protein HuR. Molecular cell *43*, 340–352.

Lee, E.K. (2012). Post-translational modifications of RNA-binding proteins and their roles in RNA granules. Current protein & peptide science *13*, 331–336.

Lee, J.H., and Skalnik, D.G. (2012). Rbm15-Mkl1 interacts with the Setd1b histone H3-Lys4 methyltransferase via a SPOC domain that is required for cytokine-independent proliferation. PLoS One *7*, e42965.

Lee, J.Y., Nakada, D., Yilmaz, O.H., Tothova, Z., Joseph, N.M., Lim, M.S., Gilliland, D.G., and Morrison, S.J. (2010). mTOR activation induces tumor suppressors that inhibit leukemogenesis and deplete hematopoietic stem cells after Pten deletion. Cell stem cell *7*, 593–605.

Lindtner, S., Zolotukhin, A.S., Uranishi, H., Bear, J., Kulkarni, V., Smulevitch, S., Samiotaki, M., Panayotou, G., Felber, B.K., and Pavlakis, G.N. (2006). RNA-binding motif protein 15 binds to the RNA transport element RTE and provides a direct link to the NXF1 export pathway. The Journal of biological chemistry *281*, 36915–36928.

Lunde, B.M., Moore, C., and Varani, G. (2007). RNA-binding proteins: modular design for efficient function. Nat Rev Mol Cell Biol *8*, 479–490.

Ma, C.T., Ghosh, G., Fu, X.D., and Adams, J.A. (2010). Mechanism of dephosphorylation of the SR protein ASF/SF2 by protein phosphatase 1. Journal of molecular biology *403*, 386–404.

Ma, C.T., Hagopian, J.C., Ghosh, G., Fu, X.D., and Adams, J.A. (2009). Regiospecific phosphorylation control of the SR protein ASF/SF2 by SRPK1. Journal of molecular biology *390*, 618–634.

Ma, W.J., Chung, S., and Furneaux, H. (1997). The Elav-like proteins bind to AU-rich elements and to the poly(A) tail of mRNA. Nucleic acids research *25*, 3564–3569.

Ma, X., Renda, M. J., Wang, L., Cheng, E.C., Niu, C., Morris, S.W., Chi, A.S., and Krause, D.S. (2007). Rbm15 modulates Notch-induced transcriptional activation and affects myeloid differentiation. Molecular and cellular biology *27*, 3056–3064.

Ma, Z., Morris, S.W., Valentine, V., Li, M., Herbrick, J.A., Cui, X., Bouman, D., Li, Y., Mehta, P.K., Nizetic, D., *et al.* (2001). Fusion of two novel genes, RBM15 and MKL1, in the t(1;22)(p13;q13) of acute megakaryoblastic leukemia. Nature genetics *28*, 220–221.

Makishima, H., Visconte, V., Sakaguchi, H., Jankowska, A. M., Abu Kar, S., Jerez, A., Przychodzen, B., Bupathi, M., Guinta, K., Afable, M.G., *et al.* (2012). Mutations in the spliceosome machinery, a novel and ubiquitous pathway in leukemogenesis. Blood *119*, 3203–3210.

Malcovati, L., Papaemmanuil, E., Bowen, D.T., Boultwood, J., Della Porta, M.G., Pascutto, C., Travaglino, E., Groves, M.J., Godfrey, A.L., Ambaglio, I., *et al.* (2011). Clinical significance of SF3B1 mutations in myelodysplastic syndromes and myelodysplastic/myeloproliferative neoplasms. Blood *118*, 6239–6246.

Mariano, A.R., Colombo, E., Luzi, L., Martinelli, P., Volorio, S., Bernard, L., Meani, N., Bergomas, R., Alcalay, M., and Pelicci, P.G. (2006). Cytoplasmic localization of NPM in myeloid leukemias is dictated by gain-of-function mutations that create a functional nuclear export signal. Oncogene *25*, 4376–4380.

Martelli, A.M., Evangelisti, C., Chappell, W., Abrams, S.L., Basecke, J., Stivala, F., Donia, M., Fagone, P., Nicoletti, F., Libra, M., *et al.* (2011). Targeting the translational apparatus to improve leukemia therapy: roles of the PI3K/PTEN/Akt/mTOR pathway. Leukemia: official journal of the Leukemia Society of America, Leukemia Research Fund, UK *25*, 1064–1079.

Matsunawa, M., Sanada, M., Yamamoto, R., Yaoshida, K., Nagata, Y., Kon, A., Yoshizato, T., Otsu, M., Nakauchi, H., and Ogawa, S. (2012). Biological Analysis of SRSF2 Mutations in Leukemogenesis. In: 54th ASH meeting.

Meggendorfer, M., Roller, A., Haferlach, T., Eder, C., Dicker, F., Grossmann, V., Kohlmann, A., Alpermann, T., Yoshida, K., Ogawa, S., *et al.* (2012). SRSF2 mutations in 275 cases with chronic myelomonocytic leukemia (CMML). Blood *120*, 3080–3088.

Melton, C., Judson, R.L., and Blelloch, R. (2010). Opposing microRNA families regulate self-renewal in mouse embryonic stem cells. Nature *463*, 621–626.

Mercher, T., Raffel, G.D., Moore, S.A., Cornejo, M.G., Baudry-Bluteau, D., Cagnard, N., Jesneck, J.L., Pikman, Y., Cullen, D., Williams, I.R., *et al.* (2009). The OTT-MAL fusion oncogene activates RBPJ-mediated transcription and induces acute megakaryoblastic leukemia in a knockin mouse model. The Journal of clinical investigation *119*, 852–864.

Moerke, N.J., Aktas, H., Chen, H., Cantel, S., Reibarkh, M.Y., Fahmy, A., Gross, J. D., Degterev, A., Yuan, J., Chorev, M., *et al.* (2007). Small-molecule inhibition of the interaction between the translation initiation factors eIF4E and eIF4G. Cell *128*, 257–267.

Mukherjee, N., Corcoran, D.L., Nusbaum, J.D., Reid, D.W., Georgiev, S., Hafner, M., Ascano, M., Jr., Tuschl, T., Ohler, U., and Keene, J.D. (2011). Integrative regulatory mapping indicates that the RNA-binding protein HuR couples pre-mRNA processing and mRNA stability. Molecular cell *43*, 327–339.

Nardella, C., Chen, Z., Salmena, L., Carracedo, A., Alimonti, A., Egia, A., Carver, B., Gerald, W., Cordon-Cardo, C., and Pandolfi, P.P. (2008). Aberrant Rheb-mediated mTORC1 activation and Pten haploinsufficiency are cooperative oncogenic events. Genes & development *22*, 2172–2177.

Newman, M.A., Mani, V., and Hammond, S.M. (2011). Deep sequencing of microRNA precursors reveals extensive 3′ end modification. RNA (New York, NY) *17*, 1795–1803.

Papaemmanuil, E., Cazzola, M., Boultwood, J., Malcovati, L., Vyas, P., Bowen, D., Pellagatti, A., Wainscoat, J.S., Hellstrom-Lindberg, E., Gambacorti-Passerini, C., *et al.* (2011). Somatic SF3B1 mutation in myelodysplasia with ring sideroblasts. The New England journal of medicine *365*, 1384–1395.

Park, S.M., Deering, R., Lu, Y., Tivnan, P., Shenoy, V., Lianoglou, S., Al-Shahrour, F., Leslie, C., Ebert, B.L., Lengner, C., Daley, G.Q., and Kharas, M.G. (2014). Musashi-2 controls self fate, lineage restriction and TGF-b signalling in HSCs. Journal of experimental medicine (in press).

Pereira, J.K., Traina, F., Machado-Neto, J.A., Duarte Ada, S., Lopes, M.R., Saad, S.T., and Favaro, P. (2012). Distinct expression profiles of MSI2 and NUMB genes in myelodysplastic syndromes and acute myeloid leukemia patients. Leukemia research *36*, 1300–1303.

Perrotti, D., Bonatti, S., Trotta, R., Martinez, R., Skorski, T., Salomoni, P., Grassilli, E., Lozzo, R.V., Cooper, D.R., and Calabretta, B. (1998). TLS/FUS, a pro-oncogene involved in multiple chromosomal translocations, is a novel regulator of BCR/ABL mediated leukemogenesis. The EMBO journal *17*, 4442–4455.

Perrotti, D., and Neviani, P. (2007). From mRNA metabolism to cancer therapy: chronic myelogenous leukemia shows the way. Clin Cancer Res *13*, 1638–1642.

Prabhu, S., Saadat, D., Zhang, M., Halbur, L., Fruehuaf, J.P., and Ong, S.T. (2007) A novel mechanism for Bcr-abl action: Bcr-abl-mediated induction of the eIF4F translation initiation complex and mRNA translation. Oncogene *8*, 1188–1200.

Quesada, V., Ramsay, A.J., and Lopez-Otin, C. (2012). Chronic lymphocytic leukemia with SF3B1 mutation. The New England journal of medicine *366*, 2530.

Raffel, G.D., Mercher, T., Shigematsu, H., Williams, I.R., Cullen, D.E., Akashi, K., Bernard, O.A., and Gilliland, D.G. (2007). Ott1(Rbm15) has pleiotropic roles in hematopoietic development. Proceedings of the National Academy of Sciences of the United States of America *104*, 6001–6006.

Raval, A., Kusler, B., Pang, W.W., Weissman, I.L., Mitchell, B.S., and Park, C.Y. (2012). Effect of nucleophosmin1 haploinsufficiency on hematopoietic stem cells. Leukemia: official journal of the Leukemia Society of America, Leukemia Research Fund, UK *26*, 853–855.

Rousseau, D., Gingras, A.C., Pause, A., and Sonenberg, N. (1996). The eIF4E-binding proteins 1 and 2 are negative regulators of cell growth. Oncogene *13*, 2415–2420.

Sattler, M., Mohi, M.G., Pride, Y.B., Quinnan, L.F., Malouf, N.A., Podar, K., Gesbert, K., Iwasaki, H., Li, S., Van Etten, R.A., Gu, H., Griffin, J.D. and Neel, B.G. (2002). Critical role for Gab2 in transformation by BCR/ABL. Cancer cell *5*, 479–492.

Somervaille, T.C., Matheny, C.J., Spencer, G.J., Iwasaki, M., Rinn, J.L., Witten, D.M., Chang, H.Y., Shurtleff, S.A., Downing, J.R., and Cleary, M.L. (2009). Hierarchical maintenance of MLL myeloid leukemia stem cells employs a transcriptional program shared with embryonic rather than adult stem cells. Cell stem cell *4*, 129–140

Sonenberg, N., and Gingras, A.C. (1998). The mRNA 5′ cap-binding protein eIF4E and control of cell growth. Curr Opin Cell Biol *10*, 268–275.

St. Johnston, D. (2005). Moving messages: the intracellular localization of mRNAs. Nat Rev Mol Cell Biol *6*, 363–375.

Stumpo, D.J., Broxmeyer, H.E., Ward, T., Cooper, S., Hangoc, G., Chung, Y.J., Shelley, W.C., Richfield, E.K., Ray, M.K., Yoder, M.C., *et al.* (2009). Targeted disruption of Zfp36l2, encoding a CCCH tandem zinc finger RNA-binding protein, results in defective hematopoiesis. Blood *114*, 2401–2410.

Szebeni, A., and Olson, M.O. (1999). Nucleolar protein B23 has molecular chaperone activities. Protein science: a publication of the Protein Society *8*, 905–912.

Thornton, J.E., and Gregory, R.I. (2012). How does Lin28 let-7 control development and disease? Trends in cell biology *22*, 474–482.

Topisirovic, I., Guzman, M.L., McConnell, M.J., Licht, J.D., Culjkovic, B., Neering, S.J., Jordan, C.T., and Borden, K.L. (2003). Aberrant eukaryotic translation initiation factor 4E-dependent mRNA transport impedes hematopoietic differentiation and contributes to leukemogenesis. Molecular and cellular biology *23*, 8992–9002.

Topisirovic, I., Siddiqui, N., Orolicki, S., Skrabanek, L. A., Tremblay, M., Hoang, T., and Borden, K.L. (2009). Stability of eukaryotic translation initiation factor 4E mRNA is regulated by HuR, and this activity is dysregulated in cancer. Molecular and cellular biology *29*, 1152–1162.

Venables, J.P. (2004). Aberrant and alternative splicing in cancer. Cancer research *64*, 7647–7654.

Visconte, V., Makishima, H., Jankowska, A., Szpurka, H., Traina, F., Jerez, A., O'Keefe, C., Rogers, H.J., Sekeres, M.A., Maciejewski, J.P., and Tiu, R.V. (2012a). SF3B1, a splicing factor is frequently mutated in refractory anemia with ring sideroblasts. Leukemia: official journal of the Leukemia Society of America, Leukemia Research Fund, UK *26*, 542–545.

Visconte, V., Makishima, H., Maciejewski, J.P., and Tiu, R.V. (2012b). Emerging roles of the spliceosomal machinery in myelodysplastic syndromes and other hematological disorders. Leukemia: official journal of the Leukemia Society of America, Leukemia Research Fund, UK.

Visconte, V., Rogers, H.J., Singh, J., Barnard, J., Bupathi, M., Traina, F., McMahon, J., Makishima, H., Szpurka, H., Jankowska, A., *et al.* (2012c). SF3B1 haploinsufficiency leads to formation of ring sideroblasts in myelodysplastic syndromes. Blood *120*, 3173–3186.

Viswanathan, S.R., Powers, J.T., Einhorn, W., Hoshida, Y., Ng, T.L., Toffanin, S., O'Sullivan, M., Lu, J., Phillips, L.A., Lockhart, V.L., *et al.* (2009). Lin28 promotes transformation and is associated with advanced human malignancies. Nature genetics *41*, 843–848.

Wang, H.Y., Xu, X., Ding, J.H., Bermingham, J.R., Jr., and Fu, X.D. (2001). SC35 plays a role in T cell development and alternative splicing of CD45. Molecular cell *7*, 331–342.

Wang, L., Lawrence, M.S., Wan, Y., Stojanov, P., Sougnez, C., Stevenson, K., Werner, L., Sivachenko, A., DeLuca, D.S., Zhang, L., *et al.* (2011). SF3B1 and other novel cancer genes in chronic lymphocytic leukemia. The New England journal of medicine *365*, 2497–2506.

Warner, J.R., Knopf, P.M., and Rich, A. (1963). A multiple ribosomal structure in protein synthesis. Proceedings of the National Academy of Sciences of the United States of America *49*, 122–129.

Wendel, H.G., De Stanchina, E., Fridman, J.S., Malina, A., Ray, S., Kogan, S., Cordon-Cardo, C., Pelletier, J., and Lowe, S.W. (2004). Survival signalling by Akt and eIF4E in oncogenesis and cancer therapy. Nature *428*, 332–337.

Wilbert, M.L., Huelga, S.C., Kapeli, K., Stark, T.J., Liang, T.Y., Chen, S.X., Yan, B.Y., Nathanson, J.L., Hutt, K.R., Lovci, M.T., *et al.* (2012). LIN28 Binds Messenger RNAs at GGAGA Motifs and Regulates Splicing Factor Abundance. Molecular cell *48*, 195–206.

Wilson, A., Ardiet, D.L., Saner, C., Vilain, N., Beermann, F., Aguet, M., Macdonald, H.R., and Zilian, O. (2007). Normal hemopoiesis and lymphopoiesis in the combined absence of numb and numblike. Journal of immunology (Baltimore, MD: 1950) *178*, 6746–6751.

Xiao, N., Jani, K., Morgan, K., Okabe, R., Cullen, D.E., Jesneck, J.L., and Raffel, G.D. (2012). Hematopoietic stem cells lacking Ott1 display aspects associated with aging and are unable to maintain quiescence during proliferative stress. Blood *119*, 4898–4907.

Yilmaz, O.H., Valdez, R., Theisen, B.K., Guo, W., Ferguson, D.O., Wu, H., and Morrison, S.J. (2006). Pten dependence distinguishes haematopoietic stem cells from leukaemia-initiating cells. Nature *441*, 475–482.

Yoshida, K., Sanada, M., Shiraishi, Y., Nowak, D., Nagata, Y., Yamamoto, R., Sato, Y., Sato-Otsubo, A., Kon, A., Nagasaki, M., *et al.* (2011). Frequent pathway mutations of splicing machinery in myelodysplasia. Nature *478*, 64–69.

Yu, J., Vodyanik, M.A., Smuga-Otto, K., Antosiewicz-Bourget, J., Frane, J.L., Tian, S., Nie, J., Jonsdottir, G.A., Ruotti, V., Stewart, R., *et al.* (2007). Induced pluripotent stem cell lines derived from human somatic cells. Science (New York, NY) *318*, 1917–1920.

Yuan, J., Nguyen, C.K., Liu, X., Kanellopoulou, C., and Muljo, S.A. (2012). Lin28b reprograms adult bone marrow hematopoietic progenitors to mediate fetal-like lymphopoiesis. Science (New York, NY) *335*, 1195–1200.

Zaharieva, E., Chipman, J.K., and Soller, M. (2012). Alternative splicing interference by xenobiotics. Toxicology *296*, 1–12.

Zhang, J., Grindley, J.C., Yin, T., Jayasinghe, S., He, X. C., Ross, J.T., Haug, J.S., Rupp, D., Porter-Westpfahl, K.S., Wiedemann, L.M., *et al.* (2006). PTEN maintains haematopoietic stem cells and acts in lineage choice and leukaemia prevention. Nature *441*, 518–522.

Zolotukhin, A.S., Uranishi, H., Lindtner, S., Bear, J., Pavlakis, G.N., and Felber, B.K. (2009). Nuclear export factor RBM15 facilitates the access of DBP5 to mRNA. Nucleic acids research *37*, 7151–7162.

Section VI

Clinical Relevance of Cancer Stem Cells in Patients

Chapter 32

Targeting Different States of Breast Cancer Stem Cells

Sean P. McDermott and Max S. Wicha
Department of Internal Medicine, Division of Hematology and Oncology, University of Michigan, Comprehensive Cancer Center, Ann Arbor, MI, USA

What Is a Cancer Stem Cell?

The term "cancer stem cell" has led to a degree of confusion within the cancer research field (Clarke et al., 2006; Dick, 2008; Valent et al., 2012). The definition of "cancer stem cell" is based on three functional characteristics: (1) the ability to initiate tumors in immunocompromised mice; (2) self-renewal capacity as measured by tumor formation in secondary mice; and (3) the capacity to differentiate into the non self-renewing bulk tumor cells. Within this review, we use 'CSC' to refer to presence of stem cells within a cancer without regard to the cell-of-origin, which might be a stem or progenitor cell (Visvader, 2011).

Technical advances, such as the development of NOD/SCID (non-obese diabetic/severe combined immune deficiency) immunocompromised mouse models (Shultz et al., 2007; Shultz et al., 1995), monoclonal antibodies, and increasing complex flow cytometers (Perfetto et al., 2004) have been instrumental in the isolation and study of CSCs (Al-Hajj et al., 2003; Bonnet and Dick, 1997; Hope et al., 2004; Lapidot et al., 1994). The isolation and characterization is complicated by the significant genetic (Notta et al., 2011) and functional (Eppert et al., 2011) heterogeneity within the CSC populations. Based on limiting dilution analysis (Hu and Smyth, 2009), the frequency of breast CSCs from primary patient samples, as defined by CD44+CD24−EpCAM+(Al-Hajj et al., 2003) or ALDH+(Ginestier et al., 2007), is still low (1 in 156 CD44+CD24−EpCAM+ cells) suggesting that not every cell within those fractions are bona fide stem cells.

CSCs can exhibit multiple fates: division, quiescence, death, or differentiation (Dick, 2008). A hallmark of stem cells is their ability to self renew whereby cell fate (i.e., tumor-initiation) is segregated asymmetrically or symmetrically to daughter cells, which must be functionally defined by secondary transplantation (Dick, 2003; McKenzie et al., 2006). The existence of multiple stem cell populations with varying self-renewal capacity has been demonstrated in the hematopoietic system (Hope et al., 2004; McKenzie et al., 2006). The widely reported in vitro sphere formation assay, as detailed by our lab (Dontu et al., 2003), is typically used as a surrogate assay to measure self-renewal. This assay is rarely performed at the single cell level and measures self-renewal on the aggregate level since spheres are combined, digested, and

Cancer Stem Cells, First Edition. Edited by Vinagolu K. Rajasekhar.

then plated above single-cell level. Given the variable self-renewal capacity of normal and leukemic stem cells when measured in vivo, it is possible that clonally derived spheres may exhibit variable self-renewal capacity.

The third property of CSC is the ability to differentiate into the bulk, non-tumorigenic cells. Cellular heterogeneity is typically measured using differential expression of protein differentiation markers (i.e., CD44, CD24, CD34, CD38, etc.) and is generated in xenograft models upon serial transplantation (Al-Hajj et al., 2003; Bonnet and Dick, 1997; Eppert et al., 2011). Traditionally, in vitro experiments to determine the differentiation potential of breast CSCs are based on sorting or selecting CSC (CD44+CD24− or ALDH+) and non-CSC fractions of cells, culturing these CSC enriched fractions either in adherent (Pollard et al., 2009) or sphere-promoting conditions (Dontu et al., 2003), and monitoring expression of the markers used to isolate the CSCs. These studies have typically determined that non-CSCs do not generate CSCs in measurable numbers, and CSCs eventually differentiate to reach equilibrium. The persistence of the CSC marker phenotype (i.e., CD44+CD24−) over time is interpreted to indicate that CSCs self-renewed in vitro, which is a common mistake without in vivo functional data showing CSCs (i.e., tumor-initiation) is maintained during culture.

Existence of Multiple CSC States

The original description of breast CSCs (Al-Hajj et al., 2003) showed that CSCs reside in the CD44+CD24− fraction and more specifically in the CD44+CD24−EpCAM+fraction. However, there is significant heterogeneity in the frequency of CD44+CD24− between patients (Al-Hajj et al., 2003; Li et al., 2008), as is seen with the leukemia stem cell (LSC) enriched CD34+CD38− fraction in AML (Eppert et al., 2011). By using limiting dilution analysis (Hu and Smyth, 2009), the frequency of CSCs in the CD44+CD24− fraction, using data from Al-Hajj and colleagues (2003) is overestimated (Table 32.1) due to a lack of mice that did not form a tumor. When EpCAM staining was used to split the CD44+CD24− fraction, then it was dramatically apparent that breast CSCs are CD44+CD24−EpCAM+since CD44+CD24−EpCAM− cells did not form tumors. The estimated CSC frequency in the CD44+CD24−EpCAM+fraction is 1 in 156 cells (Table 32.1).

Although our group was not the first to prospectively isolate stem cells based on expression of aldehyde dehydrogenase (ALDH) (Pearce et al., 2005; Storms et al., 1999), we showed that normal and malignant breast stem cells can be isolated based on the Aldefluor assay, which detects cells based on ALDH activity on a fluorescent substrate (Ginestier et al., 2007). Once again, the estimated CSC frequency in the ALDH+fraction of tumor xenografts was overestimated by LDA (Table 32.1), but showed ALDH- cells are mostly devoid of tumor initiation. Extension of these results to breast cancer cell lines showed a distinct heterogeneity of ALDH+ frequency between cell lines (Charafe-Jauffret et al., 2009). Within these cell lines, there is limited overlap between CD44+CD24−EpCAM+and ALDH+(McDermott SP, unpublished observations), indicating these CSC fractions might be mutually exclusive in cell lines.

Unlike leukemia, where it is routine to obtain millions and potential billions of leukemia blasts in a subset of patients with advanced disease, the ability to obtain sufficient primary breast cancer tissue, especially from metastatic lesions in lymph nodes or the liver, for flow cytometry examination of CSC markers, is severely limited. Therefore, numerous groups have used immunohistochemical (IHC) staining of human breast tumors for CD44+CD24−cells to evaluate the prevalence of this phenotype across patient subtypes. Honeth and colleagues (2008) reported that only 31% of samples contained a CD44+CD24− fraction, which was positively correlated with a basal cell phenotype; Idowu and colleagues (2012) also found 37% of tumors contained CD44+CD24− cells; and Mylona

Table 32.1. Limiting dilution analysis of breast CSC separated by CD44, CD24, EpCAM, or ALDH

Tumor	Histological subtype	Source	Fraction	Tumor-initiating frequency 1 in # cells*	Range (95% confidence interval)
Mouse passage #1	NA	Multiple	Unsorted	28,769	17,548 to 47,167
			CD44+CD24+	Infinity	126,847 to Infinity
			CD44+CD24−/low	1	1 to 737
			CD44+CD24−EpCAM+	156	74 to 326
			CD44+CD24−EpCAM−	Infinity	7,811 to Infinity
MC1	ER− PR HER2−	Pleural effusion	Unseparated	1	1 to 1084
			ALDH (+)	1	1 to 780
			ALDH (−)	78,359	19,414 to 316,279
UM1	ER− PR- HER2−	Primary tumor	Unseparated	455	106 to 1,951
			ALDH (+)	1	1 to 1,084
			ALDH (−)	Infinity	68931 to Infinity
UM2	ER+ PR+ HER2−	Metastasis	Unseparated	1	1 to 1084
			ALDH (+)	1	1 to 1084
			ALDH (−)	155,664	22,133 to 1,094,783
UM3	ER− PR− HER2−	Primary tumor	Unseparated	1	1 to 1084
			ALDH (+)	1	1 to 780
			ALDH (−)	125,342	17,479 to 898,836

*Calculated by extreme limiting dilution analysis (ELDA) using data from Al-Hajj et al. (2003) or Ginestier et al. (2007)

and colleagues (2008) found that 58% of tumors had at least 1% CD44+CD24− cells. Although IHC staining of primary tumors for CD44 and CD24 can be technically challenging, this work has shown that not all primary breast cancers have a detectable CD44+CD24− fraction.

As is the case with staining primary breast cancers for CD44, CD24, and EpCAM, the ability to determine the frequency of ALDH+ cells in a large patient cohort can be limited by the amount of cells. Thus, our group and others have analyzed expression of ALDH1, a major contributor to the Aldefluor phenotype (Marcato et al., 2011), in large breast cancer tissue microarrays (Ginestier et al., 2007; Gong et al., 2010; Ricardo et al., 2011). Between 7 and 30% of tumors expressed ALDH1, and the frequency of ALDH1+ cells ranged from between 5 and 20% depending on the study. Together, these IHC data indicate that some breast cancers may not express detectable levels of known CSC markers (i.e., CD44+ CD24− or ALDH+), the CSC may reside in another fraction, or there are technical limits to using IHC to measure a potentially rare (i.e., <1%) population in tissue sections.

The current dogma and hot topic in the breast CSC community is that CD44+CD24− CSCs are in fact "EMT" CSCs. Numerous studies have shown these cells (patient derived, cancer cell lines, or experimentally induced) express mesenchymal markers (i.e., VIM, SNAI1, SNAI2) and lack epithelial markers (i.e., CDH1) (Creighton et al., 2009; Mani et al., 2008; Taube et al., 2010). Furthermore, breast cancer cell lines highly enriched for CD44+CD24− (i.e., SUM159, MDA-MB-231, BT549) are typically mesenchymal (Fillmore and Kuperwasser, 2008), which has been extrapolated to suggest CD44+CD24− CSCs have undergone EMT (Mani et al., 2008). However, this hypothesis appears in direct contradiction with the experimental evidence that breast CSCs are epithelial based on expression of EpCAM (Al-Hajj et al., 2003).

The simplicity of CSCs only residing in one fraction is attractive, as put forth in the seminal work by John Dick's lab (Lapidot et al., 1994) and our group (Al-Hajj et al., 2003), but advancements in xenograft techniques (Mazurier et al., 2003; Quintana et al., 2008; Shultz et al., 2007) have led to recent work showing that AML CSCs can be identified, albeit at lower frequencies, in other fractions (Eppert et al., 2011). Furthermore, work by Barbara Vonderhaar's group has shown that breast CSCs can reside in the CD44+CD24+ fraction in breast cancer pleural effusions and xenografts (Meyer et al., 2010). Thus, a simple model of breast CSCs residing in only one fraction (i.e., CD44+CD24−) may be overly simplistic.

In contrast to a fixed CSC model, there is emerging evidence that suggests that CSC may exist in multiple states (Liu et al., 2012a; Liu et al., 2012b). In support of the concept of multiple states of breast CSCs is the association between CD44+CD24− expression and distant metastasis (Abraham et al., 2005), which is not seen with ALDH1+ expression (Ginestier et al., 2007). This is in contrast to known prognostic association between CD44+CD24− and ALDH+ to clinical outcome (Ginestier et al., 2007; Liu et al., 2007). CSCs are generally invoked as the seed of distant metastases since EMT is needed for cancer cell invasion and subsequent dissemination (Polyak and Weinberg, 2009; Rhim et al., 2012). Furthermore, CD44+CD24− cells generally display EMT properties (Mani et al., 2008) and have been associated with greater invasive capacity (Sheridan et al., 2006). Furthermore, circulating tumor cells (CTC) with a CD44+CD24− phenotype have been detected in early breast cancer patients (Balic et al., 2006) and bone marrow micrometastases whereas ALDH1+ CTCs are detected in patients with larger, clinically significant metastases (Theodoropoulos et al., 2010).

In light of the conflicting evidence regarding EMT and CSCs, we propose a model involving distinct CSC states. At the apex are CD44+CD24−EpCAM+ ALDH+cells based on our tumor-initiation data (Ginestier et al., 2007), which showed ALDH+/not CD44+CD24− and CD44+ CD24−ALDH+ cells initiated tumors, but CD44+CD24−ALDH- cells did not initiate tumors. Combined with our earlier data (Al-Hajj et al., 2003) showing CD44+ CD24−EpCAM+ are most enriched for tumor initiation, the cells at the apex most likely are EpCAM+ cells. Our data (Figure 32.1) show a subset of primary breast cancer (pleural effusions, liver biopsies) CD44+CD24− cells are ALDH+ and greater enrichment of ALDH+ within the CD44+CD24−EpCAM+fraction compared to CD44+CD24−EpCAM−. We propose that these cells can give rise to multiple fractions: CD44+CD24−(EMT), CD44+ CD24−ALDH+ (transition), and ALDH+ EpCAM+ (MET) (Figure 32.2). The ability of each fraction to give rise to other fractions may be regulated by microRNAs (Liu et al., 2012a; Liu et al., 2012b), TGF-beta, and various transcription factors (SNAI, TWIST, ZEB1) (Taube et al., 2010). Future experiments are required to test this hypothesis across molecular subtypes (luminal, basal, HER2+, claudin-low) and from multiple tumor sources (primary tumors, pleural effusions, metastatic lesions, etc.).

"De-differentiation" of Non-CSC to CSC States

Within normal tissues, especially the hematopoietic, intestinal, and skin tissues, there is a distinct hierarchy of differentiation from stem cells to progenitors to terminally differentiated cells (Doulatov et al., 2012; Li and Clevers, 2010; Munoz et al., 2012). As cells

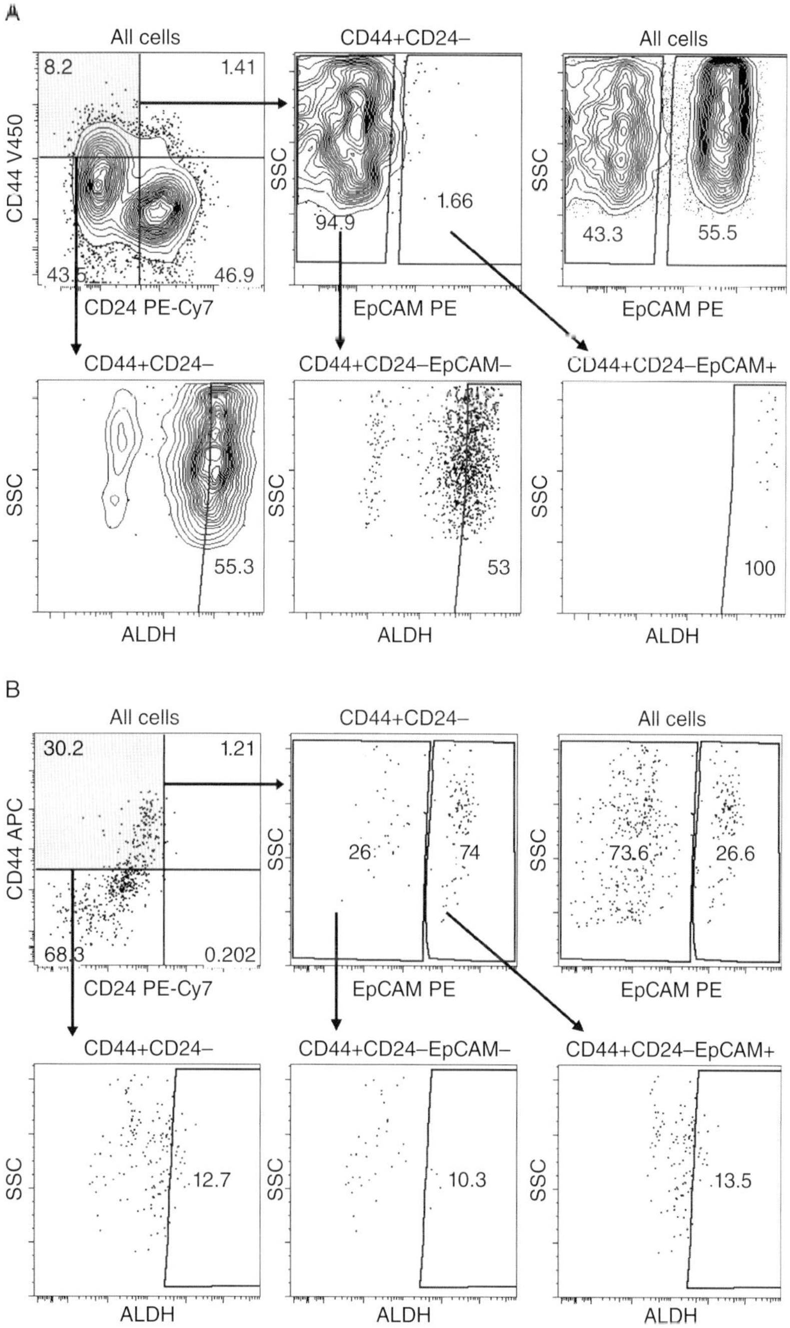

Figure 32.1. FACS analysis of EMT and MET CSC in primary malignant samples. Cells from malignant pleural effusion **A**. or needle core biopsy **B**. from the liver were processed to single cells, and stained with biotinylated lineage markers (CD45, CD31, CD140b, CD3, CD14, CD16, CD19, CD41, CD56, CD235a, HLA-DR), CD44 (APC or V450), CD24−PE-Cy7, EpCAM-PE, Aldefluor (ALDH), Streptavidin-Alexa 750, and Live/Dead Alexa750. Background gating was determined using Isotype matched and DEAB (ALDH1 inhibitor) controls. Live, lineage negative, single cells were gated (not shown). CD44+CD24− cells (*gray box*) were gated to examine EpCAM and ALDH staining. Within the CD44+CD24−EpCAM+fraction, there was enrichment for ALDH+cells compared to the CD44+CD24−EpCAM− fraction.

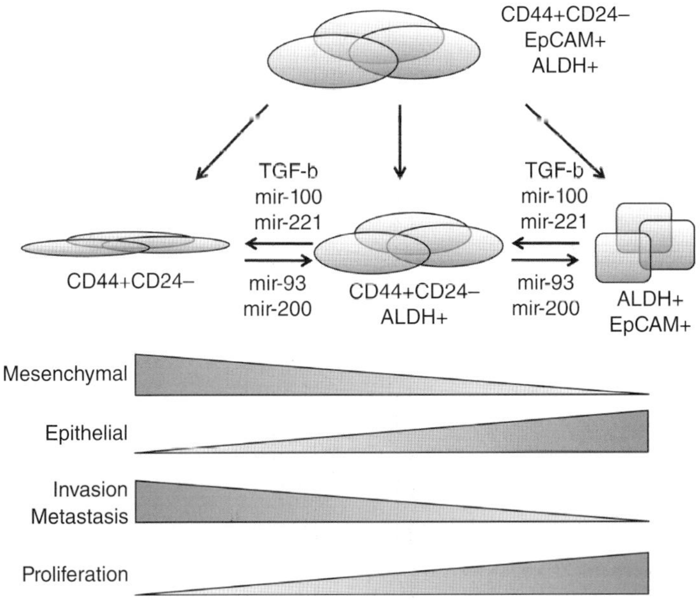

Figure 32.2. Proposed model of EMT and MET CSC states.

transit from stem cells to progenitors, albeit as multipotent or unipotent, there is a progressive loss of self-renewal (Dick, 2008; He et al., 2009). The reacquisition of self-renewal properties by normal progenitor and/or differentiated cells is typically associated with early steps of transformation (Doulatov et al., 2012; Krivtsov et al., 2006). Based on these observations in the hematopoietic, intestinal, and skin systems, it has been assumed that non-stem cells would not be able to acquire stem cell characteristics, as compared to progenitor cells "de-differentiating" into stem cells. Furthermore, the ability of prospectively isolated normal progenitor cells to reacquire self-renewal capacity without genetic manipulation or subsequent transformation has not been conclusively demonstrated. However, recent provocative work from Weinberg and colleagues (Chaffer et al., 2011) suggest the ability of normal and neoplastic non-stem cells to acquire characteristics of stem cells. Within their in vitro culture conditions of normal HME cells, there was a rare floating fraction composed of CD44+CD24−EpCAM− and $CD44^{lo}$CD24+EpCAM−cells. Over time, the floating $CD44^{lo}$ fraction gradually became CD44+CD24+ and had the ability to form normal gland structures in NOD/SCID mice. The sequential addition of oncogenic hTERT, SV40-ER, and Ras increased the switching from $CD44^{lo}$CD24+ to CD44+CD24−. These results suggest that non-CSC may switch to a CSC phenotype (CD44+CD24−), albeit at a low frequency, as the cell acquires more oncogenic hits. However, it will be critical to determine whether acquisition of EMT associated CSC markers truly represents "de-differentiation" or rather reflects inter-conversion between CSC states, which are characterized by expression of a different set of markers.

These results are highly provocative, however, there are a number of issues with the conclusion that non-CSC cells may switch to CSC. First, CSC marker expression is highly dependent on culture conditions (S.P. McDermott, unpublished observations). Secondly, if CSCs exist in alternative states then this may be reflected in marker expression. If CSCs are generated from non-CSC in vivo then the original xenograft models (Al-Hajj et al., 2003) for tumor-initiating capacity wouldn't work. Furthermore, our data with overexpression of miR-93 suggests that if CSCs are knocked out in the early tumor setting, then tumors don't recur (Liu et al., 2012b). This suggests that if CSCs are generated from non-CSCs, this must occur with low frequency, which is similar to the low frequency of re-programming to generate iPS cells (Yamanaka, 2012). The important question is whether the frequency of this type of "de-differentiation" event is clinically relevant.

Drug Discovery Efforts to Target Multiple CSC States

To date, numerous groups have used high-throughput screening to identify anti-CSC agents in various cancers, including breast cancer (Diamandis et al., 2007; Eberhard et al., 2009; Grinshtein et al., 2011; Gupta et al., 2009; McDermott et al., 2012; Sachlos et al., 2012; Smith et al., 2010). Based on the power of EMT to induce immortalized normal breast cells (HMLE) to acquire CSC markers (CD44+CD24−) (Mani et al., 2008), Gupta and colleagues (2009) used a relatively simple high-throughput cell viability screen with cells engineered to maintain an EMT state, HMLE-shEcad (anti-E-cadherin shRNA), and the immortalized parental cells to identify small molecules with anti-CSC properties. The majority of these HMLE-shEcad cells are CD44+ CD24− and enriched for tumor-initiating capacity. Using a collection of approximately 16,000 small molecules, 32 targeted the EMT cells but not the parental cells. Of the 4 compounds (salinomycin, abamectin, nigericin, and etoposide) confirmed in follow-up studies, salinomycin, a potassium ionophore, showed the greatest effect in vitro by reducing mammosphere formation as well as the CD44+CD24− fraction. In vitro treatment of breast cancer cells with salinomycin dramatically reduced tumor-initiation and lung metastases in NOD/SCID mice indicating salinomycin was able to target breast CSC. This approach has greatly advanced the field of developing anti-CSC therapeutics. However, as described above, recent data have suggested the existence of distinct CSC states (i.e., 'EMT' and 'MET') (Liu et al., 2012a; Liu et al., 2012b) and the ability of non-CSC cells to acquire CSC markers (Chaffer et al., 2011). Salinomycin may only target the EMT-CSC (i.e., CD44+CD24−) state and not MET-CSCs (ALDH+), and it is unknown if salinomycin is able to prevent the "de-differentiation" of non-CSC into CSC cells.

High-throughput screening (HTS) has the power to not only identify agents that kill CSC but also induce terminal differentiation of CSC into non-CSC. The ability of all-trans retinoic acid (ATRA) to terminally differentiate and provide robust cure of acute promyelocytic leukemia (APL) (Nowak et al., 2009) highlights the clinical importance of identifying differentiation agents in high-throughput screens. However, ATRA is unable to induce differentiation in non-APL AML cells due to the inability to induce transcriptional activation of retinoic acid receptor target genes, which is associated with a reduction in lysine methylation of histone H3 (H3K4me2) (Glasow et al., 2008). Combination of ATRA with tranylcypromine, an inhibitor of lysine-specific demethylase 1 (LSD1) was able to target AML LSC (Schenk et al., 2012) indicating that differentiation therapies have great potential to target CSC.

Incorporation of differentiation outcomes in HTS has typically focused on modulating differentiation of embryonic stem cells to specific lineages (Andrews, 2011). An early HTS cancer differentiation approach by Todd Golub and colleagues (Stegmaier et al., 2005; Stegmaier et al., 2004) utilized RT-PCR to monitor the differentiation of AML cells treated with small molecule libraries. This approach identified 8 compounds (of 1,739) that induced differentiation of AML cells, of which one was 4,5-dianilinophthalimide, a previously identified EGF inhibitor (Stegmaier et al., 2005). These results were extended to another EGFR inhibitor, gefitinib (Iressa), that was able to induce differentiation of AML cells suggesting a potential new clinical use of gefitinib (Stegmaier et al., 2005).

The ability to screen for inducers of breast CSC differentiation is a formidable goal since endogenous breast CSCs are a rare population. One possible approach is the utilization of flow cytometry in HTS (Krutzik and Nolan, 2006) to monitor the acquisition of differentiation markers (i.e., CD24) or loss of CSC markers (i.e., ALDH+) in a rare population. High-content imaging also has the power to identify modulators of stem cell differentiation as demonstrated by the use of immunostaining for CD34 and CD133 acquisition in primary human HSC to identify StemRegenin-1 (Boitano et al., 2010). Furthermore, use of an engineered cell population highly enriched for CSCs, such as the CD44+CD24− HMLE-shEcad cells, provides a rich opportunity to monitor differentiation via flow cytometry or by gene expression profiling (Krutzik and Nolan, 2006; Stegmaier et al., 2004). However, since the CD44+CD24−HMLE cells used by Gupta and colleagues (2009) fail to differentiate while in 2D culture as compared to 3D mammospheres (Chaffer et al., 2011), the 2D screening approach might have failed to identify agents that can terminally differentiate CSC. Incorporation of a screening platform with mammospheres formed from HMLE-shEcad and high-content imaging (Boitano et al., 2010) or overall viability (Diamandis et al., 2007) might identify novel breast CSC differentiation agents.

The ability to expand drug-discovery efforts to target multiple CSC states could be greatly facilitated by gene expression studies of the various CSC states. Identification of shared signaling pathways, especially druggable targets like kinases, G-coupled protein receptors, and ion-channels (Rask-Andersen et al., 2011), may identify new therapeutic targets and uses for previously approved drugs (Eberhard et al., 2009; Sukhai et al., 2011). To this end, Creighton and colleagues (2009) have identified a molecular signature of CD44+CD24− cells from primary breast cancers that was highly enriched for EMT-associated genes. We have performed microarray analyses of ALDH+cells from breast cancer cell lines (Charafe-Jauffret et al., 2009), which are highly enriched for MET-associated genes. However, the distinct sources of CSCs (i.e., cell lines and primary tumors) preclude a direct comparison between these two data sets. Future microarray or RNA sequencing studies to examine these distinct CSC states in primary tumors is greatly needed and could help identify new targets for drug discovery efforts.

Most of the recent anti-CSC drug discovery efforts have used small molecule libraries comprised mainly of agents with known activity and protein targets in the belief that repurposing old drugs for new uses may hold clinical benefit. Recent efforts by the U.S. National Institutes for Health (NIH) and the U.K. Medical Research Council (MRC) to provide academic researchers access to approximately 60 "abandoned" therapeutics from pharmaceutical partners highlight the potential to identify new uses for these therapeutics (Mullard, 2012). Importantly, all of the therapeutics available through the NIH have been tested in humans and have extensive safety and pharmacokinetic data. It remains to be seen if any of these therapeutics target pathways in breast CSC. Considering the success of ciclopirox olamine (Loprox) being identified by HTS as active against AML CSC (Eberhard et al., 2009) and rapidly repurposed from a topical antifungal into an oral formulation for a phase I AML trial (Sukhai et al., 2011), these drug "repurposing" pilot programs may provide a new source clinically active agents.

The presence of multiple CSC states and "de-differentiation" potential highlight the inherent problem of screening against one population and provide additional opportunities to identify anti-CSC therapeutics. Furthermore, it may not be possible to selectively target multiple breast CSC states with one small molecule, and the use of multiple agents might be needed. This is akin to the need of multiple anti-HIV drugs to effectively manage HIV infection to minimize mutational "escape" when only one or two drugs are used (Peters and Conway, 2011). The combination of traditional lethality screens, differentiation screens, molecular analysis of CSC states, and repurposing of previously tested agents holds great promise for identifying breast CSC agents that target multiple CSC states and can be tested in the clinical setting.

Testing Anti-CSC Agents in Clinical Trials: What Marker and Approaches to Use?

Clinically, gene signatures from CSCs have been shown to predict prognosis in large patient cohorts of AML (Eppert et al., 2011; Gentles et al., 2010) and breast cancer (Liu et al., 2007). As stated above, standard chemotherapy and radiation treatment is known to increase cells expressing CSC markers as determined in preclinical models as well as in clinical trials (Bao et al., 2006; Creighton et al., 2009; Li et al., 2008). Of particular note, the clinical studies in breast cancer typically only examined one CSC fraction, CD44+CD24− (sometimes in conjunction with EpCAM status), suggesting that critical information is being missed. Furthermore, evaluation of multiple CSC states from clinical samples can be technically difficult due to the small sample size and exclusion of contaminating hematopoietic, endothelial, and stromal cells. As such, incorporation of CSC quantification into breast cancer clinical trials has not yet been successfully accomplished. Of the 1,344 active breast cancer trials registered at ClinicalTrials.gov (as of August 2012), only 5 (0.4%) studies explicitly mention that CSCs will be measured as a primary or secondary outcome. However difficult, these issues can be overcome, as we have successfully been able to measure and isolate EMT and MET CSCs from liver, lymph node, or breast punch biopsies from patients enrolled in breast cancer clinical trials (Figure 32.1).

The overall lack of evaluating breast CSC in clinical trials indicates that additional approaches are needed to address the following critical questions: (1) whether a particular therapy targets CSC states, and (2) if targeting CSC is beneficial to the patient and prolongs survival. One such approach is the use of measuring circulating tumor cells (CTC) during treatment. The ability of breast cancer cells to disseminate into the circulation is dependent on the acquisition of EMT and CSC characteristics (Valastyan and Weinberg, 2011). Recent work has shown that breast CTCs are enriched in cells with CSC markers (Balic et al., 2006; Theodoropoulos et al., 2010). However, the only FDA-approved method to measure CTCs, CellSearch (Riethdorf et al., 2007), uses EpCAM as a capture method, which will fail to measure the EMT-CSC (i.e., EpCAM negative) (Gorges et al., 2012). Unbiased and easy-to-use methods, such as microfluidic chips, to isolate CTCs from peripheral blood are greatly needed not only for quantifying tumor burden and CSCs, but for molecular analyses such as single-cell sphere formation (Dontu et al., 2003), exome sequencing (Ramskold et al., 2012), or SNP analysis (Treff et al., 2011). Microfluidic technology to efficiently separate cancer cells from blood originally utilized EpCAM-based capture methods (Nagrath et al., 2007), which are hampered by the same issues arising from using the CellSearch method. Recent microfluidic advances, such as non-contact and label-free methods with high efficiency (Augustsson et al., 2012), could be instrumental in evaluating CSCs in the clinical setting since CTCs are highly enriched in cells with CSC markers. These approaches would provide an opportunity to conclusively determine if circulating CSCs are functionally and transcriptionally similar to CSC present in the primary tumor or metastatic lesion.

Based on the lack of clinical trials evaluating multiple breast CSC states and relative early development of non-contact and label-free microfluidic devices, there is an obvious need to address these important issues. We propose that future clinical trials include measuring CSC frequency by flow cytometry in neoadjuvant and advanced studies. The I-SPY2 neoadjuvant trial design (Barker et al., 2009) is an ideal study design where patients undergo an initial biopsy and then follow-up biopsies at 3 and 12 weeks. While I-SPY2 combines chemotherapy regimens (weekly paclitaxel for 12 weeks followed by four cycles of doxorubicin and cyclophosphamide), which are known to increase breast CSC (Bao et al., 2006; Creighton et al., 2009; Li et al., 2008), it is likely that traditional "de-bulking" chemotherapeutics will be needed for a complete cure (Zhou et al., 2009).

Conclusions and Future Directions

The last 10 years has seen a relative explosion of research dedicated to studying the existence, molecular pathways, generation, and treatment of CSCs in many tumor types including breast cancer. The field is at an exciting time where recent research has shown that CD44+CD24− CSCs can be targeted in pre-clinical models, which is currently being examined in clinical trials. Obviously, there is ample opportunity for improvement based on the emerging evidence suggesting the existence of multiple CSC states, which must be targeted to obtain successful treatment in the clinical setting. Based on the collective work from the breast CSC community, we are optimistic that successful targeting of all breast CSC is achievable in the near future and hopefully this will result in improved patient outcomes.

References

Abraham, B.K., Fritz, P., McClellan, M., Hauptvogel, P., Athelogou, M., and Brauch, H. (2005). Prevalence of CD44+/CD24−/low cells in breast cancer may not be associated with clinical outcome but may favor distant metastasis. Clin Cancer Res *11*, 1154–1159.

Al-Hajj, M., Wicha, M.S., Benito-Hernandez, A., Morrison, S.J., and Clarke, M.F. (2003). Prospective identification of tumorigenic breast cancer cells. Proc Natl Acad Sci USA *100*, 3983–3988.

Andrews, P.D. (2011). Discovering small molecules to control stem cell fate. Future Med Chem *3*, 1539–1549.

Augustsson, P., Magnusson, C., Nordin, M., Lilja, H., and Laurell, T. (2012). Microfluidic, Label-Free Enrichment of Prostate Cancer Cells in Blood Based on Acoustophoresis. Anal Chem.

Balic, M., Lin, H., Young, L., Hawes, D., Giuliano, A., McNamara, G., Datar, R.H., and Cote, R.J. (2006). Most early disseminated cancer cells detected in bone marrow of breast cancer patients have a putative breast cancer stem cell phenotype. Clin Cancer Res *12*, 5615–5621.

Bao, S., Wu, Q., McLendon, R.E., Hao, Y., Shi, Q., Hjelmeland, A.B., Dewhirst, M.W., Bigner, D.D., and Rich, J.N. (2006). Glioma stem cells promote radioresistance by preferential activation of the DNA damage response. Nature *444*, 756–760.

Barker, A.D., Sigman, C.C., Kelloff, G.J., Hylton, N.M., Berry, D.A., and Esserman, L.J. (2009). I-SPY 2: an adaptive breast cancer trial design in the setting of neoadjuvant chemotherapy. Clin Pharmacol Ther *86*, 97–100.

Boitano, A.E., Wang, J., Romeo, R., Bouchez, L.C., Parker, A.E., Sutton, S.E., Walker, J.R., Flaveny, C.A., Perdew, G.H., Denison, M.S., *et al.* (2010). Aryl hydrocarbon receptor antagonists promote the expansion of human hematopoietic stem cells. Science *329*, 1345–1348.

Bonnet, D., and Dick, J.E. (1997). Human acute myeloid leukemia is organized as a hierarchy that originates from a primitive hematopoietic cell. Nature Medicine *3*, 730–737.

Chaffer, C.L., Brueckmann, I., Scheel, C., Kaestli, A.J., Wiggins, P.A., Rodrigues, L.O., Brooks, M., Reinhardt, F., Su, Y., Polyak, K., *et al.* (2011). Normal and neoplastic nonstem cells can spontaneously convert to a stem-like state. Proc Natl Acad Sci USA *108*, 7950–7955.

Charafe-Jauffret, E., Ginestier, C., Iovino, F., Wicinski, J., Cervera, N., Finetti, P., Hur, M.H., Diebel, M.E., Monville, F., Dutcher, J., *et al.* (2009). Breast cancer cell lines contain functional cancer stem cells with metastatic capacity and a distinct molecular signature. Cancer Res *69*, 1302–1313.

Clarke, M.F., Dick, J.E., Dirks, P.B., Eaves, C.J., Jamieson, C.H., Jones, D.L., Visvader, J., Weissman, I.L., and Wahl, G.M. (2006). Cancer Stem Cells—Perspectives on Current Status and Future Directions: AACR Workshop on Cancer Stem Cells. Cancer Res *66*, 9339–9344.

Creighton, C.J., Li, X., Landis, M., Dixon, J.M., Neumeister, V.M., Sjolund, A., Rimm, D.L., Wong, H., Rodriguez, A., Herschkowitz, J.I., *et al.* (2009). Residual breast cancers after conventional therapy display mesenchymal as well as tumor-initiating features. Proc Natl Acad Sci USA *106*, 13820–13825.

Diamandis, P., Wildenhain, J., Clarke, I.D., Sacher, A.G., Graham, J., Bellows, D.S., Ling, E.K., Ward, R.J., Jamieson, L.G., Tyers, M., *et al.* (2007). Chemical genetics reveals a complex functional ground state of neural stem cells. Nat Chem Biol *3*, 268–273.

Dick, J.E. (2003). Stem cells: Self-renewal writ in blood. Nature *423*, 231–233.

Dick, J.E. (2008). Stem cell concepts renew cancer research. Blood *112*, 4793–4807.

Dontu, G., Abdallah, W.M., Foley, J.M., Jackson, K.W., Clarke, M.F., Kawamura, M.J., and Wicha, M.S. (2003). In vitro propagation and transcriptional profiling of human mammary stem/progenitor cells. Genes Dev *17*, 1253–1270.

Doulatov, S., Notta, F., Laurenti, E., and Dick, J.E. (2012). Hematopoiesis: a human perspective. Cell Stem Cell *10*, 120–136.

Eberhard, Y., McDermott, S.P., Wang, X., Gronda, M., Venugopal, A., Wood, T.E., Hurren, R., Datti, A., Batey, R.A., Wrana, J., *et al.* (2009). Chelation of intracellular iron with the antifungal agent ciclopirox olamine induces cell death in leukemia and myeloma cells. Blood *114*, 3064–3073.

Eppert, K., Takenaka, K., Lechman, E.R., Waldron, L., Nilsson, B., van Galen, P., Metzeler, K.H., Poeppl, A., Ling, V., Beyene, J., *et al.* (2011). Stem cell gene expression programs influence clinical outcome in human leukemia. Nat Med *17*, 1086–1093.

Fillmore, C.M., and Kuperwasser, C. (2008). Human breast cancer cell lines contain stem-like cells that self-renew, give rise to phenotypically diverse progeny and survive chemotherapy. Breast Cancer Res *10*, R25.

Gentles, A.J., Plevritis, S.K., Majeti, R., and Alizadeh, A.A. (2010). Association of a leukemic stem cell gene expression signature with clinical outcomes in acute myeloid leukemia. JAMA *304*, 2706–2715.

Ginestier, C., Hur, M.H., Charafe Jauffret, E., Monville, F., Dutcher, J., Brown, M., Jacquemier, J., Viens, P., Kleer, C.G., Liu, S., *et al.* (2007). ALDH1 is a marker of normal and malignant human mammary stem cells and a predictor of poor clinical outcome. Cell Stem Cell *1*, 555–567.

Glasow, A., Barrett, A., Petrie, K., Gupta, R., Boix-Chornet, M., Zhou, D.C., Grimwade, D., Gallagher, R., von Lindern, M., Waxman, S., *et al.* (2008). DNA methylation-independent loss of RARA gene expression in acute myeloid leukemia. Blood *111*, 2374–2377.

Gong, C., Yao, H., Liu, Q., Chen, J., Shi, J., Su, F., and Song, E. (2010). Markers of tumor-initiating cells predict chemoresistance in breast cancer. PLoS One *5*, e15630.

Gorges, T.M., Tinhofer, I., Drosch, M., Roese, L., Zollner, T.M., Krahn, T., and von Ahsen, O. (2012). Circulating tumour cells escape from EpCAM-based detection due to epithelial-to-mesenchymal transition. BMC Cancer *12*, 178.

Grinshtein, N., Datti, A., Fujitani, M., Uehling, D., Prakesch, M., Isaac, M., Irwin, M.S., Wrana, J.L., Al-Awar, R., and Kaplan, D.R. (2011). Small molecule kinase inhibitor screen identifies polo-like kinase 1 as a target for neuroblastoma tumor-initiating cells. Cancer Res *71*, 1385–1395.

Gupta, P.B., Onder, T.T., Jiang, G., Tao, K., Kuperwasser, C., Weinberg, R.A., and Lander, E.S. (2009). Identification of selective inhibitors of cancer stem cells by high-throughput screening. Cell *138*, 645–659.

He, S., Nakada, D., and Morrison, S.J. (2009). Mechanisms of stem cell self-renewal. Annu Rev Cell Dev Biol *25*, 377–406.

Honeth, G., Bendahl, P.O., Ringner, M., Saal, L.H., Gruvberger-Saal, S.K., Lovgren, K., Grabau, D., Ferno, M., Borg, A., and Hegardt, C. (2008). The CD44+/CD24− phenotype is enriched in basal-like breast tumors. Breast Cancer Res *10*, R53.

Hope, K.J., Jin, L., and Dick, J.E. (2004). Acute myeloid leukemia originates from a hierarchy of leukemic stem cell classes that differ in self-renewal capacity. Nature Immunology *5*, 738–743.

Hu, Y., and Smyth, G.K. (2009). ELDA: extreme limiting dilution analysis for comparing depleted and enriched populations in stem cell and other assays. J Immunol Methods *347*, 70–78.

Idowu, M.O., Kmieciak, M., Dumur, C., Burton, R.S., Grimes, M.M., Powers, C.N., and Manjili, M.H. (2012). CD44(+)/CD24(−/low) cancer stem/progenitor cells are more abundant in triple-negative invasive breast carcinoma phenotype and are associated with poor outcome. Hum Pathol *43*, 364–373.

Krivtsov, A.V., Twomey, D., Feng, Z., Stubbs, M.C., Wang, Y., Faber, J., Levine, J.E., Wang, J.,

Hahn, W.C., Gilliland, D.G., *et al.* (2006). Transformation from committed progenitor to leukaemia stem cell initiated by MLL-AF9. Nature *442*, 818–822.

Krutzik, P.O., and Nolan, G.P. (2006). Fluorescent cell barcoding in flow cytometry allows high throughput drug screening and signaling profiling. Nat Methods *3*, 361–368.

Lapidot, T., Sirard, C., Vormoor, J., Murdoch, B., Hoang, T., Caceres-Cortes, J., Minden, M., Paterson, B., Caligiuri, M.A., and Dick, J.E. (1994). A cell initiating human acute myeloid leukaemia after transplantation into SCID mice. Nature *367*, 645–648.

Li, L., and Clevers, H. (2010). Coexistence of quiescent and active adult stem cells in mammals. Science *327*, 542–545.

Li, X., Lewis, M.T., Huang, J., Gutierrez, C., Osborne, C.K., Wu, M.F., Hilsenbeck, S.G., Pavlick, A., Zhang, X., Chamness, G.C., *et al.* (2008). Intrinsic resistance of tumorigenic breast cancer cells to chemotherapy. J Natl Cancer Inst *100*, 672–679.

Liu, R., Wang, X., Chen, G.Y., Dalerba, P., Gurney, A., Hoey, T., Sherlock, G., Lewicki, J., Shedden, K., and Clarke, M.F. (2007). The prognostic role of a gene signature from tumorigenic breast-cancer cells. N Engl J Med *356*, 217–226.

Liu, S., Clouthier, S.G., and Wicha, M.S. (2012a). Role of microRNAs in the regulation of breast cancer stem cells. J Mammary Gland Biol Neoplasia *17*, 15–21.

Liu, S., Patel, S.H., Ginestier, C., Ibarra, I., Martin-Trevino, R., Bai, S., McDermott, S.P., Shang, L., Ke, J., Ou, S.J., *et al.* (2012b). MicroRNA93 Regulates Proliferation and Differentiation of Normal and Malignant Breast Stem Cells. PLoS Genet *8*, e1002751.

Mani, S.A., Guo, W., Liao, M.J., Eaton, E.N., Ayyanan, A., Zhou, A.Y., Brooks, M., Reinhard, F., Zhang, C.C., Shipitsin, M., *et al.* (2008). The epithelial-mesenchymal transition generates cells with properties of stem cells. Cell *133*, 704–715.

Marcato, P., Dean, C.A., Pan, D., Araslanova, R., Gillis, M., Joshi, M., Helyer, L., Pan, L., Leidal, A., Gujar, S., *et al.* (2011). Aldehyde dehydrogenase activity of breast cancer stem cells is primarily due to isoform ALDH1A3 and its expression is predictive of metastasis. Stem Cells *29*, 32–45.

Mazurier, F., Doedens, M., Gan, O.I., and Dick, J.E. (2003). Rapid myeloerythroid repopulation after intrafemoral transplantation of NOD-SCID mice reveals a new class of human stem cells. Nature Medicine *9*, 959–963.

McDermott, S.P., Eppert, K., Notta, F., Isaac, M., Datti, A., Al-Awar, R., Wrana, J., Minden, M.D., and Dick, J.E. (2012). A small molecule screening strategy with validation on human leukemia stem cells uncovers the therapeutic efficacy of kinetin riboside. Blood *119*, 1200–1207.

McKenzie, J.L., Gan, O.I., Doedens, M., Wang, J.C., and Dick, J.E. (2006). Individual stem cells with highly variable proliferation and self-renewal properties comprise the human hematopoietic stem cell compartment. Nat Immunol *7*, 1225–1233.

Meyer, M.J., Fleming, J.M., Lin, A.F., Hussnain, S.A., Ginsburg, E., and Vonderhaar, B.K. (2010). CD44posCD49fhiCD133/2hi defines xenograft-initiating cells in estrogen receptor-negative breast cancer. Cancer Res *70*, 4624–4633.

Mullard, A. (2012). Drug repurposing programmes get lift off. Nat Rev Drug Discov *11*, 505–506.

Munoz, J., Stange, D.E., Schepers, A.G., van de Wetering, M., Koo, B.K., Itzkovitz, S., Volckmann, R., Kung, K.S., Koster, J., Radulescu, S., *et al.* (2012). The Lgr5 intestinal stem cell signature: robust expression of proposed quiescent '+4' cell markers. EMBO J *31*, 3079–3091.

Mylona, E., Giannopoulou, I., Fasomytakis, E., Nomikos, A., Magkou, C., Bakarakos, P., and Nakopoulou, L. (2008). The clinicopathologic and prognostic significance of CD44+/CD24(−/low) and CD44−/CD24+ tumor cells in invasive breast carcinomas. Hum Pathol *39*, 1096–1102.

Nagrath, S., Sequist, L.V., Maheswaran, S., Bell, D.W., Irimia, D., Ulkus, L., Smith, M.R., Kwak, E.L., Digumarthy, S., Muzikansky, A., *et al.* (2007). Isolation of rare circulating tumour cells in cancer patients by microchip technology. Nature *450*, 1235–1239.

Notta, F., Mullighan, C.G., Wang, J.C., Poeppl, A., Doulatov, S., Phillips, L.A., Ma, J., Minden, M.D., Downing, J.R., and Dick, J.E. (2011). Evolution of human BCR-ABL1 lymphoblastic leukaemia-initiating cells. Nature *469*, 362–367.

Nowak, D., Stewart, D., and Koeffler, H.P. (2009). Differentiation therapy of leukemia: 3 decades of development. Blood *113*, 3655–3665.

Pearce, D.J., Taussig, D., Simpson, C., Allen, K., Rohatiner, A.Z., Lister, T.A., and Bonnet, D.

(2005). Characterization of cells with a high aldehyde dehydrogenase activity from cord blood and acute myeloid leukemia samples. Stem Cells *23*, 752–760.

Perfetto, S.P., Chattopadhyay, P.K., and Roederer, M. (2004). Seventeen-colour flow cytometry: unravelling the immune system. Nat Rev Immunol *4*, 648–655.

Peters, B.S., and Conway, K. (2011). Therapy for HIV: past, present, and future. Adv Dent Res *23*, 23–27.

Pollard, S.M., Yoshikawa, K., Clarke, I.D., Danovi, D., Stricker, S., Russell, R., Bayani, J., Head, R., Lee, M., Bernstein, M., *et al.* (2009). Glioma stem cell lines expanded in adherent culture have tumor-specific phenotypes and are suitable for chemical and genetic screens. Cell Stem Cell *4*, 568–580.

Polyak, K., and Weinberg, R.A. (2009). Transitions between epithelial and mesenchymal states: acquisition of malignant and stem cell traits. Nat Rev Cancer *9*, 265–273.

Quintana, E., Shackleton, M., Sabel, M.S., Fullen, D.R., Johnson, T.M., and Morrison, S.J. (2008). Efficient tumour formation by single human melanoma cells. Nature *456*, 593–598.

Ramskold, D., Luo, S., Wang, Y.C., Li, R., Deng, Q., Faridani, O.R., Daniels, G.A., Khrebtukova, I., Loring, J.F., Laurent, L.C., *et al.* (2012). Full-length mRNA-Seq from single-cell levels of RNA and individual circulating tumor cells. Nat Biotechnol.

Rask-Andersen, M., Almen, M.S., and Schioth, H.B. (2011). Trends in the exploitation of novel drug targets. Nat Rev Drug Discov *10*, 579–590.

Rhim, A.D., Mirek, E.T., Aiello, N.M., Maitra, A., Bailey, J.M., McAllister, F., Reichert, M., Beatty, G.L., Rustgi, A.K., Vonderheide, R.H., *et al.* (2012). EMT and dissemination precede pancreatic tumor formation. Cell *148*, 349–361.

Ricardo, S., Vieira, A.F., Gerhard, R., Leitao, D., Pinto, R., Cameselle-Teijeiro, J.F., Milanezi, F., Schmitt, F., and Paredes, J. (2011). Breast cancer stem cell markers CD44, CD24 and ALDH1: expression distribution within intrinsic molecular subtype. J Clin Pathol *64*, 937–946.

Riethdorf, S., Fritsche, H., Muller, V., Rau, T., Schindlbeck, C., Rack, B., Janni, W., Coith, C., Beck, K., Janicke, F., *et al.* (2007). Detection of circulating tumor cells in peripheral blood of patients with metastatic breast cancer: a validation study of the CellSearch system. Clin Cancer Res *13*, 920–928.

Sachlos, E., Risueno, R.M., Laronde, S., Shapovalova, Z., Lee, J.H., Russell, J., Malig, M., McNicol, J.D., Fiebig-Comyn, A., Graham, M., *et al.* (2012). Identification of drugs including a dopamine receptor antagonist that selectively target cancer stem cells. Cell *149*, 1284–1297.

Schenk, T., Chen, W.C., Gollner, S., Howell, L., Jin, L., Hebestreit, K., Klein, H.U., Popescu, A.C., Burnett, A., Mills, K., *et al.* (2012). Inhibition of the LSD1 (KDM1A) demethylase reactivates the all-trans-retinoic acid differentiation pathway in acute myeloid leukemia. Nat Med *18*, 605–611.

Sheridan, C., Kishimoto, H., Fuchs, R.K., Mehrotra, S., Bhat-Nakshatri, P., Turner, C.H., Goulet, R., Jr., Badve, S., and Nakshatri, H. (2006). CD44+/CD24− breast cancer cells exhibit enhanced invasive properties: an early step necessary for metastasis. Breast Cancer Res *8*, R59.

Shultz, L.D., Ishikawa, F., and Greiner, D.L. (2007). Humanized mice in translational biomedical research. Nat Rev Immunol *7*, 118–130.

Shultz, L.D., Schweitzer, P.A., Christianson, S.W., Gott, B., Schweitzer, I.B., Tennent, B., McKenna, S., Mobraaten, L., Rajan, T.V., and Greiner, D.L. (1995). Multiple defects in innate and adaptive immunologic function in NOD/LtSz-scid mice. J Immunol *154*, 180–191.

Smith, K.M., Datti, A., Fujitani, M., Grinshtein, N., Zhang, L., Morozova, O., Blakely, K.M., Rotenberg, S.A., Hansford, L.M., Miller, F.D., *et al.* (2010). Selective targeting of neuroblastoma tumour-initiating cells by compounds identified in stem cell-based small molecule screens. EMBO Mol Med *2*, 371–384.

Stegmaier, K., Corsello, S.M., Ross, K.N., Wong, J.S., Deangelo, D.J., and Golub, T.R. (2005). Gefitinib induces myeloid differentiation of acute myeloid leukemia. Blood *106*, 2841–2848.

Stegmaier, K., Ross, K.N., Colavito, S.A., O'Malley, S., Stockwell, B.R., and Golub, T.R. (2004). Gene expression-based high-throughput screening(GE-HTS) and application to leukemia differentiation. Nat Genet *36*, 257–263.

Storms, R.W., Trujillo, A.P., Springer, J.B., Shah, L., Colvin, O.M., Ludeman, S.M., and Smith, C. (1999). Isolation of primitive human hematopoietic progenitors on the basis of aldehyde dehydrogenase activity. Proc Natl Acad Sci USA *96*, 9118–9123.

Sukhai, M.A., Spagnuolo, P.A., Weir, S., Kasper, J., Patton, L., and Schimmer, A.D. (2011). New sources of drugs for hematologic malignancies. Blood *117*, 6747–6755.

Taube, J.H., Herschkowitz, J.I., Komurov, K., Zhou, A.Y., Gupta, S., Yang, J., Hartwell, K., Onder, T.T., Gupta, P.B., Evans, K.W., *et al.* (2010). Core epithelial-to-mesenchymal transition interactome gene-expression signature is associated with claudin-low and metaplastic breast cancer subtypes. Proc Natl Acad Sci USA *107*, 15449–15454.

Theodoropoulos, P.A., Polioudaki, H., Agelaki, S., Kallergi, G., Saridaki, Z., Mavroudis, D., and Georgoulias, V. (2010). Circulating tumor cells with a putative stem cell phenotype in peripheral blood of patients with breast cancer. Cancer Lett *288*, 99–106.

Treff, N.R., Su, J., Tao, X., Northrop, L.E., and Scott, R.T., Jr. (2011). Single-cell whole-genome amplification technique impacts the accuracy of SNP microarray-based genotyping and copy number analyses. Mol Hum Reprod *17*, 335–343.

Valastyan, S., and Weinberg, R.A. (2011). Tumor metastasis: molecular insights and evolving paradigms. Cell *147*, 275–292.

Valent, P., Bonnet, D., De Maria, R., Lapidot, T., Copland, M., Melo, J.V., Chomienne, C., Ishikawa, F., Schuringa, J.J., Stassi, G., *et al.* (2012). Cancer stem cell definitions and terminology: the devil is in the details. Nat Rev Cancer *12*, 767–775.

Visvader, J.E. (2011). Cells of origin in cancer. Nature *469*, 314–322.

Yamanaka, S. (2012). Induced pluripotent stem cells: past, present, and future. Cell Stem Cell *10*, 678–684.

Zhou, B.B., Zhang, H., Damelin, M., Geles, K.G., Grindley, J.C., and Dirks, P.B. (2009). Tumour-initiating cells: challenges and opportunities for anticancer drug discovery. Nat Rev Drug Discov *8*, 806–823.

Chapter 33

Difficulties in Targeting the Beating Heart: Therapeutic Implications of the Cancer Stem Cell Hypothesis in Melanoma

Jennifer Makalowski and Hinrich Abken
Center for Molecular Medicine Cologne, University of Cologne, and Department I Internal Medicine, University Hospital Cologne, Cologne, Germany

Abbreviations

ABC	adenosine triphosphate-binding cassette
ALDH	aldehyde dehydrogenase
AML	acute myeloic leukemia
BRAF	v-raf homolog B1
CSC	cancer stem cell
CAR	chimeric antigen receptor
HMW-MAA	high molecular weight-melanoma associated antigen
JARID1B	jumonji AT-rich interactive domain 1B
MCSP	melanoma-associated chondroitin sulfate proteoglycan
NK	natural killer
NOD/SCID	non-obese diabetic/severe combined immunodeficiency
NSG	NOD scid gamma
PTEN	phosphatase and tensin homolog
TCR	T cell receptor

Evidence of Cancer Stem Cells in Melanoma

Primary melanoma can be cured by surgery; therapy of metastatic melanoma, however, is still challenging. Patients often develop drug resistance and relapse after therapy with chemotherapeutics or even small-molecule inhibitors aimed at blocking the most frequently mutated oncogene $BRAF^{V600E}$ (Flaherty et al., 2010; Chapman et al., 2011). Substantial observations indicate that cells with stem cell-like properties contribute to both tumor relapse and therapy resistance.

Melanoma lesions, like many other solid cancer lesions, are composed of cells with vast phenotypic and functional heterogeneity; a subset of these cells obviously displays poly-potency

Cancer Stem Cells, First Edition. Edited by Vinagolu K. Rajasekhar.

characteristics, including self-renewal, asymmetric differentiation into the variety of cell types found in the primary melanoma lesion from which the cells were isolated, and tumor initiation and progression. These observations lead to the hypothesis that melanoma, like other cancer entities, may be initiated and maintained by so-called "cancer stem cells" (CSC). Primarily established by deciphering the hierarchical organization of cells in hematological malignancies (Lapidot et al., 1994; Bonnet and Dick, 1997) and subsequently confirmed for a number of solid cancers, including carcinomas of the mammary gland, prostate, pancreas, and colon (Al-Hajj et al., 2003; Singh et al., 2003; Dalerba et al., 2007; Ricci-Vitiani et al., 2007; Li et al., 2007), the postulated properties of cancer stem cells are thought to be fulfilled for melanoma as well.

A particular subset of melanoma cells can induce tumors in immune-deficient mice upon transplantation with a higher frequency than other melanoma cells of the same lesion. Using various cell surface markers, several melanoma cell subsets were defined to produce melanoma in this model (Fang et al., 2005; Monzani et al., 2007; Schatton et al., 2008). A first study identified a melanoma cell subset that exhibits stem-cell-like capacities and expresses CD20, a marker generally associated with B lymphocytes, along with melanoma-associated markers (Fang et al., 2005). The induction of progressively growing melanomas that recapitulate the parental tumor in histological type and diversity reflects the potential of this subset to differentiate; induction of secondary tumors by serial transplantation of re-isolated melanoma cell subsets confirmed the long-term self-renewal capacity of those cells.

Great efforts have been undertaken to attribute a specific phenotype to those "melanoma stem cells". For instance, melanoma cells with high tumor-inducing capacity express the adenosine triphosphate-binding cassette (ABC) transporter protein-B5, which belongs to a family of efflux pumps whose members are expressed by normal stem cells (Schatton et al., 2008). Those melanoma cells show increased doxorubicin resistance (Frank et al., 2005). Matricellular proteins cooperate with ABCB5, as disruption of tenascin C decreased both ABCB5 expression and tumor formation (Fukunaga-Kalabis et al., 2010). Similarly, the subset cells expressing the low-affinity nerve growth factor receptor (NGFR) CD271, which constitute approximately 20% of melanoma cells, induce melanoma and form metastases (Boiko et al., 2010). Melanoma stem cells were moreover identified based on their aldehyde dehydrogenase (ALDH) activity; 86% of ALDH$^+$ melanoma cells induced tumors compared with 10% of ALDH$^-$ melanoma cells (Luo et al., 2012). These observations suggest that melanoma is organized in a hierarchical manner, following an initiator cancer stem cell through subsequent differentiation levels to progenitor cells in a top-down direction, without reversion back to stem cells (Figure 33.1). The conclusion, however, is based on assumptions that have led to controversial discussions and question the existence of melanoma stem cells in general (Shackleton and Quintana, 2010; Marotta and Polyak, 2009; Rosen and Jordan, 2009; Shackleton et al., 2009; Hill, 2006).

Transplantation of a Single Melanoma Cell Questions the Cancer Stem Cell Hypothesis in Melanoma

In previous transplantation assays, Weissman's and Frank's groups estimated that the tumor-initiating potential of melanomas is rare, i.e., in the range of 1 in 2,000 cells (Boiko et al., 2010; Schatton et al., 2008). Testing by Morrison's group under more rigorous experimental conditions revealed that approximately 1 of 4 (1 in 2 to 1 in 15) melanoma cells induces tumors (Quintana et al., 2008; Quintana et al., 2010). No differences in the tumorigenic potential of CD271$^+$ versus CD271$^-$, or of ABCB5$^+$ versus ABCB5$^-$ melanoma cells were found. In particular, transplantation of single cells from BRAF-mutated, PTEN-negative melanomas produced tumor

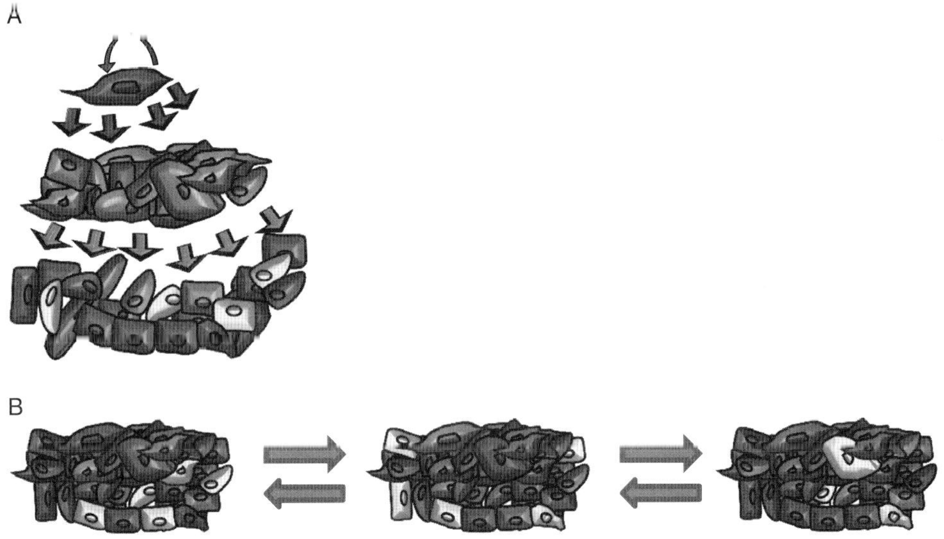

Figure 33.1. Models describing the role of putative cancer stem cells in melanoma. **A**. The hierarchic model predicts a top-down pathway in the differentiation of cancer stem cells to more differentiated cells that make up the melanoma cell mass and which lost tumor initiating capacities. **B**. The dynamic model assumes that nearly every melanoma cell has the capability of initiating cancer; the phenotype and functional capacity of each individual cell is flexible and expressed in a stochastic manner orchestrated by cell-intrinsic and/or micro-environmental conditions. For color detail, please see color plate section.

formation with high incidence (Held et al., 2010). The conclusion drawn from these and a number of subsequent assays by several other groups is that the potential to induce melanoma is a common feature of nearly all melanoma cells, independent of identified surface markers.

Minor differences in the experimental design to test for melanoma stem cells may have major effects, as recently outlined (Magee et al., 2012; Fukunaga-Kalabis et al., 2011). For instance, the processing of the tumor sample, the degree of immune deficiency of the mouse, and the environment into which the melanoma cells are transplanted can substantially impact the outcome of the transplantation assay. The NSG mice used by the Morrison laboratory are more immune compromised than the NOD/SCID mice, which have NK cells, and were used by the Frank laboratory. The difference became even more apparent when transplanting CD271-negative melanoma cells, which formed tumors in the NSG mouse but not in the NOD/SCID mouse (Civenni et al., 2011). Human cells have to overcome substantial species barriers when transplanted into the murine environment and must adapt to a non-physiologic environment, particularly the subcutaneous tissue, to reconstruct their own growth promoting niche, which in various parameters is different from the environment in the sporadic tumor. As a consequence, the transplanted cell is more challenged for functional flexibility upon transplantation than the same cell when resident in the progressing tumor tissue.

Targeted Elimination Revealed a Minor Melanoma Cell Subset Sustaining Established Melanoma

The transplantation assay does not address whether or not an established melanoma lesion is organized in a functional hierarchy to orchestrate the different cancer cell subsets within the tumor tissue. The assay does not address "tumor stemness", which describes the maintenance of tumor progression once the tumor is established (Schmidt and Abken, 2011; Adams and Strasser, 2008). Although equally capable to induce melanoma, not every melanoma cell may equally contribute to progression of an established melanoma lesion. Evidence for the latter hypothesis was provided by serial xeno-transplantation of established melanoma cells of a distinct phenotype (Roesch et al., 2010). Slow cycling cells, which are present at frequencies of 1–5% in established melanoma cell lines, induce melanomas of heterogeneous histology upon serial transfer, exhibit stem cell-like properties, and are characterized by the expression of JARID1B, a histone-3 lysine-4 demethylase that is an epigenetic marker for embryonic stem cells. Stable JARID1B knockdown resulted in the exhaustion of tumor induction by serial transplantation, indicating that JARID1B is essential for the maintenance of melanoma growth. This notion is remarkable with respect to the observation that JARID1B-positive and JARID1B-negative cells equally induce tumors in transplantation assays (Roesch et al., 2010).

In an alternative approach, Schmidt and colleagues (2011) addressed the question via specific elimination of melanoma cell subsets from an established melanoma lesion. The rationale is that if there is a clearly defined hierarchy of cancer cells in an established melanoma lesion, specific ablation of the melanoma-sustaining cell(s) will inevitably lead to growth arrest or the lasting decay of the tumor independently of targeting the cancer cell mass. Human melanoma lesions were established in immune-deficient mice by transplantation of a bulk of melanoma biopsy cells giving rise to melanoma of the same histological phenotype as the parental tumor. Defined melanoma cell subsets from those established melanoma lesions were selectively eliminated by adoptive transfer of cytotoxic T cells that were redirected toward a pre-defined biobsy antigen by engineering with a chimeric antigen receptor (CAR), which binds to a cognate antigen on the surface of the target cell. CAR binding produced T cell activation with specific cytolysis of target cells displaying the cognate antigen, whereas cells lacking the antigen remained untouched (Eshhar et al., 1993; Gilham et al., 2012). Because intravenously transferred T cells evade circulation and penetrate virtually all tissues, the T cell-based immunologic approach provides a strategy to selectively eliminate antigen-defined melanoma cell subsets without manipulating the established tumor tissue.

Targeted elimination of the minor subset of melanoma cells expressing CD20 eradicated melanoma lesions, whereas targeting of any random minor melanoma cell population in the same lesion did not (Schmidt et al., 2011). $CD20^+$ melanoma cells are rare, comprising approximately 1–2% of the melanoma, independent of histological type and the stage of the disease; approximately 80% of melanoma samples harbored $CD20^+$ melanoma cells, as detected by histological screening. CD20-negative melanoma lesions could not be eliminated by treatment with CD20-specific T cells. CD20 expression induced by genetic modification in a random subpopulation of those tumor cells did not render the tumor lesion sensitive to eradication by CD20-specific T cells, indicating that CD20 expression per se is insufficient to maintain melanoma progression. The phenotype of $CD20^+$ melanoma cells, however, likely associates with additional cell autonomous capacities that mediate tumor progression.

The first clinical evidence sustaining the dominant role of the $CD20^+$ subset of melanoma cells was recently provided by a case report (Schlaak et al., 2012). A patient with stage III/IV refractory melanoma received repetitive intra-lesional injections of the anti-CD20 therapeutic antibody rituximab with concomitant dacarbazine treatment, the latter of which was ineffective as a monotherapy. Although the melanoma lesion harbored $CD20^+$ melanoma cells at a frequency of approximately 2%, the treatment produced lasting complete remissions; one metastasis showed partial remission and was surgically removed in stable disease. The report is anecdotic and requires confirmation using a larger patient cohort, however, the data provide the first clinical evidence that established melanoma lesions may be hierarchically organized with a subset of melanoma cells sustaining melanoma progression. The caveat remains that eliminating B cells may also contribute to the effect. Several major issues need to be addressed, as discussed below.

Do CD20⁺ Melanoma Cells Have the Characteristics of Cancer Stem Cells?

Several pieces of evidence sustain the notion that $CD20^+$ melanoma cells exhibit stem cell properties, i.e., inducing melanoma upon transplantation, which is conserved upon serial transplantation, recapitulating the cellular heterogeneity of the parental tumor and indicating the capacity of these cells for self-renewal and tumor induction (Fang et al., 2005). Other melanoma cell subsets with different phenotypes but with similar tumor-maintaining capacities might, however, coexist in the same melanoma lesion. $CD20^+$ melanoma cells in an established tumor lesion are rare; the frequency is 2% or less by immune histology screen (Schmidt et al., 2011). A caveat in this respect is that approximately 20% of melanomas do not obviously harbor $CD20^+$ melanoma cells. Like others (Magee et al., 2012; Clevers, 2011; Fukunaga-Kalabis et al., 2010), we assume that melanoma-initiating cells, as estimated by transplantation assays, may not be rare, however, their frequency may be less relevant in the context of a dynamic functional equilibrium of cells in an established lesion.

Do CD20⁺ Melanoma Cells Represent a Transient Phenotype?

$CD20^+$ melanoma cells co-express markers, such as HMW-MAA (MCSP), βHMW-MAA (MCS), melanoma cell adhesion molecule (MCAM), CD133, CD44, CD271 and other markers commonly associated with CSCs in other malignancies (Schmidt et al., 2011, Boiko et al., 2010; Monzani et al., 2007; Fang et al., 2005). The remarkable phenotypic plasticity of these melanoma cells became obvious by transplantation of sorted CD133-positive and CD133-negative cells, both subsets similarly able to reconstitute a tumor with the same heterogeneous pattern of cells as the parental tumor (Quintana et al., 2008, 2010; Shackleton and Quintana, 2010; Shackleton et al., 2009). Individual melanoma cells can transiently acquire or lose markers independently of their tumor-inducing capacities. For instance, melanoma cells expressing the H3K4 demethylase JARID1B, which are more tumorigenic than cells lacking JARID1B, can convert to JARID1B-negative cells, which in turn can produce $JARID1B^+$ cells after transplantation (Roesch et al., 2010). Individual melanoma cells can also transiently acquire or lose JARID1A-mediated drug resistance (Sharma et al., 2010). The expression of a specific set of surface markers by melanoma cells may not be heritable fixed, but rather be determined as a result of a number of signals provided by the tumor environment, which has substantial consequences for therapy, as discussed below.

Can Stem-Like Cells be Therapeutic Targets for the Treatment of Melanoma?

Currently, there are two views of tumor initiation and maintenance in melanoma, which are not necessarily exclusive (Schmidt and Abken, 2011; Shackleton and Quintana, 2010). Although some experimental evidence supports the hierarchical stem cell model (Figure 33.1) in the maintenance of an established melanoma (Schmidt et al., 2011), melanoma initiation by transplantation of single melanoma cells sustains the stochastic model (Quintana et al., 2008; Quintana et al., 2010). For the development of therapeutic strategies, several aspects need to be taken into account.

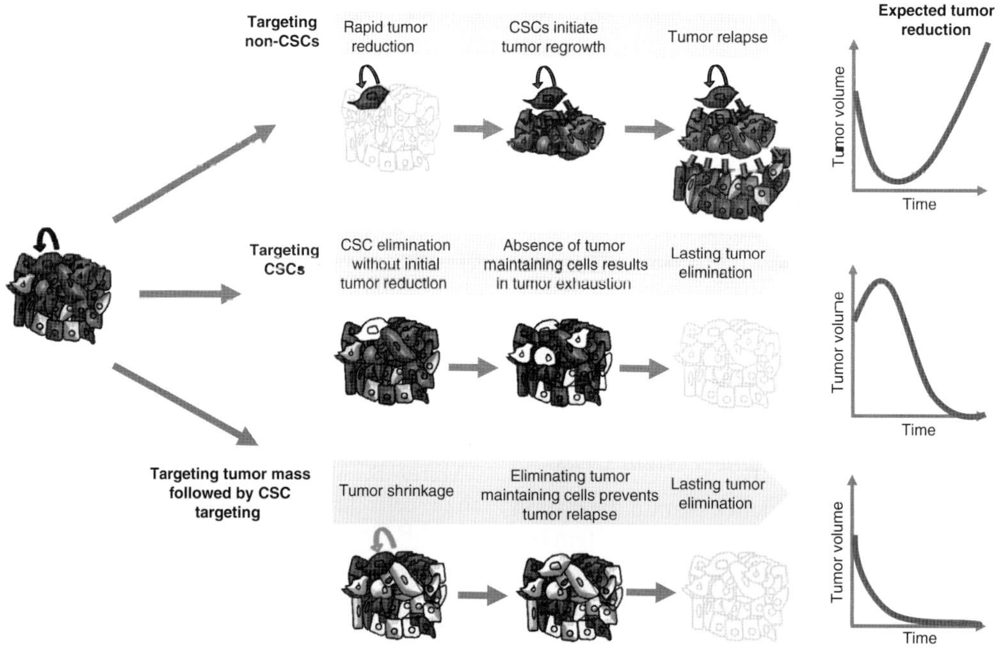

Figure 33.2. Cancer stem cell- and non-stem-cell-directed therapies will generate typical patterns of tumor responses. Targeting the tumor cell mass without targeting the melanoma sustaining cells will result in rapid tumor regression; however, the tumor will relapse fueled by drug-resistant melanoma stem cells. The tumor will initially progress when targeting CSCs due to the low number of CSCs in the tumor mass and will finally shrink due to exhaustion. This process may occur in small tumors more rapidly than in major tumor lesions. Combined strategies that "debulk" the tumor mass followed by targeting stem cells are expected to cause rapid tumor reduction and lasting tumor elimination. For color detail, please see color plate section.

First, selective targeting cancer cell subsets will produce different therapeutic outcomes depending on their functional hierarchy in the tumor lesion. For instance, targeting a cancer cell subset, which is small in number but drives tumor progression might not initially slow tumor progression while targeting every cancer cell will result in an immediate reduction of tumor volume (Figure 33.2). In contrast, if there is a fixed hierarchy in established melanoma lesions, strategies eliminating tumor-repopulating cells will be more successful than those increasing the death rate of any tumor cell in a more random fashion. Combining these two strategies to de-bulk the tumor mass and eliminate the melanoma-maintaining subset are assumed to comprise a more successful strategy in the long term while avoiding the risk of a relapse.

Second, targeting of melanoma stem cells needs to overcome the pronounced chemoresistance of these cells; some molecular mechanisms of drug resistance have recently been elucidated (Johannessen et al., 2010; Roesch et al., 2010; Villanueva et al., 2010). Chemotherapy may select for cells with efficient export of toxic drugs, resulting in remaining melanoma stem cells, which produce tumor relapse within a short period of time. This notion is supported by clinical observations in other cancer entities (Diehn et al., 2009; Li et al., 2008). As a consequence, alternative strategies to induce cell death of these stem cells are required.

Third, therapeutic strategies must take into account that the melanoma-maintaining cells need to be completely eliminated, which cannot be achieved by randomly targeting every cell. The prevalence of $CD20^+$ melanoma maintaining cells is rare, making specific targeting strategies necessary, such as therapeutic antibodies or the adoptive transfer of cytotoxic T cells. Compared with therapeutic drugs, redirected T cells have an advantage in that they can actively penetrate tissues to scan for target cells, which they eliminate upon engagement of cognate antigen. In contrast to adoptive cell therapy, therapeutic antibodies have been in clinical use for nearly a decade. The therapeutic efficacy of antibodies is not impaired by drug transporters like conventional chemotherapy, however, depends on several other prerequisites, including tissue penetration, targeting a low number of target cells, and their sensitivity to antibody-mediated killing. An explorative phase I trial applying rituximab (rituxan) as adjuvant therapy for low-dose IL-2, however, was without improvement in anti-tumor response (Aklilu et al., 2004). Schlaak and colleagues (2012) successfully used rituximab in the treatment of a melanoma patient. Finally, patients may develop resistance due to CD20 repression, as observed in lymphoma patients, rendering the targeted treatment ineffective (Jazirehi and Bonavida, 2005).

Fourth, functional plasticity of melanoma cells may result in the production of "new" stem cells after therapeutic elimination of the targeted subset cells. Melanoma cells, which initially lack CD20, may acquire stem cell properties and express CD20 under therapeutic pressure. Therapeutic targeting might therefore require a continued process and may be successful in small tumor lesions where the stochastic frequency of newly reverted cells remains low. If a large tumor mass produces sufficient numbers of reverted melanoma cells after initial treatment, reduction in melanoma cell mass before eliminating the melanoma-sustaining cells may be required (Figure 33.2). Targeting stem cells with high plasticity will require the presence of the therapeutic agent over a prolonged period of time to continuously eliminate those cells that freshly gained stem cell capacities. Blocking the conversion to the stem cell stage may improve therapeutic efficacy. Conversely, if melanoma growth depends on a fixed tumor cell subpopulation, the specific elimination of that subpopulation will effectively eradicate the tumor lesion without the need for targeting the bulk tumor.

Fifth, proliferative quiescent ("dormant") melanoma stem cells escape anti-proliferative chemotherapy, favoring tumor relapse even after decades. Proliferative quiescence of $CD20^+$ melanoma cells results in resistance to conventional therapy; inducing these cells to enter into the cell cycle, as has been successfully shown for AML (Saito et al., 2010), will be required to sensitize the cells for anti-proliferative therapy.

Targeting a Melanoma Stem Cell: Some Thoughts on Open Questions

The assumption that cancer is statically and hierarchically organized may not sufficiently describe the situation in melanoma; the dynamic development of melanoma stem cells will plague stem cell-directed therapies. The balanced, dynamic equilibrium of functional capacities and individual cell types will substantially impact the therapeutic response and prognosis of melanoma. Targeted elimination of melanoma sustaining cells may lead to an abrogation of melanoma maintenance and melanoma exhaustion, and targeting other melanoma cell subsets will not be of benefit for the patients (Figure 33.2).

Therapeutic targeting of melanoma cells with stem cell properties requires targeting the critical survival and self-renewal pathways that are active during these stages of function. The functional similarity between healthy tissue stem cells and cancer stem cells suggests that many pathways may be expressed in both cells, making selective targeting more difficult. Potential therapeutic targets include the Wnt, Notch, or Hedgehog pathway; toxicity to healthy tissue stem cells needs to be explored.

Subpopulations may fluctuate depending on cell-intrinsic signals and/or on signals from the tumor microenvironment (Hoek and Goding, 2010). The control of such a phenotypic and functional switch has not been understood in detail yet. Genetic heterogeneity may moreover contribute and arise from epigenetic differences; genetic instability will promote clonal selection, making the situation even more complex.

At least three substantial consequences for long-term therapy of melanoma need to be addressed:

- First, whether functional plasticity is a random process equally occupying each melanoma cell and which cell intrinsic and/or environmental signals trigger the process.
- Second, how the frequency of melanoma stem cells is established, and whether the number of these cells can serve as a prognostic marker and/or a surrogate for therapeutic efficacy.
- Third, which signals drive melanoma cells into a "dormant" state and which "awake" those cells to initiate melanoma relapse, even after decades.

Acknowledgments

Work in the authors' laboratory is supported by the Deutsche Krebshilfe, the European Union (European Regional Development Fund—Investing in Your Future), and the German federal state North Rhine-Westphalia (NRW), Germany.

The authors wish to apologize to authors whose work was not discussed due to space limitations.

References

Adams, J.M., and Strasser, A. (2008). Is Tumor Growth Sustained by Rare Cancer Stem Cells or Dominant Clones? Cancer Res *68*, 4018–4021.

Aklilu, M., Stadler, W.M., Markiewicz, M., Vogelzang, N.J., Mahowald, M., Johnson, M., and Gajewski, T.F. (2004). Depletion of normal B cells with rituximab as an adjunct to IL-2 therapy for renal cell carcinoma and melanoma. Ann Oncol *15*, 1109–1114.

Al-Hajj, M., Wicha, M.S., Benito-Hernandez, A., Morrison, S.J., and Clarke, M.F. (2003). Prospective identification of tumorigenic breast cancer cells. Proc. Natl. Acad. Sci. U.S.A. *100*, 3983–3988.

Boiko, A.D., Razorenova, O.V., van de Rijn, M., Swetter, S.M., Johnson, D.L., Ly, D.P., Butler, P.D., Yang, G.P., Joshua, B., Kaplan, M.J., et al. (2010). Human melanoma-initiating cells express neural crest nerve growth factor receptor CD271. Nature *466*, 133–137.

Bonnet, D., and Dick, J.E. (1997). Human acute myeloid leukemia is organized as a hierarchy that originates from a primitive hematopoietic cell. Nat. Med. *3*, 730–737.

Chapman, P.B., Hauschild, A., Robert, C., Haanen, J.B., Ascierto, P., Larkin, J., Dummer, R., Garbe, C., Testori, A., Maio, M., et al. (2011). Improved Survival with Vemurafenib in Melanoma with BRAF V600E Mutation. New England Journal of Medicine *364*, 2507–2516.

Civenni, G., Walter, A., Kobert, N., Mihic-Probst, D., Zipser, M., Belloni, B., Seifert, B., Moch, H., Dummer, R., van den Broek, M., et al. (2011). Human CD271-positive melanoma stem cells associated with metastasis establish tumor heterogeneity and long-term growth. Cancer Res. *71*, 3098–3109.

Clevers, H. (2011). The cancer stem cell: premises, promises and challenges. Nat. Med. *17*, 313–319.

Dalerba, P., Dylla, S.J., Park, I.-K., Liu, R., Wang, X., Cho, R.W., Hoey, T., Gurney, A., Huang, E.H., Simeone, D.M., et al. (2007). Phenotypic characterization of human colorectal cancer stem cells. Proc. Natl. Acad. Sci. U.S.A. *104*, 10158–10163.

Diehn, M., Cho, R.W., Lobo, N.A., Kalisky, T., Dorie, M.J., Kulp, A.N., Qian, D., Lam, J.S., Ailles, L.E., Wong, M., et al. (2009). Association of reactive oxygen species levels and radioresistance in cancer stem cells. Nature *458*, 780–783.

Eshhar, Z., Waks, T., Gross, G., and Schindler, D.G. (1993). Specific activation and targeting of cytotoxic lymphocytes through chimeric single chains consisting of antibody-binding domains and the gamma or zeta subunits of the immunoglobulin and T-cell receptors. Proc Natl Acad Sci USA *90*, 720–724.

Fang, D., Nguyen, T.K., Leishear, K., Finko, R., Kulp, A.N., Hotz, S., Belle, P.A.V., Xu, X., Elder, D.E., and Herlyn, M. (2005). A Tumorigenic Subpopulation with Stem Cell Properties in Melanomas. Cancer Res *65*, 9328–9337.

Flaherty, K.T., Hodi, F.S., and Bastian, B.C. (2010). Mutation-driven drug development in melanoma. Curr Opin Oncol *22*, 178–183.

Frank, N.Y., Margaryan, A., Huang, Y., Schatton, T., Waaga-Gasser, A.M., Gasser, M., Sayegh, M.H., Sadee, W., and Frank, M.H. (2005). ABCB5-Mediated Doxorubicin Transport and Chemoresistance in Human Malignant Melanoma. Cancer Res *65*, 4320–4333.

Fukunaga-Kalabis, M., Martinez, G., Nguyen, T.K., Kim, D., Santiago-Walker, A., Roesch, A., and Herlyn, M. (2010). Tenascin-C promotes melanoma progression by maintaining the ABCB5-positive side population. Oncogene *29*, 6115–6124.

Fukunaga-Kalabis, M., Roesch, A., and Herlyn, M. (2011). From cancer stem cells to tumor maintenance in melanoma. J. Invest. Dermatol. *131*, 1600–1604.

Gilham, D.E., Debets, R., Pule, M., Hawkins, R.E., and Abken, H. (2012). CAR-T cells and solid tumors: tuning T cells to challenge an inveterate foe. Trends Mol Med *18*, 377–384.

Held, M.A., Curley, D.P., Dankort, D., McMahon, M., Muthusamy, V., and Bosenberg, M.W. (2010). Characterization of Melanoma Cells Capable of Propagating Tumors from a Single Cell. Cancer Res *70*, 388–397.

Hill, R.P. (2006). Identifying cancer stem cells in solid tumors: case not proven. Cancer Res. *66*, 1891–1895; discussion 1890.

Hoek, K.S., and Goding, C.R. (2010). Cancer stem cells versus phenotype-switching in melanoma. Pigment Cell Melanoma Res *23*, 746–759.

Jazirehi, A.R., and Bonavida, B. (2005). Cellular and molecular signal transduction pathways modulated by rituximab (rituxan, anti-CD20 mAb) in non-Hodgkin's lymphoma: implications in chemosensitization and therapeutic intervention. Oncogene *24*, 2121–2143.

Johannessen, C.M., Boehm, J.S., Kim, S.Y., Thomas, S.R., Wardwell, L., Johnson, L.A., Emery, C.M., Stransky, N., Cogdill, A.P., Barretina, J., et al. (2010). COT drives resistance to RAF inhibition through MAP kinase pathway reactivation. Nature *468*, 968–972.

Lapidot, T., Sirard, C., Vormoor, J., Murdoch, B., Hoang, T., Caceres-Cortes, J., Minden, M., Paterson, B., Caligiuri, M.A., and Dick, J.E. (1994). A cell initiating human acute myeloid leukaemia after transplantation into SCID mice. Nature *367*, 645–648.

Li, C., Heidt, D.G., Dalerba, P., Burant, C.F., Zhang, L., Adsay, V., Wicha, M., Clarke, M.F., and Simeone, D.M. (2007). Identification of Pancreatic Cancer Stem Cells. Cancer Res *67*, 1030–1037.

Li, X., Lewis, M.T., Huang, J., Gutierrez, C., Osborne, C.K., Wu, M.-F., Hilsenbeck, S.G., Pavlick, A., Zhang, X., Chamness, G.C., et al. (2008). Intrinsic resistance of tumorigenic breast cancer cells to chemotherapy. J. Natl. Cancer Inst. *100*, 672–679.

Luo, Y., Dallaglio, K., Chen, Y., Robinson, W.A., Robinson, S.E., McCarter, M.D., Wang, J., Gonzalez, R., Thompson, D.C., Norris, D.A., et al. (2012). ALDH1A Isozymes are Markers of Human Melanoma Stem Cells and Potential Therapeutic Targets. Stem Cells.

Magee, J.A., Piskounova, E., and Morrison, S.J. (2012). Cancer stem cells: impact, heterogeneity, and uncertainty. Cancer Cell *21*, 283–296.

Marotta, L.L.C., and Polyak, K. (2009). Cancer stem cells: a model in the making. Curr. Opin. Genet. Dev. *19*, 44–50.

Monzani, E., Facchetti, F., Galmozzi, E., Corsini, E., Benetti, A., Cavazzin, C., Gritti, A., Piccinini, A., Porro, D., Santinami, M., et al. (2007). Melanoma contains CD133 and ABCG2 positive cells with enhanced tumourigenic potential. Eur. J. Cancer *43*, 935–946.

Quintana, E., Shackleton, M., Foster, H.R., Fullen, D.R., Sabel, M.S., Johnson, T.M., and Morrison, S.J. (2010). Phenotypic heterogeneity among tumorigenic melanoma cells from patients that is reversible and not hierarchically organized. Cancer Cell *18*, 510–523.

Quintana, E., Shackleton, M., Sabel, M.S., Fullen, D.R., Johnson, T.M., and Morrison, S.J. (2008). Efficient tumour formation by single human melanoma cells. Nature *456*, 593–598.

Ricci-Vitiani, L., Lombardi, D.G., Pilozzi, E., Biffoni, M., Todaro, M., Peschle, C., and Maria, R.D. (2007). Identification and expansion of human colon-cancer-initiating cells. Nature *445*, 111–115.

Roesch, A., Fukunaga-Kalabis, M., Schmidt, E.C., Zabierowski, S.E., Brafford, P.A., Vultur, A., Basu, D., Gimotty, P., Vogt, T., and Herlyn, M. (2010). A Temporarily Distinct Subpopulation of Slow-Cycling Melanoma Cells Is Required for Continuous Tumor Growth. Cell *141*, 583–594.

Rosen, J.M., and Jordan, C.T. (2009). The Increasing Complexity of the Cancer Stem Cell Paradigm. Science *324*, 1670–1673.

Saito, Y., Uchida, N., Tanaka, S., Suzuki, N., Tomizawa-Murasawa, M., Sone, A., Najima, Y., Takagi, S., Aoki, Y., Wake, A., et al. (2010). Induction of cell cycle entry eliminates human leukemia stem cells in a mouse model of AML. Nat. Biotechnol. *28*, 275–280.

Schatton, T., Murphy, G.F., Frank, N.Y., Yamaura, K., Waaga-Gasser, A.M., Gasser, M., Zhan, Q., Jordan, S., Duncan, L.M., Weishaupt, C., et al. (2008). Identification of cells initiating human melanomas. Nature *451*, 345–349.

Schlaak, M., Schmidt, P., Bangard, C., Kurschat, P., Mauch, C., and Abken, H. (2012). Regression of metastatic melanoma in a patient by antibody targeting of cancer stem cells. Oncotarget *3*, 22–30.

Schmidt, P., and Abken, H. (2011). The beating heart of melanomas: a minor subset of cancer cells sustains tumor growth. Oncotarget *2*, 313–320.

Schmidt, P., Kopecky, C., Hombach, A., Zigrino, P., Mauch, C., and Abken, H. (2011). Eradication of melanomas by targeted elimination of a minor subset of tumor cells. PNAS *108*, 2474–2479.

Shackleton, M., and Quintana, E. (2010). Progress in understanding melanoma propagation. Mol Oncol *4*, 451–457.

Shackleton, M., Quintana, E., Fearon, E.R., and Morrison, S.J. (2009). Heterogeneity in Cancer: Cancer Stem Cells versus Clonal Evolution. Cell *138*, 822–829.

Sharma, S.V., Lee, D.Y., Li, B., Quinlan, M.P., Takahashi, F., Maheswaran, S., McDermott, U., Azizian, N., Zou, L., Fischbach, M.A., et al. (2010). A chromatin-mediated reversible drug-tolerant state in cancer cell subpopulations. Cell *141*, 69–80.

Singh, S.K., Clarke, I.D., Terasaki, M., Bonn, V.E., Hawkins, C., Squire, J., and Dirks, P.B. (2003). Identification of a cancer stem cell in human brain tumors. Cancer Res. *63*, 5821–5828.

Villanueva, J., Vultur, A., Lee, J.T., Somasundaram, R., Fukunaga-Kalabis, M., Cipolla, A.K., Wubbenhorst, B., Xu, X., Gimotty, P.A., Kee, D., et al. (2010). Acquired resistance to BRAF inhibitors mediated by a RAF kinase switch in melanoma can be overcome by cotargeting MEK and IGF-1R/PI3K. Cancer Cell *18*, 683–695.

Chapter 34

Targeting Cancer Stem Cells for Overcoming Drug Resistance and Cancer Progression

Yiwei Li[1], Dejuan Kong[1], Aamir Ahmad[1], Bin Bao[1], and Fazlul H. Sarkar[2]
[1] *Department of Pathology, Barbara Ann Karmanos Cancer Institute, Wayne State University School of Medicine, Detroit, MI, USA*
[2] *Departments of Pathology and Oncology, Barbara Ann Karmanos Cancer Institute, Wayne State University School of Medicine, Hudson Webber Cancer Research Center, Detroit, MI, USA*

Introduction

Cancer cell resistance to conventional chemotherapy and radiotherapy has been known as the major cause of cancer treatment failure. The resistance of cancer cell to cancer therapy also leads to the recurrence of cancer from small amount of survival and self-renewal cancer cells after anti-cancer therapy. The drug resistance and recurrence nature of malignancies have puzzled cancer researchers for many decades. The growing studies have focused on the investigation of drug resistance and recurrence of cancers in many research directions. One of the important findings is that cancer stem cells (CSC) play critical roles in the drug resistance and recurrence of cancers. Although the debate regarding the existence of CSCs are going on for several years, it has now been experimentally documented and widely accepted that CSCs or in some context defined as cancer stem-like cells (CSLC) do exist in most types of hematopoietic and solid tumors including brain tumor, breast, head and neck, colon, lung, prostate, ovarian, pancreatic, and other cancers. The CSCs are defined as those cells that are a subset of cancer cells and possess the characteristics of extensive self-renewal capacity, multi-lineage differentiation potential, high proliferation rate, and tumor-inducing capacity. In recent years, significant advances have been made in the field of CSC research. Importantly, it has been found that CSCs are resistant to conventional cancer therapies including radiotherapy and chemotherapy. The survival of CSCs after conventional anti-cancer treatment and the rapid self-renewal of CSCs result in cancer recurrence (relapse). In addition, in vitro and in vivo studies have shown that CSCs have a high tendency of invasiveness and metastasis. It has been found that most early, disseminated cancer cells from solid tumors have a putative CSC phenotype, suggesting the metastatic nature of CSCs. Therefore, CSCs are responsible for cancer recurrence and progression including invasion and metastasis. Hence, it is important to elucidate the molecular mechanisms underlying drug resistance and cancer progression with respect to CSCs in cancers. The gain of such knowledge would allow the development of novel therapeutic strategies targeting CSCs. This would be very useful for designing effective agents for increasing drug sensitivity, which will inhibit invasion and metastasis, and thereby will lead to the successful treatment of cancers with better treatment outcome.

Cancer Stem Cells, First Edition. Edited by Vinagolu K. Rajasekhar.

Cancer Stem Cells

The first reported cancer stem cells were isolated from leukemic cells. In 1997, Bonnet and Dick (1997) reported that there was a subpopulation of leukemic cells with specific cell surface markers, $CD34^{++}$ and $CD38^{-}$. These cells had the abilities of differentiation, proliferation, and self-renewal. From then on, more hematopoietic and solid tumor stem cells were isolated by using cell surface markers.

Molecular Markers for CSC Isolation and Identification

Stem cells have their specific molecular markers that can be used for distinguishing stem cells from other cells. Because the biological characters of CSCs have some similarities with normal stem cells, CSC surface markers and stem cell surface markers overlap in many epithelia. These similarities provide some clues and opportunities to find CSCs. To identify and isolate CSCs from the whole cancer cell population, cell surface markers such as CD24, CD34, CD44, CD133, CD139, CD166, ESA, and EpCAM have been used together with immunolabeling and cell sorting by fluorescence-activated cell sorting (FACS) analysis. In addition to cell surface markers, other stem cell markers existing inside the cells have also been used as CSC markers for CSC identification. These markers include ALDH, c-Met, Bmi1, Nanog, Olfm4, Lgr5, Sox-2, Oct-4, and others. Among the cell surface markers, CD133, CD44, CD24, and ESA are more frequently used for the isolation of CSCs from different types of cancers, but these markers vary among different solid tumors. It is clear that combined use of several cell surface markers could increase the purity of isolated CSCs. The cell surface markers combined with stem cell markers inside the cells would likely be more reliable for the identification of cancer stem cells.

CD133 expression is one of the most commonly used cell surface markers for CSCs. CD133 has been used for the isolation and identification of CSCs from glioblastoma, bone sarcomas, and lung, ovarian, colon, uterine, pancreatic, breast, prostate, hepatic, thyroid, and other cancers. It was found that $CD133^{+}$ CSCs were highly tumorigenetic (Immervoll et al., 2008). CD133 combined with other CSC markers such as epithelial-specific antigen (ESA) and CD44 has been used to identify CSCs of lung, colon, and hepatic cancers. $CD133^{+}/ESA^{+}$ cells from non-small cell lung cancer (NSCLC) have higher tumorigenic potential in SCID mice and express genes that control stemness, adhesion, and motility (Bertolini et al., 2009). All of these findings suggest that CD133 is an important CSC marker, and its expression is correlated with cancer progression.

CD44 is another most commonly used CSC surface markers while CD24 is a cell surface marker that is frequently used together with CD44. $CD44^{+}/CD24^{lo/-}$ has been used for the isolation and identification of CSCs from breast, head and neck, gastric, prostate, and hepatic cancers. In addition, $CD44^{+}$ and $CD24^{lo/-}$ combined with another CSC marker $ALDH1^{+}$ could be more specific for the isolation of CSCs. CSCs with $CD44^{+}/CD24^{+}/ESA^{+}$ were isolated from pancreatic cancer. These cells have the stem cell properties including self-renewal, differentiation capacity, hedgehog signaling activation, and increased tumorigenic potential (Li et al., 2009a), suggesting that CD44 combined with CD24 and ESA could be better surface markers for the isolation and characterization of CSCs.

As mentioned above, ALDH is another CSC marker. It has been used for the isolation and identification of CSCs in leukemia, pancreatic, head and neck, colon, ovarian, breast, lung, and other cancers. $ALDH1^{+}$ cells isolated from head and neck squamous cell carcinoma patients represented a reservoir for generating tumors, showed radioresistance, and had epithelial to mesenchymal transition (EMT) characteristics (Chen et al., 2009). $ALDH^{+}$ marker combined with $CD44^{+}$/

$CD24^{lo/-}$ has also been used for the isolation and identification CSCs from breast cancer. In ovarian cancer, $ALDH^{+}$ cells were found to be chemoresistant, and were capable of initiating tumor growth. Using ALDH in combination with CD133 to further enrich ovarian CSCs, it was found that as few as 11 $ALDH1^{+}/CD133^{+}$ cells isolated from ovarian tumors were sufficient to initiate tumors in mice (Silva et al., 2011).

In addition, several other stem cell markers inside cells including c-Met, Bmi-1, Nanog, Sox-2, Oct-4, Olfm4, Lgr5, etc., have been found to be overexpressed in CSCs of various cancers. However, the expression of these stem cell markers is not CSC-specific. It is also worth mentioning that the expression of stem cell surface markers CD44, CD133, or CD24 in cancer cells could be altered within the tumor microenvironment (Wei et al., 2011). Therefore, it is obvious that the combination of several stem cell markers including cell surface markers and markers inside the cell should be used for the identification and molecular characterization of CSCs.

Molecular Signaling in CSCs

Because CSCs are a subclone of cancer cells with more progressive and renewable characteristics, CSCs have shown alterations in multiple signaling pathways, which are also observed in all cancer cells. The involved cellular signaling pathways in cancer cells include Akt, NF-κB, MAPK, Wnt, Notch, Hedgehog, TGF-β, EGF, etc. However, the most relevant pathways that control CSCs are Hedgehog, Notch, and Wnt signaling pathways because these signaling pathways are critically involved in embryonic development and differentiation. The molecules in these signaling pathways cross talk with each other, which leads to the CSC self-renewal causing tumor progression (Figure 34.1).

Hedgehog (Hh) signaling is an important signaling pathway regulating normal embryonic development, but it is silenced in adult tissues and organs. However, activation of Hh pathway has been found in various cancers. Overexpression of sonic hedgehog (Shh) could promote tumorigenesis. In most of the normal tissues, there is no detectable Shh expression; however, CSCs show significantly increased expression of sonic hedgehog (Li et al., 2009a). Downregulation of Hh signaling by Hh inhibitors could inhibit CSC and EMT-type cell proliferation and spheroid formation through the regulation of Hh and EMT-related molecules such as E-cadherin, Snail, Slug, ZEB1, and TCF/LEF, suggesting the importance of Hh signaling in the development and maintenance of CSCs.

Notch signaling is another important signaling pathway in the regulation of normal tissue development and differentiation (Miele, 2006). However, it was found that activation of Notch signaling could induce EMT through the regulation of EMT-related molecules including Snail, Slug, and TGF-β. Moreover, the expression of Notch-2 and Jagged-1 were found to be significantly increased in gemcitabine-resistant EMT-type cells (Katoh and Katoh, 2006). We have also found that overexpression of Notch-1 increased the formation of pancreatospheres, which expressed high levels of CSC markers CD44 and EpCAM (Bao et al., 2011). All of these findings clearly suggest that Notch signaling plays an important role in CSCs and EMT-type cells.

Wnt signaling is also critically involved in the regulation of development and differentiation. It has been found that the activation of Wnt signaling promoted the formation of CSCs and EMT-type cells. Inhibition of Wnt signaling could increase the expression of epithelial markers and suppress EMT-related transcription factor Slug and Twist. In addition, TGF-β, IGF, and EGF also stimulate and maintain the growth of CSCs. Therefore, Hh, Notch, and Wnt signaling pathways together with Akt, NF-κB, MAPK, TGF-β, IGF, and EGF signaling are critically involved in the development and progression of cancer mediated through the maintenance and functioning of CSCs (Figure 34.1).

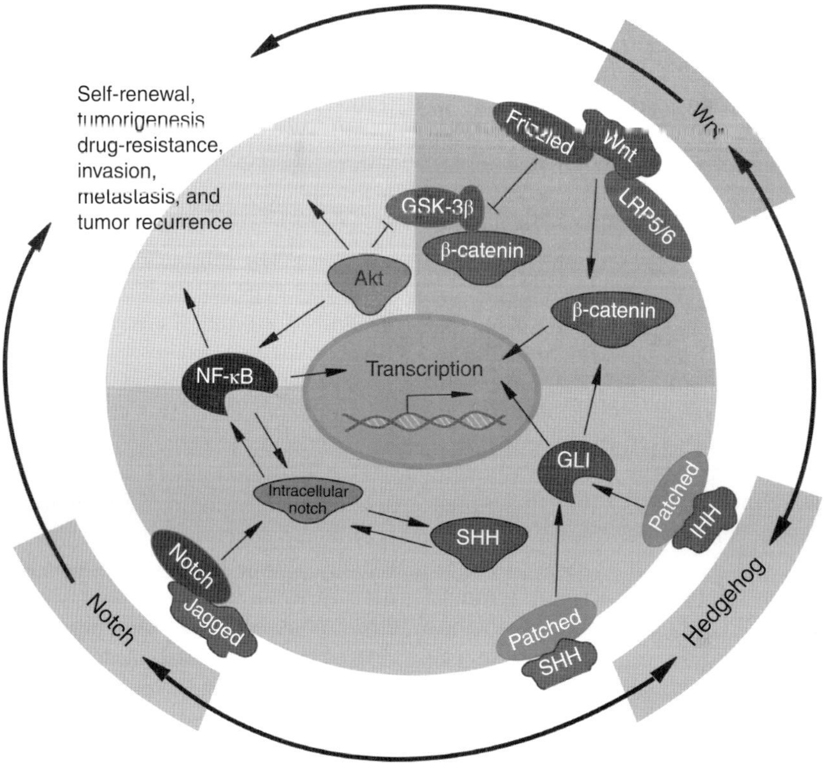

Figure 34.1. Signaling network altered in CSCs and its role in cancer development and progression. For color detail, please see color plate section.

Therapeutic Resistant and Aggressive Natures of CSCs

In an effort for the isolation and identification of CSCs, it has been reported that almost all CSCs are resistant to conventional chemotherapeutics and radiotherapy. Experimental studies have also shown that CSCs possess higher potential for tumor cell migration, invasion, and metastasis.

The Role of CSCs in Therapeutic Resistance

In recent years, emerging evidence from experimental studies has demonstrated that CSCs are resistant to conventional chemotherapy and radiotherapy. The resistance of breast CSCs to cancer therapy has been investigated in vitro and in vivo. An animal study showed that CSCs in mouse mammary tumors were cisplatin resistant (Shafee et al., 2008). Another study using biopsy tissues from patients with breast cancer have also shown that chemotherapy increased the percentage of $CD44^{+}/CD24^{-}$ cells (enrichment of CSCs) and the mammosphere formation efficiency (Li et al., 2008), suggesting the drug-resistance characteristics of breast CSCs. In a clinical and in vitro study, breast CSCs with ALDH1 expression played a significant role in resistance to chemotherapy (Tanei et al., 2009). Experimental studies with gene expression profiling demonstrated that the therapeutic resistance of breast CSCs was due to the aberrant expression of miR-128 and p16 (Zhu et al., 2011; Arima et al., 2012), suggesting the importance of miRNAs and tumor suppressor genes in the regulation of CSC drug resistance.

In colorectal cancer, CSCs have also been found to be resistant to chemotherapeutic drugs. Using the stem cell marker

CD133, colon CSCs were isolated, and these cells accounted for approximately 2% of colon cancer cells (Todaro et al., 2007). These CD133⁺ cells produced and utilized IL-4 to protect themselves from apoptosis. Importantly, treatment with IL-4Rα antagonist or anti-IL-4 neutralizing antibody significantly increased the efficacy of chemotherapeutic drugs, suggesting that colon CSCs are drug-resistant due to an increased IL-4 expression (Todaro et al., 2007). In $CD133^+/CD44^+$ colorectal CSCs, the increased phosphorylated and total IGF-IR levels were found to be correlated with resistance to 5-Fluorouracil and oxaliplatin, suggesting the importance of IGF signaling in drug-resistance (Dallas et al., 2009).

In lung cancer, cisplatin treatment could cause the enrichment of $CD133^+$ cancer cells. Furthermore, it was found that subpopulations of $CD133^+/ABCG2^+$ and $CD133^+/CXCR4^+$ cells are chemoresistant with highly tumorigenic and stem-like features (Bertolini et al., 2009). Lung CSCs also exhibited significantly decreased sensitivity to gemcitabine or a combination of cisplatin and gemcitabine through the inactivation of caspase-9 and caspase-3, and the activation of MAPK and HSP27 (Hsu et al., 2011). It was also found that CD133 positivity of lung CSCs was correlated with the expression of drug resistance-related proteins such as glutathione S-transferase, thymidylate synthase, catalase, and methylguanine-DNA methyltransferase (Salnikov et al., 2010), suggesting that drug resistance in lung cancer is in part due to the enrichment of CSCs.

CSCs from hepatocellular carcinoma exhibited strong gefitinib resistance. The CD133 expression in the hepatic CSCs was dramatically increased after gefitinib treatment. Moreover, the gefitinib treated cancer cells showed increased phosphorylation and nuclear translocation of IGF-1R, suggesting that activated IGF-1R and upregulated CD133 contribute to the acquisition of drug resistance (Bodzin et al., 2012). It was also found that the hepatic CSCs with $CD133^+/CD44^+$ exhibited resistance to chemotherapeutics due to the upregulation of ABCB1, ABCC1, and ABCG2 (Zhu et al., 2010), suggesting that multiple signaling pathways are involved in the acquisition of drug resistance, which is in part due to enrichment of CSCs.

In other cancers, CSCs also exhibit therapeutic resistance feature. In head and neck cancer, ALDH1 positive cells displayed radioresistance (Chen et al., 2009). Putative cancer stem cells in malignant pleural mesothelioma showed resistance to cisplatin and pemetrexed (Cortes-Dericks et al., 2010). The $CD133^+$ CSCs in uterine carcinosarcoma were also more resistant to cisplatin/paclitaxel-induced cytotoxicity (Choijamts et al., 2011). In glioblastoma, CD133 positive cells were significantly resistant to chemotherapeutic agents including temozolomide, carboplatin, paclitaxel, and etoposide compared to CD133 negative cells (Liu et al., 2006). Moreover, studies have also shown that the pancreatic cancers are more resistant to conventional chemotherapy if the tumor tissues contain a high number of $CD133^+$ CSCs (Hermann et al., 2007). All of these findings clearly suggest the importance of CSCs in therapeutic resistance.

The Biology of CSCs Is Consistent with Features of Tumor Aggressiveness

Another crucial feature of CSC is its aggressiveness. Most CSCs possess their tendency to invasion and metastasis. In breast cancer, it was found that a small population of cancer cells with molecular signature of CSCs had higher metastatic capacity. Both in vitro and in xenograft experiments showed that invasion and metastasis of inflammatory breast cancer were mediated by CSC marker ALDH activity. The expression of ALDH1 in inflammatory breast cancer was a predictive factor for early metastasis and poor survival in patients diagnosed with inflammatory breast cancer (Charafe-Jauffret et al., 2010). Experimental studies have also shown that breast CSCs with $ALDH1^+$ contain high level of KLF4 expression consistent with increased tumor cell migration and invasion capacity.

Downregulation of KLF4 significantly inhibited the migration and invasion of breast CSCs, suggesting that KLF4 is required for breast CSC maintenance, migration, and invasion (Yu et al., 2011).

In pancreatic cancer, $CD133^+$ cells showed enhanced migration and invasive activity. An in vivo study has demonstrated that $CD133^+/CXCR4^+$ pancreatic CSCs are located in the invasive front of pancreatic tumors, suggesting the higher invasive and metastatic abilities of pancreatic $CD133^+/CXCR4^+$ CSCs. Further studies have shown that removing these $CD133^+/CXCR4^+$ pancreatic CSCs significantly decreased the metastatic ability of pancreatic tumors cells (Hermann et al., 2007). Moreover, pancreatic $CD44^+/CD24^+$ CSCs also possess higher migration ability (Yin et al., 2011), suggesting the importance of pancreatic CSCs in cancer invasion and metastasis.

Metastatic colon cancer also contained CSCs with cancer stem cell characteristics including self-renewal and increased tumor-forming ability in animal model. A subpopulation of $CD26^+$ CSCs had been isolated from both primary and metastatic tumors from colorectal cancer patients with liver metastasis (Pang et al., 2010). The isolated $CD26^+$ CSCs induced development of distant metastasis when injected into the mice. $CD26^+$ CSCs also showed enhanced invasiveness and chemoresistance. In patients without distant metastasis at the time of experiment, the presence of $CD26^+$ CSCs in their primary tumors predicted distant metastasis on follow up, suggesting the critical role of colorectal CSCs in colorectal cancer progression.

The CSCs from other cancers have also been found to play important roles in cancer progression with differential molecular mechanisms. A recent report has shown that Matrix metalloproteinase-10 (MMP-10) was highly expressed in oncosphere cultures enriched in lung CSCs and that MMP-9 was required for lung CSC maintenance, tumor initiation, and high metastatic potential (Justilien et al., 2012). Rac1-GTP, Oct4, Nanog, Sox-2, MMP-2, integrin alphaV, and Slug signaling also contributed to self-renewal, invasion, and metastatic characteristics of CSCs in various cancers. Overexpression of miR-124 and knockdown of its target SNAI2 could significantly inhibit glioma cell invasion by reduction of neurosphere formation, $CD133^+$ cell subpopulation, and stem cell marker (Bmi-1, Nanog, and Nestin) expression. These observations demonstrate that CSCs are critically involved in cancer invasion and metastasis.

Targeting CSCs to Increase Drug Sensitivity and Inhibit Cancer Progression

Since CSCs are drug resistant and possess the capacity of increased invasion and metastasis, the novel therapeutic strategies targeting CSCs could be an effective approach for the treatment of cancers with better survival outcome. With growing knowledge gained about CSC markers and molecular signaling involved in CSCs, more investigations have focused on finding novel agents or molecules that could target CSCs, which is expected to increase drug sensitivity and inhibit cancer cell invasion and metastasis. These novel therapeutic strategies could have great benefits for the elimination of cancer cells by effective eradication of CSCs.

In a study to screen compounds specifically targeting CSCs, salinomycin showed very high CSC-specific toxicity. Salinomycin significantly inhibited the growth of CSCs at least 100 times more effectively than the commonly used anti-cancer drug paclitaxel. It was also found that salinomycin inhibited mammary tumor growth and induced epithelial differentiation of cancer cells in mice (Gupta et al., 2009). Another recent study has shown that trastuzumab, which targets HER2, did not regulate the expression of Sox-2, one of the CSC markers, in HER2-positive mammospheres. However, treatment of the mammospheres with salinomycin decreased the expression of Sox-2, suggesting a selective targeting of breast CSCs by salinomycin. Importantly, treatment of the mammospheres with combination of trastuzumab and salinomycin was effective in inhibiting the growth of HER2-positive cancer cells and cancer stem cells (Oak et al., 2012), suggesting the beneficial effects of salinomycin in combination treatment of breast cancers, which appears to be mediated via targeted elimination of CSCs. In pancreatic cancer, salinomycin combined with gemcitabine also showed significant inhibition of cell growth of pancreatic cancer cells. It is known that gemcitabine mainly inhibits non-CSC pancreatic cancer cells. Therefore, salinomycin combined with gemcitabine was able to eliminate human pancreatic cancer cells more effectively by targeting both CSCs and non-CSC cancer cells (Zhang et al., 2011). Osteosarcoma is the most common primary bone tumor, with poor prognosis in children. Bone tumor stem cells are responsible for tumor initiation, tumor relapse, and metastasis. It was found that in vitro, bone tumor stem cells formed sarcospheres, which overexpressed OCT4 and Sox-2 stem cell markers and were chemotherapy resistance (Tang et al., 2011). Importantly, salinomycin inhibited osteosarcoma by selectively targeting bone tumor stem cells both in vitro and in vivo without severe side effects. It was also found that the inhibition of bone tumor stem cells by salinomycin was mediated by the regulation of Wnt signaling pathway. All of these findings clearly suggest that salinomycin could selectively target CSCs; however, the implication of salinomycin in the clinical setting needs further in-depth investigations.

As described in the previous sections, CD133^{+} is a CSC marker commonly found in CSCs from various cancers. These CD133^{+} CSCs are responsible for the initiation, propagation, and recurrence of different types of cancers. Therefore, effective targeting of CD133^{+} tumor cells could efficiently inhibit the growth of cancers. Arsenic trioxide has been used for the treatment of solid tumors, which is believed to be mediated by the induction of apoptosis. It was found that arsenic trioxide could effectively induce apoptosis of enriched CD133^{+} gallbladder cancer cells, which were resistant to conventional chemotherapy. Arsenic trioxide decreased the expression of CD133. Moreover, the ectopic expression of CD133 attenuated the apoptotic effect of Arsenic trioxide in gallbladder cancer cells through the activation of Akt signaling (Ai et al., 2011). Furthermore, studies have also shown that arsenic disulfide together with phosphoinositide 3-kinase inhibitor PI-103 could suppress the growth of acute promyelocytic leukemia (APL) and induce the differentiation of acute myeloid leukemia stem cells. The effect of arsenic disulfide synergized with the phosphoinositide 3-kinase inhibitor PI-103 through elimination of acute myeloid leukemia stem cells by inducing differentiation (Hong et al., 2011), suggesting the selective CSC targeting effect of Arsenic disulfide.

Inhibitors targeting CSC-related signaling were also used to target CSCs. Hedgehog signaling antagonist GDC-0449 (Vismodegib) showed its inhibitory effects on pancreatic CSCs. GDC-0449 inhibits Hedgehog signaling by blocking the activities of Smoothened. Since the Hedgehog signaling is critically involved in the formation and maintenance of CSCs, the inactivation of Hedgehog signaling by Hedgehog signaling antagonist could also suppress CSCs. The GDC-0449 induced apoptosis in pancreatic CSCs with increased Fas and TRAIL expression, and decreased the expression of PDGFRα and Bcl-2. Suppression of cell viability and spheroid formation was also observed in GDC-0449-treated pancreatic CSCs (Singh et al., 2011), suggesting that Hedgehog signaling inhibitor could be useful for targeting pancreatic CSCs; however further in vivo studies are warranted prior to test this concept in clinical setting especially when GDC-0449 is used in combination with conventional chemotherapeutics. Activation of Notch signaling was also observed in CSCs and EMT-type cells. The inactivation of Notch signaling could partially reverse the EMT by downregulation of vimentin, ZEB1, Slug, Snail, and NF-κB, suggesting that the inhibition of Notch signaling by Notch inhibitors could target CSCs and EMT-type cells.

Immunotherapy has also been used in targeting CSCs. In a recent study, $ALDH^{bright}$ cells were isolated by flow cytometry and ALDH1A1-specific $CD8^+$ T cells were generated in vitro. It was found that the $ALDH^{bright}$ cells at low numbers (500 cells) were tumorigenic in SCID mice. Importantly, ALDH1A1-specific $CD8^+$ T cells inhibited tumor growth and metastases, and improved survival of tumor-bearing mice by eliminating $ALDH^{bright}$ cells (Visus et al., 2011), suggesting the potential of ALDH1A1-based immunotherapy by targeting CSCs. EpCAM is another CSC marker, and it has been used in immunotherapy. The results of a recent study have shown that a targeted immunotherapy using EpCAM/CD3-bispecific T-cell-engaging antibody MT110 was able to eliminate pancreatic CSCs. The antibody MT110 could selectively target highly tumorigenic pancreatic CSCs through EpCAM binding and T-cell killing in vitro and in vivo, leading to the inhibition of tumorigenicity of pancreatic CSCs (Cioffi et al., 2012). Another immunotherapy was designed to target $CD24^+/CD44^+/ESA^+$ CSCs. It was found that RON was highly expressed in pancreatic $CD24^+/CD44^+/ESA^+$ CSCs. An anti-RON antibody with doxorubicin (anti-RON antibody Zt/c9-directing doxorubicin-immunoliposomes [Zt/c9-Dox-IL]) was engineered to target $CD24^+/CD44^+/ESA^+$ CSCs (Padhye et al., 2011). This antibody recognized RON and rapidly caused RON internalization, leading to the uptake of liposome-coated doxorubicin. It was found that the antibody significantly decreased viability of pancreatic $CD24^+/CD44^+/ESA^+$ CSCs. These findings open newer avenues for designing novel therapeutic strategies for the treatment of various cancers through targeting CSCs.

Several natural agents have been used for targeting CSCs and EMT-type cells because of the non-toxic and multi-targeting attributes of natural agents. Sulforaphane is a natural agent, which showed its inhibitory effect on CSCs in combination treatment. We have also found that natural agents including isoflavone, 3,3′-diindolylmethane (DIM), and CDF (curcumin analogue with greater bioavailability) could inhibit cancer stem-like cells (CSLC) and EMT-type cells through the regulation of cell signaling and miRNAs (Li et al., 2009b; Bao et al., 2012). Other investigators also reported inhibitory effects of isoflavone on CSCs. In addition, δ12-prostaglandin J3, a novel and naturally produced cyclopentenone prostaglandins (CyPG) from the dietary fish-oil omega-3 fatty acid-derived metabolite, also selectively ablated leukemia stem cells in mice (Hegde et al., 2011). All of these findings demonstrate the potent inhibitory effects of natural agents on CSCs and EMT-type cells; however, further in-depth animal experimentation is warranted prior to the application of this concept in the clinical setting for the elimination of CSCs by natural agents.

Conclusions

Emerging evidence has demonstrated that CSCs exist in various malignant tumors including hematopoietic and solid tumors. The CSCs contribute to drug resistance, tumor progression, and recurrence of malignant tumors. The molecular signaling pathways that are altered in malignant tumors also exist in CSCs; however, Hedgehog, Notch, and Wnt are main signaling pathways that are altered in CSCs because of their roles in self-renewal capacity and tumorigenesis. With the discovery of CSC markers and understanding of the molecular mechanisms for the development and maintenance of CSCs, it is our expectation that novel therapeutic strategies could be designed for targeted elimination of CSCs. This strategy will allow induction of drug sensitivity and inhibition of cancer progression and recurrence. However, the complexities of CSC markers in various cancers and the low sensitivity of these cells to existing therapeutics are challenging for designing novel therapeutic strategy. Nevertheless, emerging rapid growth of investigations, focused on targeting CSCs and EMT-type cells by novel synthetic and natural agents, would likely become novel therapeutic strategies for cancer treatment mediated through the elimination of CSCs and EMT-type cells. To that end, more in vivo animal studies and clinical trials are needed to evaluate these novel agents; however, it is becoming clear that specific targeting of CSCs would lead to the opening of newer avenues for the treatment of human malignancies with better survival outcome.

Acknowledgments

The authors' work cited in this review article was funded by grants from the National Cancer Institute, NIH (R01CA083695, R01CA108535, R01CA132794, R01CA131151, R01CA154321 and R01CA164318 awarded to FHS). We also thank Puschelberg and Guido foundations for their generous contribution to support our research.

References

Ai, Z., Pan, H., Suo, T., Lv, C., Wang, Y., Tong, S., and Liu, H. (2011). Arsenic oxide targets stem cell marker CD133/prominin-1 in gallbladder carcinoma. Cancer Lett. *310*, 181–187.

Arima, Y., Hayashi, N., Hayashi, H., Sasaki, M., Kai, K., Sugihara, E., Abe, E., Yoshida, A., Mikami, S., Nakamura, S., and Saya, H. (2012). Loss of p16 expression is associated with the stem cell characteristics of surface markers and therapeutic resistance in estrogen receptor-negative breast cancer. Int J Cancer *130*, 2568–2579.

Bao, B., Ali, S., Banerjee, S., Wang, Z., Logna, F., Azmi, A.S., Kong, D., Ahmad, A., Li, Y., Padhye, S., and Sarkar, F.H. (2012). Curcumin Analogue CDF Inhibits Pancreatic Tumor Growth by Switching on Suppressor microRNAs and Attenuating EZH2 Expression. Cancer Res *72*, 335–345.

Bao, B., Wang, Z., Ali, S., Kong, D., Li, Y., Ahmad, A., Banerjee, S., Azmi, A.S., Miele, L., and Sarkar, F.H. (2011). Notch-1 induces epithelial-mesenchymal transition consistent with cancer stem cell phenotype in pancreatic cancer cells. Cancer Lett. *307*, 26–36.

Bertolini, G., Roz, L., Perego, P., Tortoreto, M., Fontanella, E., Gatti, L., Pratesi, G., Fabbri, A., Andriani, F., Tinelli, S., Roz, E., Caserini, R., Lo, V.S., Camerini, T., Mariani, L., Delia, D., Calabro, E., Pastorino, U., and Sozzi, G. (2009). Highly tumorigenic lung cancer CD133+ cells display stem-like features and are spared by cisplatin treatment. Proc Natl. Acad. Sci. USA *106*, 16281–16286.

Bodzin, A.S., Wei, Z., Hurtt, R., Gu, T., and Doria, C. (2012). Gefitinib resistance in HCC mahlavu cells: upregulation of CD133 expression, activation of IGF-1R signaling pathway, and enhancement of IGF-1R nuclear translocation. J Cell Physiol *227*, 2947–2952.

Bonnet, D. and Dick, J.E. (1997). Human acute myeloid leukemia is organized as a hierarchy that originates from a primitive hematopoietic cell. Nat. Med *3*, 730–737.

Charafe-Jauffret, E., Ginestier, C., Iovino, F., Tarpin, C., Diebel, M., Esterni, B., Houvenaeghel, G., Extra, J.M., Bertucci, F., Jacquemier, J., Xerri, L., Dontu, G., Stassi, G., Xiao, Y., Barsky, S.H., Birnbaum, D., Viens, P., and Wicha, M.S. (2010). Aldehyde dehydrogenase 1-positive cancer stem cells mediate metastasis and poor clinical outcome in inflammatory breast cancer. Clin Cancer Res *16*, 45–55.

Chen, Y.C., Chen, Y.W., Hsu, H.S., Tseng, L.M., Huang, P.I., Lu, K.H., Chen, D.T., Tai, L.K., Yung, M.C., Chang, S.C., Ku, H.H., Chiou, S.H., and Lo, W.L. (2009). Aldehyde dehydrogenase 1 is a putative marker for cancer stem cells in head and neck squamous cancer. Biochem. Biophys. Res Commun. *385*, 307–313.

Choijamts, B., Jimi, S., Kondo, T., Naganuma, Y., Matsumoto, T., Kuroki, M., Iwasaki, H., and Emoto, M. (2011). CD133+ cancer stem cell-like cells derived from uterine carcinosarcoma (malignant mixed Mullerian tumor). Stem Cells *29*, 1485–1495.

Cioffi, M., Dorado, J., Baeuerle, P.A., and Heeschen, C. (2012). EpCAM/CD3-Bispecific T-cell Engaging Antibody MT110 Eliminates Primary Human Pancreatic Cancer Stem Cells. Clin Cancer Res.

Cortes-Dericks, L., Carboni, G.L., Schmid, R.A., and Karoubi, G. (2010). Putative cancer stem cells in malignant pleural mesothelioma show resistance to cisplatin and pemetrexed. Int J Oncol *37*, 437–444.

Dallas, N.A., Xia, L., Fan, F., Gray, M.J., Gaur, P., van, B.G., Samuel, S., Kim, M.P., Lim, S.J., and Ellis, L.M. (2009). Chemoresistant colorectal cancer cells, the cancer stem cell phenotype, and increased sensitivity to insulin-like growth factor-I receptor inhibition. Cancer Res *69*, 1951–1957.

Gupta, P.B., Onder, T.T., Jiang, G., Tao, K., Kuperwasser, C., Weinberg, R.A., and Lander, E.S. (2009). Identification of selective inhibitors of cancer stem cells by high-throughput screening. Cell *138*, 645–659.

Hegde, S., Kaushal, N., Ravindra, K.C., Chiaro, C., Hafer, K.T., Gandhi, U.H., Thompson, J.T., van den Heuvel, J.P., Kennett, M.J., Hankey, P., Paulson, R.F., and Prabhu, K.S. (2011). Delta12-prostaglandin J3, an omega-3 fatty acid-derived metabolite, selectively ablates leukemia stem cells in mice. Blood *118*, 6909–6919.

Hermann, P.C., Huber, S.L., Herrler, T., Aicher, A., Ellwart, J.W., Guba, M., Bruns, C.J., and Heeschen, C. (2007). Distinct populations of cancer stem cells determine tumor growth and metastatic activity in human pancreatic cancer. Cell Stem Cell *1*, 313–323.

Hong, Z., Xiao, M., Yang, Y., Han, Z., Cao, Y., Li, C., Wu, Y., Gong, Q., Zhou, X., Xu, D., Meng, L., Ma, D., and Zhou, J. (2011). Arsenic disulfide synergizes with the phosphoinositide 3-kinase inhibitor PI-103 to eradicate acute myeloid leukemia stem cells by inducing differentiation. Carcinogenesis *32*, 1550–1558.

Hsu, H.S., Lin, J.H., Huang, W.C., Hsu, T.W., Su, K., Chiou, S.H., Tsai, Y.T., and Hung, S.C. (2011). Chemoresistance of lung cancer stem-like cells depends on activation of Hsp27. Cancer *117*, 1516–1528.

Immervoll, H., Hoem, D., Sakariassen, P.O., Steffensen, O.J., and Molven, A. (2008). Expression of the "stem cell marker" CD133 in pancreas and pancreatic ductal adenocarcinomas. BMC. Cancer *8*, 48.

Justilien, V., Regala, R.P., Tseng, I.C., Walsh, M.P., Batra, J., Radisky, E.S., Murray, N.R., and Fields, A.P. (2012). Matrix metalloproteinase-10 is required for lung cancer stem cell maintenance, tumor initiation and metastatic potential. PLoS. One. *7*, e35040.

Katoh, M. and Katoh, M. (2006). Notch ligand, JAG1, is evolutionarily conserved target of canonical WNT signaling pathway in progenitor cells. Int J Mol Med *17*, 681–685.

Li, C., Lee, C.J., and Simeone, D.M. (2009a). Identification of human pancreatic cancer stem cells. Methods Mol Biol *568*, 161–173.

Li, X., Lewis, M.T., Huang, J., Gutierrez, C., Osborne, C.K., Wu, M.F., Hilsenbeck, S.G., Pavlick, A., Zhang, X., Chamness, G.C., Wong, H., Rosen, J., and Chang, J.C. (2008). Intrinsic resistance of tumorigenic breast cancer cells to chemotherapy. J. Natl. Cancer Inst. *100*, 672–679.

Li, Y., VandenBoom, T.G., Kong, D., Wang, Z., Ali, S., Philip, P.A., and Sarkar, F.H. (2009b). Up-regulation of miR-200 and let-7 by natural agents leads to the reversal of epithelial-to-mesenchymal transition in gemcitabine-resistant pancreatic cancer cells. Cancer Res *69*, 6704–6712.

Liu, G., Yuan, X., Zeng, Z., Tunici, P., Ng, H., Abdulkadir, I.R., Lu, L., Irvin, D., Black, K.L., and Yu, J.S. (2006). Analysis of gene expression and chemoresistance of CD133+ cancer stem cells in glioblastoma. Mol Cancer *5*, 67.

Miele, L. (2006). Notch signaling. Clin. Cancer Res *12*, 1074–1079.

Oak, P.S., Kopp, F., Thakur, C., Ellwart, J.W., Rapp, U.R., Ullrich, A., Wagner, E., Knyazev, P., and Roidl, A. (2012). Combinatorial treatment of mammospheres with trastuzumab and salinomycin efficiently targets HER2-positive cancer cells and cancer stem cells. Int J Cancer.

Padhye, S.S., Guin, S., Yao, H.P., Zhou, Y.Q., Zhang, R., and Wang, M.H. (2011). Sustained Expression of the RON Receptor Tyrosine Kinase by Pancreatic Cancer Stem Cells as a Potential Targeting Moiety for Antibody-Directed Chemotherapeutics. Mol Pharm. *8*, 2310–2319.

Pang, R., Law, W.L., Chu, A.C., Poon, J.T., Lam, C.S., Chow, A.K., Ng, L., Cheung, L.W., Lan, X.R., Lan, H.Y., Tan, V.P., Yau, T.C., Poon, R.T., and Wong, B.C. (2010). A subpopulation of CD26+ cancer stem cells with metastatic capacity in human colorectal cancer. Cell Stem Cell *6*, 603–615.

Salnikov, A.V., Gladkich, J., Moldenhauer, G., Volm, M., Mattern, J., and Herr, I. (2010). CD133 is indicative for a resistance phenotype but does not represent a prognostic marker for survival of non-small cell lung cancer patients. Int J Cancer *126*, 950–958.

Shafee, N., Smith, C.R., Wei, S., Kim, Y., Mills, G.B., Hortobagyi, G.N., Stanbridge, E.J., and Lee, E.Y. (2008). Cancer stem cells contribute to cisplatin resistance in Brca1/p53-mediated mouse mammary tumors. Cancer Res *68*, 3243–3250.

Silva,I.A., Bai,S., McLean,K., Yang,K., Griffith, K–, Thomas, D., Ginestier, C., Johnston, C., Kueck, A., Reynolds, R.K., Wicha, M.S., and Buckanovich, R.J. (2011). Aldehyde dehydrogenase in combination with CD133 defines angiogenic ovarian cancer stem cells that portend poor patient survival. Cancer Res *71*, 3991–4001.

Singh, B.N., Fu, J., Srivastava, R.K., and Shankar, S. (2011). Hedgehog signaling antagonist GDC-0449 (Vismodegib) inhibits pancreatic cancer stem cell characteristics: molecular mechanisms. PLoS. One. *6*, e27306.

Tanei, T., Morimoto, K., Shimazu, K., Kim, S.J., Tanji, Y., Taguchi, T., Tamaki, Y., and Noguchi, S. (2009). Association of breast cancer stem cells identified by aldehyde dehydrogenase 1 expression with resistance to sequential Paclitaxel and epirubicin-based chemotherapy for breast cancers. Clin Cancer Res *15*, 4234–4241.

Tang, Q.L., Zhao, Z.Q., Li, J.C., Liang, Y., Yin, J.Q., Zou, C.Y., Xie, X.B., Zeng, Y.X., Shen, J.N., Kang, T., and Wang, J. (2011). Salinomycin inhibits osteosarcoma by targeting its tumor stem cells. Cancer Lett. *311*, 113–121.

Todaro, M., Alea, M.P., Di Stefano, A.B., Cammareri, P., Vermeulen, L., Iovino, F., Tripodo, C., Russo, A., Gulotta, G., Medema, J.P., and Stassi, G. (2007). Colon cancer stem cells dictate tumor growth and resist cell death by production of interleukin-4. Cell Stem Cell *1*, 389–402.

Visus, C., Wang, Y., Lozano-Leon, A., Ferris, R.L., Silver, S., Szczepanski, M.J., Brand, R.E., Ferrone, C.R., Whiteside, T.L., Ferrone, S., Deleo, A.B., and Wang, X. (2011). Targeting ALDH(bright) human carcinoma-initiating cells with ALDH1A1-specific CD8(+) T cells. Clin Cancer Res *17*, 6174–6184.

Wei,H.J., Yin,T., Zhu,Z., Shi,P.F., Tian, Y., and Wang, C.Y. (2011). Expression of CD44, CD24 and ESA in pancreatic adenocarcinoma cell lines varies with local microenvironment. Hepatobiliary. Pancreat. Dis. Int *10*, 428–434.

Yin, T., Wei, H., Gou, S., Shi, P., Yang, Z., Zhao, G., and Wang, C. (2011). Cancer stem-like cells enriched in panc-1 spheres possess increased migration ability and resistance to gemcitabine. Int J Mol Sci. *12*, 1595–1604.

Yu, F., Li, J., Chen, H., Fu, J., Ray, S., Huang, S., Zheng, H., and Ai, W. (2011). Kruppel-like factor 4 (KLF4) is required for maintenance of breast cancer stem cells and for cell migration and invasion. Oncogene *30*, 2161–2172.

Zhang, G.N., Liang, Y., Zhou, L.J., Chen, S.P., Chen, G., Zhang, T.P., Kang, T., and Zhao, Y.P. (2011). Combination of salinomycin and gemcitabine eliminates pancreatic cancer cells. Cancer Lett. *313*, 137–144.

Zhu, Y., Yu, F., Jiao, Y., Feng, J., Tang, W., Yao, H., Gong, C., Chen, J., Su, F., Zhang, Y., and Song, E. (2011). Reduced miR-128 in breast tumor-initiating cells induces chemotherapeutic resistance via Bmi-1 and ABCC5. Clin Cancer Res *17*, 7105–7115.

Zhu, Z., Hao, X., Yan, M., Yao, M., Ge, C., Gu, J., and Li, J. (2010). Cancer stem/progenitor cells are highly enriched in CD133+CD44+ population in hepatocellular carcinoma. Int J Cancer *126*, 2067–2078.

Chapter 35

The Role of Cancer Stem Cells in Tumor Radioresistance

I. Kurth*, C. Peitzsch*, M. Baumann, and A. Dubrovska
OncoRay, National Center for Radiation Research in Oncology, Faculty of Medicine Carl Gustav Carus, Dresden, Germany
* *Equal Contribution*

Abbreviations

ALDH	aldehyde dehydrogenase
ATM	ataxia telangiectasia mutated protein kinase
ATR	Ataxia telangiectasia and rad-3-related protein kinase
CD	cluster of differentiation
CHK1/2	checkpoint protein kinase 1 and 2
CSC	cancer stem cell
CXCR4	C-X-C chemokine receptor type 4
CXCL12	chemokine (C-X-C motif) ligand 12
DSB	double strand break
ECM	extracellular matrix
EGF	epidermal growth factor
FGF	fibroblast growth factor
HIF	hypoxia-inducible factor
HNSCC	head and neck squamous cell carcinoma
HR	homology-directed recombination
IFC	immunofluorescence
IHC	immunohistochemistry
IL	interleukin
JAK/STAT	Janus kinase signal transducer and activator of transcription
MMP	matrix metalloprotease
NF-kB	nuclear factor kappa-light-chain-enhancer of activated B cells
NHEJ	non-homologous end joining
NSCLC	non-small cell lung cancer
PI3K	phosphoinositide 3-kinase
PDGF	plateled derived growth factor

(*Continued*)

Cancer Stem Cells, First Edition. Edited by Vinagolu K. Rajasekhar.

(*Continued*)

ROS	reactive oxygen species
RT-PCR	reverse transcription polymerase chain reaction
SF2	surviving fraction at 2Gy
SHH	sonic hedgehog
TCD_{50}	tumor control dose 50%
TD_{50}	median transplantation dose
TGFβ	transforming growth factor β
TMA	tissue microarray
VEGF	vascular endothelial growth factor
WNT	wingless-type MMTV integration site family

Cancer Stem Cell Response to Radiotherapy

Radiotherapy is the backbone approach of the treatment of many types of cancer and is typically given in multiple fractions to optimize the therapeutic ratio between tumor and normal tissue responses. In many cases radiotherapy or radiochemotherapy can completely destroy the primary tumor, i.e. obtain local control, without clinically important damage to surrounding normal tissues. Nevertheless, particularly in larger tumors or in tumors located close to critical normal tissues, local tumor control is often impeded by tumor radioresistance. Accelerated tumor cell repopulation, tumor hypoxia, or inherent radioresistance of tumor cells have been identified as major reasons for the failure of conventional fractionated course of radiotherapy (Withers, 1975; Kim and Tannock, 2005; Harada, 2011; Weichselbaum et al., 1985). Furthermore, radiobiological studies demonstrated that radioresistance of experimental tumors correlates with their transplantability in vivo and with the tumor volume or number of tumor cells (Suit et al., 1965; Baumann et al., 1990; Hill and Milas, 1989; Dubben et al., 1998; Bentzen and Thames, 1998; Johnson et al., 1995). As suggested by preclinical studies and clinical observation, the dose necessary to control irradiated tumors increases with tumor volume that can be explained by an increase of the number of CSCs as the size of tumor increases. Taken together, these data suggest that tumor recurrence is caused by surviving CSCs (Baumann et al., 2008). Based on these findings, the radioresistant phenotype of tumors has been attributed to the cancer stem cell compartment, and is considered to be an inherent property of CSCs. Resistance to radiation therapy has been reported to be a defining characteristic of CSCs from various tumor types, including glioma, medulloblastoma, breast, colon, and atypical teratoid/rhabdoid tumors. CSC markers were linked with poor clinical outcome of head and neck squamous cell carcinomas (HNSCC), glioma, uterine cervical squamous cell carcinoma, and rectal carcinoma patients (Koukourakis et al., 2012; Pallini et al., 2008; de Jong et al., 2011; Kwon et al., 2007; Metellus et al., 2011; Saigusa et al., 2009) (Table 35.1). Experimental evidence suggests that CSCs are not uniform in their response to radiation, and differences may exist between the intrinsic radioresistance of CSCs in tumors even of the same histopatological type (Hill and Milas, 1989; Gerweck et al., 1994; Taghian et al., 1995, Yaromina et al., 2007). Many intrinsic determinants were attributed to the radioresistance of CSCs including quiescence, activated radiation response mechanisms, increase in intracellular defense against reactive oxygen species (ROS), and activation of cell survival pathways (Table 35.2). Within the tumor, physiological conditions, including hypoxia and interaction with stromal elements, can also modulate the radiosensitivity of CSCs (Figure 35.1).

Table 35.1. Cancer stem cell markers related to radiotherapy or radio-chemotherapy outcome (retrospective clinical studies)

Cancer	Treatment	Markers	Method	Clinical Parameters	Reference
Glioblastoma	Complete or partial tumorectomy followed by radiotherapy combined with temozolomide treatment	CD133/ Ki67	IHC, IFC	Shorter overall and progression-free survival	Pallini et al., 2008
Glioblastoma	Complete or partial tumorectomy followed by radiotherapy combined with temozolomide treatment	CD133	IHC	Shorter overall and progression-free survival	Metellus et al., 2011
Larynx cancer	Radiotherapy	CD44	RT-PCR	Local relapse	De Jong et al., 2010
HNSCC	Accelerated platinum-based radiotherapy	CD44, Oct4, integrin β1	IHC	Local relapse, Distant metastasis, Shorter overall survival	Koukourakis et al., 2012
Uterine cervical squamous cell carcinoma	Radiotherapy alone or with concurrent chemoradiotherapy	CD24	IHC	Distant metastasis	Kwon et al., 2007
Rectal carcinoma	5-fluorouracil (5FU) and tegafur-uracil (UFT) with concurrent radiotherapy followed by resection	CD133, OCT4, SOX2	RT-PCR	Shorter progression free survival, Distant metastasis	Saigusa et al., 2009

immunohistochemistry (ICH); immunofluorescence (IFC); reverse-transcription polymerase chain reaction (RT-PCT).

Table 35.2. Intrinsic determinants of cancer stem cell radioresistance

Mechanism	Assays	CSC Population	Reference
Enhanced DNA repair capability	Measurement of the kinetic of γH2AX foci, DNA comet assay, CHK1 and CHK2 phosphorylation, NHEJ assay, ATM phosphorylation	CD133$^+$ human glioma, Lin$^-$CD29^{high}CD24^{high} mouse mammary tumor, CD44$^+$CD24$^{-/low}$ human breast cancer	Phillips et al., 2006 Bao et al., 2006 Zhang et al., 2007 Phillips et al., 2006 Bao et al., 2006; Ropolo et al., 2009 Yin and Glass, 2011
Protection from DNA damage	Measurement of ROS level	CD44$^+$CD24$^{-/low}$ human breast cancer, Thy1$^+$CD24$^+$ mouse mammary tumor, CD44$^+$ human gastric cancer	Diehn et al., 2009 Ishimoto et al., 2011
Activation of cell survival pathways	Analysis of the activation of PI3K/Akt, WNT/β-catenin, Notch signaling pathways	CD133$^+$ human glioma, Lin$^-$CD29^{high}CD24^{high} mouse mammary tumor	Wang et al., 2010 Zhang et al., 2010

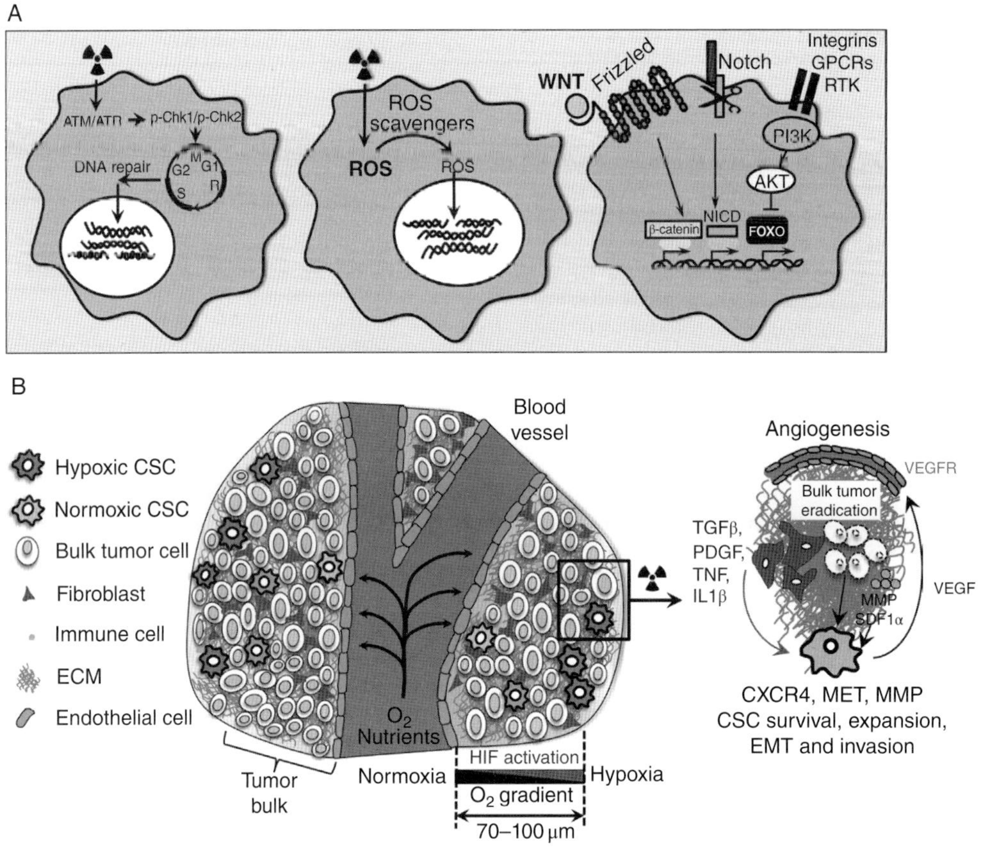

Figure 35.1. The mechanisms underlying CSC radioresistance. **A**. Intrinsic cancer stem cells radioresistance is attributed to a more efficient activation of the DNA damage response, intracellular defense against reactive oxygen species and activation of the cell survival pathways, including PI3K/AKT, WNT/β catenin, and Notch. **B**. The putative niche may protect CSCs from radiation damage. Eradication of the bulk tumor cells by conventional therapy could induce accelerated repopulation of CSCs. Hypoxic environment activates HIF signaling pathway and promotes CSC expansion and radioresistance. Ionizing radiation induces the production of the growth factors and cytokines, such as PDGF, IL1β, TNF, TGF β by irradiated fibroblasts. Irradiated immune cells, such as tumor-associated macrophages, produce CXCL12 and matrix metalloproteases (MMP) and create a microenvironment that promotes the survival of tumor and endothelial cells. In summary, an irradiated tumor bed promotes carcinogenesis and accelerates the development of highly invasive CSC phenotypes. For color detail, please see color plate section.

Retrospective clinical studies suggest that estimation of the number of CSCs in pre-therapeutic biopsies might be important for prediction of clinical outcome and selection of the optimal treatment strategy (Koukourakis et al., 2012; de Jong et al., 2011; Baumann and Krause, 2010). However, the radioresistant properties of CSCs may be dynamic in nature and can be differently employed throughout the course of radiotherapy. Therefore, various therapeutic strategies and diverse predictive biomarkers may be required at the distinct stages of tumor treatment (Baumann et al., 2008; Begg et al., 2011; Brunner et al., 2012; Koch et al., 2010). Understanding the molecular pathways regulating CSC properties during the course of radiotherapy is essential for optimization and individualization of treatment strategies.

Determinants of Cancer Stem Cell Radioresistance

Intrinsic Survival Mechanisms

The concept of the "4R's" to characterize the mechanisms of tumor resistance to fractionated radiotherapy was summarized by Withers in 1975. These "4R" parameters include (i) **r**epair, (ii) **r**epopulation, (iii) **r**edistribution, and (iv) **r**eoxygenation, and they define subpopulations of tumor cells surviving radiotherapy (i.e., quiescent in the setting of the developed tumor, fast repopulating during the course of radiotherapy, hypoxic, and having high potential of recovery or repair after sublethal damage). These radiobiological processes can be distinct for different tumor cell populations. However, they can determine tumor curability only if attributed to the tumorigenic, i.e., CSC population (Baumann et al., 2008). A growing body of experimental evidence suggests a number of different intrinsic and extrinsic adaptations that confer tumor radioresistance and may occur in CSC populations (Table 35.2).

Radiation-induced cell death is mediated through induction of DNA lesions, which are lethal to cells if not repaired (Seymour et al., 1986). Recent studies demonstrated that CSCs have accelerated resistance to radiation-induced DNA damage. This resistance can be explained by more efficient scavenging of reactive oxygen species (ROS), activation of the radiation-induced DNA damage response, and enhanced DNA repair capability. Transcriptional profiling of murine mammary gland CSCs revealed an increased expression of many genes associated with DNA damage response and DNA reparation, e.g., *Nek1, Brca1, Chek1, Hus1, Ung, Xrcc5, Sfpq, and Uhrf1* (Zhang et al., 2008), suggesting that mammary gland CSCs might be more resistant to radiotherapy than bulk tumor cells.

One of the potential mechanisms of radioresistance of CSC population is the intracellular defense against ROS. Radiation-induced excessive production of ROS, which damages cells by interactions with critical macromolecules including DNA, lipids, and proteins, leading to cell death (Cook et al., 2004). Recent studies showed that subsets of CSCs in some tumors, including breast and gastrointestinal carcinomas, have lower levels of ROS and enhanced levels of ROS scavengers compared to the non-tumorigenic populations. In support of the role of ROS scavenging in CSC radioprotection, pharmacological depletion of ROS scavengers in tumor progenitors markedly decreased their radioresistance (Diehn et al., 2009, Ishimoto et al., 2011). Ishimoto and colleagues (2011) showed that expression of the putative stem cell marker CD44 correlates with ROS resistance in cancer cells as a result of the increased synthesis of the cellular ROS scavenger glutathione. This study demonstrated that expression of CD44 might serve as a functional marker for ROS-resistant tumor progenitor cells (Ishimoto et al., 2011). These findings are consistent with previous studies and showed that overexpression of CD44 in tumors correlates with an increased capacity of resistance to radiation therapy and an increased risk of local recurrence (Situ et al., 2010; de Jong et al., 2010; Kokko et al., 2011).

Another important mechanism of CSC radioprotection is their high DNA repair capabilities (Figure 35.1). Tumor and normal cells can repair radiation-induced DNA double strand breaks (DSB) by either homology-directed recombination (HR) or non-homologous end joining (NHEJ) (Thompson et al., 2001; Karran et al., 2000; van Gent et al., 2001; Kasten-Pisula et al., 2005; Jeggo and Löbrich, 2006; Mladenov and Iliakis, 2011). HR removes DNA damage in an error-free manner, using the undamaged sister chromatid as a template, and is performed by the Rad52 group of proteins including Rad50, Rad51, Rad52,

Rad54, and MRE11. One of the best-characterized chromatin modification events in DNA DSB responses is histone H2A.X phosphorylation by ATM or ATR serine/threonine protein kinase. The phosphorylated form of H2A.X (γ-H2A.X) forms nuclear foci on the sites of DNA damage. Radiation induces significantly more Rad51 and less γ-H2A.X foci in human breast CSCs, and γ-H2A.X foci resolved faster in murine mammary gland CSCs than in non-CSC population that is perhaps reflective of more efficient DSB repair in CSCs as well as high capacity of CSCs for HR (Al-Assar, 2011; Diehn, 2006; Woodward, 2007). The $CD44^+CD24^{-/low}$ subsets from breast cancer cell lines and primary human breast cancer cells showed increased resistance to radiation compared to the $CD44^-CD24^+$ subset attributed to activation of ATM signaling. Application of an ATM inhibitor restored radiosensitivity of the $CD44^+CD24^{-/low}$ cell population, suggesting that targeting ATM signaling may provide a new tool to eradicate radioresistant breast cancer cell populations (Yin and Glass, 2011).

Important modulators of cell response to radiotherapy are the checkpoint kinase 1 and 2 (CHK1/2), which become activated after genotoxic stress and arrest cell cycle progression to allow DNA reparation. CHK1 is preferentially activated in $CD133^+$ glioblastoma progenitor cells, but not in $CD133^-$ cells in response to DNA damage caused by ionizing radiation. Moreover, radioresponsiveness of $CD133^+$ brain tumor stem cells can be restored by pharmacological inhibition of CHK1/2 (Bao et al., 2006; Ropolo et al., 2009).

Resistance to radiotherapy and accelerated repopulation of CSCs after or during radiotherapy treatment might be also conferred to the activation of the cell survival pathways, including phosphoinositide 3-kinase (PI3K), canonical wingless-type MMTV integration site family (WNT), and Notch signaling pathways (Zhang et al., 2007; Wang et al., 2012; Martelli et al., 2011). In support of this hypothesis, several groups have reported an increase in CSC phenotype after a clinically relevant dose of fractionated radiation (Phillips et al., 2006; Woodward et al., 2007; Bao et al., 2006). Accelerated growth of surviving tumor cells between irradiation fractions is a critical determinant of local tumor control and one of the key causes of treatment failure (Kim and Tannock, 2005; Baumann et al., 2003; Petersen et al., 2001; Bauman and Gregoire, 2009).

It is important to note that the experimental data may depend on the selected markers, bringing some controversy to the issue. For example, despite the fact that Bao and colleagues (2006) demonstrated more efficient DNA reparation in $CD133^+$ glioma cells, that was challenged by Ropolo and coworkers (2009) who showed that neither DNA base excision or single-strand break repair nor resolving of γ-H2A.X nuclear foci were changed in $CD133^+$ compared to $CD133^-$ cells. Currently, it is generally accepted that no single marker is absolutely informative, and the use of CD133 expression as a marker for brain tumor stem cells is intensively disputed (Prestegarden and Enger, 2010; Grosse-Gehling et al., 2012).

This controversy could also suggest that intrinsic radioresistance of CSCs is oversimplified because of their plasticity and heterogeneity. Thereby, the radioresistant properties and radioprotective mechanisms of CSCs may not only reflect the inherent features but rather the characteristics of the marker-defined population at that particular time point. In addition to this, the radioresistant properties of CSCs may be quite dynamic depending on extrinsic signals from the CSC niche, including access to the nutrients and oxygen and direct interactions between different cell types, as well as between cells and the extracellular proteins.

Extrinsic Survival Mechanisms

Beside the previously described intrinsic mechanisms of radioresistance, another important factor contributing to resistance of CSCs to radiotherapy is the protection by the microenvironmental niche. Solid tumors resemble an abnormal organ-like entity and are comprised of hierarchically organized malignant and non-transformed cells embedded in the extracellular matrix (ECM). Within this tumor mass, CSC populations are enriched in specific anatomic locations defined by various factors such as specific paracrine and autocrine signaling, direct contact with surrounding stromal fibroblasts, immune cells, endothelial cells, and components of the ECM (e.g., fibronectin, laminin, proteoglycans, and collagen fibers) as well as gas and nutrient supply, and tissue acidification and bioenergetic status (Borovski et al., 2011). The concept of a CSC niche was derived from the similarities between normal stem cells and CSCs. However, in contrast to normal stem cell niches that are characterized by fine-tuned interactive microenvironment and specific localization within the tissue, CSC niches have more disordered interactions and could occupy different areas, including hypoxic, perivascular, and invasive regions, that could be dynamically changed as the disease progresses (Lander et al., 2012) (Figure 35.1). It can be hypothesized that CSCs in the distinct niches may exist in the different stages. The hypoxic conditions may define a quiescent radioresistant CSC subset, while the more oxygenated environment defines activated CSCs enabled to proliferate and migrate. In support of this hypothesis, preclinical and clinical data demonstrated that a CSC population can be protected from the effect of radiation by hypoxic conditions (Li et al., 2009; Mohyeldin et al., 2010; Li and Rich, 2010).

The crucial role of oxygen in cell radiosensitizing has been extensively studied during the last century since 1909, when Gottwald Schwarz reported that a reduction of blood flow protected the human skin from X-ray-induced damage (Schwarz, 1909). More recent studies confirmed that the oxygenation status of cancer cells determined by tumor blood flow is an important determinant of tumor radiocurability. Preclinical and clinical data demonstrated that tumor cells can be protected from the effect of radiation by hypoxia and determination of microenvironmental parameters such as tumor hypoxic fraction, vasculature and perfusion may have a prognostic value for response to radiotherapy (Overgaard, 2011; Yaromina et al., 2011; Zips et al., 2011; Eschmann et al., 2005). Although the mechanisms of the oxygen effect on cell radiosensitivity are not completely understood, it is widely assumed that it is mediated by the production of free radical species with high stability and long lifetime (ROS). The transfer of the free radicals to the biological molecules such as DNA, proteins, and lipids can break chemical bonds and inactivate key molecular functions (Harada, 2011; Mikkelsen and Wardman, 2003). Some hopeful strategies have been developed to overcome hypoxia including direct delivery of molecular oxygen to hypoxic regions, mimicking the effect of oxygenation by using radiosensitizing drugs such as nitroimidazole derivatives, and reoxygenation during fractionated radiation therapy (Overgaard, 2007; Ansiaux et al., 2005; Harada et al., 2011). Moreover, hypoxia is a critical microenvironmental factor in regulating CSC self-renewal and differentiation (Hill et al., 2009; Li and Rich, 2010). Low oxygen tension promotes expansion of glioma and colon CSCs and converts nonstem cancer cells into a CSC-like status with increased self-renewal capacity (Zhang et al., 2008; Heddleston et al., 2009; Soeda et al., 2009). The effects of hypoxia on CSCs are mediated at least in part through the activation of the hypoxia-inducible factor (HIF) signaling pathway (Li and Rich, 2010). Li and colleagues (2009) found that HIF1α is expressed in both CSCs and non-CSCs upon induction of hypoxia. In contrast, HIF2α is only induced in glioblastoma and neuroblastoma CSCs and loss of HIF2α leads to decrease in CSC self-renewal and tumorigenicity (Li et al., 2009).

Low tumor oxygenation and other signals from the tumor microenvironment such as growth factors collaborate to promote epithelial-mesenchymal transition (EMT) associated with high invasiveness and radioresistance (Hill et al., 2009). In hypoxia, the lack of oxygen leads to the HIF1α-dependent activation of the metastasis-associated marker CXCR4, a chemokine receptor involved in CSC trafficking to distant organs (Ishikawa et al., 2009; Zagzag at al., 2006). The overlapping signaling mechanisms that govern tumor radioresistance and invasiveness could explain why in some cases cancer relapsing after radiotherapy develops into more aggressive tumors with the potential to metastasize, which are difficult to treat and associated with poor clinical prognosis (Fagundes et al., 1992; Anderson et al., 1981; Fuks et al., 1991).

However, accumulating experimental evidence indicates that, in addition to the inherent cancer cell properties, the irradiated tumor bed may also contribute to such aggressive phenotype of the recurrent tumors. Recent studies revealed multiple signaling mechanisms that promote tumor growth in the irradiated microenvironments through formation of hypoxia, induction of angiogenesis, activation and differentiation of stromal cells, and recruitment of immune cells leading to tumor growth, invasion, and metastasis (Nguyen et al., 2011; Sofia Vala et al., 2010; Kuonen et al., 2012). It has been shown that the irradiated tumor microenvironment may provide various growth factors such as transforming growth factor β (TGFβ), chemokine CXCL12, KIT ligand (stem cell factor/SCF), plateled derived growth factor (PDGF), epidermal growth factor (EGF), vascular endothelial growth factor (VEGF), fibroblast growth factor (FGF), interleukin (IL) 1β, which induces radioprotection and dissemination of tumor cells (Kuonen et al., 2012; Chung et al., 2006; Sofia Vala et al., 2010) (Figure 35.1). These findings highlight the importance of the microenvironment in modulating tumor radiocurability and open new opportunities for treatment of tumor recurrence after radiotherapy.

Clinical Relevance of Cancer Stem Cell radioresistance: the Promises and Challenges for Translation Studies

Given all above mentioned unique functional properties, CSC hypothesis provides a potential explanation of discordance between response of the bulk tumor cells and permanent local tumor control. The impact of the number of CSCs within a given tumor on local tumor control after radiotherapy was first demonstrated more than 30 years ago and has been shown by numerous laboratories (Suit et al., 1965; Baumann et al., 1990; Hill and Milas, 1989; Dubben et al., 1998; Bentzen and Thames, 1998; Johnson et al., 1995). The studies by Hill and Milas (1989) also indicate that in tumor models, tumor transplantability (TD_{50}) inversely correlate with tumor cure rate (TCD_{50}) after single dose irradiation under homogeneous hypoxia.

Nevertheless, some tumor models with equal TD_{50} show significant discordance in TCD_{50} values, suggesting that not only the absolute number of CSCs, but also their intrinsic radiosensitivity may impact local tumor control after irradiation (Hill and Milas, 1989; Baumann et al., 2008). Moreover, CSC radiosensitivity features may be dynamic in nature and can be differently employed throughout the course of radiotherapy. For instance, hypoxia and quiescence that convert CSCs into a radioresistant state could play an important role at the initial stage of radiotherapy followed by reoxygenation and recruitment of CSCs into the proliferation pool (Hittelman et al., 2010). Accelerated tumor repopulation is one of the major reasons for the failure of conventional fractionated course of radiotherapy. Similarly to normal stem cells, CSCs may undergo a switch from asymmetric to symmetric type of cell division after ionizing radiation, employing activation of the developmental signaling pathways like Notch, Wnt, and Sonic hedgehog (SHH) (Pajonk et al., 2010). Indeed, accelerated radiotherapy resulted in improved treatment efficacy, as shown in randomized trials, for example, in HNSCC and non-small cell lung cancer (NSCLC) patients (Bentzen and Baumann, 2002; Baumann et al., 2011; Tao et al., 2010; Mauguen et al., 2012; Fu et al., 2000, Bauman and Gregoire, 2009; Bourhis et al., 2006).

Because repopulation of surviving tumor cells between irradiation fractions is one of the major causes of radiotherapy failure, especially if prolonged radiotherapy schedules are applied, it may be suggested that CSCs undergoing rapid repopulation in response to irradiation, and could be more effectively targeted by accelerated radiotherapy schedules (Suwinski et al., 1998; Bauman and Gregoire, 2009). Nevertheless, eradication of CSCs by accelerated radiotherapy may be challenged by high intrinsic radioresistance and dynamic nature of their proliferative status, as well as by radiation effects on the tumor microenvironment (Figure 35.1B).

A strong clinical rationale necessitates the identification of CSC specific markers to be able to predict a group of patients with maximal benefit from radiotherapy. During the last decades, a large number of CSC-specific antigens have been characterized. Analysis of the efficacy of radiotherapy and radio-chemotherapy showed that an extensive expression of the stem cell markers is associated with poor prognosis for HNSCC, glioma, uterine cervical squamous cell carcinoma, and rectal carcinoma patients (Koukourakis et al., 2012; Pallini et al., 2008; de Jong et al., 2011; Kwon et al., 2007; Metellus et al., 2011; Saigusa et al., 2009) (Table 35.1). Proliferating glioblastoma CSCs with $CD133^+/Ki67^+$ phenotype were found to represent a highly significant prognostic factor for poor clinical outcome, which may be interpreted as an effect of radiation-induced repopulation of CSCs (Pallini et al., 2008). In a study by Koukourakis and colleagues (2012), an extensive expression of CSC markers including integrin-β1, Oct4, and CD44 by tumor cells was associated with reduced radiotherapy efficacy in HNSCC patients. These results are in accordance with a previous clinical study, which defined CD44 expression as the most significant factor that correlates with response to radiotherapy in early stage larynx cancer patients (de Jong et al., 2010).

An important finding that emerged from the study by Koukourakis and coauthors (2012) is that some tumors have extensive expression of putative stem cell markers that could not be restricted only to stem cell compartment. One of the explanations is that such stem cell markers are also functional proteins that may play a role in cell metabolism, like ALDH or cell motility, such as CD44 and integrins. The induced expression of these CSC markers may occur as a stress response under hypoxic or acidic conditions, which resemble important environmental changes in many solid cancers. Moreover, because the extensive expression of these proteins is important for diverse biological processes such as cell survival, proliferation, and migration, it may have a radioprotective effect not only restricted to the CSC compartment. Therefore, the association between these proteins and the clinical outcome can be explained by their radioprotective effect on the entire tumor cell population rather than on the CSCs themselves.

Nevertheless, despite the high inter-tumor heterogeneity of CSC marker expression, the current findings suggest that analysis of the number of CSCs in pre-therapeutic biopsies may prove important for the prediction of tumor radiocurability, estimation of the total radiation dose required, and the selection of the optimal therapeutic strategy (Baumann et al., 2008; Baumann and Krause, 2010). Therefore, the identification of new biomarkers for the early detection of CSCs is essential for the optimization of treatment strategy and individualization of radiotherapy.

As suggested by current preclinical and clinical observations, CSCs may play a role in tumor resistance to radiotherapy, and specific targeting of CSCs may render cancer treatment more efficient and less toxic to the patients. Currently, a number of drugs and biological substances are developed that specifically inhibit the developmental signaling pathways Notch, WNT, and SHH and the oncogenic pathways PI3K/Akt, NF-kB, and JAK/STAT that are also strongly implicated as CSC regulators (Shigdar et al., 2012; Martelli et al., 2011). Currently, some target-specific drugs against these signaling pathways are in clinical trials for cancers, and more are in the pipeline awaiting approval to enter human testing, but their efficacy against CSC function in cancer patients remains to be determined.

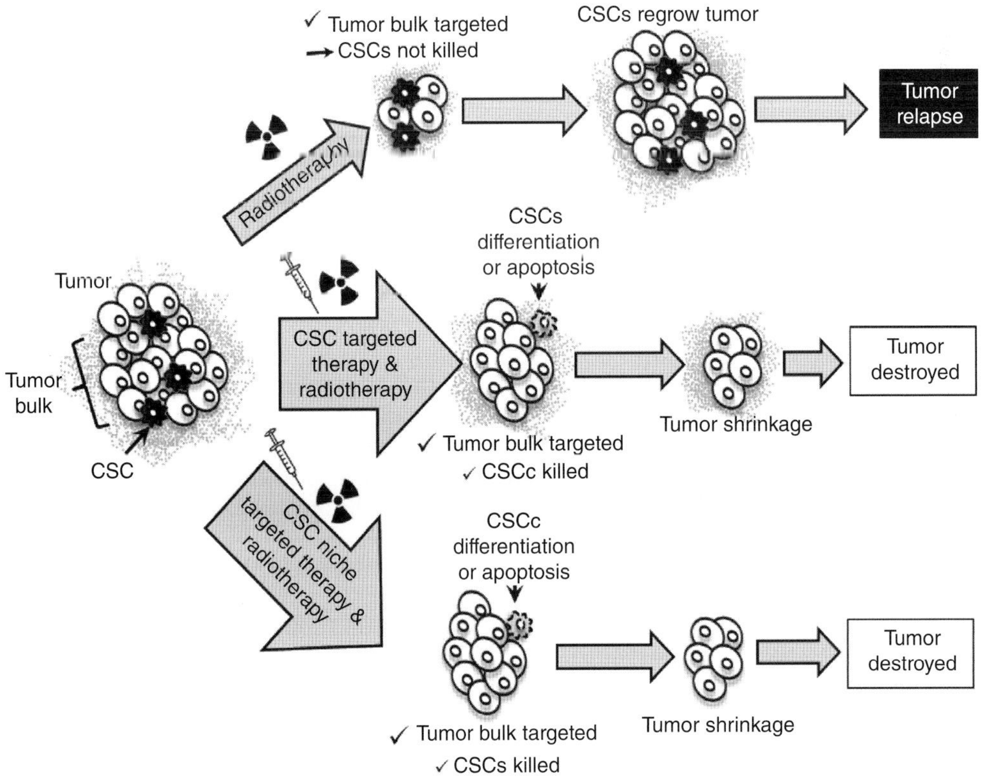

Figure 35.2. Cancer stem cells and radiotherapy. Standard radiotherapy fails to eradicate CSCs that results in tumor regrowth and disease relapse. However, combination of radiotherapy with administration of therapeutic agents targeting CSCs or their niche may eliminate radioresistant population and lead to tumor regression.

The current findings support a hypothesis that targeted therapies aiming to eliminate CSC population or to induce differentiation of progenitor cells to a more radioresistant cell population can provide a lasting clinical response. These treatment modalities may be especially beneficial in aggressive radioresistant cancers with a high percentage of CSCs.

Accumulating experimental evidence suggests that combinatorial strategies based on bulk tumor reduction and CSC specific pathway inhibition offer a promising treatment modality and are predicted to have greater efficacy in tumor reduction and prevention of relapse than monotherapies (Baumann et al., 2008; Linkous and Yazlovitskaya, 2012; Shigdar et al., 2012) (Figure 35.2). Therefore, radiation therapy is expected to undergo biological refinement by combination with agents that specifically target or radiosensitize CSC population.

However, given the fact that CSCs share features with normal stem cells, it remains challenging to identify a new drug that is specific for CSCs only. A major factor that can prevent a successful introduction of CSC-targeting therapy into the clinic is a potential lack of specificity and toxic side effects on the stem cell compartment in normal tissues while targeting CSCs. This may be especially important when this treatment is combined with radiotherapy associated with depletion of normal tissue progenitors, to avoid increased radiation toxicity.

Targeting the tumor microenvironment can be another promising strategy for the effective control of CSCs. Preclinical and clinical data demonstrated that hypoxic

CSC niches directly protect CSC by lack of oxygen and affect CSC maintenance and expansion through the activation of HIF signaling pathway. Expression of HIF2α is shown to be specific for glioblastoma CSCs and not a general stem cell feature (Li et al., 2009). Inhibition of such CSC-specific targets may have important clinical application and therapeutic translation.

Another important aspect, which is critical for the clinical implementation of CSC targeted therapies, is the investigation of sensitive molecular modalities for in vivo CSC imaging. Non-invasive methods of CSC-specific imaging will enable more accurate staging, treatment planning, and evaluation of the effectiveness of CSC-targeted therapies in vivo (Brunner et al., 2012; Shigdar et al., 2012). Currently, no CSC-specific imaging is available in the clinic; however, the preclinical studies in animal models demonstrated the ability to detect specific subset of $CXCR4^+$ and $CD133^+$ cells using ^{64}Cu-labeled peptide antagonist or fluorophore conjugated antibody, respectively (Jacobson et al., 2011; Tsurumi et al., 2010). However, high accumulation in the healthy organs and nonspecific binding to the blood cells may impede the use of these imaging tools in the clinic. Emphasis should be placed on the improvement of spatial resolution and sensitivity of the imaging modalities because of the low abundance of CSC in the tumor mass.

A better understanding of the molecular mechanisms underlying CSC pathology will be essential in the optimal development of CSC-targeted therapies for the successful eradication of cancer.

Validation Studies: Cancer Stem Cell-Based Radiobiological Assays In Vitro and in Preclinical Models

Clonogenic cell survival assays remain one of the gold standard methods to assess tumor cell survival after irradiation. This assay is based on the ability of a single cell (clonogen) to generate a colony of greater than 50 cells that is equivalent to five to six cell divisions. Results of this assay reflect the reproductive survival of the committed progenitor cells and CSCs after radiation treatment (Baumann et al., 2009). In the early 1990s, a few translational studies demonstrated that clonogenic assays ex vivo based on the pretreated tumor biopsies may predict clonogenic survival in vivo following fractionated radiotherapy treatment (West et al., 1993; Girinsky et al., 1993). However, these findings were not supported in a number of other studies (Brock et al., 1990; Allalunis Turner et al., 1992; Eschwege et al., 1997; Stausbol-Gron and Overgaard, 1999). This controversy could be due, in part to the lack of the microenvironment-derived signals that regulate CSC properties and cannot be offered in vitro.

Indeed, a growing body of evidence documents that the tumor microenvironment and direct interactions between cells as well as between cells and the ECM may regulate CSC fate and edit CSC phenotype and properties, including sensitivity to radiation and cytotoxic drugs (Medema and Vermeulen, 2011; Mohyeldin et al., 2010). Conventional cell culture on two-dimensional (2D) substrates utilized in clonogenic survival assay roughly resembles the environment that tumor cells experience in living tissues. In contrast, three-dimensional (3D) culture models that mimic tumor environment allow CSC expansion in vitro and provide an important addition to both complex in vivo models and 2D approaches with its spatial limitations. Tumor cells grown under 3D conditions form tumor-like structures called spheres or spheroids that more accurately reflect in vivo mechanisms of action of the treatment modalities and the diffusion properties of nutrients and oxygen than cell monolayers. An effect of 3D growth conditions on cell radioresistance was first demonstrated in early 1970s. Sutherland and colleagues (1970) found that cell-cell interaction in spheroid cultures dramatically increased cell survival after irradiation (Sutherland at al., 1970; Sutherland at al. 1971). Since then, this finding has quickly gained recognition as an important tool in proving the efficacy of treatment techniques (Friedrich et al., 2007). Over the last decade,

3D cell culture models were modified by using an extract of basement membrane proteins, such as Matrigel, for 3D cell culture, where inter-cell interactions are complemented by the interaction of tumor cells with ECM proteins (Eke and Cordes, 2011). These studies demonstrated that cell radioresistance depends on cell-cell communication and cell-matrix interaction and is largely determined by cell growth conditions. The different approaches are aimed at embodying a complexity of the tumor molecular architecture in 3D culture models, including (i) reconstruction of ECM structural components, (ii) providing the endocrine, paracrine, and extracellular influences by co-culturing of cancer cells with fibroblasts or endothelial monolayers, and (iii) nutrient and oxygen restriction mimicking the tumor microenvironment (Yamada and Cukierman, 2007). It is important to note that some of the 3D models explored so far provide an increase in CSC properties of the cultured cells as compared to monolayer conditions, including cell self-renewal, migration, release of exocrine and paracrine factors regulating tumor development and angiogenesis, formation of hypoxic niches, and therapy resistance (Dontu et al., 2003; Ponti et al., 2005; Yamada and Cukierman, 2007; Schwartz et al., 2010). Some of these 3D culture models are robust enough to be utilized for the development of new radiobiology assays that may better reflect CSC survival after irradiation. For example, Lomonaco and colleagues (2009) used neurosphere formation assay to demonstrate that radiation-induced authophagy is one of the radioprotective mechanisms for $CD133^+$ glioma stem cells. The isolation of brain tumor progenitors and growing them as spheres in anchorage-independent conditions and serum-free media was the first breakthrough to culture adult stem cells (Reynolds and Weiss, 1992). These cancer progenitors were further functionally defined by increased self-renewal capacity and tumorigenicity. The isolation and culture of cancer progenitor populations as spheres represent a convenient radiobiological assay to analyze clonogenicity estimated by the number of spheres and self-renewing capacity defined by the size of the spheres.

The first encouraging study that utilized a 3D clonogenic assay for the prediction of the intrinsic radiosensitivity of tumor biopsies from the patients with cervical carcinoma was published about 30 years ago. In this study, West and colleagues (1993) used a soft agar clonogenic assay to demonstrate that in vitro measurement of cell surviving fraction at 2Gy (SF2) can predict clinical outcome for patients treated with radiotherapy. However, it has not yet been proven directly that clonogens, which determine tumor recurrence after radiotherapy are the same as the cells with the CSC phenotype. For this, in vivo radiobiological assays are needed to be established where tumor control probabilities after different radiation doses are analyzed in parallel with the frequency of tumorigenic CSCs (Figure 35.3). This study would help to address a few key questions about radioresistance of CSCs, such as CSC properties of the surviving radioresistant populations and potential transdifferentiation of cancer cells during the course of radiotherapy. Moreover, taking into consideration the heterogeneity and phenotypic plasticity of CSCs, CSC-specific biomarkers of radioresistance should be validated in significant numbers of patients before they are assumed to be informative. Future clinical trials that effectively test these unique biomarkers to predict patient outcome will be critical.

In conclusion, accumulating experimental and clinical evidence supports the notion that a small subset of the cancer cell population with stem cell properties may affect tumor radiocurability and determine clinical outcome. A growing body of evidence suggests that CSC radioresistance depends on a complex interaction between inherent cellular features and the microenvironmental niche. A highly dynamic nature of the intrinsic and extrinsic determinants of radiosensitivity makes the CSC population a target that may be hard to treat. Further genomic and

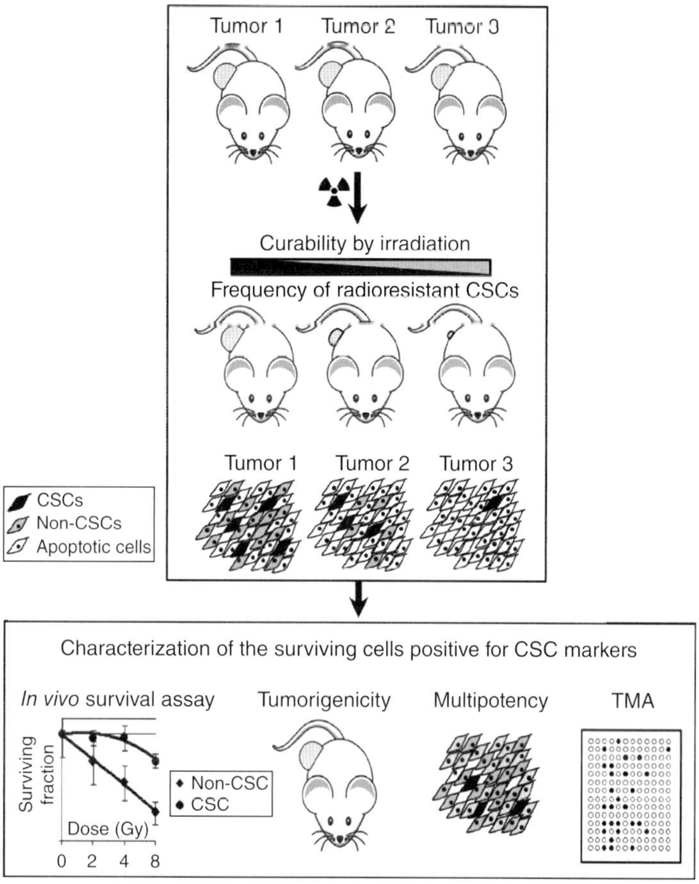

Figure 35.3. Design of CSC specific in vivo radiobiological assay. Xenograft tumors established in mice are treated with different doses of radiation. A frequency of CSCs in the residual tumors can be analyzed by IHC or flow cytometry and compared with tumor curability (TCD_{50}). Taking into consideration the phenotypic plasticity of CSCs, stem cell properties of survived radioresistant cell populations, such as tumorigenicity and multipotency need to be reconfirmed. Relevance of CSC-specific markers to radiotherapy outcome will be validated by tissue microarray (TMA) or PCR-based techniques.

proteomic studies accompanied by functional radiobiological assays and clinical observations are expected to provide more information about new CSC-associated biomarkers in order to better understand the molecular mechanisms of radioresistance and to predict clinical outcome after therapy.

References

Al-Assar, O., Mantoni, T., Lunardi, S., Kingham, G., Helleday, T., Brunner, T.B. (2011). Breast cancer stem-like cells show dominant homologous recombination due to a larger S-G2 fraction. Cancer Biol Ther. 11(12):1028–1035.

Allalunis-Turner, M.J., Barron, G.M., Day, R.S. 3rd, Fulton, D.S., Urtasun, R.C. (1992). Radiosensitivity testing of human primary brain tumor specimens. Int J Radiat Oncol Biol Phys. 23(2):339-343.

Anderson, P., Dische, S. (1981). Local tumor control and the subsequent incidence of distant metastatic disease. Int J Radiat Oncol Biol Phys. 7(12):1645-1648.

Ansiaux, R., Baudelet, C., Jordan, B.F., Beghein, N., Sonveaux, P., De Wever, J., Martinive, P., Grégoire, V., Feron, O., Gallez, B. (2005).

Thalidomide radiosensitizes tumors through early changes in the tumor microenvironment. Clin Cancer Res. 11(2 Pt 1):743–750.

Bao, S., Wu, Q., McLendon, R.E., Hao, Y., Shi, Q., Hjelmeland, A.B., Dewhirst, M.W., Bigner, D.D., Rich, J.N. (2006). Glioma stem cells promoteradioresistance by preferential activation of the DNA damage response. Nature 444 (7120):756–760.

Baumann, M., Dubois, W., Suit, H.D. (1990). Response of human squamous cell carcinoma xenografts of different sizes to irradiation: relationship of clonogenic cells, cellular radiationsensitivity in vivo, and tumor rescuing units. Radiat Res. 123(3):325–330.

Baumann, M., Dörr, W., Petersen, C., Krause, M. (2003). Repopulation during fractionated radiotherapy: much has been learned, even more is open. Int J Radiat Biol. 79(7):465–467.

Bauman, M., Gregoire, V. (2009). "Modified fractionation." In: Joiner, M., van der Kogel, A. (Eds.): Basic Clinical Radiobiology, pp. 135–148. London, United Kingdom, Arnold.

Baumann, M., Krause, M., Hill, R. (2008). Exploring the role of cancer stem cells in radioresistance. Nat Rev Cancer. 8(7):545–554.

Baumann, M., Krause, M., Thames, H., Trott, K., Zips, D. (2009). Cancer stem cells and radiotherapy. Int J Radiat Biol. 85(5):391–402.

Baumann, M., Krause, M. (2010). CD44: a cancer stem cell-related biomarker with predictive potential for radiotherapy. Clin Cancer Res. 16(21):5091–5093.

Baumann, M., Herrmann, T., Koch, R., Matthiessen, W., Appold, S., Wahlers, B., Kepka, L., Marschke, G., Feltl, D., Fietkau, R., Budach, V., Dunst, J., Dziadziuszko, R., Krause, M., Zips, D.; CHARTWEL-Bronchus study group. (2011). Final results of the randomized phase III CHARTWEL-trial (ARO 97-1) comparing hyperfractionated-accelerated versus conventionally fractionated radiotherapy in non-small cell lung cancer (NSCLC). Radiother Oncol. 100(1):76–85.

Begg, A.C., Stewart, F.A., Vens, C. (2011). Strategies to improve radiotherapy with targeted drugs. Nat Rev Cancer. 11(4):239–253.

Bentzen, S.M., Thames, H.D. (1996). Tumor volume and local control probability: clinical data and radiobiological interpretations. Int J Radiat Oncol Biol Phys. 36(1):247–251.

Bentzen, S.M., Baumann, M. (2002). "The linear-quadratic model in clinical practice". In: Steel GG (Ed.): Basic Clinical Radiobiology, pp. 134–146. London, United Kingdom, Arnold.

Bourhis, J., Overgaard, J., Audry, H., Ang, K.K., Saunders, M., Bernier, J., Horiot, J.C., Le Maître, A., Pajak, T.F., Poulsen, M.G., O'Sullivan, B., Dobrowsky, W., Hliniak, A., Skladowski, K., Hay, J.H., Pinto, L.H., Fallai, C., Fu, K.K., Sylvester, R., Pignon, J.P. (2006). Meta-Analysis of Radiotherapy in Carcinomas of Head and neck (MARCH) Collaborative Group. Hyperfractionated or accelerated radiotherapy in head and neck cancer: a meta-analysis. Lancet. 368(9538):843–854.

Borovski, T., De Sousa, E., Melo, F., Vermeulen, L., Medema, J.P. (2011). Cancer stem cell niche: the place to be. Cancer Res. 71(3):634–639.

Brock, W.A., Baker, F.L., Wike, J.L., Sivon, S.L., Peters, L.J. (1990). Cellular radiosensitivity of primary head and neck squamous cell carcinomas and local tumor control. Int J Radiat Oncol Biol Phys. 18(6):1283–1286.

Brunner, T.B., Kunz-Schughart, L.A., Grosse-Gehling, P., Baumann, M. (2012). Cancer stem cells as a predictive factor in radiotherapy. Semin Radiat Oncol. 22(2):151–174.

Chung, Y.L., Jian, J.J., Cheng, S.H., Tsai, S.Y., Chuang, V.P., Soong, T., Lin, Y.M., Horng, C.F. (2006). Sublethal irradiation induces vascular endothelial growth factor and promotes growth of hepatoma cells: implications for radiotherapy of hepatocellular carcinoma. Clin Cancer Res. 12(9):2706–2715.

Cook, J.A., Gius, D., Wink, D.A., Krishna, M.C., Russo, A., Mitchell, J.B. (2004). Oxidative stress, redox, and the tumor microenvironment. Semin Radiat Oncol. 14(3):259–266.

Diehn, M., Cho, R.W., Lobo, N.A., Kalisky, T., Dorie, M.J., Kulp, A.N., Qian, D., Lam, J.S., Ailles, L.E., Wong, M., Joshua, B., Kaplan, M.J., Wapnir, I., Dirbas, F.M., Somlo, G., Garberoglio, C., Paz, B., Shen, J., Lau, S.K., Quake, S.R., Brown, J.M., Weissman, I.L., Clarke, M.F. (2009). Association of reactive oxygen species levels and radioresistance in cancer stem cells. Nature 58 (7239):780–783.

Dische, S., Saunders, M., Barrett, A., Harvey, A., Gibson, D., Parmar, M. (1997). A randomised-multicentre trial of CHART versus conventional radiotherapy in head and neck cancer. Radiother Oncol. 44(2):123–136.

Dontu, G., Abdallah, W.M., Foley, J.M., Jackson, K.W., Clarke, M.F., Kawamura, M.J., Wicha, M.S. (2003). In vitro propagation and transcriptional profiling of human mammary stem/progenitor cells. Genes Dev 17(10):1253–1270.

Dubben, H.H., Thames, H.D., Beck-Bornholdt, H.P. (1998). Tumor volume: a basic and specific response predictor in radiotherapy. Radiother Oncol. 47(2):167–174.

Eke, I., Cordes, N. (2011). Radiobiology goes 3D: how ECM and cell morphology impact on cell survival after irradiation. Radiother Oncol. 99 (3):271–278.

Eschmann, S.M., Paulsen, F., Reimold, M., Dittmann, H., Welz, S., Reischl, G., Machulla, H.J., Bares, R. (2005). Prognostic impact of hypoxia imaging with 18 F-misonidazole PET in non-small cell lung cancer and head and neck cancer before radiotherapy. J Nucl Med 46(2):253–260.

Eschwege, F., Bourhis, J., Girinski, T., Lartigau, E., Guichard, M., Deblé, D., Kepta, L., Wilson, G.D., Luboinski, B. (1997). Predictive assays of radiation response in patients with head and neck squamous cell carcinoma: a review of the Institute Gustave Roussy experience. Int J Radiat Oncol Biol Phys. 39(4):849–853.

Fagundes, H., Perez, C.A., Grigsby, P.W., Lockett, M.A. (1992). Distant metastases after irradiation alone in carcinoma of the uterine cervix. Int J Radiat Oncol Biol Phys. 24(2):197–204.

Friedrich, J., Ebner, R., Kunz-Schughart LA. (2007). Experimental anti-tumor therapy in 3-D: spheroids—old hat or new challenge? Int J Radiat Biol. 83(11–12):849–871.

Fu, K.K., Pajak, T.F., Trotti, A., Jones, C.U., Spencer, S.A., Phillips, T.L., Garden, A.S., Ridge, J.A., Cooper, J.S., Ang, K.K. (2000). A Radiation Therapy Oncology Group (RTOG) phase III randomized study to compare hyperfractionation and two variants of accelerated fractionation to standard fractionation radiotherapy for head and neck squamous cell carcinomas: first report of RTOG 9003. Int J Radiat Oncol Biol Phys. 48(1):7–16.

Fuks, Z., Leibel, S.A., Wallner, K.E., Begg, C.B., Fair, W.R., Anderson, L.L., Hilaris, B.S., Whitmore, W.F. (1991). The effect of local control on metastatic dissemination in carcinoma of the prostate: long-term results in patients treated with 125I implantation. Int J Radiat Oncol Biol Phys. 21(3):537–547.

van Gent, D.C., Hoeijmakers, J.H., Kanaar, R. (2001). Chromosomal stability and the DNA double-stranded break connection. Nat Rev Genet 2:196–206.

Gerweck, L.E., Zaidi, S.T., Zietman, A. (1994). Multivariate determinants of radiocurability. I: Prediction of single fraction tumor control doses. Int J Radiat Oncol Biol 30;29(1):57–66.

Girinsky, T., Lubin, R., Pignon, J.P., Chavaudra, N., Gazeau, J., Dubray, B., Cosset, J.M., Socie, G., Fertil, B. (1993). Predictive value of in vitro radiosensitivity parameters in head and neck cancers and cervical carcinomas: preliminary correlations with local control and overall survival. Int J Radiat Oncol Biol Phys. 25(1):3–7.

Grosse-Gehling, P., Fargeas, C.A., Dittfeld, C., Garbe, Y., Alison, M.R., Corbeil, D., Kunz-Schughart, L.A. (2012). CD133 as a biomarker for putative cancer stem cells in solid tumours: limitations, problems and challenges. J Pathol. doi: 10.1002/path.4086.

Harada, H. (2011). How can we overcome tumor hypoxia in radiation therapy? J Radiat Res. 52(5):545–556.

Heddleston, J.M., Li, Z., McLendon, R.E., Hjelmeland, A.B., Rich, J.N. (2009). The hypoxic microenvironment maintains glioblastoma stem cells and promotes reprogramming towards a cancer stem cell phenotype. Cell Cycle. 8(20):3274–3284.

Hittelman, W.N., Liao, Y., Wang, L., Milas, L. (2010). Are cancer stem cells radioresistant? Future Oncol. 6(10):1563–1576.

Hill, R.P., Marie-Egyptienne, D.T., Hedley, D.W. (2009). Cancer stem cells, hypoxia and metastasis. Semin Radiat Oncol. 19(2):106–111.

Hill, R.P., Milas, L. (1989). The proportion of stem cells in murine tumors. Int J Radiat Oncol Biol Phys.16(2):513–518.

Ishikawa, T., Nakashiro, K., Klosek, S.K., Goda, H., Hara, S., Uchida, D., Hamakawa, H. (2009). Hypoxia enhances CXCR4 expression by activating HIF-1 in oral squamous cell carcinoma. Oncol Rep. 21(3):707–712.

Ishimoto, T., Nagano, O., Yae, T., Tamada, M., Motohara, T., Oshima, H., Oshima, M., Ikeda, T., Asaba, R., Yagi, H., Masuko, T., Shimizu, T., Ishikawa, T., Kai, K., Takahashi, E., Imamura, Y., Baba, Y., Ohmura, M., Suematsu, M., Baba, H., Saya, H. (2011). CD44 variant regulates redox status in cancer cells by stabilizing the xCT subunit of system xc(−) and thereby promotes tumor growth. Cancer Cell 19(3):387–400.

Jacobson, O., Weiss, I.D., Szajek, L.P., Niu, G., Ma, Y., Kiesewetter, D.O., Farber, J.M., Chen X. (2011). PET imaging of CXCR4 using copper-64 labeled peptide antagonist. Theranostics 1:251–262.

Jeggo, P.A., Löbrich, M. (2006). Contribution of DNA repair and cell cycle checkpoint arrest to the maintenance of genomic stability. DNA Repair (Amst). 5(9–10):1192–1198.

Jeter, C.R., Liu, B., Liu, X., Chen, X., Liu, C., Calhoun-Davis, T., Repass, J., Zaehres, H., Shen, J.J., Tang, D.G. (2011). NANOG promotes cancer stem cell characteristics and prostate cancer resistance to androgen deprivation. Oncogene 30(36):3833–3845.

Johnson, C.R., Thames, H.D., Huang, D.T., Schmidt-Ullrich, R.K. (1995). The tumor volume and clonogen number relationship: tumor control predictions based upon tumor volume estimates derived from computed tomography. Int J Radiat Oncol Biol Phys. 33(2):281–287.

de Jong, M.C., Pramana, J., van der Wal, J.E., Lacko, M., Peutz-Kootstra, C.J., de Jong, J.M., Takes, R.P., Kaanders, J.H., van der Laan, B.F., Wachters, J., Jansen, J.C., Rasch, C.R., van Velthuysen, M.L., Grénman, R., Hoebers, F.J., Schuuring, E., van den Brekel, M.W., Begg, A.C. (2010). CD44 expression predicts local recurrence after radiotherapy in larynx cancer. Clin Cancer Res. 16(21):5329–5338.

Karran, P. DNA double strand break repair in mammalian cells. (2000). Curr Opin Genet Dev. 10:144–150.

Kasten-Pisula, U., Tastan, H., Dikomey, E. (2005). Huge differences in cellular radiosensitivity due to only very small variations in double-strand break repair capacity. Int J Radiat Biol. 81(6):409–419.

Kim, J.I., Tannock, I.F. (2005). Repopulation of cancer cells during therapy: an important cause of treatment failure. Nature Reviews Cancer 5(7):516–525.

Koch, U., Krause, M., Baumann, M. (2010). Cancer stem cells at the crossroads of current cancer therapy failures—radiation oncology perspective. Semin Cancer Biol. 20(2):116–124.

Kokko, L.L., Hurme, S., Maula, S.M., Alanen, K., Grénman, R., Kinnunen, I., Ventelä, S. (2011). Significance of site-specific prognosis of cancer stem cell marker CD44 in head and neck squamous-cell carcinoma. Oral Oncol. 47 (6):510–516.

Koukourakis, M.I., Giatromanolaki, A., Tsakmaki, V., Danielidis, V., Sivridis, E. (2012). Cancer stem cell phenotype relates to radio-chemotherapy outcome in locally advanced squamous cell head-neck cancer. Br J Cancer 106(5):846–853.

Kuonen, F., Secondini, C., Ruegg, C. (2012). Molecular Pathways: Emerging pathways mediating growth, invasion and metastasis of tumors progressing in an irradiated microenvironment. Clin Cancer Res. PMID 22730447.

Kwon, G.Y., Ha, H., Ahn, G., Park, S.Y., Huh, S.J., Park, W. (2007). Role of CD24 protein in predicting metastatic potential of uterine cervical squamous cell carcinoma in patients treated with radiotherapy. Int J Radiat Oncol Biol Phys. 69(4):1150–1156.

Lander, A.D., Kimble, J., Clevers, H., Fuchs, E., Montarras, D., Buckingham, M., Calof, A.L., Trumpp, A., Oskarsson, T. (2012). What does the concept of the stem cell niche really mean today? BMC Biol. 10:19.

Li, Z., Bao, S., Wu, Q., Wang, H., Eyler, C., Sathornsumetee, S., Shi, Q., Cao, Y., Lathia, J., McLendon, R.E., Hjelmeland, A.B., Rich, J.N. (2009). Hypoxia-inducible factors regulate tumorigenic capacity of glioma stem cells. Cancer Cell 15(6):501–513.

Li, Z., Rich, J.N. (2010). Hypoxia and hypoxia inducible factors in cancer stem cell maintenance. Curr Top Microbiol Immunol. 345:21–30.

Linkous, A.G., Yazlovitskaya, E.M. (2012). Novel radiosensitizing anticancer therapeutics. Anticancer Res. 32(7):2487–2499.

Lisanti, M.P., Martinez-Outschoorn, U.E., Chiavarina, B., Pavlides, S., Whitaker-Menezes, D., Tsirigos, A., Witkiewicz, A., Lin, Z., Balliet, R., Howell, A., Sotgia, F. (2010). Understanding the "lethal" drivers of tumor-stroma co-evolution: emerging role(s) for hypoxia, oxidative stress and autophagy/mitophagy in the tumor micro-environment. Cancer Biol Ther. 10(6):537–542.

Lomonaco, S.L., Finniss, S., Xiang, C., Decarvalho, A., Umansky, F., Kalkanis, S.N., Mikkelsen, T., Brodie, C. (2009). The induction of autophagy by gamma-radiation contributes to the radioresistance of glioma stem cells. Int J Cancer. 125(3):717–722.

Martelli, A.M., Evangelisti, C., Follo, M.Y., Ramazzotti, G., Fini, M., Giardino, R., Manzoli, L., McCubrey, J.A., Cocco, L. (2011). Targeting the phosphatidylinositol 3-kinase/Akt/mammalian target of rapamycin signaling network incancer stem cells. Curr Med Chem. 18(18):2715–2726.

Mauguen, A., Le Péchoux, C., Saunders, M.I., Schild, S.E., Turrisi, A.T., Baumann, M., Sause, W.T., Ball, D., Belani, C.P., Bonner, J.A., Zajusz, A., Dahlberg, S.E., Nankivell, M., Mandrekar, S.J., Paulus, R., Behrendt, K., Koch, R., Bishop, J.F., Dische, S., Arriagada, R., De Ruysscher, D., Pignon, J.P. (2012). Hyperfractionated or accelerated radiotherapy in lung cancer: an individual patient data meta-analysis. J Clin Oncol. 30(22):2788–2797.

Medema, J.P., Vermeulen, L. (2011). Microenvironmental regulation of stem cells in intestinal homeostasis and cancer. Nature. 474(7351): 318–326.

Metellus, P., Nanni-Metellus, I., Delfino, C., Colin, C., Tchogandjian, A., Coulibaly, B., Fina, F., Loundou, A., Barrie, M., Chinot, O., Ouafik, L., Figarella-Branger, D. (2011). Prognostic impact of CD133 mRNA expression in 48 glioblastoma patients treated with concomitant radiochemotherapy: a prospective patient cohort at a single institution. Ann Surg Oncol. (10):2937–2945.

Mikkelsen, R.B., Wardman, P.(2003). Biological chemistry of reactive oxygen and nitrogen and radiation-induced signal transduction mechanisms. Oncogene. 22(37):5734-5754.

Mladenov, E., Iliakis, G. (2011). Induction and repair of DNA double strand breaks: the increasing spectrum of non-homologous end joining pathways. Mutat Res. 711(1–2):61–72.

Mohyeldin, A., Garzón-Muvdi, T., Quiñones-Hinojosa, A. (2010). Oxygen in stem cell biology: a critical component of the stem cell niche. Cell Stem Cell 7(2):150–161.

Nguyen, D.H., Oketch-Rabah, H.A., Illa-Bochaca, I., Geyer, F.C., Reis-Filho, J.S., Mao, J.H., Ravani, S.A., Zavadil, J., Borowsky, A.D., Jerry, D.J., Dunphy, K.A., Seo, J.H., Haslam, S., Medina, D., Barcellos-Hoff, M.H. (2011). Radiation acts on the microenvironment to affect breast carcinogenesis by distinct mechanisms that decrease cancer latency and affect tumor type. Cancer Cell. 19(5):640–651.

Overgaard, J. (2007). Hypoxic radiosensitization: adored and ignored. J Clin Oncol. 25(26): 4066–4074.

Overgaard, J. (2011). Hypoxic modification of radiotherapy in squamous cell carcinoma of the head and neck—a systematicreview and meta-analysis. Radiother Oncol. 100(1):22–32.

Pajonk, F., Vlashi, E., McBride, W.H. (2010). Radiation resistance of cancer stem cells: the 4 R's of radiobiology revisited. Stem Cells. 28(4): 639–648.

Pallini, R., Ricci-Vitiani, L., Banna, G.L., Signore, M., Lombardi, D., Todaro, M., Stassi, G., Martini, M., Maira, G., Larocca, L.M., De Maria, R. (2008). Cancer stem cell analysis and clinical outcome in patients with glioblastomamultiforme. Clin. Cancer Res. 14(24):8205–8212.

Petersen, C., Zips, D., Krause, M., Schöne, K., Eicheler, W., Hoinkis, C., Thames, H.D., Baumann, M. (2001). Repopulation of FaDu human squamous cell carcinoma during fractionated radiotherapy correlates with reoxygenation. Int J Radiat Oncol Biol Phys. 51(2):483–493.

Phillips, T.M., McBride, W.H., Pajonk, F. (2006). The response of CD24(−/low)/CD44+ breast cancer-initiating cells to radiation. J Natl Cancer Inst 98:1777–1785.

Pietras, A., Hansford, L.M., Johnsson, A.S., Bridges, E., Sjölund, J., Gisselsson, D., Rehn, M., Beckman, S., Noguera, R., Navarro, S., Cammenga, J., Fredlund, E., Kaplan, D.R., Påhlman, S. (2009). HIF-2alpha maintains an undifferentiated state in neural crest-like human neuroblastoma tumor-initiating cells. Proc Natl Acad Sci USA. 106(39):16805–16810.

Ponti, D., Costa, A., Zaffaroni, N., Pratesi, G., Petrangolini, G., Coradini, D., Pilotti, S., Pierotti, M.A., Daidone, M.G. (2005). Isolation and in vitro propagation of tumorigenic breast cancer cells with stem/progenitor cell properties. Cancer Res. 65(13):5506–5511.

Prestegarden, L., Enger, P. ø. (2010). Cancer stem cells in the central nervous system—a critical review. Cancer Res. 70(21):8255–8258.

Reynolds, B.A., Weiss, S. (1992). Generation of neurons and astrocytes from isolated cells of the adult mammalian central nervous system. Science. 255(5052):1707–1710.

Ross, G.M. (1999). Induction of cell death by radiotherapy. Endocr Relat Cancer. 6(1):41–44.

Saigusa, S., Tanaka, K., Toiyama, Y., Yokoe, T., Okugawa, Y., Ioue, Y., Miki, C., Kusunoki, M. (2009). Correlation of CD133, OCT4, and SOX2 in rectal cancer and their association with distant recurrence after chemoradiotherapy. Ann Surg Oncol. 16(12):3488–3498.

Schwartz, J.D., Rowinsky, E.K., Youssoufian, H., Pytowski, B., Wu, Y. (2010). Vascular endothelial growth factor receptor-1 in human cancer: concise review and rationale for development of IMC-18F1 (Human antibody targeting vascular endothelial growth factor receptor-1). Cancer. 15;116(4 Suppl):1027–1032.

Schwarz, G. (1909). Ueber Desensibilisierung gegen Roentgen- und Radiumstrahlen. Muenchner Medizinische Wochenschrift. 56:1217–1218.

Seymour, C.B., Mothersill, C., Alper, T. (1986). High yields of lethal mutations in somatic mammalian cells that survive ionizing radiation. Int J Radiat Biol Relat Stud Phys Chem Med. Jul;50(1):167–179.

Shigdar, S., Lin, J., Li, Y., Yang, C.J., Wei, M., Zhus, Y., Liu, H., Duan, W. (2012). Cancer stem cell targeting: the next generation of cancer therapy and molecular imaging. Ther Deliv. 3(2):227–244.

Situ, D., Long, H., Lin, P., Zhu, Z., Wang, J., Zhang, X., Xie, Z., Rong, T. (2010). Expression and prognostic relevance of CD44v6 in stage I non-small cell lung carcinoma. J Cancer Res Clin Oncol. 136(8):1213–1219.

Soeda, A., Park, M., Lee, D., Mintz, A., Androutsellis-Theotokis, A., McKay, R.D., Engh, J., Iwama, T., Kunisada, T., Kassam, A.B., Pollack, I.F., Park, D.M. (2009). Hypoxia promotes expansion of the CD133-positive glioma stem cells through activation of HIF 1alpha. Oncogene. 28(45):3949–3959.

Suit, H.D., Shalek, R.J., Wette, R. (1965). Radiation response of C3H mouse mammary carcinoma evaluated in terms of cellular radiation sensitivity. In: Cellular radiation biology. Baltimore: Williams and Wilkins; 514–530.

Sofia Vala, I., Martins, L.R., Imaizumi, N., Nunes, R.J., Rino, J., Kuonen, F., Carvalho, L.M., Rüegg, C., Grillo, I.M., Barata, J.T., Mareel, M., Santos, S.C. (2010). Low doses of ionizing radiation promote tumor growth and metastasis by enhancing angiogenesis. PLoS One 2010;5: e11222.

Sutherland, R.M., Inch, W.R., McCredie, J.A., Kruuv, J. (1970). A multi-component radiation survival curve using an in vitro tumour model. Int J Radiat Biol Relat Stud Phys Chem Med. 18(5):491–495.

Stausbøl-Grøn, B., Overgaard, J. (1999). Relationship between tumour cell in vitro radiosensitivity and clinical outcome after curative radiotherapy for squamous cell carcinoma of the head and neck. Radiother Oncol. 50(1):47–55.

Sutherland, R.M., McCredie, J.A., Inch, W.R. (1971). Growth of multicell spheroids in tissue culture as a model of nodular carcinomas. Journal of the National Cancer Institute 46(1):113–120.

Suwinski, R., Taylor, J.M., Withers, H.R. (1998). Rapid growth of microscopic rectal cancer as a determinant of response to preoperative radiation therapy. Int J Radiat Oncol Biol Phys. 1998 Dec 1;42(5):943–951.

Taghian, A., DuBois, W., Budach, W., Baumann, M., Freeman, J., Suit, H. (1995). In vivo radiation sensitivity of glioblastoma multiforme. Int J Radiat Oncol Biol Phys. 32(1):99–104.

Tao, Y., Daly-Schveitzer, N., Lusinchi, A., Bourhis, J. (2010). Advances in radiotherapy of head and neck cancers. Curr Opin Oncol. 22(3):194–199.

Thompson, L.H., Schild, D. (2001). Homologous recombinational repair of DNA ensures mammalian chromosome stability. Mutat Res. 477:131–153.

Tsurumi, C., Esser, N., Firat, E., Gaedicke, S., Follo, M., Behe, M., Elsässer-Beile, U., Grosu, A.L., Graeser, R., Niedermann, G. (2010). Non invasive in vivo imaging of tumor-associated CD133/prominin. PLoS One. 5(12):e15605.

Wang, J., Sullenger, B.A., Rich, J.N. (2012). Notch signaling in cancer stem cells. Adv Exp Med Biol. 727:174–185.

Wang, J., Wakeman, T.P., Lathia, J.D., Hjelmeland, A.B., Wang, X.F., White, R.R., Rich, J.N., Sullenger, B.A. (2010). Notch promotes radioresistance of glioma stem cells. Stem Cells. 28(1):17–28.

Weichselbaum, R.R., Dahlberg, W., Little, J.B. (1985). Inherently radioresistant cells exist in some human tumors. Proc Natl Acad Sci USA. 82(14):4732–4735.

West, C.M., Davidson, S.E., Roberts, S.A., Hunter, R.D. (1993). Intrinsic radiosensitivity and prediction of patient response to radiotherapy for carcinoma of the cervix. Br J Cancer. 68(4):819–823.

Withers, H.R. (1975). The four R's of radiotherapy. New York: Academic Press. Lett JTAH, editor. Advances in radiation biology. 5.

Woodward, W.A., Chen, M.S., Behbod, F., Alfaro, M.P., Buchholz, T.A., Rosen, J.M. (2007). WNT/beta-catenin mediates radiation resistance of mouse mammary progenitor cells. Proc Natl Acad Sci USA. 104(2):618–623.

Yamada, K.M., Cukierman, E. (2007). Modeling tissue morphogenesis and cancer in 3D. Cell 130(4):601–610.

Yaromina, A., Krause, M., Thames, H., Rosner, A., Krause, M., Hessel, F., Grenman, R., Zips, D., Baumann, M. (2007). Pre-treatment number of clonogenic cells and their radiosensitivity are major determinants of local tumour control after fractionated irradiation. Radiother Oncol. 83(3):304–310.

Yaromina, A., Kroeber, T., Meinzer, A., Boeke, S., Thames, H., Baumann, M., Zips, D. (2011). Exploratory study of the prognostic value of microenvironmental parameters during fractionated irradiation in human squamous cell carcinoma xenografts. Int J Radiat Oncol Biol Phys. 80(4):1205–1213.

Yin, H., Glass, J. (2011). The Phenotypic Radiation Resistance of $CD44^{+}/CD24^{-\text{or low}}$

Breast Cancer Cells Is Mediated through the Enhanced Activation of ATM Signaling. PLoS One. 6(9): e24080.

Zagzag, D., Lukyanov, Y., Lan, L., Ali, M.A., Esencay, M., Mendez, O., Yee, H., Voura, E.B., Newcomb, E.W. (2006). Hypoxia-inducible factor 1 and VEGF upregulate CXCR4 in glioblastoma: implications for angiogenesis and glioma cell invasion. Lab Invest. 86(12): 1221–1232.

Zhang, M., Behbod, F., Atkinson, R.L., Landis, M.D., Kittrell, F., Edwards, D., Medina, D., Tsimelzon, A., Hilsenbeck, S., Green, J.E., Michalowska, A.M., Rosen, J.M. (2008). Identification of tumor-initiating cells in a p53-null mouse model of breast cancer. Cancer Res. 68(12):4674–4682.

Zhang, X.P., Zheng, G., Zou, L., Liu, H.L., Hou, L.H., Zhou, P., Yin, D.D., Zheng, Q.J., Liang, L., Zhang, S.Z., Feng, L., Yao, L.B., Yang, A.G., Han, H., Chen, J.Y. (2008). Notch activation promotes cell proliferation and the formation of neural stem cell-like colonies in human glioma cells. Mol Cell Biochem. 307(1–2):101–108.

Zhang, M., Atkinson, R.L., Rosen, J.M. (2010). Selective targeting of radiation-resistant tumor-initiating cells. Proc Natl Acad Sci USA 107:3522–3527.

Zips, D., Böke, S., Kroeber, T., Meinzer, A., Brüchner, K., Thames, H.D., Baumann, M., Yaromina, A. (2011). Prognostic value of radiobiological hypoxia during fractionated irradiation for local tumor control. Strahlenther Onkol. 187(5):306–310.

Index

Page references followed by f denote figures. Page references followed by t denote tables.

Cancer Stem Cells, First Edition. Edited by Vinagolu K. Rajasekhar.